International Conference on Mechanical Engineering and Applied Composite Materials (MEACM2017)

IOP Conference Series: Materials Science and Engineering
Volume 307

Hong Kong, China
23 – 24 November 2017

ISBN: 978-1-5108-5888-6
ISSN: 1757-8981

Printed from e-media with permission by:

Curran Associates, Inc.
57 Morehouse Lane
Red Hook, NY 12571

CURRAN ASSOCIATES INC. proceedings.com

Some format issues inherent in the e-media version may also appear in this print version.

Printed by Curran Associates, Inc. (2018)

For permission requests, please contact the Institute of Physics
at the address below.

Institute of Physics
Dirac House, Temple Back
Bristol BS1 6BE UK

Phone: 44 1 17 929 7481
Fax: 44 1 17 920 0979

techtracking@iop.org

Additional copies of this publication are available from:

Curran Associates, Inc.
57 Morehouse Lane
Red Hook, NY 12571 USA
Phone: 845-758-0400
Fax: 845-758-2633
Email: curran@proceedings.com
Web: www.proceedings.com

Table of contents

Volume 307

International Conference on Mechanical Engineering and Applied Composite Materials

23–24 November 2017, Hong Kong

Accepted papers received: 24 January 2018
Published online: 19 February 2018

Preface

Papers

Design and Simulation

Manufacturing and Automation

Mechanical Engineering

IOP Conf. Series: Materials Science and Engineering **307** (2018) 011001 doi:10.1088/1757-899X/307/1/011001

Preface

Dear Distinguished Authors and Guests,
The Organizing Committee warmly welcomes you to 2017 International Conference on Mechanical Engineering and Applied Composite Materials (MEACM2017), held on 23th-24^{th,} November, 2017, Hong Kong, China.

The aim of MEACM2017 is to present the latest research and results of scientists (professors, students, PhD Students, engineers, and post-doc scientist) related to Mechanical Engineering and Materials topics. This conference provides opportunities for the different areas delegates to exchange new ideas and application experiences face to face, to establish business or research relations and to find global partners for future collaboration.

After the peer-review process, the submitted papers were selected on the basis of originality, significance, and clarity for the purpose of the conference. The selected papers and additional late-breaking contributions to be presented as lectures will make an exciting technical program. The conference program is extremely rich, featuring high-impact presentation. We hope that the conference results constituted significant contribution to the knowledge in these up to date scientific field.

The proceeding records the fully refereed papers presented at the conference. The main conference themes and tracks are Materials Engineering and Processing Technologies, Physical Chemistry, Surface chemistry, Mining Engineering, Advanced Materials Science, Materials Processing Technology, Electrical Engineering, Vehicle and Motor, Virtual Manufacturing and Simulation. Hopefully, all participants and other interested readers benefit scientifically from the proceedings and also find it stimulating in the process.

On behalf of the organizing committee, I would like to especially thank all the editors from IOP Conference Series for their great support to MEACM2017. Without their excellent editorial work, MEACM2017 will not be published so timely and successfully.

Finally we wish all the authors and attendees of MEACAM2017 a unique, rewarding and enjoyable memory at MEACM2017 in Hong Kong, China. We look forward to your participation in the 2thMEACM in 2018.
With our warmest regards,
Peng-Sheng Wei
Conference Organizing Chair
November 30th 2017

IOP Conf. Series: Materials Science and Engineering **307** (2018) 011002 doi:10.1088/1757-899X/307/1/011002

MEACM2017 Committee

Conference Chair

Professor Peng-Sheng Wei, National Sun Yat-Sen University, Kaohsiung, Taiwan

Program Committees

Prof. Steven Y. Liang, Georgia Institute of Technology, USA

Prof. Dr. Farid Nasir bin Ani, Universiti Teknologi Malaysia, Malaysia

Prof. Jacob Opadeyi, The University of the West Indies, Trinidad and Tobago

Prof. Dr. Mohammad Nazri bin Mohd. Ja'afar, Universiti Teknologi Malaysia, Malaysia

Prof. Dr. Md Mustafizur Rahman, Universiti Malaysia Pahang, Malaysia

Assoc. Prof. Jiyun Zhao, City University of Hong Kong, Hong Kong

Assoc. Prof. Yongsheng Chen, The Chinese University of Hong Kong, China

Asst. Prof. Yu (Will) Wang, Washington State University, USA

Asst. Prof. Zhiquan Shu, Washington State University, USA

Asst. Prof. Mohammad H. Rahman, University of Wisconsin-Milwaukee, USA

Asst. Prof. Kow Kien Who, University College Sedaya International, Malaysia

Dr. Mohd Farid bin Muhamad Said, Universiti Teknologi Malaysia,

IOP Conf. Series: Materials Science and Engineering **307** (2018) 011003 doi:10.1088/1757-899X/307/1/011003

Peer review statement

All papers published in this volume of *IOP Conference Series: Materials Science and Engineering* have been peer reviewed through processes administered by the proceedings Editors. Reviews were conducted by expert referees to the professional and scientific standards expected of a proceedings journal published by IOP Publishing.

IOP Conf. Series: Materials Science and Engineering **307** (2018) 012001 doi:10.1088/1757-899X/307/1/012001

Simulation based optimization on automated fibre placement process

Lei Shi

Autodesk (China) Software Research and Development Co. Ltd.
399 Pudian Road, Pudong New District, Shanghai 200122, China
E-mail: forrest.shi@autodesk.com

Abstract. In this paper, a software simulation (Autodesk TruPlan & TruFiber) based method is proposed to optimize the automate fibre placement (AFP) process. Different types of manufacturability analysis are introduced to predict potential defects. Advanced fibre path generation algorithms are compared with respect to geometrically different parts. Major manufacturing data have been taken into consideration prior to the tool paths generation to achieve high success rate of manufacturing.

1. Introduction

Automated fibre placement (AFP) technique is widely applied on composites automated manufacturing field. The problems followed, however, are critical when compared to traditional handcraft. Experimental studies[1] proved the necessity of the application of the manufacturing system upgraded with simulation tools. Autodesk TruPlan is based on multiple advance optimization algorithms and assesses the cost-impact of unique design features and surface definitions across multiple manufacturing processes, such as: Automated Fiber Placement (AFP), Automated Tape Laying (ATL), Hand Layup (HLU) or any combination of these strategies. This study focuses on the AFP process and the statistical data necessary to evaluate the impact of various design concepts against various manufacturing processes. Course trajectory propagation algorithms are compared between different shapes of products. Manufacturing feasibility analysis have been carried out to catch common fabrication issues at the early step. Down to the manufacturing stage, numbers of advanced features such as 'Operation Strategy' and 'Process Zone' have been involved to optimize tool paths.

2. Manufacturability Study

2.1. Propagation Mode

Automated fiber placement technology was invented originally to solve the manufacturing issue of geometrically complex parts due to the presence of tight radii, double curvatures and surface joggles. As a result, various fiber path generation algorithms (i.e., Propagation Mode) need to be taken into consideration and compared to achieve high-quality products. Autodesk TruPlan supports three types of propagation modes, known as 'Constant Angle', 'Constant Angle Offset', and 'Guide Curve Offset', respectively as shown in Figure 1.

IOP Conf. Series: Materials Science and Engineering **307** (2018) 012001 doi:10.1088/1757-899X/307/1/012001

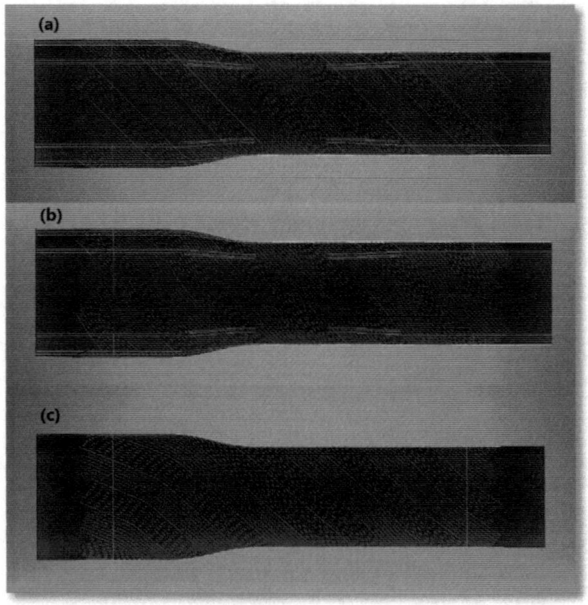

Figure 1. Three types of propagation modes applied on the same spar part: (a) Constant Angle, (b) Constant Angle Offset and (c) Guide Curve Offset.

2.1.1. Constant angle. The Constant Angle propagation mode consists of creating courses whose centreline follows the direction of material specified for the ply as shown in the Figure 2, where 1 is to transfer the rosette onto the surface at the point and 2 is to rotate the projected vector to the ply orientation. In other words, the tangent to the centreline at any point along the centreline is equivalent to the reference vector for the ply projected onto the tangent plane of the support and rotated around the normal to the specified orientation. Constant Angle is propagated to minimize material angle deviation, but cannot control the gap/overlap in between adjacent courses (Figure 1 (a)) very well. So it is recommended to be used on geometrically simple parts.

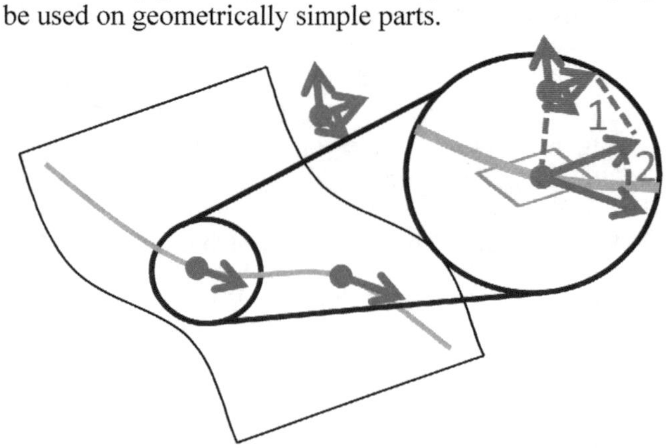

Figure 2. Schematic diagram of the tangent to the centerline at one point.

2.1.2. Constant Angle Offset. The first course generated has a constant angle centreline, following the rule of Constant Angle propagation mode, and other courses are adjacent to the previous course and are parallel offsets of the centreline of the first course. Figure 1 (b) shows the fibre paths generated on spar part, where steering tows will be laid. In spite of this, the tow-steered bands decrease the occurrence of gap or overlap defects, which improves the mechanical properties of the laminates.

IOP Conf. Series: Materials Science and Engineering **307** (2018) 012001 doi:10.1088/1757-899X/307/1/012001

2.1.3. Guide Curve Offset: The Guide Curve Offset propagation mode uses curve geometry (See the short blue line in Figure 1(c)) to create the centreline for the first course. The supplied guide curve is extrapolated using a geodesic curve to the ply or region boundary if it does not span the entire domain of the ply. The remaining courses are generated as parallel offsets of the first course. A well-defined guide curve geometry is a good compensation of the potential deviations of the Constant Angle mode, and meantime the Guide Curve Offset also holds the advantage of Constant Angle Offset in terms of 'offset' to eliminate most gap and overlap defects.

2.2. Manufacturability Analysis

Manufacturability or producibility analysis makes it possible to let composite manufacturers anticipate the costs of fabricating a certain part and also optimize their entire process to avoid re-engineering and downtime on the shop floor. The data collected enables engineers to perform comparative cost-benefit analyses to determine optimum materials, processes, and equipment. Figure 3 shows the six different types of producibility analysis implemented in Autodesk TruPlan to comprehensively predict potential defects during automated fiber placement operation.

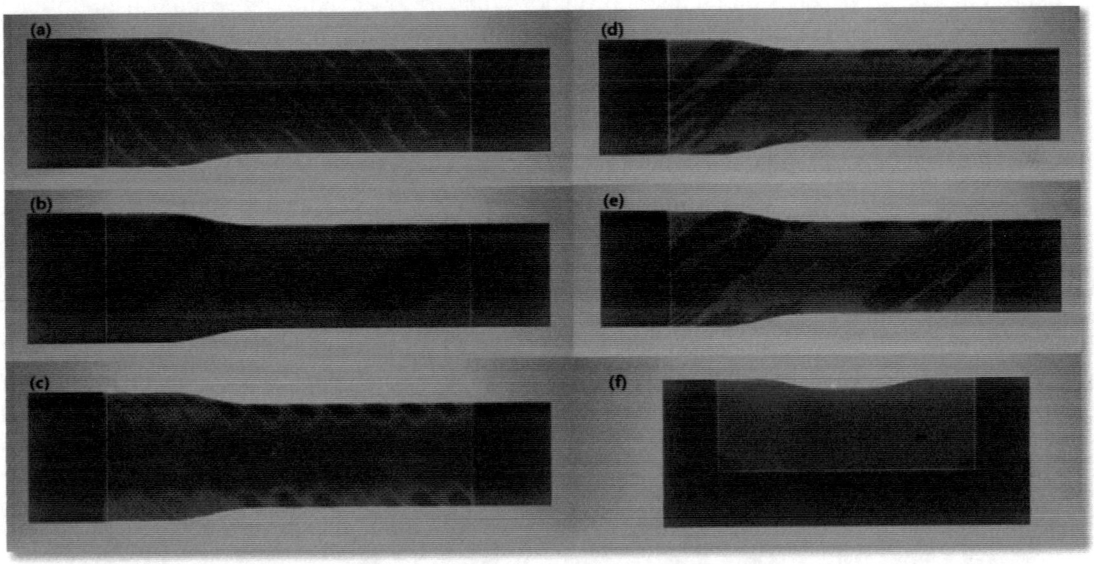

Figure 3. Six types of manufacturability analysis applied on the same spar part: (a) Gap/Overlap, (b) Angle (c) Compaction, (d) Wrinkle, (e) Steering and (f) Tow Length.

2.2.1. Steering. Automated fibre placement was used to optimise the tow steering of carbon fibre reinforced plastic/polymer whereby the minimum defect-free steering radius achievable[2]. In this instance, 1000 mm and 500 mm are set as steering radius warning and limit values, respectively. The most risky area in terms of tow steering is depicted in red in Figure 3 (e). With the simulation results, the laminates designers or the manufacturers can be well guided to fix the defects prior to the physical processing.

2.2.2. Gap/Overlap. Figure 3 (a) shows the results of Gap/Overlap analysis with the settings where the values of gap and overlap warning are ± 0.0254 mm, respectively, and the ones of those limit are ± 0.254 mm.

2.2.3. Angle. Figure 3 (b) simulates the fibre angle deviation results. The warning and limit are set to 1 and 5 degree, respectively.

2.2.4. Compaction. Figure 3 (c) displays analysis results of compacting distance of AFP roller with respect to the substrate. Here the compaction warning and limit are 2 mm and 5 mm, respectively.

2.2.5. Wrinkle. Material wrinkle analysis counts two major parameters: One is shearing angle along width direction and the other is material spreading along length direction. Figure 4 (d) indicates the wrinkle analysis results of this spar part with the settings, where maximum shearing angle is 1 degree and maximum spreading is 3%.

2.2.6. Tow Length. An AFP equipment always has specified allowable tow cut length value due to the limitations of the mechanical structure of the machine head. Tow length analysis consumes the value to find those tows that are too short to cut. As shown in Figure 3 (f), all the tows shorter than 100 mm are displayed in red.

3. Kinematics Optimization

Fiber placement operation is not straightforward to interpret, although the design information has been well collected, since it depends on numbers of factors such as tows laying direction, off-part motion, compacting pressure of AFP roller and feed rates of machine head. Autodesk TruFiber, using data from TruPlan, is designed to optimize multiple processes and generate tool paths with their associated NC codes for manufacturing.

3.1. Operation Strategy

3.1.1. Layup Direction. Typically, there are two types of machine movements need to be eliminated for higher fabrication efficiency: Off-part motion and Head reversing motion. For small parts, unidirectional layup (As shown in Figure 4 (a)) is recommended as the head rotation is much more time-consuming. For big-size parts, however, the off-part motion becomes a big concern due to the long linking path, so the bidirectional strategy (As shown in Figure 4 (b)) is a preferable choice to reverse machine head after completing each fibre course. An optimal 'combined' solution is provided by TruFiber tool to nominate the 'Maximum Unidirectional Length', meaning that any course shorter than the defined value is not reversed, otherwise it goes with bidirectional way. Figure 4 (c) shows the tool paths computed using the 'combined' layup strategy, where the 'Maximum Unidirectional Length' is set to 1000 mm.

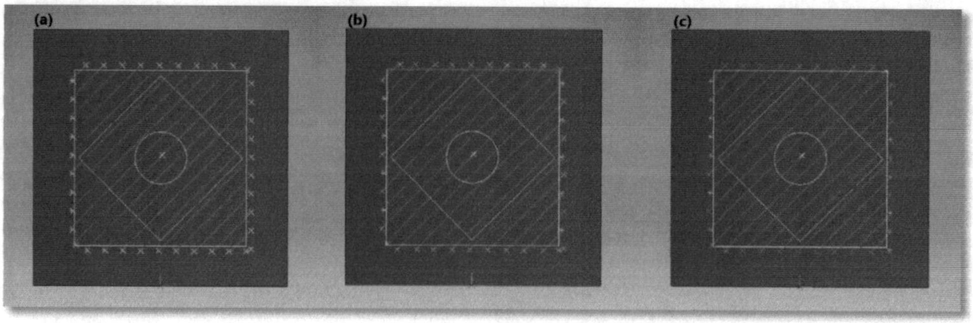

Figure 4. Tool paths computed results using three types of layup direction strategies on the same circle-diamond-square part: (a) Unidirectional layup, (b) Bidirectional layup and (c) Bidirectional with 'Maximum Unidirectional Length'.

3.1.2. Automatically Break Courses. In most cases, the holes have to be first filled with bands (As shown in the centre of Figure 5 (a)) and then be drilled in the trimming process. The best idea is that the courses can be broken when it passes through the holes. Figure 5 (b) illustrates the situation that TruFiber determines to 'Automatically Break Courses' to fly through the hole with the diameter of 400 mm, where the 'minimum course break distance' is set to 200 mm.

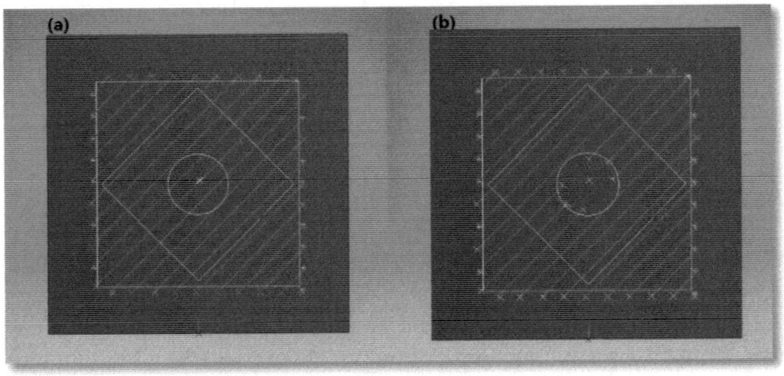

Figure 5. The comparison of tool paths computed results between (a) 'Automatically Break Courses' disabled and (b) 'Automatically Break Courses' enabled.

3.2. Process Zone

A Process Zone is an area to use specific manufacturing parameters appropriate for the characteristics of that area. For example, when a roller is being applied to a ramped surface, a lighter compaction force is desired so as not to crush the underlying surface at the start of the ramp, or to use over-feeding to make sure there are no gaps between the material and the surface. In this case, a zone for the ramp can be created as shown in the Figure 6 (The region in between the two blue contours).

Figure 6. Process Zone defined on the ramp surface of the engine cover part.

4. Conclusion

Current research status concerning automated fibre placement process and related equipment has been reviewed. The results imply the requirement of the advanced system equipped with simulation software to optimize the design and manufacturing lifecycle. Autodesk TruPlan, as a result, is introduced to conduct course propagation and analysis to assess the manufacturability of unique design features and surface definitions. Another software tool (Autodesk TruFiber) is proposed to optimize the fibre placement operation from kinematics point of view.

References

[1] Kaven Croft. Experimental Characterization of AFP Process Defects in Composite Structures. Thesis submitted to McGill University in partial fulfillment of the requirements of the degree of Master of Engineering, August 2010.

[2] RP Smith, Z Qureshi, RJ Scaife and HM El-Dessouky. Limitations of processing carbon fibre reinforced plastic/polymer material using automated fibre placement technology. Journal of Reinforced Plastics and Composites 2016, Vol. 35(21) 1527–1542.

IOP Conf. Series: Materials Science and Engineering **307** (2018) 012002 doi:10.1088/1757-899X/307/1/012002

Research on dynamic balancing simulation of rotary shaft based on ADAMS

Weiqiang Zheng[1], Chengjie Rui[1], Jie Yang[1] and Pingyi Liu[1]
[1]College of Engineering, China Agricultural University, Beijing 100083, China
E-mail: liupingyi@cau.edu.cn

Abstract. Due to the design and processing technology of rotary shaft, the mass center of it does not coincide with the rotating axis of the rotary shaft and there is an unbalanced mass. The unbalanced mass can have some disadvantages, such as the centrifugal force, the vibration and so on. Those disadvantages could reduce the accuracy and service life of the equipment.In this paper, the dynamic balance of the rotary shaft is analysed by the theory analysis combined with the dynamic simulation software. This method ensures that the rotary shaft meets the dynamic balancing requirements during the design stage. It effectively supports the structural design of the rotary shift, and provides a way of thinking and method for the design and development of the same type of products.

1. Introduction

When using dynamic balancing machine to correct the unbalance of rotary components, it is necessary to install a load cell between the rotor and supporting bearing to measure the bearing force during the rotation of the rotor[1], so as to measure the magnitude and direction of unbalance force[2-3]. It is also possible to arrange the sensor horizontally and vertically on the bearing housing supporting the rotor. First, the vibration value of the original unbalance at the measuring point is measured, and then the rotor is weighted and the vibration displacement of the bearing housing after the test weighting measured. The imbalance is corrected by calculating the influence coefficient[1,4-5].

In this paper, theoretical analysis combined with dynamic simulation software is used in the product design of the dynamic balance of crankshaft virtual test. It provides a solution to the problem of balancing the parts where it is difficult to find an unbalanced position.

2. Theories and methods of dynamic balancing

When the rotor rotates, the centrifugal inertia force exists because the part has the eccentric mass, and the eccentric mass is not on the axis of rotation. That is means that the mass diameter product is not zero. Since balancing is generally carried out on both sides of the balance, the balance mass can be decomposed into two sections offset from the center of mass and perpendicular to the axis of rotation, and the imbalance mass is then balanced in both sections[1]. As shown in Figure 1:

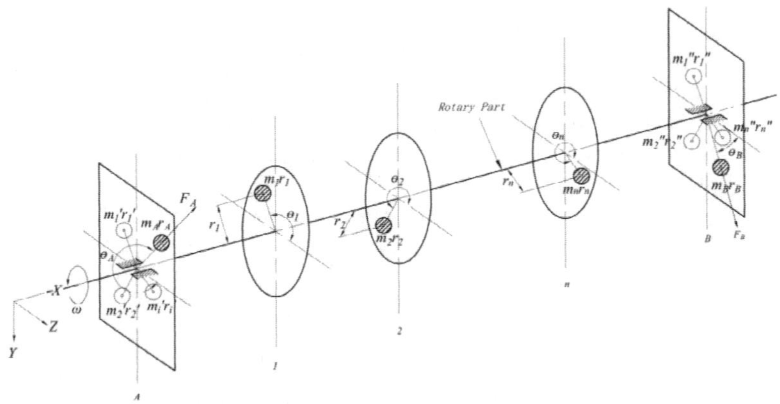

Figure 1. Dynamic balance principle

Because this method only obtains the unbalance and the phase by the measuring force, it only needs to balance the unbalanced amount obtained in the corresponding phase pair. It is not necessary to give the magnitude and phase of the unbalanced mass of each section, which is suitable for the motion of the irregularly shaped rotating part Balance design issues. The position of the rotation pair is the balance cross section, and the measurement of the maximum force along the Y direction and its phase can be processed by means of virtual simulation analysis. A description will be given with examples.

3. Experimental Design of Crankshaft Dynamic Balancing

In this paper, the ADAMS software is used to monitor the centrifugal force and phase angle of a crankshaft. In ADAMS, imbalance balanced simulation tests are carried out to balance the existing imbalances.

3.1. Model establishment

The established three-dimensional model is saved as .x_t format, imported into ADAMS. The crankshaft material is No.45 steel, and the density is 7850kg / m^3.And the mass m = 6.5098kg. The gravity is along the Y-axis positive direction in ADAMS.

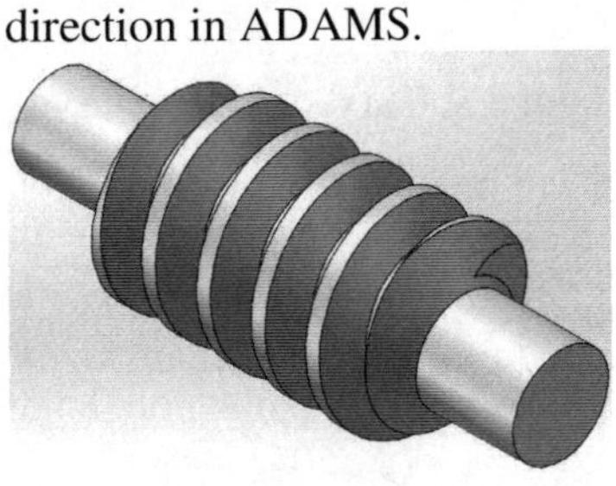

Figure 2. Crankshaft 3D model

3.2. Dynamic balance simulation analysis

Two sections are selected for dynamic balancing. The distance from section 1 to plane A is 56.5 mm, the distance from section 2 to plane A is 69.5 mm, and rotary pair 1 and rotary pair 2 are respectively added in section 1 and section 2. Drive 1 is added in the position of the drive 1, the size of its speed is 157rad / s, the direction from section 2 to section 1 as clockwise.

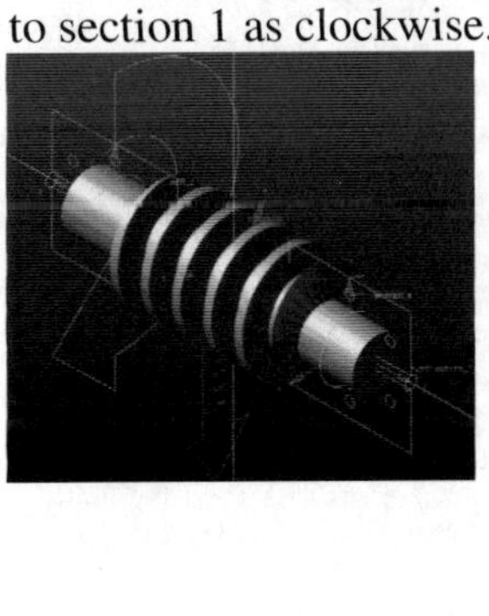

Figure 3. Simulation analysis

Establish two measurements: measuring the magnitude of the force F_{Y1} in the positive direction of the Y-axis at the rotation pair 1, and measuring the magnitude of the force F_{Y2} that it receives in the positive direction of the Y-axis at the rotation pair 2. The simulation time is set to 0.08s (2 cycles) and the step size is set to 4000. After setting up, we carry on the simulation analysis. It can be seen that the crankshaft rotates about the X axis and the two measurements of forces F_{Y1} and F_{Y2} change periodically. The change rules are shown in Figure 4 and Figure 5.

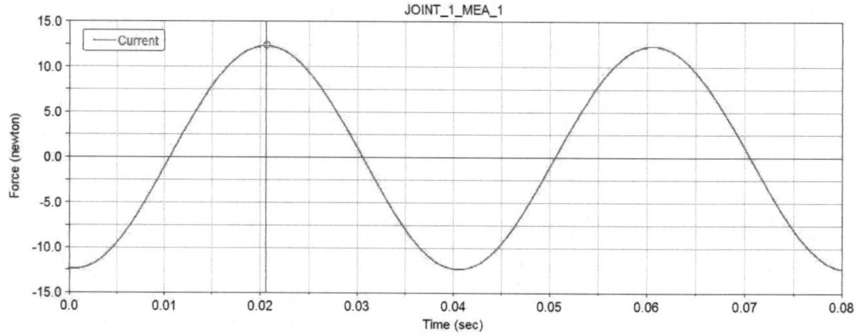

Figure 4. Curve of the force F_{Y1}

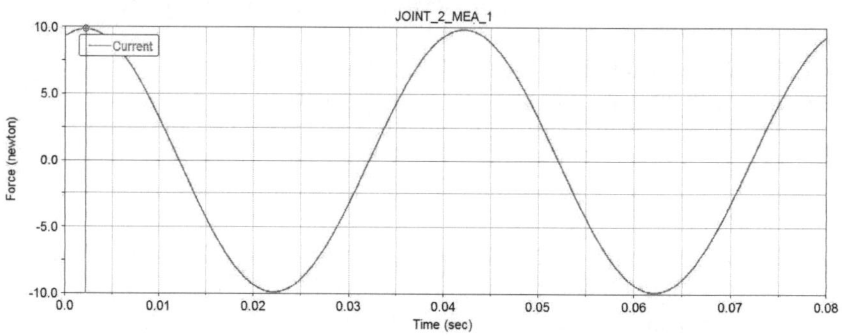

Figure 5. Curve of the force F_{Y2}

Through the analysis of the curve, the maximum force max(F_{Y1})= 12.3505N in section 1, the phase angle is 185.625 °; the maximum force in section 2 max(F_{Y2})= 9.8763N, the phase angle is 19.665 °. Set the magnitude of the unbalance in section 1 to m_1r_1 and the magnitude of the unbalance in section 2 to m_2r_2.

In section 1, the magnitude of residual unbalance m_1r_1 is:

$$m_A r_A = \frac{\max(F_{Y1})}{\omega^2} = \frac{12.3505}{157^2} \times 1000 g \cdot mm = 0.5011 g \cdot mm \quad (1)$$

The phase angle $\varphi = \theta_A$ = 360 ° -185.625 ° = 174.375 °;

In section 2, the magnitude of residual unbalance m_2r_2 is:

$$m_B r_B = \frac{\max(F_{Y2})}{\omega^2} = \frac{9.8763}{157^2} \times 1000 g \cdot mm = 0.4007 g \cdot mm \quad (2)$$

The phase angle $\varphi = \theta_A$ = 360 ° -19.665 ° = 340.335 °;

The overall balance accuracy A is calculated as follows:

$$A = \frac{m_A r_A + m_B r_B}{m}\omega = \frac{0.5011 + 0.4007}{6.5098 \times 1000} \times 157 mm/s = 0.02176 mm/s \quad (3)$$

3.3. Balance method of unbalance amount

First, balance the amount of unbalance in section 1. In the direction of the end plate (clockwise), remove 8.56 mm depth material in this direction at a distance of 180 ° + 5.625 ° = 185.625 ° from the end plate. The distance from the position where the material is removed to the origin is X = 53 mm. In this case, the maximum force max(F_{Y1}) = 0.0854N in section 1 and the maximum force max(F_{Y2}) = 9.7915N in section 2, both of which have been reduced. The phase angle θ_2 = 0.002185 * 9000° = 19.665° remains unchanged.

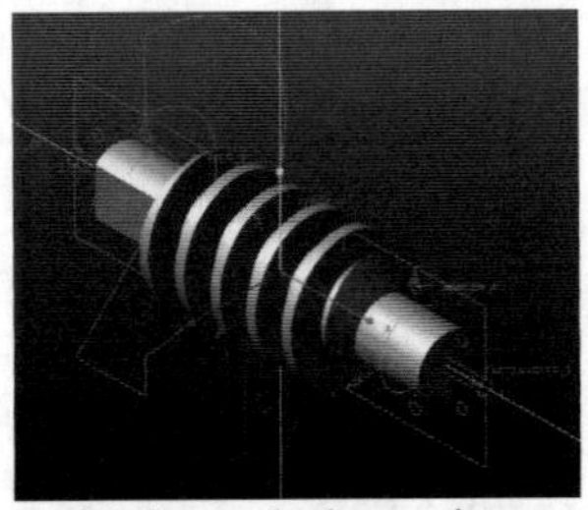

Figure 6. The unbalance in section 1

Then, the amount of unbalance in section 2 is balanced, and the position of mass removal is shown in FIG. 7. Seen from the end plate direction (clockwise), 5.54 mm deep material is removed in this direction at 19.665 ° from the end plateau. The distance from the side where the material is removed to the origin is X = 64 mm. In this case, the maximum force in section 2 max(F_{Y2})= 0.0274N, and the maximum unbalance force max(F_{Y1})= 0.0907N in section 1, which is improved but not increased slightly.

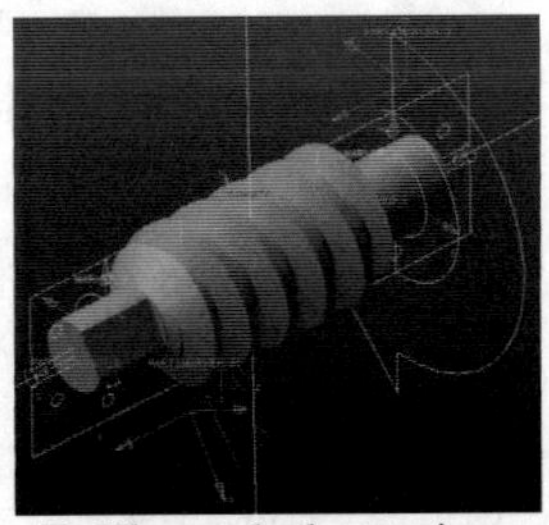

Figure 7. The unbalance in section 2

After the balance, the curve of internal force of the two sections is shown in Figure 8 and Figure 9:

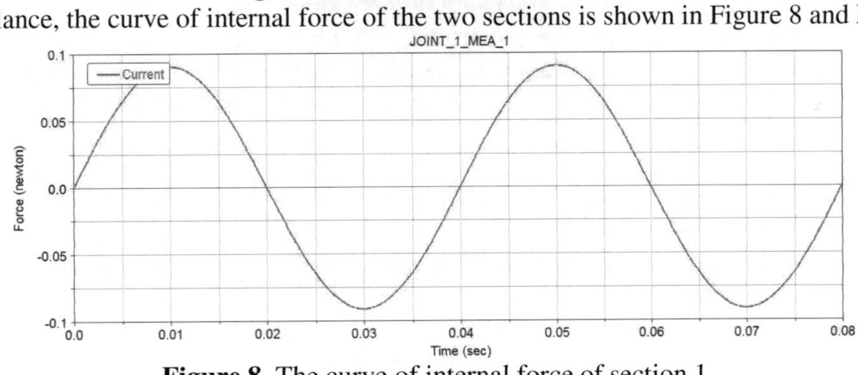

Figure 8. The curve of internal force of section 1

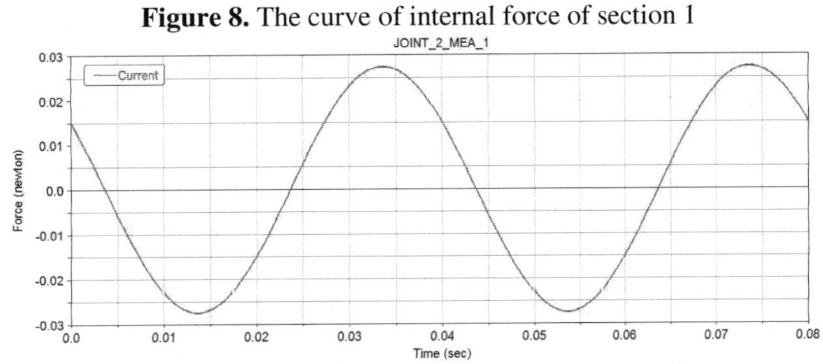

Figure 9. The curve of internal force of section 2

After dynamic balancing, the size of residual unbalance m_1r_1 in section 1 is:

$$m_A r_A = \frac{\max(F_{Y1})}{\omega^2} = \frac{0.0907}{157^2} \times 1000 g \cdot mm = 0.00368 g \cdot mm \quad (4)$$

The phase angle $\varphi = \theta_A = 360\,° - 185.625\,° = 174.375\,°$;
In section 2, the size of the residual unbalance $m_2 r_2$ is:

$$m_B r_B = \frac{\max(F_{Y2})}{\omega^2} = \frac{0.0274}{157^2} \times 1000 g \cdot mm = 0.00111 g \cdot mm \quad (5)$$

The phase angle $\varphi = \theta_A = 360\,° - 19.665\,° = 340.335\,°$;
The overall balance accuracy A is calculated as follows:

$$A = \frac{m_A r_A + m_B r_B}{m}\omega = \frac{0.5011 + 0.4007}{6.5098 \times 1000} \times 157 mm/s = 0.02176 mm/s \quad (6)$$

After the balance is completed, the three-dimensional solid model of the crankshaft is shown in Figure 10:

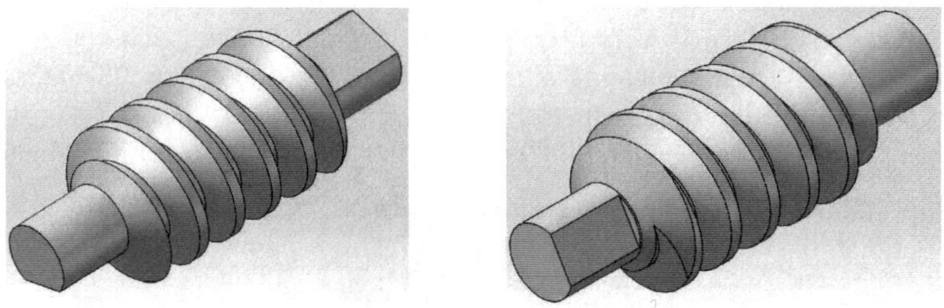

(a) section 1 **(b)** section 2

Figure 10. After being balanced three-dimensional solid model

4. Conclusion

(1) In this paper, theoretical analysis combined with dynamic simulation software is used in the product design of the dynamic balance of crankshaft virtual test, effectively support the structure of the crankshaft design, and provides a design method for the same type of product.

(2) After the crankshaft is dynamically balanced by this method, the magnitude of the residual unbalance in section 1 is 0.00368 g · mm, the residual unbalance in section 2 is 0.00111 g · mm, and the overall balance accuracy is 0.02176 mm / s, the result meets the design requirements.

5. References

[1] Zhou S. Modeling Estimation and Active Balancing of Speed Varying Rotor System. Ann Arbor: University of Michigan. 2001:15～16

[2] A Gusarov VIS, U I Susanin Automatic Balancing of Rotors by the Application of Servo Systems Moscow: Izdatel'stvo Nauka, 1974.177-172

[3] C.H.Chen,K.W.Wang. An Integrated Approach Toward the Dynamic Analysis of High-Speed Spindles[J].Vibration and Acoustics,1994,116:514-522.

[4] Kim S M, ee S K, ee K J. Effect of Bearing Surroundings on the High-Speed Spindle-Bearing Compliance[J]. International Journal of Advanced Manufacturing Technology, 002, 9: 551-557.

[5] Shin Y C. Bearing Non-linearity and Stability Analysis in High Speed Machining[J]. Journal of Engineering for Industry, 992, 14: 23-30.

Acknowledgments

The work in this paper was fully supported by National Natural Science Foundation of China under Grant No. 51475460.

IOP Conf. Series: Materials Science and Engineering **307** (2018) 012003 doi:10.1088/1757-899X/307/1/012003

FEM simulation of the effect of cooling rate on prediction of surface deviation of compression molded glass lenses

Guanjun Chen, Bo Tao*, Shuang xu, Gongfa Li

Key Laboratory of Metallurgical Equipment and Control Technology of Ministry of Education, Wuhan University of Science and Technology, Wuhan, Hubei 430081, China

*Corresponding author e-mail:taoboq@wust.edu.cn

Abstract. For precision compression glass molded lenses, surface deviation is vital for the quality check. In this research, the effects of cooling rates on the prediction of surface deviation in glass lenses by precision compression molding were studied. Two types of cooling method with many different cooling rates were investigated. Based on the simulation results, two steps cooling method is better. And slower the cooling rate cause larger surface deviation.

1. Introduction

Compression glass molding is an ideal manufacturing technique for precision glass optical elements because of its net shape forming and high production manufacturing capabilities [1-4]. Due to the hot pressing and cooling, surface deviation was inevitable introduce to the molded glass lenses. The surface deviation will cause the lens's unnecessary refraction and lead to a decline in quality. Finite element method (FEM) has been applied to simulate and investigate the influence of the process parameters in compression molding method on the quality of glass lenses [2-4]. The Tool-Narayanaswamy-Moynihan (TNM) model [5, 6] was used to simulate the glass structural relaxation behavior in cooling stage. In this research, the FEM method with TNM model was used to study the surface deviations of compression molded glass lenses. Based on different cooling methods and different cooling rates, the surface deviations of aspherical and spherical surface studied. Based on the results, a two steps cooling method was chosen. However, the cooling rate at the first step should be considered carefully, because of the side effect for residual stresses.

2. Numerical simulation

The compression molding proceeds in two stages called heating and hot pressing of glass blank, and cooling. The glass blank was first heated to the molding temperature $T_{molding}$, higher than the glass's glass transition temperature T_g, in the heating and molding stage. In $T_{molding}$, the glass was simulated as a Newtonian fluid. Because of the non-linear properties of the glass material associated with its current temperature and temperature history in cooling stage, the glass was molded by using the TNM model.

Figure 1 (a) illustrates the design drawing of the glass lens in mm. The convex side is spherical surface and the concave side is aspherical surface.

Figure 1(b) illustrates an example of two steps cooling procedure for the compression molding process. Hot pressing was executed at 684 °C, and maintaining the temperature for 4 minutes to release stresses due to pressing. The next step is to cool the molds and glass lens to room temperature. In Figure 1(b), the cooling procedure was divided into two parts: from the pressing temperature to

520 °C was implemented on a cooling rate of 0.8 °C/s, then from 520 °C to 200 °C was carried out on a cooling rate of 1.6 °C/s. Then the molded glass lens was taken out for room temperature cooling when the temperature achieved 200 °C [2, 7].

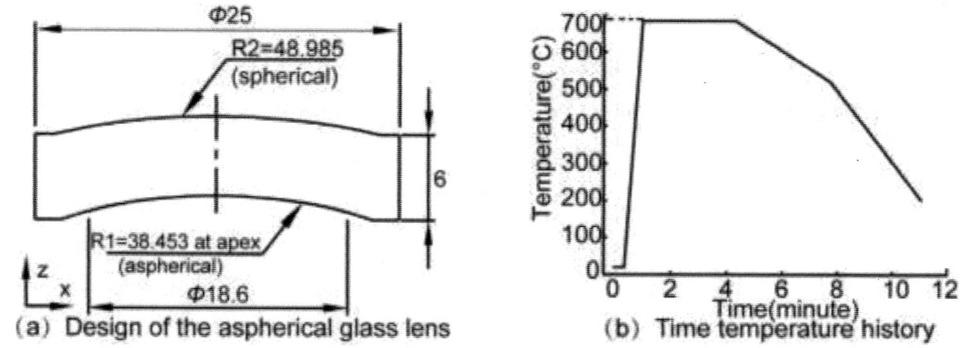

Figure 1. (a) Design drawing of the aspherical glass lens in mm, (b) An example of two steps cooling procedure

In FEM simulation, a two dimensional (2D) axisymmetric model was applied in this research based on the axisymmetric properties of glass lens and experimental environment. And a four-node isoparametric quadrilateral elements was used to mesh the glass and molds[7-9]. The upper mold, glass lens and lower molds were meshed into 381, 3,500 and 365 elements, respectively. Figure 2 illustrates the moulded lens and molds model after the molding procedure.

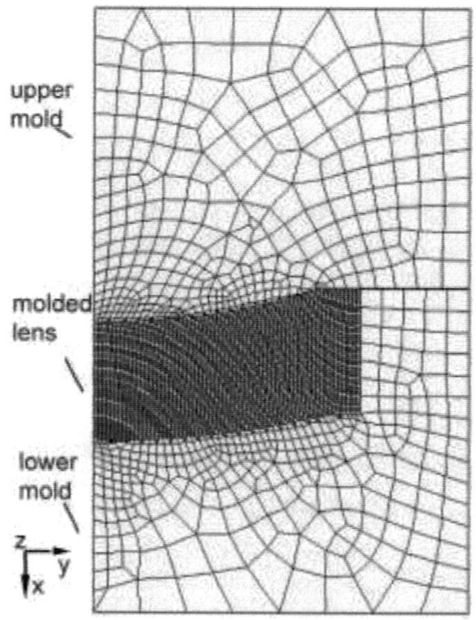

Figure 2. FEM model of the moulded glass lens and molds after molding procedure

3. Simulation results and discussion

Figure 3 illustrates the deviations of aspherical surface and spherical surface at different temperature stage during the cooling process. The deviation means difference between the lens's surface and its designed surface. The cooling process was started with a cooling rate of 0.8 °C/s from molding temperature to 520 °C, then the cooling rate was changed to 1.6 °C/s until the temperature reached 200 °C. Then a room temperature natural cooling was followed by taking out the glass lens. In Figure 3, black dot denotes the deviations at molding temperature. Red diamonds denotes the deviations at 200 °C. Green square means the deviations under room temperature.

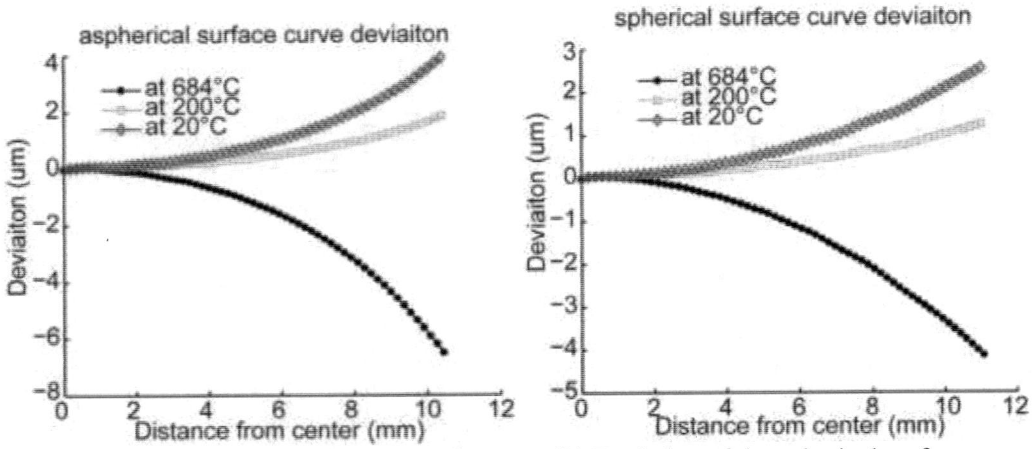

(a) Deviation of the aspherical surface (b) Deviation of the spherical surface

Figure 3. Deviations of the glass lens at aspherical and spherical surface at different temperature stage during cooling process

Figure 4 shows the deviations of aspherical and spherical surfaces under different cooling rates. The lenses were cooled to 200 °C with a constant cooling rate. The cooling rates were set as 0.03 °C/s, 0.13 °C/s, 0.43 °C/s, 1 °C/s, 2 °C/s, 3 °C/s, 5 °C/s, respectively. It can be seen that slower cooling rate causes larger deviations on both aspherical and spherical surfaces.

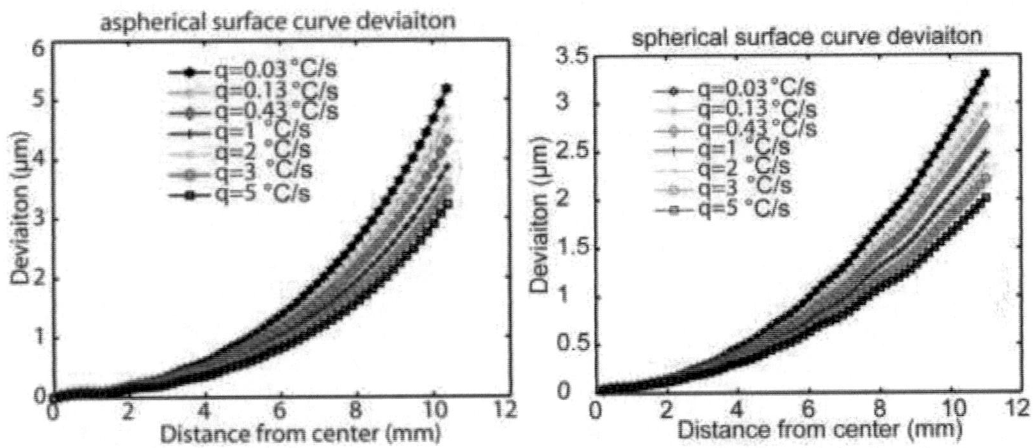

(a) Deviation of the aspherical surface (b) Deviation of the spherical surface

Figure 4. Deviation of the glass lens at room temperature by using different cooling rates

Figure 5 shows the deviations of aspherical and spherical surfaces by using two steps cooling method with different first step cooling rates. The first step cooling rates were set as 0.13 °C/s, 0.43 °C/s, 0.8 °C/s, 1.25 °C/s, respectively. The lens was cooled with the first step cooling rate from molding temperature to 520 °C. Then a second cooling rate was followed untill 200 °C. The second cooling rate was set as 1.6 °C/s. Then the lens was removed from molding machine for room temperature cooling. It can been seen that the largest difference of deviations at both aspherical and spherical surfaces by different first step cooling rates were about 1μm.

IOP Conf. Series: Materials Science and Engineering **307** (2018) 012003 doi:10.1088/1757-899X/307/1/012003

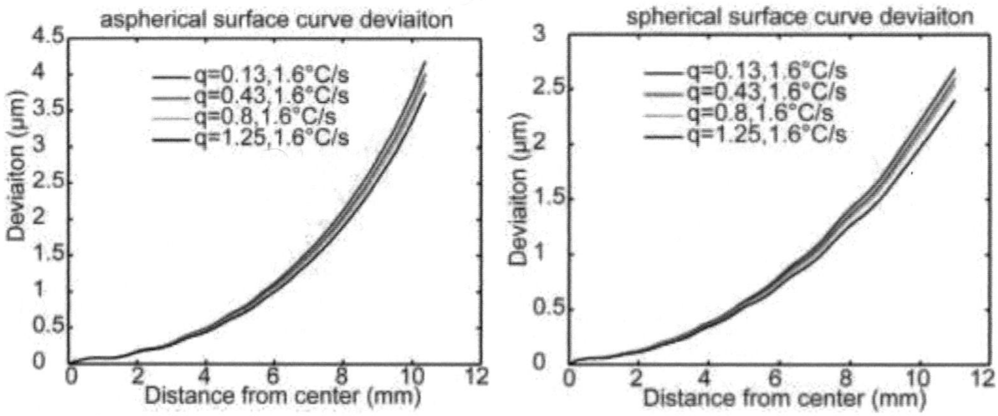

(a) Deviation of the aspherical surface (b) Deviation of the spherical surface

Figure 5. Deviation of the glass lens at room temperature by using two steps cooling method with different first step cooling rates

Figure 6 shows the deviations of aspherical and spherical surfaces by using two steps cooling method with different second step cooling rates. First, the lens was cooled from molding temperature to 520 °C with the first step cooling rate. And the first step cooling rate was set as 0.8 °C/s. Then the cooling rate was changed to the second step cooling rate until the temperature reached 200 °C. The second step cooling rates were set as 0.8 °C/s, 1.6 °C/s, 3 °C/s, respectively. Then the lens was taken out for natural cooling. It can been seen taht there is no significant changes in surface deviation at both aspherical and spherical surfaces by different second step cooling rates.

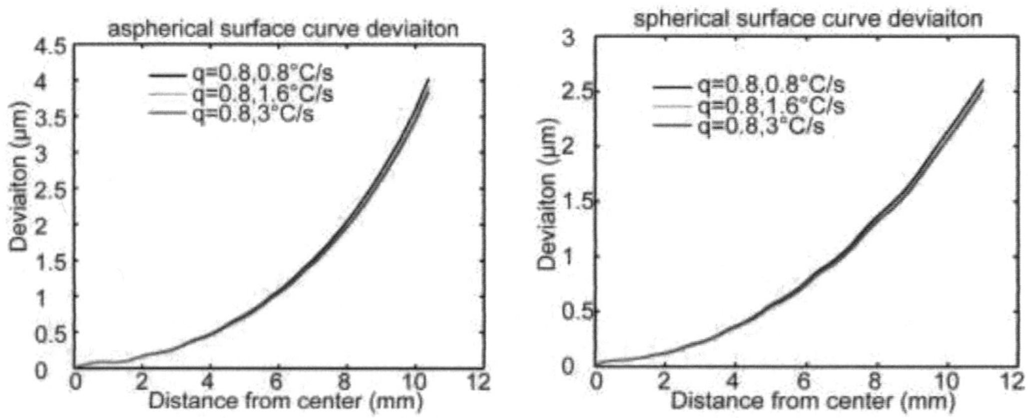

(a) Deviation of the aspherical surface (b) Deviation of the spherical surface

Figure 6. Deviation of the glass lens at room temperature by using two steps cooling with different second step cooling rates

Based on the results in Figures 4 and 5, the cooling rate between the molding temperature and 520 °C shows a big influence on surface deviation. The slower cooling rate, the larger deviation. However, the cooling rate at the second step gives smaller influence. There is only 0.1 μm changes in deviation when the cooling rate at the second step raised from 0.8 °C/s to 3 °C/s.

4. Conclusions

The surface deviation of compression molded glass lenses were investigated by using FEM. The results showed that cooling rate before the transition temperature has larger effect on surface deviation and should be chosen carefully. And the slower cooling rate cause larger surface deviation. When the temperature cooled under the transition temperature, the cooling rate shows smaller effect. Therefore, two steps cooling method is a way to keep small surface deviation while decrease cooling time.

IOP Conf. Series: Materials Science and Engineering **307** (2018) 012003 doi:10.1088/1757-899X/307/1/012003

Acknowledgment
This research reported in the paper is supported by National Natural Science Foundation of China (51505349, 51575407)

References
[1] Allen Y and Jain A 2005 Compression molding of aspherical glass lenses - a combined experimental and numerical analysis, *J. Am. Ceram. Soc.* **88**(3) 579-586.
[2] Yang C, Allen Y, Lijuan S, Klocke F, and Pongs, G 2008 Numerical simulation and experimental study of residual stresses in compression molding of precision glass optical components, *J. Manuf. Sci. E.* **130**(5) 051012
[3] Jain A, Firestone G, and Allen Y 2005 Viscosity measurement by cylindrical compression for numerical modeling of precision lens molding process, *J. Am. Ceram. Soc.* **88**(9) 2409-2414
[4] Jain A 2006 *Experimental study and numerical analysis of compression molding process for manufacturing precision aspherical glass lenses* (Doctoral dissertation, The Ohio State University)
[5] Narayanaswamy O 1971 A model of structural relaxation in glass, *J. Am. Ceram. Soc.* **54**(10) 491-498
[6] Soules T, Busbey R,, Rekhson S, Markovsky A and Burkem A 1987 Finite-element calculation of stresses in glass parts undergoing viscous relaxation, *J. Am. Ceram. Soc.* **70**(2) 90-95
[7] Bo T, Peng H, Lianguan S and Allen Y 2014 Quantitatively measurement and analysis of residual stresses in molded aspherical glass lenses, *Int. J. Adv. Manuf. Tech.* **74**(9-12) 1167-1174
[8] Bo T, Peng H and Lianguan S 2014. Measurement of Residual Stresses in Molded Glass Lenses, *Adv. Mat. Res.* **902** 144-147
[9] Bo T, Peng H, Lianguan S and Allen Y 2014 Annealing of compression molded aspherical glass lenses, *J . Manuf . Sci . E.-T . Asme* **136**(1) 011008

IOP Conf. Series: Materials Science and Engineering **307** (2018) 012004 doi:10.1088/1757-899X/307/1/012004

Design and simulation analysis of a novel pressure sensor based on graphene film

M Nie, Y H Xia and A Q Guo

Key Laboratory of MEMS of Ministry of Education, Southeast University, Nanjing, China
E-mail: mnie1980@163.com

Abstract. A novel pressure sensor structure based on graphene film as the sensitive membrane was proposed in this paper, which solved the problem to measure low and minor pressure with high sensitivity. Moreover, the fabrication process was designed which can be compatible with CMOS IC fabrication technology. Finite element analysis has been used to simulate the displacement distribution of the thin movable graphene film of the designed pressure sensor under the different pressures with different dimensions. From the simulation results, the optimized structure has been obtained which can be applied in the low measurement range from 10hPa to 60hPa. The length and thickness of the graphene film could be designed as 100μm and 0.2μm, respectively. The maximum mechanical stress on the edge of the sensitive membrane was 1.84kPa, which was far below the breaking strength of the silicon nitride and graphene film.

1. Introduction

Pressure sensor is widely used in many fields, such as industry, agriculture, medical treatment, space flight, meteorology. At present, the mainstream device is the silicon piezoresistive pressure sensor, whose sensitive movable part is made of silicon thin film. The piezoresistive pressure sensor has a lot of advantages such as high sensitivity, fast dynamic response, high accuracy, good stability, wide working temperature range, small and easy to miniaturization[1-5]. However, the fabrication technology limits the size of traditional MEMS pressure sensor structures within micro scales, and according to the material characteristics, the silicon based pressure sensor is more suitable for medium and high measurement range. Once it reaches a low measurement range, the sensitivity of such sensor will decline. However, in many application fields such as breathing, small flow, unmanned aerial vehicle, the pressure sensor needs the high sensitivity to measure low pressure, the effective method to solve this problem is to reduce the thickness of the movable thin film of the sensor. But in actually, it is difficult to manufacture the movable film to nano scale by using silicon materials. Moreover, the electrical and material properties would be changed as the reduction of dimensions of the film structure.

Graphene can be fabricated to be a very thin film, which can reduce the size of the sensor structure to one tenth of the traditional one's. Graphene is arranged in SP2 hybrid, and C=C bond endows graphene with excellent mechanical properties. Each carbon atom and around three carbon atoms are combined by using aσ bond, and the interaction of C - C bond is very strong. The connection between each carbon atom is very flexible which make it has high flexibility, and its elastic extension can achieve 20%. So, graphene film is one of the world's most hard and thin material with excellent mechanical properties, such as high coefficient of strain (about 1.8*104), which determines that it can has high sensitivity when it is used in devices[6-12].

IOP Conf. Series: Materials Science and Engineering **307** (2018) 012004 doi:10.1088/1757-899X/307/1/012004

This paper proposed a novel pressure structure, graphene film and silicon nitride were used as movable sensing film and supported film, respectively. The proposed graphene based pressure sensor has small dimension and high sensitivity in very low measurement range.

2. Process flow design of the pressure sensor

The schematic diagram of the sensor structure is shown in figure 1(h), and the fabrication process is described in two parts:

Firstly, the main processes of the silicon substrate and the supported silicon nitride film are shown in figure 1(a)-(d):

(a) A thin layer of silicon nitride film was grown on silicon substrate;

(b) Etch holes were etched;

(c) Silicon dioxide was etched;

(d) Silicon substrate was etched by anisotropic etching to create the cavity in the front side of the substrate, and completed the support structure part of the silicon nitride film.

Secondly, the main process flow of graphene film transfer to silicon substrate is shown in figure 1(e)-(g):

(e) Put graphene film onto the substrate, such as slide, and taped the heat stripping on it, then coated with a layer of PDMS and made it solidified;

(f) Cut the graphene film into appropriate size and loaded a force on the surface of the substrate, then heated up to 80 ℃ to make the graphene film fully fitted to the substrate;

(g) Took away the adhesive tape at the peeling temperature (135℃), and left the transferred graphene film on the supported silicon nitride film. After the transfer, the whole sensor structure profile was completed, which was shown in figure 1(h).

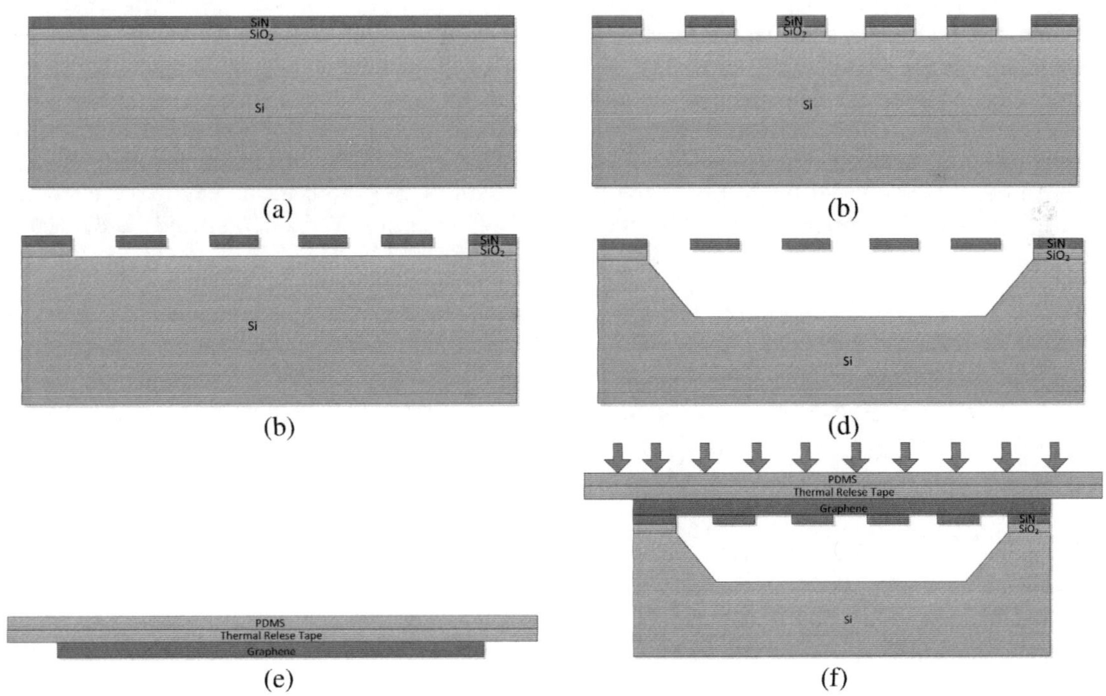

IOP Conf. Series: Materials Science and Engineering **307** (2018) 012004 doi:10.1088/1757-899X/307/1/012004

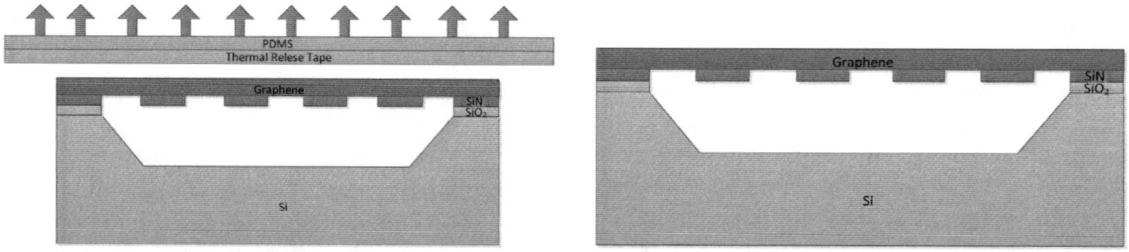

(g) (h)

Figure 1. Overview of the fabrication process flow for the proposed graphene-based pressure sensor.

Compared with the traditional processing method of silicon based pressure sensor, the proposed fabrication process used the front etching method instead of the back etching method to get the cavity of the structure, which can reduce the chip area and save the cost. It is very convenient to use the hot stripping technology to transfer the graphene film to the silicon substrate, which is commonly used in microelectronics fabrication technology.

3. Simulation and analysis of the pressure sensor

ANSYS software was used to simulate and analyze the performance of the pressure sensor structure, the movable sensitive graphene film was square, the boundary conditions were fixed around the membrane, the Young's modulus of graphene was 280GPa, Poisson's ratio was 0.288. The thickness t of the film was 0.1μm, 0.2μm, 0.3μm, 0.4μm, 0.5μm, respectively. The length L of the film was 50μm, 100μm, 150μm, respectively. The pressure was loaded from 10hPa to 60hPa.The simulated results are shown in figure 2-5, which illustrate the relationships between the maximum displacement of movable film and the loaded pressure. It can be analyzed that when the movable film length was 50μm, the displacement of movable film was too small, which would get very low sensitivity; when the movable film length was 150μm, the displacement of movable film was too large, which would make the movable membrane too fragile; when the movable film length was 100μm, the displacement of movable film was suitable for low pressure measurement. Obviously, it can be obtained from figure 5 when the thickness of the film was 0.2μm, the maximum displacement of the movable film changed between several microns, which is quite suitable dimension to be designed for the structure. Furthermore, it is shown in figure 6 that the maximum stress generated in the edge area is 1.84kPa under 60hPa loaded pressure, which is far below the fracture strength of silicon nitride and graphene film.

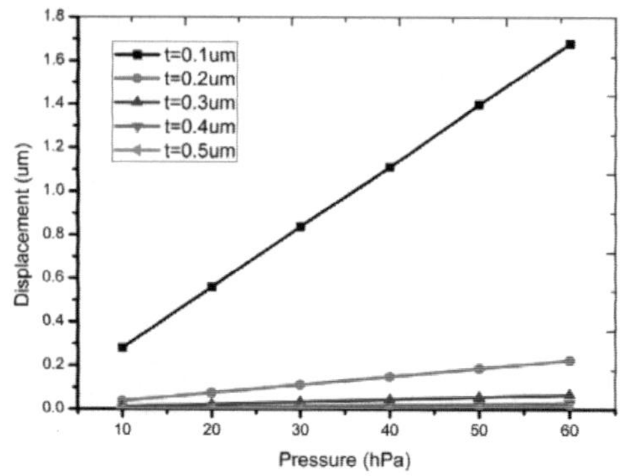

Figure 2. The relationship between the maximum displacement of movable film and the loaded pressure(L=50μm).

IOP Conf. Series: Materials Science and Engineering **307** (2018) 012004 doi:10.1088/1757-899X/307/1/012004

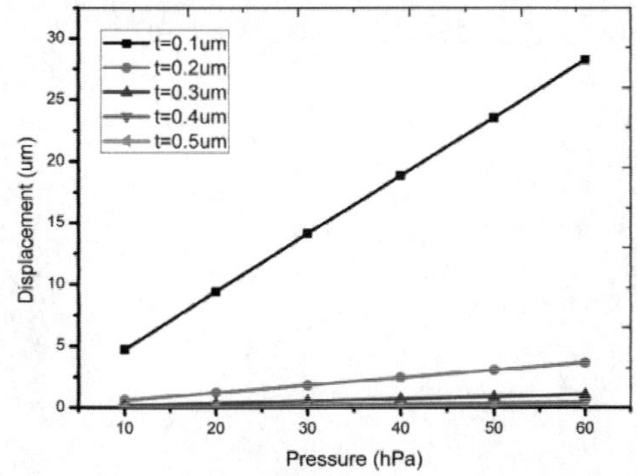

Figure 3. The relationship between the maximum displacement of the movable film and the loaded pressure(L=100μm).

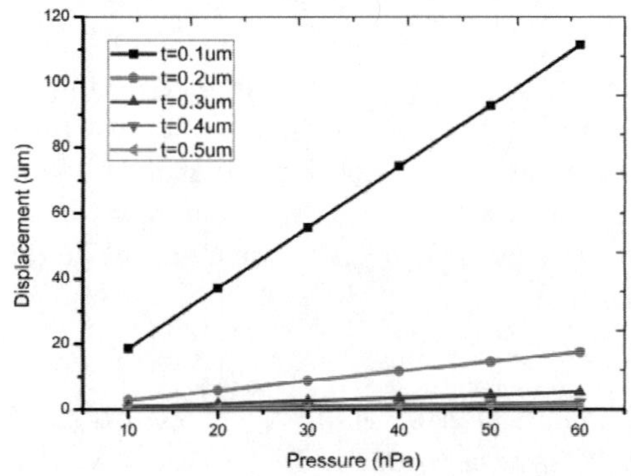

Figure 4. The relationship between the maximum displacement of the movable film and the loaded pressure(L=150μm).

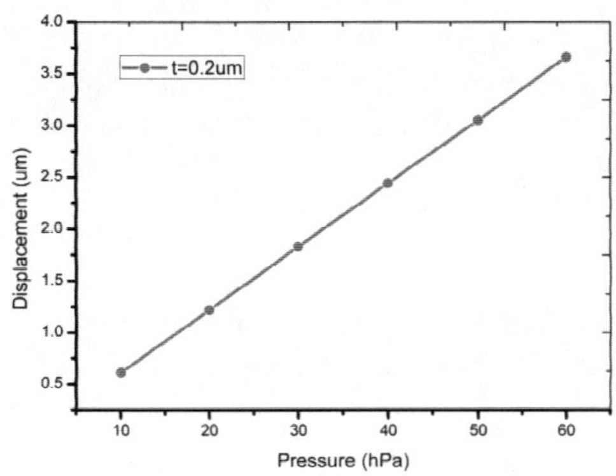

Figure 5. The relationship between the maximum displacement of the movable film and the loaded pressure(L=50μm, t=0.2μm).

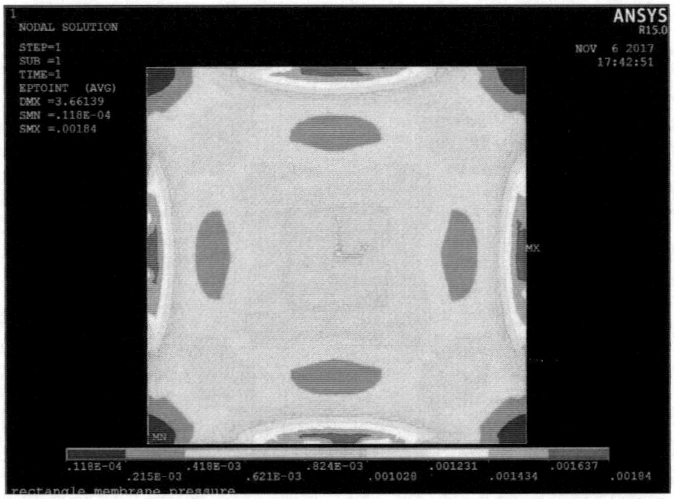

Figure 6. The stress distribution of the movable film(L=50μm, t=0.2μm).

4. Conclusion

In this paper, a novel pressure sensor structure based on graphene film as the sensitive membrane was proposed, and the fabrication processes was designed which was compatible with CMOS IC fabrication technology. The simulations have been given to analyze the maximum displacement of the sensitive film under the pressure from 10hPa to 60hPa, and an optimized structure dimension has been obtained. The design of the pressure sensor can give a meaningful guide to the low pressure measure field.

5. References

[1] Kumar S S, Ojha A K, Paut B D 2016 *Microsyst. Technol.* **22** 83
[2] Nie M, Gao Y 2017 *AIP Advances* **7** 035120
[3] Nie M, Bao H Q 2016 *AIP Advances* **6** 105302
[4] Niu Z, Zhao Y L 2014 *Review of Scientific Instruments* **85** 015001
[5] Okojie R S, Lukco D, Nguyen V, et al. 2015 *IEEE Electron Device Letters* **36** 174
[6] Tadakaluru S, Thongsuwan W, Singjai P 2014 *Sensors* **14** 868
[7] Tang Y C, Zhao Z B, Hu H, et al. 2015 *ACS Appl. Mater. Interfaces* **7** 27432
[8] Sheng L Z, Liang Y, Jiang L L, et al. 2015 *Adv. Funct. Mater.* **25** 6545
[9] Yan C Y, Wang J X, Kang W B, et al. 2014 *Adv. Mater.* **26** 2022
[10] Wang Y, Wang L, Yang T T, et al. 2014 *Adv. Funct. Mater.* **24** 4666
[11] Sakhaee-Pour A, Ahmadian M T, Vafai A 2008 *Solid State Commun.* **14 7**336
[12] Li X, Zhang R J, Yu W J, et al. 2012 *Sci. Rep.* **2**8 70

Acknowledgments

This work was supported by National Natural Science Foundation of China (Grand No. 61474023) and National Key Technology Support Program of China (Grand No.2015BAF16B01).

Simulation Analysis of 25A-Size Corrugated Metal Gasket Coated Copper to Increase Its Performance

Didik Nurhadiyanto[1], Mujiyono[2], Sutopo[3], Febrianto Amri Ristadi[4]

1 Associate Professor, Yogyakarta State University, Indonesia
2, 3, 4 Assistant Professor, Yogyakarta State University, Indonesia
E-mail: didiknur@uny.ac.id

Abstract. Previous study, researchers investigated on 25A-size corrugated metal gasket using material SUS304 and it modified. The elastic model design of corrugated metal gasket still leakage and need high axial force for tightening process. The purpose of this study is to analyze coated corrugated metal gasket by FEM simulation. Experimental result used for validating the data. The characteristic of coating material should softer than base material. The coating material will stick and fill in the surface roughness of gasket. The simulation result shows that contact width increase and contact stress decrease after the gasket coated by copper. It mean that the gasket performance increase to stop the leakage.
Keywords: *corrugated, metal gasket, coating, copper, simulation*

1. Introduction

The need of gasket for sealing in oil and mining industries is very high. Asbestos-containing sealing materials is superior in preventing leakage in piping, and it used in several industries. However, it is totally dangerous because chemicals effect and it can cause serious illness. In the world, the production and usage of asbestos materials were banned [1]. But, the need of gasket material to replacement of asbestos material is expensive and have short period of live [2, 3, 4].

Saeed and friends [5] investigated super seal gasket of 25A size metal gasket. The gasket formed the circumferential annular lips. The gasket performance optimization based on the contact width and contact stress. It is an important design parameters. The contact width limit that the leakage did not occur studied by Haruyama et al [6]. They used water pressure test to measure the leakage rate.

Nurhadiyanto et al [7] designed the optimum metal gasket 25A size by FEM simulation based on contact width and contact stress effect considering forming effect. Plastic design of gasket performs better than elastic design. Haruyama et al [8] investigated the effect of surface roughness on leakage. The leakage performance examined by flange surface roughness. Also the axial force examined by flange surface roughness. The low surface roughness is better than high surface roughness for leakage performance.The real contact width result using helium leakage measurement proposed by Haruyama et al [9] and the real simulation result of contact width and contact stress proposed by Nurhadiyanto et al [10]. Both of study considered of surface roughness of flange. The local contact stress has been studied by FEM analysis and they clarified the distribution of contact stress.

The result of previous research that used a single material SUS304 was clarified. The corrugated metal gasket elastic design contact with high surface roughness flange still leaking because of only partial contact occurs. Gasket elastic design and plastic design which contact with surface roughness of flange 1.5 and 2.5 μm is good gasket, but it still used need high axial force. Haruyama et al. [11] studied the three layers corrugated metal gasket. The layers consist of softer-base-softer materials. The softer material are Nickel, Cupper or Aluminum.The average contact stress for three layers gasket lower than single layer and the contact width for three layers gasket was wider than single layer. Karohika et al. [12] investigated the contact width and contact stress of three layers corrugated metal gasket. They concluded that thickness ratio decreased will increase contact width but the average contact stress will decrease. Both article [11] and [12] used simulation analysis, they did not compare the simulation and experimental result. Also, the leakage performance did not clear yet. The springback effect and the deformation every peak of the three layers gasket did not discuss clearly in this research. The influence of thickness layer investigated by simulation in research but it did not investigated by experimentally.

In this light, this study aims to reengineering of corrugated metal gasket by coating using Copper (Cu) material. The Cu material is softer than base material (SUS304). The coating material will stick on the outer surface of metal gasket. The characteristic of coating material should softer than SUS304. When metal gasket contact with flange, after tightening process, coating material fill up the rough surface of flange.

2. Material and Methods

The SUS304 used as gasket material used in present research. This material is very effective in high temperature and pressure. The property of this SUS304 determined using test of tensile based on JISZ2241 standard [13]. Ultimate stress (σ) is 210GPa, elasticity modulus (E) is 398.83MPa and tangential modulus respectively is 1900.53MPa. The stress-strain diagram of SUS304 shown Figure 1. Table 1 shows the characteristic of SUS304 material.

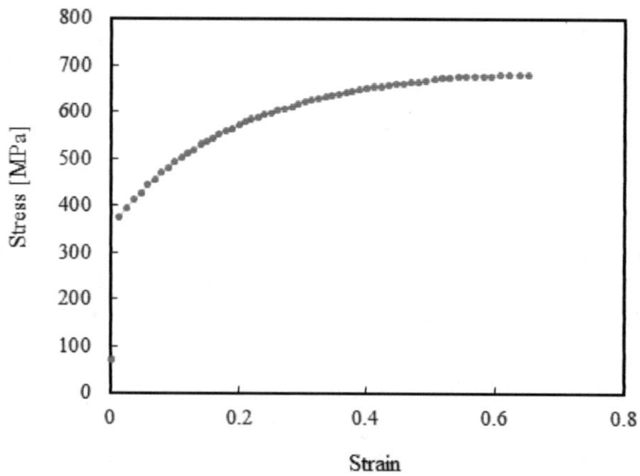

Figure 1. SUS304's stress-strain diagram

Table 1. The material of SUS304

Properties	Value
Yield stress	398.83MPa
Elasticity modulus	210GPa
Tangential modulus	1900.53MPa
Poisson ratio	0.3

The coating material was Cu. Copper is softer than SUS304 and resistant to temperature and chemical material. The characteristic of copper and SUS304 shown in Figure 2.

The simulation process is tightening simulation. The simulation using MSC Marc software [14]. The various design of gasket model generated using four basic steps. Firstly, the 2-D models paramerization using Solid Work software (IGES file). Secondly, using Hypermesh for automatic meshing by batch command to produced NAS file. The next step was computation of preprocessing and running the model. Model run on MSC Marc software. The last step is post-processing in batch mode optimization. The analysis result could be generated in TXT file after the completely of FEM analysis. Each convex position, the stress value, contact status and body force at every time are the output result.

IOP Conf. Series: Materials Science and Engineering **307** (2018) 012005 doi:10.1088/1757-899X/307/1/012005

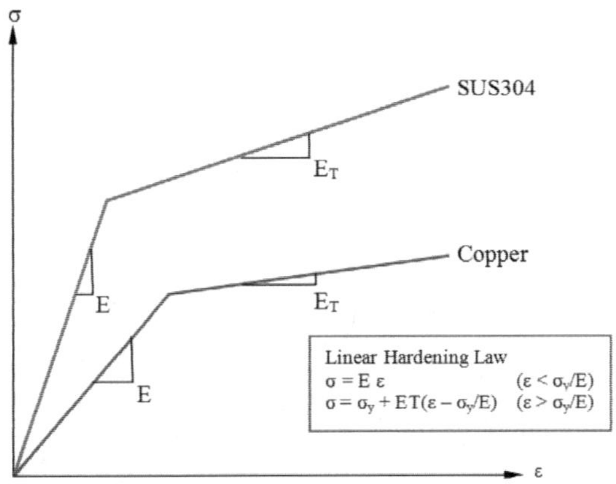

Figure 2. Linear strain hardening model for SUS304 and copper material

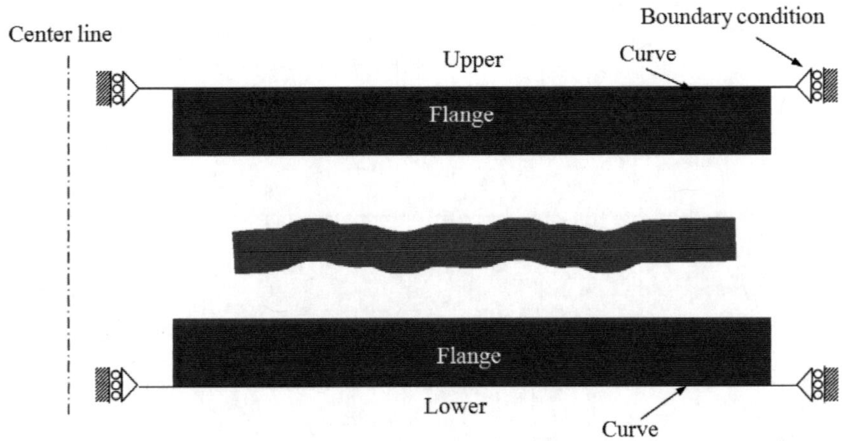

Figure 3. Simulation setting up for coating gasket

Figure 3 shows the set up simulation for coating gasket. The flange, gasket material and coating material assumed deformable bodies. The curve assumed as rigid body. The gasket did not move, but the lower dies moved up and upper gasket moved down.

3. Result and Discussion

Figure 4 shows the contact width and average contact stress result of gasket coated 3μm Cu and standard gasket. The thicness ratio Cu-base-Cu is 3μm-1.5mm-3μm. The contact width increase when the axial force increase. In low axial force (0-70kN) contact width for both type of gasket are similar even there is different cotact width when the axial force in 40-65kN. Contact width is very different when the axial force start 70-120kN. The contact stress also litle bit different when the axial force 40-120kN. This condition caused by the outer layer of gasket. The outer gasket coated by 3μm thickness of Cu, it stick and filled the surface roughness of flange. So, the contact width will increase. The disadvantage is contact strees will decrease. It because of the relationship between stress, force and contact area. If the contact witdh increase so the contact stress will decrease, see equation (1).

$$\sigma = \frac{F}{A} \qquad (1)$$

where σ = stress ($\frac{N}{m^2}$), F = force (N) and A = contact area (m^2)

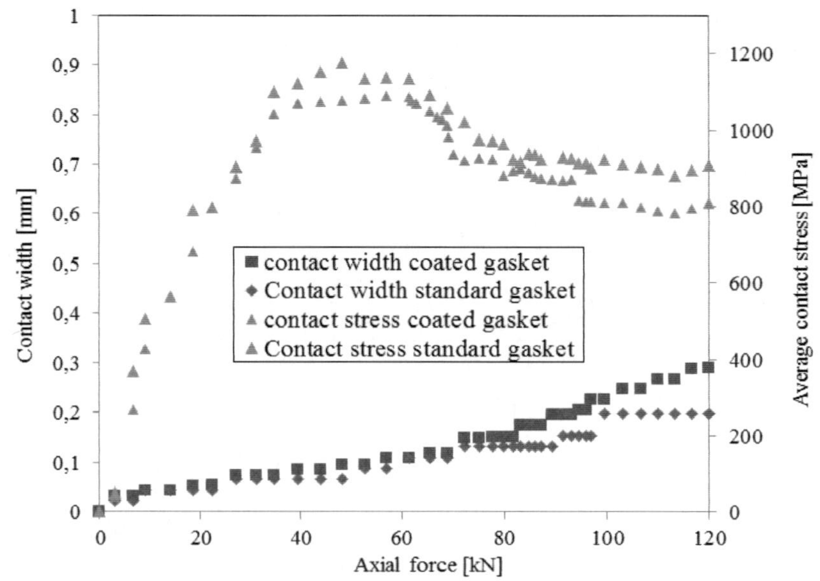

Figure 4. Simulation result of coating tickness 3µm

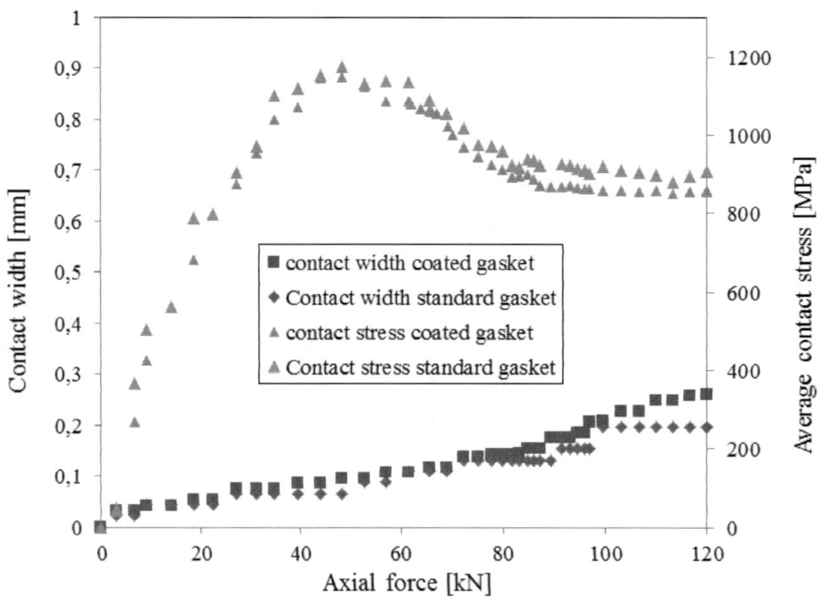

Figure 5. Simulation result of coating thickness 2µm

Figure 5 shows the simulation result of modified gasket which thicness ratio Cu-base-Cu is 2µm-1.5mm-2µm compared standard gasket. In low axial force (0-70kN) contact width for both type of gasket are similar. Contact width is very different when the axial force start 70-120kN. The contact sress also litle bit different when the axial force 40-120kN. This condition caused by the outer layer of gasket. The outer gasket coated by 2µm thickness of Cu, it stick and filled the surface roughness of flange. So, the contact width will increase.

There is a differences of contact width and contact stress between 2μm and 3μm thickness of Cu coated. The contact width of gasket Cu coated 3μm is wider than 2μm when contact with flange. The contact stress of gasket Cu coated 3μm than 2μm when contact with flange.

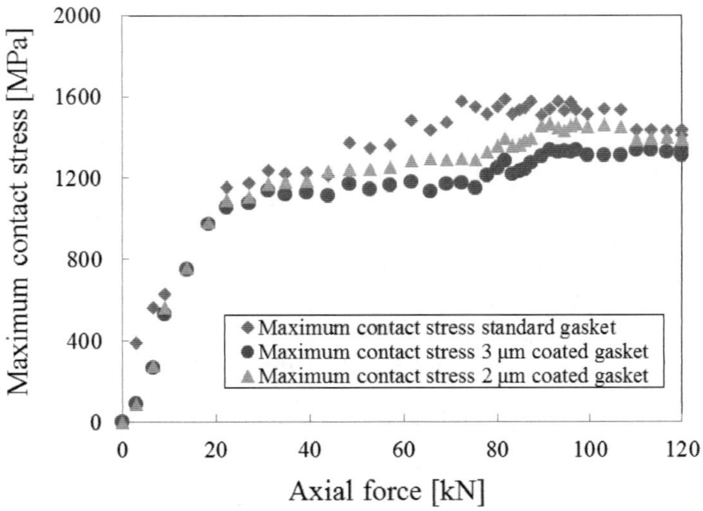

Figure 6. Maximum contact stress for all type of gasket

Figure 6 shows the relationship between axial force-maximum contact stress for standard gasket and coating gasket which has thickness 2μm and 3μm. The axial force start at 0kN until 120kN, it because the experimental axial force was taken 40kN, 60kN, 80kN, 100kN and maximum 120kN. The maximum increase very significant when the axial force 0-30kN, after that the maximum contact stress increase slowly or tend to be constant. The maximum contact stress shows the similary increase for gasket standard, 2μm and 3μm Cu coated when the axal force 0-30kN. When the axial force 30kN above, there is differences maximum contact stress value for gasket standard, 2μm and 3μm Cu coated. The maximum contact stress of gasket standard has highest maximum contact stress than coating gasket. The maximum contact stress of gasket 2μm Cu has higher 3μm Cu. The reason is contact width of gasket standard is lower than 2μm and 3μm Cu coated, using equation (1) if the contact width increase so the contact stress will decrease.

Table 2. Leakage measurement result

Axial force (kN)	Time (second)	Pressure (MPa)	Standard gasket	Cu Coated	
				2μm	3μm
40	600	10	Leaked	Leak	Leak
60	600	10	Leaked	No Leaked	No Leaked
80	600	10	No Leaked	No Leaked	No Leaked
100	600	10	No Leaked	No Leaked	No Leaked
120	600	10	No Leaked	No Leaked	No Leaked

Table 2 shows the experimental result for standard gasket and coated gasket for compared simulation and experimental result. The experimetal using water pressure test. The standard gasket did not leakage when axial force 80kN, 100kN and 120kN. Coated gasket 2μm and 3μm thickness of Cu did not leak when the axial force 60kN, 80kN, 100kN and 120kN. Coated gasket with Cu both thickness 2μm and 3μm has higher performance than standard gasket. This result a good agreement with simulation result that contact width increase when the gasket coated by Cu 2μm and 3μm thickness.

4. Conclussion

The conclussion as follow:

a. Contact width of coated gasket is wider than standard gasket and contact stress of coated gasket is lower than standard gasket.
b. The simulation and experimental result is a good agreement.
c. A coated gasket is better than standar gasket based on simulation and experimental result.

References

[1] Asbestos is hard to replace as Japan heads for ban, *Sealing Technology* 1 2006 1.
[2] Estrada H, Parsons I D 1999 Strength and Leakage Finite Element Analysis of GFRP Flange Joint *International Journal of Pressure Vessel and Piping* Vol 76 p 543-550.
[3] Krishna M M, Shunmugam M S, Prasad N S 2007 A Study on the Sealing Performance of Bolted Flange Joints with Gaskets using Finite Element Analysis *International Journal of Pressure Vessel and Piping* 84 p 349-357.
[4] Ross E, Kockelmann H, and Hann R 2002 Gasket Characteristics for Design of Bolt Flange Connections of Metal-to-metal Contact Type *International Journal of Pressure Vessel and Piping* 79 p 45-52.
[5] Saeed H A, Izumi S, Sakai S, Haruyama S, Nagawa M, and Noda H 2008 Development of New Metallic Gasket and its Optimum Design for Leakage Performance, *Journal of Solid Mechanics and Material Engineering* Vol 2: 1 p 105-114.
[6] Haruyama S, Choiron M A, and Kaminishi K 2009 A Study of Design Standard and Performance Evaluation on New Metallic Gasket, Proceeding of the 2nd International Symposium on Digital Manufacturing, p 107-113.
[7] Nurhadiyanto D, Choiron M A, Haruyama S, and Kaminishi K. 2012 Optimization of New 25A-size Metal Gasket Design Based on Contact Width Considering Forming and Contact Stress Effect *World Academy of Science, Engineering and Technology* Vol 6 p 661-665.
[8] Haruyama S, Nurhadiyanto D, Choiron MA, and Kaminishi, K. 2013 Influence of Surface Roughness on Leakage of New Metal Gasket *International Journal of Pressure Vessels and Piping*, Vol 111-112 p 146-154.
[9] Haruyama S, Nurhadiyanto D, and Kaminishi K. 2014 Contact Width Analysis of Corrugated Metal Gasket based on Surface Roughness, *Advanced Material Research*, Vol. 856 p 92-97.
[10] Nurhadiyanto D, Haruyama S, Kaminishi K, Karohika IMG, Mujiyono 2015 Contact Stress and Contact Width Analysis of Corrugated Metal Gasket *Applied Mechanics and Materials*, Vol 799-800 p 765-769.
[11] Haruyama S, Karohika, IMG Sato, A, Nurhadiyanto, D, and Kaminishi K.2016 *Development of 25A-Size Three-Layer Metal Gasket by Using FEM Simulation, International Journal of Mechanical, Aerospace, Industrial, Mechatronics and Manufacturing Engineering*, Vol 10 p 527-533.
[12] Karohika IGM, Haruyama S, Kaminishi K, Oktavianty O., Nurhadiyanto D, 2017 Analysis of Contact Width and Contact Stress of Three-Layer Corrugated Metal Gasket, *International Journal of Mechanical, Aerospace, Industrial, Mechatronics and Manufacturing Engineering*, Vol 11 p 870-879.
[13] JIS Z2241 1998 *Method of Tensile Test for Metallic Materials*, Japanese Standards Association.
[14] MSC Marc. 2007 *User manual.*

Acknowledgments

The author wishing to acknomledge Ministry of Research, Technology and Higher Education of the Republic of Indonesia and Yogyakarta State University for giving us a change to pursue the research grant. The research financial supported by Director General of Research, Technology and Higher Education and Yogyakarta State University.

IOP Conf. Series: Materials Science and Engineering **307** (2018) 012006 doi:10.1088/1757-899X/307/1/012006

The research of hourglass worm dynamic balancing simulation based on SolidWorks motion

Wang Zhuangzhuang[1], Yang Jie[1], Liu Pingyi[1] and Zhao Junpeng[2]
[1]College of Engineering, China Agricultural University Beijing 100083, China
[2]Beijing Zhongzhuo Fire Fighting Equipment Co.,Ltd Beijing 101300,China

E-mail: liupingyi@cau.edu.cn
Abstract. Hourglass worm is extensively used in industry due to its characteristic of heavy-load and a large reduction ratio. Varying sizes of unbalanced mass distribution appeared in the design of a single head worm. With machines developing towards higher speed and precision, the vibration and shock caused by the unbalanced mass distribution of rotating parts must be considered. Therefore, the balance grade of these parts must meet higher requirements. A method based on theoretical analysis and SolidWorks motion software simulation is presented in this paper; the virtual dynamic balance simulation test of the hourglass worm was carried out during the design of the product, so as to ensure that the hourglass worm meet the requirements of dynamic balance in the design process. This can effectively support the structural design of the hourglass worm and provide a way of thinking and designing the same type of products.

1. Introduction

There are two main methods for correcting the dynamic unbalances of rotate parts: One way is to measure the dynamic force of the support to correct the amount of unbalances. Another way is to measure the vibration of the support to correct the amount of unbalance [1]. The above two methods, both using a certain test system to detect the existence dynamic unbalance of rotate parts after the completion of manufacture, are unable to provide guidance for the design of the dynamic balance of the parts during the design phase. In addition, for single-headed worms, single-headed screws, single-head conveyor screws and other non-axisymmetric complex surface rotary parts, the size and phase of the unbalanced mass calculation is extremely complex. What has been the focus of designers' attention is how to effectively solve the dynamic unbalance of such parts.

Therefore, in order to effectively solve the problem of dynamic unbalance of non-axisymmetric complex surface rotating parts in the design stage, a virtual dynamic balance test platform is designed and a dynamic balance design method based on SolidWorks motion is proposed in this paper. A single-head hourglass worm is used as the experimental object as one of the non-axisymmetric complex surface rotating parts, and a simulation experiment is carried out to verify and analyze the method.

2. Design and test method of virtual test platform

2.1. Composition and structure of virtual test platform

In three-dimensional software, the frame, spring, swing-rod, column and the motor are constructed, as shown in Figure 1.

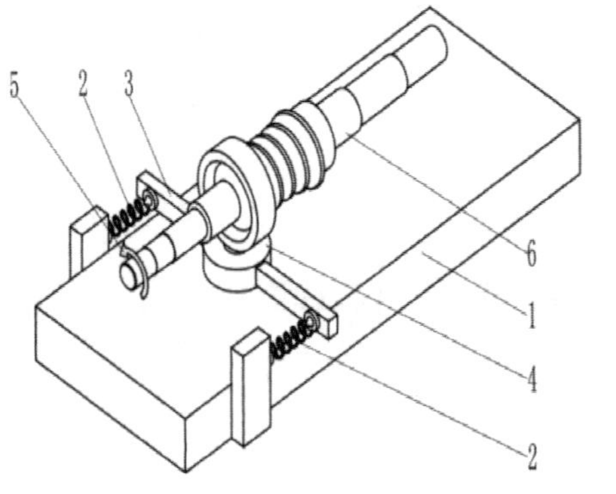

Figure 1. The structure of the test platform [2]
1—frame 2—spring 3—swing-rod 4—column
5—motor 6—rotary parts

The midpoint of the swing-rod is fixedly connected to the middle of the column. The lower end of the column is connected to the frame and the column axis is perpendicular to the horizontal plane of the frame. The upper end of the column is set with a revolute joint and the axis of revolute joint is perpendicular to the swing-rod. Two identical sets of springs are both fixed to the frame on the one end and the other end is mounted vertically at both ends of the swing-rod. The length of the swing-rod is s and the stiffness of the spring is k. The two springs are parallel and the distance is s. The motor is located on the revolute joint of the column. The motor drives the unbalance of hourglass worm to rotate around the axis of revolute joint and the speed is ω. Establish a fixed coordinate system σ (C; X, Y, Z) on the frame, with column axis being the Z-axis and intersect to the horizontal plane of the frame at point C. The XCY plane is on the horizontal plane of the frame. The Y-axis is collinear with the initial installation of the axis of revolute joint; the X-axis direction is the vertical direction of the C-point and Y-axis on the horizontal plane of the frame [2].

2.2. Experiment method

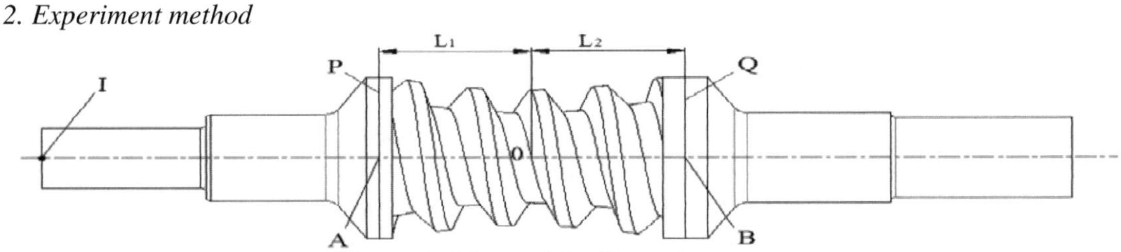

Figure 2. The model of hourglass worm

The three-dimensional model of the hourglass worm is established in the SolidWorks as the test part, as shown in Figure 2. Two planes are selected in the axial direction of the rotating parts suitable for increasing and reducing the weights. The left balance plane P is on the left side of middle of the throat and the distance is L_1, and the right balance plane Q is on the right side of middle of the throat and the distance is L_2. Plane P and plane Q are perpendicular to the axis of the rotating part and intersects the axis at point A and B respectively. Optionally select a shoulder and set a mark line along its radial direction as I.

The first step in the hourglass worm simulation test is to balance the right balance plane Q. The following three conditions are required to ensure the hourglass worm to be balanced during the simulation test. First of all, the axis of the unbalanced hourglass worm is aligned with the axis of revolute joint. Secondly, the left balance plane P is overlap with the axis of column. Finally, line I has to be made parallel to the plane of the frame at the beginning of installation.

The motor drives the hourglass worm with a rotation speed of ω, X_1 is the maximum amplitude of the point B in the X-axis direction, φ_1 is the phase angle corresponding to the maximum amplitude.

The mass-radius product G_1 can be calculated by equation (1).

$$G_1 = \frac{ks^2X_1}{2(L_1+L_2)^2 w^2} \tag{1}$$

According to the calculation results, there are two ways to balance the right balance plane Q. The first way is to add a counterweight to the phase angle of φ_1+180° on the right balance plane Q and the mass-radius product is G_1. Another way is to reduce a counterweight to the phase angle of φ_1 on the right balance plane Q and the mass-radius product is $-G_1$ [2].

The next step after balancing the right balance plane P is to balance the left balance plane Q. Just like balancing the right balance plane Q, the three conditions are also required to install the hourglass worm to be balanced during the simulation test. First of all, the axis of the unbalanced hourglass worm is aligned with the axis of revolute joint. Secondly, the right balance plane Q is overlap with the axis of column. Finally, line I has to be make parallel to the plane of the frame at the beginning of installation.

The motor drives the hourglass worm with a rotation speed of ω; X_2 is the maximum amplitude of the point A in the X-axis direction; φ_2 is the phase angle corresponding to the maximum amplitude.

Use the same way as to balance the right balance plane Q to measure the phase angle φ_2 and the maximum amplitude of the point A in the X-axis direction. Then, calculate the value of mass-radius product G_2. Reduce or add a counterweight to the left balance plane P based on the calculation.

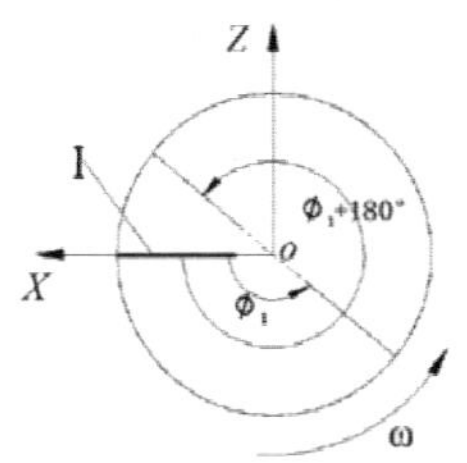

Figure 3. The phase diagram of the right balance plane Q[2]

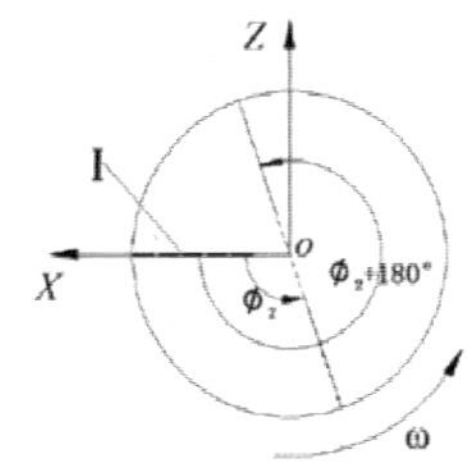

Figure 4. The phase diagram of the left balance plane P[2]

3. Dynamic balance simulation analysis

Set the parameters for the simulation: the motor speed ω as 60r/min, spring stiffness k as 1N /mm, the frames per second as 200, simulation time as 5s (five cycles), L_1=48mm, L_2=50mm, s=270mm.

First of all, ensure that the axis of the unbalanced hourglass worm is aligned with the axis of revolute joint. The left balance plane P must overlap with the axis of column. Line I was parallel to the plane of the frame at the beginning of installation. Measure the maximum amplitude X_1 of the point B in the X-axis direction. The result of X_1 is illustrated in the Figure 5.

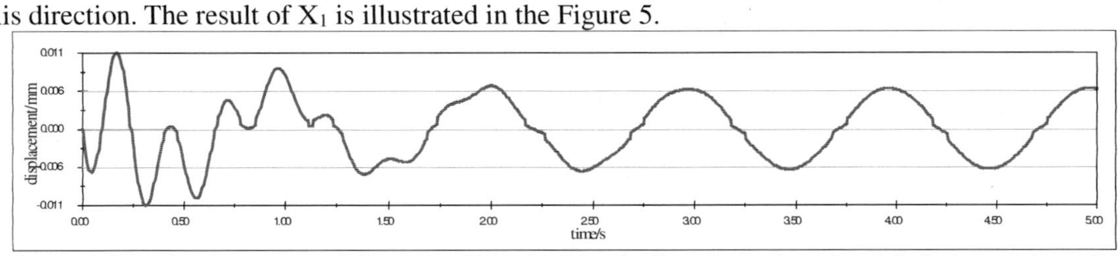

Figure 5. The displacement of the point B in the X-axis direction when the right balance plane Q is not balanced

Select the last stable period to calculate the maximum amplitude X_1. With the assistance of SolidWorks, it is easy to know that the maximum value of displacement is 0.006019mm at 4.965S, and the minimum value of displacement is -0.006011mm at 4.46S. The maximum amplitude is X_1=0.01203mm, the phase angle $\varphi_1=360°\times0.965=347.4°$, and the mass-radius product is $G_1 = -1157.69\,g\cdot mm$.

Remove the counterweight in the right balance plane Q based on the calculated value to balance the right balance plane Q and keep the installation position unchanged. Measure the displacement of the point B in the X-axis direction again. The results are illustrated in Figure 6.

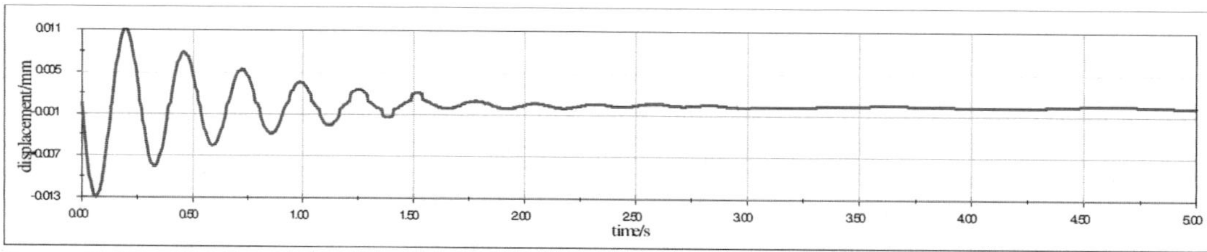

Figure 6. The displacement of the point B in the X-axis direction when the right balance plane Q is balanced

The maximum amplitude is 0.00024mm. It can be easily found that the maximum amplitude was down by 98% compared to the maximum amplitude before the right balance plane Q was balanced. Therefore, it was concluded that the effect of balancing the right balance plane Q is very significant.
Secondly, it becomes the right balance plane Q overlap with the axis of column while other installation requirements are unchanged. Measure the maximum amplitude X_2 of the point A in the X-axis direction. The results of X_2 were illustrated in the Figure 7.

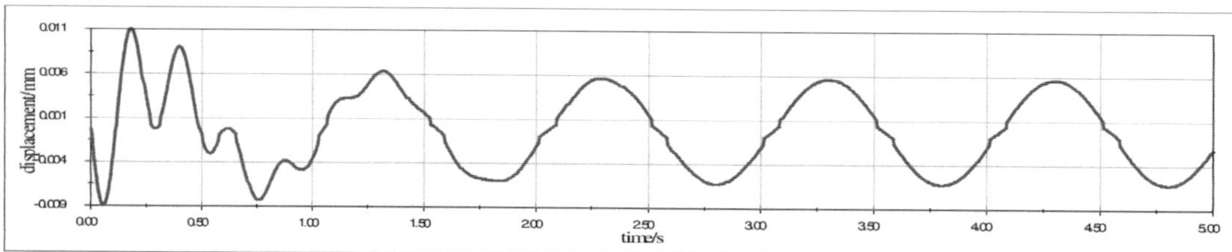

Figure 7. The displacement of the point A in the X-axis direction when the right balance plane Q is not balanced

Select the last stable period to calculate the maximum amplitude X_2 as well. It is easy to know that the maximum value of displacement is 0.005994mm at 4.305S, and the minimum value of displacement is -0.005993mm at 4.805S. The result of maximum amplitude is X_2=0.01187mm, the phase angle $\varphi_1=360°\times0.3055=109.8°$, and the mass-radius product is $G_1 = -1153.5 g\cdot mm$.
Remove the counterweight in the left balance plane P based on the calculated value and keep the installation position unchanged. Measure the displacement of the point A in the X-axis direction again. The results were illustrated in the Figure 8.

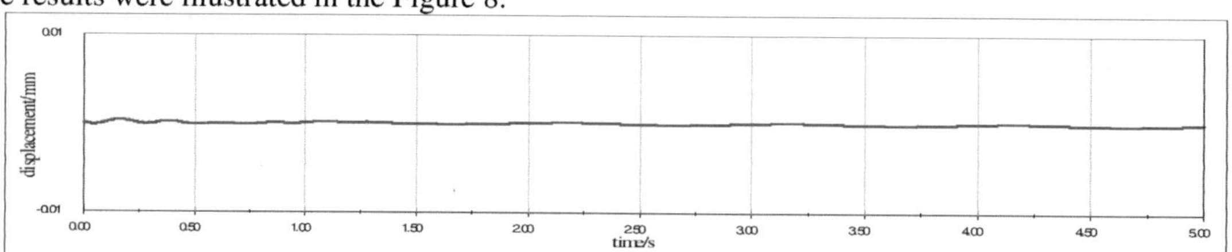

Figure 8. The displacement of the point A in the X-axis direction when the right balance plane Q is balanced

The maximum amplitude is 0.000198mm. It can be easily found that the maximum amplitude was down by 98.3% compared with the maximum amplitude before the left balance plane P was balanced. Thus, it was concluded that the effect of balancing the left balance plane P is very significant as well.

4. Balance check calculation for the balanced worm

m=5353g is the mass of the balanced hourglass worm and the coordinates of the centroid is (-0.0011, 0.0063, 16.4102). The distance between the centroid and the left balance plane P is $a=L_1-16.4102=31.5898$mm. $b=L_2+16.4102$mm=66.4102mm is the distance between the centroid and the right balance plane Q. Some other parameter information is as follows: the motor speed ω is 6.28rad/s; the dynamic balance grade of hourglass worm G is 6.3; the dynamic balance precision of hourglass worm G is A=6.3mm/s [3].
The total permissible residual unbalance was calculated by equation (2).

$$U = \frac{Am}{\omega} = \frac{6.3 \times 5353}{6.28} = 5370 g \cdot mm \quad (2)$$

The permissible residual unbalance of left balance plane P was calculated by equation (3).

$$U_1 = \frac{b}{(a+b)} U = \frac{66.4102}{98} \times 5370 = 3639 g \cdot mm \quad (3)$$

The permissible residual unbalance of right balance plane Q was calculated by equation (4).

$$U_2 = \frac{a}{(a+b)} U = \frac{31.5898}{98} \times 5370 = 1730.99 g \cdot mm \quad (4)$$

The amount of unbalance of the left balance plane P is U_1' after the hourglass worm is balanced. Equation (5) shows the relationship between U_1' and U_1.

$$U_1' = \frac{3639}{6.28^2} = 92 g \cdot mm < U_1 \quad (5)$$

The amount of unbalance of the right balance plane Q is U_2' after the hourglass worm is balanced. Equation (6) indicates the relationship between U_2' and U_2.

$$U_2' = \frac{1730}{6.28^2} = 43.8 g \cdot mm < U_2 \quad (6)$$

According to the calculation of the above check, the balanced hourglass worm satisfied the permissible unbalance requirements.

5. Conclusion

（1）A virtual dynamic balance simulation test method is proposed. It can effectively solve the problem of single-headed worm, single-headed screw, single-head extrusion screw, single-head conveyor spiral and other non-axisymmetric complex surface rotary parts dynamic balance at design stage.

（2）A virtual dynamic balance test platform was established. A platform consisting of a frame, spring, swing rod, column and a motor was built in three-dimensional software. This platform can measure and calculate the mass-radius product and corresponding phase of the simulation experiment.

（3）The hourglass worm is used to simulate experiment in the virtual dynamic balance test platform. The results show that the maximum amplitude of point B and point A decrease by 98% and 98.3% respectively after the virtual dynamic balance simulation test method is used to balance the worm. Therefore, this method meets the design requirements and can provide guidance for the design of similar products.

6. References

[1] Li X F, Zheng L X, and Liu Z X 2013 *J.Vibration.measurement&Diagnosis.***33** 565-70 & 721

[2] Li H T, Yang J, Rui C J, Wei W J and Liu P Y 2017 *Non-axisymmetric complex curved surface rotating parts virtual dynamic balancing machine and dynamic balancing design method.* China Patent.10030636.4

[3] Wang H Y 1988 *Rotor Balance Technology and Balancing Machine* (Bei Jing: China Machine Press)

[4] Shen Y S 2015 *Theory of Machines and Mechanisms* (Bei Jing: Tsinghua University Press)

Acknowledgments

The work in this paper was fully supported by National Natural Science Foundation of China under Grant No. 51475460.

IOP Conf. Series: Materials Science and Engineering **307** (2018) 012007 doi:10.1088/1757-899X/307/1/012007

Design of a Novel Electro-hydraulic Drive Downhole Tractor

Delei Fang[1*], Jianzhong Shang[1], Junhong Yang[1], Zhuo Wang[1] and Wei Wu[1]

[1] National University of Defense Technology, College of Mechatronic Engineering and Automation, Changsha, 410073, China
Corresponding author e-mail: fangdelei365 @126.com

Abstract. In order to improve the traction ability and the work efficiency of downhole tractor in oil field, a novel electro-hydraulic drive downhole tractor was designed. The tractor's supporting mechanism and moving mechanism were analyzed based on the tractor mechanical structure. Through the introduction of hydraulic system, the hydraulic drive mechanism and the implementation process were researched. Based on software, analysis of tractor hydraulic drive characteristic and movement performance were simulated, which provide theoretical basis for the development of tractor prototype.

1. Introduction

With the rapid development of horizontal well in petroleum exploration, it's becoming more and more difficult to carry down logging and other instrument to well [1-2]. As the tool for transporting the underground instrument, the downhole tractor has been paid more and more attention and developed gradually [3].

The downhole tractor is also called horizontal well crawling robot, which can carry logging and drilling instrument to move in horizontal well, and the tractor power usually comes from itself or the drilling fluid[4]. In 1996, the cable wheel tractor was developed by Statoil, Maritime well services and Welltec. As the two most advanced tractors, SmarTract and Sondex are playing an important role in horizontal well development[5-6].

However, in the face of the complex oil well environment, the tractor has two key problems, how to raise the traction ability and how to move fast and steadily inside the pipe. Currently most tractors are driven by motors, which simplify the design, but reduce the drag ability and the speed. Some tractors are hydraulic driven, which selects the servo valve and proportional valve as the control component. Due to the stringent requirements of the hydraulic medium purity, this type increases the tractor operating costs and the accidents in the hydraulic system. At the same time, the small opening area of traditional valve usually causes the pressure loss and produces so much heat in the system to reduce the reliability of the tractor. Therefore, how to improve the traction ability, realize the steady walking and ensure operation security have become the key problems to develop the tractor.

In this paper, a new type of electro-hydraulic drive downhole tractor is proposed based on the analysis of the product and research before. The paper firstly introduces the mechanical structure of the tractor, analyzes the locking principle of the supporting mechanism and the moving principle of walking mechanism. Then the hydraulic system of the tractor is introduced and its hydraulic drive and control mechanism were analyzed. Finally, the walking process of the tractor is studied based on software, which researched the characteristics of hydraulic drive and the performance of the whole machine.

2. Mechanical Structure and Moving Principle

IOP Conf. Series: Materials Science and Engineering **307** (2018) 012007 doi:10.1088/1757-899X/307/1/012007

The mechanical structure of the horizontal well tractor is shown in **figure 1**, including the main body of the tractor and two moving mechanism modules which contain reciprocating cylinder 1, reciprocating cylinder 2 and supporting cylinder respectively. The cylinder body of supporting cylinder is connected to the piston rod of reciprocating cylinder 1. The piston rod of reciprocating cylinder 2 and the piston rod of the supporting cylinder are hinged to the supporting arm respectively.

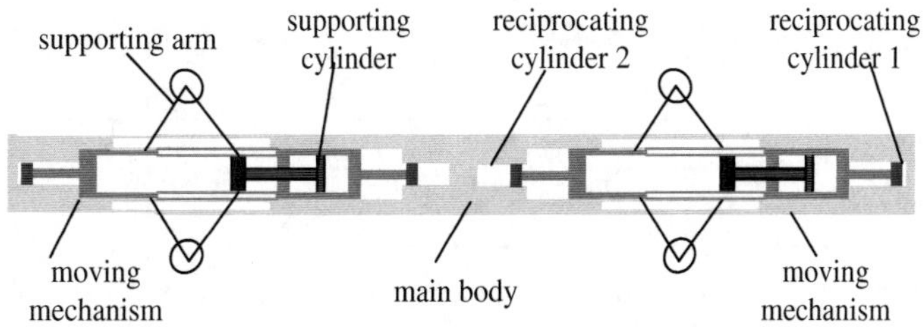

Figure 1. Schematic diagram of tractor structure

2.1. Locking Principle of Supporting Arm

The mechanical locking principle of the supporting arm is shown in **figure 2**. The *CAM* is in contact with the pipe wall at point *B* under the torsional spring pretension force, *O* is the rotation axis of the *CAM*, *F* is the positive pressure of the *CAM* on the wall, α is the angle between *OB* and positive pressure *F*, β is the equivalent friction coefficient of the roller and the inner wall of the pipe. As $\tan\alpha < \beta$ is satisfied, when the mechanism is under the action of F_1, the *CAM* turns away from the pipe wall, and the mechanism can move to the right overcoming the friction resistance. When the mechanism is under the action of F_2, *CAM* is rotating in the direction of the wall of extrusion, and the positive pressure and friction between the *CAM* and the wall is increasing, then the mechanism locks the inner wall of the pipe.

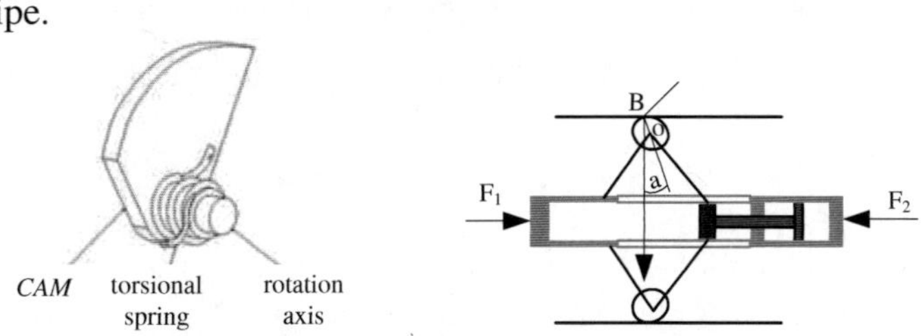

（a）locking structure （b）locking principle

Figure 2. Diagram of supporting arm locking principle

2.2. Continuous Propulsion Principle

Figure 3 shows the schematic diagram of the continuous propulsion principle of downhole tractor. In initial state of **figure 3(a)**, supporting arms are located at both ends of main body.

Figure 3 (b) is the work step 1, the both supporting arms move toward the middle. As the right supporting arm is pulled by the left force, its *CAM* turns clockwise and locks the wall. Under the reaction force, the main body of the tractor is pulled by the right reciprocating cylinder 1 and moving forward ΔL. As the left supporting arm is pulled by the right force, its *CAM* turns counterclockwise and releases the wall, the left supporting arm moves towards the right ΔL relative to body.

Figure 3 (c) is the work step 2, the both supporting arms move toward the opposite. As the left supporting arm is pulled by the left force, its *CAM* turns clockwise and locks the wall. Under the reaction force, the main body of the tractor is pulled by the right reciprocating cylinder 2 and moving .As the right supporting arm is pulled by the right force, its *CAM* turns counterclockwise and releases the wall, the right supporting arm moves towards the right relative to body.

IOP Conf. Series: Materials Science and Engineering **307** (2018) 012007 doi:10.1088/1757-899X/307/1/012007

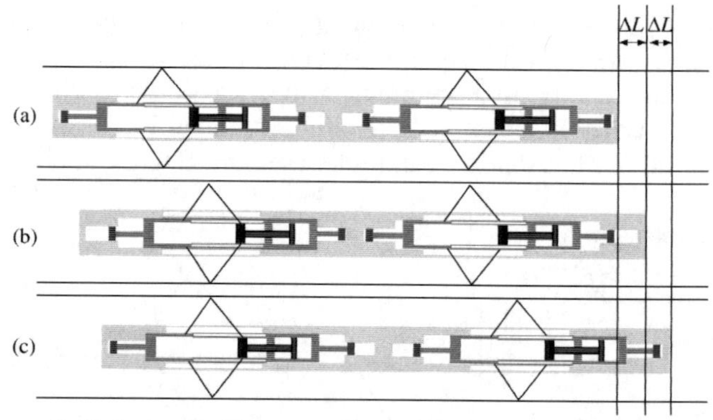

Figure 3. Schematic diagram of continuous propulsion principle

It can be seen that the main body of the tractor moves forward two ΔL at work step 1 and step 2, so that the rapid and efficient movement of the tractor can be realized.

3. Hydraulic System and Control Scheme

The hydraulic system is shown in **figure 4**, including pump 4, the overflow valve 41 and several high frequency switch valves, *SV* 5i (i=1,2,3…8). The switch valve loop unit consists of two reciprocating hydraulic branches #1, #2 and two supporting hydraulic branches #3, # 4.

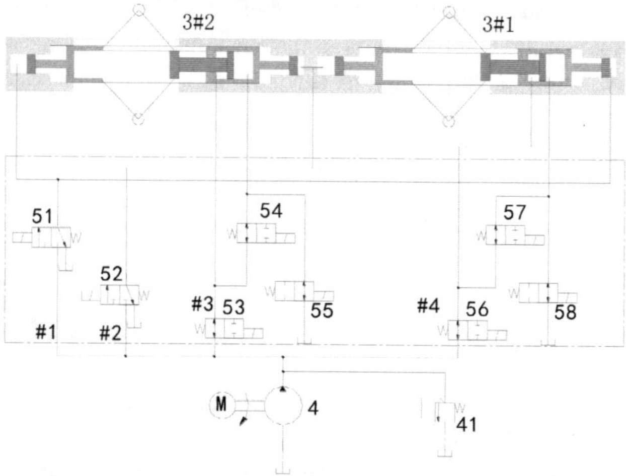

Figure 4. Hydraulic system of the downhole tractor

The specific implementation steps are as follows.

1) After the hydraulic system is connected to the oil circuit, the *SV* 51 is disconnected and the *SV* 52 is connected, so that the two supporting arms are pushed away. When *SV* 55 and *SV* 58 are off, *SV* 53, *SV* 54, *SV* 56 and *SV* 57 are connected with circuit, allowing two supporting arms to expand. Mark the right supporting arm is 3#1 and the left is 3#2.

2) The supporting arm 3#1 realizes locking by turning off *SV* 56 and *SV* 58, turning on *SV* 57. When turning on *SV* 53 and *SV* 55, turning off *SV* 54, the supporting arm 3#1 releases.

3) Turning off *SV* 52, turning on *SV* 51, the supporting arm 3#1 and 3#2 move close to each other, the tractor moves forward.

4) Turning off *SV* 55, turning on *SV* 53 and *SV* 54, the supporting arm t 3#2 can realize locking. Turning on *SV* 56 and *SV* 58, turning off *SV* 57, the supporting arm 3#1 releases.

5) Turning off *SV* 51, turning on *SV* 52, the supporting arm 3#1 and 3#2 move away to each other, the tractor moves forward.

By repeating the steps above, the tractor can achieve continuous motion.

4. Simulation and Discussion

IOP Conf. Series: Materials Science and Engineering **307** (2018) 012007 doi:10.1088/1757-899X/307/1/012007

Based on the analysis of mechanical structure and hydraulic drive mechanism of the tractor, the moving process is simulated by software. Setting the load of the tractor is 3000N, and the frictional force of each supporting mechanism is 300N. The piston diameter of the reciprocating cylinder is 50mm, the piston rod is 15mm, the piston diameter of supporting cylinder is 30mm, and the diameter of piston rod is 10mm.

According to the principle of hydraulic control of the tractor, the signals of switch valves in supporting module are set up, as shown in **figure 5**. According to signal period of the switch valves in two supporting modules, the action sequence diagram of the switch valves in reciprocating hydraulic branch are set, as shown in **figure 6**.

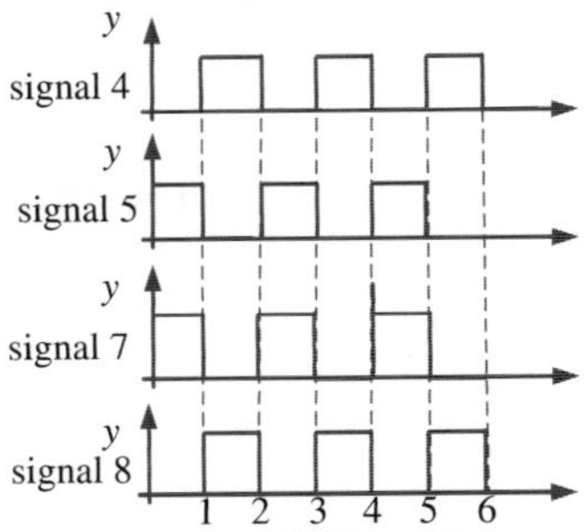

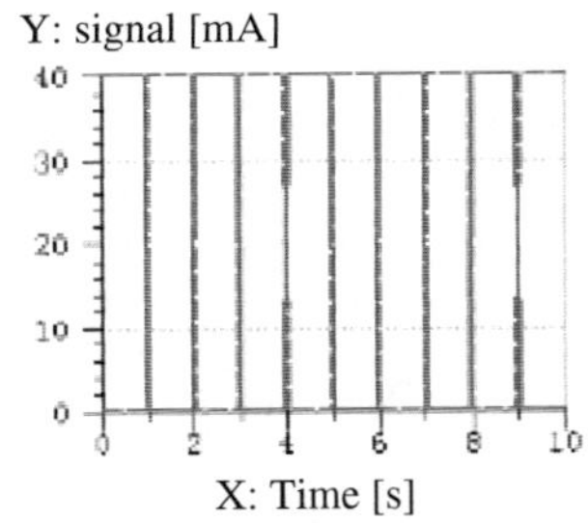

Figure 5. Signals of supporting module valves **Figure 6.** Signals of supporting module valves

As can be seen from **figure 5** and **figure 6**, the signal of each switch valve is the change of the form period of the rectangular wave, and the period is 2s. The signal levels are also switched according to the hydraulic drive principle of the tractor. Since all are the high frequency switch valves, the valve core response frequency are also large enough to ensure the continuity of the traction movement.

The kinematics characteristics of tractor reciprocating mechanism are simulated, as shown in **figure 7**. The response curve of the switch valves are shown in **figure 8**.

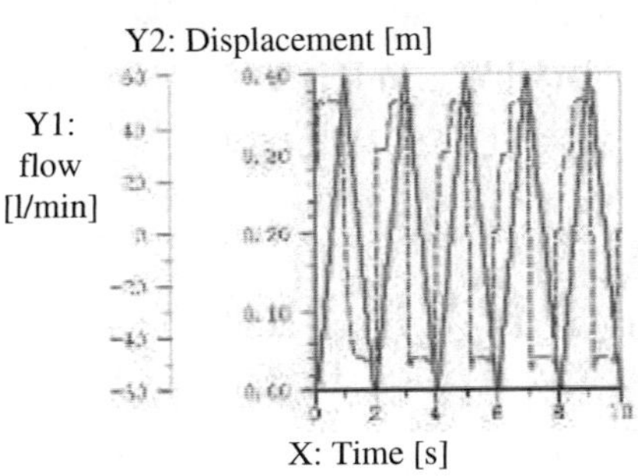

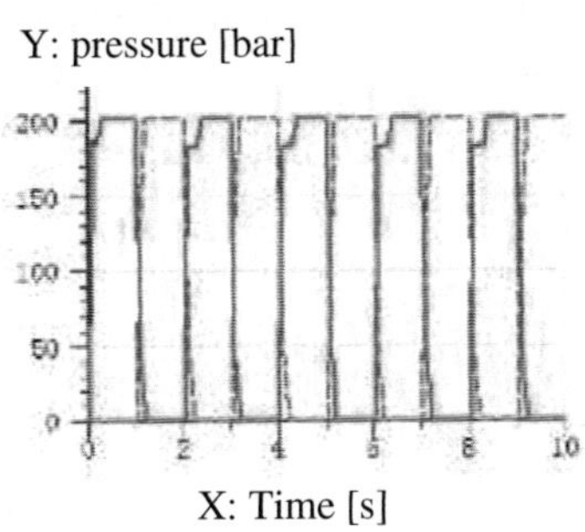

Figure 7. Curve of reciprocating motion **Figure 8.** Curve of switch valves action

As shown in **figure 7**, the displacement curve and flow curve of the right reciprocating cylinder are obtained. The two curve cycles are also 2s, corresponding to the switch valve cycles. The red solid line indicates the reciprocating motion of the piston, and the maximum stroke is 0.4m, then it returns to the initial position at the end of the motion. The internal flow of the cylinder changes periodically and rises rapidly. **Figure 8** shows the pressure change in cavities of reciprocating cylinder. As can be seen from the picture, the pressure of the two chambers changes periodically. Because of the function of the high frequency switch valve, the switching speed is fast, ensuring the continuous and steady operation of the tractor.

It is known by the tractor driving principle that the left reciprocating cylinder has the same characteristics. From the above simulation, it can be seen that the proposed electro-hydraulic drive tractor can realize continuous motion, and has better stability and reliability.

5. Conclusion

In this paper, a novel electro-hydraulic drive horizontal well tractor is presented. The embedded mechanical structure is applied to make the whole electro-hydraulic drive tractor more simple and compact, which can increase the pipe crossing performance. The motion principle and the continuous propulsion principle of the tractor are analyzed.

The hydraulic drive system is designed whose supporting mechanisms adopt hydraulic differential control to improve the tractor speed and efficiency. The inlet and outlet of all hydraulic cylinders are independently controlled with high frequency switch valves, which can enhance anti-pollution ability, reduce loss of valve body throttling and improving the reliability. Finally, the kinematics characteristics and hydraulic characteristics of the continuous moving are analyzed based on simulation, which can lay theoretical basis for the development of tractor.

Acknowledgments
The work was supported by the National Natural Science Foundation of China (51675527).

References

[1] S. Gao， W. Sun， H. Ni, "Status and understandings of downhole tractor design" , Journal of Machine Design, vol. 31, 2014, pp. 1-8.
[2] A. Shukla, H. Karki, Application of Robotics in Onshore Oil and Gas industry, A review Part I, Robotics and Autonomous Systems, vol. 75, 2016, pp. 490–507.
[3] J. S. Cao, L. Lin, Y. Li, Research and Development Progress in Oil and Gas Pipeline Robotics, Oil & Gas Storage and Transportation, vol. 32, 2013, pp. 1-7.
[4] A. Shukla, H. Karki, Application of robotics in onshore oil and gas industry—a review part i. Shukla A, Karki H. Application of robotics in onshore oil and gas industry—A review Part I. Robotics & Autonomous Systems, vol. 75, 2016, pp. 490-507.
[5] L. Q. Li, D. K. Zhou, G. Y. Zhang, Y. Q. Li, W. P. Song, Optimal design of horizontal well traction robot traction force regulating mechanism. Advanced Materials Research, vol. 1033, 2014, pp. 1350-1353.
[6] Q. Liu, Y. Li, T. Ren, Y. Chen, An active helical drive in-pipe robot. Robot. vol. 36, 2014, pp. 711-718.

IOP Conf. Series: Materials Science and Engineering **307** (2018) 012008 doi:10.1088/1757-899X/307/1/012008

Research on Pin Tumbler Locks and the Characteristics of Surface Traces Formed by Unlocking Guns

Gao Yi [1] **Jin Yifeng** [2,*] **Bai Yanping** [2,*] **Mei Hongcheng** [2] **Xu Zhen** [2]

1 Key Laboratory of Evidence Examination and Identification Technology, Ministry of Public Security, Criminal Investigation Police University of China
2 Department of Public Security Document Identification Center
*E-mail: jinyifeng5685 @163.com, yp0916 @163.com

Abstract. In this paper, taking the structure of the pin tumbler locks and the principle of opening and closing locks as breakthrough point, we have studied the machining traces of pin tumbler components and trace characters of pin surface during normal operating. At the same time, taking the unlocking gun as an example, the paper analyzes the unlocking technology and the formation of traces ,and summarizes the changing rules of the surface traces of elastic bead locks by comparing the positions and the characteristics of the traces, which provides technical support for the future study of traces of pin tumbler locks.

1. Itroduction

The pin tumbler lock is also known as the marble lock, the spring lock, or the pin lock, is one of the most common lock structures. The principle is to use a number of cylindrical parts of different heights(called the lock spring, pin or beads)to lock the cylinder. When the correct key is inserted, the round -headed pins are pushed to the same height and the cylinder is released. In this paper, taking the unlocking gun as an example, we focus on the trace characteristics of the pin tumbler's part structure and summarize the characteristics of surface trace changes when the pin tumbler lock is unlocked.

2. The opening and closing principle of the pin tumbler lock

The pin tumbler lock is usually used in a cylindrical lock. The center of the cylindrical lock is a cylindrical hole with a cylindrical cylinder. When opened, the cylinder should be able to rotate.
The middle of the cylinder is for the slot into the key, the end of the slot open to insert the key, the end is a lever. When the key is fully inserted, the key bits cooperate with the pins, so that the bottoms of round -headed pins flush with the outer surface of the lock cylinder ,and the lock cylinder has the opportunity to rotate. There is a long protrusion in the same direction inside the key slot, and the projection cooperate with the key slot to fix the key, which is also used to prevent the round -headed pin(also known as the ball, Beads) from falling ,and increase the difficulty to unlock doors with unlocking tools. The end of the round -headed pin which contacts the key is spherical, so that the key can be easily inserted through . On the top of each round -headed pin,there is a corresponding flat -headed pin (also known as the marbles, the upper spring or the beads). The flat -headed pin is pressed down by the spring. Each round -headed pin isl only equipped with a flat pin. There will be two and more flat -headed pins on each round -headed pin in a more complex lock. (For example,a lock can be locked by multiple keys) .
When the pin tumbler lock is assembled, the pins are pressed down into the lock cylinder by the spring. The lock cylinder and the lock body connect at a place called cut-off point. After the appropriate key is inserted into the lock cylinder, the work part of the key(key bits) will push the round -headed pin upwards so that the gap between the round -headed pin and the flat -headed pin is exactly at the cut-off

IOP Conf. Series: Materials Science and Engineering **307** (2018) 012008 doi:10.1088/1757-899X/307/1/012008

point, and the lock cylinder can be rotated, then turned on. After the key is pulled out, the pins fall, the flat -headed pin stop the cut-off point, and the lock cylinder can not rotate, as shown in Figure1.

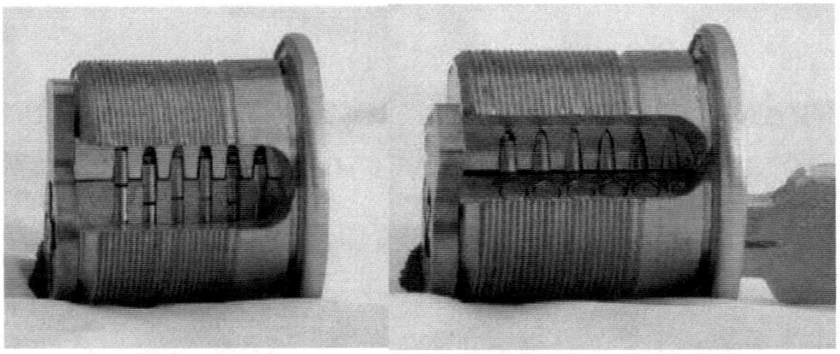

Fig. 1. Opening and closing states of the lock

3. Study on pin tumbler lock components

When unlocking with technical methods, the main parts of the lock which the tool contacts are the pin, the cylinder surface and the keyhole. Because the lock can not be observed into the keyhole without anatomy, the project did not take pictures inside the keyhole before research ,but photographed and fixed the keyhole after unlocking with technical methods and sawing.The following describes the shapes of each component processing trace.

Machining traces on pins: The pin surface without uses is relatively clean, and there is no dust, grease or dirt and other attached substances. The traces on the pin spheres are fresh and concentric round sometimes with mild wear ;There are longitudinal strip traces on the surface of the pin cylinder or it's smooth without any traces. as shown in Figure2.

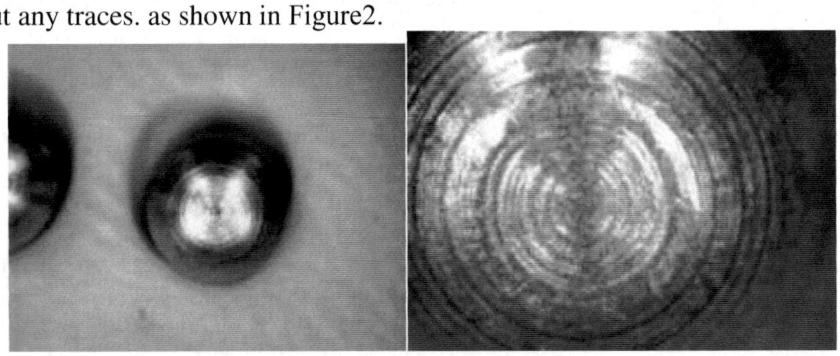

Fig. 2 Traces on new pins

Machining traces on the surface of the lock cylinder: the new lock cylinder surface is relatively clean, and occasionally there are scratches, depression, dirt or wear traces.The machining traces on the cylinder surface vary according to different surface shapes. Under normal circumstances ,the shapes can be concentric rounds or parallel lines.

Lock cylinder slot traces: the new lock cylinder surface is relatively clean,and there are very few scratches, depression, dirt or wear traces,machining traces of the lock keyhole are generally vertical strip traces.

4. Study on the normal use traces of the pin tumbler lock components

The shape of the original key and the shape of the keyhole are coordinated with each other so that the original key can only be oriented in the space defined by the lock cylinder slot. When unlocking, the key enters the slot along the lock cylinder axis, when the key enters the lock cylinder positioning ,at the same time, the round -headed pin finishes cooperating with the key bits. At this point ,the key is applied to torque, so that the lock cylinder rotates to complete the unlock. After unlocking, the key is pulled out. So in the process of the whole unlocking, the key works on the pin twice. This form of movement makes the key surface and the top of the key bits is always in a sliding friction state with the key slot and the pin sphere surface.,in which the top of the key bits and there is the most friction effect on the pin sphere surface. This is also the main reason for the formation of the using traces of pins.

The wear of the lock components is mainly related to the material of the lock and key.The most commonly used material for the components (including the pin, shrapnel, etc.) and the key is copper.The cylinder and the components are sometimes made of nickel brass, steel and other materials. The materials of machining keys are a lot,, in addition to copper, but also nickel brass, aluminum, iron, steel and various alloys. At the same time, the wear of the lock components is also related to the design of the key and components.However,this project is not enumerated.

In order to better study the situations of each pin tumbler lock component, we simulate the normal use process of the key , and respectively observed a key used 50 times(simulating using for a month), 1000 times (simulating using for a year) and 3000 times (simulating using for 3 years) . We found that normally plugging the key 50 times in the ball on the spherical surface can form a bunch of relatively slight scratches through the center of the ball. Around the scratches,the sporadic distribution of scratched line traces; And there are clear wear traces sporadically around the scratches.There are wear traces on the outer surface of the lock cylinder caused because of the friction between the flat -headed pins and the outer surface of the lock cylinder when rotating,as shown in Figure3. When normally plugging 1000 times, the ball surface appears very obvious scratches traces, scratches to the vertex of the spherical surface as the center radially outward;The lock cylinder outer surface's wear traces increased, as shown in Figure 4. When normally plugging about 3000 times, the ball surface appeared serious wear, and the original machining traces on the ball are almost abrasive.The original using traces have been polished.Slight scratch traces are occasionally seen such as Figure 5 shows. The pin close to the surface of the ball gets the most heavy wear.The pin away from the surface is worn gradually slightly. The lock surface gets a certain degree of wear,and the constant use of the keyhole in the lock slot (key bit side) will cause pressure traces, which should be distinguished from the traces caused by bump unlocking.

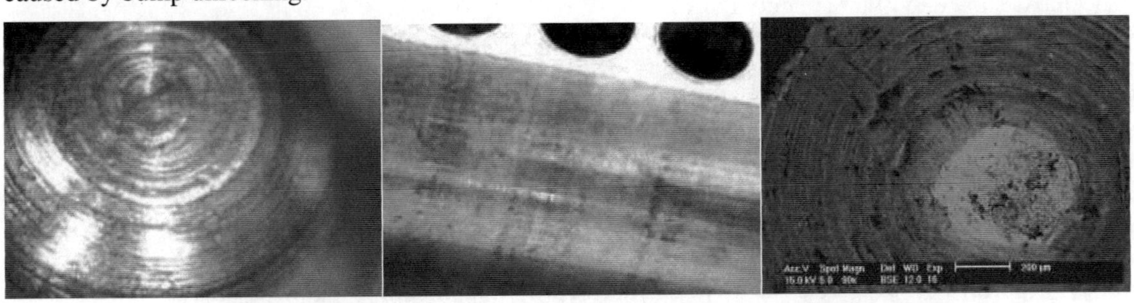

Fig. 3 Traces formed of inserting and removing the key 50 times

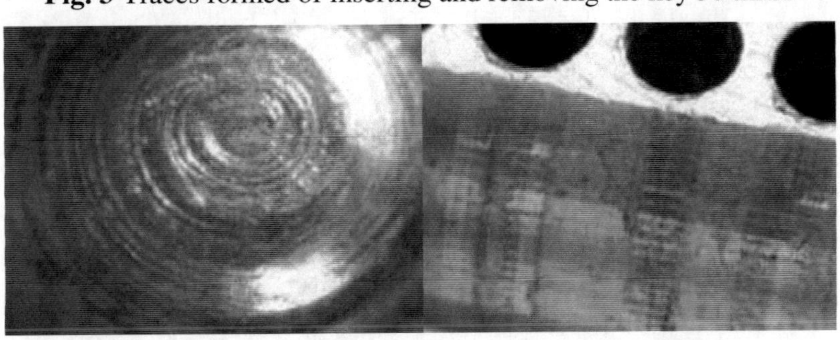

Fig. 4 Traces formed of inserting and removing the key 1000 times

IOP Conf. Series: Materials Science and Engineering **307** (2018) 012008 doi:10.1088/1757-899X/307/1/012008

Fig. 5 Traces formed of inserting and removing the key 3000 times

5. Research on technology of unlocking with unlocking guns and its traces

The unlocking gun is a tool to open the pin tumbler lock, making use of the shake of the swiftach in the front of the gun,so that the round -headed pins and the flat -headed pins separate immediately and a achieve the purpose of unlocking. Unlocking guns are divided into manual and automatic ones,.Manual unlocking guns are equipped with a swiftach in front of the toy guns,which can be replaced. As shown in Figure 6 and Figure 7.

Using unlocking gun locks is also a better way to unlock, especially suitable for opening a lock with a shaped pin. When using manual unlocking guns, the push rod should be set first to form a push dislocation, and then push the swiftach into the keyhole, so that the swiftach can be placed under all the pins.Then pull the trigger to flick round -headed pins,so that round -headed pins can pass the kinetic energy to flat -headed pins to make flat -headed pins shake up and down in the pin hole. When all the flat-headed pins jump into the pin hole of the lock, the round-headed pins are all in the lock cylinder pin hole, and the top round-headed pins separate.Then the unlocking is completed. Electric unlocking guns' and manual unlocking guns'principle is similar.However the difference is that manual unlocking guns need to shock the swiftach backwards and forwards to make the swiftach shake. Electric unlocking guns are to control the shake frequency of the swiftach to fit the shake frequency of flat-headed pins, making the flat-headed pins all jump to the lock body pin hole to achieve the purpose of unlocking, as shown in Figure 8.

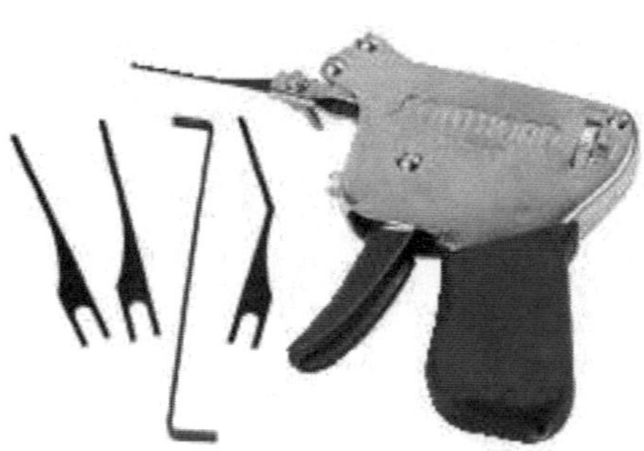

Fig.6 Manual unlocking gun

Fig. 7 Unlocking demo with the unlocking gun

IOP Conf. Series: Materials Science and Engineering **307** (2018) 012008 doi:10.1088/1757-899X/307/1/012008

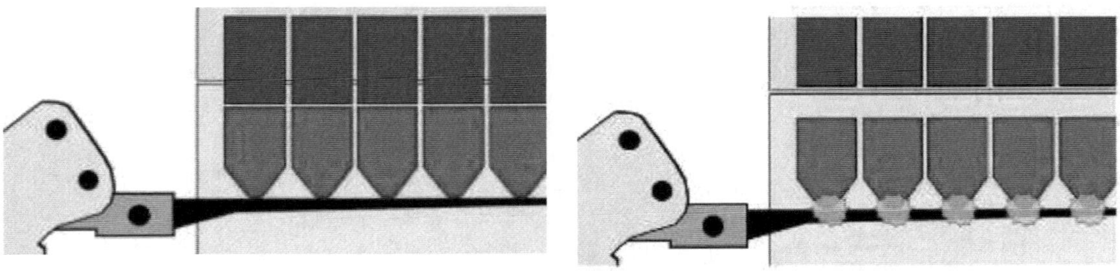

Fig. 8 Unlocking gun's principle

Traces caused by unlocking guns mainly concentrate in the pin sphere parts, that is, the parts of the swiftach hitting, as shown in Figure 9. Traces on the round-headed pin sphere caused by the unlocking guns are different from the scratch traces caused by single hook unlocking .Unlocking guns form pressing traces, and sometimes traces caused by vibration are similar to the arrangement of bicycle wheel spokes, as shown in Figure 10 . Each shock of the unlocking will make the pin a slight rotation, so the number of unlocking gun vibration can be determined through the number of traces on the pin head. As with other techniques, the swiftach will form pressing traces on the inner wall of the key slot and the scratches are formed on the cam at the forefront of the keyhole. If the scratches are formed deep enough, the traces can be identified as shown in Figure 11. Electric unlocking guns sometimes form a relatively dense shades of different pit-like traces on pins. If the suspicious unlocking gun is found, whether there is residue on the swiftach should be paid attention to. When using the unlocking gun , it will form a long stripe, intermittent or indentation mark on the inner wall of the keyway, as shown in Figure 12, as shown in Figure 13.

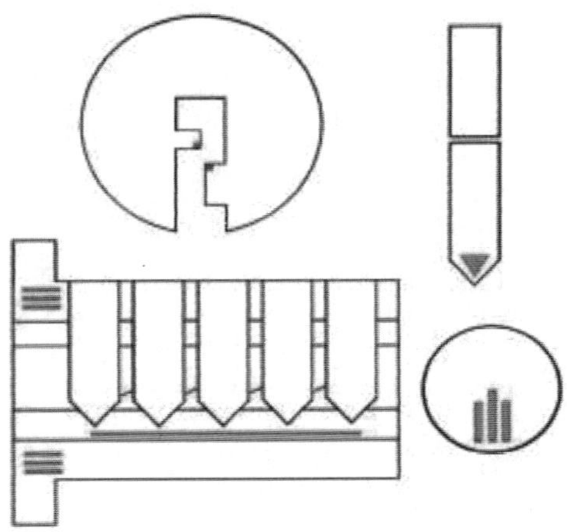

Fig.9 The location picture caused by single hook unlocking

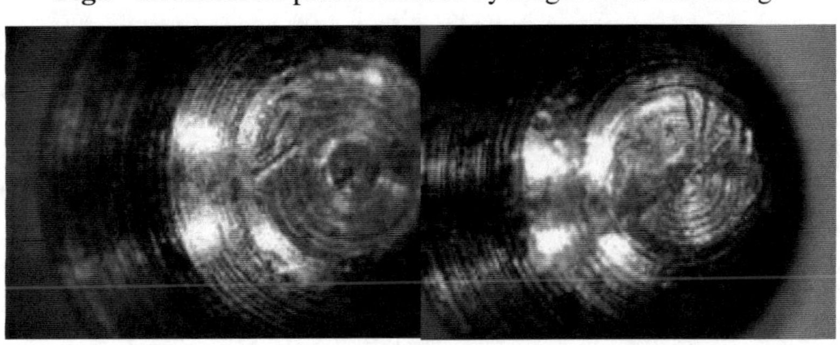

Fig.10 Traces on the pins caused by unlocking guns

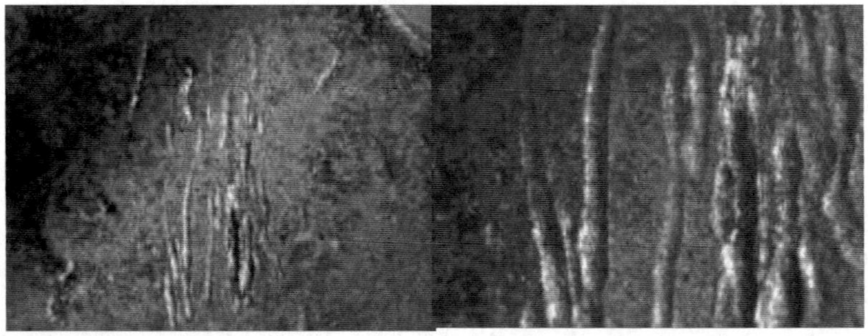

Fig.11 The traces on the keyhole forefront components caused by unlocking guns

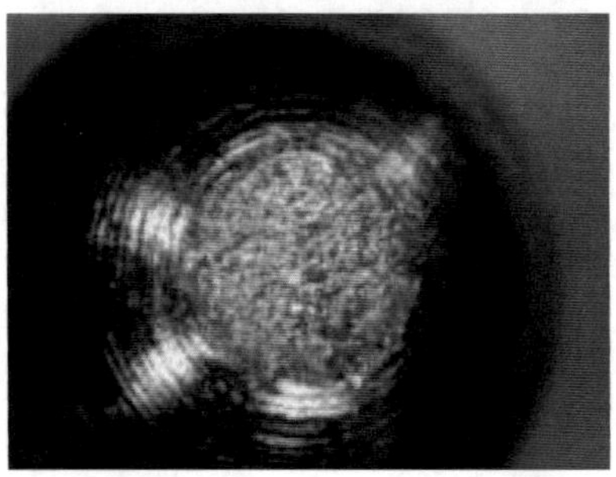

Fig.12 Pits formed on the pins. **Fig.13** Traces formed on the inner wall of the keyhole

References:

1. Single Hui. Engineering Mechanics and Mechanics of Materials [M]. Beijing: Higher Education Press, 2004.
2.Liu Qingmin, Lin Chao, Yang Xin, Design of a Ballistic Separation Feed Device for Padlocks Automatic Assembly System [J]. Mechatronics, 2017 (5): 460-464
3. Zhao Xiao, Wei Machuan, Bi Jing, Study on the Technological unlocking traces of Axial Pin Tumbler Locks [J]. Police Technology, 2017 (2): 49-52
4. GUO Wei, ZHAO Yong-feng,Elementary Discovery of the Pin Tumbler Lock UnlockingTraces with Electrical Unlocking Guns [J]. Chinese People's Public Security University, 2007 (2): 12-16
5. Ma Yuan, Ao Qi, Experiment and Research on Unlocking Traces with Selected Keys[J]. Journal of Chinese People's Public Security University, 2009 (4): 35-36
6. Zhang Xuanguang, Research on Traces of Unlocking and Breaking Billiard Padlocks[D]. Gansu Political Science and Law, 2015
7. Zhu Honghui, the Impact of the Cylinder Material Hardness on the Single Hook Unlocking Traces [J]. Journal of Yunnan Police Officer Academy, 2014 (6): 76-79

Author: Gao Yi (1978-), male, Jilin Jilin people, master, professor in Criminal Investigation Police University of China, mainly engaged in traces inspection technology work, E-mail:396406005@ qq.com.

Corresponding Author: Jin Yifeng (1988--), male, Zhejiang Wenzhou, MA, Ministry of Public Security Evidence Identification Center at the assistant researcher traces, traces of inspection techniques in the work. E-mail: jinyifeng5685 @ 163.com;

Corresponding Author: Bai Yanping (1973-), female, Liaoning Tieling, Master, researcher at the Ministry of Public Security Evidence Identification Center traces of traces of inspection techniques in the work. E-mail: byp0916 @ 163.com

Mei Hongcheng, male, Hubei Suizhou people, 1984.12, associate researcher, trace material certificate, meihongcheng @ 163.com, the Ministry of Public Security evidence identification center

Xu Zhen, female, June 1986, Hubei Xiaogan, deputy chief physician, biochemical and molecular, forensic genetics, xuzhen@cifs.gov.cn, Ministry of Public Security evidence center

Fund Project: This paper is a staged achievement of "Research on Comprehensive Test Methods of Pin-type Mortise Lock" Support by Key Laboratory of Impression Evammination and Identification Technology,Ministry of Public Security (Project No. HJ2014003).

IOP Conf. Series: Materials Science and Engineering **307** (2018) 012009 doi:10.1088/1757-899X/307/1/012009

Research on Efficiency of a Wave Energy Conversion System

Zhongyue Lu[1], Jianzhong Shang[1], Zirong Luo[1], Chongfei Sun[1] and Gewei Chen[1]

1 National University of Defense Technology, Changsha, China
E-mail: jz_shang_nudt@163.com

Abstract. The oceans are rich in wave energy that is green energy, and the wave energy are now being used to generate electricity on a massive scale. It can also be used as a single generator for beacon, buoy or underwater vehicle. Micro small wave energy power generation device is a kind of wave energy power generation devices, main characteristic is mobility is good, and can be directly assembled on various kinds of equipment for the power supply, with good prospects for development. . The research object of the paper is a new adaptive reversing wave energy generating device belongs to micro-sized wave energy generating device. Using the upper and lower absorber blade groups, the low speed and large torque wave energy can be converted into electric energy which can be used for load and lithium battery charging.

1. Introduction

The wave energy is caused by wind energy, and it is influenced by the geographical environment [1, 2]. Its energy change amplitude and frequency fluctuates greatly, which is an unstable and discontinuous energy form. To convert this energy into other forms of energy and convert to electrical energy for storage, a variety of energy conversion devices need to be designed and researched to maximize the ability of converting complex motion forms of waves into stable electrical energy that can be utilized and stored [3]. At present, the energy absorption conversion principle and structural form of wave energy generating devices are different [4]. For example, It can be divided into large-scale grid power generation, special equipment power supply and other direct energy supply by functional purposes [8]; According to the workplace and technology, it can be divided into shore-based, low efficiency and offshore type; According to its working principle, it can be divided into the over topping, the oscillating column and the oscillation float type; And according to the way of energy conversion type of back-end system, it can be classified into pneumatic, hydraulic and mechanical type [5, 6, 7]. The classifications like these can help researchers to understand the energy conversion ways of wave energy devices.

2. Model analysis

The absorption of wave energy is the source of wave energy generation systems. As the wave in reality is a random reciprocating oscillation, it is not conducive to direct absorption. The paper presents a kind of adaptive reversing wave energy generating device. As a new way of absorbing wave energy, it can convert the irregular motion of wave energy into a continuous rotary motion for generator, which reduced the complex transmission link, and can achieve high efficiency absorption and conversion of wave energy. In order to study the energy characteristics of the wave energy generating device, and to verify the feasibility of the method of stable charging, this paper designed a physical prototype, as shown in figure 1. The characteristics of the generator and the power generation and charging effect of the device are verified experimentally. The material used for blade is PE plate,

IOP Conf. Series: Materials Science and Engineering **307** (2018) 012009 doi:10.1088/1757-899X/307/1/012009

the blade ring is made of aluminum alloy, the connecting sleeve of blade and upper and lower absorber is carbon fiber, and the upper and lower absorber joint shaft is stainless steel.

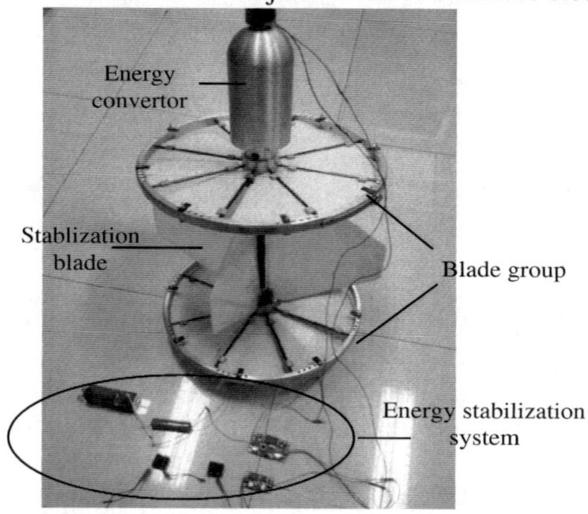

Figure 1. The structure of the prototype

Divide any blade on the device into small elements. Assume that, there is no interference between each element.

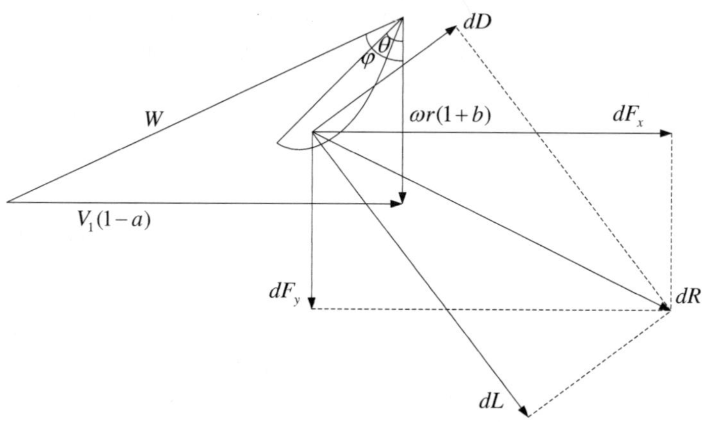

Figure 2. Force analysis of any element

α is the angle of attack, θ is the torsion angle of blade, φ is the angle relative to water velocity. Under axial induction and tangential induction, the water flow produced a relative water speed W, and the water flow also produced a hydraulic resultant force R. The resultant force R is decomposed into the vertical force (lift force L)and horizontal force(drag force D) relative to the water flow's direction. The expression of lift force element and drag force element is:

$$dL = \frac{1}{2}\rho W^2 C C_L dr \qquad dD = \frac{1}{2}\rho W^2 C C_D dr \tag{1}$$

$$W = \frac{V}{\sin\varphi} \tag{2}$$

L is lift force, C is chord length, C_L is Lift coefficient, C_D is drag coefficient.

Decompose the lift force and drag force in the horizontal and vertical direction respectively, the horizontal and vertical resultant forces are calculated as:

$$dF_X = dL\cos\varphi + dD\sin\varphi = \frac{1}{2}\rho W^2 C dr C_X \qquad dF_Y = dL\sin\varphi - dD\cos\varphi = \frac{1}{2}\rho W^2 C dr C_Y \tag{3}$$

Among them, $C_X = C_L \cos\varphi + C_D \sin\varphi$
$C_Y = C_L \sin\varphi - C_D \cos\varphi$

F_X is the axial thrust on blade, the thrust on the ring of the width dr at radius r is calculated as:

$$dT = NdF_X = \frac{1}{2}\rho W^2 NCdrC_X \tag{4}$$

And the torque is $dM = NdF_Y r = \frac{1}{2}\rho W^2 NCrdrC_Y$

N is the number of blades.

Combining the previous analysis, a and b is calculated as:

$$\frac{\alpha}{1-\alpha} = \frac{NCC_X}{8\pi r \sin^2\varphi} \tag{5}$$

$$\frac{b}{1-b} = \frac{NCC_Y}{4\pi r \sin 2\varphi} \tag{6}$$

If ignore the resistance generated by blade's shape.

$$\begin{aligned} C_X &\cong C_L \cos\varphi \\ C_Y &\cong C_L \sin\varphi \end{aligned} \tag{7}$$

So, $b(b+1)\lambda^2 = a(1-a)$

$\lambda = \frac{\omega r}{V_1}$ is the speed ratio at radius r

$$\tan\varphi = \frac{V_1(1-a)}{\omega r(1+b)} = \frac{1}{\lambda}\cdot\frac{1-a}{1+b} \tag{8}$$

$$\frac{BCC_L}{r} = \frac{8\pi a}{(1-a)}\frac{\sin^2\varphi}{\cos\varphi} \tag{9}$$

It can be see that, when absorber's basic design parameter determined, the flow angle is only relative to axial Interference coefficient a and tangential interference coefficient b.

3. Results and discussion

Build the experiment platform as shown in figure 3. The platform contains differential wave energy power generation device, linear heaving type electric cylinder, pool, Data Acquisition Card and PC etc.

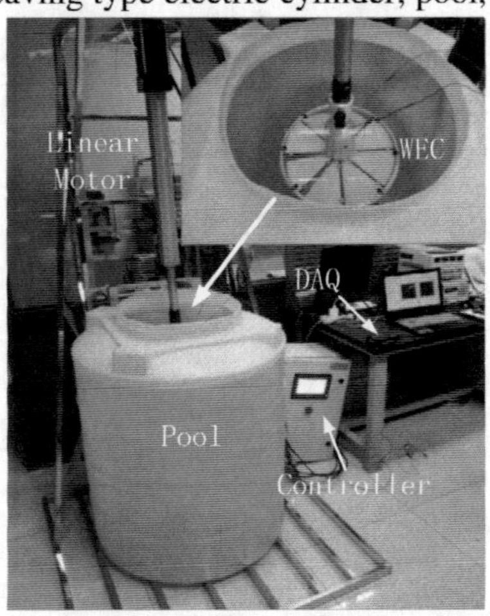

Figure 3. The platform for experiment

In consideration of the Amplitude and period of the absorber's motion are relate to the wave speed relative to device, so the second experiment verified the oscillating period's affection to the output voltage. Keeping the other parameters constant, and set the amplitude is 150mm, the electrical load is

10Ω, and set the maximum blade angle is ±25°. Set the periodic variation in the range of 2~4s, the interval is 0.5s, the results are shown in figure 12.

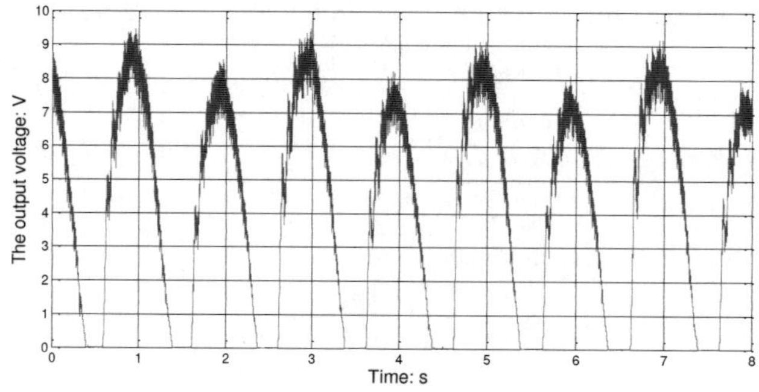

(a) The period is 2*s*

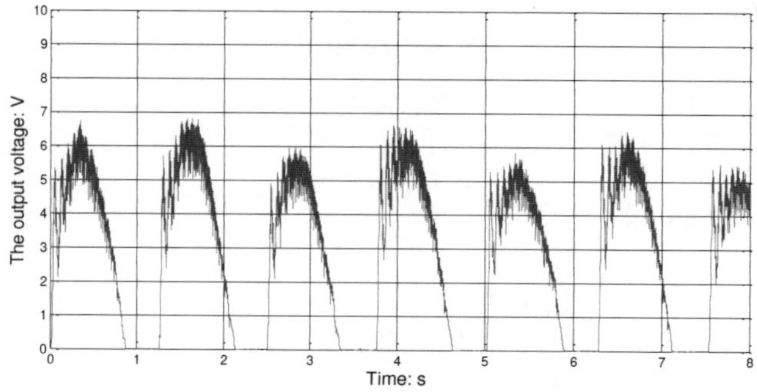

(b) The period is 2.5s

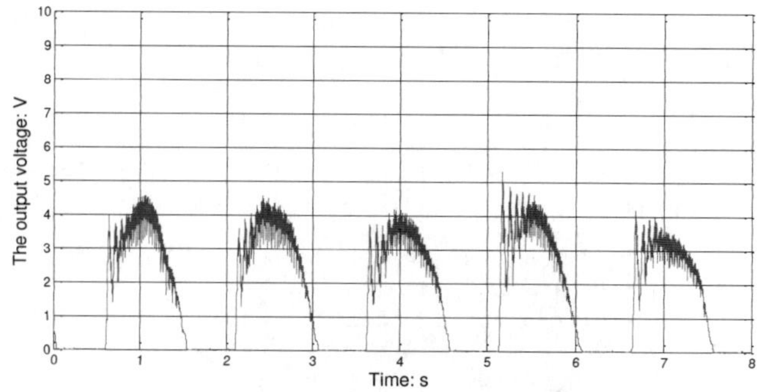

(c) The period is 3s

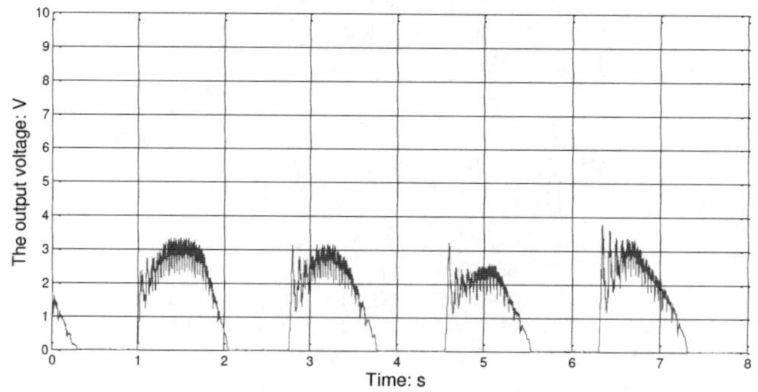

(d) The period is 3.5s

IOP Conf. Series: Materials Science and Engineering **307** (2018) 012009 doi:10.1088/1757-899X/307/1/012009

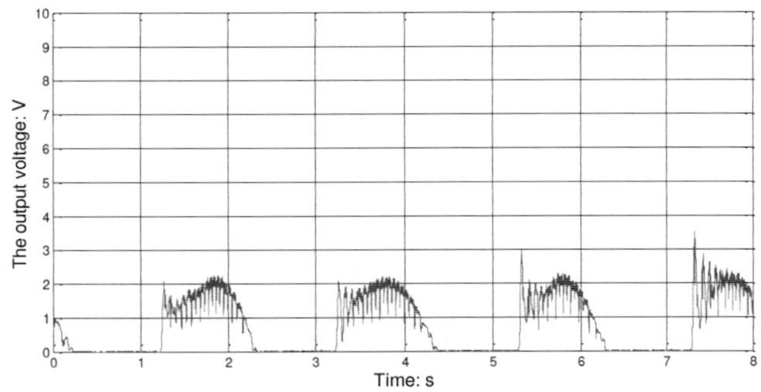

(e) The period is 4s

Figure 4. The output voltage corresponding to different period

From the results, it can be seen that, the longer the period, the more severe oscillation on the rising voltage waveform. It is because under the low speed, nonlinear generating voltage will increase the oscillation.

4. Conclusion

In this paper, a finite element analysis of the blades is carried out for the characteristics of the wave energy generating device, and the experimental platform is built and verified by experiments. According to the wave condition of 150mm wave height in different periods, the actual output voltage curve of the wave energy generating device is obtained. The periodic oscillation of the device causes the periodic oscillation of the output voltage. In one period of the device's movement, the output voltage varies by two periods. Between the device upward and the downward oscillation, there will be a time when the blades are not rotating, the output value on the voltage value is 0. This is also caused by the water resistance of the blade and the torque of the generator itself. To improve this situation, it can increase the speed of the wave relative to blade, or add quality to the outside wheel to increase the rotation inertia, and under the action of inertial force to ensure a sustainable blade rotation, and a continuous power generation process.

Acknowledgments

This work was financially supported by NSFC(Natural Science Foundation of China) E050701.

References

[1] Falcão AFDO. Wave energy utilization: A review of the technologies[C].Renewable & Sustainable Energy Reviews. 2010:899-918.

[2] Muetze A, Vining J G. Ocean Wave Energy Conversion - A Survey[J]. IEEE Industry Applications Conference, 2006, 144(1):1410 - 1417.

[3] González Andreu A, Piza Señas A, Muñoz Arjona E, et al. Device for generating electricity from wave energy,WO/2015/144956[P]. 2015.

[4] Yang S H, He G Y, Zhang H R. Design and Development of Wave Energy -Wind Energy Hybrid Power Generation System[J]. Applied Mechanics & Materials,2014, 535:17-21.

[5] Lafsah, M.D, Ibrahim M Z, Albani A. The Development of Wave Energy Conversion Device to Generate Electricity[J]. Applied Mechanics & Materials, 2015, 773-774:460-464.

[6] Strati F M, Malara G, Laface V, et al. A Control Strategy for PTO Systems in aU-OWC Device[C]// ASME 2015, International Conference on Ocean, Offshore andArctic Engineering. 2015:493-497.

[7] Clement A, McCullen P, Falcao A, et al. Wave energy in Europe: current status and perspectives[J]. Renewable & Sustainable Energy Reviews, 2002, volume 6(2):405-431(27).

[8] Hiramatsu T, Huang X, Kato M, et al. Capacity Design of Supercapacitor-Battery Hybrid Energy System with Repetitive Charging[J]. Ieej Transactions on Industry Applications, 2016, 197(3):58 - 66.

IOP Conf. Series: Materials Science and Engineering **307** (2018) 012010 doi:10.1088/1757-899X/307/1/012010

Wa-Chair: A concept for development of economical stair-climbing wheelchair

Nayan Jyoti Baishya and Harutoshi Ogai

Graduate School of Information, Production and Systems (IPS), Waseda University, Production Engineering, 2-7 Hibikino, Fukuoka, Japan

E-mail: nayan2385@gmail.com

Abstract. In this paper, a concept for development of cost effective and reliable stair climbing wheelchair is being proposed. Slider-crank mechanism is being used to compensate for any variation in inclination angle of the wheelchair during ascent or descent on stairs. Controlling wheelchair's inclination angle can reduce risk for the rider as it prevents the wheelchair from toppling. A prototype is being developed to validate proposed mechanism. Proposed mechanism allows rider to view in the direction of progress which adds additional safety to the rider.

1. Introduction

Physical disability bound to occur in most people if they live longer, as strength of bones and muscle deteriorate after certain age. Unfortunately, for many it happens earlier due to accidents or decreases. Moreover, some are affected by birth. Also, population ageing is one of the concerned issues through out the world. Hence, researches are working continuously towards the development of assistive device for elderly and disabled people. One of the most popular field in development of assistive technology is assistance in commuting people. In this field, wheelchairs had served for a very long period of time and has being evolving continuously. Although electric wheelchairs was one of the benchmark development, climbing up stair is one of the major concerned area.

Researchers had developed track based stair climbing wheelchair which consists of both wheels as well as track belt. Tracks were used for climbing up stairs, whereas wheels were used in plain [1, 2, 3]. Another design was based on attaching a separate module. This is attached to the wheelchair, whenever the rider needs to climb up the stairs [4]. Cluster based stair climbing wheelchair were developed where wheel were replaced by a cluster of three wheels[5, 6, 7, 8, 9]. Moreover, many advance wheelchairs were developed which could balance on two wheels [10].

Most of stair climbing wheel chair available in the market are expensive, complex or bulky. Hence, development of cost effective stair climbing wheelchair is need of present. This paper is organized as follows: Section 2 illustrates about working principle of proposed mechanism; Section 3 explains sensor and filters used; Section 4 presents overview of PID control; Section 5 results of developed prototype and experiments is being discussed. Conclusion and future works are mentioned in section 6.

IOP Conf. Series: Materials Science and Engineering **307** (2018) 012010 doi:10.1088/1757-899X/307/1/012010

2. Working Principle

In this section, working principle of proposed mechanism is being explained. Figure 1 shows a schematic of methodologies followed by the system to climb up or down on the stairs. The system comprises of two front and a rear wheel. Diameter of front wheel is bigger as compared to rear wheel. Chair is mounted on front axle and rear wheel is connected to the chair by means of slider-crank mechanism. This ensures controllability of wheelchair's inclination angle.

In slider-crank mechanism, mechanical parts are assembled in such a way that linear motion is converted to rotary motion or vise-versa. The link which rotate is termed as crank and reciprocating link as slider. With little consideration, it can be confirmed that any wheelchair fabricated using rigid structures will tilt as it climbs up or down the steps. This is because of change in elevation of front and rear wheel. During climbing motion wheelchair will tend to tilt backward whereas it will tilt forward while climbing down the steps. This arises risk for the rider as wheel chair may topple during descent on steps. Although, wheelchair will not topple during ascent motion, but it can tilt to an extent where rider will not be able to see the steps. Hence, use of control system is being proposed to compensate for any change in inclination angle of the wheelchair. This can be achieved by use of slider-crank mechanism. In proposed system, control of inclination angle can be achieved by three methods, i.e., varying actuator's length or by rotating crank shaft or by controlling speed of rear wheel. Also, for better reliability combination of method can be used.

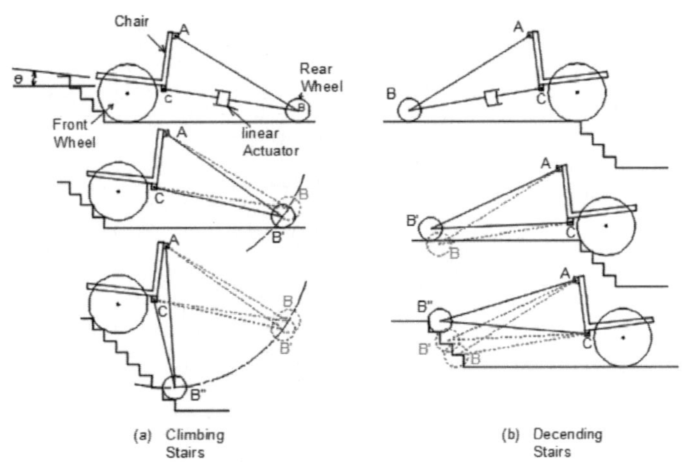

Figure 1: Schematic of mechanism

Figure 1(a) describes about motion of wheelchair during climbing up the steps. Initially, the inclination angle of the chair is set as θ with respect to horizontal plane. As the wheelchair climbs 1st step, elevation of front wheel's center increase. Red dotted lines represents previous position of rear wheel from which it moves to new position. Final image describes about the situation when both front and rear wheels are on steps. From diagrams it can be noticed that in order to maintain initial angle θ, length of link BC gets reduced whereas link AB rotates as a crank.

Similarly, in figure 1(b), motion of wheelchair during descent is being illustrated. It can be seen that inclination angle of the wheel chair can be controlled by either adjusting length of link BC or by rotating link AC or by varying speed of rear wheel.

For successful operation of proposed wheelchair, it is important to keep track of wheelchair's inclination angle continuously. For this purpose MPU6050 is being used. In following section

IOP Conf. Series: Materials Science and Engineering **307** (2018) 012010 doi:10.1088/1757-899X/307/1/012010

details of MPU6050 is being discussed.

3. Sensor

3.1. Kalman Filter

Accuracy in sensing is one of the key factor for proper operation of a control system. To make systems cost effective a low cost sensor is being used, but these sensor is found to be less accurate. Kalman filter is one of the most popular stochastic estimation method to estimate correct data from inaccurate sensor [11]. This filter have gained popularity in many research field, especially in the field of autonomous or assisted navigation. Prediction are performed based on three equations of kalman gain, current estimate and new error in estimate. These equations for the Kalman filter is given as:

$$
\begin{aligned}
KG &= \frac{E_{est_{t-1}}}{E_{est} + E_{mea}} \\
EST_t &= Est_{t-1} + KG(MEA - EST_{t-1}) \\
E_{est_t} &= (1 - KG)E_{est_{t-1}}
\end{aligned}
$$

where,
$KG =$ Kalman Gain
$E_{est} =$ Error in estimated value
$E_{mea} =$ Error in Sensor
$EST =$ Sensor's estimate Value
$MEA =$ Measured sensor value
$t =$ Iteration number

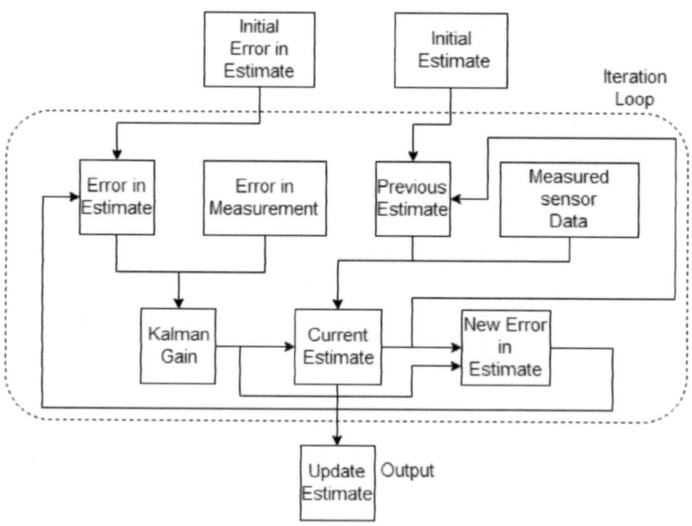

Figure 2: Flow chart of Kalman filter

Figure 2 illustrates the flow process of kalman filter. *Initial error*, *initial estimate* and *Error in Measurement* is provided manually. Value of *Error in Measurement* remains unchanged throughout the iteration process. Iterations are performed till data variation falls under predefined limits.

IOP Conf. Series: Materials Science and Engineering **307** (2018) 012010 doi:10.1088/1757-899X/307/1/012010

3.2. MPU 6050

Tilt sensors plays an important role in proposed stair climbing wheelchair. The sensor continuously sends attitude of the robot to micro-controller which processes the signals and controls speed of rear wheel accordingly. MPU 6050 sensor consists of a 3 axis accelerometer and a 3 axis gyroscope. Accelerometers and gyroscopes are arranged mutually perpendicular axis. This enables to measure change in three degree of linear and rotation motion, respectively. However, in prototype, two axis of accelerometer and single axis of gyroscope is being used. Figure 3 shows image of the sensor. MPU 6050 uses I^2C communication to communicate between sensors and micro-controller.

Figure 3: MPU 6050

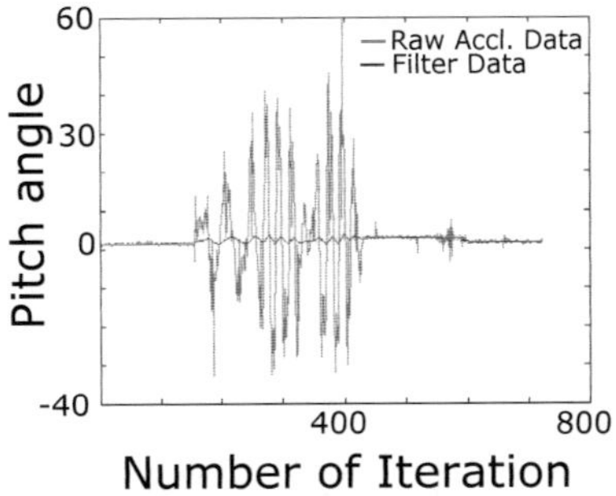

Figure 4: Accelerometer raw data Vs comp. filter

Accelerometer sensors can determine force in terms of gravity in particular direction. Hence, if any axis of the sensor is aligned to true vertical then an output of 1g can be seen on that sensor if it is held fixed. If sensor is tilted, components of 1g will be seen in other axis axis as well, however resultant of all the components will be 1g. This components can be resolved to find the tilt angle. Although, accelerometers are efficient in measuring angles, but these sensors are sensitive enough to pick up small vibrations also. This will cause noise in accelerometer readings.

On the other hand, gyroscopes provides the information of angular velocity. This angular velocities are multiplied by time to derive angular displacement and then integrated to get attitude of the robot. Since, the data is integrated any error in data will also be added to subsequent result. This causes drift in gyroscopic measurements.

Another problem associates with this sensors is that if initially sensor is not perfectly horizontal, gyroscope will assume that angle as zero. Hence, these sensors are incapable of sensing initial attitude. To solve this problem, initially accelerometer data is used to compute initial inclination of prototype. Later, output from complementary filter is considered to compute inclination of the prototype.

3.3. Complementary Filter

To compensated for the error form accelerometers, complementary filter is being used. A large fraction of gyroscope's data is added to a small portion of accelerometer data to get the final value of tilt angle. Comparison of data generated by accelerometer and using complementary filter is shown in figure 4. The spikes seen in accelerometer data represents the situation when sensor was subjected to movement in horizontal plane only. Since, tilt angle calculation is based on accelerometer data in all direction, results shows change in tilt angle. It can be seen that using complementary filter such noises can be rectified. Although some ripples can be seen on those situations, but those are too small to effect control system of the robot.

IOP Conf. Series: Materials Science and Engineering **307** (2018) 012010 doi:10.1088/1757-899X/307/1/012010

4. Proportional Integral Derivative (PID) control

In developed prototype, proportional integral derivative (PID) controller is being used. Figure 5 shows the flow chart of the control system. Feedback signal generated by tilt sensors are fed to micro-controller, which is continuously computes any deviation from desired inclination angle value. Each terms of PID control represents different types of errors. While deviation from desired output represents proportional part, Integral part represents summation of all the errors in iterations. Derivative parts accounts for errors with respect to previous error. Gains K_P, K_I and K_D are multiplied to the proportional, integral and derivative errors respectively. Finally the combined signal is sent as an output to actuators. Following equation are used in PID control system.

$$\begin{aligned} E_p &= \text{Current Value - Destextired Value} \\ E_I &= \sum E_p \\ E_D &= E_{P_K} - E_{P_{K-1}} \\ E &= K_p \cdot E_p + K_I \cdot E_I + K_D \cdot E_D \end{aligned}$$

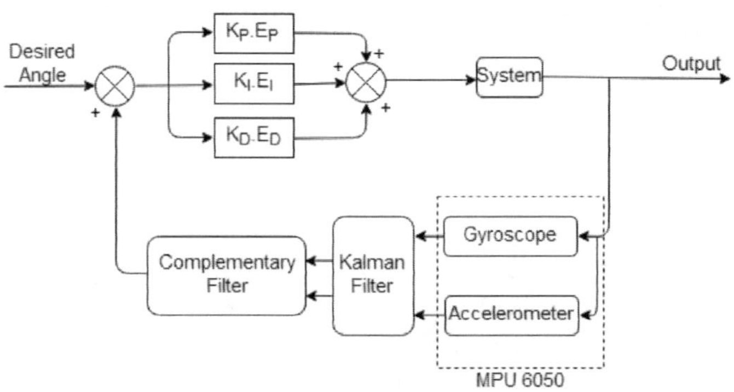

Figure 5: Block Diagram of PID control used in the robot

5. Results

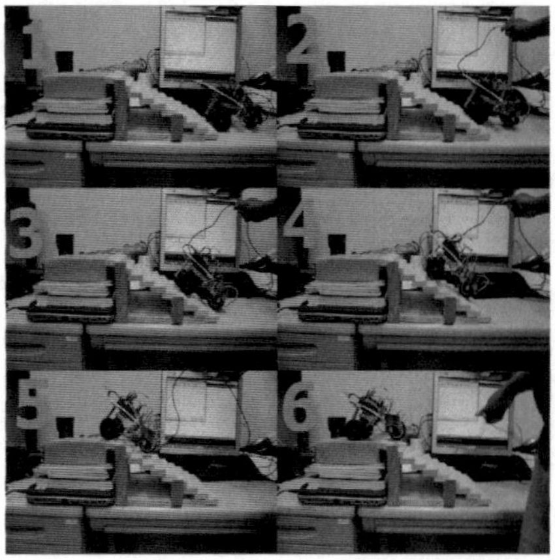

Figure 6: Ascending on steps

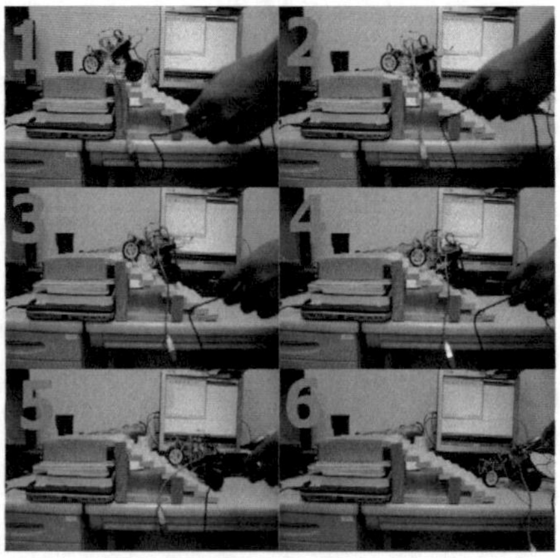

Figure 7: Descending on steps

A prototype has been developed to conduct experiments. It consist of two front wheel and a rear wheel driven by geared DC motor. MPU 6050 is being used as a sensor for gathering information of prototype's inclination angle. To vary inclination angle of the prototype a servo motor was used. Rotation of all motors has been controlled by *ARDUINO UNO*.

Experiments were conducted with and without control. Figure 6 & 7 are snap shots of experiments. These shows that by controlling inclination angle of the wheelchair, stair climbing can be successfully achieved. It was found that when controls were not used wheelchair was unable to climb up in some instant. In that case a slight push was required for the motion to continue. Also, it was found that by using PID control variation of inclination angle had reduced by 7 and 10 degree while ascending and descending respectively. Moreover, in case of descending without control at one instant the angle crossed true vertical. Such situation should be avoided as this may lead to toppling of wheelchair.

6. Conclusion and Future Work

A novel concept for step climbing wheelchair having two front wheel and a rear wheel was proposed. For safe operation of wheelchair inclination angle is one of the important factor. Hence, variation of inclination angle was controlled with slider-crank mechanism. A low cost tilt sensor was used in combination with kalman and complementary filter to sense inclination angle of the robot. PID controller was used to process feedback signal and fed to actuators accordingly. A prototype was developed and experiments were performed. It was found that proposed system can climb up and down maintaining the initial inclination angle. In future a 3-D model will be developed to evaluate kinematics, dynamics and stability of proposed wheelchair. Finally, a bigger prototype will be developed and experiments will be performed.

References

[1] Shigeo Hirose, Edwardo F. Fukushima, Riichiro Damoto, and Hideichi Nakamoto. Design of terrain adaptive versatile crawler vehicle HELIOS-VI. *IEEE International Conference on Intelligent Robots and Systems*, 3:1540–1545, 2001.

[2] Murray J Lawn, Toshihide Sakai, Megumu Kuroiwa, and Takakazu Ishimatsu. Development and practical application of a stairclimbing wheelchair in Nagasaki. *Int. J. Human-Friendly Welfare Robot System*, pages 33–39, 2001.

[3] Suyang Yu, Ting Wang, Xiaofan Li, Chen Yao, Zhong Wang, and Di Zhi. Configuration and tip-over stability analysis for stair-climbing of a new-style wheelchair robot. *2010 IEEE International Conference on Mechatronics and Automation, ICMA 2010*, pages 1387–1392, 2010.

[4] Rintaro Misawa. Stair-Climbing Wheelchair Carrier, 1986.

[5] Giuseppe Quaglia, Walter Franco, and Riccardo Oderio. Wheelchair.q, a mechanical concept for a stair climbing wheelchair. In *2009 IEEE International Conference on Robotics and Biomimetics (ROBIO)*, pages 800–805. IEEE, dec 2009.

[6] Lixin Fang, Tao Lu, Wenhao He, and Kui Yuan. Dynamic and Tip-Over Stability Analysis of A Planetary Wheeled Stair-Climbing Wheelchair. pages 2541–2546, 2012.

[7] Giuseppe Quaglia and Walter Franco. Design of a reconfiguration mechanism for an electric stair-climbing wheelchair. pages 1–10, 2014.

[8] Luis A.M. Riascos. A low cost stair climbing wheelchair. *IEEE International Symposium on Industrial Electronics*, 2015-Septe:627–632, 2015.

[9] Giuseppe Quaglia and Matteo Nisi. Design of a self-leveling cam mechanism for a stair climbing wheelchair. *Mechanism and Machine Theory*, 112:84–104, 2017.

[10] R. Morales, V. Feliu, A. Gonzalez, and P. Pintado. Kinematic Model of a New Staircase Climbing Wheelchair and its Experimental Validation. *The International Journal of Robotics Research*, 25(9):825–841, 2006.

[11] Gary Bishop and Greg Welch. *An introduction to the kalman filter*. Proc of SIGGRAPH, Course, 2001.

IOP Conf. Series: Materials Science and Engineering **307** (2018) 012011 doi:10.1088/1757-899X/307/1/012011

Increasing of the lifetime of large forging dies by repairwelding

M Duchek[1], Koukolikova M[2], J Kotous[3], M Majer[4]

1 Head of Department Metallurgical Technologies, COMTES FHT a.s., Dobřany, CZ
2 R&D engineer Mechanical Testing, COMTES FHT a.s., Dobřany, CZ
3 R&D engineer Metallurgical Technologies, COMTES FHT a.s., Dobřany, CZ
4 Head of Technical Department, CZECH PRECISION FORGE a.s., Plzeň, CZ
E-mail: Michal.duchek@comtesfht.cz, martina.koukolikova@comtesfht.cz, jakub.kotous @comtesfht.cz, miroslav.majer@cpforge.com

Abstract. Repair welding is often used for rebuilding discarded or failed forging dies. It saves the cost of new tools. Increased useful life of repaired dies is another motivation for repair welding. This article focuses on the development of new filler materials for this purpose. The main goal was to prolong the life of tools of DIN 1.2714 material. Filler wires of two chemistries were made and several samples were experimentally welded. Metallographic and tribological analyses were carried out.

1. Introduction

Most forge shops face the issue of worn forging dies and their repair welding or rebuilding. However, those dies are typically small-size tools which are easy to pre-heat for better outcomes of repair welding. Pre-heating of large forging dies is a rather complicated operation with greater demands on the process route and, consequently, on the experience of the engineers involved. [1]

Hot-forming dies are most often scrapped after reaching the maximum permitted wear (approx. 70 % of cases), followed by those affected by mechanical fatigue (25 %) [2]. The factors which play a major role in the die life include the die design, the heat treatment, and the materials used (i.e. both the base material and the filler material). The design of a forging die is dictated by the shape of the forged part and by the capabilities of the forming machine. Special software tools are available for optimizing the impression to achieve the best possible cavity filling [8]. There are many heat treatment and thermochemical treatment routes for extending the die life. In each case, the method chosen must be appropriate for the material in question. For instance, plasma nitriding improves wear resistance but the product becomes more prone to cracking. Such cracks are caused by thermal shock and mechanical loads [3]. Recently, deep cryogenic treatment has been receiving renewed attention – as a process which can provide full martensitic transformation, and therefore eliminates retained austenite from the product. Deep-cryogenic-treated dies show approximately 30% longer life than conventionally-treated tools [4].

Base materials with the right chemistry can provide improved utility properties. Finding the optimal chemical composition is a demanding endeavour that entails numerous trial heats and testing of properties. Modification of chemical composition can deliver the life improvement of 10-20% at considerably increased costs of the material [5]. However, there is also a more affordable variant which involves hardfacing the base material with special materials. Various ordinary filler materials and special hardfacing materials can be used. Another advantage of this method is that it enables discarded dies to be rebuilt or repaired [2], [9]. In practice, all these methods are applied to small and medium-sized dies, i.e. those with weights of around 100 kg. By contrast, the dimensions of large dies that weigh several tonnes make it almost impossible to use and combine heat treatment and thermochemical treatment. Their hardfacing is complicated as well because they are difficult to pre-heat. Repair welding of sizable forging dies has been embarked upon by CZECH PRECISION FORGE, a forging company which intends to use it to rebuild damaged dies. One advantage is that repair-welded dies have longer life than new dies. A consortium of companies has been formed to pursue this research. The consortium members are COMTES FHT, a research organisation, DG Weld, an Italian company engaged in welding which has extensive experience in die rebuilding, and CZECH PRECISION FORGE, a forge shop and the main user of the results. The main objective is to develop a new material for welding large dies.

2. Experimental programme

The basic principle of the experiment was to vary the levels of the key alloying elements in such a way that the effects of such variation could be identified. Two materials were developed in this manner: CPF and N31 materials. In CPF, the carbon and chromium levels were reduced, whereas the nickel content was increased. In N31, the amount of manganese was reduced and more chromium was added. Details of their chemical compositions cannot be published at this time because their legal protection is under discussion.

Table 1. Material: DIN 1.2714, chemical composition [%]

C	Si	Mn	P	S	Cr	Mo	Ni	V	Fe
0.50 0.60	0.30 0.60	0.50 0.90	0.030	0.030	0.90 1.30	0.30 0.50	1.50 1.90	0.10 0.25	balance

Table 2. Material: N3, chemical composition [%]

C	Mn	Si	Cr	Ni	Mo	V	P	S	Fe
0.106	1.162	0.5	0.131	4.3	1.87	0.128	0.01	<0.15	91.4

Samples were cut from a discarded die. Their shapes were chosen to enable the same hardfacing process as the actual die. Having the dimensions of 180×90×150 mm, the samples were provided with a V-notch of 80 mm depth, Figure 1. Each sample was hardfaced according to specified welding procedure, table 3. The samples were then examined by metallographic techniques, their hardness was measured and their wear resistance tested by the pin-on-disc method.

Figure 1. In Macrophotograph of weld, N31 – Etching revealed the heat-affected zone (HAZ) and the boundary of the weld

IOP Conf. Series: Materials Science and Engineering **307** (2018) 012011 doi:10.1088/1757-899X/307/1/012011

Table 3. Welding parameters

Welding position	Welding parameters	Heat treatment
Horizontal	290 A 28 V	Stress-relieving 575°C

3. Metallographic examination

Metallographic examination was carried out on all samples to verify the quality of the weld joint. HV10 hardness numbers were determined using STRUERS DuraScan 50 laboratory tester. The indents were spaced at 1 mm. The locations of the indents are indicated in the macrographs. Metallographic specimens were ground. Their microstructures were brought out by etching with 3% nital and documented using NIKON Eclipse MA200 optical microscope. Macrographs of the locations with indents were taken using a CARL ZEISS Observer Z1m microscope. The macrostructure was revealed by etching with aqua regia. A close-up view of a weld joint and a hardness profile for the CPF material is shown in Figure 2. In the heat-affected zone (HAZ), adjacent to the weld metal, there was coarse tempered martensite and bainite. The base material consisted of bainite (tempered lower and upper bainite). The variation within the hardness profile was acceptable: in the range of 125 HV10.

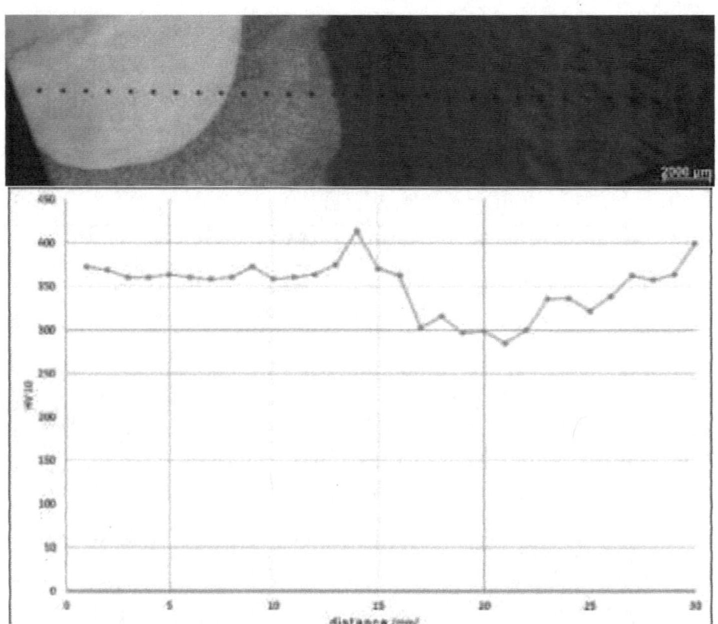

Figure 2. A close-up view of the weld joint and a hardness profile for the CPF material.

The weld metal of the N31 material in the region under examination consisted of a highly-tempered hardening microstructure with minute oxide inclusions. In the heat affected zone, adjacent to the base material, there was fine tempered bainite. The variation in hardness was in the range of 150 HV10.

IOP Conf. Series: Materials Science and Engineering **307** (2018) 012011 doi:10.1088/1757-899X/307/1/012011

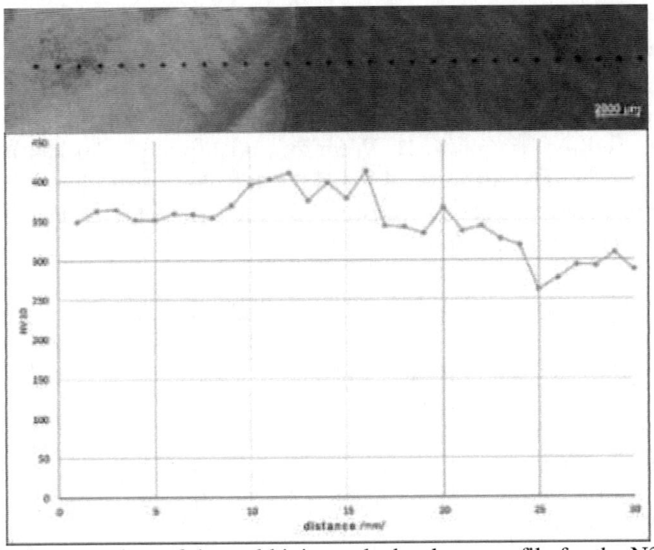

Figure 3. A close-up view of the weld joint and a hardness profile for the N31 material.

4. Mechanical testing

Impact toughness testing was chosen for comparing fundamental mechanical properties of the specimens. Samples for making impact toughness test pieces were taken from four locations, Figure 1. The test pieces had the standard size of 10×10×55 mm and contained a V-notch. The results achieved with optimum welding parameters for both newly-developed materials are given in table 4.

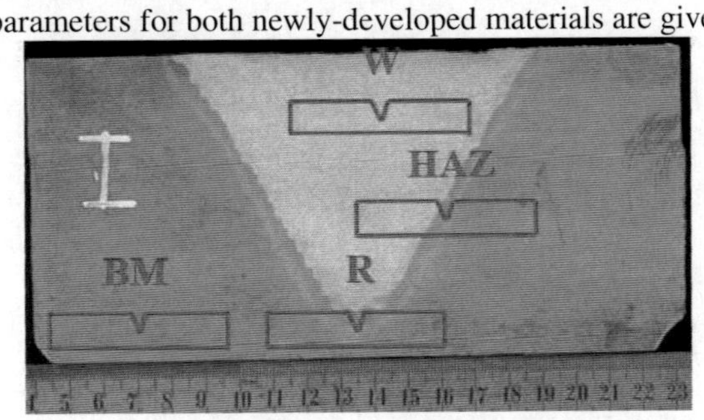

Figure 5. Locations of samples for impact toughness testing (W–Weld, HAZ-Heat Affected Zone, R-Root, BM–Basic Material)

Table 4. Impact toughness test values

Specimen	KCV J/cm^2	Note
N3	13.2	HAZ
	8.6	R
	8.5	W
CPF	47.0	HAZ
	8.0	R
	31.6	W
N31	12.2	HAZ
	15.5	R
	6.8	W
1.2714	20.8	BM

5. Tribological testing– PIN-ON-DISC test

To study the life of the newly-created surfaces on specimens, the PIN-ON-DISC test was chosen. As the materials were intended for hot-working operations, the tests were carried out at 150 °C and 300 °C.
The purpose of the test was to determine the wear resistance of materials under operating conditions. The test consisted in forcing a ball indenter into the surface of a rotating flat specimen. The indenter was loaded by a prescribed force (exerted by a weight) and attached to a rigid arm. Strain gauges fitted to the arm recorded the resulting friction force. Wear data was found by measuring the resulting tracks in test specimens. The measured parameters included the actual amount of wear (material loss) and the impressions (deformation) in the test specimen. The readings are given in figure 5 and table 5. Average values obtained from three readings are shown.

Specimens taken from the welds were tested under the following conditions:

Test temperature: 150°C and 300 °C
Ball material: Si_3N_4, diameter 6 mm
Load: 10 N
Wear track diameter: 3 mm
Number of cycles: 5000
Linear travel speed: 2.5 cm/s
Rotation speed: 80 rev/mi

Table 5. Results of pin-on-disc testing

Designation	Wear rate (10^{-6} mm^3/Nm)	Mean friction coefficient
N1 150°C	47	0.768
N31 150°C	88	0.721
CPF 150°C	74	0.768
N1 300°C	246	0.496
N31 300°C	175	0.490
CPF 300°C	181	0.504

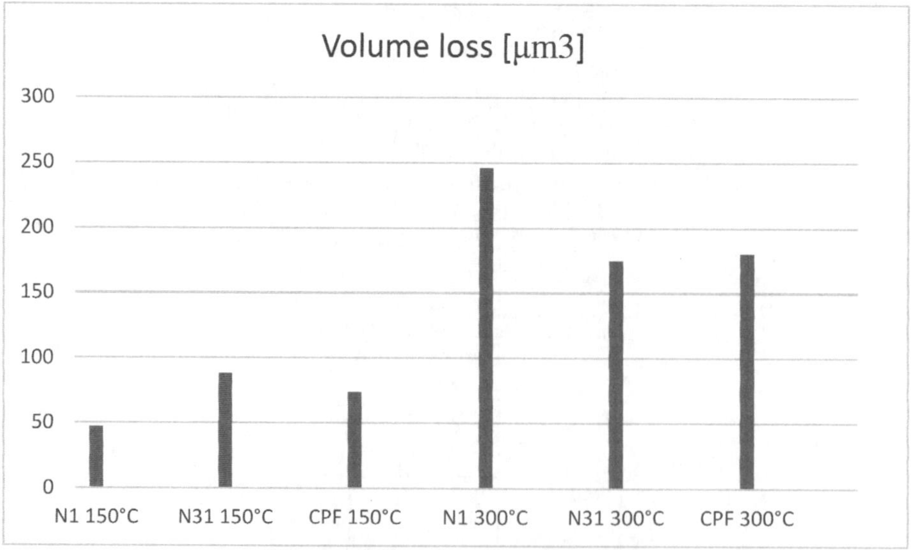

Figure 5. Results of wear testing

6. Conclusion

In the project, two variants of filler material were proposed and compared with the N3 material. At COMTES FHT, all materials were cast and converted into filler wires. Test samples were manufactured at DG Weld. Upon metallographic examination, the weld joints were found to exhibit good dilution. The microstructures were similar in both materials: tempered hardening microstructure. The variation in the hardness profile was examined. The CPF material performed better because its difference in hardness levels was 125 HV10. In the N31 material, the variation was 150 HV10. Both profiles can be deemed very good. The comparison of impact notch toughness levels shows that the CPF material is tougher than N3 and N31. In the HAZ, these values were 47 J/cm^2 and 13.2 J/cm^2. The toughness of the N31 material is similar to that of N3. Only the toughness in the vicinity of the weld root is twice as high.

The tribological test focused on wear at elevated temperatures to which dies are subjected during forging. At 150°C, the performance of both newly-proposed materials was poorer than that of the reference material N3. In N31, the wear rate was $88*10^{-6}$ mm3/Nm, in CPF it was $74*10^{-6}$ mm3/Nm and in N3 it was $47*10^{-6}$ mm3/Nm. By contrast, at 300°C, both newly-proposed materials performed better than the reference material. The performance of the N31 material was 29% better and that of the CPF material was 27% better than in the reference material. These materials proved to be more effective at higher operating temperatures of dies. At temperatures around 150°C, the N3 material is a more suitable choice.

7. References

[1] Duchek M, Šuchmann P, Majer M.: Development of New Welding Materials for Repair Welding of Sizable Forging Dies. In: *Kovárenství, 53/2015*. Brno, 2015. p. 4/81-84. ISSN 1213-9289

[2] Dong-Gyu A: Hardfacing Technologies for Improvement of Wear Characteristics of Hot Working Tools: A Review, *International Journal Of Precision Engineering And Manufacturing* Vol. **14**, No. 7, pp. 1271-1283,

[3] Paschkea H, Weber M, Braeuer G, Yilkiran T, Behrens B A, Brand H: Optimized plasma nitriding processes for efficient wear reduction of forging dies, *Archives Of Civil And Mechanical Engineering 12*, pp. 407–412, 2012

[4] Hauserova D, Dlouhy J, Novy Z, Zrnik J & Duchek M, Behrens B A, Brand H: Forming of C45 Steel at Critical Temperature, *11th Int. Conf. On the Mechanical Behavior of Materials* (icm11), 2011, 10

[5] Duchek M, Suchmann P, Reznar J, Krejcik J & Marek R, Behrens B A.; Brand, H.: Development of Modified Versions of 1.2343 Tool Steel, *21st Int. Conf. On Metallurgy and Materials (metal 2012)*, 2012, TANGER; VSB - TU, OSTRAVA

[6] Dzugan J, Konopik P., Rund M and Prochazka R 2015 ASME Pressure vessels and piping conf. - 2015 Determination of local tensile and fatigue properties with the use of sub-sized specimens, Volume 1A: Codes and Standards (New York: Amer. Soc. Mech. Eng.).

[7] Konopik P, Dzugan J, Rund M: Dynamic Tensile And Micro-Tensile Testing Using DIC Method, *Metal 2014*, May 21st – 23rd 2014, Brno, Czech Republic, ISBN 978-80-87294-52-9

[8] Podaný P, Zemko M, Balcar M: Development of forging technology for microalloyed steel for oil industry, (2010) *Metal 2010 - 19th Int. Conf. on Metallurgy and Materials, Conf. Proc.*, pp. 287-292

[9] Duchek M, Koukolíková M, Nižňanská J, Majer M: Optimization Of Filler Materials For Large Forging Dies, *Materials Science and Technology Conf. and Exhibition 2016*, MS and T 2016, Association for Iron and Steel Technology, Warrendale (PA), USA

Acknowledgments

These results were created under the EUREKA - WELDWIRE project entitled "Development of special welding wires for repair welding of large forging dies" financed by the Ministry of Education, Youth and Sports of the Czech Republic and "Development of West-Bohemian Centre of Materials and Metallurgy", No.: LO1412, which financed by the Ministry of Education of the Czech Republic.

IOP Conf. Series: Materials Science and Engineering **307** (2018) 012012 doi:10.1088/1757-899X/307/1/012012

Experimental Studies of Heat-Transfer Behavior at a Casting/Water-Cooled-Mold Interface and Solution of the Heat-Transfer Coefficient

Y D Zeng , F Wang

College of Materials Science and Engineering, Inner Mongolia University of Technology, Huhehaote,010051, China
E-mail: zyidan@sina.com

Abstract. In this paper, we propose an experimental model for forming an air gap at the casting/mold interface during the solidification process of the casting, with the size and formation time of the air gap able to be precisely and manually controlled. Based on this model, experiments of gravity casting were performed, and on the basis of the measured temperatures at different locations inside the casting and the mold, the inverse analysis method of heat transfer was applied to solve for the heat-transfer coefficient at the casting/mold interface during the solidification process. Furthermore, the impacts of the width and formation time of the air gap on the interface heat-transfer coefficient (IHTC) were analyzed. The results indicate that the experimental model succeeds in forming an air gap having a certain width at any moment during solidification of the casting, thus allowing us to conveniently and accurately study the impact of the air gap on IHTC using the model. In addition, the casting/mold IHTC is found to first rapidly decrease as the air gap forms and then slowly decrease as the solidification process continues. Moreover, as the width of the air gap and the formation time of the air gap increase, the IHTC decreases.

1. Introduction

A water-cooled mold has a very strong cooling capability, and is often used in cast molding of non-ferrous-metal casting. Generally speaking, if the heat-transfer capability of the casting is strong, then the cooling rate of the casting is high, the obtained casting microstructure is fine, and the corresponding mechanical performance is high. However, during the solidification process of the casting, due to casting shrinkage and casting expansion, the casting and mold no longer closely contact each other and an air gap is formed between them. The air gap inhibits the heat exchange between the casting and the water-cooled mold, which greatly lowers the original cooling ability of the water-cooled mold. As a result, the expected cooling rate can hardly be achieved. Thus, the impact of the air gap on heat exchange at the interface has drawn intensive and wide attention from researchers.

Often, the heat transfer at the casting/mold interface is described using the IHTC. The theoretical studies have already confirmed that once an air gap has formed at the interface, the IHTC will be changed dramatically[1]. Approaches of studying the interface air gap and its impact on heat transfer mainly include the actual measurement method and the heat-transfer coefficient inverse analysis method. Early-stage investigations were focused primarily on actual measurement of the size of the air gap at the interface[2–4] but as it is affected by various factors, the formation time of the air gap and the its width can hardly be measured precisely, and the rules indicating how the air gap influences heat transfer cannot be obtained. Given the fact that insurmountable issues existed in the experimental measurements, subsequent research focused more on using the nonlinear estimation method to obtain the IHTC between the casting and mold[4–8]. However, because the impact of the air gap was not taken into consideration, the deviation of the result could be large. Current studies are focused on establishing mathematical models with higher accuracy in which the air gap is taken into consideration[9–12]. It can be seen that no matter whether the actual measurement method or the heat-transfer coefficient inverse analysis method is used, both methods are based on the accurate measurement data of the size and formation time of the air gap. The challenge of the research lies in the fact that the air gap formation time and the size of the air gap still cannot be accurately measured. Directed towards the above issues, in this paper we first propose a new experimental model that is able to accurately and manually control the size and the formation time of the air gap at the interface during the solidification process, thus overcoming the difficulties of direct measurement. Further, using this model, pure aluminum was adopted as the experimental material to perform casting experiments, and the cooling curves of the casting under different conditions of the size and formation time of the air gap were measured, and on this basis, the influence of the air gap on the heat transfer coefficient was analyzed.

2. Experimental design

2.1. Experimental Model

As shown in Figure 1, the experimental model is a combined casting with the inner chamber being rectangular and the top, bottom, and three sidewalls made of an insulating material. Additionally, the rest of the sidewall of the casting is a metallic water-cooled mold. There is a water channel inside the water-cooled mold, which is cooled by the circulating water. The water-cooled mold may move with respect to the other five casting walls. When the metallic liquid is poured into the casting, the water-cooled mold moves in a direction toward the casting to closely contact the casting, such that no air gap is formed. When the solidification thickness reaches a preset value, the movement of the water-cooled mold is reversed to form an interface air gap. Thus, the size and formation time of the air gap can be precisely controlled. A pressure sensor is installed on the water-cooled puncher, which can accurately capture the contact status between the water-cooled mold and the casting.

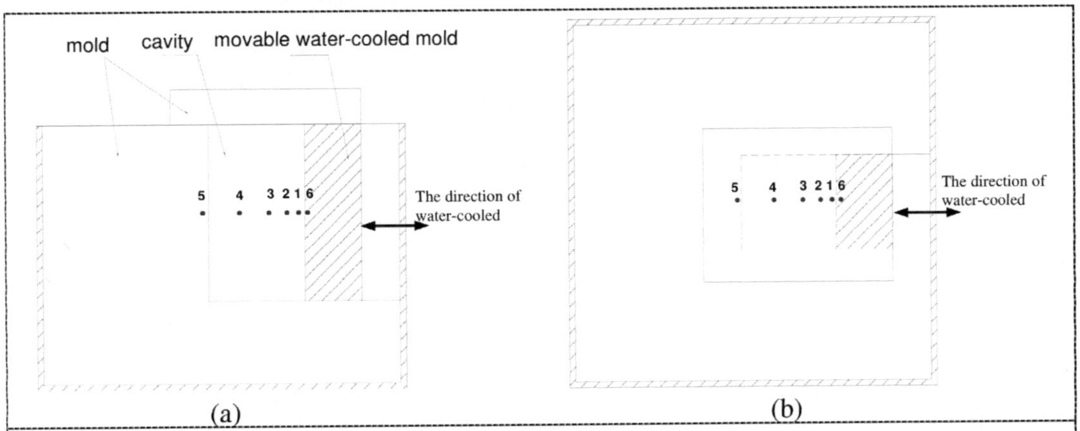

Figure 1. Experimental apparatus: (a) Front sectional view (b) Horizontal sectional view

2.2. Experimental Strategy

To study the impact of the air gap on the IHTC, two factors—the width of the air gap and its formation time—are used as the controlling factors to carry out the experiments, and, in particular, the formation time of the air gap is controlled by the thickness of the solidification layer, which is converted into time. The width of the air gap is configured to be 0.5, 1, and 2 mm, and the formation time of the air gap is configured to be the time when the solidification thickness of the casting reaches 30 and 50 mm..

2.3. Temperature Measurement and Formation of Air Gap
The temperature measurement points of the thermocouple are distributed as shown in Figure 1, where points 1–4 show a distance of 5, 10, 30, and 50 mm, respectively, to the casting/water-cooled-mold interface. Furthermore, point 6 measures the temperature of the water-cooled-mold wall, which is inside the water-cooled-mold wall and displays a distance of 1 mm towards its surface. Point 5 measures the temperature of the casting, which is 5 mm from the surface of the casting. The K-type thermocouple is applied to measure the temperature; the diameter of the thermocouple wire is 0.1 mm. To have a faster response to the temperature, the exposed thermocouple may be applied. Moreover, data collection is realized by utilizing a data acquisition module (DAQ-USB-2401, Omega Engineering, USA), in which the sampling frequency is 100Hz. The temperature variance at each temperature measurement point is recorded during the experimental process.
The air-gap-controlling method refers to the fact that when the metallic liquid is poured into the casting, the water-cooled mold moves in a direction toward the casting, thus allowing the water-cooled mold to closely contact the casting to avoid any air gap. When the formation time of the air gap reaches a preset value, the movement of the water-cooled mold is reversed to form an interface air gap with a preset width.

2.4. Experimental Material
The material of the experimental casting is commercially pure aluminum, the casting temperature is controlled between 780 and 790°C, and top casting is applied. For the casting and the mold material, their thermophysical parameters, the solid fraction, etc. vary as the temperature varies, which may be calculated using PROCAST software

3. Solution of the heat-transfer coefficient
The interface heat-transfer coefficient is calculated using the nonlinear estimation method proposed by Beck[13] where the minimization of F(h) is applied as the determination base, as shown in the equation (1):

$$F(h) = \sum_{j=1}^{K_1} \sum_{i=0}^{K_2} (T_{j,i} - \theta_{j,i})^2 \tag{1}$$

where $T_{j,i}$ and $\theta_{j,i}$ are the simulated temperature and the experimental measurement temperature, respectively, of the measurement point *j* during the time period *i*; K_1 and K_2 are the total number of temperature measurement points and the total number of time periods, respectively; and h is the IHTC. By mathematical deduction, the IHTC is eventually obtained.

4. Experimental results and analysis

4.1. Actual Measurement Curves
Figure 2 illustrates the temperature variance curves of measurement points 1–6 when the width of the air gap is 0, 0.5, 1 and 2 mm., respectively. It is found that when the air gap width is 0 mm [Fig. 2(a)], the curves of temperature measurement points 1–4 inside the casting do not exhibit the phenomenon that the temperature first decreases and then increases, and the decreasing rate of the temperature at each point is relatively high. When the the air gap is formed [Fig. 2(b,c,d)], the curves of temperature measurement points 1–4 inside the casting first decrease rapidly and then increases, and the rate of temperature decrease of each point is significantly lower than that in Fig. 2(a). This indicates that the heat transfer between the casting/water-cooled-mold interface before formation of the air gap occurs very quickly, and the temperature at the casting wall decreases rapidly. However, after the air gap is formed, the heat transfer between the casting/water-cooled-mold interface is inhibited, and the heat-transfer rate decreases, which leads to the increase of the temperature at the casting surface. Thus, the formation of the air gap has a great impact on the interface heat transfer.
As can be seen from the [Fig. 2(b,c,d)], the wider the air gap width, the higher the temperature increase at the casting surface. This indicates that an increase in the width of the air gap results in greater heat-transfer resistance of the casting/mold interface. The width of the air gap in castings is

often smaller than 3 mm, and as a result the air gap is a factor that must be considered during study of the interface heat transfer and actual casting solidification process.

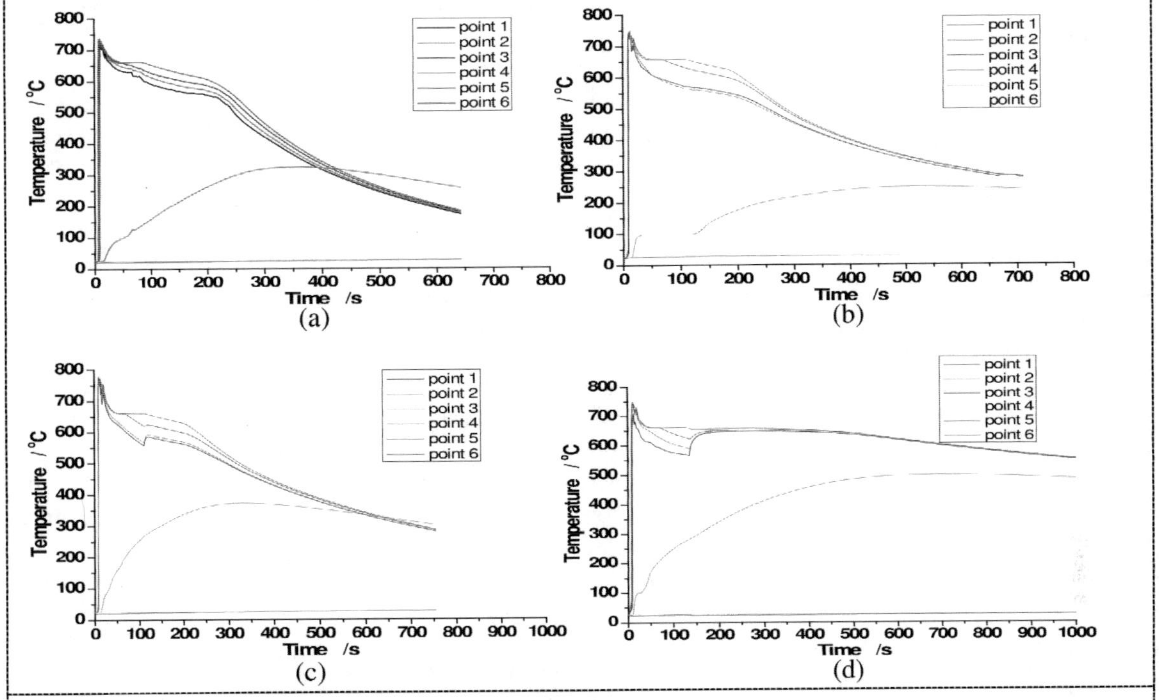

Figure 2. Measured temperature curves at different air gap widths.

(a) air gap width: 0mm (b) air gap width: 0.5mm (c) air gap width: 1mm (d) air gap width: 2mm

Figure 3 presents the temperature variance curves of each measurement point when the formation time of the air gap increases from 90 to 140 s and the width of the air gap is established to be 1 mm. The temperature curves all indicate that the temperature first decreases rapidly, then increases, and, later, decreases further. As the formation time of the air gap increases, the rising point of the temperature is postponed correspondingly. Specifically, when the formation time of the air gap is 90 s, the temperature at measurement point 1 increases by 18°C, and when the formation time of the air gap is 140 s, the temperature at measurement point 1 increases by 12°C. Similar rules are found for the temperature variance measured at point 2; however, with a decreased amplitude of the increase in the temperature, and, furthermore, the temperatures at the same measurement point are different. This is

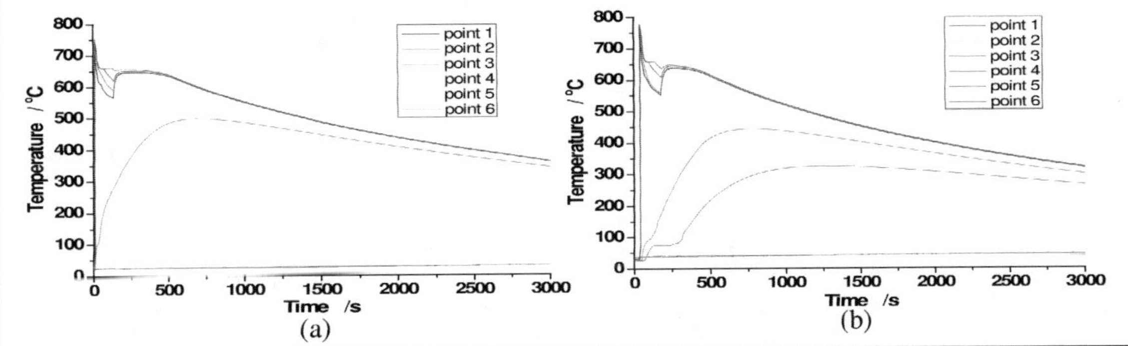

Figure 3 Measured temperature curves at different durations of air gap formation
(a) duration of air gap formation :90s (b) duration of air gap formation :140s

because when the formation time of the air gap varies, the time that the interface contacts the casting varies, and the heat-transfer condition changes correspondingly, which results in the change of the cooling rate of the casting. Clearly, the formation time of the air gap influences the interface heat transfer.

4.2. Solution of Interface Heat-Transfer Coefficient

Taking Eq. (1), the PROCAST software is applied to calculate the IHTC under conditions of different air gap widths and air gap formation times, and the results are shown in Figs. 4–5.

Figure 4 shows the calculation results of the heat-transfer coefficients when the air gap width is 0, 0.5, 1, and 2 mm, and the formation time of the air gap is 90 s. As the solidification time increases, the aforementioned heat-transfer coefficients all show a decreasing trend. The heat-transfer coefficient corresponding to an air gap width of 0 mm is far greater than that of other three situations. When the air gap width is 0.5 mm, the decrease of the heat-transfer coefficient is relatively slow, and when the air gap width is 2 mm, the heat-transfer coefficient decreases rapidly [Fig. 4(d)]. Under the condition that an air gap exists, by comparing the heat-transfer coefficients it can be found that when the air gap width increases, the heat-transfer coefficient decreases. Furthermore, the greater the air gap width, the faster the decrease in the heat-transfer coefficient. Under the present experimental conditions, the cooling time is 90 s, the air gap width increases from 0.5 to 2 mm, and the heat-transfer coefficient decreases from 559 to 128 $W/m^2{\cdot}K$. As the cooling time increases, the difference between heat-transfer coefficients for different air gap widths decreases.

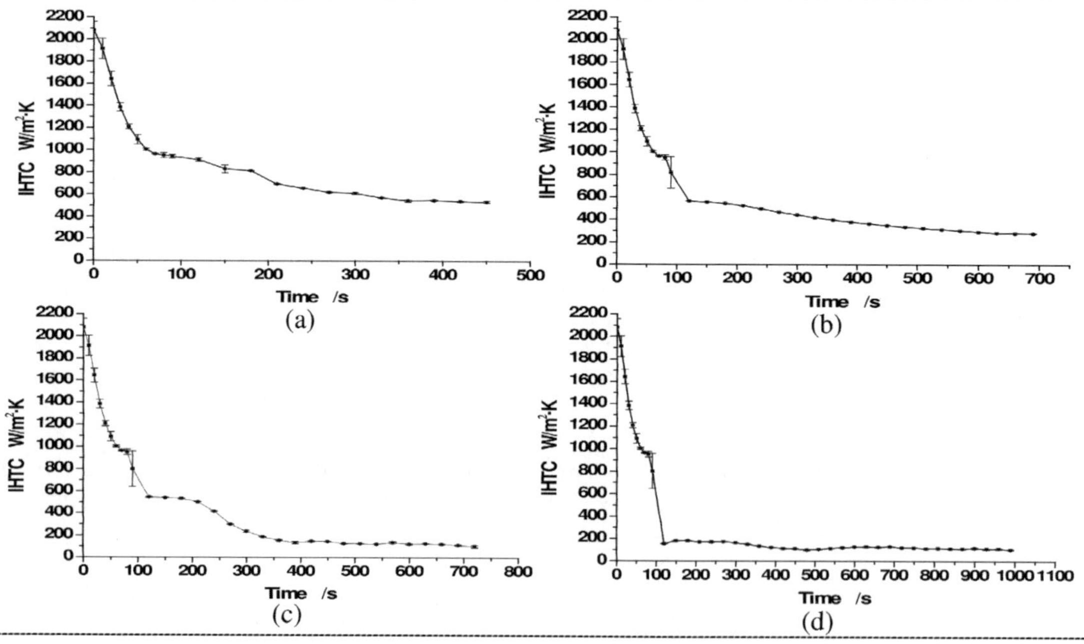

Figure 4. The calculated IHTC values at different air gap widths

(a) air gap width: 0mm (b) air gap width: 0.5mm (c) air gap width: 1mm (d) air gap width:2mm

Figure 5 shows the calculation results of the heat-transfer coefficient when the air gap width is 2 mm, and the formation time is 90 and 140 s. As the air gap formation time increases, the interface heat-transfer coefficient decreases. When the cooling time is 300 s, the heat-transfer coefficients of the above two formation times are 156 and 123 W/m2·K, respectively. This because when the air gap formation time increases, the contact time between the casting and the water-cooled mold elongates, and the surface temperature of the casting thus decreases rapidly. By then, the air gap is formed, and the radiation effect of the casting interface is weakened. Accordingly, the heat-transfer coefficient becomes relatively low. As the cooling time increases, the difference between heat-transfer coefficients calculated using different air gap formation times decreases. The heat-transfer coefficients corresponding to different air gap formation times basically all show the phenomenon that the IHTC first decreases rapidly and then decreases slowly.

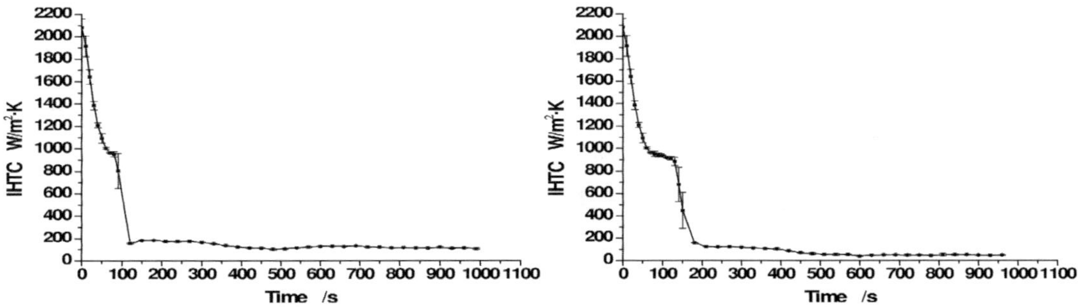

Figure 5. The calculated IHTC values at different durations of air gap formation
(a) duration of air gap formation :90s (b) duration of air gap formation :140s

5. Conclusions

The experimental model for precisely controlling the size and formation time of the air gap at a casting/mold interface is designed, and is applied for studies of casting experiments and of the heat-transfer coefficient. The experimental results indicate that the usage of such a model yields a more accurate and a more effective understanding of the impact of the size and the formation time of the model on the heat-transfer coefficient.

The temperature variance curves of the experimental casting are measured, and the heat-transfer coefficients are obtained via inverse analysis. After the air gap is formed, the temperature inside the casting displays an increasing phenomenon. The wider the air gap, the higher the temperature increases. The calculation results of the heat-transfer coefficient indicate that the impact of the air gap on the heat-transfer coefficient is significant. In the present experiments, by comparing a situation where no air gap is formed and another situation where a 0.5mm air gap is formed, the maximum heat-transfer coefficient decreases from 2157 to 559 W/m2·K. Moreover, the heat-transfer coefficient decreases as the air gap width and formation time increase.

6. References

[1] Sarjant R J and Slack M R 1954 *Heat transfer*(Iron and Steel Institute)p 428
[2] Nishida Y and Matsubara H 1976 *British Foundryman* **69** 274-8
[3] Li T J, Ren Z M and Ren X F 1990 *Journal of DALIAN University of Technology* **30** 297-300
[4] Trovant M and Argyropoulos S 2000 *Metal. Trans. B.* **31** 75-86
[5] Zhu L P and Chen L 2009 *Engineering Journal of Wuhan University* **42** 268-72
[6] Zhang L Q, Li L X, Zu B W 2010 *Journal of Hunan University(Natural Sciences)* **37** 22-6
[7] Guo Z P and Xiong S M 2007 *Acta Metallurgicai Sinica* **43** 1149-54
[8] Sahin H M, Kocatepe K, Kayikci R *2006 Energy conversion and management* **47** 19-34
[9] Florio B J , Vynnycky M, Mitchell S L and O'Brien S B G 2015 *Applied Mathematics and Computation* **268** 1122-39
[10] Bertelli F , Faria J D, Goulart P R, Brito C and Cheung N 2016 *Applied Thermal Engineering* **96** 454-62
[11] Chen L P, Wang Y X, Peng L M,Peng H F and Jiang H Y 2014 *Experimental Thermal and Fluid Science.* **54** 196-203
[12] Szeliga D, Kubiak K, Ziaja W, Cygan R, Suchy J Sz, Burbelko A, Nowak W J and Sieniawski J 2017 *Experimental Thermal and Fluid Science* **87** 149-60

IOP Conf. Series: Materials Science and Engineering **307** (2018) 012013 doi:10.1088/1757-899X/307/1/012013

Adhesion of *Candida albicans* to Vanillin Incorporated Self-Curing Orthodontic PMMA Resin.

K Zam [1], P Sawaengkit [1], S Thaweboon [2], B Thaweboon [2]

[1]Department of Orthodontics, [2]Department of Oral Microbiology
Faculty of Dentistry, Mahidol University, Bangkok 10400, Thailand

Kuenzam05@gmail.com
pornrachanee.saw@mahidol.ac.th

Abstract. It has been observed that there is an increase in *Candida* carriers during the treatment with orthodontic removable appliance. Vanillin is flavouring agent, which is known to have antioxidant and antimicrobial properties. The aim of this study was to evaluate the effect of vanillin incorporated PMMA on adhesion of *Candida albicans.* A total of 36 orthodontic self-curing PMMA resin samples were fabricated. The samples were divided into 3 groups depending on percentage of vanillin incorporated (0.1%, 0.5% and PMMA without vanillin as control). PMMA samples were coated with saliva. The adhesion assay was performed with *C. albicans* (ATCC 10231). The adherent yeast cells were stained with crystal violet and counted under microscope by random selection of 3 fields at 10X magnification. The statistical analyses performed by Kruskal Wallis and Mann Whitney non-parametric test. It was found that the PMMA resin samples with vanillin incorporation significantly reduced the adhesion of *C. albicans* as compared to the control group. This study indicates that vanillin incorporated resin can impede the adhesion of *C. albicans* to about 45 - 56 %. With further testing and development, vanillin can be employed as an antifungal agent to prevent adhesion of *C. albicans* to orthodontic self-curing PMMA resin.

1. Introduction

Vanillin is an important component of natural vanilla, extracted from the seeds of tropical orchid, *Vanilla planifolia*. Vanillin is mainly used as a flavouring agent for food, beverages and also used for intermediates for pharmaceuticals [1]. Additional to flavouring agent, vanillin being phenolic compound with low molecular weight, has antioxidant and antimicrobial properties [2]. Several studies reported the effectiveness of vanillin against certain food spoilage bacteria, yeasts and molds [3-5].

Candida albicans is a commensal pathogen that lives on various surfaces of body like genital and intestinal tracts, oral mucosa as well as skin [6-8]. The increased growth of *C. albicans* can cause soreness, unpleasant taste, bad breath with bleeding and inflammation of oral mucosa. Also, it can penetrate the blood stream causing candidemia which will infect the internal organs causing significant morbidity and mortality [9, 10]. For efficient colonization and development of pathogenesis, the adhesion of *C. albicans* to host cells or biomaterials is necessary [10, 11].

PMMA is a transparent acrylic based resin, it is used for the fabrication of oral appliances and retainers in Orthodontics. It has been reported that with orthodontic appliances and other oral devices leads to change in oral microbiota resulting in rise of Candidal density at oral mucosa [12, 13]. The microbial plaque biofilm can be controlled by mechanical and chemical aids which demand stringent patient compliance which unfortunately is unobtainable in patients with disabilities. The objective of this study is to investigate the inhibitory effect of vanillin on *C. albicans* adhesion to vanillin incorporated Orthodontic self-curing PMMA resin surface.

2. Method and materials

2.1. *Collection of saliva*
Unstimulated saliva was collected from a normal healthy individual (non-smoker, no medication). The subject was asked to refrain from using oral hygiene products like antimicrobial mouthwashes and the saliva was collected 2 h after consumption of meals.

2.2. *Candida suspension preparation.*
C. albicans (ATCC 10231) was obtained from Department of Oral Microbiology, Faculty of Dentistry Mahidol University, Thailand. They were maintained in Sabouraud dextrose agar at 4°C. Few colonies from the agar were inoculated in phosphate buffer saline (PBS) to yield a concentration of approximately 10^7CFU/ml using McFarland standard No.1.

2.3. *Preparation of PMMA Samples*
The Orthodontic self-curing PMMA resin powder (Siam cement group, SCG) was mixed with 0.1% and 0.5 % vanillin powder. Thirty-six orthodontic self-curing PMMA resin samples were prepared with dimension of 15 mm x 15 mm x 1.5 mm. The samples were divided into 3 groups depending on percentage of vanillin incorporated (0.1%, 0.5% and PMMA without vanillin as control). The tissue surfaces of these samples were unpolished to replicate an exact impression of tissue surface of retainers. The resin samples were placed in distilled water for 3 days to leach out excess monomer and sterilized with UV light for 30 min. The PMMA samples were place in sterile saliva collected from a healthy individual for 30 min.

2.4. *Adhesion Assay*
The self–curing orthodontic PMMA samples were placed in 24-wellplate. To each well containing PMMA sample, 2 mL of candida suspension was added. The 24-wellplate with PMMA sample and candida suspension was placed in shaking incubator (120r /min) at 37°C for 90 min. The PMMA samples were then washed twice with PBS to remove the non -adherent cells. The remaining adherent cells on PMMA resin surface was fixed with 95% ethanol and stained with crystal violet. They were counted under microscope by random selection of 3 fields at 10X magnification. The mean number of adherent yeast cells were expressed as cells/mm^2.

2.5. *Statistical analyses*
All tests were performed triplicate on 4 separate occasions. The analysis of data was done using SPSS statistics. Kruskal Wallis and Mann Whitney U test. The significance level of 0.05 was used.

3. Results

Table 1 presents the number of adherent cells on orthodontic self–curing PMMA surface. Both the vanillin incorporated resin samples had significant reduction of candida adhesion as compared to the control. It was found that there was a reduction of about 45 to 56 % as illustrated in figure 2. Although the number of adherent yeast cells was more reduced in 0.5% than 0.1% vanillin group, there was no statistical significant difference found between these two groups.

Table 1. The number of *C. albicans* cells adhered on PMMA resin surface (cells/mm^2)

Vanillin		
0%	0.1%	0.5%
18.9183±**1.22431**	10.3233±1.32726	8.4117±1.30667

values are expressed as mean±SD

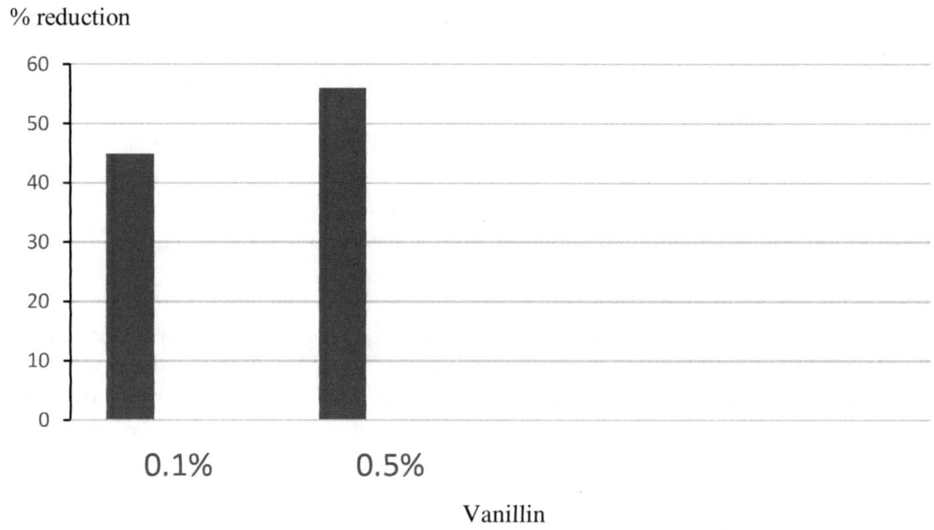

Figure 1 Percentage of reduction of *C. albicans* adhesion on vanillin incorporated self-curing orthodontic PMMA resin

4. Discussion:

The adhesion of *C. albicans* to host cells or polymers such as acrylic resin is a necessary step which is required for efficient colonization and progression of pathogenesis and infection [6, 14].The adherence of the candida strain is important to prevail over the host clearance mechanisms and for colonization, it depends on the effectiveness of the adhesion mechanism and the yeast growth rate [6]. Factors affecting the adhesion of *C. albicans* to PMMA depends on the substrate surface (type, chemical nature and surface properties like free surface energy, surface roughness), the presence of salivary proteins and serum, the presence of other adherent microorganisms, strain variability, consumption of carbohydrate-rich diet, and culture conditions [15].

A study conducted on orthodontic patients undergoing removable appliance therapy indicated that the *Candida* carriers was increased from 39% before treatment to 79% during treatment in 9 months. However, following the completion of treatment, it was reduced significantly to 14% [16]. Both mechanical and chemical methods can be used to control plaque accumulation on orthodontic appliances. Despite their effectiveness, these procedures demand good patient compliance, which may not be readily achievable in those patients with disabilities. Hence, the addition of antimicrobial agents in orthodontic acrylic resin for plaque biofilm reduction is highly recommendable [9, 12].

So, in this study, the effect of vanillin on adhesion of a standard strain of *C. albicans* to self-curing orthodontic PMMA was investigated. The results showed that candida adhesion to 0.1 % and 0.5% vanillin incorporated orthodontic PMMA resin was significantly reduced as compared to control.

Vanillin, a major component of natural vanilla is not only an important flavour but also has antioxidant and antimicrobial properties. A study conducted using food bacteria like *Escherichi coli, Lactobacillus plantarum* and *Listeria innocua* have shown that the antimicrobial activity of vanillin is mainly due to its ability to affect the integrity of the cytoplasmic membrane with the resultant loss of ion gradients, pH homeostasis and inhibition of respiratory activity [3]. Vanillin can inhibit several strains of *Cryptococcus neoformans* and *C. albicans* .The study conducted showed that vanillin is effective against *C. albicans* in vitro and it can be used for treatment of infection caused by non-dermatophyte fungus[17]. It was suggested that the antifungal property of the vanillin could be due to aldehyde moiety its structure [3].

5. Conclusion

The vanillin incorporated orthodontic PMMA resin exhibited significant inhibitory effect against candida adhesion to the surface about 45-56%. Hence vanillin can be incorporated in self curing orthodontic PMMA resin to reduce candida adhesion for fabrication of intraoral orthodontic appliance.

Acknowledgment

First of all, I would like to thank Thailand international Cooperation Agency (TICA) and Faculty of Dentistry, Mahidol University for funding this research. I would also like to convey my heart filled thanks to the staffs of Department of Oral Microbiology, Mahidol University, for their help, guidance, their support extended during the research. Lastly, I would like to thank the staffs from Siam cement groups for providing the materials.

References

1. Walton NJ, Mayer MJ, Narbad A 2003 *Phytochemistry.* **63** 505
2. Priefert H, Rabenhorst J, Steinbüchel A 2001 *Appl. Microbiol. Biotechnol.* **56** 296
3. Fitzgerald DJ, Stratford M, Gasson MJ, Ueckert J, Bos A, Narbad A 2004 *J. Appl. Microbiol.* **97** 104
4. Fitzgerald DJ, Stratford M, Gasson MJ, Narbad A 2005 *J. Agri. Food. Chem.* **53** 1769

IOP Conf. Series: Materials Science and Engineering **307** (2018) 012013 doi:10.1088/1757-899X/307/1/012013

5. Cerrutti P, Alzamora SM, Vidales SL 1977 *J. Food. Sci.* **62** 608
6. Cannon RD, Chaffin WL 1999 *Crit. Rev. Oral. Biol. Med.* **10** 359
7. Calderone RA, Fonzi WA 2001 *Trends. Microbiol.* **9** 327
8. Henriques M, Azeredo J, Oliveira R. 2004 *Colloids Surf. B.* **33** 235
9. Akpan A, Morgan R 2002 *Postgrad. Med. J.* **78** 455
10. Thaweboon B, Thaweboon S 2011 *Asian. Pac. J. Trop. Med.* **4** 41
11. Ramage G, Saville SP, Thomas DP, Lopez-Ribot JL 2005 *Eukaryot. Cell.* **4** 633
12. Hagg U, Kaveewatcharanont P, Samaranayake YH, Samaranayake LP 2004 *Eur. J. Orthod.* **26** 623
13. Gristina AG, Giridhar G, Gabriel BL, Naylor PT, Myrvik QN 1993 *Int. J. Artif. Organs.* **16** 755
14. Verheyen C, Dhert W, de Blieck-Hogervorst J, Van Der Reijden T, Petit P, De Groot K 1993 *Biomaterials.* **14** 383
15. Radford DR, Challacombe SJ, Walter JD 1999 *Crit. Rev. Oral. Biol. Med.* **10** 99
16. Addy M, Shaw WC, Hansford P, Hopkins M 1982 *Br. J. Orthod.* **9** 158
17. Boonchird C, Flegel TW 1982 *Can. J. Microbiol.* **28** 1235

IOP Conf. Series: Materials Science and Engineering **307** (2018) 012014 doi:10.1088/1757-899X/307/1/012014

The Control of Welding Deformation of the Three-Section Arm of Placing Boom of HB48B Pump Truck

Zhi-ling Wang

Jiangsu province xuzhou technician institute
Xuzhou, Jiangsu 221151, China
379713268@qq.com

Abstract: The concrete pump truck is the construction equipment of conveying concrete with self contained base plate and distributing boom. It integrates the pump transport mechanism of the concrete pump, and the hydraulic roll-folding type distributing boom used to distribute materials, and the supporting mechanism into the automobile chassis, and it is the concrete conveying equipment with high efficient and the functions of driving, pumping, and distributing materials. The placing boom of the concrete pump truck is the main force member in the pump parts with bearing great pressure, and its stress condition is complex. Taking the HB48B placing boom as an example, this paper analyzes and studies the deformation produced by placing boom of pump truck, and then obtains some main factors affecting the welding deformation. Through the riveter "joint" size, we controlled the process parameters, post-welding processing, and other aspects. These measures had some practical significance to prevent, control, and reduce the deformation of welding.

1. Use Background of Concrete Pump Truck

The concrete pump truck is a special vehicle which integrates the pump transport mechanism used for pumping concrete and the boom used for distributing materials into the automobile chassis. When under working, it uses the power of the diesel engine in the automobile chassis; passes the power to the hydraulic pump through the transfer case; and then drives the concrete pumping mechanism and the boom system; the pumping mechanism adds pressure on the concretes in the hopper and pumps them into the pipe; the pipe is attached to the boom; the boom can be moved; the concrete pumped out by the pumping mechanism will be directly sent to the pouring point. Compared with other pumps, the pump truck has the following advantages. 1. The pipe is attached to the boom. When the pump truck is driven to the workplace, it can soon open the boom and work. The preparation time is short, and usually the the pump truck is ready for work within half an hour. 2. The pump truck is equipped with hydraulic roll-folding type boom, which can flexibly rotate within the scope of work. The distributing material is convenient with fast pumping speed generally within 903-1503m/h, and the work efficiency is high. 3. The degree of automation is high. Only one person can operate the entire pump from the pumping to the distributing material. The pump truck is generally equipped with wireless remote control system, easy to operate. 4. The maneuverability is good. After the completion of one project, the pump truck can quickly come to another project for continuous operation, and the pump truck can simultaneously pump concrete to several projects, so the equipment utilization rate is high. The pump truck was first used in 1979 in China. Then, the pump trucks were introduced from Japan, and then they were constructed in Shanghai Baosteel. In 1982, the Hubei Construction

Machinery Factory introduced the boom production technology from Japan IHI Corproation and began to produce pump truck, so the Hubei Construction Machinery Factory became the first concrete pump truck production plant. With the development of the construction industry, the pump manufacturers are gradually increased, but the booms were mostly imported in the beginning. Now, the production supporting mode of the booms is mainly homemade with auxiliary import. In 1999, the SANY Heavy Industry Co., Ltd began to develop its own 37-meter-long concrete pump truck, becoming the earliest enterprise to self develop the long boom concrete pump truck.

2. The Main Description of the Three-Section Arm

The main structures of the three-section arm are composed of base plates, side plates, cover plates, bar plates, reinforced plates, and axle sleeves. The axle sleeves in the front and rear end of the three-section arm have the role to connect the second and the fourth arm. In the welding process, the front and rear end of the three-section arm are deviated in the horizontal direction. In the welding process, the welding parts of the undertaking axle sleeves produce the angular deformation and distortion deformation, mainly resulting into this problem. The welding deformation caused by the welding seam in the front and rear ends of the side plates makes the dimension deviation appeared in the front and rear ends exceed the technical requirement of 2mm. This does not conform to the production process requirement, and the product quality is not guaranteed, resulting in unnecessary rework and repair as well as reducing the production efficiency.

3. The Deformation Cause of HB48B Three-Section Arm

3.1 Dimensional Deviations Caused in Splicing

The axle sleeves in the two ends of the three-section arm and the side plates formed an annular welding seam; the side plates and the cover plate lap of base plate formed the main welding pass; the welding seam had the V-shaped groove. The material of the plate parts was the WELDOX700 low-alloy high-strength steel; the groove angle was 40°; the thickness of plate was 12mm. The joint type was the butt joint; the butt welding seam was formed after welding; the flat welding was used in the welding position, as shown in Figure 1 and 2. Because some steel plates were bended in the material-discharging, this led to overall not too flat the cover plate of the base plate. Moreover, in the splicing process of riveters, the process was not strictly controlled with assembly error, resulting in the uneven gap of welding seam of red arrow. The gap exceeded the 0-1mm stipulated by the process. The welding pass size of the annular welding seam in the axle sleeves was uneven, and a gap was appeared among the base plate, the cover plate, and the side plate. This caused the too big partial hot input, so the heat was too concentrated, thus producing the welding deformation after welding by welder.

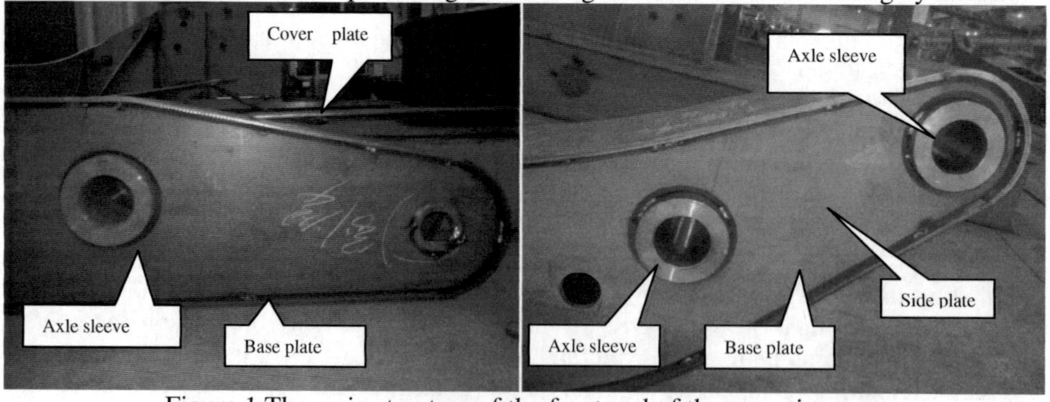

Figure 1 The main structure of the front end of three-section arm
Figure 2 The main structure of the rear end of three-section arm

3.2 Wrong Welding Order

The welding order has a great influence on the welding residual stress and deformation, and the distribution law of residual stress can be changed when different welding sequence is adopted, which has a great effect on controlling the welding deformation. Especially in the welding lap plate, the

IOP Conf. Series: Materials Science and Engineering **307** (2018) 012014 doi:10.1088/1757-899X/307/1/012014

welding seam distribution is more concentrated and the function is more obvious. In order to prevent leakage welding, the welder often first conducted bottom welding on the two welding seams in the side plate and the base plate, and then turned over the welding parts. The welder then welded the main welding pass in the side plate and the cover plate. It is shown in Figure 3 below.

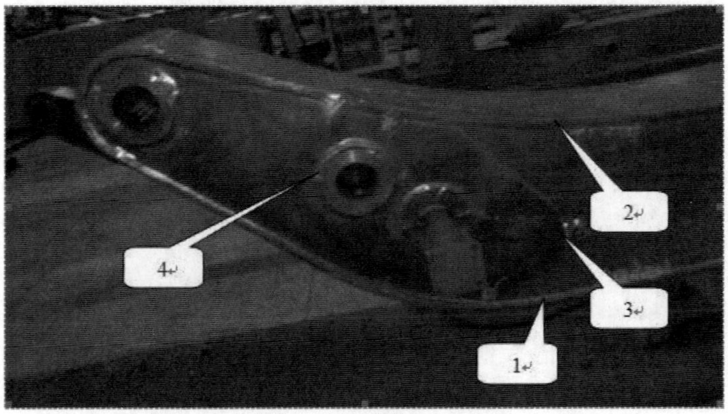

Figure 3 Wrong welding order

If we first finished welding the four main welding seams, and then if we welded the inner welding seams and the annular welding seams in the axle sleeve, this welding method could not make the cover plate, the axle sleeve and the side plate effectively rigid fixed, resulting in the welding deformation. Furthermore, in the actual welding process of the main welding pass, the welders did not use decentralized symmetrical welding, but they often directly used continuous straight-through welding, so that the heat of the whole welding seam was more concentrated and uneven distribution, which made the components produce greater internal stress and increased the amount of deformation.

3.3 Unreasonable Selection of Welding Parameters

Because the HB48B three-section arm uses the 12mm WELDOX700 high-strength steel plate, the process guidance book requires that we adopt the 3-layer 3-way welding, and it also requires that we preheat the welding seam 100-150 ℃ before welding. Meanwhile, it also requires that we control the temperature among layers within 150-170 ℃ in the welding process. In actual production, due to the heavy production task, in order to complete the daily production task as soon as possible, the workers often adopted incorrect welding parameters, as shown in table 1. The workers increased the current and voltage without preheating, and also increased the thickness of the bottom welding. Meanwhile, they did not pay attention to control the temperature among the layers, resulting in increased welding heat input, heat concentration, and increased deformation.

Table 1 The actually used welding parameters

Welding layers	Welding method	Welding position	Method of moving welding rod	Welding materials	Welding current (A)	Arc voltage (V)	Welding speed (cm/min)	Gasflow (L/min)	Welding torch angle (°)
The first layer					260~280	26~28	36~40		
The second layer	CO_2 gas shielded welding	Flat welding	Diagonal circle reciprocating	SLD-70 diameter 1.2	260~280	26~28	32~36	15~25	40~55
The third layer					280~300	28~32	30~32		

3.4 Unreasonable Display of Welding Parts

Because the three-section arm was a Z-type structure. When welding both ends, it required that backing plates were used to make completely flat two ends and parallel both ends of the arm. But in

IOP Conf. Series: Materials Science and Engineering **307** (2018) 012014 doi:10.1088/1757-899X/307/1/012014

the actual welding, in order to improve production efficiency, the welders would often ignored the parallel, resulting in unnecessary welding deformation as shown in Figure 4.

Figure 4 Wrong placement when welding

4. Measures for Controlling Deformation of HB48B

4.1 Guaranteeing the Quality When the Arm is in Splicing

In the bended plate connected by the base plate, the cover plate, and the side plate, for the positions with gaps, we used hammer to hit them and made the gaps reduced to the technical requirements of the 0-1mm, and we ensured that the surrounding gap of the annular welding seam connected by the side plate and axle sleeve was even. After completing the splicing, we reasonably reinforced the entire arm, and we welded 40mm every 30cm, guaranteeing the rigidity strength of steel plate connection, in order to control the welding distortion.

4.2 Reasonable Selection of Welding Order

According to the degree of contraction, we first welded the welding seam with small shrinkage, and then we welded the long welding seam with large shrinkage. After this, we conducted bottom welding on the axle sleeve and the base plate of the side plate near the axle sleeve. Finally, we welded the inner welding seam. After all welding seams were completed, we welded the cover plate. The welding order could effectively connect the axle sleeve, the side plate with the base plate. For the welding seam of the side plate near the welding axle sleeve, the welding order could prevent the welding deformation, could effectively fix welding, and could reduce the welding deformation.

4.3 Reasonable Selection of Welding Parameters

The welding parameter requirements for the three-section arm in the HB48B Welding Operation Guide are shown in table 2.

Table 2 Requirements on welding parameters from the process guidance book

Welding layers	Welding method	Welding position	Method of moving welding rod	Welding materials	Welding current (A)	Arc voltage (V)	Welding speed (cm/min)	Gas flow (L/min)	Welding torch angle (°)
The first layer	MAG welding	Flat welding	Diagonal circle reciprocating	SLD-70 diameter 1.2	240~250	23~26	32~36		
The second layer					250~260	24~26	30~32	15~25	40~55
The third layer					240~250	26~28	28~30		

And it requires the welding pieces to be preheated to 100-150 ℃ before welding. The welding torch is perpendicular to the base material, and the angle between the welding torch and welding seam is 40-55°. The welding wire grade is SLD-70 wire. We require that we use the welding parameters in strict accordance with the requirements of the welding operation guidance book in the welding process, and we require that we can conduct welding after we preheat the welding seam within 100-150 ℃.

IOP Conf. Series: Materials Science and Engineering **307** (2018) 012014 doi:10.1088/1757-899X/307/1/012014

During the welding process, we pay attention to control the temperature among the layers within 150-75 ℃. Using this welding method can not only improve the weld quality, but also make the welding seam heat even. It also reduces the concentration of heat, and effectively reduces the welding deformation.

4.4 Reasonable Placement of Welding Parts

According to the drawings as well as the actual size of the three-section arm, we developed backing plate with suitable height, which could guarantee the parallelism of the three-section arm in the production. The appropriate use of backing plate avoided workpiece hit the ground in the turning and reduced unnecessary deformation. It is shown in Figure 5 below ,in the figure 5, h=570mm.

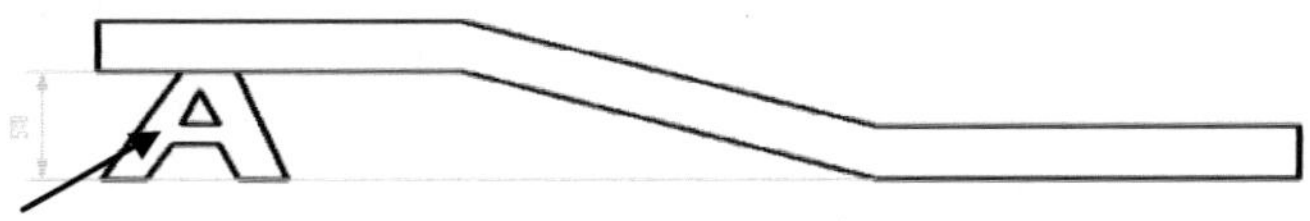

Figure 5, the three-section arm

4.5 Flame Correction

After the welding of the three-section arm was finished, we measured the three-section arm. If the measured results were different to those in the process sizes of the drawings, we conducted flame correction on the arm. The flame correction used linear heating method, using 700-800 ℃. We used infrared thermometer for real-time measurement to ensure the accuracy of the temperature. After the correction was completed, the welding parts were measured again.

4.6 Control Results

After we applied the series of measures mentioned above to measure the deformed parts, the deformation errors were all within 2mm, which conformed to the requirements of the process technology completely. The problem of assembly interference caused by welding deformation was avoided. The comparison before and after controlling deformation is shown in table 3.

Table 3 The results comparison before and after controlling deformation

Numbers of welding groups	The deformation before improvement (mm)	The deformation after improvement (mm)
The first group	1.5	0.8
The second group	1.8	0.9
The third group	3	1.2
The fourth group	2.2	1.1
The fifth group	5	1
The sixth group	4.5	0.9
Mean	3	0.98

Conclusion:

In the manufacturing process of the arm supports, we ensured the quality of the arm in splicing, reasonably selected the welding orders and the welding parameters, and we reasonably placed the welding parts and corrected flame. After we applied the series of measures mentioned above to measure the deformed parts, the deformation errors were all within 2mm, which conformed to the requirements of the process technology completely.

References:

[1] Ba Lianwen compiled. The Quality Control and Inspection Textbook in the Welding Process [M]. Standards Press of China, 2006.

[2] Ma Xianmang. Fault Maintenance of Arm Support System of Concrete Pump Truck [J]. Plant Maintenance Engineering, 2017 (10).

IOP Conf. Series: Materials Science and Engineering **307** (2018) 012014 doi:10.1088/1757-899X/307/1/012014

[3] Zhou Fulong. Defender Sticking to Change-The operation debate on Arm Support Type Concrete Pump Truck in 2014 [J]. Construction Machinery & Maintenance, 2014 (01).
[4] Liu Yonghui. Common Faults and Maintenance of Concrete Pump Truck [J]. Technology and Market, 2013 (02).

IOP Conf. Series: Materials Science and Engineering **307** (2018) 012015 doi:10.1088/1757-899X/307/1/012015

Deformation of products cut on AWJ x-y tables and its suppression

L M Hlaváč[1,4], I M Hlaváčová[1], Š Plančár[2], T Krenický[3], V Geryk[1]

[1]VŠB - Technical University of Ostrava, Department of Physics, 17.listopadu 15/2172, 70833 Ostrava – Poruba, Czech Republic

[2]Watting, s.r.o., Budovateľská 38, 080 01 Prešov, Slovak Republic

[3]Technical University of Košice, Faculty of Manufacturing Technologies, Štúrova 31, 080 01 Prešov, Slovak Republic

libor.hlavac@vsb.cz

Abstract. The aim of this study is namely investigation of the abrasive water jet (AWJ) cutting of column pieces on commercial x-y cutting machines with AWJ. The shape deformation in curved and/or stepped parts of cutting trajectories caused by both the trailback (declination angle) and the taper (inclination of cut walls) can be calculated from submitted analytical model. Some of the results were compared with data measured on samples cut on two types of commercial tables. The main motivation of this investigation is determination of the percentage difference between predicted and real distortion of cutting product, i.e. accuracy of prepared analytical model. Subsequently, the possibility of reduction of the distortion can be studied through implementation of the theoretical model into the control systems of the cutting machines with the system for cutting head tilting. Despite some limitations of the used AWJ machines the comparison of calculated dimensions with the real ones shows very good correlation of model and experimental data lying within the range of measurement uncertainty. Results on special device demonstrated that the shape deformation in curved parts of the cutting trajectory can be substantially reduced through tilting of the cutting head.

Nomenclature

a_m average size of material structural units [m]
C_A coefficient for adjusting jet power with respect to variations in abrasive content below and above the saturation level [-]
C_Q coefficient enabling calculation of the traverse speed ensuring achievement of desired cut wall quality [-]
d_o water nozzle diameter [m]
D_{Ie} experimentally determined inlet diameter of the column sample cut by AWJ [mm]
D_{It} theoretically determined inlet diameter of the column sample cut by AWJ [mm]
D_{Oe} experimentally determined outlet diameter of the column sample cut by AWJ [mm]
D_{Ot} theoretically determined outlet diameter of the column sample cut by AWJ [mm]
h actual cut depth [m]
h_{lim} limit cut depth at a given traverse speed [m]
h_Q cut depth at which the desired wall quality is achieved [m]
H material thickness [m]

IOP Conf. Series: Materials Science and Engineering **307** (2018) 012015 doi:10.1088/1757-899X/307/1/012015

k_m attribute characterizing resistance of material to AWJ machining [-]
K material resistance to abrasive (e.g. hardness) [-]
L stand-off distance [m]
p_j pressure determined from density and speed of abrasive jet [Pa]
Q cutting quality indicator - it is expressed as the value of the average angle of striations determined at the bottom of the cut material (measured in degrees)
S_P coefficient of particle integrity (ratio of number of intact particles to the total number of particles) [-]
t_i interaction time [s]
v_P traverse speed (speed of jet movement across material surface) [$m.s^{-1}$]
v_{Plim} limit traverse speed calculated for cutting material of thickness H [$m.s^{-1}$]
v_{Pmin} minimum traverse speed determined from abrasive particles size [$m.s^{-1}$]
v_{PQ} traverse speed ensuring desired cut wall quality on the thickness H of material [$m.s^{-1}$]
α coefficient of jet speed loss in the interaction process with material [-]
ϑ angle between the tangent to the striation curve at depth h and the jet symmetry axis at the point of impact [°]
ϑ_{lim} angle between the tangent to the striation curve at depth h_{lim} and the jet symmetry axis at the point of impact [°]
ξ_j coefficient of attenuation of abrasive jet between focusing tube outlet and material surface [m^{-1}]
ρ_j density of abrasive jet (conversion to homogeneous liquid) [$kg.m^{-3}$]
ρ_m density of machined material [$kg.m^{-3}$]
σ tensile strength (or compressive strength) of processed material [Pa]
σ_s shear strength of processed material [Pa]

1. Introduction

Abrasive water jets (AWJ) can be used for machining of a broad range of materials, provided that suitable parameters are set for respective processes. Today, AWJ is tested as a tool in many machining processes assigned to the rigid tools in past: turning, milling, grinding and polishing. Nevertheless, the most frequent application remains the cutting of shapes from plate-type materials, mainly metals, rocks, glass and plastics. Actually, this method can achieve very good standards of quality selecting lower traverse speeds. The process could be more economical with higher traverse speeds, but the product distortion increases over the acceptable tolerances. Compensation techniques implemented into modern machines act automatically without necessity to set and control them by operator. Nevertheless, some operations cannot be performed. This contribution is aimed at potential improvement of quality through implementation of the controllable device for tilting of the cutting head based on presented cutting model.

The quality of cutting, with a particular focus on the role of material parameters and AWJ parameters, has been studied in past. Some basic models were prepared by Hashish [1], Zeng and Kim [2], Kovacevic and Yong [3, 4], Paul et al. [5, 6], and many other researchers [7-10]. A specific approach to the explanation of several phenomena occurring during AWJ material cutting was presented by Hlaváč [11]. This model is based on the laws of conservation of energy and momentum and on the geometric analysis [12] of phenomena caused by AWJ on cut walls, namely the declination angle and the respective product distortion [13]. Further studies of AWJ cutting phenomena focus on the kerf taper [14] or surface roughness [15]. Nevertheless, the declination angle seems to be a sensitive parameter reflecting the setting of parameters for AWJ processes and material properties (even more sensitive than roughness or the taper). Therefore, this study focuses primarily on comparing the predicted results calculated from theory presented in more detail in [13] and data measured on steel "column" samples cut from plates of various thickness using commercial AWJ machines. Steels were selected due to requirement of relatively homogeneous structure of tested

IOP Conf. Series: Materials Science and Engineering **307** (2018) 012015 doi:10.1088/1757-899X/307/1/012015

material, because laminar and honeycomb structures or significantly non-homogeneous materials destroy the penetrating jet. Therefore, the samples are destroyed rather randomly than deterministically.

2. Theoretical background

The theoretical results were calculated from the model presented in [13] and derived for limit cutting parameters defined in previous publications [13, 14]. One of these significant limit parameters is the limit traverse speed. It is the maximum attainable traverse speed for the given thickness and type of material at which neither non-cut-through parts are more frequent than once per 50 mm, nor the lengths of them are greater than 25% of the cut width. The modified Eq. (1) is used for calculation of the limit traverse speed v_{Plim} belonging to a given material thickness H, i.e. the maximum traverse speed with a minimum quality at the bottom edge (with even some small uncut parts, according to the definition reminded above). The coefficient α is defined by Eq. (2) derived from the law of momentum conservation on the micro-level (grain of abrasive – grain of material) [16]. It determines the effectiveness of the disintegration process regarding selected material properties and several characteristics of the liquid jet. Due to a very difficult getting of all necessary variables in practice (K, t_i, a_m and σ_s), the combination of them present in Eq. (2) is often substituted by material parameter k_m characterizing resistance of material to AWJ penetration. This material characteristic can be calculated from an experimental cut performed on specimen of tested material. Subsequently, the parameter k_m is used for calculation of the coefficient α (substituting the respective combination of variables K, t, a_m and σ_s for k_m) – Eq. (3).

$$v_{Plim} = \left(\frac{C_A S_P \pi d_o \sqrt{2\rho_j p_j^3 e^{-5\xi_j L}} (1-\alpha^2)}{8H(p_j \rho_m \alpha^2 e^{-2\xi_j L} + \sigma \rho_j)} \right)^{\frac{2}{3}} - v_{Pmin} \tag{1}$$

$$\alpha = 1 - \frac{\sqrt{2p_j^3} K t_i}{8\sqrt{\rho_j} a_m \sigma_s} \tag{2}$$

$$\alpha = 1 - \frac{\sqrt{2p_j^3} k_m}{8\sqrt{\rho_j}} \tag{3}$$

For the purpose of determining speed v_{PQ}, which ensures the achievement of the desired quality of the cut walls, the basic relation for determining the limit traverse speed was complemented by the coefficient C_Q (see [11]) – the speed v_{PQ} is determined by Eq. (4). It is also possible to determine the value C_Q from the normalized average angle of striations at the bottom of the cut material (then the Eq. (5) is to be respected).

$$v_{PQ} = C_Q v_{Plim} \tag{4}$$

Speed v_{Plim} (or v_{PQ}) and the dependency of the declination angle on the material thickness [11] can be used to determine the jet delay, which causes deformations of the work-piece (see line segment BC in figure 1). The declination angle is calculated at a selected traverse speed from the Eq. (5) being derived from equation for traverse speed presented in [11]:

$$\vartheta = \vartheta_{lim} \left(\frac{v_P}{v_{Plim}} \right)^{\frac{3}{2}} \tag{5}$$

Plane λ in figure 1 is determined by points ABC; A is the intersection of the jet axis with the inlet surface of material, B is the intersection of the jet entry axis with the outlet surface of material, and C is the intersection of the momentary axis of the curved cutting jet with the outlet surface of material. The red line is the centerline of the trajectory of the jet penetrating the material and lying in plane λ.

IOP Conf. Series: Materials Science and Engineering **307** (2018) 012015 doi:10.1088/1757-899X/307/1/012015

The blue dashed line indicates the cross-section of the jet flowing out of cut material with the bottom plane of the plane-parallel material plate. The dotted line marks the resulting cut shape of the sample (product of the column cutting without use of the compensating tilting). It is necessary to realize that plane λ is "rolling" along the column surface drawn by impinging jet axis.

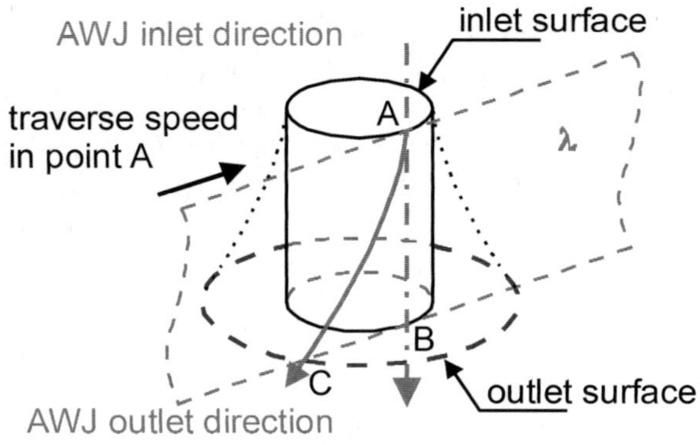

Figure 1. Scheme of the column cutting from a plane-parallel plate of material without tilted cutting head (jet perpendicular to the inlet material surface).

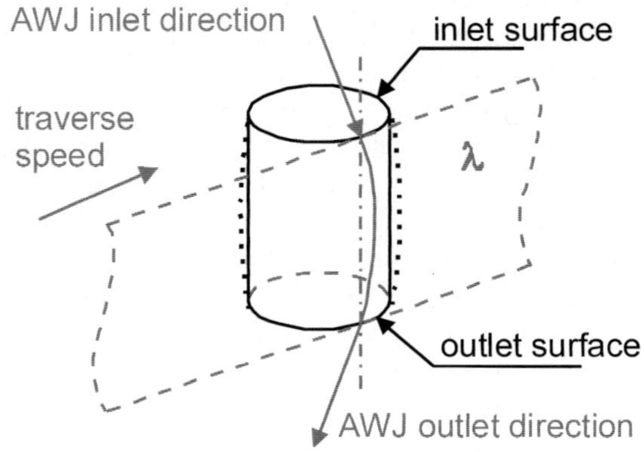

Figure 2. Cutting of column with inclined cutting head.

The taper, studied in [14], can add some deformation described by Eq. (6). The total deformation of the column shape on its bottom can be then calculated from the Eq. (7).

$$\varphi = \varphi_{lim}\left(\frac{h}{h_{lim}}\right)^{\frac{2}{5}} + q \qquad (6)$$

$$D_{bc} = 2\left[\sqrt{\left(\frac{2}{5}H\tan\vartheta\right)^2 + R^2} + \frac{2}{5}H\tan\varphi\right] + d_a \qquad (7)$$

Compensation of the "column" deformation can be accomplished by jet tilting (see figure 2). The attention has been focused namely on the first part of the deformation component in this study, i.e. on the deformation caused by declination angle (retardation of jet in cut). Some calculations of the taper are also presented.

3. Experimental set-up

The preparation and evaluation of the experiments are based on these two assumptions:

- the actual jet axis inside kerf is parallel with the tangent to the front line of the curved cutting edge (see drawing in figure 3),
- the character of the curves of striations produced on the cut walls is identical to the character of the kerf cutting edge front line (shape of the front wall of the channel formed by penetrating jet).

The following process was applied in experiment: The respective traverse speed v_{PQ} was calculated for each tested material and selected declination angle from Eq. (4). Subsequently, cuts were performed at this speed along a circular trajectory in a plane-parallel metal plate using non-tilted jets (perpendicular to the inlet surface of material).

The experiments were carried out at the machines located in the Laboratory of Liquid Jet at the VŠB-Technical University of Ostrava (Czech Republic) and in the company Watting s.r.o. in Prešov (Slovak Republic). The equipment in each case consisted of a pump (enabling working pressure and flow to be set within a certain range) and a CNC table. The table is without controlled tilting in Ostrava and with an automated tilting that can be switched off in Prešov. The selected parameters at the individual workplaces were identical except type of the abrasive and respective mean grain size of the inlet particles, determined by laser analyzer of particles Mastersizer 2000. The conditions are summarized in table 1.

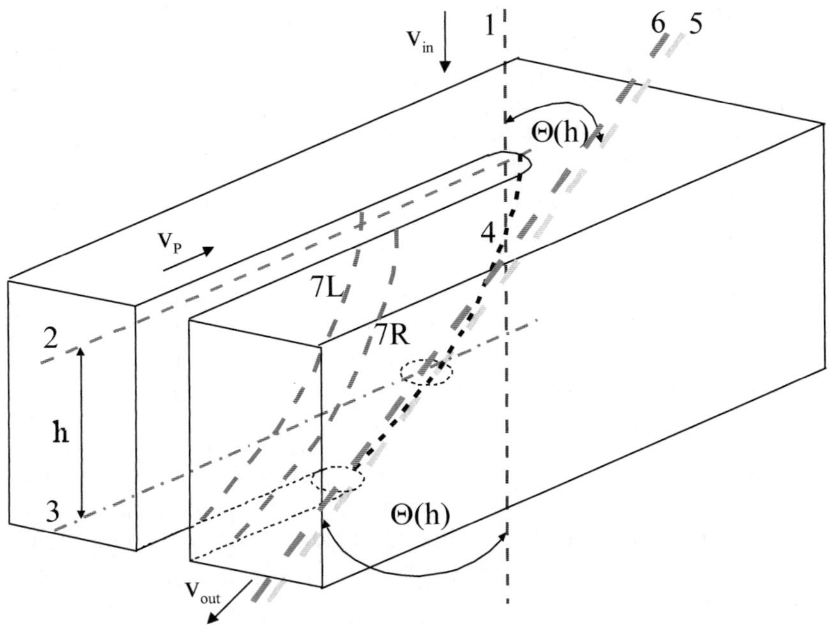

Figure 3. Kerf cutting: 1 – jet axis, 2 – kerf width axis on the top, 3 – kerf width axis in the depth *h*, 4 – kerf cutting edge front line, 5 – tangent to the front line in the depth *h*, 6 - actual jet "axis" parallel to the tangent of the front line in the depth *h*, 7L, 7R – left and right kerf wall striations with shapes copying the shape of the kerf cutting edge front line.

Table 1. Constant parameters used for experiments and calculations

	Constant cutting parameters on testing workplaces	
Factor	Ostrava	Prešov
Pressure in pump	380 MPa	380 MPa
Water jet diameter	0.254 mm	0.254 mm
Focusing tube diameter	1.02 mm	1.02 mm
Focusing tube length	76 mm	76 mm
Abrasive mass flow rate	250 g/min	250 g/min
Mean abrasive grain size[a]	0.275 mm (80 mesh)[b]	0.225 mm (80 mesh)[b]
Abrasive type	Australian garnet (almandine)	Indian garnet (almandine)
Stand-off distance	2 mm	2 mm

[a] Mean grain size is determined on the commercial particles size analyzer.
[b] The "mesh" specification is commercial indication provided by suppliers.

Several kinds of steel samples with thickness 10 mm were used in experiments. One type of steel – WNR 1.0038 (RSt 37-2) – was tested with three thicknesses: 10, 15 and 20 mm. All the cut shapes were the cylindrical "columns" with the diameter 10 mm drawn by the intersection of the jet axis with the selected material surface. The angle was 0° (between jet axis and normal to the material surface), i.e. the perpendicular impact of the jet on the surface. The traverse speeds were calculated for the expected declination of the cutting jet at the exit from material 20°. So, the declination angle 20° is expected to be measured as declination of the striation at the bottom edge of the cut wall. Selected angle (20°) corresponds with the medium cutting quality, which is considered to be the best economical solution for most applications. However, the jet delay causes visible deformations of the sample at these traverse speeds, overcuts and undercuts in stepped parts of jet trajectories and, therefore, improvement is needed. The declination angle 30° has been also tested, as it corresponds with quality sufficient for cutting of products prepared for subsequent finishing processes. The traverse speeds necessary to achieve the selected declination angles were also calculated in the first series of experiments for each material separately.

4. Experimental results

The results of calculations and experiments are presented in tables 2 – 4. Table 2 summarizes results of the preparation phase of investigation – calculation of limit traverse speed and respective traverse speeds for expected declination angles 20° and 30° and calculation of the inlet and outlet diameters of column samples cut in steel plates.

Table 2. Steel plate thicknesses H, calculated limit traverse speed v_{Plim}, predicted traverse speeds v_{P20} and v_{P30} for expected declination angles 20° and 30° respectively and predicted inlet (D_{I20t}) and outlet (D_{O20t}) diameters of "column" samples cut without compensations (like on machine in Ostrava).

Material type	H	v_{Plim}	v_{P20}	v_{P30}	D_{I20t}	D_{O20t}
WNR norm (DIN norm)	mm	mm/min	mm/min	mm/min	mm	mm
1.057 (St 52-3)	10	250	146	191	8.96	9.40
1.0503 (C 45)	10	200	116	153	8.96	9.40
1.7131 (16 MnCr 5)	10	220	128	168	8.96	9.40
1.7225 (42 CrMo 4)	10	180	105	137	8.96	9.40
1.4541 (X6 CrNiTi 18 10)	10	200	116	153	8.96	9.40
1.2436 (X210 CrW 12)	10	160	93	122	8.96	9.40
1.0038 (RSt 37-2)	10	300	175	229	8.96	9.40
1.0038 (RSt 37-2)	15	200	116	153	8.96	9.88
1.0038 (RSt 37-2)	20	140	82	107	8.96	10.58

Table 3 contains comparison of calculated and measured results for technological equipment without tilting (in Ostrava) and evaluation of the respective relative uncertainties. The equipment in company Watting s.r.o. in Prešov, unfortunately, prevented operators to switch off functions of automated compensation of the jet diameter and the taper. This fact influenced comparison of results from Ostrava and Prešov (although the conditions were set identical except the abrasive material type and respective grain size). Results calculated for samples with compensating functions switched on and respective data measured on samples are summarized in table 4. All results presented in tables indicate that differences between values calculated from the model and the ones determined from experiment are small for both the input and the output diameters of "column" samples prepared on both commercial machines. The results from Prešov proved that input of corrections into the theoretical model during calculations yields results differing from the experimental ones within the uncertainty of measurement (see table 4, columns with theoretical and experimental diameters).

Table 3. Comparison of calculated and experimentally determined inlet (D_{It}, D_{Ie}) and outlet (D_{Ot}, D_{Oe}) diameters of "column" samples cut from 10 mm thick steel plates by a non-tilted jet with traverse speeds v_{P20} including differences ΔD_I and ΔD_O of the diameters.

Material type WNR norm (DIN norm)	v_{P20} mm/min	D_{It} [a] mm	D_{Ie} mm	D_{Ot} mm	D_{Oe} mm	ΔD_I %	ΔD_O %
1.057 (St 52-3)	146	8.96	8.96	9.40	9.53	0.22	1.36
1.0503 (C 45)	116	8.96	9.00	9.40	9.52	0.44	1.37
1.7131 (16 MnCr 5)	128	8.96	8.92	9.40	9.54	0.45	1.57
1.7225 (42 CrMo 4)	105	8.96	8.97	9.40	9.55	0.11	1.57
1.4541 (X6 CrNiTi 18 10)	116	8.96	8.98	9.40	9.53	0.22	1.47
1.2436 (X210 CrW 12)	93	8.96	8.92	9.40	9.61	0.22	2.29

[a] Theoretical inlet diameter is determined from the circle diameter set on the cutting machine and the diameter of the abrasive water jet (regarding the stand-off distance).

Table 4. Measurement in Prešov – comparison of theoretical (D_{It}, D_{Ot}) and experimental (D_{Ie}, D_{Oe}) inlet and outlet diameters of "column" samples for traverse speed v_{P20} (from table 1) with non-tilted jet and respective percentage differences ΔD_{I20} and ΔD_{O20} of these diameters. The samples were prepared with automatically implemented compensation of jet diameter.

Material type WNR norm (DIN norm)	H mm	D_{It} mm	D_{Ie} mm	D_{Ot} mm	D_{Oe} mm	ΔD_{I20} %	ΔD_{O20} %
1.057 (St 52-3)	10	10.01	10.03	10.41	10.44	0.20	0.19
1.0503 (C 45)	10	10.01	10.03	10.41	10.33	0.20	0.77
1.7131 (16 MnCr 5)	10	10.01	10.04	10.41	10.48	0.30	0.67
1.7225 (42 CrMo 4)	10	10.01	10.02	10.41	10.26	0.10	1.56
1.4541 (X6 CrNiTi 18 10)	10	10.01	10.03	10.41	10.34	0.20	0.68
1.2436 (X210 CrW 12)	10	10.01	10.04	10.41	10.44	0.30	0.29
1.0038 (RSt 37-2)	10	10.01	10.05	10.41	10.55	0.40	1.42
1.0038 (RSt 37-2)	15	10.01	10.04	10.91	11.08	0.30	1.53
1.0038 (RSt 37-2)	20	10.01	10.06	11.57	11.46	0.50	0.96

All measurements of samples were carried out using a digital Vernier caliper. Micrometer measurements and evaluation of diameter from the surface of "bases" transformed into circles with the same area were also tested and detectable differences were in the scope of the uncertainties for each method. Therefore, the most simple and quick method was selected. An average combined expanded uncertainty of 0.4 mm was determined for all measurements. It represents a relative uncertainty about 4 % that is usual for this type of experiment. Nevertheless, individual values determined for differences between inlet and outlet bases of cut column samples are usually lower, as it is evident from the results summarized in tables. Therefore, presented results lay in the scope of the overall measurement accuracy.

Table 5 summarizes results calculated for expected declination angle 30° and data measured on cut "column" samples. This declination angle corresponds to the most common traverse speeds used in practice for rough semi-finished products preparation. Therefore, it was necessary to investigate, whether the model yields sufficiently exact results also for these traverse speeds.

Table 5. Measurement in Prešov – comparison of theoretical (D_{It}, D_{Ot}) and experimental (D_{Ie}, D_{Oe}) inlet and outlet diameters of "column" samples for traverse speed v_{P30} (from table 1) with non-tilted jet and respective percentage differences ΔD_{I30} and ΔD_{O30} of these diameters. The samples were prepared with automatically implemented compensation of jet diameter.

		v_{P30}					
Material type	H	D_{It}	D_{Ie}	D_{Ot}	D_{Oe}	ΔD_{I30}	ΔD_{O30}
WNR norm (DIN norm)	mm	mm	mm	mm	mm	%	%
1.057 (St 52-3)	10	10.01	10.04	10.78	10.72	0.30	0.65
1.0503 (C 45)	10	10.01	10.03	10.78	10.58	0.20	1.89
1.7131 (16 MnCr 5)	10	10.01	10.08	10.78	10.62	0.70	1.51
1.7225 (42 CrMo 4)	10	10.01	10.06	10.78	10.55	0.50	2.27
1.4541 (X6 CrNiTi 18 10)	10	10.01	10.04	10.78	10.54	0.30	2.28
1.2436 (X210 CrW 12)	10	10.01	10.04	10.78	10.57	0.30	1.99
1.0038 (RSt 37-2)	10	10.01	10.05	10.78	10.68	0.40	0.84
1.0038 (RSt 37-2)	15	10.01	10.08	11.47	11.72	0.70	2.13
1.0038 (RSt 37-2)	20	10.01	10.04	12.32	11.80	0.30	4.41

The last group of experiments was intended for verification of benefits yield by tilted-jet cutting according to proposals published in [13]. This experiment, however, requires compliance with the proper geometry of cutting around the whole column circumference like it is shown in figure 2.

This geometry is not easily feasible on classical x-y tables with integrated software automatically tilting the cutting head regarding internal procedures reflecting projected cutting trajectory and set traverse speeds. Therefore, it was necessary to design and construct a special device (figure 4) facilitating the desired geometry sketched in figure 2.

Figure 4. Special device for sample rotation.

This device has been used for carrying out the experiments most exactly complying requirements stated for compensation of distortion caused by the declination angle on column samples (without compensation of the taper). The device is used on x-y table with stable cutting head tilted to the selected angle and it rotates with the cut material. Therefore, the tilted jet is properly situated in the plane tangential to the column surface around the whole column circumference and this configuration makes possible to study compensation of the declination angle very precisely and independently to the taper. Measured results are summarized in table 6.

Table 6. Ostrava – experiments with rotating cut material 10 mm thick and stable jet: inlet D_I and outlet D_O diameters of "columns" are reported. Speed v_{P20} was applied for expected declination angle of 20° for non-tilted jet and tilting 10°was applied for expected elimination of distortion effects caused by trailback. The percentage differences between the theoretical and experimental results at the inlet and the outlet column bases are compared (ΔD_I for inlet, ΔD_O for outlet). All results are rounded to two decimal places.

		tilted – predicted		tilted – measured		difference	
Material type	v_{P20}	D_{It}	D_{Ot}	D_{Ie}	D_{Oe}	ΔD_I	ΔD_O
WNR norm (DIN norm)	mm/min	mm	mm	mm	mm	%	%
1.057 (St 52-3)	146	9.33	9.68	9.33	9.73	0.01	0.48
1.0503 (C 45)	116	9.30	9.66	9.33	9.66	0.28	0.06
1.7131 (16 MnCr 5)	128	9.29	9.65	9.32	9.73	0.36	0.84
1.7225 (42 CrMo 4)	105	9.30	9.66	9.32	9.63	0.22	0.24
1.4541 (X6 CrNiTi 18 10)	116	9.28	9.64	9.31	9.62	0.32	0.25
1.2436 (X210 CrW 12)	93	9.31	9.67	9.34	9.66	0.29	0.03

The relative impact of declination angle and the taper on the cut part shape depends on the radius of cutting trajectory – the smaller the radius the major relative deformation. Detailed explanation of this fact was presented in [13]. In view of the uncertainties mentioned above and the measurement uncertainties, it was not possible to determine clearly the amount of deformation caused by the convergence or divergence of the cut walls (the taper), as described in [14]. This issue will be examined in subsequent studies of jet – material investigations. It is also necessary to evaluate influence of other phenomena affecting the process, as documented in [16, 17, 18], i.e. material refining and heat treatment, vibrations of the cutting system or quality of abrasive material and non-homogeneity of its feeding rate. In certain systems, it can be also problem with pulsing of the water due to stopping and starting of the system, like described in [19]. Quality of cutting on certain materials, namely aluminium alloys, can be influenced also by size of abrasive grains [20]. Therefore, further research will be focused also on these factors. Nevertheless, contemporary results show a very good agreement of predicted product dimensions with the real ones measured after cutting.

5. Discussion

This contribution is aimed at proof of predictions derived from the own theoretical models on the real machines for water jetting used in practice. The predicted results were fully confirmed as it can be seen not only from results presented in Tabs 3 – 5, but namely from results presented in Tab 6, where the taper calculated for respective traverse speeds has been incorporated. It can be seen, that respective relative differences can be caused by some defects in tested material, operation of manipulating technique, abrasive feed rate etc. Further investigations will be focused also on the extent of these factors and their impact on product shape and accuracy to make the tool abrasive water jet even more precise.

6. Conclusions

Presented investigations were focused on distortion of "column"-shaped samples during abrasive water jet cutting on equipment in commercial firms. The expected results were calculated from presented model and compared with results measured on produces samples. Two machines were used for this investigation. A special device constructed so that it can eliminate limitations introduced by manufacturers into software controlling movement of cutting machines was produced. The device rotates with material under stable jet (either non-tilted or tilted). Therefore, no machine software is used during experiment. Results of this investigation reliably proved that theoretical model is powerful tool for calculations of the predicted product shapes both for non-tilted and tilted jets. The results also proved that tilting of the cutting head (and thus AWJ) can substantially suppress an undesirable shape damage caused by AWJ trailback and taper.

Acknowledgements
This paper was created with the support of the Student grant No.44/2017 funded by Ministry of Education, Youth and Sports of the Czech Republic.

References
[1] Hashish M 1989 *J. Eng. Mater.-T. ASME* **111** 154
[2] Zeng J, Kim T J 1996 *Wear* 193, 207
[3] Kovacevic R, Yong Z 1996 *Proc. 13th Int. Conf. on Jetting Technology (Cagliari)* (Bury St Edmunds & London: Mech. Eng. Pub. Ltd.) p 73
[4] Yong Z, Kovacevic R 1996 *Proc. 13th Int. Conf. on Jetting Technology (Cagliari)* (Bury St Edmunds & London: Mech. Eng. Pub. Ltd.) p 83
[5] Paul S, Hoogstrate A M, van Luttervelt C A, Kals H J J 1998 *J. Mater. Process. Technol.* **73** 189
[6] Paul S, Hoogstrate A M, van Luttervelt C A, Kals H J J 1998 *J. Mater. Process. Technol.* **73** 206
[7] Vikram G, Babu N R 2002 *Int. J. Mach. Tools Manuf.* **42** 1345
[8] Chen F L, Wang J, Lemma E, Siores E 2003 *J. Mater. Process. Technol.* **141** 213
[9] Deam R T, Lemma E, Ahmed D H 2004 *Wear* **257** 877
[10] Orbanic H, Junkar M 2008 *Wear* **265** 821
[11] Hlaváč L M 1998 *Proc. 14th Int. Conf. on Jetting Technology (Brugge)* (Bury St Edmunds & London: Prof. Eng. Pub. Ltd.) p 25
[12] Hlaváč L M 2009 *J. Mater. Process. Technol.* **209** 4154
[13] Hlaváč L M, Strnadel B, Kaličinský J, Gembalová L 2012 *Int. J. Adv. Manuf. Technol.* **62** 157
[14] Hlaváč L M, Hlaváčová I M, Geryk V, Plančár Š 2014 *Int. J. Adv. Manuf. Technol.* **77** 1811
[15] Caydas U, Hascalik A 2008 J. Mater. Process. Technol. **202** 574
[16] Strnadel B, Hlaváč L M, Gembalová L 2013 *Int. J. Mach. Tools Manuf.* **64** 12
[17] Salokyová Š 2016 *Key Engineering Materials* **669** 243
[18] Hlaváčová I M, Geryk V 2017 *Int. J. Adv. Manuf. Technol.* **90** 1217
[19] Hlaváčová I M, Čikovová M, Wicher K 2014 *Int. J. Heat Mass Transf.* **77** 704
[20] Yuvaraj N, Kumar M P 2017 *Mach. Sci. Technol.* **21** 385

IOP Conf. Series: Materials Science and Engineering **307** (2018) 012016 doi:10.1088/1757-899X/307/1/012016

Micro-mirror with hybrid photoelectric-electrostatic driving of PLZT ceramic

Y J TANG [1], Y S CHEN [2], K QIAO[2]and X J WANG [2]

[1]School of Intelligent Science and Control Engineering, Jinling Institute of Technology,Nanjing 211169, China
[2]School of Mechanical Engineering, Nanjing University of Science and Technology, Nanjing 210094,China

Email: yjtang@jit.edu.cn

Abstract. A novel micro-mirror driven by hybrid photoelectric-electrostatic actuation of PLZT ceramic is proposed based on the anomalous photovoltaic effect of PLZT ceramic. On the basis of analysis of coupling relationships of multi-physics fields of PLZT ceramic, the photovoltage of PLZT ceramic is established during illumination and light off phases. Furthermore, the relationship between the driving displacement and the photovoltage of PLZT ceramics is present. In addition, the feasibility of micro-mirror with hybrid photoelectric-electrostatic driving is verified via closed-loop control for photovoltage of PLZT ceramic. The experimental results show that the photovoltage of PLZT ceramics has good dynamic control precision using on-off closed-loop control method.

1. Introduction

Micro-Opto-Electro-Mechanical System (MOEMS) micro-mirror is widely applied in optical precision measurement and communication, biomedical imaging and other military and civil fields because of the merits of small size, light weight, low power consumption, etc. There are some driving methods used most for micro-mirror, such as piezoelectric, electromagnetic, and electrostatic driving, etc. [1]. The above-mentioned driving methods can be used in different operating environment and technical fields according to their own advantages and disadvantages. However, additional external power should be supplied for above-mentioned driving methods. It not only produces electromagnetic interference, but also has more limitations working in the clean operating space, vacuum environment and other independent working environment. Compared with the above-mentioned driving mode, optical driving technology has the advantages of driving cleaning, such as no electromagnetic interference, non-contact remote optical control and wireless energy transmission, which is an ideal driving mode for MOEMS micro mirror [2]. Actuation using PLZT ceramic is one of optical driving methods. When illuminated by UV light, PLZT ceramic can produce the voltage of several kV/cm and photo-induced deformation in the polarization direction based on anomalous photovoltaic effect and photostrictive effect, Therefore, the photovoltage and photo-induced deformation of PLZT ceramics can be applied in the driving mode for MOEMS micro-mirror, which can offset the drawbacks of traditional micro-mirror driving method.

The mathematical model of PLZT ceramic with coupled multi-physics fields is a basis for the theoretical analysis of MOEMS micro-mirror with PLZT ceramics. Up to now, the mathematical model of PLZT ceramic has been studied by some scholars in recent years. In 1983, RC charging circuit model was proposed by Brody et al [3], but did not analyze the change of the electric field after the light turned off. In 1993, Fukuda et al [4] experimentally confirmed that the photostrictive effect of PLZT ceramic is the result of multi-physical fields coupling. However, the coupling relationship of different physical fields has not been modeled. In 2005, Shih et al [5] proposed a photostrictive constitutive model of PLZT ceramic. However, the impacts of thermal expansion on the electric field are ignored in the photostrictive constitutive model. Considering the effect of thermal expansion on the photoelectric fields, Wang and Huang et al [6, 7] proposed a mathematical model of photovoltage and photo-induced deformation based on coupled multi-physical fields in 2014. Furthermore, they experimentally verified the established mathematical model. Additionally, Zheng et al [8] establish the constitutive equations for 0-1 and 0-3 polarized PLZT ceramic via finite element method. Jiang et al [9] established an electrical model of hybrid photostrictive/piezoelectric actuation mechanism. The publications mentioned above provide a theoretical basis for active control for micro-driving with PLZT ceramic.

In this paper, a micro-mirror driven by hybrid photoelectric-electrostatic driving of PLZT ceramic is proposed. Working principle of the micro-mirror based on hybrid photoelectric-electrostatic driving mechanism is present. The relationship between the driving displacement of micro-mirror and the photovoltage of PLZT ceramic is established. The feasibility of micro-mirror driven by hybrid photoelectric-electrostatic mechanism is experimentally verified.

2. Working principle of the micro-mirror with photoelectric-electrostatic driving

A novel micro-mirror with hybrid photoelectric-electrostatic driving based on the anomalous photovoltage effect of PLZT ceramic is shown in figure 1. The micro-mirror is located at the free end of the flexible cantilever beam. One copper foil laminated on the lower surface of the beam, the other copper foil is correspondingly fixed on the base. The two copper foils are respectively connected to the electrodes of the PLZT ceramic. When the PLZT ceramic is irradiated by the high-energy ultraviolet light, the photovoltage produced by PLZT ceramic is applied on the two copper foils. Based on the electrostatic driving force between the two copper foils, the flexible cantilever beam will produce a deflection at the free end. Finally, the micro-mirror is driven down.

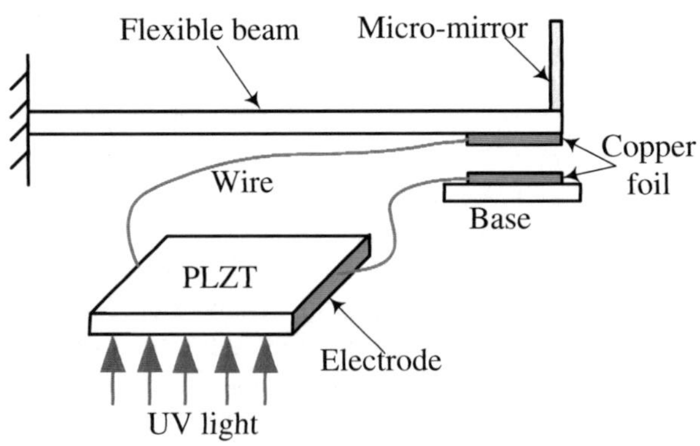

Figure 1. Micro-mirror with hybrid photoelectric-electrostatic driving

Since the electrostatic twisting mechanism has lots of advantages, such as simple structure, fast response, no assembly error, small size and easy control [10], which has been widely used in MOEMS micro-mirror.

3. Multi-physics fields coupling model of photovoltage of PLZT ceramics

When the PLZT ceramic is illuminated by UV light, photo-induced voltage is generated in the poling direction based on the anomalous photovoltaic effect (light energy to electric energy conversion); A part of the light energy is converted to thermal energy by the photothermal effect (light energy to thermal energy conversion), which causes the rising temperature. Meanwhile, the rising temperature triggers the pyroelectric effect (thermal energy to electric energy conversion) and thermal expansion effect. Based on thermal expansion and the direct piezoelectric effect, some electric energy is generated (mechanical energy to electric energy conversion). It is noted that the electric field generated by the anomalous photovoltaic effect and pyroelectric effect are in the same direction as the residual polarization and enhance the internal electric field in PLZT ceramic. However, a portion of electric field produced by the thermal dilatation deformation is in opposite direction with the poling direction and weakens the electric field in PLZT ceramic [11]. Figure 2 shows the coupling relationships of multi-physics fields of PLZT ceramic illuminated by the UV light.

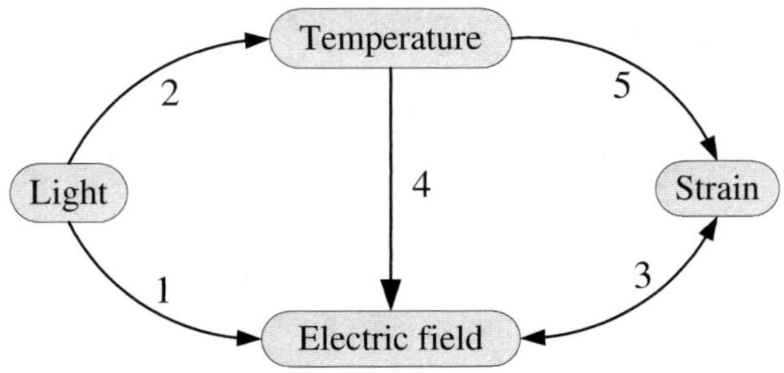

(1)Anomalous photovoltaic effect, (2) Photothermal effect, (3) Piezoelectric effect, (4) Pyroelectric effect, (5) Thermal expansion effect

Figure 2. The coupling relationships of multi-physics fields of PLZT ceramic

According to the coupling mechanism analysis of photovoltage effect, the total photovoltage expression of PLZT ceramic can be written as [12, 13]:

$$\begin{aligned} V(t) &= V_p(t) + V_\theta(t) - \beta V_e(t) \\ &= V_s(1-e^{-\frac{t}{\tau_1}}) + (\frac{AP}{C_p} - \beta\frac{\lambda D_e}{d_{3i}Y_a})\Delta T_s(1-e^{-\frac{t}{\tau_\theta}}) \end{aligned} \tag{1}$$

where d_{3i} indicates the piezoelectric constant of the PLZT ceramic; V_p is photovoltage and $V_p(t)=V_S(1-e^{-\frac{t}{\tau_1}})$; V_s is the saturated photovoltage; τ_1 is the illumination time constant; V_θ is pyroelectric during light on stage and $V_\theta(t)=AP\Delta T_S/C_p(1-e^{-\frac{t}{\tau_\theta}})$; A is the electrode area of PLZT ceramic; P is the pyroelectric coefficient of PLZT ceramic; ΔT_S is the maximal temperature variation, τ_θ is the thermal time constant; λ is the thermal stress coefficient of PLZT ceramic; V_e is voltage caused by thermal deformation during light on stage and $V_e(t)=\lambda D_e/(d_{3i}Y_a)\Delta T_S(1-e^{-\frac{t}{\tau_\theta}})$; Y_a is the elastic modulus of PLZT ceramic; D_e is the distance between the two electrodes of PLZT ceramic and β is the conversion coefficient of thermal deformation and electric field.

After UV light is switched off, the residual voltage and deformation of PLZT ceramics gradually decrease to the initial state of the PLZT before illuminated. Investigating the trend of PLZT ceramics in the light-off phase can help further study of photo-deformation characteristics of PLZT ceramics and promote the engineering application of PLZT ceramics. The residual voltage after turning off the UV light can be obtained as:

$$\begin{aligned} V_d(t) &= V(t_0) + V_{\theta-d}(t) - \beta_2 V_{e-d}(t) \\ &= V(t_0) - (\frac{AP}{C_p} - \frac{\beta_2\lambda D_e}{d_{3i}Y_a})\Delta T_{s-d} + (\frac{AP}{C_p} - \frac{\beta_2\lambda D_e}{d_{3i}Y_a})\Delta T_{s-d}e^{-\frac{t}{\tau_d}} \end{aligned} \tag{2}$$

where $V_{\theta\text{-}d}$ is pyroelectric during light off stage and $V_{\theta-d}(t)=-(\frac{AP}{C_p}-\frac{\beta_2\lambda D_e}{d_{3i}Y_a})\Delta T_{s-d}$; ΔT_{s-d} is the maximal temperature decrement during light-off phase and $\Delta T_{s-d}=T(t_0) - T_0$; $T(t_0)$ is the temperature of PLZT ceramic when the ultraviolet light is switched off; τ_d is the dark time constant. β_2 is the conversion coefficient of thermal deformation and electric field during light off stage. $V_{e\text{-}d}$ is voltage caused by thermal deformation during light off stage.

4. Driving displacement of the micro-mirror with hybrid photoelectric-electrostatic driving

As shown in figure 3, it is a deflection analysis of flexible beam with hybrid photoelectric-electrostatic driving. l is the length of the flexible beam; b is the width of the flexible beam; d is the initial distance between the two cooper foil; l_{cu} and b_{cu} is the length and width of the cooper foil respectively, and $b_{cu}=b$. According to figure 1, the driving displacement of the micro-mirror in Y direction is approximately equal to the deflection of the free tail of the flexible beam.

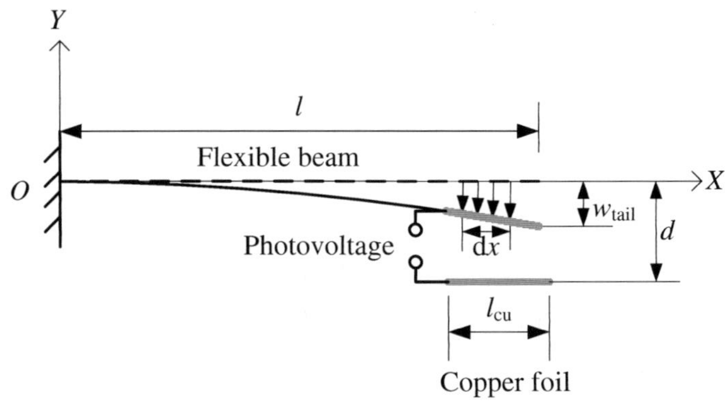

Figure 3. Deflection analysis of flexible beam with hybrid photoelectric-electrostatic driving

When the PLZT ceramic is irradiated by the UV light, the photovoltage is generated and applied to the two copper foils via the lead wire. Micro-mirror will be driven due to electrostatic force between the two copper foils. Herein, taking an infinitesimal unit dx along the X direction in the two copper foils is analysed. It is assumed that the two infinitesimal units in the upper and lower copper foil are parallel. Hence, the two infinitesimal units can be taken as a parallel plate capacitor. The formula of capacitance energy W can be written as

$$W = \frac{1}{2} C U^2 \tag{3}$$

where C is the capacitance of the capacitor; U is the potential difference between the two plates of the capacitor.

According to the principle of virtual work, the electrostatic force between the two plates of the capacitor can be defined as

$$\mathrm{d}F = \frac{\partial W}{\partial(\mathrm{d}w)} = \frac{1}{2} \frac{\varepsilon_{\mathrm{a}} b_{\mathrm{cu}} U^2}{(d - \mathrm{d}w)^2} \mathrm{d}x \tag{4}$$

where ε_a is the dielectric constant of the air; dw is the deflection of the beam at the location of the infinitesimal unit dx.

It is noted that the electrostatic force dF is a concentrated force and located in the region of $[l\text{-}l_{cu}, l]$ in X direction. So the deflection of the free tail of the flexible beam $\mathrm{d}w_{\mathrm{tail}}$ caused by dF can be expressed as

$$\mathrm{d}w_{\mathrm{tail}} = \frac{\mathrm{d}F x^2}{6EI}(3l - x) \quad (l - l_{\mathrm{cu}} \le x \le l) \tag{5}$$

Substituting equation (4) into equation (5), one can get the following formula

$$\mathrm{d}w_{\mathrm{tail}} = \frac{1}{2} \frac{\varepsilon_{\mathrm{a}} b_{\mathrm{cu}} U^2}{(d - \mathrm{d}w)^2} \frac{x^2}{6EI}(3l - x) dx \quad (l - l_{\mathrm{cu}} \le x \le l) \tag{6}$$

According to the working principle of the micro-mirror with photoelectric-electrostatic driving, the photovoltage generated by PLZT ceramic is applied to the two copper foils. Thus, the potential difference U between the two plates of the capacitor should be replaced by the photovoltage $V(t)$ and $V_d(t)$ expressed in equations (1) and (2). Integrating with respect to x, the relationship between the driving displacement (w_{tail}) and the photovoltage of PLZT ceramics is established.

5. Close-loop control for photovoltage of PLZT ceramic

5.1. Experimental setup of close-loop control for photovoltage of PLZT ceramic

In this section, the feasibility of micro-mirror with hybrid photoelectric-electrostatic driving is verified via closed-loop control for photo-induced voltage of PLZT ceramic. The block diagram of close-loop control for photovoltage of PLZT ceramic is shown in figure 4. When the UV light is applied to PLZT ceramic through optical shutter, the photovoltage generated by PLZT ceramic is measured by a high impedance voltmeter and the data is acquired and processed by signal processing module. Afterwards, the voltage data are transferred to the optical shutter control thread. Then the control command is inputted to the optical shutter. Finally, closed-loop servo control for photovoltage of PLZT ceramic is achieved. Experimental setup of closed-loop control for photovoltage of PLZT ceramic is illustrated in figure 5. In the actual operation, the software of the servo control system written by LabVIEW should be started firstly. Afterwards, enter the target value of voltage and initialize the light shutter to "ON".

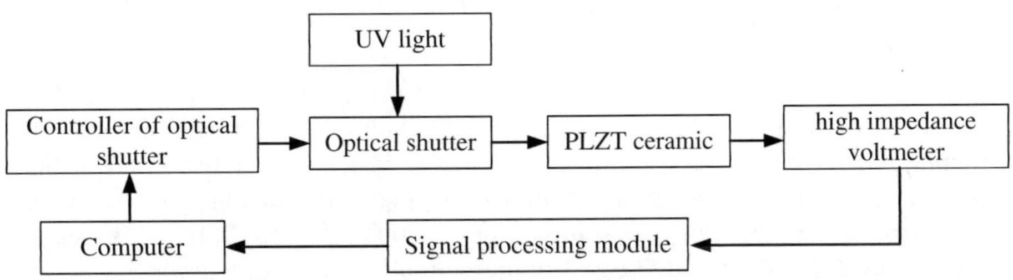

Figure 4. Block diagram of close-loop control for photovoltage of PLZT ceramic

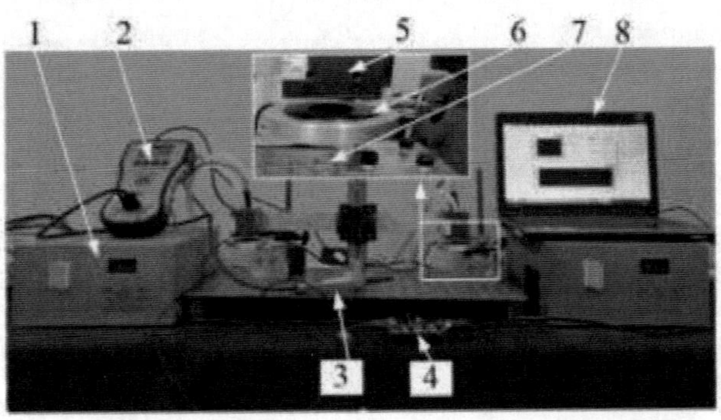

(1) UV light controller; (2) High impedance voltmeter; (3) Probe of high impedance voltmeter; (4) Controller of optical shutter; (5) UV probe; (6) Optical shutter; (7) PLZT ceramic; (8) Computer

Figure 5. Experimental setup of close-loop control for photovoltage of PLZT ceramic

Since the closed-loop control scheme is carried out based on the photovoltage of PLZT ceramic is real-time measured and feedback, so test program for the high impedance voltmeter is written and developed using LabVIEW. Figure 6 is the program front panel, which is divided into two threads: data acquisition thread and light shutter control thread.

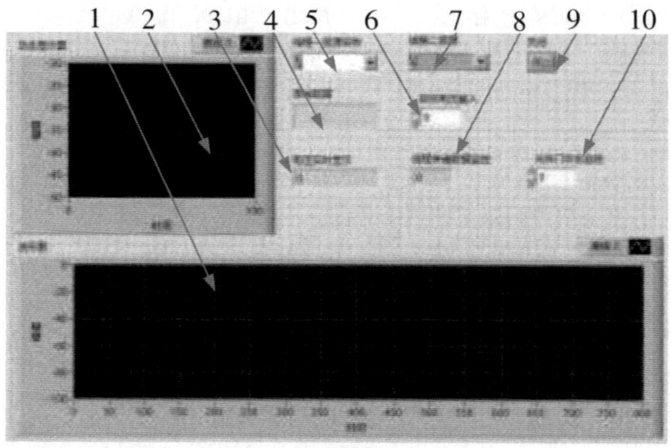

(1) Voltage curve full display window; (2) Voltage curve real-time display window; (3) Real time voltage value; (4) Original sampling date; (5) Thread of the first resource; (6) Target voltage input window; (7) Thread of the second resource; (8) Data transfer value between the two threads; (9) Stop button; (10) Status monitoring window of light shutter

Figure 6. Front panel of the high-impedance voltmeter

5.2. Experimental results of closed-loop control of PLZT ceramic photovoltage

The closed-loop photovoltage control experiment of single PLZT ceramic carried out, where the sampling period is set to 300ms. Figure 7 shows the experimental curve of the closed-loop photovoltage control of single PLZT ceramic with light intensity of 50mW/cm^2 and100mW/cm^2. As illustrated in figure 7(a), the target value of photovoltage is 500V and 1000V. The photovoltage of PLZT ceramic quickly increases to the target value and then fluctuates around the target voltage. As shown in figure 7(b), the target value of photovoltage is 500V, 800V, 1000V, 1100V and 1300V.The photovoltage of PLZT can successfully keep around the multi-target voltage repectively.

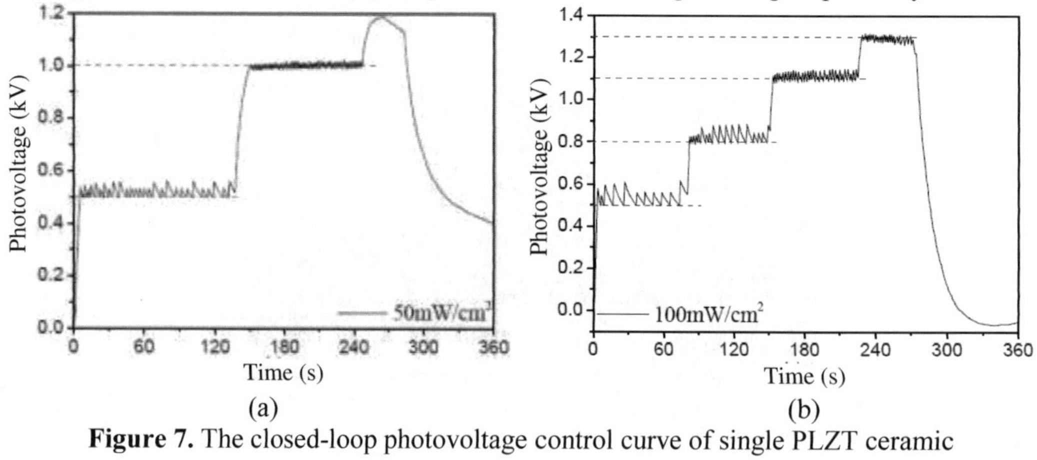

Figure 7. The closed-loop photovoltage control curve of single PLZT ceramic

The experimental results verify the feasibility of hybrid photoelectric-electrostatic driving micro-mirror. Because of the limitation of the sampling frequency of the high impedance voltmeter, the photovoltage data is acquired with certain time delay. Improving the sampling frequency of the high impedance voltmeter and the control algorithm will be carried out to achieve a better control effect in our future research.

6. Conclusions

The micro-mirror with hybrid photoelectric-electrostatic driving is proposed based on anomalous photovoltaic effect in PLZT ceramics exposed to UV light. The photovoltage equations in illumination phase and light-off phase are obtained. Based on the multi-physics coupling model of PLZT ceramics,

IOP Conf. Series: Materials Science and Engineering **307** (2018) 012016 doi:10.1088/1757-899X/307/1/012016

the relationship between the driving displacement of the micro-mirror and the photovoltage of the PLZT ceramic is established. In addition, the closed-loop experimental study on the photovoltage of PLZT ceramics under different light intensities is carried out. The experimental results show that the photovoltage can be successfully controlled around the target value, which means that the deflection of the flexible beam or the driving displacement of micro-mirror can achieve the target value via controlling the photovoltage of PLZT ceramic. Thus, the feasibility of hybrid photoelectric-electrostatic driving micro-mirror is proved.

References

[1] Lu A J, Zhang Z P, Huang Z J2016*Optics and Precision Engineering***24**400-406.
[2] Zhu Y, Liu W, Jia K, et al. 2011*Sensors and Actuators A: Physical***167** 495-501.
[3] Brody P S. 1983*Ferroelectrics***50**27-32.
[4] Fukuda T, Hattori S, Arai F, Matsuura H, Hiramatsu T, Ikeda Y and Maekawa A1993*Proceeding of IEEE Robotics and Automation Conference* 618-23.
[5] Shih H R, Tzou H S and Saypuri M 2005*Journal of Sound and Vibration***284** 361-378.
[6] Huang J H, Wang X J and Wang J 2015*Smart Materials and Structures* **25**025002.
[7] Huang J H, Wang X J and Wang J2015*Optics and Precision Engineering***23** 760-768.
[8] Zheng S J, Tong LY and Luo Q T2015*International Journal of Applied Electromagnetics and Mechanics***49**513-530.
[9] Jiang J, Li X N, Yue H H and Deng Z Q2017*Journal of Mechanical Engineering***53**65-71.
[10] Huang J M, Liu A Q, Deng Z L, Zhang Q.X.Ahna J and Asundid 2004*Sensors and Actuators A: Physical***115**159-167.
[11] Wang X J, Lu F, Liu Y F and Huang J H2016*Optics and Precision Engineering***24** 2505-14.
[12] Fridkin V M1979*Photoferroelectrics* (New York: Springer).
[13] Zhong W L1996*Ferroelectric physics* (Beijing: Science Press)487-488.

IOP Conf. Series: Materials Science and Engineering **307** (2018) 012017 doi:10.1088/1757-899X/307/1/012017

The protective ability of Camellia meal extract on the silk protein

JZ Weng[1,*], C Cai, DY Zhang, BK Dai

[1] College of Chemistry and Material Science, Fujian Normal University Fuzhou
*E-mail:jackyweng@vip.163.com

Abstract. With the enhancement of living standards, people pay more and more attention to the health. The edible oil become more and more popular, but also produced a large amount of Camellia meal which can not fully put into utilization. In this study, the extracting liquid of Camellia meal was used on the process of silk degumming. Firstly, tussah silk was treated by degumming in the Na2CO3 solution, and the preliminary condition of tussah silk degumming was obtained by orthogonal experiment: the concentration Na2CO3 was 0.1%, the degumming time was 1 hour, and the ratio of silk/water was 40:1. Then the extract of Camellia meal (GCJSY) was added before the bleaching process of tussah silk to investigate the protective ability of GCJSY on the silk protein basry on the residual ratio of the silk. While the concentration of GYJSY was 0.08%, the residual ratio of silk after degumming in the Na2CO3 solution and bleaching in the 2% H2O2 solution was 87.2%.

1. Preface

China is the hometown of silk.Silk is a kind of natural macromolecule material with good mechanical properties[1].Silk is mainly composed of the inner silk fibroin and the outer layer sericin[2]. Silk sericin, which is composed of 18 kinds of amino acids, plays a protective and adhesive role in silk fibroin, which is mainly composed of globular proteins. Sericin accounted for 20%~30%` in the silkworm silk, and accounted for about 12%~13% in tussah silk[3].Recent studies shown that sericin has great potential application in functional materials[4].Therefore, the selection of suitable additives could help to form sericin protein composite[5],which also will help to maintain the quality of silk and will expand the functional materials applications based on the silk protein composite[6].In recent years, great efforts have been made to develop Camellia meal plantation in China.

With the production of tea oil, a large amount of Camellia meal producing residue--Camellia meal has been produced[7].At present, the application of Camellia meal in our country was still not enough, most of the detoxification simple as feed or directly as cleaning agent, fuel, fertilizer use[8].So the Camellia meal was resulting in a serious waste of resources. The main components of Camellia meal were some natural organic polymer, such as tea saponin, cellulose, protein and polysaccharide, which can be exacted by biomass refining technology and their equitable applications were the current research focus[9].

In this paper, tussah silk was firstly treated by degumming in the , Na_2CO_3 solution, and the preliminary condition of tussah silk degumming was obtained by orthogonal experiment.Then, the protective ability of the extracted liquid of the Camellia meal on the silk protein were studied.

2. Experimental

2.1Main reagents

All reagents were used as received without further purification. Camellia meal were provided by factory. Tussah cocoon, modifier and catalyst were provided by research group. HCl(AR) was purchased from Quanzhou Donghai Reagent Co., Ltd. NaOH(AR) was purchased from Sinopharm Chemical Reagent Co. Ltd. H_2O_2 was purchased from Sinopharm Chemical Reagent Co., Ltd, Na_2CO_3 was purchased from Guangdong Chemical Reagent Engineering Technology Research and Development Center.

2.2 Experiment content

2.2.1 Preparation of GCJSY

A certain amount of Camellia meal was added into the three necked flask, then water, modifiers, and catalysts were also added. The reacting temperature was controlled at 60℃, and the solution was magnetically stirred with a constant rotation rate of 30 r/min. After a certain time , the mixture was filtered to obtain the desired extract.

2.2.2 Degumming process of silk

The cocoon was dried at 100 ℃ for 2H. Then silks were degumming in the Na_2CO_3 solution.

2.2.3 Bleaching process of silk

The silk was bleached in a hydrogen peroxide solution of 2% concentration. The reaction time is one hour, the reaction temperature is 60 ℃.

2.2.4 Characterization of the silk surface

Scanning electron microscopy (SEM, JSM-7500F, JEOL) examined the surface morphology of silks which were treated by different process with 15 kV electron acceleration voltage.

2.2.5 The residual ratio of silk

The residual ratio of silk reflects the weight loss during the treating process of silk. The remaining ratio of silk is calculated according to the following formula.

$$c\% = \frac{m_2}{m_1} \times 100\% \qquad (1)$$

m_1: the weight of the raw silk .

m_2: the weight of the treated raw silk.

3 Results and discussion

3.1. Orthogonal test of cocoon degumming

In order to obtain the optimum process conditions for silk degumming, an orthogonal test was carried out. Three controllable variables, initial RB concentration (A), FeSO4 concentration (B), and H_2O_2 concentration(C), were set at three levels, respectively. With reference to the experimental design theory, the orthogonal array L9 (33) was selected to arrange the exexperiments (Table 1). The results are listed in Table 2.

Table.1 Orthogonal factor level table

level	Alkali dosage (A)/%	Time(B) /min	The weight ratio of water/silk (C)
1	0.2	60	40:1
2	0.3	90	50:1
3	0.4	120	60:1

According to the R values in Table 2, the influential order of the three factors on the residual ratio of silk degumming was A>B>C. Thus H_2O_2concentration and initial RB concentration have a more important influence on the residual ratio of silk.

According to the value of k, the optimum condition was A1B1C2. However, the optimum condition of A1B1C2 did not appear in the orthogonal test. In the orthogonal test, the best condition was experiment No. (A1B1C1) and its residual ratio of silk was 92.69%.

Thus the optimum conditions were determined as follows: the concentration of , Na_2CO_3 is 0.2%, the degumming time is 1H, and the ratio of water/silk is 40:1.

Table 2 Orthogonal table and range analysia of results

Experimental serial number	A	B	C	Residual ratio of silk/%
1	1	1	1	92.69
2	1	2	2	92.01
3	1	3	3	91.73
4	2	1	2	90.95
5	2	2	3	90.14
6	2	3	1	89.78
7	3	1	3	89.43
8	3	2	1	89.09
9	3	3	2	88.86
K1	276.43	273.07	271.56	
K2	270.87	271.24	271.82	
K3	267.38	270.37	271.30	
k1	92.14	91.02	90.52	
k2	90.29	90.41	90.61	
k3	89.13	90.12	90.43	
R	3.02	0.90	0.18	

3.2 The protective ability of GCJSY on silk protein

3.2.1 The concentration of GCJSY

In this section, GCJSY was added after the degumming process of silk to investigate its protective ability on silk protein. The results were shown in Figure 1. It can be concluded from Figure 1 that the residual ratio of silk decreased slowly with the increasing of the GCJSY's amount. This was because While sericin was dissolved in water, the addition of GCJSY facilitated the separation of sericin from

IOP Conf. Series: Materials Science and Engineering **307** (2018) 012017 doi:10.1088/1757-899X/307/1/012017

silk fibroin. While the concentration of GCJSY reaches 0.1%, the GCJSY has no more influence on the residual ratio of silk.

Cocoons needed to be bleached after degumming, which was due to tannic acid, the pigment in the tussah silk can not be removed under alkaline condition. After the cocoons were degummed under different concentrations of GCJSY, the above cocoons were bleached by 2% H_2O_2. The result was shown in Figure 2.

According to Figure 2, the residual ratios of the silk treated by GCJSY were higher than that of the silk without GCJSY treating. And the residual ratio of the silk was increasing with the increasing of GCJSY's concentration. While the concentration of GCJSY reached 0.08%, the residual ratio of silk reached maximum.

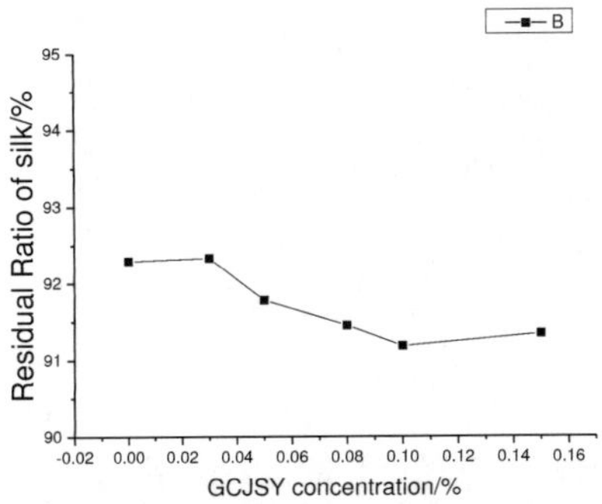

Figure.1 The residual ratio of silk by different concentration of GYCJS after degumming

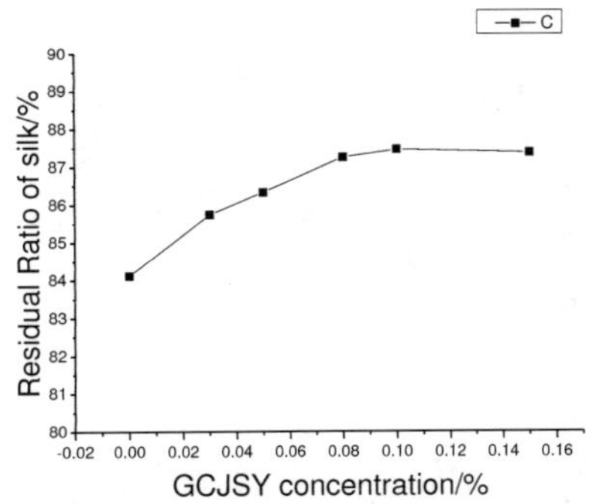

Figure.2 The residual ratio of silk by different concentration of GYCJS after bleaching

3.2.2 Extracting method of Camellia meal

Different exacts which were exacted from camellia meal with different modifiers were added after the degumming process of silk. The concentration of exact was controlled at 0.08%, and the results were shown in Figure 3. The least residue of silk was appeared while the silk was treated by GCJSY-NaOH.

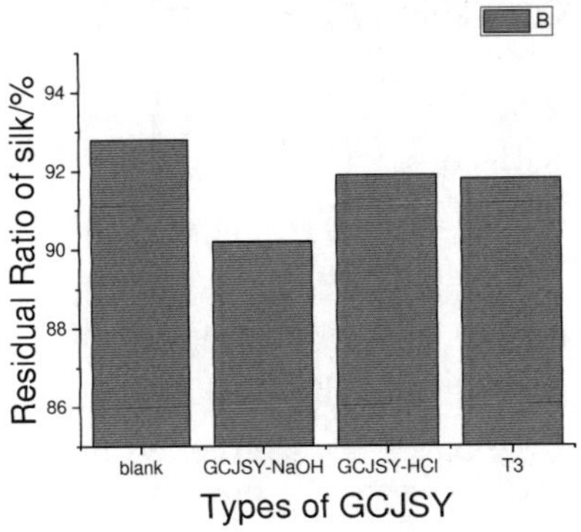

Figure.3 Effect of silk degumming of different types of GCJSY

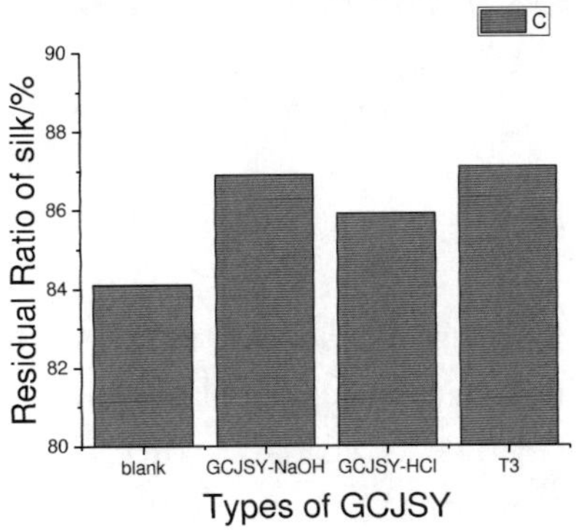

Figure.4 Effect of silk bleachof different types of GCJSY

After treated by different GCJSY , the cocoon silks were bleached at 2% hydrogen peroxide . The the residual ratio of the silks were shown in Figure 4. The residual ratio of silk treated by GCJSY-NaOH and GCJSY was 86.82% and 87.25% respectively.

3.3 Surface morphology of silk

The protective ability of GCJSY on the silk was confirmed by SEM (Figure 5). The surface appearance of silks by different treatments were different. As shown on the Figure 5 (a) , there are many impurities on the silk surface.It was because silk protein could not completely dissolved in water by Na_2CO_3.And some dents were appeared on the surface of the silk fiber(Figure 5 (c)), this was mean that the silk fiber was damaged during bleaching process.Silk surface of silk treated by GCJSY was relatively clean(Figure 5 (b)).And in Figure 5 (d), the surface of the silk fiber was smoother and more tidier than the surface of the silk fiber without GCJSY treatment. This may be due to the attachment of some active ingredient inside GCJSY to the fibroin surface and play a protective role in silk fibroin fibers.

（a）Silk Degumming with sodium carbonate

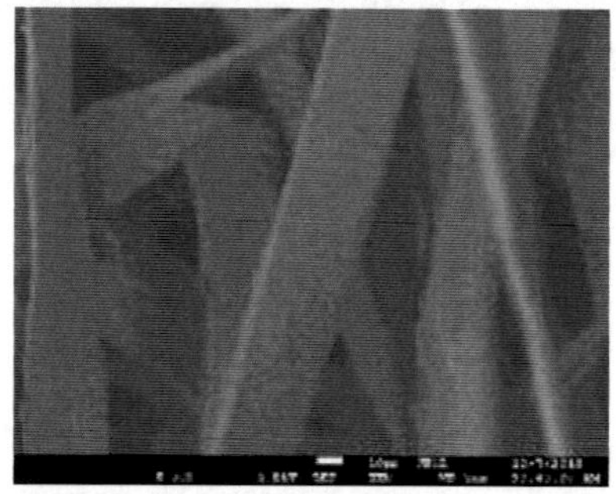

（b）After the degumming of sodium carbonate, the treated silk was treated with the extract

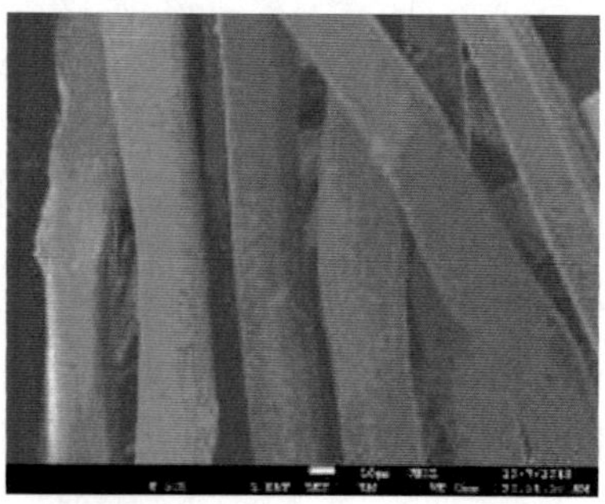

（c）Direct degumming of silk with sodium bleaching carbonate degumming

（d）Degumming with sodium carbonate and of silk after GCJS-T3 treatmen

Figure.5 SEM pictures of silk surface

4 conclution

While the concentration of GYJSY was 0.08% , the residual ratio of silk after degumming in the Na_2CO_3 solution and bleaching at 2% hydrogen peroxide was 87.2% . And the residual ratio of silk in the blank experiment was only 84.12%. The surface morphology of silk was observed by scanning electron microscope (SEM). Silk surface of silk treated by GCJSY was more smooth and neat than silk surface of silk without GCJSY treatment. It was concluded that GCJSY play protective ability during degumming process and bleaching process of silk.

References

[1] Chen H Z, Liang J,Min S J and HU G L 2002 *Journal of dong hua university 3 132-136*

[1] Md.M R K, Masuhiro T, Zhang X H and Hideaki M 2013 *J Mater Sci.*48 3731–3736

[2] Ma J, Wang X Y, L Q, Shi S L and Ma J B 2005 *Journal of Anhui Agri Sci.*33 674-675

[3] Darshil U S, David P and Fritz V 2014 *Composites Science and Technology* 101 173-178

[4] Akira M, Peggy C and David L K 2006 *The Journal of Physical Chemistry B* 110 21630-21638

[5] Subhas C.K, Biraja C D, Rupesh D and David L K 2008 *Progress in Polymer Science* 33 998–1012

[6] Ma L, Chen Y.Z, Peng S F, Wang X N, Chen L S, Wang R, Yang X H and Luo J 2013 *Journal of Central South University of Forestry & Technology* 10 34-37

[7] Li Y F, Hu L and Wang L Z 2009 *Journal of Guangxi Agricultural Sciences* 4 450-454

[8] Ma L and Chen Y Z 2009 *Chinese Agricultural Science Bulletin*.25(08):82-843.2009,40(4);450-454

Simulation of a 3D Turbulent Wavy Channel based on the High-order WENO Scheme

Bor-Jang Tsai [1], Chung-Chyi Chou [2†], Yeong-Pei Tsai [3], Ying Hung Chuang[4]

Ph.D. Program in Engineering Science,Chung Hua, University[1, 3]
Department of Fire and Safety, Da-Yeh University, [2†,4]
dustin@cloud.dyu.edu.tw, [2†]

Abstract. Passive interest turbulent drag reduction, effective means to improve air vehicle fuel consumption costs. Most turbulent problems happening to the nature and engineering applications were exactly the turbulence problem frequently caused by one or more turbulent shear flows. This study was operated with incompressible 3-D channels with cyclic wavy boundary to explore the physical properties of turbulence flow. This research measures the distribution of average velocity, instant flowing field shapes, turbulence and pressure distribution, etc. Furthermore, the systematic computation and analysis for the 3-D flow field was also implemented. It was aimed to clearly understand the turbulence fields formed by wavy boundary of tube flow. The purpose of this research is to obtain systematic structural information about the turbulent flow field and features of the turbulence structure are discussed.

Keyword：Lower drag, Navier-Stokes, Wavy wall.

1. Introduction

Recent researches on wavy channel concentrate on the areas of revelation of fluxion structure and the control and reduction of anti-obstruction mechanism. The research was aimed to theorize this phenomenon and use it in the engineering applications to reduce obstruction. However the effects of fluid structure change and the wall surface pressure gradient on the change of wall surface obstruction in the near wall area are very complex. The reduction in obstruction from the wavy channel is apparently due to the improvement of the partial In the near-wall region of turbulence structure, which has reduced the friction in the fluids and between the fluid the solid wall. Therefore in order to further the revelation of wavy channel in obstruction reduction, it is believed that the near wall flowing should be characterized.

Experimental research of flow over the wavy wall was performed as early as 1932. The theory of the surface wall variation has been developed. Zilker and Hanratty[3], Zilker et al.[4] continued to conduct studies including the experiments of gauging wall surface pressure, measuring mean velocity of stream and the wall surface shearing stress and analyzing the mobile characteristics. In 1985 Patel[5] has improved simulation for several kinds of near wall with model (low Reynolds number). Patel and Chen[6] used double-decked model to solve turbulent complex stalled flow. Since this model saves the grids, therefore it can effectively save the computation space and time and enhances the computation feasibility. Patel et al.[7] calculation of the wavy wall turbulent flow was conducted on two kinds of wave height ratios in 1991. The simulated streamlines picture in the near wall vortex area and the cross section velocity distribution, the numerical simulation is consistent with experimental results.

IOP Conf. Series: Materials Science and Engineering **307** (2018) 012018 doi:10.1088/1757-899X/307/1/012018

Comparison of friction coefficient curves with pressure coefficients was also made between the wavy wall surface and flat surface. The effects of the profile on flows in the near wall area were analyzed. Ferrira and Lopes [8] carried on the multi-group wind tunnel experiments on the unimodular flows of sinusoidal wavy wall flow field. The low Reynolds number k-ε model with the control volumetric method was used to compute wavy wall near zone stalled flow for various wave height ratios . Montalbano and McCready [9] used the wave stability theory and added small perturbation quantity to Orr-Sommerfeld equation to develop the relation between the laminar wall surface pressure and the shear stresses.

Airiau and Giovannini [10] utilized statistical simulation to obtain the stream function and vortex charts at different time for the sinusoidal wavy wall. Also the average stream function and the average wall surface pressure coefficient curves were calculated using the time average method. With these tools vorticity and variation of the pressure gradient along the wall surface were analyzed. Malamataris1 and Bontozoglou[11] Dimensional Navier-Stokes equations within the entire range of laminar flow conditions. Relatively very low Reynolds numerical and experimental data. Focus of discussion in the results is due to the free surface contour, flow lines, velocity and pressure distribution along the free surface and the wall of the exhibit. Number of dimensionless flow interactions were studied for flow inversion condition is established, and the resonance phenomenon in the high Reynolds number is studied. Boersma[12]. Release particles in the flow field waveform observed flow trajectories of particles in the flow field. This will show on channel bottom small waves can generate large longitudinal vortex is similar to that observed at the free surface vortex flow Langmuir. The simulation results show that the particle has the highest concentration in the trough. Nakagawa et al.[13], Turbulence using laser Doppler velocimetry (LDV) measurements were compared, for the turbulence over one flat surface, and with the small wavelength of the sine wave surface. Wavy border, the height of the flow separation roughened. Reynolds set to 46,000.

2. NUMERICAL METHOD

The numerical code in the current work is intended for the direct calculation of solving 3D Navier-Stokes equations with no subgrid scale turbulence model. Flow field contains multiple regions, with large gradients, we need to be carefully distinguished. Grid near the wall in the vertical direction requires encryption, to resolve the boundary layer. The separated shear layer, which is located the separation bubble, also need special attention. Because of these requirements, a fine resolution should be maintained until the middle of the channel, in order to resolve the separated shear layer normal. The present flow domain consists of three wavelengths compared with four in A. Nakayama and Maass & Schumann. For the present configuration of the flow, one of the coordinates is taken to conform to the wavy surface on the bottom and the flat upper boundary and the second coordinate is taken so that it is close to orthogonal to the first as close as possible without much distortion in the main flow domain. Approximate orthogonally near the solid boundary is important to capture the motion accurately in this region. Boundary conditions are set: no slip on the bottom of the wave-shaped walls, no slip boundary, and regularly flow direction and cross flow. In order to shorten the development time of initial flow, the current calculation first began with our previous DNS straight open channel flow as a result of the calculation of the initial flow of a cycle of big waves.

2.1. Grid setup

*The (x, y, z) direction was discretized into 210125 elements. The number of grid points:125*125*41. The turbulent flow occurs mostly in the flow field nearby the wavy wall surface. In order to better reveal the turbulence movement, this study densifies the grid at both sides in the physical space Figure1.*

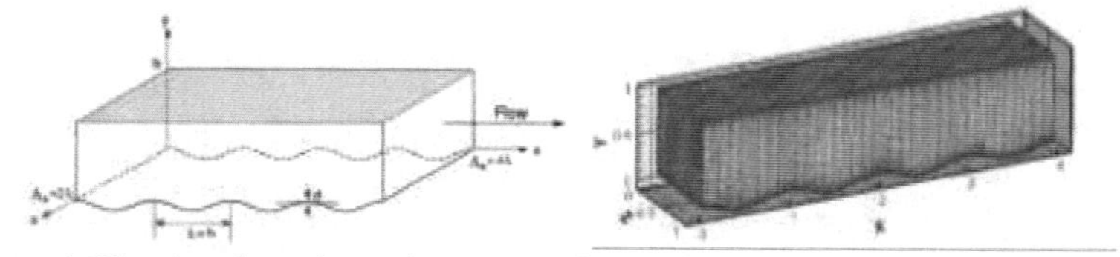

Figure.1 Physics domain and computational domain of flow over a wavy wall.

2.2. Initial and boundary conditions

The inflow speed is at Mach number 0.6, and the Reynolds number is 3.66x10^5. The pressure is in the static atmosph- eric condition and the density is 1.2kg/m^3. Boundary value of the flow field is assumed to be the initial flow field exit condition and kept constant.

$$\frac{\partial p}{\partial y}=0,\quad \frac{\partial u}{\partial y}=0,\quad v=0,\quad \frac{\partial w}{\partial y}=0$$

The boundary condition is subsonic of sound and expressed as follows:

$$\rho u = C_1 \quad , \quad \frac{\gamma}{\gamma-1}\cdot\frac{p}{\rho}+\frac{1}{2}(u^2+v^2)=C_2$$

3. RESULTS

Turbulent Flow Field in this study can be obtained from the sinusoidal wall having a lower wavy channel, however, mainly the production of turbulent shear layer is used to separate the back of the associated wave. Turbulent flow generation and maintenance guide whirlpool, found the flat wall seems to be unimportant wave wall of the channel. This is important, but little is known about many of flow structure, because it is a priority for the complex flow of applicability of the reference flow. Early numerical simulation and analysis of flow through the infinitely small amplitude wave surface wave think. The numerical simulation results and Cherukat [1] for comparison. Flow field contains multiple regions, with large gradients, we need to be carefully distinguished. Grid encryption requires the flow field in the vertical direction, in order to resolve the boundary layer. The separated shear layer, which is located the separation bubble, also need special attention. Because of these requirements, a fine resolution should be maintained until the middle of the channel, in order to resolve the separated shear layer normal.

Comparisons of streamlines, ψ , of the mean flow fields are presented in Figure 2. For computing grid test. (A) Mesh: 100 * 20 * 20; (B) Mesh: 125 * 41 * 41; (C) Grid: 251 * 41 * 41. B and C at the downstream of the vortex can correctly observed phenomena, and then to B Streamwise velocity distribution grid computing. There are obvious vortex phenomenon downstream of middle layer and the average speed is higher than the waveform wall.

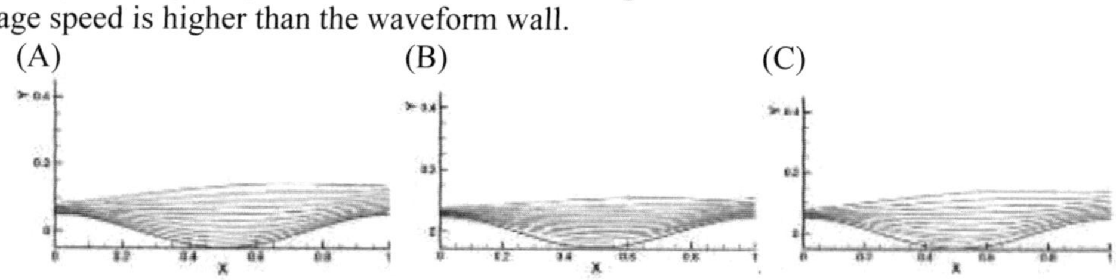

Figure. 2 Computing grid test. (A) Mesh: 100 * 20 * 20; (B) Mesh: 125 * 41 * 41; (C) Grid: 251 * 41 * 41. Computational domain of flow over a wavy wall and mean streamwise velocity and streamline for channel flow over wavy wall at Re=3460.

Comparisons of Cherukat like, and Hudson (1993) present results are compared both flow position means both peaks and troughs velocity curve, showing good agreement in Figure 3.The flat boundary, single wavy wall and bilateral wavy wall, three different instantaneous streamwise velocity chart are shown in Figure4. At the flat boundary, both upper and lower parts of inner flow field were flat. it was impossible for bilateral flat wall to trigger vortex happening between tube walls and flowing bodies. Thus, the apparent separation phenomenon no happening nearby flowing fields. And Figure5 through Figure6 shows three different instantaneous cross-stream and spanwise velocity contour chart. At the single wavy wall, because the upper wall was flat, it was unavailable to trigger vortex happening between flowing bodies and tube wall. The separation phenomenon could be clearly seen. The lower wall part was a sinusoidal wavy one with apparent vortex happening to the wave trough. At the bilateral wavy wall, both upper and lower wall parts of inner flow field were sinusoidal wavy walls with the amplitude(A) rated at 2.54mm. Because both upper and lower wall parts were sinusoidal ones, there were vortex and pressure changes could be seen nearby the wave trough.

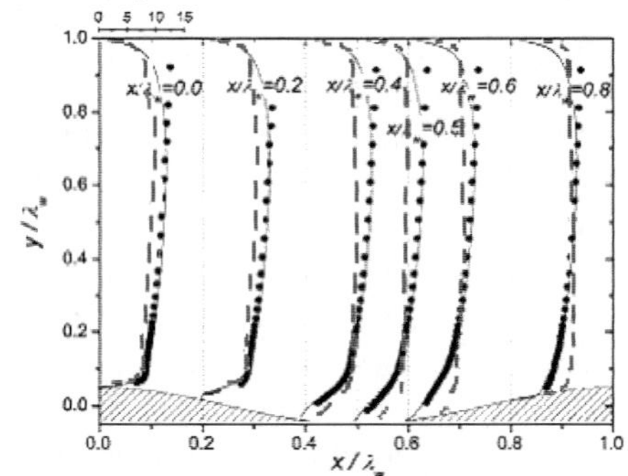

Figure.3 Comparison with measured profiles of mean streamwise velocities at locations for x=0.0, 0.2, 0.5, 0.6 and 0.8: Nakagawa(2003) LDV measurement; •, Nakagawa(2003) DNS; —, Present mean streamwise;---.

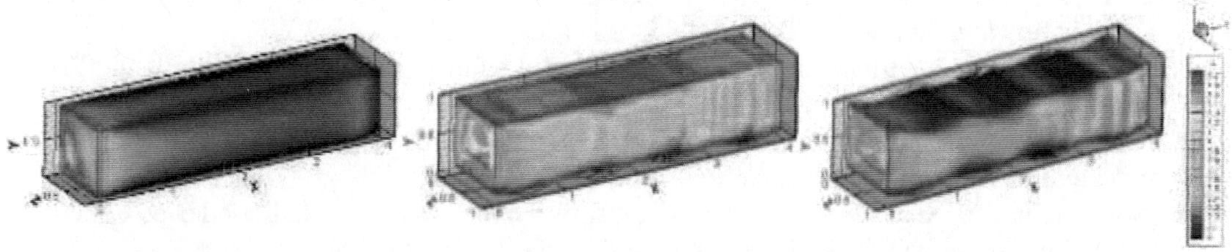

Figure 4 Three-dimensional instantaneous velocity chart for (A)Flat boundary; (B)Single wavy wall; (C)Bilateral wavy wall in the mean streamwise velocity.

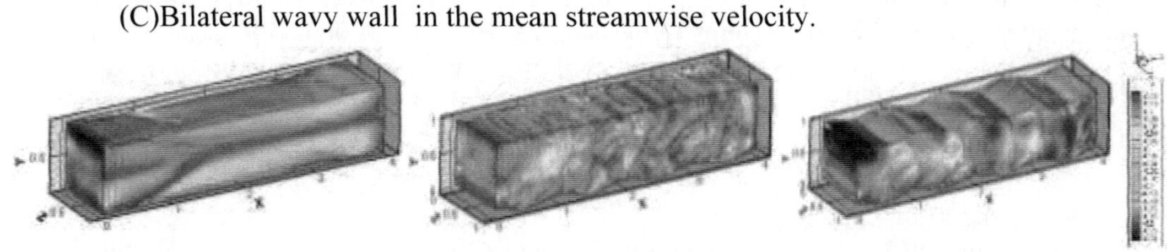

Figure5 Three-dimensional instantaneous velocity chart for (A)Flat boundary; (B)Single wavy wall; (C)Bilateral wavy wall in the cross-stream direction.

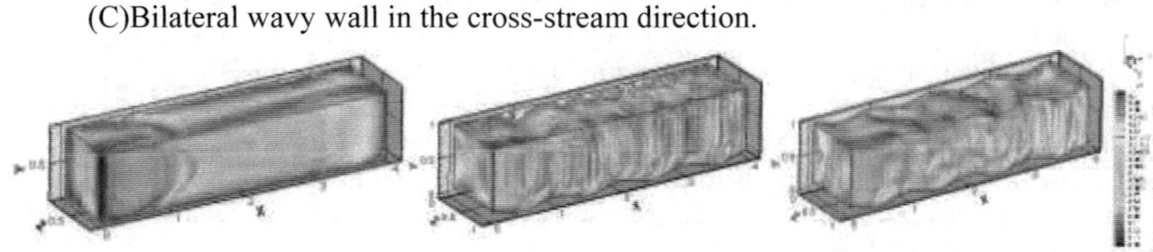

Figure 6 Three-dimensional instantaneous velocity chart for (A)Flat boundary; (B)Single wavy wall; (C)Bilateral wavy wall in the spanwise direction.

IOP Conf. Series: Materials Science and Engineering **307** (2018) 012018 doi:10.1088/1757-899X/307/1/012018

4. Conclusion

This study successfully simulates the flow field characteristics of the two-dimensional profile wall using the Upwind and LU-SSOR schemes for space-time discretization and a 3D grid of 251x41x41 with both sides densified, the numerical results agree well with the experimental results, and this agreement can be used to verify the numerical codes and numerical results. Due to the strong mixing of the core flow and turbulence intensities, Reynolds stresses, turbulent energy production, instantaneous and mean flow field, and the pressure distribution of the turbulent channel flow with a wavy wall are calculated and investigated. The purpose of this research is to obtain systematic structural information about the turbulent flow field. Features of the turbulence structure are discussed.

References

[1] S. Nakagawa, Y. Na, T.J. Hanratty., "Influence of a wavy boundary on turbulence I", Highly rough surface Experiments in Fluids. 2003, 35. 422–436

[2] N. A. Malamataris1 and V. Bontozoglou. "Computer Aided Analysis of Viscous Film Flow along an Inclined Wavy Wall", Journal of Computational Physics. 1999. 154, 372–392

[3] E. D. Montalbano and M. J. McCready. Laminar channel flow over long and moderate waves, J. Comp. Phys.1998. 6. 1-23.

[4] A. D. Ferreira etc., "Experimental and numerical simulation of flow around two-dimensional hills", J. of Wind Eng. and Industrial Aerodynamics. 1995, (54/55), 173-181.

[5] Liu, et al., "Weighted Essentially Non-oscillatory Schemes", Journal of Computational Physics. 1994. 115. 200-212.

[6] V. C. Patel, etc. "Turbulent flow in a channel with a wavy wall", J. of Fluids Eng. 1991, 113, 579-586.

[7] V. C. Patel and H. C. Chen. "Near-wall turbulence models for complex flows including separation". AIAA J. 1988. 26. 641-648.

[8] V. C. Patel, W. Rodi and G. Scheuerer. "Turbulence models for near-wall and low Reynolds number flows: a review", AIAA J. 1985. 23. 1308-1319.

[9] P.Zilker, J.Hanratty. "Influence of the amplitude of a solid wavy wall on a turbulence flow Part 2, separated flows", J. Fluid Mech,. 1979. 90, 257-271.

[10] P.Zilker, W.Cook, and J. Hanratty. "Influence of the amplitude of a solid wavy wall on a turbulent flow. Part 1, Non-separated flows". 1977. 82 , 29-51.

[11] A. Nakayama and K. Sakio "Simulation of flows over wavy rough boundaries", Center for Turbulence Research Annual Research Briefs, 2002. pp.313-324.

[12] Bendiks Jan Boersma, "Particle distributions in the flow over a wavy wall", Center for Turbulence Research Proceedings of the Summer Program., 2000. 109.

[13] C. Airiau and A. Giovannini. "Vorticity evolution on a separated wavy wall flow", Third International Workshop on Vortex Flows and Related Numerical Methods. 1999. 7. 1-11.

[14] P. Cherukat. "Direct Numerical Simulation of a Fully Developed Turbulent Flow over a Wavy Wall" , Theoretical and Computational Fluid Dynamics, 1998. 11, 109 – 134. Reports, Theses, and Individual Papers

[15] Marie Denison (2015) "A Brief Study of the Effect of Rigid Swept Surface Waves on Turbulent Drag" Advanced Modeling & Simulation (AMS) Seminar Series, August 27, 2015

[16] G. Lodato, etc. (2016) Direct numerical simulation of shock wavy-wall interaction: Analysis of cellular shock structures and flow patterns, J. Fluid Mech, 789: 221-258.

[17] B. Denet, etc. (2015) Model equation for the dynamics of wrinkled shock waves. Comparison with DNS and experiments. Combust. Sci. Tech., 187(1-2): 296-323.

[18] O. Zanotti and M. Dumbser, A high order special relativistic hydrodynamic and magnetohydrodynamic code with space-time adaptive mesh refinement, Computer Physics Communications, 188 (2015), 110-127.

[19] Y. Yang, M. Wan, Y. Shi, K. Yang and S. Chen, A hybrid scheme for compressible magnetohydrodynamic turbulence, Journal of Computational Physics, 306 (2016), 73- 91.

IOP Conf. Series: Materials Science and Engineering **307** (2018) 012019 doi:10.1088/1757-899X/307/1/012019

Feasibility Study on Manufacturing Lightweight Aggregates from Water Purification Sludge

Ching-Fang Peng [1] and How-Ji Chen [1]

[1] Department of Civil Engineering, National Chung-Hsing University, Taiwan;
hojichen@nchu.edu.tw

Abstract. This study mainly discussed the feasibility of manufacturing lightweight aggregates from water purification sludge in Taiwan. They were analysed for the physical and chemical composition before the sintering test for lightweight aggregates in a laboratory. Then the physical and mechanical properties of the synthesized aggregates were assessed. The result showed that the chemical composition of sludge in the water purification plants was within the appropriate range for manufacturing lightweight aggregate as proposed in the literature. The sintering test demonstrated that the particle density of aggregates from the ten types of water purification sludge were mostly less than 1.8 g/cm3. In addition, the dry unit weight, the organic impurity, the ignition loss, and other characteristics of synthesized aggregates met the requirement of CNS standards, while its water absorption and crushing strength also fulfilled the general commercial specifications. Therefore, reclamation of water purification sludge for production of lightweight aggregate is indeed feasible.

1. Introduction

Lightweight aggregate (LWA) is a general term for natural or artificial aggregate with a bulk density of less than 1200 kg/m3 [1]. LWA can be used to replace ordinary aggregate to produce lightweight aggregate concrete (LWAC), which has the advantages of light weight, heat insulation, strong fire and seismic resistance [2-4]. As a result, LWA has long been used by many countries as building materials [1,5,6]. In recent years, the development of LWA has focused on industrial waste or municipal solid waste as a raw material to reduce the use of natural resources. A number of studies have been proposed on the production of LWA. For instance, the feasibility of the use of paper sludge, textile sludge, sewage sludge, incineration fly ashes, reaction ashes, and reservoir sediments in manufacturing LWA has been confirmed. In other words, industrial waste or municipal solid waste can be reused as a sustainable resource in the manufacturing process of artificial LWA.

2. Experimental Procedure

In this study, the overall procedure involves granulation, drying, and sintering. Sintering was further divided into a preheating phase and a sintering phase. A series of different conditions (i.e., soaking time at preheating phase and sintering temperature at sintering phase) were planned. Then the sintering test was conducted in a laboratory to analyze the thermal expansion range and the range of attainable particle density for mass production reference. Physical and mechanical properties of the synthesized aggregates were subsequently assessed.

The sludge was thoroughly blended with a controlled amount of water to produce a mix consistency, and was grained by an extrusion pelletizer that was manually chopped into cylinder-shaped pellets as raw pellets. The cylindrical pellets were approximately 10 mm in diameter and 12 mm in length. After sintering, the green pellets were dried in an oven at 105 ± 10 °C for 24 hours to obtain sufficient particle strength, and it could prevent possible fragmentation from high temperature rupture or steam explosion.

3. Results and Discussion

The physical test of water purification sludge included grain size analysis, soil classification, and specific gravity, in which two samples from the low flow period and the high flow period were used for particle size distribution test. The specific gravity, plasticity index, medium particle size, and composition of water purification sludge from the ten purification plants are shown in Table 1.

Table 1. Physical properties of water purification sludge.

Name of purification plant	Specific gravity	Plasticity Index P.I. (%)	Medium particle size D_{50} (mm)		Composition (% wt.) Gravel > 4.75 mm		Sand 4.75-0.075 mm		Silt 0.075-0.005 mm		Clay < 0.005 mm	
			High flow period	Low flow period	High flow period	Low flow period	High flow period	Low flow period	High flow period	Low flow period	High flow period	Low flow period
Hsinsan	2.60	9	0.009	0.002	0	0	9	1	58	25	33	74
Banxin	2.62	10	0.007	0.005	0	0	6	1	52	46	42	53
Pingzhen	2.75	13	0.014	0.003	0	0	2	3	67	30	31	67
Hsinchu	2.69	17	0.006	0.005	0	0	3	2	49	53	48	55
Fengyuan	2.60	13	0.010	0.009	0	0	12	3	56	68	32	29
Linnei	2.75	11	0.005	0.004	0	0	1	1	38	47	61	52
Nanhua	2.63	13	0.010	0.004	0	0	1	4	65	38	34	58
Pingding	2.7	27	0.008	0.007	0	0	4	2	78	68	18	30
Kaotan	2.64	26	0.007	0.003	0	0	5	1	55	31	40	68
Chengcing Lake	2.63	14	0.011	0.008	0	0	8	3	75	58	17	39

The chemical composition of the water purification sludge from the ten purification plants is shown in Table 2. It can be seen that the water purification sludge from the ten purification plants satisfied Riley's recommended requirements [7]. In other words, the chemical composition of sludge in the ten water purification plants was within the appropriate range as proposed in the literature. Especially, the mineral composition of the clean water sludge sample shows that all samples contained quartz and clay mineral compounds such as chlorite, illite, feldspar and others. The result showed two characteristics that the mineral compositions of all water purification sludge samples conformed with striking similarity and water purification sludge indeed show manufacturing potential of LWA.

Table 2. Chemical composition ratio of water purification sludge.

Chemical composition	Name of purification plant Hsinsan	Banxin	Pingzhen	Hsinchu	Fengyuan	Linnei	Nanhua	Pingding	Kaotan	Chengcing Lake	C.M. Riley Recommended Range

SiO_2 (%)	64.3	63.6	63.8	64	62.3	66.9	62.3	54.8	65.6	63.7	53-79
Al_2O_3 (%)	21.2	22.2	20.6	23	22	20.6	21.4	23	22.4	19.6	12-26
Fluxing (%)	14.5	14.1	15.5	12.9	15.6	12.4	16.3	22.1	12.0	16.4	8-24
Total (%)	100	100	100	100	100	100	100	100	100	100	—

Note: Fluxing=Fe_2O_3+CaO+MgO+K_2O+Na_2O

Table 3. Physical properties of aggregates.

Name of purification plant	Preheating phase		Sintering phase		Particle density (g/cm^3)	Water absorption (%)	Ignition loss (%)	Bloating index (%)
	Preheating temperature (°C)	Soaking time (min.)	Sintering temperature (°C)	Soaking time (min.)				
Shinshan	500	0	1275	10	1.29	11.3	14.5	104.4
		7.5			1.33	10.6	14.0	101.4
		15			1.40	9.6	14.7	96.5
Banxin	500	0	1275	10	1.29	11.3	15.2	102.9
		7.5			1.39	7.3	15.0	95.7
		15			1.65	3.7	15.0	80.5
Pingzhen	500	0	1275	10	1.50	5.0	14.3	106.5
		7.5			1.57	4.5	14.5	102.0
		15			1.89	1.9	14.6	84.5
Hsinchu	500	0	1250	10	0.65	8.0	10.1	224.9
		7.5			1.74	1.7	10.7	83.8
		15			1.88	0.6	10.6	77.5
Fengyuan	500	0	1150	10	1.31	19.0	14.6	110.7
		7.5			1.43	15.0	15.2	101.4
		15			1.52	11.0	14.5	95.4
Linnei	500	0	1200	10	0.78	1.2	8.4	162.2
		7.5			1.46	1.0	8.2	86.2
		15			2.38	0.3	7.9	53.0
Nanhua	500	0	1275	10	1.26	7.4	15.1	104.0
		7.5			1.29	4.9	15.3	101.6
		15			1.58	2.7	15.1	82.8
Pingding	500	0	1200	10	0.74	3.7	7.5	228.1
		7.5			1.67	0.9	7.7	100.6
		15			2.05	0.1	7.1	82.0
Kaotan	500	0	1200	10	0.75	4.8	6.5	223.4
		7.5			1.28	1.2	6.1	130.5
		15			2.03	0.3	6.6	82.1
Chengcing Lake	500	0	1200	10	0.73	6.6	6.3	181.8
		7.5			0.79	4.2	6.8	166.2
		15			1.97	0.7	6.4	66.9

Therefore, the samples of water purification sludge should be a suitable expanded material that can develop gases at high temperature. Moreover, they can produce a highly viscous liquid phase at the temperature that could entrap the gases. In short, the specific gravity of sludge was between 2.60 and 2.75, and all of the sediments of silt or clay were suitable as raw materials for the manufacture of artificial LWA.

Name of purification plant	Soaking time at preheating phase/Bloating index		
	0 min.	7.5 min.	15 min.

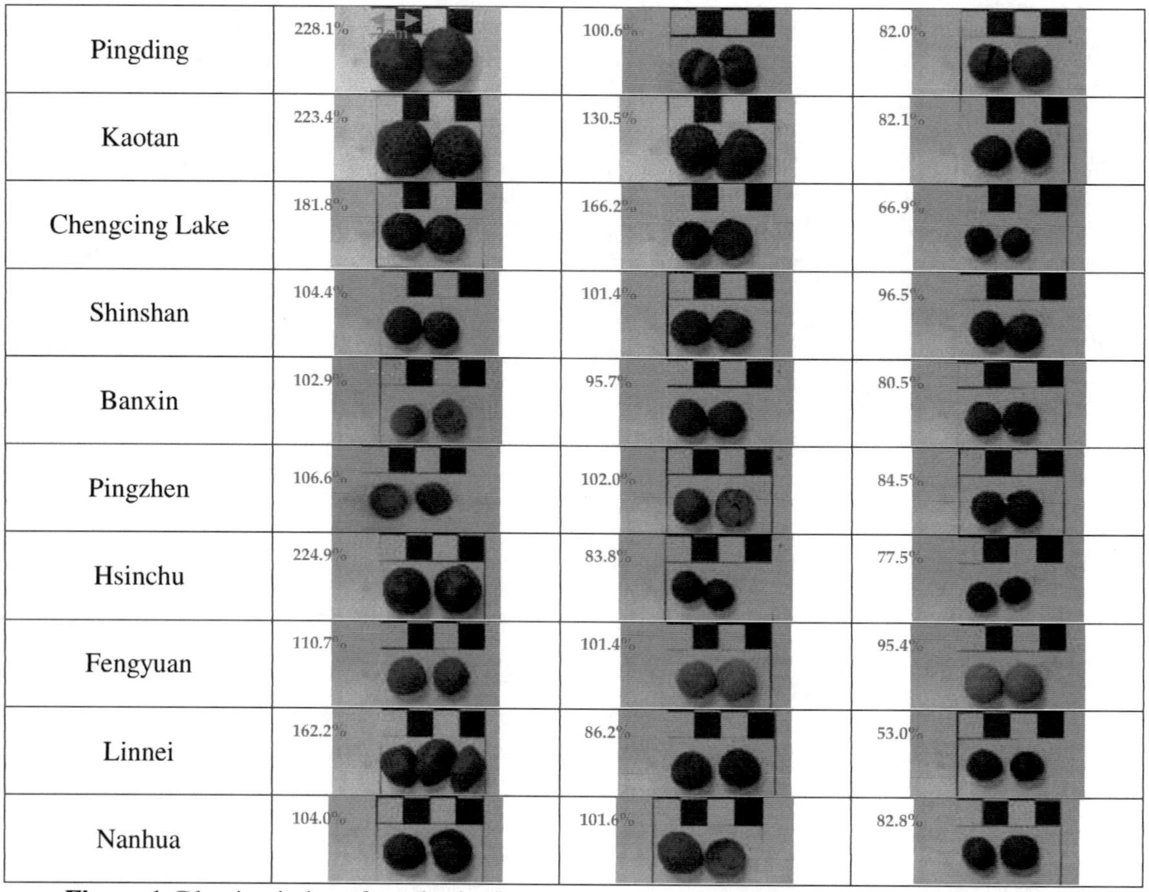

Figure 1. Bloating index of synthesized aggregates versus soaking time at preheating phase.

Taking into account previous findings, the preheating temperature was set at 500 °C and the soaking times were 0, 7.5, and 15 minutes, respectively. On the other hand, based on the preliminary trial results, the sintering temperature was the optimal temperature in each group and the soaking time was fixed at 10 minutes. The sintering conditions for water purification sludge of the ten purification plants are shown in Table 3. In addition, the particle density, water absorption, ignition loss, and bloating index of the synthesized aggregates are also shown in Table 3.

The particle density of the synthesized aggregates ranged from 0.65 to 2.38 g/cm3, most of which was less than 1.8 g/cm3. It also illustrated that the soaking time at preheating phase could affect the particle density of aggregate. For any water purification sludge from the ten purification plants, a longer soaking time would result in aggregates of a higher particle density.

As can be seen in Table 3, the 24-hour water absorption values of the synthesized aggregates ranged from 0.1% to 19%, most of which was less than 8%. It demonstrated contrastive effect of lower absorption and higher particle density with increased soaking time. For structural concrete applications, the water absorption of LWA is usually less than 20%. The results indicate that the synthesized LWA in this study meet with regulations.

Table 3 shows that for any water purification sludge from the ten purification plants, the soaking time at preheating phase did not significantly affect the ignition loss of aggregate. On the whole, a higher sintering temperature (1275 °C) would result in aggregates of a higher ignition loss (> 14%). With exception of 1150 °C, a lower sintering temperature (1200 °C) would result in aggregates of a lower ignition loss (between about 6.4% and 8.2%). The results demonstrated a positive correlation between sintering temperature and ignition loss that higher sintering temperature increased the loss on ignition.

IOP Conf. Series: Materials Science and Engineering **307** (2018) 012019 doi:10.1088/1757-899X/307/1/012019

From Table 3, it can be seen, a longer soaking time would result in aggregates of a lower bloating index, even a petty volume contraction. Overall, there was a clear inverse relationship between bloating index and particle density/soaking time (see Table 3 and Figure. 1).

4. Conclusions

Based on the above results and discussion, the following conclusions can be drawn:

(1) The specific gravity of sludge was between 2.60 and 2.75, and all of the sediments of silt or clay were suitable as raw materials for the manufacture of lightweight aggregate.

(2) The chemical composition of sludge in the ten water purification plants was within the appropriate range as proposed in the literature. The sintering test demonstrated that the particle density of aggregates from the ten types of water purification sludge were mostly less than 1.8 g/cm^3.

(3) The dry unit weight, the organic impurity, the ignition loss, and other characteristics of synthesized aggregates met the requirement of Chinese National Standards, while its water absorption and crushing strength also fulfilled the general commercial specifications. Therefore, reclamation of water purification sludge for production of lightweight aggregate is indeed feasible.

References

[1] Somayaji, S. Civil engineering materials; Prentice Hall, Upper Siddle River, New Jersey, 2001; 45.

[2] Tang, C.W. Uniaxial bond stress-slip behavior of reinforcing bars embedded in lightweight aggregate concrete. Structural Engineering and Mechanics 2017, 62(5), 651-661.

[3] Chandra, S.; Berntsson, L. Lightweight Aggregate Concrete. Noyes Publications 2002, New York, USA.

[4] Tang, C.W. Effect of presoaking degree of lightweight aggregate on the properties of lightweight aggregate concrete. Computers & Concrete 2017, 19(1), 069-078.

[5] Liu, R.; Coffman, R. Lightweight aggregate made from dredged material in green roof construction for stormwater management. Materials 2016, 9, 611, doi:10.3390/ma9080611.

[6] Lee, H.S.; Ismail, M.A.; Woo, Y.J.; Min, T.B.; Choi, H.K. Fundamental study on the development of structural lightweight concrete by using normal coarse aggregate and foaming agent. Materials 2014, 7, 4536–4554.

[7] Riley, C.M. Relation of chemical properties to the bloating of clays. ACI Journal 1951, 34, 121–128.

IOP Conf. Series: Materials Science and Engineering **307** (2018) 012020 doi:10.1088/1757-899X/307/1/012020

Effect of Dimension and Shape of Magnet on the Performance AC Generator with Translation Motion

A. Indriani[1]*, Dimas, S.[1], Hendra[2]

1 Electrical Engineering Dept., University of Bengkulu, Indonesia
2 Mechanical Engineering Dept., University of Bengkulu, Indonesia
E-mail: aniz_raimin@yahoo.com, h7f1973@yahoo.com

Abstract. The development of power plants using the renewable energy sources is very rapid. Renewable energy sources used solar energy, wind energy, ocean wave energy and other energy. All of these renewable energy sources require a processing device or a change of motion system to become electrical energy. One processing device is a generator which have work principle of converting motion (mechanical) energy into electrical energy with rotary shaft, blade and other motion components. Generator consists of several types of rotation motion and linear motion (translational). The generator have components such as rotor, stator and anchor. In the rotor and stator having magnet and winding coil as an electric generating part of the electric motion force. Working principle of AC generator with linear motion (translation) also apply the principle of Faraday that is using magnetic induction which change iron magnet to produce magnetic flux. Magnetic flux is captured by the stator to be converted into electrical energy. Linear motion generators consist of linear induction machine, wound synchronous machine field, and permanent magnet synchronous [1]. Performance of synchronous generator of translation motion is influenced by magnet type, magnetic shape, coil winding, magnetic and coil spacing and others. In this paper focus on the neodymium magnet with varying shapes, number of coil windings and gap of magnetic distances. This generator work by using pneumatic mechanism (PLTGL) for power plants system. Result testing of performance AC generator translation motion obtained that maximum voltage, current and power are 63 Volt for diameter winding coil 0.15 mm, number of winding coil 13000 and distance of magnet 20 mm. For effect shape of magnet, maximum voltage happen on rectangle magnet 30x20x5 mm with 4.64 Volt. Voltage and power on effect of diameter winding coil is 14.63 V and 17.82 W at the diameter winding coil 0.7 and number of winding coil is 1260 with the distance of magnet 25 mm.

1. Introduction

The development of power plants from renewable energy sources is currently needed. The availability of fossil and petroleum power generation resources from petroleum depletion, the scarcity of petroleum availability and uncertain price fluctuations causes the demand for the development of power plants with renewable energy sources is higher. This demand is fulfilled with the utilization of wind energy sources, solar energy, ocean wave energy and other energy. The power plant with energy source requires several components so that the existing potential can be utilized properly. The components are shafts, generators, inverters, batteries, drive components or heat receivers. The axis serves to continue the rotation of the driving components such as blades, piston and other components to the generator. In the generator, the rotating motion of the shaft is converted into electrical energy through the induction force

that occurs inside the generator. Induction force occurs because on the inside of the generator there is a rotor and stator that interaction moves to produce an electric motion force.
Induced motion in the generator can occur linear and rotation. Motion generator linear contains magnetic element. The linear motion generator works by translating and using magnetic induction to produce the magnetic flux processed by the stator to produce electrical energy. Linear motion generators consist of linear induction machines, field-wound synchronous machines, and permanent magnet synchronous [1]. Linear induction machine is a kind of conventional generator that simple and widely used by industries. The generator induction machine has a larger air resistance than the rotary induction and has reactance to the low excitation coil. This causes the generator to require greater excitation current than the rotating rotor design. Field-wound synchronous machine is a high-speed rotary generator that is rarely used in industry. Field-wound synchronous machines have a disadvantages that is small power and requires a large spin to generate electricity. The advantage is the excitation current can be controlled and the resulting voltage is stable. Permanent magnetic synchronous is a generator that can be used for power generation. The magnitude of the output produced by Permanent magnet synchronous depends on the magnet. The principle of permanent magnetism works using Lorentz force and its performance is influenced by the direction of flux and the translational movement [1].
Magnets are used in permanent magnet generator is made of a magnetic material such as neodymium (NdFeB), Samarium-Cobalt magnets, ceramic magnets, plastic magnets and alnico magnet [2]. Magnets on the generator can create their own magnetic field, where the magnetic field is an area that has a magnetic force (flux). Magnetic flux [3] flows from the North Pole into the South Pole and never intersecting. The magnetic field is getting stronger with a dense magnetic flux. The movement that occurs in the magnetic flux generated and the amount will cause an electromotive force (EMF). The amount of voltage coming out of the generator depending on the magnetic field strength, number of windings coil, the addition of the iron core and motion generator (speed). The strength of the magnetic field depends on the type of magnet in the generator.
Application of generators can be seen in hydroelectric power plants, wind power plants, ocean wave power plants, diesel power plants, and etc [4-12]. The principle works of generator can move in rotation and translation. For generator with rotational motion is often found in hydroelectric power plants, diesel turbine and wind turbine. Micro hydro power plants also use generators with rotational work system [1]. For the kinetic energy of ocean waves can be seen on the piston mechanism that has been done in previous studies [4-6].
The previous study used generator for sea wave power plant is a DC generator which utilizes rotational move pneumatic motion and has been able to generate electrical energy (able to turn the incandescent and LED) [4-6]. The weakness that occurred in previous research is that the resulting voltage is not maximal, less stable and also rotational motion requires the initial trigger for pneumatic motion can run well. To overcome this problem is done by replacement rotation motion system into translational motion. In this paper focused on the manufacture and experimental testing of performance generators with a translational motion system using neodymium magnets. Neodymium magnet shape, number of winding coil, dimension of winding coil and gap of the magnet are used as a reference to being investigated. The resulting generator is an AC generator.
These generators are used for fishing in the sea to replenish the batteries [7] and the electrical energy needs while at sea. By doing experimental testing of performance AC generator with translational motion is obtained a better voltage and power supply. Experimental testing was done by varying the number of windings coil and distance of magnetic. From experimental results obtained maximum voltage, current and power for AC generators.

2. Methods
AC generator with the translational motion mechanism is made of PVC tubing. AC generator has a rotor and stator components. The rotor as moving parts on the generator which contains shaft as a piston and winding coil. The stator having magnet component is mounted on the generator tube wall. The magnet used is neodymium magnet (NdFeB) with variations of round shape, trapezium and rectangle magnet

2.1. Stage of Testing

Experimental testing of AC generator with the translational motion mechanism can be seen in Figure 1.

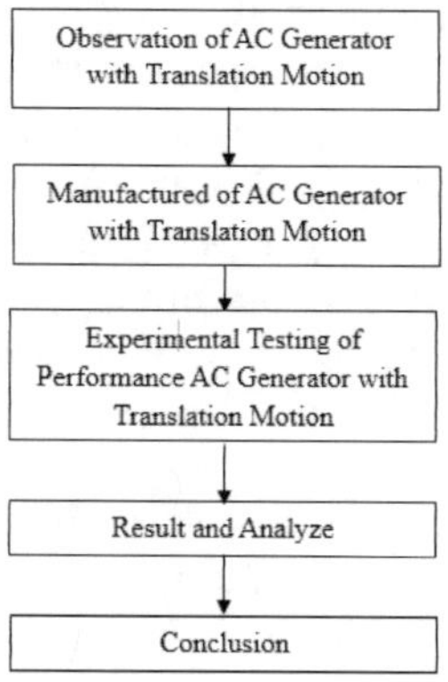

Figure 1. Stages of Experimental Test Performance of AC Generator

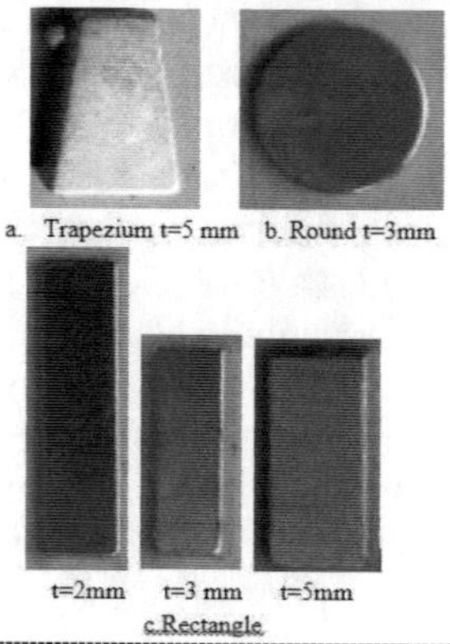

Figure 2. Dimension and Shape of Magnet for AC Generator

Figure 3. AC Generator with Translation Motion

Figure 4. Koker

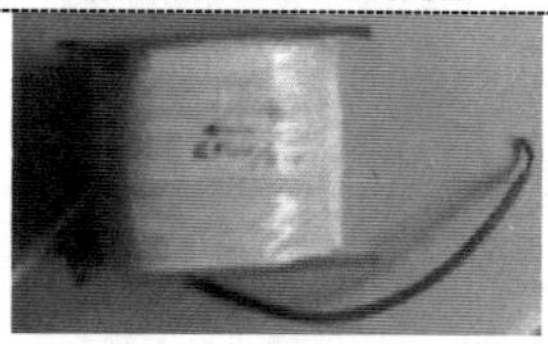

Figure 5. Diameter Coil of Winding 0.7 mm

Figure 6. AC Generator Translational Motion Mechanism

2.2. Tools and Materials

Tools and materials are used in the AC generator have mechanical and electrical components. Mechanical components include: PVC tube d = 8 in a house with a generator mounting length of 600 mm; piston shaft 12 mm with a length of 1000 mm; rectangle coil with size 45 mm x 45 mm and t = 4 mm; and hexagon-shaped coil holder containers with t = 4 mm. Electrical components include: neodymium magnet rectangle, trapezium-shaped magnet with a thickness of 5 mm and round-shaped magnet with a thickness of 3 mm; winding coil with size 0.5-0.7 mm with the number of winding coil 920 to 13000. The shape and dimensions of the magnet can be seen in Figure 2.

For dimensional measurements, the test data is done by using a caliper, multimeter, tachometer, and other measuring equipment. The shape of the mechanical component of AC generator with translation motion can be seen in Figure 3 which consists of piston tube, piston, koker (see Fig.4) and buoy. Electrical component can be seen in Figure 5 includes a magnet and windings coil (diameter).

3. Results

The design of generator can be seen in Figure 6 and the experimental testing were conducted using one side of the tube with the variation number of windings coil and distance of magnetic between 10-25 mm. Shape of magnets used in AC generator are rectangle, trapezoid and round. The results of performance testing of AC generator with translation motion using various shape of neodymium magnets, the number of coil windings 920 and 0.7 mm diameter of coil windings can be seen in Table 1. Table 1 shows that the rectangle-shaped magnet neodymium with the dimension 30x20x5 mm having maximum a voltage is 4.64 V, the lower value of voltage on shape of rectangle magnet happen with the dimension 40x10x2 mm. From experimental testing by using various shape of magnet neodymium show that voltage by the rectangle-shaped magnet is higher than the round shape and the trapezium shape of magnet. In other hand, magnetic dimensions such as length, width and magnetic thickness will be improve the performance of generators such as voltage, current and power.

Table 1. Performance of Generator with Variation Shape of Magnet

Dia Coil (mm)	Shape of Magnet	Size of Magnet/ LxWxt (mm)	Number of Coil	Number of Pole (P)	Distance of magnet (mm)	Voltage (V)	Current (A)	Power (W)
0.7	Rectangle	30x10x5	920	5	20	1.40	0.241	0.337
0.7	Rectangle	30x20x5	920	5	20	4.64	0.725	3.364
0.7	Rectangle	40x10x2	920	5	20	0.58	0.108	0.062
0.7	Round (dxt)	2.5x3	920	5	20	1.85	0.228	0.421
0.7	Trapezium	5	920	5	20	3.44	0.535	1.841

Table 2. Performance of Generator with Shape of Rectangle Magnet with Number of Coil

Dia. Coil (mm)	Number of Winding Coil	Number of Pole (P)	Distance of magnet (mm)	Voltage. (V)	Current. (A)	Power (W)
0.7	460	5	20	2.23	0.627	0.925
0.7	920	5	20	4.64	0.725	3.364
0.7	1380	5	20	4.00	0.506	2.024

Table 3. Performance of Generator with Shape of Rectangle Magnet with Number of Coil

Dia. Coil (mm)	Number of Winding Coil	Number of Pole (P)	Distance of magnet (mm)	Voltage. (V)	Current. (A)	Power (W)
0.12	13000	5	15	53	0.024	1.251
0.12	13000	5	20	57	0.028	1.590
0.15	13000	5	20	63	0.039	2.470
0.18	9600	5	20	59	0.070	4.124
0.20	6000	5	20	34	0.092	3.135

Based on the number of windings coil with the diameter 0.7 mm is obtained that performance of power generator is 3.364 W, voltage 4.64 V and current 0.725 A on the amount of number of winding coil 920 as shown in Table 2. Table 2 shows that an increase in the number of windings will increase the voltage and current is generated. On the difference in coil diameter, the number of winding coil decreases due to the ability of the koker with dimensions 45x45x4 mm can accommodate a certain number of windings as indicated by Table 3. As shown in Table 3, for winding coil diameter 0.2 mm the koker is able to accommodate 6000 windings coil and for the diameter 0.12 mm the number of windings coil is 13000. The resulting voltage for diameter winding coil 0.12 mm with the number of coil 13000 is 53 V, current 0.024 A and power 1.25 W at magnetic distance 15 mm. The addition of magnetic distances to 20 mm will increase the voltage, current and power as shown in Table 3. For the diameter winding coil 15 mm, the koker is still capable having a number of coil 13000 where the resulting voltage is 63 V, the current is 0.039 A and power 2.47 W. This phenomenon shows that the power increases with increasing winding coil diameter and distance of magnet. The addition of the winding coil diameter will increase the current and power but the voltage decreases as the number of coil becomes small. To increase the power generator can be done by increasing the current value by enlarging the diameter of winding coil, the number of winding coil and the installation of the generator on a parallel basis.

Results of experimental testing of AC generator for variation diameter of coil 0.55 mm and 0.7 mm, a distance of magnet 15-25 mm using a rectangle magnet with dimensions 40x18x5 (mm) and 40x21x10 (mm) can be seen in Table 4. Table 4 shows that for diameter coil 0.55 mm with a magnet dimensions 40x18x5 (mm) provides a lower voltage, current and power value than the magnetic dimension

40x21x10 (mm). The effect of distance of magnet shows the increase of the greatest voltage value and it's found that in 20 mm. The same phenomenon occurred in coil diameter 0.7 mm but the highest voltage happen at distance of magnetic 25 mm.

Table 4. Performance of Generator with Shape Rectangle of Magnet

Dia. Coil (mm)	Number of Winding Coil	Number of Pole (P)	Core distance anchor (mm)	Distance of magnet (mm)	Voltage. (V)	Current. (A)	Freq. (Hz)	Power (W)
Dimension of Magnet 40x18x5 (mm)								
	2000	5	30	15	3.914	0.351	7.62	1.374
0.55	2000	5	30	20	3.894	0.313	10.07	1.219
	2000	5	30	25	1.517	0.275	7.61	0.417
Dimension of Magnet 40x21x10 (mm)								
	2000	5	30	15	9.19	0.739	13.78	6.791
0.55	2000	5	30	20	14.27	0.666	19.29	9.504
	2000	5	30	25	12.78	0.666	10.44	8.511
Dimension of Magnet 40x18x5 (mm)								
	1260	5	30	15	3.21	0.322	9.86	1.034
0.7	1260	5	30	20	2.886	0.508	7.4	1.466
	1260	5	30	25	3.459	0.589	7.47	2.037
Dimension of Magnet 40x21x10 (mm)								
	1260	5	30	15	8.15	0.878	28.2	7.1557
0.7	1260	5	30	20	10.67	1.3	11.54	13.871
	1260	5	30	25	14.63	1.218	14.63	17.819

4. Conclusions

From the experimental testing of performance AC generator with translation motion we get conclusion is:

1. Variations shape of magnet neodymium trapezium, round and rectangle has maximal value for voltage, current and power generated at 20 mm of gap magnetic distances.
2. The maximum voltage and power obtained by AC generator with translation motion using the shape of rectangle magnet (46x21x10mm) is 14.63 V and 17,82W.
3. From effect diameter of winding coil show that the number of winding coil is very large to influence performance of AC generator such increasing diameter and number of winding coil made voltage, current and power of AC generator become increase.

5. References

[1] Wahyudianto Bagus Nugroho, Indra Ranu Kusuma, dan Sardono Sarwitto, Kajian Teknis Gejala Magnetisasi pada Linear Generator untuk Alternatif Pembangkit Listrik, Jurnal Teknik POMITS Vol. 3, No. 1, (2014) ISSN: 2337-3539 (2301-9271 Print)

[2] Aris Budiman, Hasyim Asy'ari, Arief Rahman Hakim, Desain Generator Magnet Permanen Untuk Sepeda Listrik, Jurnal Emitor Vol. 12 No. 01 ISSN 1411-8890.

[3] Higuchi, K., Guan Y., Yokomizu Y., and Matsumura T., Observation of Transient Behavior of Magnetic Flux in Inductive-type Fault Current Limiter with YBCO Thin Film Disc, Physics Procedia 36 (2012) 1254 – 1257.

[4] Hendra, Indriani, A., and Hernadewita, Applying of Piston Mechanism Design used in the Wavelength Electrical Generating of Ocean for Fishing Communities, Advanced Materials Research, Vol. 918, pp. 73-78, ISSN: 1662-8985, Trans Tech Publications, Switzerland

[5] Indriani, A., Hendra, Suhartini, Y., Tanjung, A., Performance of Ocean Wave Power Plant Using Pneumatic System With Variety Of Position Of Exhaust Valves Tube Piston And The Number Of Buoys, Proceeding of International Symposium on Material Science and Engineering, Kuala Lumpur, Malaysia, 2017 .

[6] Indriani, A, Jonianto, H., Hendra, Optimasi Desain Piston Pembangkit Listrik Tenaga Gelombang Laut Sistem Pneumatik Untuk Output Daya Listrik, Prosiding Seminar Nasional Teknik Mesin XV, ITB Bandung, 2016

[7] Eka Putri R. L., Sarwoko, M., Rusdinar, A., Adam, K. B., Perancangan Dan Implementasi Pembangkit Listrik Tenaga Ombak Laut Menggunakan Sistem Generator Dc Untuk Pengisian Baterai Di Perahu Nelayan, e-Proceeding of Engineering : Vol.3, No.1 April 2016.

[8] Baileya W., Wena H., Yanga Y., Forsythb A., Jiab C., A Cryogenic Dc-Dc Power Converter For A 100 Kw Synchronous HTS Generator At Liquid Nitrogen Temperatures, Physics Procedia 36 (2012) 1002 – 1007.

[9] Anand, S., Turbines for Wave Energy Plants, Proceedings of the 8th International Symposium on Experimental and Computational Aerothermodynamics of Internal Flows, Lyon, 2007.

[10] Amundarain, M., Alberdi, M., Garrido, J., and Garido, I., Modeling and Simulation of Wave Energy Generation Plants: Output Power Control, IEEE Transactions on Industrial Electronics, Vol. 58, and No.1, 2011.

[11] Falcao, A. F., Wave Energy Utilization: A Review of the Technologies, Renewable and Sustainable Energy Reviews, 14, pp. 899-918, 2010.

[12] Casman, D. P., Sullivan, D. L., Egan, M.,M., and Hayes, J. G., *Modeling and Analysis of an Offshore Oscillating Water Column Wave Energy Converter,* Proceedings of the 8th European Wave and Tidal Energy Conference, 2009, pp. 924-933, Sweden.

Acknowledgements:

Thank you to the Government of the Republic of Indonesia through the Ministry of Research, Technology and Higher Education of the Republic of Indonesia and Institute for Research and Community Service of University of Bengkulu who has funded this research, at the Higher Research Grants of Higher Education in 2017.

IOP Conf. Series: Materials Science and Engineering **307** (2018) 012021 doi:10.1088/1757-899X/307/1/012021

Control Systems of Rubber Dryer Machinery Components Using Programmable Logic Control (PLC)

Hendra[1]*, A. S. Yulianto[1], A. Indriani[2], Hernadewita[3] and Hermiyetti[4]

1 Mechanical Engineering Dept., University of Bengkulu, Indonesia
2 Electrical Engineering Dept., University of Bengkulu, Indonesia
3 Industrial Engineering Dept., University of Mercubuana, Jakarta, Indonesia
4 Economics Faculty, University of Bakrie, Jakarta, Indonesia
E-mail: h7f1973@yahoo.com, aniz_raimin@yahoo.com, hadeita@yahoo.com, hermi_yetti@yahoo.com

Abstract. Application of programmable logic control (PLC) is widely used on the control systems in the many field engineering such as automotive, aviation, food processing and other industries [1-2]. PLC is simply program to control many automatic activity, easy to use, flexible and others. PLC using the ladder program to solve and regulated the control system component. In previous research, PLC was used for control system of rotary dryer machine. In this paper PLC are used for control system of motion component in the rubber dryer machinery. Component of rubber dryer machine is motors, gearbox, sprocket, heater, drying chamber and bearing. Principle working of rubber dryer machinery is wet rubber moving into the drying chamber by sprocket. Sprocket is driven by motors that conducted by PLC to moving and set of wet rubber on the drying chamber. Drying system uses greenhouse effect by making hanger dryer design in the form of line path. In this paper focused on motion control system motors and sensors drying rubber using PLC. The results show that control system of rubber dryer machinery can work in accordance control input and the time required to dry the rubber.

1. Introduction

Application of control systems are commonly found in the engineering field such as automotive, manufacturing, food processing and other industries [1-2]. The control system is used to manage and arrange how to make the product by automatically processing. To control of motion component, control machine, control equipment, control of manufacturing process and quality product must using control system mechanism. Control of motion component can be seen at the conveyor, motors, servo, pump, gearbox, blower and sprocket. This control system has been used for manufacturing process, CNC machine, cutting process, assembly and joining by robot. In the food processing industry, the control system can be seen in the process of separation of the initial product, washing, transfer of food, packing process and quality control of product. In rubber industry control system are used for regulate of production processing like cleaning rubber processing.

Control system have open and close loop systems, which a system have some advantages and disadvantages. Open loop system is easily to use, simply program, easy to maintenance, cheaper and no need feedback information from output system. For close loop system the program is complex, high accuracy, more expensive and feedback from output is required. Control system can be create by using microcontroller, PLC and others programming [1-2]. The language program for control system used assembler language, C ++, ladder program and other programming languages. In this paper we

IOP Conf. Series: Materials Science and Engineering **307** (2018) 012021 doi:10.1088/1757-899X/307/1/012021

will focus for control system on the drying rubber processing. This control are used to get good quality of rubber dry, short time processing and smaller area for drying. Smart Relay Zelio PLC type and ladder languages programming is applied to control system mechanism by close loop system. Ladder language programs are used because simpler programming, easier to operate, flexible and can be changed faster without stopping running processes.
PLC system for rubber processing is applying by using greenhouse effect to drying a rubber. Greenhouse effect to drying rubber have some component such as drying chamber, wet rubber hanger, hanger tracking rail, sprocket and shaft, heater, blower, drive motor and sensor. Temperature sensor, motion wet/dry rubber sensor and heater sensor is applied in this system. All sensors activity and motion component of dryer machine is control by Smart Relay Zelio PLC using ladder programming language. In previous based on the processing of rubber especially crepe drying is done naturally by dried in the open space/area where as drying time required 12-14 days depend on the climate/weather. If the weather conditions is rain so drying time will be longer and rubber quality become decreases.
To solve a problem of rubber dry processing so in this paper we designed a rubber dryer machinery using PLC system. The drying process utilizes the greenhouse effect which heat from solar is enclosed in drying chamber made of glass.
The working process of control system of drying rubber is a wet rubber hung on a hanger installed outside drying chamber. Furthermore the PLC system will instruct the motor to move the hanger running on the track rail to enter wet rubber into the drying chamber. By setting time for enter of wet rubber into the drying chamber then starting of drying process using solar heat and heater. If solar heat condition is low due to cloudy/rainy/night so heater will be active to support heat into the drying chamber. The movement of rubber and heat into the drying chamber is arranged and control according to the condition of the rubber. If temperature inside of drying chamber is high then the blower be active to support the air to decrease of temperature.

2. Methods

Programmable Logic Control (PLC) [3-6] is a programmable for control system be able to communicate with the networking and easier to operation. In automation systems PLC playing important rule for control system, which can monitor the state of the system through signals input, and determine the action control of the output equipment. PLC have good ability to control system, simple tasks over and control more complex processes.
PLC consists of several models or types power supply, number of I/O terminals, and output circuit. Type of PLC power supply is AC and DC, the number of I/O depending on PLC brand. One PLC unit sometimes has 10 into 60 I/O terminals or up to 100 I/O. The I/O ratio is 3: 2 for PLC with 10 I/O terminals having 6 input and 4 output terminals. For output circuit in PLC there are 3 types: relay output, output of sinking transistor and output saucing transistor.

PLC has several advantages, namely:

1. Use wire for PLC control system is less.
2. Easy in modifying the control system without changing the wiring.
3. Does not require a control component such as timer for connecting equipment system output to the power source.
4. Operating speed of PLC system is very fast and high productivity.
5. PLC costs are cheaper.
6. PLC control system is more reliable.
7. PLC control program can be printed quickly.

PLC system in the rubber dryer machinery is control some component namely;

1. Temperature in the drying chamber.
2. Motion of drying machine components such as motor, sprocket, gearbox, chain, blower and heater.
3. Air exhaust and color of dry rubber.

Rubber dryer machinery consist of:

1. Drying chamber from acrylic.
2. Frame hanger and chain holder of wood, iron and steel.

3. Hanger rubber clamp from iron material.
4. Shaft of the gear holder of steel material.
5. Gear.
6. Hanger rubber clamping rail.
7. Temperature sensor.
8. Motion sensor.
9. Motor.
10. Chain.
11. PLC.

Stage manufacture and testing of control systems on rubber drying machines include:

1. Design and manufacture of control system using PLC for greenhouse effect drying machine (See Fig. 1).
2. Preparation of motion component of greenhouse effect drying machine.
3. Modeling of component control system with ladder programming language and PLC.
4. Testing of drying control system using PLC includes drying time, drying temperature, drying capacity, efficiency of rubber motion control system.

Dimension of rubber for experimental performance control system of rubber dryer machinery is thickness 10 mm, length 100 mm with width 50 mm and 150 mm with width 75 mm as shown in Fig. 2. The ladder programing language before using PLC has been tried in PC as shown in Fig. 3. Ladder program language consist on the motor, motion wet rubber sensor, temperature sensor, blower and heater. Setting time and number of hanger inside of dryer chamber is 3 second and 12 hanger with hang of wet rubber. After the wet rubber put in the hanger, the motor become active to moving chain insert of wet rubber.

3. Results

The result of manufacture and testing of control system of rubber drying machine by using PLC can be seen in Fig. 1. Figure 1 show that the dryer process and control system components have been installed on the rubber dryer machinery working in accordance with the input. The ladder program for this motion control system can be seen in Fig.2. After the system is functional well than it is connected to the rubber dryer machinery. The result show that the drier engine actuator component works according to the command input to the PLC.

The result of experimental of rubber drying with dimension of 100 x 50 mm and thickness of 10 mm can be seen in Table 1. Temperature setting on drying chamber is 32^0C-70^0C which the resources from the solar heat and heater. When the wet rubber is hung on the clamp hanger, the motion sensor give input signal to PLC to activate the motor to move the running chains on the guide rail. And PLC will activate the heater if the room temperature is higher than 70^0C or temperature is less than 32^0C. If the temperature more than 70^0C the blower will activate to dropout the hot air in the drying chamber.

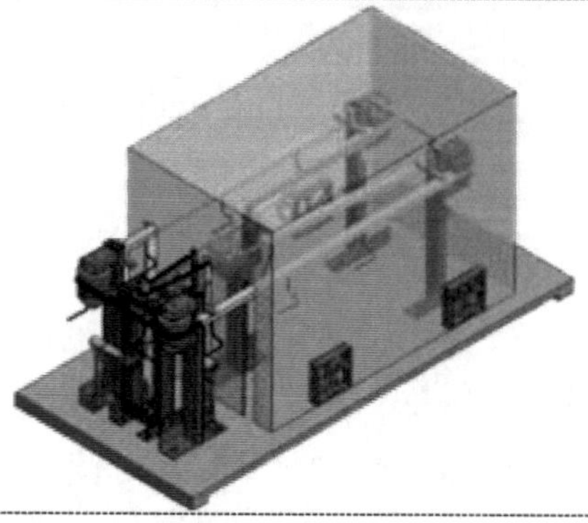

Figure 1. Control System Components Dryer Rubber Machinery

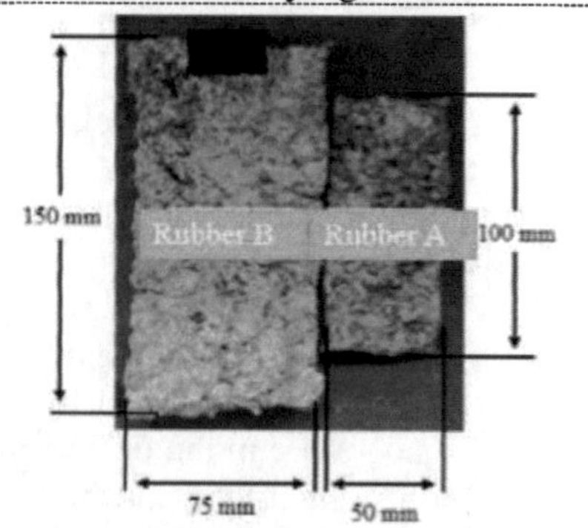

Figure 2. Dimension of Rubber

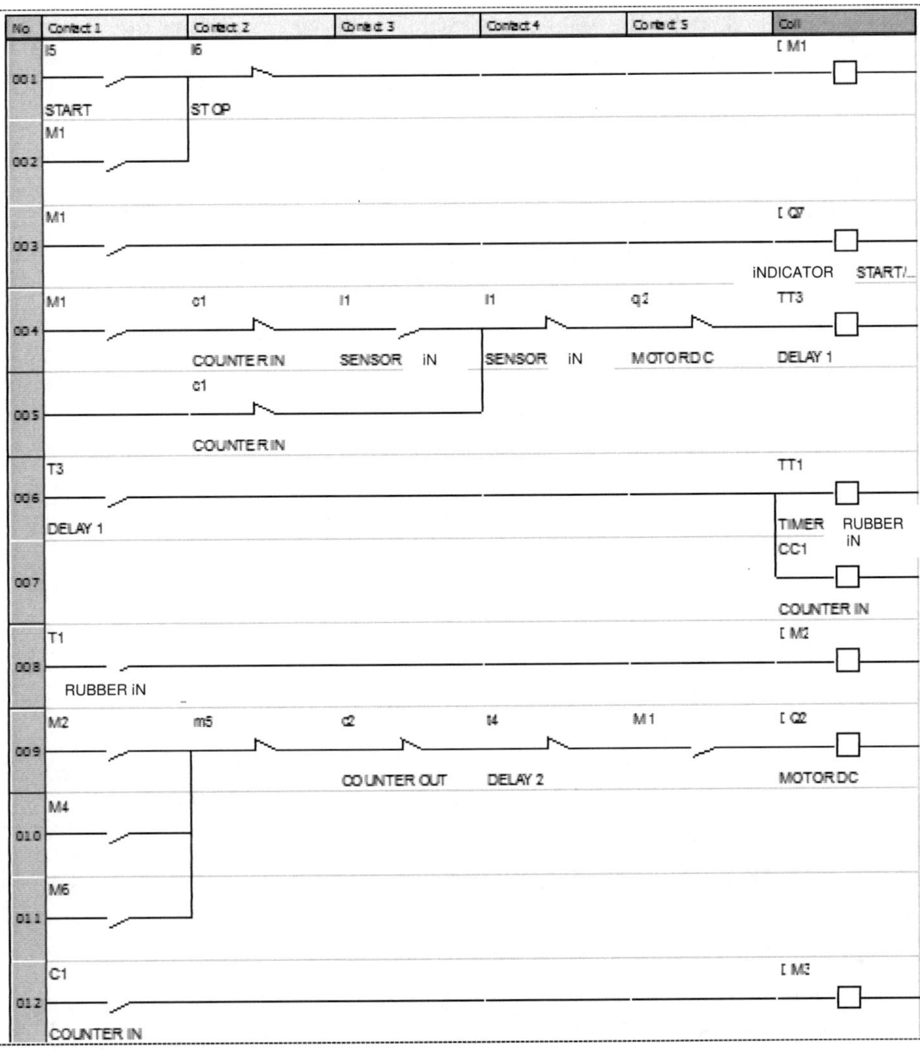

Figure 3. Ladder Program Control System Components Dryer Rubber Machinery

Table 1. Water Quality and Weight of Rubber before Drying Processing

Number Hanger of Rubber	Ave. Water Quality (%)	Weight (gram)
1	23,40	40
2	24,85	40
3	26,00	40
4	23,64	40
5	25,42	40
6	23,07	40
7	24,78	40
8	22,71	40
9	23,07	40
10	23,42	40
11	23,21	40
12	21,92	40

IOP Conf. Series: Materials Science and Engineering **307** (2018) 012021 doi:10.1088/1757-899X/307/1/012021

The amount of time required to dry the 12 piece of rubber with the dimensions of 100 x 50 mm from the initial temperature of 32 ^{0}C to the final temperature of 70.2^{0}C is 109 minutes. For wet rubber the moisture content 23-26% and weight 40 grams after drying processing become 30 grams and 0%. These data show that in rubber testing the weight is reduced 10 grams and in other words the motion control system on the rubber dryer machine can function well to dry the rubber. The amount moisture content of dry rubber from beginning until the end of drying process can be seen in Table 1 and 2 with the amount of dried rubber 12 pieces.

Table 2. Water Quality and Weight of Rubber after Drying Processing

Number Hanger of Rubber	Ave. Water Quality (%)	Weight (gram)
1	0	30
2	0	30
3	0	30
4	0	30
5	0	30
6	0	30
7	0	30
8	0	30
9	0	30
10	0	30
11	0	30
12	0	30

4. Conclusions

From the results of control system and performance of rubber drying machinery using PLC was concluded that:

1. The motion control system and the temperature of the rubber drying machine the greenhouse effect by using a PLC can work well to drying the rubber.
2. Time required for drying rubber is 109 minutes from initial temperature of 32^{0}C to 70.2^{0}C.
3. The moisture content and the final weight of dry rubber is 0% and 30 grams while the initial conditions of wet rubber is 23-26% for the moisture content and weight of 40 grams.

5. References

[1] Indriani, A, Hendra, Witanto, Y., Error of Assembly Microcontroller Arduino Mega and ATmega in the Control of Temperature for Heating and Cooling System, *Applied Mechanics and Materials*, ISSN: 1662-7482, Vol. 842, pp 319-323, Trans Tech Publications, Switzerland.
[2] Hendra, Indriani, A., Hernadewita, Rizal, Y., Assembly Programmable Logic Control (PLC) in the Rotary Dryer Machine for Processing Waste Liquid System, *Applied Mechanics and Materials*, ISSN: 1662-7482, Vol. 842, pp 319-323.
[3] Setiawan I., Programmable Logic Controller dan Teknik Perancangan Sistem Kontrol, ISBN 979-763-099-4, Penerbit Andi Yogyakarta, 2006.
[4] John R. Hackworth, J. R., And Hackworth, Jr., F., D., Jr., Programmable Logic Controllers: Programming Methods And Applications
[5] Boysen, A. R., Programmable Logic Controllers and Ladder Logic, Department Of Humanities South Dakota School Of Mines and Technology.
[6] Collins K, PLC Programming For Industrial Automation.

Acknowledgements:

Thank you to the Government of the Republic of Indonesia through the Ministry of Research, Technology and Higher Education of the Republic of Indonesia and Institute for Research and Community Service of University of Bengkulu who has funded this research, at the Higher Research Grants of Higher Education in 2017.

IOP Conf. Series: Materials Science and Engineering **307** (2018) 012022 doi:10.1088/1757-899X/307/1/012022

Performance of Control System Using Microcontroller for Sea Water Circulation

A. Indriani[1]*, Y. Witanto[2], A. Pratama, S.[1], Supriyadi[1], Hendra[2], A. Tanjung[3]

1 Electrical Engineering Dept. University of Bengkulu, Indonesia
2 Mechanical Engineering Dept. University of Bengkulu, Indonesia
3 Polytechnic State of Padang, Indonesia

E-mail: aniz_raimin@yahoo.com, yovanmail@gmail.com, h7f1973@yahoo.com, niniamrin@yahoo.com

Abstract. Now a day control system is very important rule for any process. Control system have been used in the automatic system. Automatic system can be seen in the industrial filed, mechanical field, electrical field and etc. In industrial and mechanical field, control system are used for control of motion component such as motor, conveyor, machine, control of process made of product, control of system and soon. In electrical field, control system can met for control of electrical system as equipment or part electrical like fan, rice cooker, refrigerator, air conditioner and etc. Control system are used for control of temperature and circulation gas, air and water. Control system of temperature and circulation of water also can be used for fisher community. Control system can be create by using microcontroller, PLC and other automatic program [1][2]. In this paper we will focus on the close loop system by using microcontroller Arduino Mega to control of temperature and circulation of sea water for fisher community. Performance control system is influenced by control equipment, sensor sensitivity, test condition, environment and others. The temperature sensor is measured using the DS18S20 and the sea water clarity sensor for circulation indicator with turbidity sensor. From the test results indicated that this control system can circulate sea water and maintain the temperature and clarity of seawater in a short time.

1. Introduction

Now a day control system having important rule which the development of the current control system is very fast. Control system have been used in the automatic system. Automatic system can be seen in the industrial filed, mechanical field, electrical field and etc. In previous, control system only is use in industry for control of part or component machine and processing to make of product. In industrial and mechanical field, control system are used for control of motion component such as motor, conveyor, machine, control of process made of product, control of system and soon During the development of equipment control system is fast and easy to use so the application of control system can be seen in home equipment such as refrigerator, rice cooker, air conditioner, heater and cooler for house, and etc. Control system are used for control of temperature and circulation gas, air and water. Control system of temperature and circulation of water also can be used for fisher community. Control system

IOP Conf. Series: Materials Science and Engineering **307** (2018) 012022 doi:10.1088/1757-899X/307/1/012022

consist an open loop system and close loop system. Open loop system are simple construction than close loop system, no need feedback from output system, easy for maintenance, cheaper but no stability problem. In closed loop system more complicated, need information feedback from output system to evaluate the process or program. Control system can be create by using microcontroller, PLC and other automatic program. In this paper we will focus on the close loop system by using microcontroller Arduino Mega to control of temperature and circulation of sea water for fisher community. Performance control system is influenced by control equipment, sensor sensitivity, test condition, environment and others.
In previous study, microcontroller has been used to control temperature on the circulation sea water in fish ponds by using ATmega and Arduino Mega [1] [3]. In this paper we will focus on the control of circulation sea water that integrated by temperature, water clarity, moisture content and acid (PH). Composition settings of control circulation of sea water are 30-33 ppt salt, oxygen ± 4 ppm, temperature 24^0 - 31^0C and acidity (pH) 7,6 - 7,8 [4-6]. The ideal water flow velocity is about 20 to 40 cm / sec where this velocity is required for the turn of water and oxygen. The temperature sensor is measured using the DS18S20 and the sea water clarity sensor for circulation indicator with turbidity sensor.

2. Research Method

Control system of temperature and circulation of sea water condition can be conducted by using Microcontroller Arduino Mega. Temperature sensor of sea water measure by sensor DS18S20 and for measure of other condition of sea water such as clarity using turbidity sensor. Testing step of this experiment can be seen in Fig.1. From Fig. 1 show that first step is observation of fish habits condition, second is manufacture of control system using Microcontroller Arduino Mega. Then continue to assembly and testing performance of control system temperature and circulation of sea water. After we get the data then continue to analyse and make the conclusion.

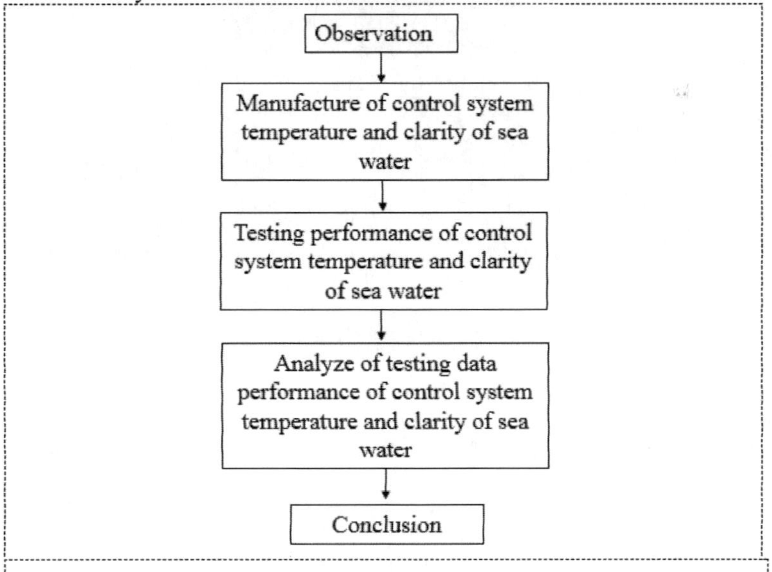

Figure 1. Testing step Control system of temperature and circulation of sea water condition

IOP Conf. Series: Materials Science and Engineering **307** (2018) 012022 doi:10.1088/1757-899X/307/1/012022

Figure 2. Microcontroller Arduino Mega for control system of Temperature and circulation of sea water

Material and equipment for experimental of control system of temperature and circulation of sea water includes:

1. Microcontroller Arduino Mega for control system of temperature and sirculation sea water (See Fig. 2).
2. Pipe for sea water circulation.
3. Pump.
4. Cooling machine and heater for sea water.
5. Pond pool.
6. Temperature sensor DS18S20, Turbidity sensor and measuring instrument (see Figure 3).
7. Multimeter to measure the voltage and current for generated of temperature and turbidity control sea water.
8. Power supply.
9. Solar panel
10. ADC.

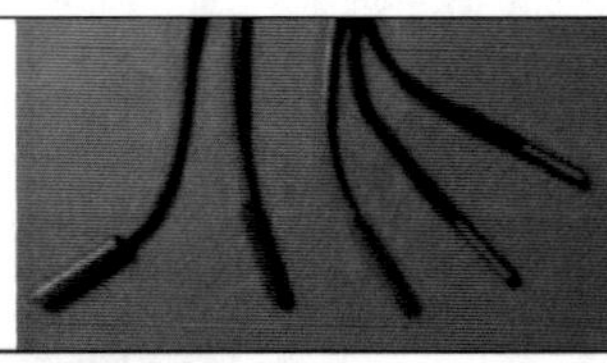

a. Temperature Sensor DS18S20

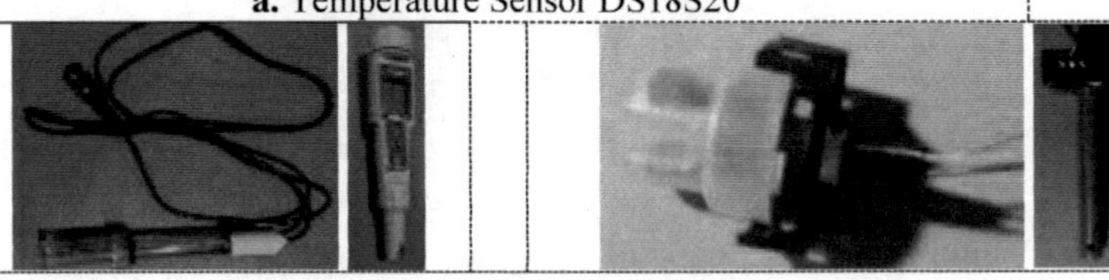

b. Circulation Sensor (Turbidity, Clarity)

Figure 3. Temperature Sensor DS18S20 and Circulation Sensor.

3. Results

Result of machine control system for temperature and circulation of sea water can be seen in Table 1 into Table 3. The control system working by circulation sea water from pond pool using the pump. When the condition temperature setting of sea water has a lower or higher than the temperatures setting 25^0C-30^0C then the pump become active to circulated clean sea water into the pond pool. The heater become on when the temperature is lower than temperature setting of sea water (<25^0C). If the temperature sea water more than 30^0C then cooling machine become on to make sea water is cool. The amount of time required to decrease temperature of sea water with a capacity of 10 l from 31.8^0C to 30.53^0C is 6 minutes 37 seconds. At a capacity of 15 l the time required to decrease temperature from 31.6^0C to 30.69^0C (0.91^0C) is 13 minute 16 seconds. The length of time required to reduce the temperature with the capacity of sea water 10 l and 15 litters can be seen in Table 1.

Table 1. Testing Data Microcontroller Arduino Mega of Cooling Time for Sea Water Circulation

Vol. (L)	Volt. (V)	Current (A)		T_s (^{0}C)		t (m:d)	T Amb. (^{0}C)	Humidity (%)
		Pump	Compressor	T_o	T_i			
10	219.9	0.234	1.068	31.8	30.53	6:37	31.5	58%
	220.1	0.234	1.068	32.4	30.54	5:24	31.5	58%
	220.4	0.234	1.068	33.5	30.58	9:15	31.5	58%
	220.1	0.234	1.068	34.5	30.46	6:56	31.5	58%
	219.7	0.234	1.068	35.5	30.57	13:24	31.5	58%
15	219	0.234	1.068	31.6	30.69	13:46	33	61%
	219.3	0.234	1.068	32.4	30.65	15:58	33.5	51%
	219.4	0.234	1.068	33.7	30.66	19:50	33.3	50%
	219.5	0.234	1.068	34.5	30.65	29:35	33.2	54%
	219.5	0.234	1.068	35.5	30.68	34:28	33.1	54%

Table 2. Testing Data Microcontroller Arduino Mega of Heating Time for Sea Water Circulation

Vol. (L)	Volt. (V)	Current (A)		T_s (^{0}C)		time (m:d)	T Amb (^{0}C)	Humidity (%)
		Pump	Heater	T_o	T_i			
10	220.2	0.234	0.72	27.5	28.35	4:50	32.4	67
	219.7	0.234	0.72	26.5	28.37	7:03	33	61
	219.9	0.234	0.72	25.5	28.37	9:43	32.9	61
	220.1	0.234	0.72	24.5	28.36	13:10	34.4	56
	219.7	0.234	0.72	23.5	28.35	16:17	34	58
15	220	0.234	0.72	27.5	28.37	5:31	33.6	50
	218.2	0.234	0.72	26.5	28.35	9:00	34.2	46
	220	0.234	0.72	25.5	28.36	17:45	33.5	49
	220.4	0.234	0.72	24.5	28.36	21:03	33.3	51
	220.8	0.234	0.72	23.5	28.35	24:04	33.2	50

For increasing temperature of sea water with a capacity of 10 l, control system by Microcontroller Arduino Mega set the heater to be active. Current needed for activation of the heater is 0.72 Ampere and time required to increasing temperature is 4 minutes 50 seconds to raise the temperature of 27.5 ^{0}C until 28.35^{0}C. For sea water with capacity 15 l in pond pool, the time required to raise the temperature from 27.5 ^{0}C to 28.37^{0}C (0.8^{0}C) is 9 minutes as shown in Table 2. Referring to Tables 1 and 2 it is seen that the time required to increase the temperature is faster compared to process decrease in temperature.

Table 3. Testing Data Performance of Microcontroller Arduino Mega for Control System Temperature and Circulation of Sea Water

Vol. (l)	pH		Temp (^{0}C)		Clarity		Salt	
	Sensor	Manual	Sensor	Manual	Sensor	Manual	Sensor	Manual
10	8.4	8.42	30.01	30.4	0	0	3.59	3.08
15	8.8	8.3	29.08	30	0	0	3.58	3.03
17,5	8.17	7.85	30.31	30.4	0.006	0	3.53	3.1
20	8.38	8.01	30.16	30.1	0.003	0	3.59	3
22,5	8.53	8.03	30.54	29.7	0	0	3.65	2.96
25	8.19	8.03	30.44	30.2	0	0	3.62	2.94
27,5	8.4	8.12	30.04	29.9	0	0	3.6	3
30	8.49	8.1	29.67	29.6	0	0	3.64	3.02

Table 3 shows performance of microcontroller arduino mega for sea water circulation and difference measurement using manual and automatic by the microcontroller. From the measurement results visible difference in pH 0.02 in 10 l of sea water, sea water temperature 0.39 and clarity 0. For the other sea water capacity fluctuation margin manual measurements and use a microcontroller that is not too large (<1).
From Table 1 until Table 3 show that the control system of temperature and circulation of sea water working well and can control temperature and circulation system of sea water easily.

4. Conclusions

From experimental of performance of control system for temperature and circulation of sea water by using microcontroller we get conclusions are:

1. The control system by using Microcontroller Arduino Mega can be work well to regulate the temperature and circulation of sea water.
2. Time for decreasing of temperature of sea water with a capacity of 10 l from 31.8^{0}C to 30.53^{0}C is 6 minutes 37 seconds. At a capacity of 15 l the time required to decrease temperature from 31.6^{0}C to 30.69^{0}C (0.91^{0}C) is 13 minute 16 seconds. For increasing temperature of sea water condition with a capacity of 10 l need current for activation of the heater is 0.72 A. and time required is 4 minutes 50 seconds to raise the temperature of 27.5 ^{0}C until 28.35^{0}C. For sea water with capacity 15 l in pond pool, the time required to raise the temperature from 27.5 ^{0}C to 28.37^{0}C (0.8^{0}C) is 9 minutes
3. From measurement by using microcontroller and manual, the results show that visible difference in pH 0.02 in 10 l of sea water, sea water temperature 0.39 and clarity 0. For the other sea water capacity fluctuation margin manual measurements and use a microcontroller that is not too large (<1).

References

[1] Indriani, A., Hendra, Witanto, Y., 2016, Error of Assembly Microcontroller Arduino Mega and ATmega in the Control of Temperature for Heating and Cooling System, *Applied Mechanics and Materials, Vol. 842, pp 324-328, doi:10.4028/www.scientific.net/ AMM.842.324 Online: 2016-06-2, Trans Tech Publications, Switzerland.*
[2] Hendra, A. Indriani, Hernadewita and Y. Rizal, 2016, Assembly Programmable Logic Control (PLC) in the Rotary Dryer Machine for Processing Waste Liquid System, *Applied Mechanics and Materials* ISSN: 1662-7482, Vol. 842, pp 319-323 doi:10.4028/www.scientific.net/AMM.842.319, Trans Tech Publications, Switzerland
[3] Indriani, A., 2015, Mesin Pengontrol Temperatur Air Aquarium otomatis Berbasis Mikrokontroller Untuk Budidaya Ikan Kerapu dengan Menggunakan Sumber Energi Terbarukan, Laporan Hibah Bersaing KEMENRISTEKDIKTI.
[4] Subachri, W., Zainuddin, Yanuarita, D., Makmur, 2011, Pamudi, Budidaya Ikan Kerapu-Sistem Karamba Jaring Apung dan Tancap, WWF-Indonesia.
[5] Suwoyo, H. S., Kajian Kualitas Air Pada Budidaya Kerapu Macan (Epinephelus fuscoguttatus) Sistem Tumpang Sari Di Areal Mangrove, Berkala Perikanan Terubuk Vol 39 No.2 Juli 2011.
[6] Risdianto, D., Amri, J., Athoo' Illah, Z., Aplikasi Probiotik Herbafarm Ikan, Udang Dan Tambak Pada Pemeliharaan Udang Vename (Litopenaeus Vannamei) Dan Ikan Kerapu Macan Untuk Meningkatkan Produksi Perikanan Nusantara, Prosiding SNST ke 6 Tahun 2015, ISBN 978-602-99334-4-4.

Acknowledgements:
Thank you to the Government of the Republic of Indonesia through the Ministry of Research, Technology and Higher Education of the Republic of Indonesia and Institute for Research and Community Service of University of Bengkulu who has funded this research, at the Higher Research Grants of Higher Education in 2017.

Research on criticality analysis method of CNC machine tools components under fault rate correlation

Shen Gui-xiang, Zhao Xian-zhuo, Zhang Ying-zhi*, Han Chen-yu

School of Mechanical Science and Engineering, Jilin University, Changchun, 130000, China

E-mail: zhangyz@jlu.edu.cn

Abstract. In order to determine the key components of CNC machine tools under fault rate correlation, a system component criticality analysis method is proposed. Based on the fault mechanism analysis, the component fault relation is determined, and the adjacency matrix is introduced to describe it. Then, the fault structure relation is hierarchical by using the interpretive structure model (ISM). Assuming that the impact of the fault obeys the Markov process, the fault association matrix is described and transformed, and the Pagerank algorithm is used to determine the relative influence values, combined component fault rate under time correlation can obtain comprehensive fault rate. Based on the fault mode frequency and fault influence, the criticality of the components under the fault rate correlation is determined, and the key components are determined to provide the correct basis for equationting the reliability assurance measures. Finally, taking machining centers as an example, the effectiveness of the method is verified.

Keywords: CNC machine tools; Fault correlation; Criticality; ISM; Pagerank

1. Introduction

CNC machine tool is a complex mechatronic system, the reliability decline of it is the result of the independent system fault and its propagation fault. Criticality is a key index to measure reliability, and it is also the basis of system reliability improvement. Therefore, it is necessary to analyze the criticality of its components under fault rate correlation.

Traditional CNC machine tool fault analysis methods mostly start with the fault mode analysis, and then determine the key components of the whole machine through the fault modes and effects and hazard analysis (FMECA) [1-3]. However, these methods do not consider the fault association among the system components. Instead, these literatures consider the fault mode as independent, so there will be some deviations in the assessment of the criticality. In [4-5], the fault tree analysis method is used to analyze the importance degree of the fault in the tool magazine by solving the importance degree of the basic event. However, when the method is used, the fault rate of the basic event needs to be known., otherwise this method is invalid. In [6], the criticality of components is calculated and ranked according to the relationship between the fault mode frequency rate, the index weight of fault rate and the fault mode.

Based on the above literatures analysis , this paper proposed a system component criticality analysis method under fault rate correlation. Firstly, according to fault mechanism analysis and matrix transformation, ISM model is used to realize fault association model hierarchy, and then Pagerank algorithm is introduced to calculate the influence degree of component fault. Finally, by calculating the component fault rate, the comprehensive fault influence and the fault mode frequency, the system component criticality under fault rate correlation is calculated. Finally, taking machining centers as an example to verify the effectiveness of the method.

2. System components criticality analysis principle under fault rate correlation

2.1. *Component comprehensive fault rate calculation based on ISM model.*

Reference [7], ISM model of CNC machine tool can be established. Based on it, combined relative influence degree CI of components and the fault rate λ_i, the comprehensive fault rate $\lambda_{i\Sigma}$ of components i can be calculated as shown in equation (1):

$$\lambda_{i\Sigma} = \lambda_i + \sum_{1}^{k} \theta_{j\to i} \lambda_{j\Sigma} \tag{1}$$

Where: λ_i is fault rate under time correlation, reference [8], which can be obtained by equation (2),

$$\lambda(t) = \frac{\beta}{\alpha^{\beta}} t^{\beta-1} \tag{2}$$

Among them: α is the scale parameter, β is the shape parameter.

$\theta_{j\to i}$ is influence coefficient of component j on component i, which can be obtained by equation $\theta_{j\to i} = \dfrac{CI(V_j)}{C(V_j)}$, where: V_j is component which has influence on i, $C(V_j)$ is the out of component V_j, which is the number of components V_j points to other components. $CI(V_j)$ is the r relative influence value of component V_j, reference literature [9], using equation (3) can calculate it:

$$CI^{(x+1)} = \frac{1-d}{n} * E + d * \left((A^T)'\right)^T * CI^{(x)} \tag{3}$$

Where d is the damping factor, which reflects the probability that the fault propagates along the propagation path, so the value is the ratio of the number of related faults to the total number of faults,

E is the $(n\times 1)$ order matrix with all elements being 1, $(A^T)'$ is the adjacency matrix after transpose, and then transform the transposed adjacency matrix into the state transition matrix.

2.2. *Calculation of component criticality under fault rate correlation*

2.2.1. *Criticality calculation of independent fault modes*

When the fault mode i of component a is independent of other fault modes, its criticality can be calculated by the following equation:

$$F_{ai} = p_{ai} s_{ai} \lambda_{a\Sigma} \quad (4)$$

Where, p_{ai} is the frequency of fault mode i in component a, s_{ai} is the severity of i, its assignment is shown in Table 1

Table1. The assignment of s_{ai}

Severity level of fault mode	very low	low	middle	high	very high
Assignment	1	3	5	7	9

2.2.2. *Criticality calculation of correlative fault modes*

When the fault mode j of component a is related to fault mode k of component b ,its criticality can be calculated by the following equation:

$$F'_{aj} = F_{aj} + e_{k\to j} F'_{bk} \quad (5)$$

Where, $e_{k\to j}$ is influence coefficient of k to j, which can be calculated $e_{k\to j} = \dfrac{n_{k\to j}}{N_{b\to a}} \cdot \theta_{b\to a}$,

where $n_{k\to j}$ is frequency of fault mode k causing fault mode j, $N_{b\to a}$ is fault frequency of component b causing component a.

2.2.3. *Calculation of component criticality under fault rate correlation*

The component criticality is equal to the sum of the criticality of each fault mode occurring within the component, therefore, which can be calculated by equation:

$$F_a = \sum_i F_{ai} + \sum_j F'_{aj} \quad (6)$$

3. Case Studies

3.1. *Calculation of comprehensive fault rate under fault rate correlation*

According to the 106 fault data of domestic machining centers, the components of the machining center are divided and coded, which is shown in Table 2

Through the modeling analysis of the fault data, component fault rate based on time correlation can be calculated by equation (2), shown in table 3.

Table 2. The component division and its code and the its fault frequency

Code	Component	Frequency	Rate	Code	Component	Frequency	Rate
M	Tool system	27	0.2547	G	Pneumatic system	4	0.0377
S	Spindle system	14	0.1321	K	Chip removal	12	0.1132
W	Cooling system	3	0.0283	L	Lubrication system	3	0.0283
J	Feeding system	31	0.2925	NC	CNC system	6	0.0566
V	Electrical system	6	0.0566				

Table 3. Component fault rate based on time correlation

Component	α	β	$\lambda_i(t)$
S	3156.143	1.136	0.0007306
M	2272.155	0.847	0.0002320
J	2002.848	0.903	0.0003372
NC	7692.308	1.12	0.0001857
V	3225.806	1.155	0.0005610
G	32786.885	0.743	0.0000195
L	2857.143	0.725	0.0001106
W	18867.925	0.888	0.0000415
K	3846.154	1.016	0.0002759

Through fault cause analysis, establish fault directed graph of machining center as shown in figure 1, and construct adjacency matrix *A* through figure 1.

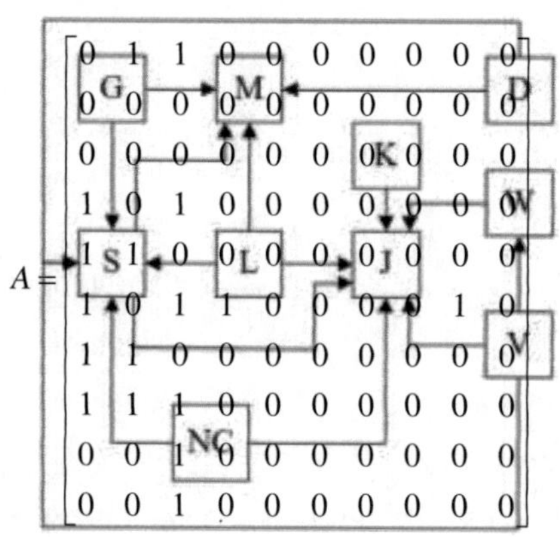

Figure 1. Fault directed graph of machining center

The probability transfer matrix $\left[\left(A^T\right)'\right]^T$ can be obtained by A, and there are 33 associated faults, therefore damping factor $d = 33/106 \approx 0.3$, according to equation (3), *CI* can be obtained as shown in figure 2, combined with ISM model of component, The model of fault propagation with weight can be obtained.

IOP Conf. Series: Materials Science and Engineering **307** (2018) 012023 doi:10.1088/1757-899X/307/1/012023

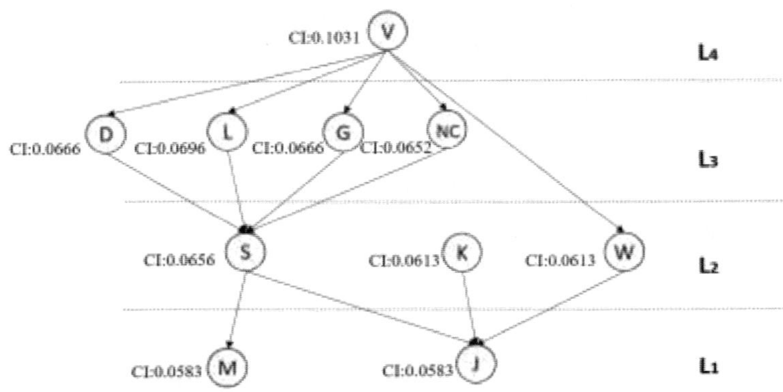

Figure 2. The model of fault propagation with weight

Combining figure 2, table 3 and equation (1), the comprehensive fault rate under fault rate correlation $\lambda_{i\Sigma}$ can be obtained as shown in table 4.

Table 4. The comprehensive fault rate under fault rate correlation $\lambda_{i\Sigma}$

Component	$\lambda_{i\Sigma}$	Component	$\lambda_{i\Sigma}$
S	0.0007548	NC	0.0001973
V	0.0005610	L	0.0001222
J	0.0003821	W	0.0000530
K	0.0002759	G	0.0000311
M	0.0002567		

3.2. *Calculation of component criticality under fault rate correlation*

According to table 4, combined fault data, the component criticality under fault rate correlation F_i can be obtained through equation (6), shown in table 5. Whereas, reference [2], the component criticality CR_i calculated by traditional FMECA is shown in table 6

Table 5. Value of F_i

Code	F_i	Code	F_i
S	0.0031270	L	0.0008554
V	0.0026181	NC	0.0005261
J	0.0020337	W	0.0003712
K	0.0017475	G	0.0002021
M	0.0015499		

Table 6. Value of CR_i

Code	CR_i	Code	CR_i
M	0.001944	V	0.000751
J	0.001795	NC	0.000495
K	0.001241	W	0.00029
S	0.000955	G	0.000127
L	0.000774		

4. Conclusions

(1) Based on traditional FMECA, component M is the most key component of machining center as shown in table 6, whereas component S is not.

(2) Based on the method proposed in this paper, component S is the most key component, although

there are not many faults in this component, there are many associated faults with the other systems, component S is in a critical position in the whole process of fault transmission, so it has a higher criticality. Whereas, although there are many faults in component M, it is in the terminal position in the fault transmission process, so it is not the most critical component.

(3) Compared to traditional FMECA, the component criticality analysis under fault rate correlation can reflect the intrinsic link between the various components, it has guiding significance for the equationtion of reliability measures, and provide right direction of improving the reliability of machining center.

Reference

[1] Xue Yu-xia, Shen Gui-xiang, Zhang Ying-zhi. Fault analysis for numerical control machine tools based on fuzzy logic [J]. Journal of Jilin University (Engi- neering and Technology Edition), 2008, (Sup.1): 115-118

[2] Jie, Jia Ya-zhou. Failure mode effect and criti- cality analysis on certain serial CNC lathes [J]. Journal of Harbin Institute of Technology, 2005(12):1725-1727.

[3] Zhang Ying-zhi, Jia Ya-zhou, Shen Gui-xiang,et al.Gray correlative analysis of failure distribution of CNC machine tool[J]. Transactions of the Chinese Society for Agricultural Machinery, 2004, 35(6):194-196.

[4] Dong Y H，Yu D T. Estimation of failure probability of oil and gas transmission pipelines by fuzzy fault tree analysis [J]. Journal of Loss Prevention in the Process Industries, 2005, 18(2): 83-88

[5] Yao C Y, Zhao J Y. Research on fuzzy fault tree analysis method of hydraulic system [J].China Mechanical Engineering, 2007, 18（14）:1656-1659,1675.

[6] Wang Xiao-feng, Shen Gui-xiang, Zhng Ying-zhi et al. Prioritizing failures of abc-axis feeding systems based on improved criticality and DEMATEL method[J], Journal of Jilin University (Engineering and Technology Edition), 2012, 42(01) :122-127.

[7] Shen GX , Sun SG, Zhan, YZ,et al.System failure analysis based on DEMATEL-ISM and FMECA[J] JOURNAL OF CENTRAL SOUTH UNIVERSITY, 2014, (21): 4518-4525.

[8] Shen Gui-xiang, Meng Shu, Zhang Ying-zhi,et at. Aplication of average rank time method in reliability modeling for subsystems[J], Journal of Jilin University (Engineering and Technology Edition), 2014,44(1):101-105.

[9] Long Zhe, Shen Gui-xiang, Zhang Ying-zhi et al. Evaluation of the fault correlation of machining center component [J]. Jouranl of Harbin Institute of Technology, 2017,49(01):133-138.

IOP Conf. Series: Materials Science and Engineering **307** (2018) 012024 doi:10.1088/1757-899X/307/1/012024

Feed rate affecting surface roughness and tool wear in dry hard turning of AISI 4140 steel automotive parts using TiN+AlCrN coated inserts

Phacharadit Paengchit[1] and Charnnarong Saikaew[1]*

1 Department of Industrial Engineering, Faculty of Engineering, Khon Kaen University, Khon Kaen 40002 Thailand
E-mail: charn_sa@kku.ac.th

Abstract. This work aims to investigate the effects of feed rate on surface roughness (Ra) and tool wear (VB) and to obtain the optimal operating condition of the feed rate in dry hard turning of AISI 4140 chromium molybdenum steel for automotive industry applications using TiN+AlCrN coated inserts. AISI 4140 steel bars were employed in order to carry out the dry hard turning experiments by varying the feed rates of 0.06, 0.08 and 0.1 mm/rev based on experimental design technique that can be analyzed by analysis of variance (ANOVA). In addition, the cutting tool inserts were examined after machining experiments by SEM to evaluate the effect of turning operations on tool wear. The results showed that averages Ra and VB were significantly affected by the feed rate at the level of significance of 0.05. Averages Ra and VB values at the feed rate of 0.06 mm/rev were lowest compared to average values at the feed rates of 0.08 and 0.1 mm/rev, based on the main effect plot.

1. Introduction

Conventionally, hard turning is one of the most widely used machining processes in various industries such as electronics, automotive, and aerospace. In the automotive industry, surface quality and accuracy characteristics of machined parts using computer numerical control has increased production costs. This is uneconomical for manufacturing industries that utilize extensive machining operations [1]. Hard turning provides relatively high accuracy for a variety of parts, but sometimes problems arise with the quality of the surface finish under different cutting operating conditions and different cutting tools [2]. The turning processes with hardness values higher than 45 HRC use various cutting tools such as cemented carbide inserts, polycrystalline cubic boron nitride (PCBN), cubic boron nitride (CBN), and ceramic tools [3,4]. Sahoo and Sahoo [5] investigated the machinability performance of AISI 4340 grade steel of different hardness using uncoated, multilayer TiN, and ZrCN coated inserts at high cutting speeds. They also studied hard turning of AISI H13 steel using CBN insert tools and predicted surface roughness and tool wear [6]. Chinchanikar and Choudhury [7] explored the effect of workpiece hardness and types of coated carbide tools on different machinability, tool life, and surface roughness during the turning of hardened AISI 4340 steel. Krolczyk et al. [8] studied the effect of cutting speed on the geometrical parameters of surface integrity after turning operation of duplex stainless steel with coated sintered carbide wedges. Lin et al. ([9] investigated the performance of TiN, TiAlN and AlCrN coated inserts on tool life and wear resistance under dry and wet turning operations at very high cutting speeds. They found that AlCrN can achieve higher performances of tool life and wear resistance than TiN and TiAlN coated inserts. Chandiran et al. [10] also performed a comparative study on tool life and wear resistance of TiN and AlCrN coated inserts under high cutting speed and feed rate in various turning conditions using design of experiments (DOE) and response surface methodology (RSM). They stated that AlCrN has better performance than TiN. The aim of this study was to investigate the effects of feed rate on average Ra and VB values in turning operation of AISI 4140 with TiN+AlCrN coated insert using DOE and main effect plot in order to obtain the appropriate feed rate with low Ra and VB.

2. Experimental methodology

In this work, workpiece materials were AISI 4140 chromium molybdenum steel bars with an average hardness of 58 HRC used for producing automotive parts. Chemical composition of workpiece material consisted of 0.47% C, 1.05% Cr, 0.2% Mo, 0.0042% V, 0.012% Al, 0.01% P, 0.01% S, 0.26% Si, 0.67% Mn and 0.19% Cu. An automotive part as a workpiece material was machined with turning operations of four different diameters. Fanuc CNC lathe machine (Takisawa: model NEX-106) was used to carry out the turning experiments with TiN+AlCrN coated inserts for the surface finish. The physical and mechanical properties of the coated insert included hardness of 94 HRA, transverse rupture strength of 0.9 GPa, modulus of elasticity of 400 GPa, and surface roughness of 0.101 μm. Turning experiments were conducted at three different feed rates of 0.06, 0.08 and 0.1 mm/rev while the cutting speed and depth of cut were kept constant at 220 m/min and 0.1 mm, respectively. Figure 1 shows the experimental setup for turning operation dealing with customer desired value of Ra less than 0.8 μm. The Ra values of workpieces produced during these experiments were assessed with a commercial surface roughness tester machine (Germany, Mahr model: MarSurfPSl) with a cut-off length of 0.8 mm and sampling length of 5 mm. A completely randomized experiment with three levels of feed rate and three replicates was run in order to investigate the effect of feed rate on Ra and VB. The statistical model for one-way ANOVA was used to investigate the effect of feed rate on average Ra and VB at the level of significance of 0.05.

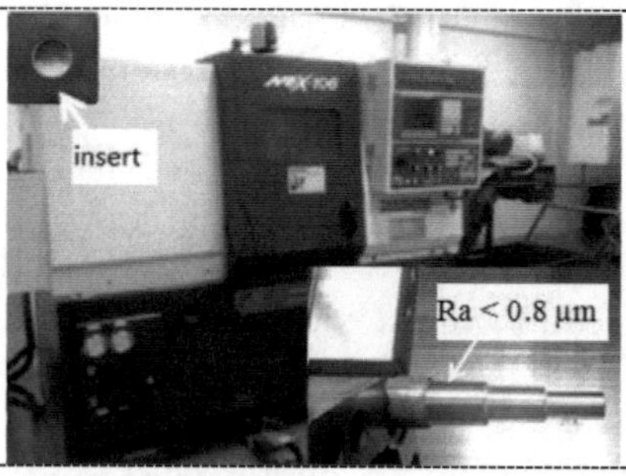

Figure 1. The experimental setup for turning operation.

3. Results and discussion

Experimental matrix and results of Ra and VB values when turning AISI 4140 steel bar with TiN+AlCrN coated inserts after machining time of 32 min are shown in Table 1.

Table 1. Experimental matrix and results for feed rate affecting Ra and VB.

Run order	Feed rate (mm/rev)	Ra (μm)	VB (μm)
8	0.06	0.768	48
2	0.06	0.788	53
6	0.06	0.747	36
9	0.08	1.025	158
1	0.08	1.079	192
3	0.08	0.886	133
7	0.10	0.896	139
4	0.10	0.909	148
5	0.01	0.946	180

IOP Conf. Series: Materials Science and Engineering **307** (2018) 012024 doi:10.1088/1757-899X/307/1/012024

According to the results of Ra and VB values from Table 1, Figures. 2(a) and (b) show the scatter plots of Ra versus VB and VB versus Ra, respectively. These revealed that there was a correlation between the two responses. This indicated that Ra and VB depended on tool performance. This also implied that Ra of the machined part increased with increasing VB of the coated insert. Hence the TiN+AlCrN coated inserts should be investigated for machined part surface and tool life improvements. Before performing ANOVA, the data set of Ra and VB values should be tested whether they distribute as normality. The probability plot was used to display the data points, fitted line of the data points and associated confidence intervals (C.I.) based on parameters estimated from the data sets along with an Anderson-Darling (AD) goodness-of-fit statistic and associated *p*-value [11]. Figures 3 (a) and 4 (a) show probability plots of Ra and VB, respectively. The probability plots reveal that the plotted points of the complete data set of Ra roughly form a straight line and fall within the 95 % confidence interval. In addition, AD statistics of both data sets were small with high *p*-values compared to the level of significance of 0.05. This confirmed that normal probability distribution fitted the data sets of the Ra and VB values well.

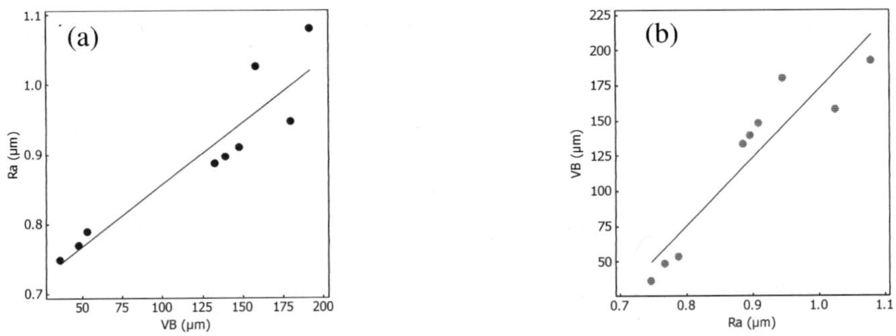

Figure 2. (a) Scatter plot of Ra versus VB values and (b) Scatter plot of VB versus Ra values.

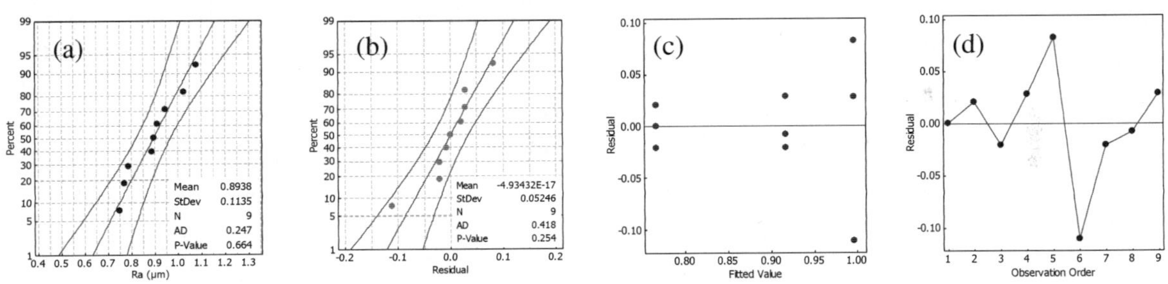

Figure 3. (a) Probability plot of Ra value and model adequacy checking for Ra (b) normal probability residual plot (c) residual versus fitted value and (d) residual versus run order.

Additionally, the adequacy of the underlying model should be checked. The primary diagnostic tool is residual analysis. Figures 3 (b) and 4 (b) illustrates normal probability plots of residuals. If the underlying error distribution is normal, this plot will resemble a straight line [11]. Clearly, the normal probability plot of residuals for Ra did not indicate anything particularly troublesome, although some residuals upward slightly on the left side. The normal probability plot of residuals for VB did not indicate problem of normality. If the model is adequate and the assumption is satisfied, the residuals should be strutureless. The residuals should be unrelated to any other factors including the fitted values. A plot of residuals versus fitted values is used to check nonconstant variance. Nonconstant variance arises if the plot look like an outward-opening funnel or megaphone. Figures. 3(c) and 4(c) b. do not show violation of the assumption of homogeneity of variances. Figures. 3 (d) and 4(d) exhibit the validity of the independence assumption of model. The model adequacy checking indicated that the one-way ANOVA model was effective for investigating the effects of feed rate on Ra and VB.

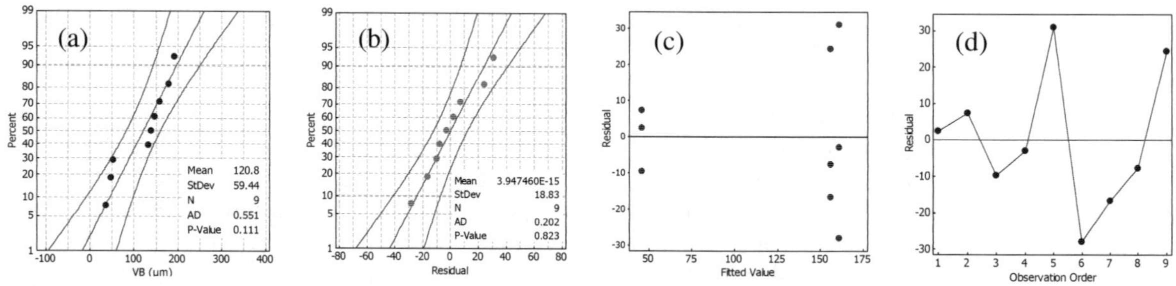

Figure 4. (a) Probability plot of VB value and model adequacy checking for VB (b) normal probability residual plot (c) residual versus fitted value and (d) residual versus run order

The results in Tables 2 and 3 were analyzed with ANOVA, which is employed for investigating the effect of factor feed rate on responses Ra and VB, respectively. Table 2 shows that the model *F*-value of 11.05 with a small *p*-value is adequate with a coefficients of determination (R^2) of 0.7865 which define as the ratio of the explained variation to the total variation and is a measure of the degree of fit. The adjusted R^2 of 0.7153 and predicted R^2 of 0.5196 are in reasonable agreement in which the difference between the two values is less than 0.2. This concluded that the feed rate significantly affected the mean Ra at the level of significance of 0.05. This result agrees with the previous research works [12,13]. Wang et al. [13] reported that the effects of feed rate and interaction between feed rate and cutting speed were statistically significant to mean Ra based on ANOVA results. Table 3 also shows that the model *F*-value of 26.91 with very small *p*-value is adequate with high R^2 of 0.8997, adjusted R^2 of 0.8663 and predicted R^2 of 0.7743. This also concluded that the feed rate significantly affected the mean VB at the level of significance of 0.05. This result also agrees with [12].

Table 2. ANOVA table for Ra during turning operation with coated insert TiN+AlCrN

Source	Df	SS	MS	*F*-value	*p*-value
Model	2	0.08129	0.04054	11.05	0.009
Feed rate	2	0.08129	0.04054	11.05	0.009
Error	6	0.02202	0.00367		
Total	8	0.10310			
	PRESS = 0.05	R^2 = 0.7865	Adj. R^2 = 0.7153	Pred. R^2 = 0.5196	

Table 3. ANOVA table for VB during turning operation with coated insert TiN+AlCrN

Source	Df	SS	MS	*F*-value	*p*-value
Model	2	25430.22	12715.11	26.91	0.001
Feed rate	2	25430.22	12715.11	26.91	0.001
Error	6	2835.33	472.56		
Total	8	28265.56			
	PRESS =	R^2 = 0.8997	Adj. R^2 = 0.8663	Pred. R^2 = 0.7743	

The main effect plot is a plot of the averages at each level of a factor. A main effect occurs when the average response changes across the levels of a factor. The main effect plot is used to compare the relative strength of the effect of the factor. Although this plot is useful to compare main effect, the ANOVA table is a primary tool used to significantly evaluate the effect of the factor. The main effect plot illustrated in Figure. 5(a) proves that the Ra value was the lowest at the feed rate of 0.06 mm/rev whereas the VB value was also the lowest at the feed rate of 0.06 mm/rev as illustrated in Fig. 5(b). Hence the appropriate feed rate of 0.06 mm/rev has the lowest Ra and VB.

Figure 6 show SEM micrographs of wear on the TiN+AlCrN coated insert during machining of AISI 4140 chromium molybdenum steel bar at the feed rate of 0.06 mm/rev after a cutting time of 32 min. Micro-wear behaviors observed during the experiments through SEM micrographs included edge chipping, micro abrasion on the TiN+AlCrN coated insert. Heat energy and workpiece temperature highly increased resulting the cutting edge lost shapes leading to plowing of chips from workpiece on shear plane and flank wear land heat sources.

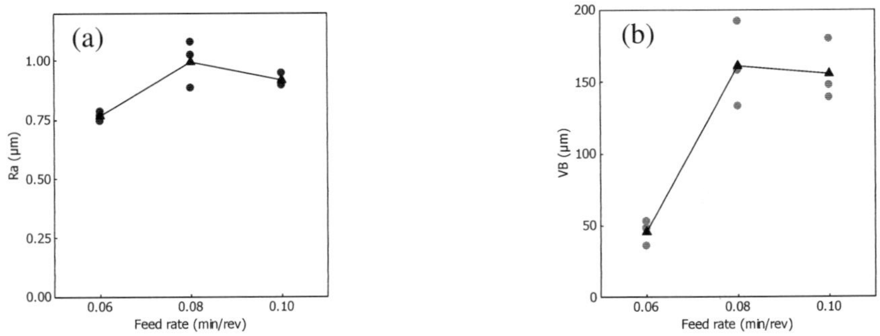

Figure 5. Ra and VB as the function of feed rate for the coated insert TiN+AlCrN.

Figure 6. SEM images of the wear of TiN+AlCrN (a) before and (b) after machining.

4. Conclusion

Feed rate significantly affected the averages Ra and VB values at the level of significance of 0.05 based on the results of analysis of variance. Ra and VB values increased when feed rate increased. The appropriate feed rate of 0.06 mm/rev has the lowest Ra and VB. SEM micrographs indicated edge chipping, micro abrasion on the TiN+AlCrN coated insert.

Acknowledgements

The authors would like to thank the Office of the Higher Education Commission of Thailand and Faculty of Engineering, Khon Kaen University for the financial supports of this research.

References

[1] Tamizharasan T, Selvaraj T, and Noorul A 2006 *Inter. J. Adv. Manuf. Technol.* **28** pp 671-679
[2] Grzesik W, and Wanat T 2006 *Inter. J. Machine Tool & Manuf.* **46** pp 1988–1995
[3] Bouach K, Yallese M A, Mabrouki T, and Rigal J F 2010 *Inter. J. Refract. Metals & Hard Mater.* **28** pp 349-361
[4] Çiçek A, Kara F, Kivak T, and Ekici E 2013 *Inter. J. Refract. Metals & Hard Mater.* **41** pp 461-469
[5] Sahoo A K, and Sahoo B 2012 *Measurement* **45** pp 2153–2165
[6] Sahoo A K, and Sahoo B 2012 *Measurement* **46** pp 2695–2704
[7] Chinchanikar S, and Choudhury S 2013 *Inter. J. Refract. Metals & Hard Mater.* **38** pp. 124-133
[8] krolczyk G, nieslony P, Legutko S, Hloch S, and Samardzic I 2015 *Metalurgija* **54** pp. 91-94
[9] Lin Y J, Agraval A and Fang Y (2007) *Wear* **264** pp 226-234
[10] Chandiran A, Anandan N and Sakthivel S 2012 *Tribo.-Mater., Surf. & Interfaces* **6** pp160-167
[11] Montgomery D C 2001 *Design and Analysis of Experiments* 6th ed John Wiley & Sons New York 2001
[12] Suresh R, Basavarajappa S, Gaitoned V N and Samuel G L 2012 *Inter. J. Refract. Metals & Hard Mater.* **33** pp 75-86
[13] Wang M, Xu B, Zhang J, Dong S 2013 *Inter. J. Adv. Manuf. Technol.* **67** pp 1537-1548

IOP Conf. Series: Materials Science and Engineering **307** (2018) 012025 doi:10.1088/1757-899X/307/1/012025

Automatic design of *conformal cooling channels* in injection molding tooling

Yingming Zhang[1], Binkui Hou[1], Qian Wang[1], Yang Li[1]* and Zhigao Huang[1]

1 State Key Laboratory of Material Processing and Die & Mold Technology, Huazhong University of Science and Technology, Wuhan 430074, Hubei, China
E-mail: liyang@hust.edu.cn

Abstract. The generation of cooling system plays an important role in injection molding design. A conformal cooling system can effectively improve molding efficiency and product quality. This paper provides a generic approach for building conformal cooling channels. The centrelines of these channels are generated in two steps. First, we extract conformal loops based on geometric information of product. Second, centrelines in spiral shape are built by blending these loops. We devise algorithms to implement the entire design process. A case study verifies the feasibility of this approach.

1. Introduction

Injection molding is widely used to produce plastic product. The plastic product needs to be cooled before mold opening. This cooling stage takes up most of time in injection molding process. Cooling channels are devised for reducing cooling time and distributing temperature uniformly in injection molding design. Effective cooling has been proved by using conformal cooling system [1, 2]. However, it's a complex process to build conformal cooling channels for product with complicated shapes.

In order to improve the design efficiency, researchers have carried out a series of studies on automated design of conformal cooling channels. Li [3] decomposed product into a lot of simple shape features based on their characteristics and built cooling system by connecting simple cooling channels established for each simple feature. However, it's difficult to decompose complex products and identify characteristics of complex shapes. Yu et al. [4, 5] offset product surface to get conformal surface and computed cooling circuit with the information of this conformal surface to get conformal cooling channels. However, offsetting free-formed surface needs complex calculations and is error-prone in actual design.

The objective of this paper is to put forward a more general approach for rapid design of conformal cooling channels. Firstly, we create conformal loops to represent the geometric information of product. Secondly, channel centerlines are built by converting these conformal loops into spiral curves. Finally, cooling system is generated by sweeping cross-section along channel centerlines. The whole design process is achieved by algorithms proposed in this study. This method can effectively avoid problems of identifying or offsetting complex surface. A case study with complex shape verifies the feasibility of this approach.

2. System process

Conformal cooling system consists of a series of channels whose centerlines maintain a certain distance with product. Typical structure of conformal cooling channels is shown in figure 1, where W is the pitch distance between adjacent channels and D is the offset distance between centerline and mold surface.

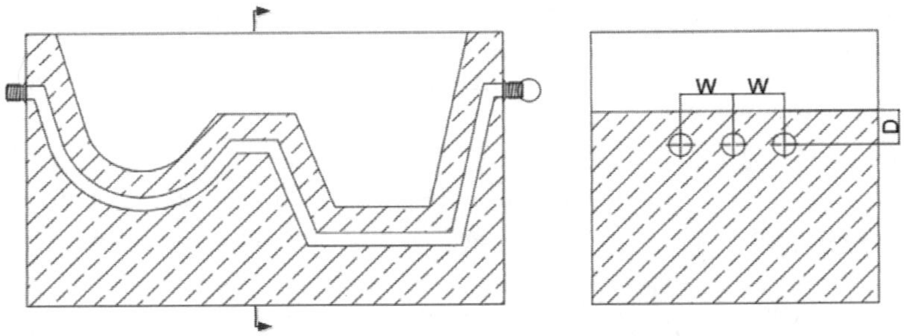

Figure 1. Conformal cooling channels

The main challenge of building conformal channels lies in creating channel centerlines. These centerlines are generated by extracting conformal loops from product and blending these loops into spiral curves. In order to achieve the automatic design of cooling channels, all the data needed in design process are stored in a knowledge base. These data include pitch distance, offset distance, spiral type and cross-section of cooling channels. When the cooling area is specified, conformal loops can be generated by extracting section loops with pitch distance and offsetting loops with offset distance. We convert conformal loops into spiral curves according to the circuit type. Cooling channels are generated by sweeping cross-section along these spiral curves. The whole process can be illustrated in figure 2.

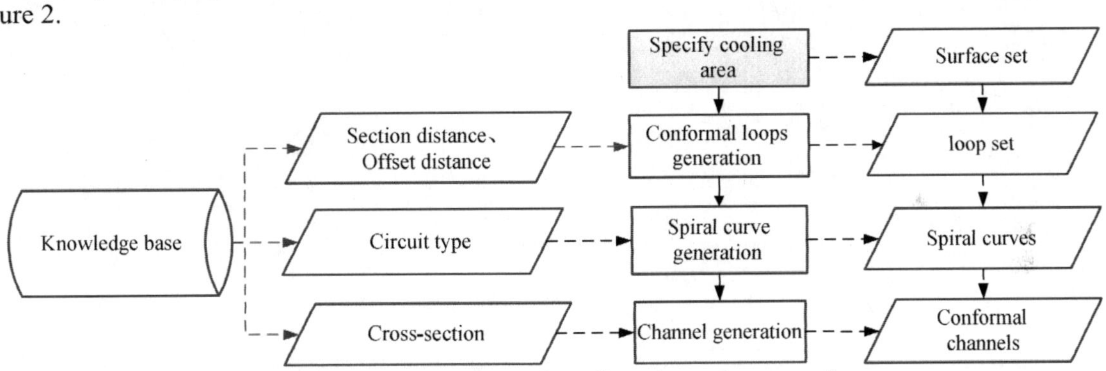

Figure 2. The process of cooling channels generation

3. Main algorithms

3.1. Conformal loops generation

Conformal loops are a set of contours which maintain certain distance with product. We create a set of reference planes and intersect them with cooling area to get intersection curves. The normal $\vec{n}$ of each reference plane and section distance d_1 between inference planes can be read from constructed knowledge base. These Curves contain geometric information of the product surface, and we can use them to create section loops. Conformal loops are built by offsetting these section loops with offset distance d_2. The input of this algorithm is cooling area S_c, section distance d_1, offset distance d_2 and parting direction $\vec{n}$. We can create conformal loops $\{L\}$ by following steps:

1. Calculate anchor points of reference planes in a range created by projecting S_c onto $\vec{n}$. All the anchor points are stored in a point set $\{P_i\}$.
2. FOR each i DO
3. Create a reference plane RE whose normal is $\vec{n}$ at point P_i
4. Calculate a curve set $\{C_k\}$ by intersecting RE with S_c

5. Create loop collection {LC} and simple curve collection {SC}
6. FOR each k DO
7.   IF C_k is a loop, add C_k into {LC}
8.   Else, add C_k into{SC}
9.   Create a loop L_{SC} by Connecting curves in {SC} and add L_{SC} into {LC}
10. Create a total loop TL by combining loops in {LC}
11. Create a conformal loop CL by offsetting TL on RE with a method proposed in [6] and add CL into {L}

A loop L_{SC} can be generated by connecting adjacent curves whose endpoints are adjoining, and a total loop is created by combining discrete loops. The combining process is achieved by splitting loops into curves and connecting curves to be a loop. The Concrete illustration is shown in figure 3.

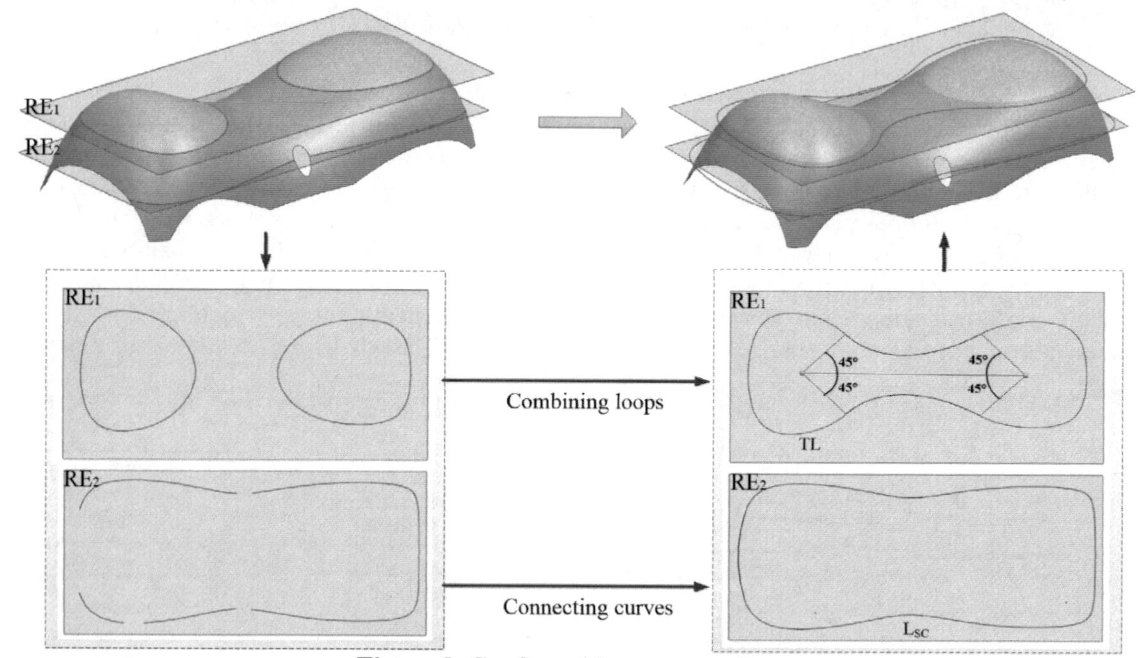

Figure 3. Conformal loops generation

3.2. Spiral curves generation

Spiral curves are generated by blending conformal loops. There are three types of cooling circuit in injection mold design. They are single spiral in series connection, double spirals in series connection and spirals in parallel connection (see the illustration in figure 4). Their generation processes are explained in following sections.

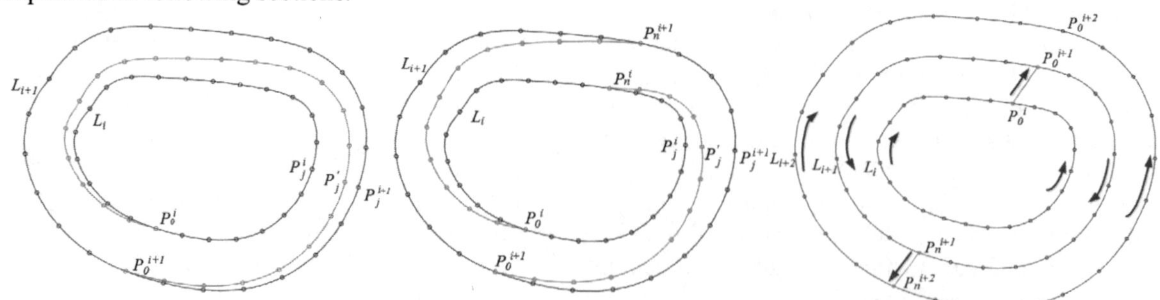

(a)Single spiral in series connection (b)Double spirals in series connection (c)Spirals in parallel connection

Figure 4. Spiral curves generation

3.2.1. Single spiral in series connection. Single spiral is generated by creating a curve between adjacent loops. We use an algorithm to convert two loops into a spiral curve. The input of this algorithm is two adjacent loops L_i and L_{i+1}. The concrete generation algorithm consists of following steps (see the illustration in figure 4(a)).

1. Set a point P_0^i in L_i as the start point of spiral curve in this section. Find a point P_0^{i+1} in L_{i+1}, which is the closest point from P_0^i. P_0^{i+1} is viewed as the end point of spiral curve in this section and the start point of start point of spiral curve in next section.
2. Discrete curves L_i and L_{i+1} into n points separately. These points are separately stored in $\{P_j^i\}$ and $\{P_j^{i+1}\}$, where $j=0, 1, \ldots, n-1$.
3. Create a point P_j^* between P_j^i and P_j^{i+1} with a function $P_j^*=(1-j/n)P_j^i+(j/n)P_j^{i+1}$. All these new points are stored in $\{P_j^*\}$.
4. Connect points consecutively in $\{P_j^*\}$ to form a spiral curve S_i between L_i and L_{i+1}.

3.2.2. Double spirals in series connection. Double spirals are generated by creating two curves between adjacent loops. We use an algorithm to convert two loops into a pair of spiral curves. The input of this algorithm is two adjacent loops L_i and L_{i+1}. The concrete generation algorithm consists of following steps (see the illustration in figure 4(b)).

1. Set a point P_0^i in L_i as start point of first spiral curve in this section. Discrete curves L_i into *2n* points. These points are stored in $\{P_j^i\}$, where $j=0, 1, \ldots, 2n-1$. P_n^i is viewed as the start point of second spiral curve in this section.
2. Find a point P_0^{i+1} in L_{i+1}, which is the closest point from P_0^i. Discrete curves L_{i+1} into *2n* points. These points are stored in $\{P_j^{i+1}\}$, where $j=0, 1, \ldots, 2n-1$. P_n^{i+1} is viewed as end point of first spiral curve in this section and start point in next section. P_0^{i+1} is viewed as end point of second spiral curve in this section and start point in next section.
3. Create a point P_j^* between P_j^i and P_j^{i+1}. If $j<n$, $P_j^*=(1-j/n)P_j^i+(j/n)P_j^{i+1}$. Else, $P_j^*=(1-(j-n)/n)P_j^i+((j-n)/n)P_j^{i+1}$. All these new points are stored in $\{P_j^*\}$.
4. Connect points consecutively in $\{P_j^*\}$ to form two spiral curves between L_i and L_{i+1}.

3.2.3. Spirals in parallel connection. Parallel spirals are generated by creating connection between adjacent loops. We use an algorithm to connect two loops with a straight line. The input of this algorithm is two adjacent loops L_i and L_{i+1}. The concrete generation algorithm consists of following steps (see the illustration in figure 4(c)).

1. Set a point P_0^i in L_i as start point. Find a point P_0^{i+1} in L_{i+1}, which is the closest point from P_0^i. Discrete curves L_i and L_{i+1} into *2n* points separately. These points are separately stored in $\{P_j^i\}$ and $\{P_j^{i+1}\}$, where $j=0, 1, \ldots, 2n-1$.
2. If i is an even number, Connect P_0^i and P_0^{i+1} with a straight line.
3. If i is an odd number, Connect P_n^i and P_n^{i+1} with a straight line.

All loops will be connected one by one with this algorithm. Figure 4(c) shows a concreate example. The straight line between P_0^i and P_0^{i+1} is an outlet for L_i and an inlet for L_{i+1}. The straight line between P_n^{i+1} and P_n^{i+2} is an outlet for L_{i+1} and an inlet for L_{i+2}. Water can be effectively conveyed in these parallel spirals.

4. Case study

The algorithms proposed in this paper has been developed with C++ and interfaced to the NX system. In this section, a case study of mouse is presented to verify the feasibility of the proposed cooling channels design process. The major steps of the design process are illustrated in figure 5.

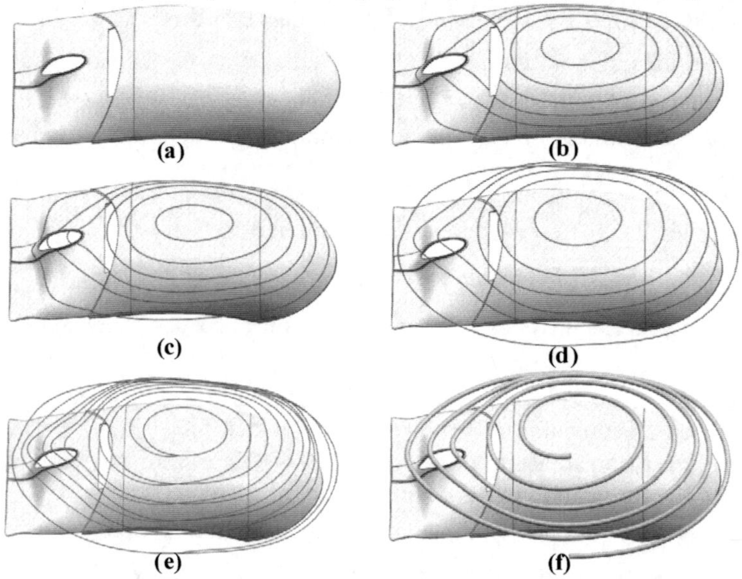

Figure 5. The case of conformal cooling channel generation (a) cooling area (b) generate intersections (c) generate section loops (d)conformal loops generation (e) spiral curve generation (f) cooling system generation

In this case, there are many defects on the surface of cooling area. The proposed algorithm of conformal loops generation can effectively convert discrete curves into loops. Cooling circuit can be created by the method of spiral curve generation. This approach can be used for rapid design of cooling channels.

5. Conclusion

In this paper, an automatic method is proposed to design cooling channels. By using algorithm of generating conformal loops, simplified information of product shape will be obtained. We convert these loops into spiral curves by an algorithm proposed in this study. These spiral curves are used for generating the final cooling system. A case study verifies the reliability of this approach.

References

[1] Xu R X and Emanuel S 2009 Rapid thermal cycling with low thermal inertia tools *Polymer Engineering & Science* **49** 305-16

[2] Park H S and Pham N H 2009 Design of conformal cooling channels for an automotive part *International Journal of Automotive Technology* **10** 87-93

[3] Li C L 2007 Part segmentation by superquadric fitting-a new approach towards automatic design of cooling system for plastic injection mould *International Journal of Advanced Manufacturing Technology* **35** 102-14

[4] Wang Y, Yu K M, Wang C C L and Zhang Y 2011 Automatic design of conformal cooling circuits for rapid tooling *Computer-Aided Design* **43** 1001-10

[5] Wang Y, Yu K M and Wang C C L 2015 Corrigendum to "Spiral and conformal cooling in plastic injection molding" [J. Comput. Aided Des. 63C (2015) 1–11] *Computer-Aided Design* **63** 1-11

[6] Farouki R and Srinathu J 2017 A real-time CNC interpolator algorithm for trimming and filling planar offset curves ☆ *Computer-Aided Design*

Preparation of photochromic silk fabrics based on thiol-halogen click chemistry

Ji Fan[1,2], Wei Wang[2] and Dan Yu[1,2]
[1] Key Laboratory of High Performance Fibers & Products, Ministry of Education, Donghua University, Shanghai, P. R. China 201620
[2] College of Chemistry, Chemical Engineering & Biotechnology, Donghua University, Shanghai, 201620, China
*Corresponding author: E-mail: yudan@dhu.edu.cn.

Abstract: In this study, we have synthesized a photochromic compound 9'-(2-bromo-2-methylethoxycarbonyl)-1,3,3-trimethyl-spiro[2H-indole-2,3'-[3H]naphtha[2,1-b][1,4]oxazine] and applied it to the silk fabric to acquire photochromic properties. First, tris (2-carboxyethyl) phosphine (TCEP) was used as a reducing agent to produce thiol groups on the surface of silk fabric. Then, these thiol groups will react with -Br groups of the spirooxazine via thiol-halogen click chemistry. The spirooxazine was characterized by FTIR and UV. And the color change properties of the as-prepared silk fabrics were evaluated by the color measurement methods. The results indicated that treated silk fabric has excellent photochromic properties.

1. Introduction

In recent decades, photochromic compounds have been extensively studied for their applications in many areas [1, 2]. Several methods of preparing photochromic textiles using spirooxazine have been reported, including conventional dyeing techniques and screen printing [3, 4]. However, the conventional preparation methods of the photochromic fabric have the disadvantages like complicated preparation process, slower color switching speed and rough handle. In addition, the spirooxazine and the fabric mainly relied on physical adsorption or hydrogen bonding, so it is easy to be removed after washing and affected the photochromic effect [5, 6]. Peng [7] et al. has synthesized N-methyl-3,3-dimethyl-9'-dichloro-s-triazine group spiro [2H-indole-2'-[3H]naphtho[2,1-b][1,4]oxazine] and applied it to cotton fabrics, using the -Cl groups on the spirooxazine derivative to react with -OH groups in the cotton fibers. The experimental results demonstrate that the cotton fabric has good photochromic properties and better washing resistance after dyeing. Currently, other photochromic textiles have not been reported, especially those could be easily prepared and with good durablity. Therefore, our target of this paper is to find a novel chemical method of preparing photochromic textiles with durable properties.

In this paper, a novel photochromic compound with a halogen end group was firstly synthesized. The silk fabric was then treated by TCEP to generate reactive thiol groups. The synthetic photochromic compound was finally grafted onto the silk fabric through the thiol-halogen click chemistry to form a covalent bond, and the reaction mechanism of the photochromic compound with silk fabric was shown in figure 1. In addition, the photochromic properties, fatigue resistance and washing fastness of photochromic silk fabrics were evaluated.

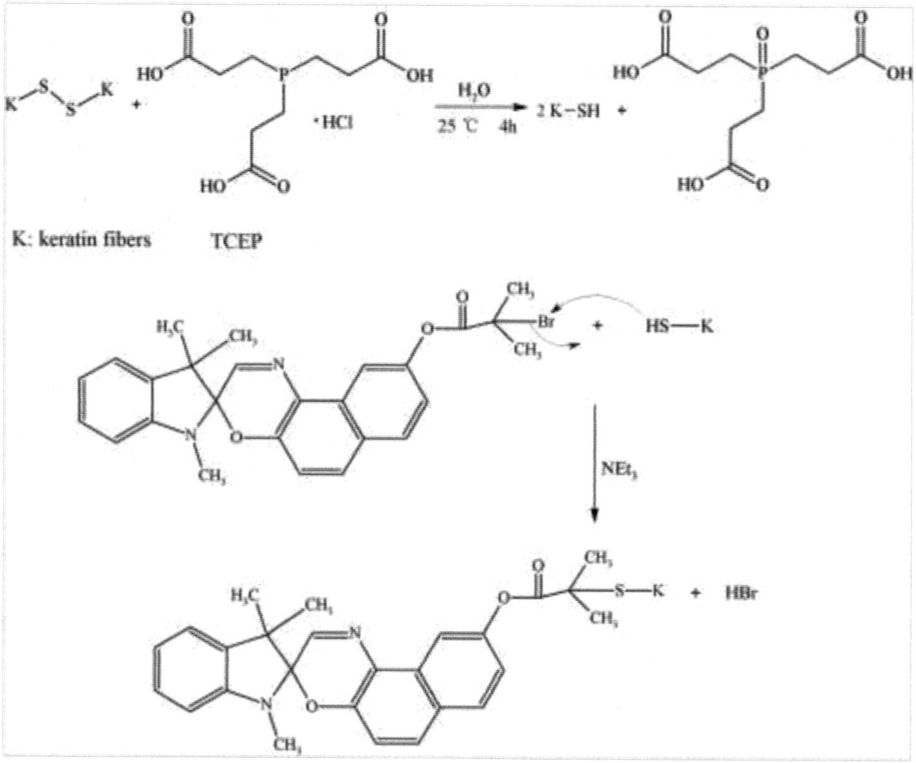

Figure 1. Mechanism of reacting wool with photochromic compound.

2. Experimental

2.1 Materials

Degummed silk fabric was purchased from Yufeng Co. Ltd. 1,3,3-trimethyl-2-methyleneindoline (97 %) was purchased from Shanghai Jiuding Chemical Technology Co., Ltd. 2-bromoisobutyryl bromide (98 %) was purchased from Beijing InnoChem Science & Technology Co., Ltd. Tris (2-carbonxyethyl) phosphine hydrochloride (TCEP) was purchased from Sigma Aldrich. Other chemicals used in the experiment were obtained from Sinopharm Chemical Reagent Co., Ltd. All chemicals were analytic grade reagents, and used without further purification.

2.2 Synthesis of spirooxazine compound

The synthesis process of spirooxazine compound was shown in figure 2. 9'-hydroxy-1,3,3-trimethyl-spiro[2H-indole-2,3'-[3H]naphtha[2,1-b][1,4]oxazine] was synthesized by Kakishita et al and carefully characterized it in the article [8]. And 9'-(2-bromo-2-methylethoxycarbonyl)-1,3,3-trimethyl-spiro[2H-indole-2,3'-[3H]naphtha[2,1-b][1,4]oxazine] was synthesized and described in detail by Such et al [9].

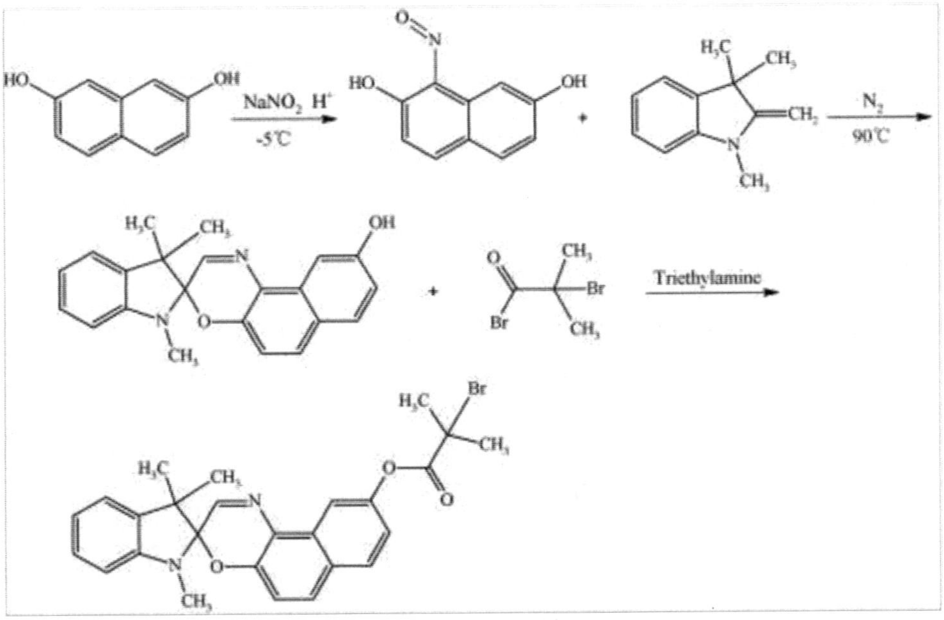

Figure 2. Synthesis process of spirooxazine.

2.3 Pretreatment of silk with TCEP

A pretreatment solution was prepared by dissolving TCEP in an alcohol-water mixture (v/v=1:1) to form a 20 mmol/l TCEP solution. The liquor ratio was 50:1 and the pH values were adjusted by the buffer ($NaHCO_3$/Na_2CO_3) pH=7.0. And pretreated with TCEP solution at 25 ℃ for 4 h [10].

2.4 Application of spirooxazine to the silk fabrics

The final synthesized spirooxazine was dissolved with dichloromethane, and the pretreated silk fabric was added into the prepared solution at 25 ℃ for 1 h with a small amount of triethylamine as a catalyst. After treatment, washed the silk fabric with standard detergent thoroughly and rinsed with clean water and dried to obtain the final product.

2.5 Characterization

The chemical structure of spirooxazine was investigated by fourier transform infrared analysis (Nicolet 6700, Thermo Fisher, America). The UV absorption spectrum of spirooxazine dissolved in ethanol solution was tested by using a UV/Vis spectrophotometer (UV-3600, Japan) with the wavelength changed from 400 to 700 nm. The color yields (K/S values) of the photochromic silk fabrics were carried out by Datacolor 650 (Datacolor, USA).

3. Results and discussion

Figure 3(a) showed the FTIR spectrum of spirooxazine. The peak at 1752 cm^{-1} is assigned to the stretching vibration of the C=O bonds. The peak at 1629 cm^{-1} is attributed to the stretching vibration of C=N, while the peaks at 1608 cm^{-1} and 1511 cm^{-1} are assigned to the stretching vibrations of C=C in the naphthalene ring. Two peaks appeared in the region of 1251 cm^{-1} and 1082 cm^{-1}, which are attributed to the stretching vibration of the functional groups C_{spiro}-O-C= [11]. At 1167 cm^{-1}, the peak belongs to the stretching vibration of Ar-O, and that at 651 cm^{-1} is assigned to the stretching vibration of C-Br. The results demonstrate the spirooxazine compounds were successfully synthesized. Figure 3(b) showed the UV spectrum of the spirooxazine compound under UV irradiation. When the spirooxazine solution was irradiated with UV light at 365 nm, it was observed that the wavelength became longer and the solution turned blue. This is because the molecular structure of spirooxazine is changed from closed loop to open loop under the UV irradiation, resulting in a significant change in color.

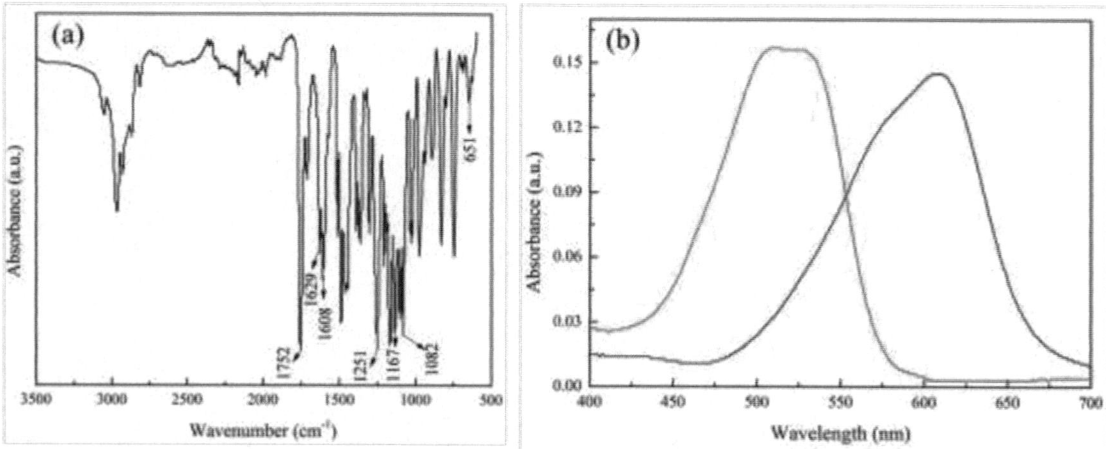

Figure 3. (a) FTIR spectra of spirooxazine and (b) UV spectra of spirooxazine.

In order to prove the mechanism of photochromism, the color change was evaluated from L*, a*, b* values by using CIE color space, and the test results are listed in table 1. It can be observed that the brightness of the original silk fabric (L*) was highest, but it decreased obviously after treated with spirooxazine. The redness (a*) of photochromic silk fabric was increased but the blueness (b*) was decreased. It is indicated that the color of the silk fabric is blue after UV irradiation.

Table 1. Color properties of the sample.

Samples		L*	a*	b*
Original silk		92.01	0.01	4.06
Pretreated silk		-5.11	2.56	-0.25
Photochromic silk	Before irradiation	-35.27	16.15	0.06
	After irradiation	-37.15	9.49	-3.44

4. Conclusion

This paper introduces a method for preparing photochromic silk fabrics by thiol-halogen click chemistry. The method comprises the following three steps: firstly synthesizing the spirooxazine compound with photochromic properties, and then the silk fabric was pretreated with TCEP. Finally, the synthetic spirooxazine compound was reacted with the reduced silk fabric via thiol-halogen click chemistry. It was confirmed by FTIR and UV that the spirooxazine compound was synthesized and it was successfully grafted onto the silk fibers. The results indicated that treated silk fabric has excellent photochromic properties.

Acknowledgements
The research was supported by National Natural Science Foundation of China (No. 51403032) and "the Fundamental Research Funds for the Central Universities (No. 16D110518)".

References

[1] Sun B, Hou Q, He Z, Liu Z and Ni Y 2014 *Carbohydr. Polym.* **111** 419-424

[2] Wang M S, Gang X, Zhang Z J and Guo G C 2009 *Chem. Commun.* **46** 361-376
[3] Little A F and Christie R M 2011 *Coloration Technology*. **127** 275-281
[4] Mokhtari J, Akbarzadeh A, Shahrestani Z and Ferdowsi P 2015 *Fibers Polym.* **16** 2299-2307
[5] Pardo R, Zayat M and Levy D 2009 *J. Mater. Chem.* **19** 6756-60
[6] Such G K, Evans R A and Davis T P 2016 *Macromolecules* **39** 1391-96
[7] Peng L, Guo R, Jiang S, Lan J, He Y and Huang X 2015 *Fibers Polym.* **16** 1312-18
[8] Kakishita T, Matsumoto K, Kiyotsukuri T, Matsumura K and Hosoda M 1992 *J. Heterocycl. Chem.* **29** 1709-15
[9] Such G K, Evans R A and Davis T P 2005 *Mol. Cryst. Liq. Cryst.* **430** 273-9
[10] Yu D, Cai J Y, Church J S and Wang L 2014 *Acs Appl. Mater. Interfaces.* **6** 1236-42
[11] Zou Q, Li X, Zhou J, Bai K and Ågren H 2014 *Dyes Pigm.* **107** 174-81

IOP Conf. Series: Materials Science and Engineering **307** (2018) 012027 doi:10.1088/1757-899X/307/1/012027

Study of the nanoporous CHAP photoluminiscence for developing the precise methods of early caries detection

D Goloshchapov[1], P Seredin[1], D Minakov[2] and E Domashevskaya[1]

[1]Department of Solid State Physics and Nanostructures,Voronezh State University, Voronezh, Russia
[2]Department of Physical and Mathematical Sciences,Voronezh State University, Voronezh, Russia

E-mail: goloshchapov@phys.vsu.ru

Abstract. This paper deals with the luminescence characteristics of an analogue of the mineral component of dental enamel of the nanocrystalline **B**-type carbonate-substituted hydroxyapatite (CHAP) with 3D defects (i.e. nanopores of ~2-5 nm) on the nanocrystalline surface. The laser-induced luminescence of the synthesized CHAP samples was in the range of ~515 nm (~2.4 eV) and is due to CO_3 groups replacing the PO_4 group. It was found that the intensity of the luminescence of the CHAP is caused by structurally incorporated CO_3 groups in the HAP structure. Furthermore, the intensity of the luminescence also decreases as the number of the above intracentre defects (CO_3) in the apatite structure declines. These results are potentially promising for developing the foundations for precise methods for the early detection of caries in human solid dental tissue.

1. Introduction

Calcium hydroxyapatite (HAP) - $Ca_{10}(PO_4)_6(OH)_2$ and its substituted forms are the major biomineral component of bone tissue, enamel and dentine of human teeth as well as prosthesis material [1]. It is becoming increasingly important to develop methods (like laser-induced luminescence (PL)) for detecting pathologies of dental enamel at the outset of their development [2,3] and thus, analysing the overall condition of the bioapatite of dental enamel. It is well-known that as people age, calcium hydroxyapatite in the dental enamel transforms to carbonate-substituted calcium hydroxyapatite (CHAP) with many impurities [4,5], leading to additional risk of caries. It has been repeatedly shown [6] that healthy, affected and caries-damaged areas have a displacement as well as additional bands in the photoluminescence spectra. This helps to visualize the PL signal and map the affected area or the oral cavity, indicating overall changes in colour from pathogenic areas. Thus, lately there have been many methods developed and a variety of tools employed for clinical use to estimate the photoluminescence signal from solid dental tissues [7,8].

Regarding the luminescence of solid dental tissues, both the organic and mineral components contribute to their photoluminescence spectrum [2]. Moreover, the type of defect in the structure of calcium hydroxyapatite can be crucial to the luminescence maximum [9]. In addition to intracentre defects, changes in the bioapatite photoluminescence spectrum can be caused by a variety of phase transformations occurring leading up to caries development in human enamel and dentine [10].

Unfortunately, as yet, modern methods are incapable of detecting such pathologies in solid dental tissue (caries) at an early stage and along its depth, due to the signal being difficult to detect at the level of intralattice replacements in a crystal lattice of an apatite and molecular local atomic surroundings (different impurities in the composition). However, addressing the synthesis of composite biomimetic materials similar to dental tissues (intact and caries-damaged) in their composition, as well as further analysis of their photoluminescence will assist in finding more effective approaches for the early detection of local areas of intact enamel and caries-damage and, more importantly, in making these tools more sensitive [7,11].

Therefore, the objective of the paper is to study emission properties of analogues of the mineral component of native dental tissue, i.e. carbonate-substituted nanocrystaline hydroxyapatite, and to identify the dependences of spectral characteristics of synthesized biomaterials on the amount of impurities in them.

2. Materials and research methods

In this study, nanocrystalline carbonate-substituted hydroxyapatite was obtained using egg shells by means of the our method detailed previously [12]. Changes in the concentration of carbonate-containing groups was controlled using a pH from 9 to 7, with a step of 0.5 (pH-meter/ionometer IPL 111-1) and accuracy of 0.01 as in [13].

The samples of hydroxyapatite were investigated by X-ray diffraction (DRON-4.07 Co $K\alpha$ = 1,790Å), Fourier infrared spectroscopy (Vertex-70, Bruker), Raman scattering spectroscopy (LabRam HR800), transmission electron microscopy (Libra-120 Carl Zeiss) and photoluminescence spectroscopy (a setup based on fibre optics spectrometer USB4000-VIS-NIR by OceanOptics) and a laser excitation source with the wavelength of 405 nm.

3. Results and discussion

The results of X-ray diffraction showed that the materials are nanocrystalline forms of hydroxyapatite, with an average crystallite size of 35 nm according to [12]. In order to specify the local microstructure and phase constancy of the samples, high-resolution transmission electron microscopy was employed (Fig. 1).

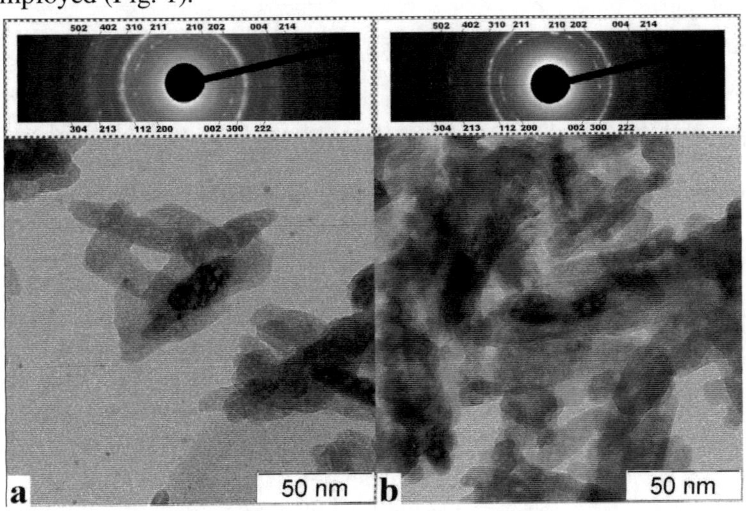

Fig. 1: Transmission electron microscopy of a microphotograph and microdiffraction of the samples obtained using the solution of a) pH = 9 and b) pH = 7.

The study of morphology of hydroxyapatite nanocrystals using nanometer resolution transmission electron microscopy showed the presence of nanocrystals with a defective porous structure on the surface (Fig. 1). The presented microphotographs suggest that as the pH of the solution decreases, the size of the nanopores increases.

IR-spectroscopy revealed that all the samples are carbonate-substituted forms of hydroxyapatite. The vibration modes in the IR-spectra at 1415 cm^{-1}, 1425 cm^{-1} and 1450 cm^{-1} indicate CHAP of B-type, that is, replacement of PO_4 group with CO_3 group as shown previously in [12,13]. The experimental data showed that in the samples investigated for the current study, the intensity of the vibration modes of CO_3 group in the IR-spectra decreases from sample to sample, while that of the major mode of PO_4 group remains unchanged. This fact refer to degree of replacement of the PO_4 group by CO_3 in the structure of apatite [13,14]. At the stage of the synthesis of the samples, as the pH of the solution decreases from 9 to 7 (introduction of a large number of PO_4 groups), so does the intensity of the vibration mode of the carbonate-anion included in the hydroxyapatite structure.

Studies of calcium hydroxyapatite suggest that when there are displacements in the apatite crystal lattice and a CO_3 anion is included in the hydroxyapatite structure, there is a shift in PO_4 bands in the spectra of Raman scattering spectroscopy. Raman spectroscopy was performed to identify the influence of a defect in the structure of the synthesized samples of calcium hydroxyapatite on their optical properties (see Fig. 2).

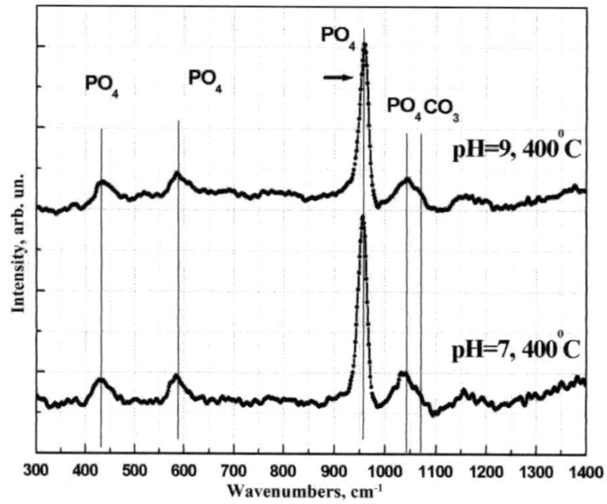

Fig. 2: Raman spectra of the samples with pH 9 and 7 annealed at 400°C.

As seen in Figure 2, except for the major modes of the phosphorus-oxygen group identified at around 961, 587 and 431 cm^{-1}, in the spectra of the samples from different pH solutions, there is an additional mode in the area of ~1070 cm^{-1}. It might be due to the carbonate anion CO_3^{2-} in the structure of calcium hydroxyapatite that replaced PO_4^{3-} in the lattice. Note that in this spectrum range, 1077, 1045, 1035 cm^{-1}, there are low-intensive modes of PO_4 group. It should also be mentioned that the ratio of the intensities of additional modes and the major phosphate mode identified at around 961 cm^{-1} in all the spectra remains identical. A detailed study of the form of the vibration of PO_4 (Fig. 2) showed that the inclusion of the CO_3 anion into the structure of hydroxyapatite caused asymmetries and a shift of the maximum by 5 cm^{-1} for the samples with pH values of 7 and 9. This correlates with the IR-spectroscopy data on the nature of carbonate-substituted samples of B-type CHAP.

Photoluminescence spectroscopy of the samples revealed a strong dependence of the intensity of the major photoluminescence maximum on the conditions of obtaining materials (CHAP), by means of the suggested method as well as on their subsequent thermal processing. Fig. 3a shows the photoluminescence spectra of all the samples of calcium hydroxyapatite obtained in this study for different (final) pH values (9, 8.5, 8, 7.5, 7) and annealed at 400°C. The maximum of the photoluminescence band is in the range of ~515 nm, corresponding with the transition energy of ~2.4 eV. The analysis of the spectra of CHAP showed that the photoluminescence intensity of the samples with a pH of 7.5 and 7 is lower than that of the samples with a pH of 9-8. This might be caused by a decrease in the concentration of the impurity luminescence centres of CO_3^{2-} in the samples obtained for smaller values of pH. The ratios of the intensities in the profile of the photoluminescence bands might be due to the element composition of the original material (egg shells) that was used to synthesize hydroxyapatite. In synthesized materials, the maximum concentration of the carbonate anion is ~5%, what is comparable with their content in the bioapatite of human dental enamel [15]. Comparing the method of synthesis described in our previous paper [12] and the data of similar studies, it appears that obtaining hydroxyapatite by means of liquid-phase methods leads to inclusion of the atmospheric gas CO_2 in the resultant material, resulting in fluctuations in the carbonate anion content, thereby rendering the photoluminescence bands of the specified intensity particularly important for observations.

An analysis of the literature shows that photoluminescence in pure hydroxyapatite might be caused by structural defects occurring as a result of intentional or random modifications of the composition [16,17].

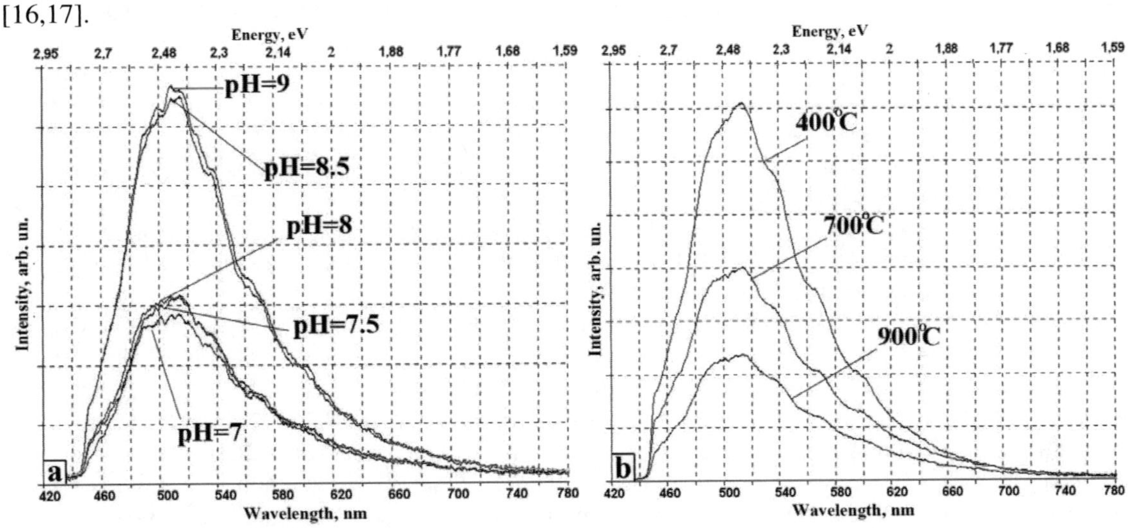

Fig. 3. Photoluminescence spectra of calcium hydroxyapatite: a - obtained at pH values of 9, 8.5, 8, 7.5, 7 and annealed at 400°C; b - with a pH = 8.5 and annealed at 400, 700, 900°C.

In [9], there are results of a study of own photoluminescence of hydroxyapatite according to which its forbidden band width is ~7,7 eV. Using this value and the obtained photoluminescence spectra with the maximum ~2.4eV (Fig. 3a), we consider a possible variant of a model of photoluminescence of carbonate-substituted hydroxyapatite whose center of luminescence is CO_3 group replacing PO_4 group in the structure of calcium hydroxyapatite. Note that in [9] the photoluminescence maximum that depends on own defects of pure hydroxyapatite was measured at ~3,8 eV. However, the energy of the source of excitation that we used does not allow photoluminescence to be observed at such energies, thus the measured laws for the intracenter photoluminescence with a maximum at ~2.4 eV.

The PL investigation of the calcium hydroxyapatite samples exposed to thermal annealing in the temperature range of 400-900°C proved the relatively impurity mechanism (intracentre defects CO_3) of the observed photoluminescence. The photoluminescence of the samples obtained at the same pH (8.5), but annealed at 400, 700, 900°C decreased as the temperature increased (Fig. 3b), due to a reduction in the carbonate ion content of the calcium hydroxyapatite. This is in good agreement with the results obtained by means of X-ray structure analysis and IR-spectroscopy, and does not contradict the photoluminescence model being investigated.

4. Conclusion

This paper investigated the luminescence characteristics of synthesized samples of carbonate-substituted hydroxyapatite of B-type (calcium-deficient hydroxyapatite), i.e. an analogue of the mineral component of dental enamel. Using the methods of X-ray diffraction and transmission electron microscopy, all the synthesized samples were confirmed as nanocrystalline calcium hydroxyapatite, with nanopores on the surface of nanocrystals. IR and Raman spectroscopy indicated that the PO_4 group was replaced by a CO_3 group during synthesis of the obtained materials. The study of the luminescence properties of the synthesized samples showed that they have a maximum luminescence at ~515 nm (~2.4 eV). The comparison of the luminescence spectra revealed that changes in the pH of the solution used in the synthesis of calcium hydroxyapatite influenced the intensity of the photoluminescence, possibly due to fewer CO_3^{2-} groups in the calcium hydroxyapatite structure obtained at a lower pH. Furthermore, the features of the photoluminescence spectra can be utilized in the development of precise methods for the early detection of caries in human solid dental tissue.

Acknowledgments
This work was supported by the grant of Russian Science Foundation, grant number 17-75-10046.

References

[1] Dorozhkin S V 2017 *Hydroxyapatite and Other Calcium Orthophosphates: Bioceramics, Coatings and Dental Applications [Hardcover]* (Nova Science Publishers, Inc New York)
[2] Seredin P V, Goloshchapov D L, Kashkarov V M, Ippolitov Y A and Prutskij T 2016 Emission properties of biomimetic composites for dentistry *Results Phys.* **6** 447–8
[3] Rocha-Cabral R M, Mendes F M, Maldonado E P and Zezell D M 2015 A simple dental caries detection system using full spectrum of laser-induced fluorescence ed C Kurachi, K Svanberg, B J Tromberg and V S Bagnato p 95311A–1–95311A–13
[4] Leventouri T, Antonakos A, Kyriacou A, Venturelli R, Liarokapis E and Perdikatsis V 2009 Crystal Structure Studies of Human Dental Apatite as a Function of Age *Int. J. Biomater.* **2009** 1–6
[5] Piga G, Goncalves D, Thompson T J U, Brunetti A, Malgosa A and Enzo S 2016 Understanding the Crystallinity Indices Behavior of Burned Bones and Teeth by ATR-IR and XRD in the Presence of Bioapatite Mixed with Other Phosphate and Carbonate Phases *Int. J. Spectrosc.* **2016** 1–9
[6] Chen Q G, Zhu H H, Xu Y, Lin B and Chen H 2015 Quantitative method to assess caries via fluorescence imaging from the perspective of autofluorescence spectral analysis *Laser Phys.* **25** 085601
[7] Panayotov I, Terrer E, Salehi H, Tassery H, Yachouh J, Cuisinier F J G and Levallois B 2012 In vitro investigation of fluorescence of carious dentin observed with a Soprolife® camera *Clin. Oral Investig.* **17** 757–63
[8] Sarycheva I, Yanushevich O, Minakov D and Shulgin V 2015 Fluorescence of intact human teeth enamel in vivo *J. Stomatol.* **68** 424–9
[9] Feldbach E, Kirm M, Kotlov H and Mägi H Luminescence Spectroscopy of Ca-apatites under VUV Excitation *DESY Photon Sci. Annu. Rep.*
[10] Bachmann L, Zezell D M, Ribeiro A da C, Gomes L and Ito A S 2006 Fluorescence Spectroscopy of Biological Tissues—A Review *Appl. Spectrosc. Rev.* **41** 575–90
[11] Seredin P V, Goloshchapov D L, Prutskij T and Ippolitov Y A 2017 Fabrication and characterisation of composites materials similar optically and in composition to native dental tissues *Results Phys.* **7** 1086–94
[12] Goloshchapov D L, Kashkarov V M, Rumyantseva N A, Seredin P V, Lenshin A S, Agapov B L and Domashevskaya E P 2013 Synthesis of nanocrystalline hydroxyapatite by precipitation using hen's eggshell *Ceram. Int.* **39** 4539–49
[13] Yusufoglu Y and Akinc M 2008 Effect of pH on the Carbonate Incorporation into the Hydroxyapatite Prepared by an Oxidative Decomposition of Calcium–EDTA Chelate *J. Am. Ceram. Soc.* **91** 77–82
[14] Komlev V S, Fadeeva I V, Gurin A N, Kovaleva E S, Smirnov V V, Gurin N A and Barinov S M 2009 Effect of the concentration of carbonate groups in a carbonate hydroxyapatite ceramic on its in vivo behavior *Inorg. Mater.* **45** 329–34
[15] Grunenwald A, Keyser C, Sautereau A M, Crubézy E, Ludes B and Drouet C 2014 Revisiting carbonate quantification in apatite (bio)minerals: a validated FTIR methodology *J. Archaeol. Sci.* **49** 134–41
[16] Liu J, Wu Q and Ding Y 2005 Self-Assembly and Fluorescent Modification of Hydroxyapatite Nanoribbon Spherulites *Eur. J. Inorg. Chem.* **2005** 4145–9
[17] Zhang C, Yang J, Quan Z, Yang P, Li C, Hou Z and Lin J 2009 Hydroxyapatite Nano- and Microcrystals with Multiform Morphologies: Controllable Synthesis and Luminescence Properties *Cryst. Growth Des.* **9** 2725–33

IOP Conf. Series: Materials Science and Engineering **307** (2018) 012028 doi:10.1088/1757-899X/307/1/012028

Fabrication of an ordered mesoporous nanoparticle SiO_2/Mc and their CMP of fused silica application

L Xu [1,2,3], C L Zou[1,2,3], X Zhang[1], C X Kang[1], G H Luo[2,3] and G S PAN*[1,2,3]

[1] State Key Laboratory of Tribology, Tsinghua University, Beijing, China
[2] Research Institute of Tsinghua University in Shenzhen, Shenzhen, China
[3] Guangdong Provincial Key Laboratory of Optomechatronics, Shenzhen, China

E-mail: pangs@tsinghua.edu.cn.

Abstract. Here we developed an inorganic SiO_2 core/ mesoporous carbon shell structured (SiO_2/Mc) nano-composite particle with an average size ~50 nm as abrasives with improved dispersibility and distribution via a hydro-thermal route to obtain reserved fused silica surface and subsurface. The obtained SiO_2/Mc were characterized by scanning electronic microscope. Atomic force microscopy was used to assess the surface before and after planarization. The results indicated the as-prepared SiO_2/Mc composite abrasives gave a much lower surface roughness as well as lower topographical variations than that of traditional colloidal silica abrasives.

1. Introduction

Fused silica optic (FS) is one of the most important optical materials in the development of high-power laser systems. The laser induced damage threshold (LIDT) of FS, used in large high power laser facilities, has been an extensive research subject [1-2]. High surface/subsurface qualities are elementary required in most above mentioned applications. Thus, the final polishing step, remove surface and subsurface defects induced during grinding and lapping processes, is very critical. However, it is the very properties of FS render it as difficult-to-machine brittle materials. Currently, ceria particles are broadly adopted as abrasives to polish siliceous oxides glass for the high material remove rate (MRR), but it inevitably induces both of Ce contaminants and subsurface damages. Core/shell structured compositions, such as polystyrene (PS)/CeO_2 [3], PS/SiO_2 [4], polymethylmethacrylate (PMMA)/CeO_2 [5], and (PMMA)/SiO_2 [6], have shown an important potential as novel abrasives in efficient and damage-free polishing for their their uniform non-rigid mechanical property and spring-like effect. Nevertheless, most of the inorganic shell is not beneficial enough to improve its dispersibility and distribution which are crucial for a desired CMP performance. Thus, we developed an inorganic SiO_2 core/ mesoporous carbon structured nano-composite particle (SiO_2/Mc) as abrasives to obtain superior FS surface and subsurface.

2. Experiments.

IOP Conf. Series: Materials Science and Engineering **307** (2018) 012028 doi:10.1088/1757-899X/307/1/012028

SiO_2/Mc synthetized via a hydro-thermal route. Phenolic resin and commercial triblock copolymer Pluronic F127 were used as carbon source and template, respectively. A field-emission scanning electron microscope (FE-SEM, TESCAN, Mira 3 Xmh) and High resolution transmission electron microscopy (HRTEM) measurement using a FEI TECNAI G^2 F30 at an acceleration voltage of 300 kV were performed to character the morphology and structure of the nano composited particles.

In this work, a FS wafer, with 100 mm in diameter and 5 mm in thickness, which had been preliminarily polished with ceria, was applied. The wafer was processed with different size of SiO_2 abrasives via a UNIPOL 1000S machine from *Kejing Co., Ltd* under the conditions of the pressure 64g/cm^2, slurry flow rate 20 mL/min, carrier plate rotating speed 50 rpm**.** After that, the wafer was cleaned by liquid cleaner and deionized water sequently, and then dried off by air spray. And the chemical mechanical polishing behaviors for FS were evaluated by atomic force microscopy (AFM).

3. Results and discussion

3.1 Characteristic of samples.

The morphology and structure of the obtained composites were characterized with FE-SEM and HRTEM, as shown in Fig. 1. Fig.1(a) displays a nano SiO_2/Mc composites with an average diameter of ~50 nm. The mesoporous structure can be detected by HRTEM shown in Fig.1(b).

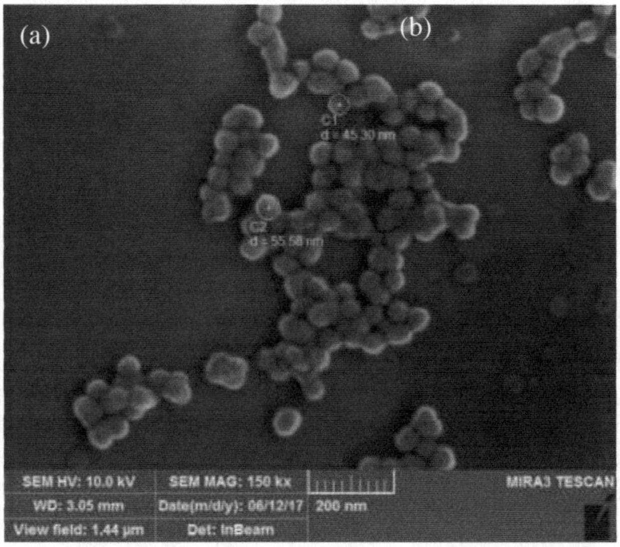

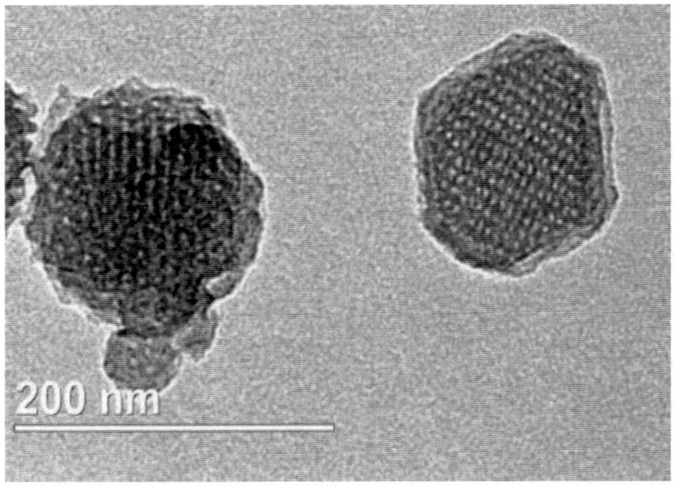

Fig.1 (a) SEM image, (b) TEM image of the SiO_2/Mc nanocomposites.

3.2 AFM measurements of polished surface

Fig.2 presented the typical 3-dimensional AFM morphologies of the FS wafer before and after polishing with SiO_2/M_c abrasive particles. As depicted in Fig.2 (a), the initial FS wafer presents a comparable rough surface with varied height hills, and the Rq roughness is 0.4 nm. While the Rq roughness is reduced to 0.08 nm after polished with SiO_2/M_c abrasive particles, as shown in Fig.2 (b). A quality of FS wafer surface polished with traditional SiO_2 abrasive as comparison, and the roughness Rq is about 0.16 nm.

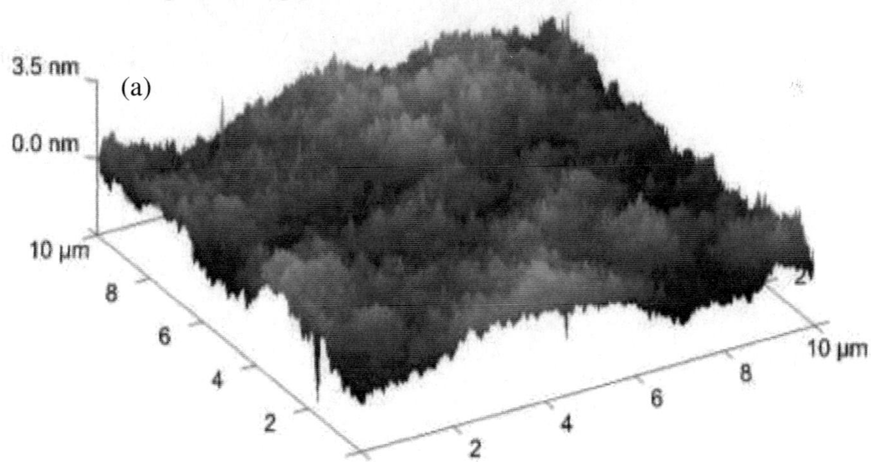

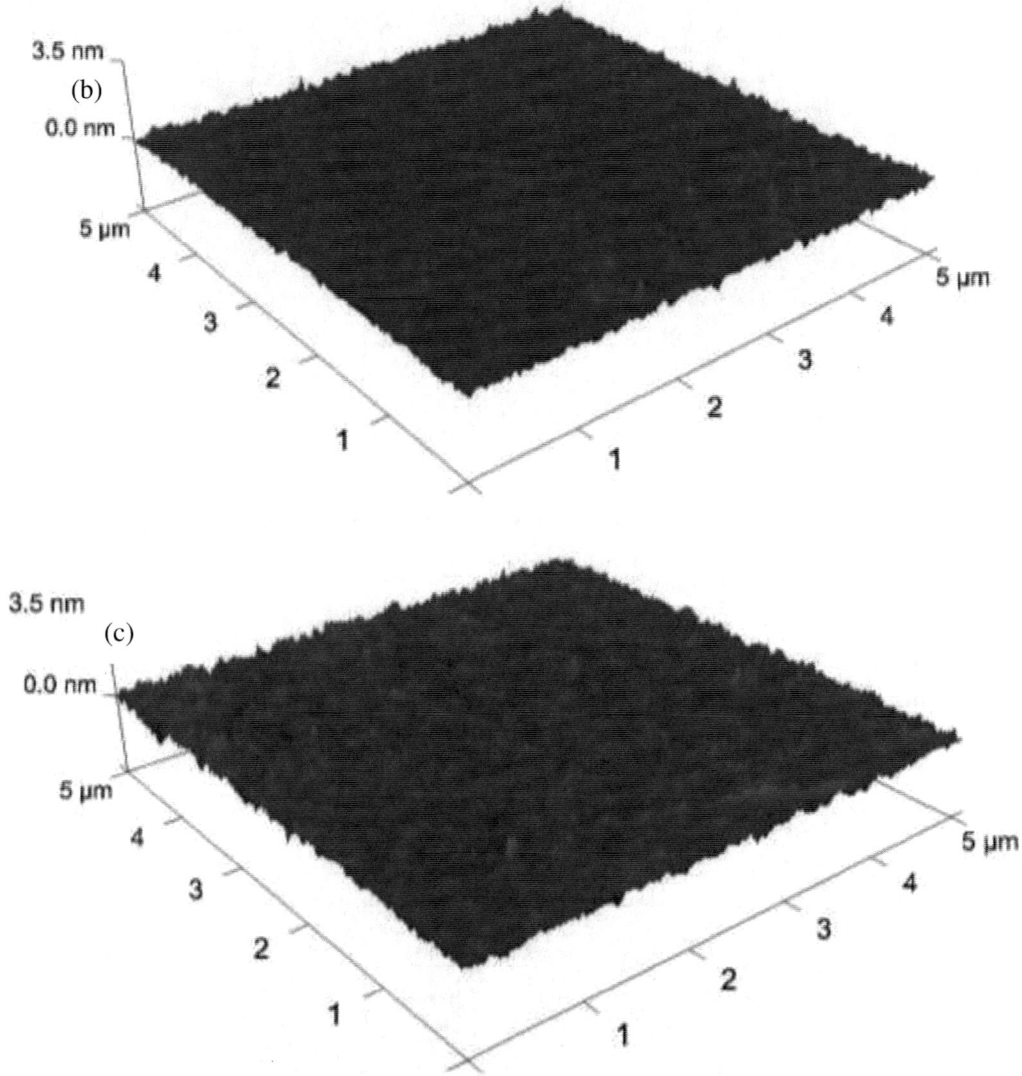

Fig.2 AFM morphologies of the FS wafer before (a) and after (b) CMP with SiO_2/M_c abrasive particles; (c) with traditional SiO_2 abrasives.

Acknowledgments
This work was supported by the Science Challenge Project (No. JCKY2016212A506-0504) and National Natural Science Foundation of China (No. 51675348).

References

[1] Wong J, Ferriera L, Lindsey F *et al* 2006 *J. Non-Cryst Solids* **352** 255.
[2] Shi F, Zhong Y, Dai Y *et al* 2016 *Optics Express* **24** 20842.
[3] Zhang L, Wang H, Zhang Z *et al* 2011 *Appl. Surf. Sci.* **258** 1217.
[4] Armini S, Whelan C M, Moinpour M *et al* 2007 *J. Electrochem. Soc.* **156** H18.

[5] Armini S, Messemaeker J, Whelan C M *et al* 2008 *J. Electrochem. Soc.* **155** pH653.
[6] Chaudhuri R G, Paria S. 2012 *Chem. Rev.* **112** p2373.

IOP Conf. Series: Materials Science and Engineering **307** (2018) 012029 doi:10.1088/1757-899X/307/1/012029

UO_2 fuel pellets fabrication via Spark Plasma Sintering using non-standard molybdenum die

E K Papynov[1,2], O O Shichalin[1,2], A Yu Mironenko[1], I G Tananaev[1,2], V A Avramenko[1,2] and V I Sergienko[1]

1 Institute of Chemistry, Far Eastern Branch of Russian Academy of Sciences, Vladivostok, 690022, Prospect 100-let Vladivostoku 159, Russia
2 Far Eastern Federal University, Vladivostok, 690920, Suhanova street 8, Russia
E-mail: papynov@mail.ru

Abstract. The article investigates spark plasma sintering (SPS) of commercial uranium dioxide (UO_2) powder of ceramic origin into highly dense fuel pellets using non-standard die instead of usual graphite die. An alternative and formerly unknown method has been suggested to fabricate UO_2 fuel pellets by SPS for excluding of typical problems related to undesirable carbon diffusion. Influence of SPS parameters on chemical composition and quality of UO_2 pellets has been studied. Also main advantages and drawbacks have been revealed for SPS consolidation of UO_2 in non-standard molybdenum die. The method is very promising due to high quality of the final product (density 97.5-98.4% from theoretical, absence of carbon traces, mean grain size below 3 μm) and mild sintering conditions (temperature 1100 °C, pressure 141.5 MPa, sintering time 25 min). The results are interesting for development and probable application of SPS in large-scale production of nuclear ceramic fuel.

1. Introduction

Fabrication of high mechanically stable compounds based on uranium dioxide (UO_2), which is a high melting oxide with melting temperature 2800 °C, is a tough technological challenge. World's practice widely uses sintering technologies with and without external load (pressurizing) at high temperatures ranging in 1700-1800 °C and are carried out in reductive medium with hours-long holding [1]. In this case conventional sintering conditions provide mechanically stable ceramic compounds in the form of UO_2 pellets with density ranging from 10.40 to 10.70 g/cm^3 (95-97.5 % from theoretical) and with various porosity and microstructure peculiarities. However, experts indicate significant drawbacks of the mentioned methods related to properties of the final product, e.g., non-uniform contraction and density of the material along the height dimension as well as distortion of geometrical shape. These increase the number of defected products, which don't meet the quality requirements.
To date as a high-tech approach to efficient synthesis of ceramic nuclear fuel electrophysical method of sintering are pointed out, in particular, the technology of spark plasma sintering (SPS). The technology is a brand new one in powder metallurgy and is actively applied for production of various ceramics [2,3]. The process is based on powder consolidation in constant electric field under the impact of high energy pulse of low voltage under constant pressure [4,5]. Obvious advantages of SPS are lower sintering temperatures due to high speed heating of consolidated powder by electric pulse current and also short process time. In this case grain growth is limited, microstructure is preserved and mechanical strength increases greatly without using any linking (reinforcing) agents that contaminate final product. Successful application of SPS to synthesis of ceramic fuel based on UO_2 is

presented in [6,7], where authors showed that the method allowed producing compounds with density of 96 % from theoretical at 1050 °C, holding time from 30 s and pressure of 40 MPa.

Despite the obvious advantages SPS technology has significant drawbacks that limit its development for synthesis of fuel products for industry. The main reason is the type of dies used that are usually such conducting materials as graphite and its composites. In this case, negative factors are active carbon diffusion from the surface of the dies and graphite surroundings into the bulk of sintered material, where it agglomerates as described in [8,9]. While known and described methods of carbon elimination by aluminum, molybdenum and tungsten foils as well as absorbing layers that interact with carbon and isolate it from the sintered material are not appropriate for fuel production because radioactive byproducts are formed after the contact of isolating materials with fuel composition and additional contamination of the material can occur. Besides that, graphite has a rather limited compression limit that does not allow higher pressures applied onto sintered powder that provide maximal stability of grain growth. Carbon diffusion into the sintered compound can reduce the degree of uranium oxidation due to its partial carbothermic reduction. The presence of carbon impurities in the fuel tablet product can be cause uncontrolled deceleration of neutrons during irradiated in the reactor. Mentioned factors impact such quality parameters as density, stability, thermal conductivity, resistance to bulk gas formation under radiation, full combustion. It is obvious that indicated drawbacks can be avoided using carbon-free dies. However, there is nothing to prove this fact because such research is absent in the literature.

Taking this into account, this work is aimed on investigating peculiarities of SPS consolidation of commercial UO_2 ceramic powder using different die instead standard graphite die. In particular, we tested molybdenum die for production of highly dense UO_2 pellets of required quality.

2. Experimental

2.1. Materials and methods

We used uranium dioxide powder with lean isotope U-235 of ceramic type produced by "PJSC MSZ" (Russia). Powders have bulk density ranging in 1.0-2.0 g/cm^3, specific surface area 3.0-5.0 m^2/g and particle size from 20 to 1000 μm.

We conducted sintering on a device SPS-515S produced by "Dr. Sinter*LAB™" (Japan) using molybdenum die (type of molybdenum is TC 48-19-203-85) 6.5 mm in diameter and 30 mm in height.

2.2. Synthesis technique

Sample of initial uranium dioxide powders was place into the die, moved into the vacuum chamber with residual pressure 6 Pa, after that sintering was conducted at optimal temperature and pressure followed by cooling and removing sample from the chamber.

Due to uranium dioxide easily undergoes oxidation in air into U_3O_8 we conducted sintering in vacuum to preserve the initial O/U ratio in the range 2.06-2.16. Sintering conditions: SPS temperature 1100 °C, applied pressure 141.54 MPa, heating rate 58 °C/min, holding time at maximal temperature 5 min and cooling time 30 min. The impulse ON/OFF regime was 12/2 frequency and duration 3.3 to 329 ms.

2.3. Characterization methods

Crystal phases of initial powders and sintered pellets based on uranium dioxide were identified by XRD on a multipurpose X-ray diffractometer D8 Advance "Bruker AXS" (Germany). Microstructure of the samples was carried out using scanning electron microscopy (SEM) on TM 3000 and S 5500 "Hitachi" (Japan) devices. Samples' elemental composition based on quantitative and qualitative determination of the elements in the investigated materials was carried out using energy dispersive microanalysis (EDX) on Bruker station within TM 3000 microscope "Hitachi" (Japan). Specific density of the samples was determined using hydrostatic weighting in various liquids on a balance AdventurerTM "OHAUS Corporation" (USA).

IOP Conf. Series: Materials Science and Engineering **307** (2018) 012029 doi:10.1088/1757-899X/307/1/012029

3. Results and discussions

The main idea of implementing molybdenum die is to fully eliminate carbon diffusion from working die surfaces into the bulk of the compound. Additional advantage is the increased upper limit of applied pressure due to higher mechanical stability of molybdenum as compared to graphite. Such approach enables one to decrease considerably process' temperature and to obtain samples of desired geometrical shape close to the practical products without loss of quality.

In this experiment we used powder of ceramic type, which microstructure is defined by small crystallites not bigger than 500 nm (Fig. 1).

Figure 1. SEM images of UO_2 powder of ceramic type.

SPS consolidation of powder was conducted stepwise with periodical holding at certain temperature and at constant pressure as shown on Fig. 2.

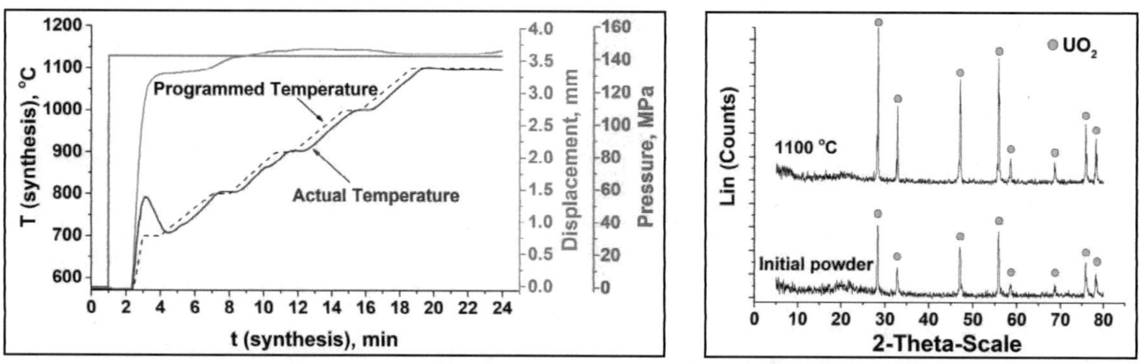

Figure 2. SPS parameters of UO_2 powder consolidation in molybdenum die and XRD patterns of initial powder and UO_2 pellets obtained via SPS.

According to presented trends the main compaction of sintered powder occurs during the first 5 min in 600-800 °C temperature range under the conditions of stepwise heating and high pressure (141.5 MPa) that is several times higher than in earlier experiments. Full densification is reached within 10 min of sintering and doesn't change significantly further. The peculiarity is proved by microstructural investigation discussed below (Fig. 3).

Initial powder according to XRD is a crystal uranium dioxide UO_2 (Fig. 2). In the course of SPS sintering at 1100 °C in molybdenum die its composition doesn't change. However, diffraction peak intensities increase indicating the improvement of crystallinity (transformation of amorphous phase into crystal one) and also possible oxide's crystallite growth.

While investigating microstructure of the sample obtained via SPS in molybdenum die (Fig. 3) we revealed that particles of the initial powder are actively sintered forming crystallites with well-defined planes and open porosity along intergranular contacts and in the bulk. The main factor deserving attention is the crystallite size below 3 µm in average.

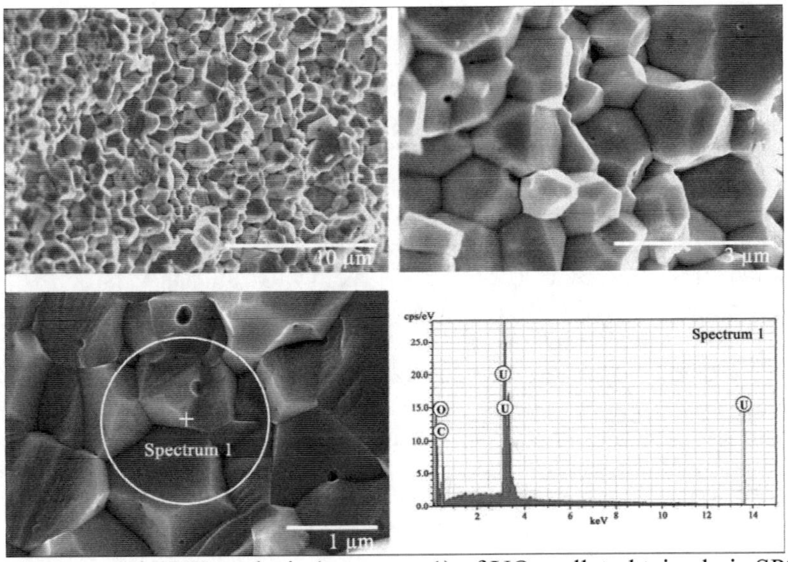

Figure 3. SEM images and EDX analysis (spectrum 1) of UO_2 pellet obtained via SPS of powder in molybdenum die at 1100 °C and under 141.5 MPa.

In addition, SEM images clearly evidence the absence of agglomerated carbon in the bulk of consolidated pellet (Fig. 3). Carbon traces are observed only in small amount on the sample's surface (Table 1) and can be explained by the EDX error introduced due to using carbon adhesive to fix the sample to the support. Carbon can also come into the sample during the treatment (polishing and smoothing) by abrasive. Despite this, it is obvious that bulk diffusion into the sample during SPS as described for previous experiments does not occur in this case.

Table 1. EDX analysis of UO_2 pellet obtained via SPS that presented at Fig.3.

Element	Spectrum 1		
	U	O	C
Content, wt. %	85.57	11.33	3.10
Content, at. %	27.12	53.42	19.45

Density measurements of obtained UO_2 pellet as one of the main quality criterion clearly indicate the efficiency of SPS process applied to sintering of UO_2 powders. According to them (Table 2) apparent density of the sample consolidated in molybdenum die varies in the range 97.5-98.4% from theoretical value depending the liquid type.
Given the data above it is obvious using molybdenum die for SPS consolidation allows consolidating uranium dioxide powder at lower temperature with high compaction due to considerable increase of pressure as compared to less mechanically stable graphite. In addition, carbon contamination is avoided. All these are crucial, when assessing the quality of produced fuel pellets based on UO_2.

Table 2. Apparent density of UO_2 pellet measured in various liquids, g/cm^3.

Water	Alcohol	Toluene	Benzene	Chloroform
0.991	0.803	0.860	0.870	1.470
10.7238	10.6947	10.7890	10.7729	10.7795

4. Conclusions

The work has investigated peculiarities of SPS consolidation of commercial uranium dioxide ceramic powder using no-standard die, in particular, the applicability of molybdenum die vs. graphite one was studied with respect to fabrication of highly dense fuel products (pellets) of required quality. Investigations have shown the SPS technology provides high-rate synthesis of UO_2 compounds with required physico-chemical characteristics. SPS consolidation of UO_2 powders in molybdenum die as alternative approach has proved to have a number of advantages. SPS consolidation of ceramic UO_2 powder is conducted at 1100 °C and 141.5 MPa during 25 min resulting in high quality product (density 97.5-98.4% from theoretical, grain size below 3 μm and carbon-free). In addition, molybdenum die prevents carbon diffusion during SPS and, therefore avoids sample's contamination. According to obtained data it is obvious that SPS technology is promising for fabrication of UO_2 pellets with various ceramic properties. Suggested alternative approach using carbon-free equipment is rational for improving the product's quality. These results can be recommended for development and possible implementation of SPS technology in the industry.

References

[1] Carrea A J 1963 Sintering of uranium dioxide in an atmosphere of controlled hydrogen content *J. Nucl. Mater.* **8** 275–7

[2] Orru` R, Licheri R, Locci A M, Cincotti A and Cao G 2009 Consolidation/synthesis of materials by electric current activated/assisted sintering *Mater. Sci. Eng. R Reports* **63** 127–287

[3] Munir Z A, Quach D V. and Ohyanagi M 2011 Electric current activation of sintering: A review of the pulsed electric current sintering process *J. Am. Ceram. Soc.* **94** 1–19

[4] Tokita M 1993 Trends in Advanced SPS Spark Plasma Sintering Systems and Technology *J. Soc. Powder Technol. Japan* **30** 790–804

[5] Mamedov V 2002 Spark plasma sintering as advanced PM sintering method *Powder Metall.* **45** 322–8

[6] Ge L, Subhash G, Baney R H, Tulenko J S and McKenna E 2013 Densification of uranium dioxide fuel pellets prepared by spark plasma sintering (SPS) *J. Nucl. Mater.* **435** 1–9

[7] Cologna M, Tyrpekl V, Ernstberger M, Stohr S and Somers J 2016 Sub-micrometre grained UO2 pellets consolidated from sol gel beads using spark plasma sintering (SPS) *Ceram. Int.* **42** 6619–23

[8] Bertrand A, Carreaud J, Delaizir G, Duclère J R, Colas M, Cornette J, Vandenhende M, Couderc V and Thomas P 2014 A comprehensive study of the carbon contamination in tellurite glasses and glass-ceramics sintered by spark plasma sintering (SPS) *J. Am. Ceram. Soc.* **97** 163–72

[9] Bokhonov B B, Ukhina A V., Dudina D V., Anisimov A G, Mali V I and Batraev I S 2015 Carbon uptake during Spark Plasma Sintering: investigation through the analysis of the carbide "footprint" in a Ni–W alloy *RSC Adv.* **5** 80228–37

Acknowledgements

The investigation was financially supported by Russian Scientific Foundation (project number 17-73-20097). Uranium dioxide powders was provided for the scientific research by Federal State Unitary Enterprise "Industrial Association "Mayak" (Ozersk city, Russia).

IOP Conf. Series: Materials Science and Engineering **307** (2018) 012030 doi:10.1088/1757-899X/307/1/012030

Manganese Oxide-Based Sorbent for Sr-90 Radionuclide Removal from Seawater

E K Papynov[1,2], A M Egorin[1,2], T A Sokolnitskaya[1,2], D V Marinin[1], Y A Azarova[1], A S Portnyagin[1,2], M I Balanov[1,2], D S Misko[2], E A Shelestiuk[2], A M Kalashnikova[2], E A Tokar[1,2], , I G Tananaev[1,2], V A Avramenko[1,2]

1 Institute of Chemistry, Far Eastern Branch of Russian Academy of Sciences, Vladivostok, 690022, Prospect 100-let Vladivostoku 159, Russia
2 Far Eastern Federal University, Vladivostok, 690920, Suhanova street 8, Russia
E-mail: papynov@mail.ru

Abstract. The morphology, specific surface area, and sorption properties toward Sr-90 radionuclides in seawater of a sorption material based on manganese oxide have been investigated. The material was fabricated through interaction of $KMnO_4$ and H_2O_2 with subsequent annealing at 500 °C. The sorbent is characterized with the mechanical strength sufficient for using under dynamic sorption conditions: here, the efficiency of Sr-90 removal at feeding of 150 bed volumes exceeds 95 %. The values of Sr-90 distribution coefficients are equal to 0.8–1.2×10^3 ml/g in the real seawater and to 1.6-1.8 ×10^3 ml/g in the simulated one.

1. Introduction

The importance of creating novel sorption materials for Sr-90 removal from seawater is caused by active construction works and operation of nuclear powers plants (NPPs) located in the coastal area – some of them use seawater for direct cooling. The latter created potential risks of the release of anthropogenic radionuclides into the World Ocean with formation of huge amounts of liquid radioactive waste (LRW) streams of complex chemical compositions in case of emergency. Presently, formation of such LRW comprising the radioactively contaminated seawater (SW) occurs in the process of operation, repair, service, and decommissioning of nuclear-powered ships.

The most hazardous radionuclides include Cs-137 and Sr-90 belonging to the group of high-toxicity isotopes. The increased content of these radionuclides in the environment, especially in natural water basins, creates risks of elevated radiation doses for human population. Whereas the task of Cs-137 removal from SW can be solved through application of ferrocyanide sorbents and phenol-aldehyde resins, the Sr-90 removal is still an urgent problem.

The developed and successfully applied technology of seawater decontamination from Sr-90 using the sorption-reagent material (barium silicate) [1–3] has an essential disadvantage – formation of significant amounts of secondary wastes in the form of spent material. That is why to develop an efficient technology of decontamination of seawater from Sr-90 with formation of minimal amounts of secondary wastes, it is necessary to create materials characterized with high selectivity to Sr and the possibility of extended used in the sorption–elution–regeneration cycle. The radionuclide concentrate obtained in the form of eluate can be evaporated down to the minimal volume or reprocessed using smaller amounts of the sorbent, which will allow reducing the volume of radioactive waste to be sent for final disposal.

To solve the task of the removal of Sr-90 from seawater, the most promising are sorption materials based on manganese oxides and characterized with an acceptable selectivity toward Sr^{2+} ions in the presence of Mg^{2+} and Ca^{2+}. The sorbed Sr can be eluted by solution of a strong mineral acid with a repeated application of the sorbent in the sorption process. However, despite a large number of works in this field, just a small part of them are devoted to strontium removal from seawater, while there are virtually no works on studying the possibility of application of manganese oxide in the sorption–elution–regeneration cycle. In our opinion, the layered modification of manganese oxide, which can be synthesized through a relatively simple procedure, is the most efficient in solution of the problem of decontamination of LRW streams comprising the radioactively contaminated seawater [4]. The present work was devoted to studies of the Sr sorption from seawater by the layered modification of manganese's oxide in the sorption–elution–regeneration mode.

2. Experimental

The sorbent was obtained through mixing of concentrated solutions of $KMnO_4$ and H_2O_2 under constant stirring at heat and reaction gaseous products removal. The produced precipitate was washed with deionized H_2O and annealed, after preliminary drying, at 500 °C for 6 h in air. The produced sorbent comprises granules of irregular shapes and black color with characteristic metallic gloss; its bulk weight is equal to 1.56 g/ml.

The strontium sorption under static conditions was carried out under constant joint stirring of the sorbent and the liquid medium – the stirring duration was 24 hours, V/m = 1000 ml/g, and the sorbent weight was 0.01 g. The initial solutions activity was 1000 Bq/ml. The distribution coefficient of Sr-90 (K_d of Sr-90) was determined upon determination of the solution residual activity.

The Sr-90 distribution coefficient was calculated according to the formula (1):

$$K_d = \frac{A_0 - A_1}{A_1} \times \frac{V}{m} \quad (1),$$

where A_0 – the liquid medium initial activity (Bq/ml), A_1 – the liquid medium residual activity after sorption (Bq/ml), V – the liquid medium volume at sorption (ml), m – the sorbent weight (g).

The sorption under dynamic conditions was studied by feeding the liquid medium through a column of an internal diameter of 0.5 cm containing 1 ml of the sorbent with grain sizes of 0.1-0.2 mm. The feeding rate was 10 bed volumes per hour (BV/h). Elution of the adsorbed strontium was made using 0.5 M HCl solution at a rate of 10 BV/h. Thereafter, the sorbent was regenerated by 100 ml of 1.0 M NaOH solution and washed with 30 ml of distilled H_2O, and the sorption cycle was repeated.

The Sr-90 desorption was calculated according to the formula (2):

$$\text{Desorption\%} = \left(\frac{\sum_1^i V \times A_1}{A_x + A_{x-1}}\right) \times 100 \quad (2)$$

where A_1 – the eluate activity (Bq/ml), V – the eluate volume (ml), A_x – the activity of Sr-90 removed at the sorption stage (Bq), A_{x-1} – the residual activity of Sr-90 from the previous cycle (Bq), x – the cycle number, i – the number of the eluate fraction.

Seawater (SW) taken at the Amursky bay (Primorsky Territory) and preliminarily filtered through two "blue ribbon" cellulose filters was used. To study sorption-selective characteristics, seawater samples containing the Sr-90 tracer of an activity of 100–300 Bq/ml were used. To prepare the simulated seawater (S-SW), the composition suggested in [5] was used.

3. Results and Discussion

The materials notations and their sorption characteristics and specific surface areas as well as Na and K contents are presented in Table 1.

Table 1. Characteristics of manganese oxide samples used in the present work.

Sample	Thermal	Treatment by	$\omega(Na^+)$,	$\omega(K^+)$,	Specific	K_d of Sr-90

No.	treatment temperature (°C)	HCl+NaOH solutions	wt.%	wt.%	surface area (m^3/g)	SW	S-SW
1	105	No	Not determined		13.4	520	950
2	500	No	4.1	4.37	9.4	890	1690
3	500	Yes	6.06	0.34	12	1240	1830

The XRD spectra of the samples 1, 2, and 3 were recorded (Figure 1). The materials are characterized with low degree of crystallinity, which complicates correct identification of the phase composition. However, the obtained XRD patterns allow assuming the presence of several phases, mainly K- and Na-birnessite.

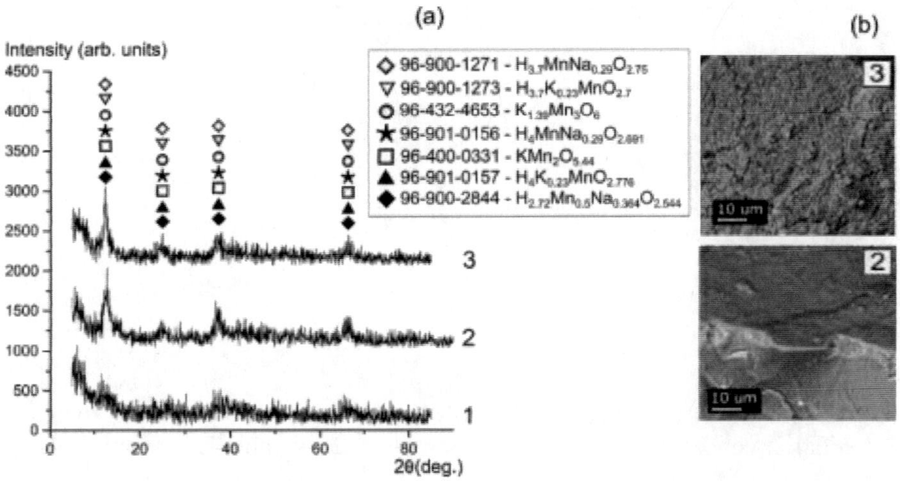

Figure 1. XRD-spectra of the synthesized samples (a) and REM images of their surfaces (b); 1 – sample 1, 2 – sample 2, 3 – sample 3.

It was revealed that the increase of the temperature of thermal treatment from 105 to 500 °C resulted in the increase of the Sr-90 K_d more than 1.5-fold. Probably, this is related to formation of a clearly expressed phase of birnessite at higher heating temperatures. Preliminary sequential treatment of the samples by HCl and NaOH solutions also results in the increase of the efficiency of Sr-90 removal. Such results can be explained by high content of potassium (Table 1) having high competing impact on the sorption (as compared to sodium) in the initial sample. That is why the potassium removal from the manganese oxide composition has a positive effect on the efficiency of Sr-90 removal. Besides, the treatment by HCl and NaOH solutions yields the increase of the specific surface area from 9.4 to 12 m^3/g, which could also increase the Sr-90 K_d values.

In the presence of Na^+, the sorbents based on manganese oxides are known to be efficient materials for strontium removal [6]. However, since it is evident that the most pronounced effect on sorption is provided by Ca^{2+} and Mg^{2+} ions, we studied their effect on the efficiency of Sr-90 removal by manganese oxide (the results are shown in Table 2). The obtained results indicate to the fact that Ca^{2+} ions decrease the strontium sorption to the highest degree, as compared to Mg^{2+} ions. However, in the simulated solution containing 0.01 mol/L of Ca^{2+}, the Sr-90 K_d values remain rather high, which indicates to the selectivity of manganese oxide that is sufficiently high to remove this radionuclide from seawater.

Table 2. Effect of Ca^{2+} and Mg^{2+} ions concentrations on the efficiency of Sr-90 sorption by manganese oxide.

Ca^{2+} concentration, mol/l	Sr-90 K_d, ml/g	Mg^{2+} concentration, mol/l	Sr-90 K_d, ml/g
0,0001	40.2×10^3	0,0001	18.9×10^4

0,001	13.7×10^3	0,001	5.4×10^4
0,01	1.9×10^3	0,01	2.2×10^4
0,01	0.7×10^3	0,01	0.6×10^4

The manganese oxide sorption characteristics under dynamic conditions were investigated. Figure 2 shows the output curves of sorption of non-active strontium from S-SW, and the calculated values of full dynamic exchange capacity were equal to 9.16, 7.82, and 6.85 mg/g in the first, second, and third cycles, respectively. Such a capacity reduction can be related to a gradual material dissolution. For example, the height of the sorption layer decreased by 20 % after three cycles. The dissolution can be explained by disproportionation of Mn^{3+} in acidic media with formation of Mn^{4+} and soluble Mn^{2+}.

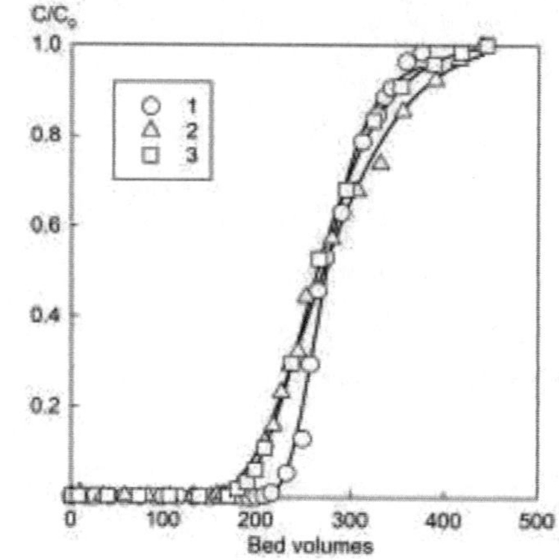

Figure 2. Sorption of non-active strontium from S-SW by manganese oxide under dynamic conditions, initial Sr concentration 50 mg/l: 1 – sample 1, 2 – sample 2, 3 – sample 3.

The sorption and desorption of Sr-90 micro-quantities from real seawater under dynamic conditions were investigated and the results are shown in Fig. 3. In the course of removal of Sr-90 micro-quantities from SW, the sorbent resource, taking into account the real strontium content, decreases and does not attain its maximal capacity. The reason of the decrease of the sorbent resource could consist in the presence of organic contaminants and organic carbon in SW, which promote Mn reduction until lowest oxidation states, and suspensions.

The Sr-90 removal from SW is characterized with low efficiency in the first cycle (Fig. 3a, curve 1), which is related to the sorbent achievement of the working mode, transition into the Na-form, and formation of a developed surface. During the second and further cycles, the sorbent resource exceeds 150 bed volumes, while no changes in the sorbent or filtrate color were observed at the sorption stage.

Figure 3b shows the curves of elution of Sr-90 sorbed from real SW under dynamic conditions. The obtained results indicate to instability of radionuclide elution, but after 4 cycles the residual activity of Sr-90 did not exceed 4 % of total activity absorbed by the sorbent.

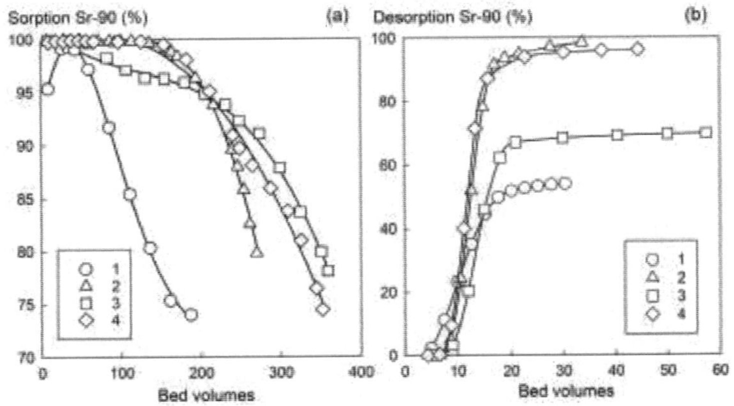

Figure 3. Sorption (a) and desorption (b) of Sr-90 from real seawater under dynamic conditions.

4. Conclusions

The interaction of $KMnO_4$ and H_2O_2 with subsequent treatment by 1 M NaOH solution and annealing at 500 °C produced the material with the structure of birnessite characterized with high selectivity toward Sr-90. In the presence of 0.4 g/l of Ca^{2+} ions, the Sr-90 distribution coefficient is equal to 1.9×10^3 mg/g, which enables one to use this material for seawater decontamination from this radionuclide. The Sr-90 distribution coefficients is equal: in real seawater – to $0.8–1.2 \times 10^3$ mg/g, in simulated seawater – to $1.6–1.8 \times 10^3$ mg/g. Under dynamic sorption conditions, the material is mechanically stable and preserves the grain structure. However, at prolonged functioning, the material dynamic exchange capacity decreases due to its dissolution, and a regular addition of new sorbent batches is required. Under dynamic sorption conditions, the average value of the volume of seawater fed until the Sr-90 breakthrough of 95 % and higher exceeds 150 bed volumes at a filtering rate of 10 BV/h. The produced material can be applied in decontamination of liquid radioactive waste streams containing seawater from Sr-90.

References

[1] Avramenko V A, Zheleznov V V, Kaplun E V, Sokol'nitskaya T A and Yukhkam A A 2001 Sorption Recovery of Strontium from Seawater *Radiochemistry* **43** 433–6

[2] Avramenko V A, Sergienko V I and Sokol'nitskaya T A 2016 Application of sorption-reagent materials in the technology of liquid radioactive waste treatment *Theor. Found. Chem. Eng.* **50** 593–7

[3] Feng Q, Kanoh H and Ooi K 1999 Manganese oxide porous crystals *J. Mater. Chem.* **9** 319–33

[4] Kester D R, Duedall I W, Connors D N and Pytkowicz R M 1967 Preparation of Artificial Seawater1 *Limnol. Oceanogr.* **12** 176–9

[5] Kirillov S A, Lisnycha T V and Pendelyuk O I 2006 Appraisal of Mixed Amorphous Manganese Oxide/Titanium Oxide Sorbents for the Removal of Strontium-90 from Solutions, with Special Reference to Savannah River Site and Chernobyl Radioactive Waste Simulants *Adsorpt. Sci. Technol.* **24** 895–906

[6] Drits V A, Silvester E, Gorshkov A I and Manceau A 1997 Structure of synthetic monoclinic Na-rich birnessite and hexagonal birnessite; I, Results from X-ray diffraction and selected-area electron diffraction *Am. Mineral.* **82** 946–61

Acknowledgments
The equipment of the Center of Collective Use of Scientific Equipment "Far Eastern Center of Structural Investigations" was used in the present work.
The work was financially supported by the Russian Science Foundation (project no. 14-13-00135).

IOP Conf. Series: Materials Science and Engineering **307** (2018) 012031 doi:10.1088/1757-899X/307/1/012031

SPS-RS technique for solid-phase "in situ" synthesis of biocompatible ZrO_2 porous ceramics

O O Shichalin[1,2], M A Medkov[1], D N Grishchenko[1], V Yu Mayorov[1], A N Fedorets[2], A A Belov[1,2], A V Golub[1], E A Gridasova[2], E K Papynov[1,2]

1 Institute of Chemistry, Far Eastern Branch of Russian Academy of Sciences, Vladivostok, 690022, Prospect 100-let Vladivostoku 159, Russia
2 Far Eastern Federal University, Vladivostok, 690920, Suhanova street 8, Russia
E-mail: oleg_shich@mail.ru

Abstract. The prospective method of spark plasma sintering-reaction synthesis (SPS-RS) for fabrication of ceramics based on ZrO_2 and biocompatible with living tissue is presented. Nanostructured ceramics has high mechanical strength (more than 400 MPa) and controlled porosity depending on specified sintering conditions. Biocompatible phases $Ca_{10}(PO_4)_6(OH)_2$ are formed "in situ" during SPS sintering of ZrO_2 powder due to chemical interaction of phosphate precursors preliminary introduced into the mixture. The effective method to improve (to develop) porous structure of bioceramics obtained by SPS or SPS-RS techniques using poreforming agent (carbon black) is proposed. Suggested original SPS-RS "in situ" technique provides fabrication of new ZrO_2 ceramics containing biocompatible phosphate components and possessing unique structural and mechanical characteristics. Such ceramics is indispensable for bone-ceramic implants that are able to activate processes of osteogenesis during bone tissue recovery.

1. Introduction

Ceramic materials for bone implants raise a great interest of scientific community due to growing demand in orthopedics. Inertness to biological medium and high mechanical properties of the ceramics based on ZrO_2 allow successful implementing it in the production of porous constructions implanted into organism [1]. ZrO_2 is referred to bioinert type of materials, which means that negative chemical reactions with tissues and intertissular liquids in organism are avoided. Nevertheless, it has a significant drawback: due to shielding of mechanical loads the gradual resorbtion of nearest bone tissues occurs [1].

There are several approaches to the problem indicated above. One of them is to increase the bioactive properties of ZrO_2 ceramics. In that case ceramics have its own ability to bioresorbtion and during implantation executes the function of building material, participates in the process of recovery of bone material. This can be made via creating on the basis of ZrO_2 ceramics composite materials with resorbable phase, in place of which the calcium-phosphate compounds, for example hydroxyapatite, which has the closest composition to the mineral part of bone tissue [2]. Secondly, stimulating activity during reparative osteogenesis is affected also by microstructure of bone implant, in particular, porosity is responsible for the depth of accretion of biological tissue into implant. In that case, during synthesis of bioactive composites such parameters as curvature, degree of crystallinity and porosity are modified in the first place and by a variety of methods [3]. Also, in order to achieve desired mechanical characteristics of ZrO_2 composites methods of powder metallurgy are used: hot/cold or isostatic pressing, sintering under pressure, slip casting, microwave synthesis and oth. [4], as results obtaining composite ceramics of high mechanical strength. However, it is known that conventional routines do not ensure preservation of material's porous structure, which collapses and forms highly dense compound with density more than 99.5% from theoretical [5]. This is caused by severe technological conditions of the process, under which the grain growth and destruction are intensified leading to distortion and demolition of porous framework in the solid body.

High level of modern technologies allows eliminating complications connected with preservation of porous framework of bulk ceramics. In that case one is advisable to use innovative technology of powder metallurgy known as Spark Plasma Sintering (SPS). Main purpose of SPS technology is high-speed sintering (consolidation) of powder materials of any origin resulting in formation of ceramics with novel compositions and properties [6]. Successful application of SPS technology for synthesis of nanostructured ceramics based on ZrO_2 and on its composite forms is shown by a number of researchers in [7]. Almost all of the works in the field contain limited description of correlations between porous structure changes and SPS conditions. Recommendations or suggestions concerning control of organization or development of porous structure of ceramic composites based on ZrO_2 during SPS synthesis are practically absent. In addition, in our point of view, more prospective direction in this scientific field is search and study of original ways of synthesis of such ceramics, for example Spark Plasma Sintering-Reaction Synthesis (SPS-RS) based on conducting chemical reaction under spark plasma exposure onto reacting mixture during SPS synthesis Information about SPS-RS preparation of biocompatible ZrO_2 ceramic composites containing bioactive phase are absent.

Present research is devoted to fabrication of multiphase biocompatible ceramics based zirconium dioxide via reactive spark plasma sintering. Given that the aim of this work is establishing capability of formation of bioactive phase within the composition of ceramics in the course of chemical reaction during SPS consolidation of ZrO_2 powder, doped with reactive mixture. Also studying of possibility of regulation of porous structure of obtained ceramics via using carbon template as a poreforming agent with conservation of high mechanical strength. The suggested in present work original method ensures fabrication of new bioceramics for medical purposes, prospective as bone-ceramic implants.

2. Experimental

2.1. Materials

To prepare initial dry mix (IM) we used ZrO_2 powder of analytical grade. Crystal structure of ZrO_2 ceramics was stabilized by introduction of magnesium oxide (chemically pure). As a reactive mixture (RM) for the formation of biocompatible phase we used a $(NH_4)_2HPO_4$ (pure) and $CaCO_3$ (chemically pure). Proportion of these components was adjusted to reach Ca/P = 1.66. Dispersed technical carbon (TC) with particle size 1-500 μm was used as a poreforming agent during ceramic synthesis.

2.2. Synthesis

Ceramic synthesis was carried out in an equipment of SPS-515S "Dr.Sinter*LAB^{TM}" (Japan) according to following scheme: 3 g of IM containing ZrO_2 powder (80 wt.%), reactive mixture (RM) of $CaCO_3$ and $(NH4)_2HPO_4$ (15 wt. %) and stabilizing agent MgO (5 wt.%) were placed into a graphite die (15.5 mm diameter), prepressed (20.7 MPa), then the stock was placed into vacuum chamber (pressure 6 Pa) and then sintered. To prevent baking of sintered powder on the die and on the plungers as well as to remove obtained compound easily we used carbon foil 200 μm thick. Sample sintering was conducted at 900, 1000 and 1300 °C with heating rate 170 °C/min. Samples were being held at the maximal temperature for 5 min and then were cooled down to room temperature. Pressure onto the sample during sintering was 22.3 MPa and remained constant throughout the process. Periodicity of generated low-voltage pulse in ON/OFF regime is 12/2, duration 3.3-329 ms.

To increase porous volume of the obtained ceramic samples the poreforming agent was additionally introduced. As an agent we used low dispersed technical carbon (TC) of 1-500 μm fraction in the amount 2, 5, 10 and 15 wt.% of total IM. After SPS-RS consolidation obtained ceramic samples were exposed to 900 °C (heating rate 5 °C/min, holding time 30 min) temperature in air atmosphere to remove TC in the furnace "Nabertherm GmbH" (Germany).

2.3. Characterization methods

Crystal phases of initial powders and sintered pellets based on uranium dioxide were identified by XRD on a multipurpose X-ray diffractometer D8 Advance "Bruker AXS" (Germany). Microstructure of the samples was carried out using scanning electron microscopy (SEM) on TM 3000 and S 5500 "Hitachi" (Japan) devices. Samples' elemental composition based on quantitative and qualitative determination of the elements in the investigated materials was carried out using energy dispersive microanalysis (EDX) on Bruker station within TM 3000 microscope "Hitachi" (Japan).

3. Results and discussions

During experiment the SPS consolidation of ZrO_2 powders, containing reactive mixture of initial compounds of bioactive phase including poreforming agent, was carried out. According to the goal of investigation, primarily it was necessary to establish a fact of formation of bioactive phases of HAP in the course of chemical reaction between $(NH_4)_2HPO_4$ (pure) and $CaCO_3$ during SPS-RS consolidation of ZrO_2 powder ("in situ"). Ratio of RM was chosen as described in [8] and corresponded to Ca/P=1.66. In that case there are two reactions possible with the first being dominating.

$$10CaCO_3 + 6(NH_4)_2HPO_4 = Ca_{10}(PO_4)_6(OH)_2 + 10CO_2 + 8H_2O + 12NH_3 \quad (1)$$

$$3CaCO_3 + 2(NH_4)_2HPO_4 = Ca_3(PO_4)_2 + 3CO_2 + 3H_2O + 4NH_3 \quad (2)$$

According to reaction (1) and (2) thoroughly described in [8] thermal treatment of the indicated reaction mixture the final products are calcium phosphates $Ca_3(PO_4)_2$ and $Ca_{10}(PO_4)_6(OH)_2$. Formation of one from the two phosphates or of their mixture is ruled by several factors, like partial interaction of the initial components, destruction of reaction products and synthesis conditions such as components' ratio, treatment temperature and holding time. As a result, the series of dense ZrO_2 ceramics with various physico-chemical characteristics have been obtained (Table 1).

We considered 1300 °C as the optimal SPS-RS temperature for the studied system, because it is enough for sintering of ZrO_2 powder and for deep chemical interaction of the components in the reaction mixture resulting in hydroxyapatite phase. As a blank experiment we conducted SPS-RS at 900 and 1000 °C to evaluate the depth of reaction between the components during SPS-RS.

Table 1. Characteristics of ceramics based on ZrO_2.

No.	Sample composition	P, MPa	$T_{sintering}$, °C	Compressive strength, MPa	$S_{spec.}$ (BET), m^2/g
1	IM (80% ZrO_2 + 15% RM)	22.3	900	9	0.68
2	IM (80% ZrO_2 + 15% RM)	22.3	1000	373	0.12
3	IM (80% ZrO_2 + 15% RM)	22.3	1300	400	0.12
4	IM (80% ZrO_2 + 15% RM)+TC 2%	22.3	1300	362	0.24
5	IM (80% ZrO_2 + 15% RM) + TC 5%	22.3	1300	242	0.25
6	IM (80% ZrO_2 + 15% RM)+ TC 10%	22.3	1300	150	0.29
7	IM (80% ZrO_2 + 15% RM) + TC 15%	22.3	1300	47	0.35

According to XRD (Fig. 1), solid composition depends on SPS temperature and includes zirconium oxide in monocline and(or) tetragonal modifications, containing calcium phosphates in its bulk. SPS treatment of reaction mixture with ZrO_2 at 900, 1000 and 1300 °C provides formation of crystalline HAP, while the phase composition of the samples is identical (Fig. 1, curves 1 and 2). However, diffraction peak intensities are obviously different, in particular, ceramics obtained at 1300 °C possesses higher crystallinity of the phases (Curve 2) compared to 900 and 1000 °C samples (Curve 1). It is noteworthy that adding the poreforming agent does not change the final ceramics' composition as found from XRD patterns of samples 3-5 that are identical and correspond to Curve 2.

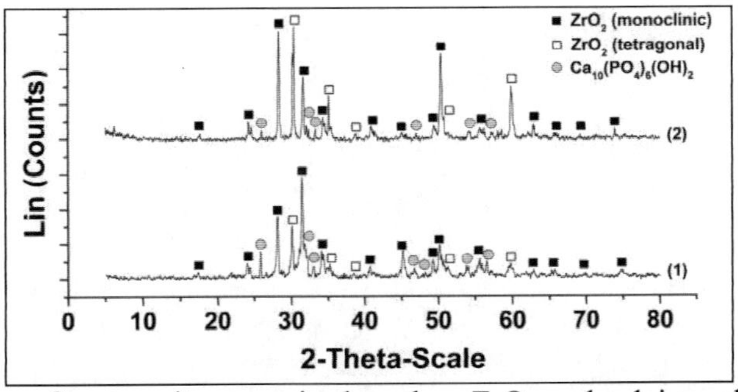

Figure 1. XRD patterns of composite ceramics based on ZrO_2 and calcium phosphates obtained via SPS-RS: (1) – samples No. 1, 2; (2) – samples No. 3, 4, 5, 6, 7. Sample description is given in Table 1.

Investigations of surface morphology showed that microstructure of consolidated ZrO_2 sample is represented by agglomerates of spherical particles with mean diameter 1-1.5 μm (Fig. 2). Also there are melted, shapeless units that somehow stick together spherical particles. According to EDX, particles consist of ZrO_2 with small amount of calcium and phosphorus (Fig. 2b, area 1, spectrum 1 and Table 2). Shapeless regions, which are observed mainly inside pores, composed of calcium phosphates with little zirconium content (Fig. 2b, area 2, spectrum 2 and Table 2). Mixed composition, obtained by energy-dispersive spectra and indicated in the table – is the result of scattering from particles that are close to the imaging area.

Table 2. Element composition of ZrO_2 composite ceramic obtained by SPS-RS using carbon template (sample No. 6, table 1). Spectrums on Fig.2.

No.	Element composition, wt. %					
	Zr	Ca	P	Mg	O	C
1	49.21	7.26	0	1.74	34.59	7.20
2	61.27	3.81	1.46	1.64	14.24	17.58
3	56.71	3.16	1.38	1.85	21.27	15.63
4	15.95	16.22	10.53	4.79	33.63	18.87
5	6.17	14.55	12.68	3.50	46.69	16.40

SEM images of obtained composite ceramics clearly show the presence of macroporous formations with pore diameter ~5-10 μm (Fig. 2b). In order to structure porous framework of the ceramics we suggested to introduce poreforming agent (TC) into the initial system subjected to sintering with its consequent removal. Prospect of such approach was confirmed by us earlier on the example of synthesis of highly porous bioinert silicate ceramics [9]. Suggested method allows controlling pore size and volume in the SPS ceramics. However, such solution requires additional stage of thermooxidative treatment of final samples. It is caused by SPS operation procedure, which is carried out in vacuum in order to avoid burning of the materials. Oxygen-free gaseous medium prevents removal of TC. Thermooxidative treatment of consolidated samples solves the mentioned problem.

Intrusion curves presented on Fig. 3 indicate the influence of carbon additive on sample's microstructure. First, the increase of porous volume depending on quantity of introduced TC is observed (2, 5, 15 wt.%). Second, material with various pore size distribution is formed. Sample obtained without carbon additive has micro- and mesopores that is reflected on intrusion curves in high pressure region (Fig. 3a, curve 1). and also on differential curve (Fig. 3b, curve 1) within the range 5-50 nm, and also a number of macropores sized ~500 nm. Formation of this type of pores is connected primarily due to packing of fine particles of sintered powder. Volume of such pores depends directly on the degree of interconnectivity of particles, shape change (deformation or grain growth), density of their mutual packing and also on the sintering parameters.

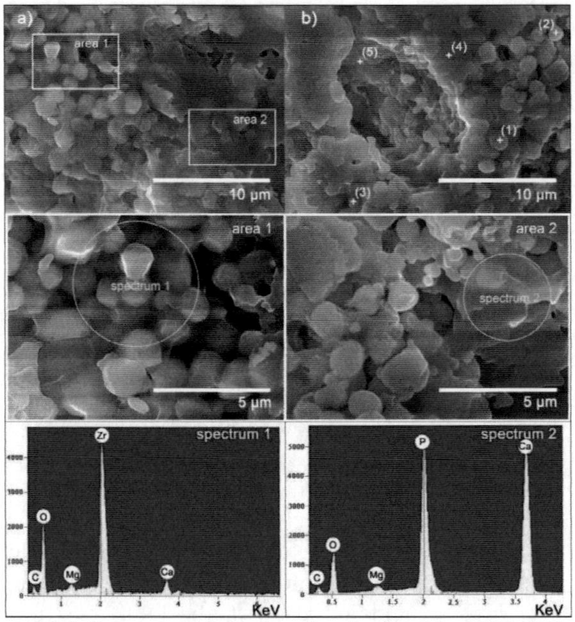

Figure 2. SEM-images of ZrO_2 composite ceramic obtained by SPS-RS using carbon template (sample No. 6, table 1).

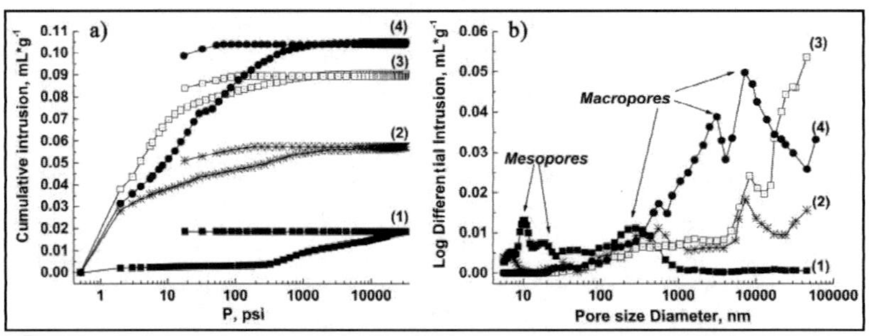

Figure 3. Cumulative and differential mercury intrusion curves of ZrO_2 ceramic samples, obtained via SPS-RS at 1300 °C: (1)-IM without carbon (sample 3); (2)-IM+2 wt.% TC (sample 5); (3)-IM+5 wt.% TC (sample 6); (4)-IM+15 wt.% TC (sample 7). Sample's description is given in Table 1.

Introducing of pore forming agent leads to eliminating of micro- and mesopores (Fig. 3b) that results in reduction of specific surface area (Table 1). However, large macropores are formed with size 1 µm and above (Fig. 3b).

Another plausible reason of variety of pore sizes in ceramic samples, in particular, synthesized without TC, is the consequences of chemical reaction according to reaction equations 1 and 2. During interaction of RM there are a lot of volatile products formed. While those gases are liberated in the process of sintering formation of defects in the volume of ceramics can possibly occur. Such defects are voids (pores) in the volume of the material and have various size and shape. One can reasonably suggest that surface of such pore type is enriched by HAP that cover pore's inner surface (Figure 2b).

In addition, one should note that quality of ceramics of any type besides structural characteristics is defined by mechanical parameters. Suggested in present work method provides formation of ZrO_2 ceramics with bioactive phosphate basis of macroporous structure with wide pore size range from 1 to 100 µm and more, high mechanical strength, compressive strength reaches 400 MPa (Table 1). While the sintering cycle does not last longer than 10 minutes of heating and holding at 1300 °C. However, the suggested SPS-RS method to control porous structure of the ceramics is plausible but at the expense of mechanical strength, which linearly depends on the amount of poreforming agent (TC).

IOP Conf. Series: Materials Science and Engineering **307** (2018) 012031 doi:10.1088/1757-899X/307/1/012031

4. Conclusions

In present work an original way of synthesis of biocompatible ceramic material based on zirconium dioxide via SPS-RS is presented. Obtained ceramics possess unique set of structural and mechanical characteristics: wide range of macropore size, from 1 to 100 μm and more, compressive strength reaches 400 MPa. We have shown for the first time that chemical interaction between $CaCO_3$ and $(NH_4)_2HPO_4$ (reaction mixture) can be caused by spark plasma with formation of biocompatible phases HAP "in situ" during sintering of ZrO_2 ceramics. It is supposed that volatile products of reaction are the reason of structural defects as macropores, on the surface of which bioactive phase is formed in a thin layer. The effectiveness of using poreforming agent (carbon additive) to organize macroporosity of ceramics along with preserving its high mechanical strength has been shown. It was established by structural analysis (BET and mercury porosimetry) that pore size and volume depend on the quantity of introduced poreforming agent, which avoids presence of micro- and mesopores, however ensures formation of large macropores when template content is higher than 2 wt.%.

Presented ZrO_2 ceramics is characterized by a number of unique properties: chemical, corrosive and mechanical stability, developed macroporous structure that is similar to bone tissue texture and is able to execute its main functions, high mechanical strength, which is responsible for durability, and also presence of biocompatibility with living tissues of the organism and stimulates the osteogenesis processes, all these make ceramics of this type very prospective solution for medical practice.

References

[1] Dubok V A 2001 Bioceramics - Yesterday, Today, Tomorrow *Powder Metall. Met. Ceram.* **39** 381–94

[2] Medkov M A, Rudnev V S, Grishchenko D N, Djuizen I V, Lukiynchuk I V, Papynov E K, Portnyagin A S and Gridasova E A 2016 Nano-Sized Calcium-Phosphate Powders and Glass-Ceramic Coatings for Medical Purposes *Bioglass: Properties, Functions and Applications* ed L Cannon (Nova Science Publisher, Inc) pp 55–96

[3] Izhizaki K, Komarneni S and Nanko M 1998 *Porous materials* vol 15, ed R Ford (Springer US) p 240

[4] Lashneva V V., Shevchenko A V. and Dudnik E V. 2009 Bioceramic based on zirconium dioxide *Glas. Ceram.* **66** 140–3

[5] Hu C, Li F, Qu D, Wang Q, Xie R, Zhang H, Peng S, Bao Y and Zhou Y 2014 Developments in hot pressing (HP) and hot isostatic pressing (HIP) of ceramic matrix composites *Advances in Ceramic Matrix Composites* ed I M Low (Woodhead Publishing Limited) pp 164–89

[6] Tokita M 2013 Spark Plasma Sintering (SPS) Method, Systems, and Applications *Handbook of Advanced Ceramics: Materials, Applications, Processing and Properties* ed S Somiya (Elsevier Inc.) pp 1149–78

[7] Li W and Gao L 2000 Rapid sintering of nanocrystalline ZrO_2(3Y) by spark plasma sintering *J. Eur. Ceram. Soc.* **20** 2441–5

[8] Safronova T V., Putlyaev V I, Ivanov V K, Knot'ko A V. and Shatalova T B 2016 Powders Mixtures Based on Ammonium Pyrophosphate and Calcium Carbonate for Preparation of Biocompatible Porous Ceramic in the $CaO–P_2O_5$ System *Refract. Ind. Ceram.* **56** 502–9

[9] Papynov E K, Mayorov V Y, Portnyagin A S, Shichalin O O, Kobylyakovt S P, Kaidalova T A, Nepomnyashiy A V., Sokol'nitskaya T A, Zub Y L and Avramenko V A 2015 Application of carbonaceous template for porous structure control of ceramic composites based on synthetic wollastonite obtained via Spark Plasma Sintering *Ceram. Int.* **41** 1171–6

Acknowledgements

This study was supported by grant for young scientists from the President of the Russian Federation (grant MK-177.2017.3).

IOP Conf. Series: Materials Science and Engineering **307** (2018) 012032 doi:10.1088/1757-899X/307/1/012032

Analysis of Deformation and Equivalent Stress during Biomass Material Compression Molding

Guiying Xu [1]*, Hetao Wei, Zhien Zhang ,Shaohui Yu[2]*, Congzhe Wang, Guowen Huang[3]*

1College of chemistry and chemical engineering of Chongqing university of Technology, Chongqing 400054,China

E-mail: xgy092500@126.com[1];yu1838249176@outlook.com[2];hgw1978@cqut.edu.cn[3]

Abstract. Ansys is adopted to analyze mold deformation and stress field distribution rule during the process of compressing biomass under pressure of 20Mpa. By means of unit selection, material property setting, mesh partition, contact pair establishment, load and constraint applying, and solver setting, the stress and strain of overall mold are analyzed. Deformation and equivalent Stress of compression structure, base, mold, and compression bar were analyzed. We can have conclusions: The distribution of stress forced on compressor is not completely uniform, where the stress at base is slightly decreased; the stress and strain of compression bar is the largest, and stress concentration my occur at top of compression bar, which goes against compression bar service life; the overall deformation of main mold is smaller; although there is slight difference between upper and lower part, the overall variation is not obvious, but the stress difference between upper and lower part of main mold is extremely large so that reaches to 10 times; the stress and strain in base decrease in circular shape, but there is still stress concentration in ledge, which goes against service life; contact stress does not distribute uniformly, there is increasing or decreasing trend in adjacent parts, which is very large in some parts.
in constructing both.

1. Introduction

Because of loose texture and small bulk density, biomass raw material contains the shortcomings of small density while high collection as well as storage and transportation cost. Besides, the difficult scale and efficient utilization and poor economic benefit are the major elements restricting the pace of turning biomass into commodity energy. Among the technologies developing and utilizing biomass energy, biomass solidification forming technology can process the low-grade biomass raw material into efficient and clean briquettes fuel, which solves the problems of high biomass storage and transportation cost and greatly enhances the utilization efficiency of biomass resources [1,2,3,4]. The use of solid after molding is so extensive that, it can be used as the raw material of living burning [5,6,7], thermal power generation[8], and gasification[9]; so it plays important supporting role in the development of entire biomass energy industry. Since biomass is the only storable renewable resource, the energy density of its raw material after compression molding is equal to that in intermediate coal, whose combustion characteristics have been obviously improved, such as none black smoke, less flying ash and complete burning. Moreover, biomass briquette fuel can be deeply processed into mechanism charcoal, which carries the advantages of regular shape, high strength, more pores, easy for combustion, burning resistance and pollution free. Based on convenient storage and transportation and less loss, briquette fuel is the renewable energy source easy for commercialized production and sales.

Existing biomass molding equipments also expose serious problems during operation, which restricts the scale development of biomass briquette industry. Therefore, to optimize existing biomass molding equipments, improve equipment productivity, reduce ton energy consumption and increase service life are the key issues need to be solved urgently. By virtue of Ansys software, this paper implements nonlinear contact statics simulation of molding process, which uncovers mold stress and strain distribution rule during molding process and provides theoretical reference for optimizing equipment structure.

1.1. Establishment of Finite Element Model

1.1.1. Establishment of Model. Based on taking distillers' grain as compression fuel, its stress and deformation distribution during compression were researched in this experiment researches. The theoretical model in this research totally adopts the shape and size that are the same as actual model. Compression process indeed is a nonlinear process; however, the process of nonlinearity is so complex that this simulation regards it as a small deformation linear issue. Due to the symmetry of structure and loaded load, axial symmetry is applied to modeling. The model in this analysis originates from Ansys Workbench modeling and the model is shown in Fig.1. Since the establishment is 1/2 of model section, it is axial symmetry model. Elastic Modulus of Distillers' Grain and Distillers' Grain 16000 and 200000 Respectively. Poisson's Ratio is 0.3.

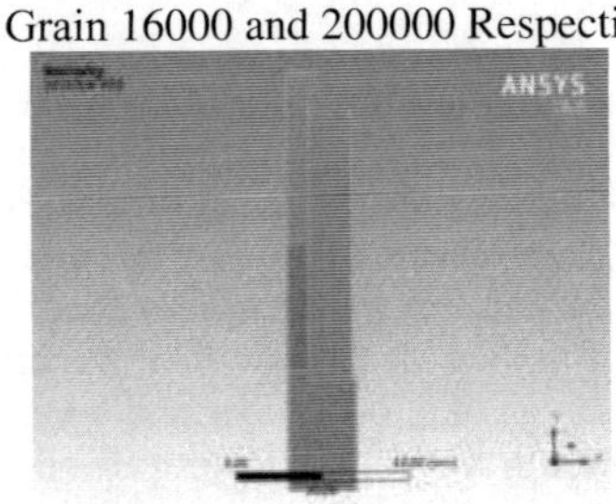

Fig 1. Modeling

Fig 2. Mesh Generation

1.1.2. Determination of Material Parameter and Property

Table 1. Material Parameter.

Material	Elastic Modulus/MPa	Poisson's Ratio
Distillers' Grain	16000	0.3
Distillers' Grain	200000	0.3

1.1.3. Unit Selection and Mesh Generation. This analysis applies Ansys Workbench and mesh module for mesh generation with adopting two-dimensional 8-node plane 183. Because PLANE183 carries quadratic displacement function, it can be better utilized in irregular mesh generation. This unit contains 8 nodes and each has 2 freedom degrees with translation toward directions x and y. This unit can be used as both plane unit (plane stress, plane strain and generalized plane strain) and axial symmetry unit. This unit possesses the abilities of plasticity, creep, stress rigidity, large deformation and large strain. In addition, its capability of force-displacement mixture formula can simulate the deformation approaching to incompressible elastoplastic material with supporting initial stress option.
The module and mesh are divided by using WB mesh so as to partition physical field as Mechanical; mesh relevance =100, and surface mapping mesh generation is adopted. The whole is shown in Fig. 2.

1.1.4. Establishment of Contact Pair, Constraint and Load. During the process of modeling analysis, because mold and distillers' brain both are solid with small displacement, it is available to hypothesize that the relative displacement is unchanged, which means there is no friction. As shown in Fig.3, 5 contact pairs are established in contact state of Frictionless, which is no frictional contact. In the entire analysis, there only is pressure forced on compression bar in the additional load, but the base is fixed, so the adopted constraint is fixed constraint. Displacement constraint is established as shown in the place A of Fig. 4. Pressure is established, whose strength at the place B in picture is 20MPa. Establishment boundary conditions and load are specifically as follows:

IOP Conf. Series: Materials Science and Engineering **307** (2018) 012032 doi:10.1088/1757-899X/307/1/012032

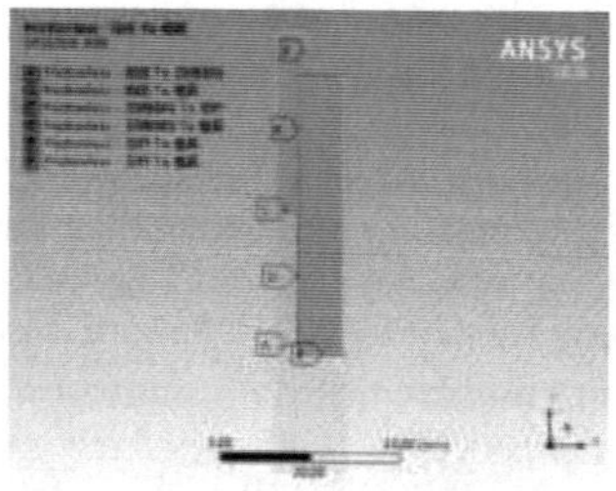

Fig 3. Contact Pair Setting Diagram

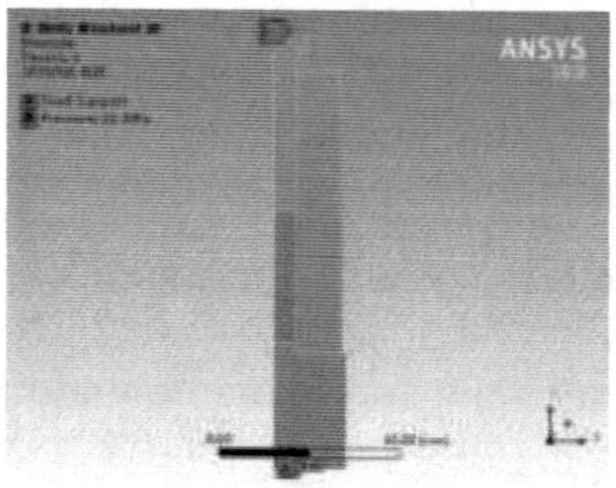

Fig 4. Constraint Setting and Pressure Applying Diagram

1.2. Statics Analysis Result

The unit selection, material property setting, mesh generation, contact pair establishment, load and constraint applying, solver setting and other earlier preparation of analysis are completed based on above settings. The examination of above set constraint, load and relevant solving parameters is required before solving. Then the following result analysis is acquired:

1.2.1. Overall Deformation and Overall Equivalent Stress. Fig. 5 is the overall deformation of mold. According to it, the mold deformation is small while the displacement of compression bar and compressed fuel is larger. Since displacement is the manifestation of overall deformation situation, the lower part of fuel will definitely be lower than that of upper part with the compaction of fuel, but the displacement of compression bar is the largest. As shown in Fig. 5, overall equivalent pressure diagram indicates that, the pressure on compression bar is the largest, followed by compressor. The stress of base is shown in the Fig.6, the pressure at the contact place of compressor and base is larger while other places is smaller so that it decreases in the form of semi circle.

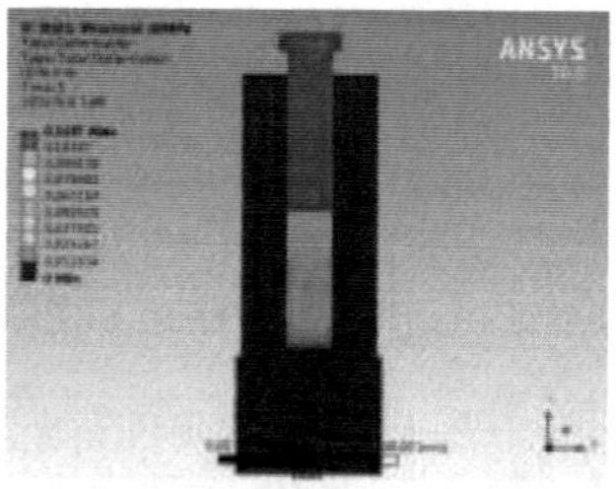

Fig 5. Overall Deformation Diagram under 20MPa

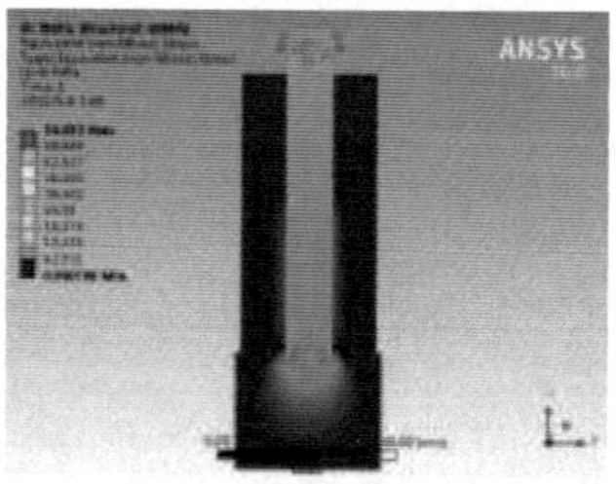

Fig 6. Overall Stress Diagram under 20MPa

1.2.2. Deformation of Compression Structure and Compression Structure Equivalent Stress. Since the distributions of stress and deformation sizes in different parts are various, the single overall cloud picture cannot display the complete variation situation of each part. Thus, it is necessary to separately analyze the stress of each part so as to obtain more accurate analysis result. The stress of each part is analyzed. Fig.7 is the deformation diagram of compression fuel under 20MPa. As shown in the fig.7, during the compression process, the deformation of the end closer to base is smaller while that in the end closer to compression bar is larger; the deformation at horizontal position is the same. Fig. 8 is the stress diagram of compression fuel under 20MPa. It shows that, the stress forced on fuel is not in uniform distribution, where the slight non-uniform appeared in the place close to base, but the overall difference is not significant without great influence on briquettes fuel; so it can be ignored in practical use.

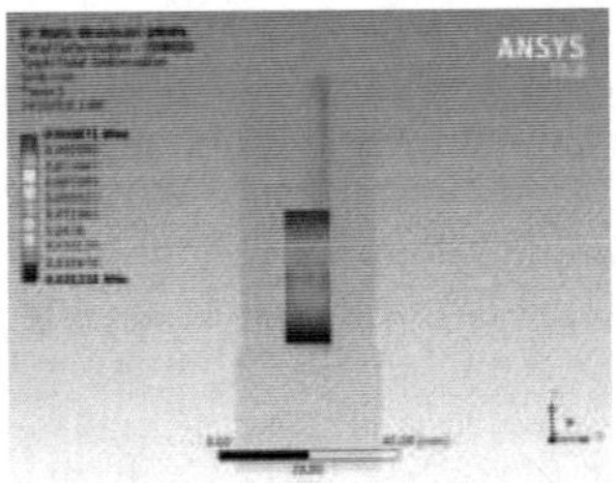

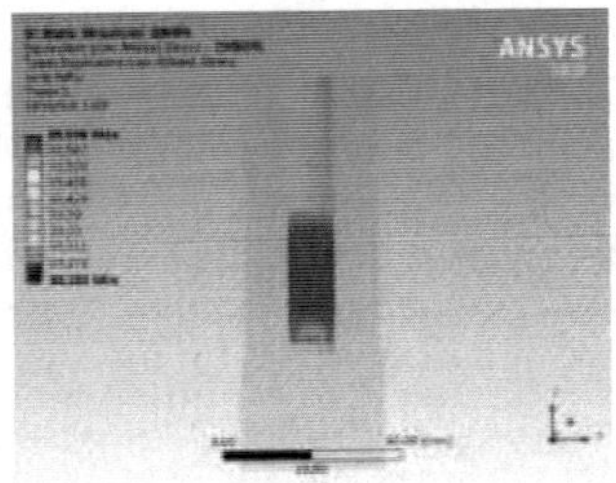

Fig 7. Compression fuel Deformation Diagram under 20MP under 20MPa **Fig 8.** Compression fuel Stress Diagram

1.2.3. Base Deformation and Base Equivalent Stress. Fig. 9 is the deformation diagram of base under 20MPa. It shows that, in base deformation that in ledge of base is obviously larger than the overall part. Meanwhile, the deformation in base decreases in the form of semi circle. Fig. 10 is the diagram of base stress under 20MPa. It shows that, for base, both stress and deformation present the largest at the part close to compressor, and then decreases in the form of semi circle. What needs to come into notice is that, the stress concentration phenomenon is generated at base ledge, which greatly influences the service life as well as reliability of this part.

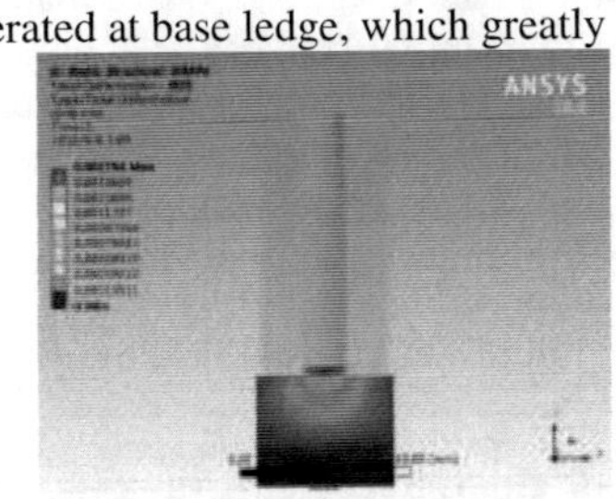

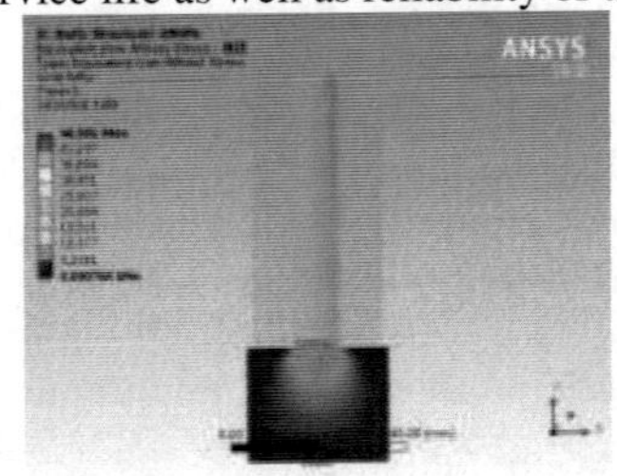

Fig 9. Base Deformation Diagram under 20MPa **Fig 10.** Base Equivalent Stress under 20MPa

1.2.4. Mold Deformation and Mold Equivalent Stress. As shown in Fig. 11, it can be obviously concluded that, the adjacent place of compression bar and compressor increases obviously. However, since mold is rigid body, the change of numerical value is not obvious. As shown in Fig. 12, the stress forced on mold at compressor part is obviously higher than that in upper part. The adjacent part of stress variation is in the contact part of compression and compression fuel. The huge variation of stress in this place has direct influence on main mold.

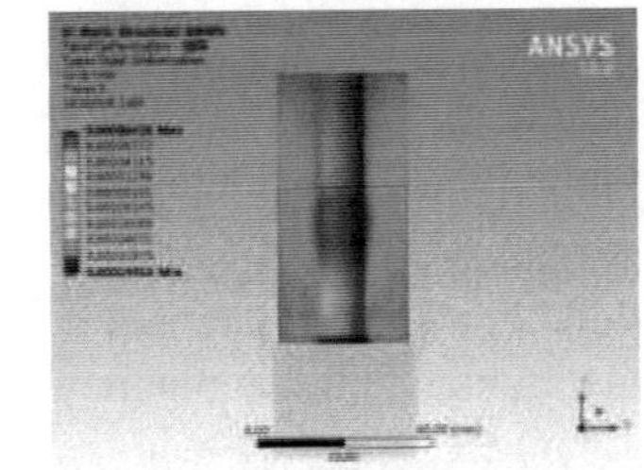

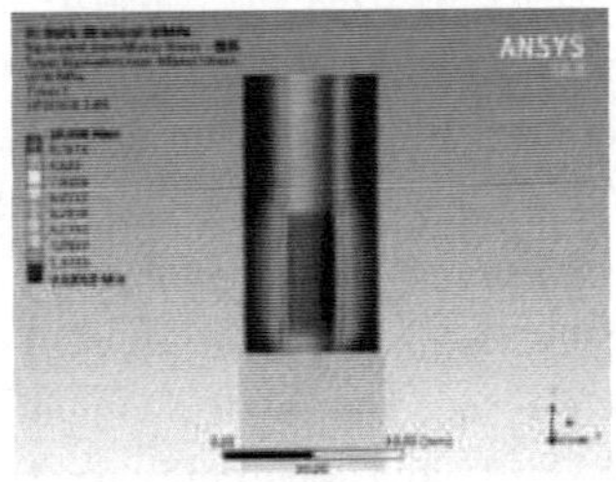

Fig 11. Main Mold Deformation Diagram under20MPa **Fig 12.** Main Mold Stress Diagram under 20MPa

2. Summary

Due to the existence of extrusion force and friction, the large stress concentration in molding mold is the major deficiency of cylinder molding equipment. The data obtained from simulating and analyzing mechanics during biomass molding, deformation process, stress and strain distribution, and comparing existing experimental conclusions are reliable. Meanwhile, it provides better data as well as method for further research of biology molding mechanism. Based on two pressures, this paper analyzes the stress of the whole and each part under different pressures. The following conclusions are obtained: the distribution of stress forced on compressor is not completely uniform, where the stress at base is slightly decreased; the stress and strain of compression bar is the largest, and stress concentration my occur at top of compression bar, which goes against compression bar service life; the overall deformation of main mold is smaller; although there is slight difference between upper and lower part, the overall variation is not obvious, but the stress difference between upper and lower part of main mold is extremely large so that reaches to 10 times; the stress and strain in base decrease in circular shape, but there is still stress concentration in ledge, which goes against service life; contact stress does not distribute uniformly, there is increasing or decreasing trend in adjacent parts, which is very large in some parts.

Since the stress sizes on mold in different parts are various, particularly that the stress concentration greatly influences mold service life, the suggestions are proposed aiming at above issues: with respect to cylinder fuel molding, because the distributions of stress and strain under different pressures are similar, this analysis can be referred under other stress; in stress concentration part, redesign of mold, heat treatment of material or adoption of better material are necessary; apply different materials to compression bar and other parts because the stress and train in compression bar are the largest in entire system, or the increase of compression bar appropriately is available; it is available to decrease circular radius or adopt other cheaper material for main mold; it is suggested to use cladding material or heat treatment to the internal contact surface of mold so as to further improve material performance, reduce pitting caused by alternating stress during continuous production, and improve reliability as well as service life.

3. Acknowledgments

This work was funded by the National Natural Research Program of China (Program No. 21406023）, Chongqing People's Livelihood Science and Technology Innovation Program（cstc2015shmszx20005, the Scientific and Technological Research Program of Chongqing Municipal Education Commission (No. KJ1400901)，and the Scientific Research Foundation and innovation team of Chongqing university of technology.

4. References

[1] M.G. Montiano , A.M. Fernández , E. Díaz-Faes. Tar from biomass/coal-containing briquettes. Evaluation of PAHs.Fuel, 154 (2015) 261-267

[2] S.C. Bhattacharya , M. Augustus Leon , Md. Mizanur Rahman． A study on improved biomass briquetting ［J］.Energy for Sustainable Development,6(2002)67-71

[3] M.Mohon Roy , K. W. Corscadden .An experimental study of combustion and emissions of biomass briquettes in a domestic wood stove. Applied Energy ,99 (2012) 206–212

[4] M.J. Stolarski, S. Szczukowski , J. Tworkowski,etal. Comparison of quality and production cost of briquettes made from agricultural and forest origin biomass. Renewable Energy 57 (2013) 20-26

[5] F. O.Okey. Evaluation of the effect of palm oil mill sludge on the properties of sawdust briquette. Renewable and Sustainable Energy Reviews,52(2015)1749-1758

[6] A.K., H. Yi, V. Puri. Fundamental mechanical properties of ground switchgrass for quality assessment of pellets. Powder Technology ,283 (2015) 48-56

[7] S. Araújo, M. A. V. Boas, D. M. Neiva, etal. Effect of a mild torrefaction for production of eucalypt wood briquettes under different compression pressures. Biomass and Bioenergy,90 (2016) 181-186

[8] C. Sakkampang, T. Wongwuttanasatian.Study of ratio of energy consumption and gained energy during briquetting process for glycerin-biomass briquette fuel. Fuel,115 (2014) 186-189

[9] S.Y. Luo, C.J. Yi, Y.M. Zhou. Direct reduction of mixed biomass-Fe_2O_3 briquettes using biomass-generated syngas.Renewable Energy, 36 (2011) 3332-3336

IOP Conf. Series: Materials Science and Engineering **307** (2018) 012033 doi:10.1088/1757-899X/307/1/012033

Interaction of chlorine adatom with lithium adatom on opposite sides of graphene

Meifeng Dong[1] and Xinxiang Song[1]

1 Colllege of Science, China University of Petroleum (East China), Qingdao 266555, China
Email: dongmf@upc.edu.cn

Abstract. Using first principles density functional theory, we study chlorine (Cl) and lithium (Li) adatoms co-adsorption properties on different sides of single-layer graphene sheet. Cl adatom gathers more charge and the adsorption system become more stable. But due to the effects of graphene, the interaction between Li and Cl adatoms is weakened and the lost charge of Li adatom is not all transferred from graphene to Cl adatom.

1. Introduction

Having superior properties, graphene can be used in various carbon based nanodevices [1-3]. The 2s, $2p_x$ and $2p_y$ atomic orbitals on each carbon of graphene hybridize to form strong covalent sp^2 bonds with its three neighbouring carbons and the remaining $2p_z$ orbital on each carbon overlaps to form a big delocalized π bond. Due to this unique structure, graphene has ultrafast carrier mobility and extremely high electrical conductivity [4]. But the forming of the big π bond reduces the activity of graphene, meaning that pure graphene has inertia in some way. Graphene has other outstanding features such as the wide open double-side surface. Therefore, it can be functionalized by adsorbing atoms to produce novel nanomaterials [5-7].

In our paper, Li and Cl adatoms co-adsorbed on different sides of graphene are studied. And the adsorption geometric structure, electronic properties are reported. The effects of graphene on the interaction between Li and Cl adatoms are discussed.

2. Methods

All calculations have been performed within DFT under the generalized gradient approximation [5]. The kinetic energy cutoff was taken as 550 eV. The reciprocal space was sampled using a 9×9×1 k-point grid during the relaxation process. The standard of convergence during the relaxation process is that the force on each atom was less than 0.01 eV/ Å. The perpendicular distance between the neighbouring graphene layers was taken 25Å.

Figure 1 represents a 5×5 supercell of graphene model with a single adatom adsorbing at three different sites: the T-site, B-site and H-site.

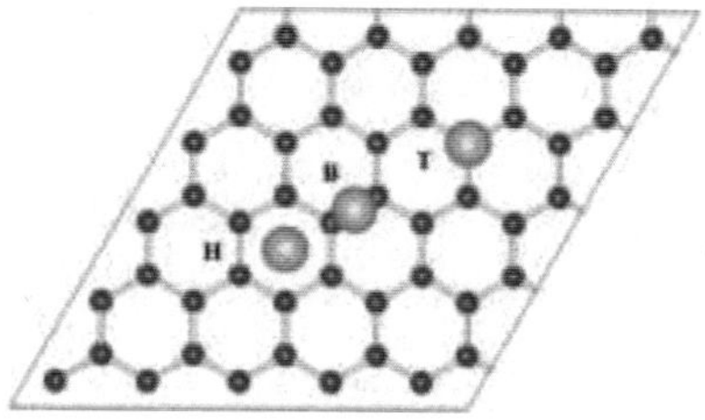

Figure 1. (Color online) Top view of the three adsorption sites.

3. Results

Table 1 lists the adsorption results of a single Li and Cl adatoms.

Table 1. Adsorption results of a single Li and Cl atom adsorbed on graphene.

adatom	site	ΔE(eV)	h(Å)	Δh_c(Å)	Δq(e)
	H	-1.52	1.74	0.01	-0.92
Li	B	-1.22	1.92	0.07	-0.92
	T	-1.19	1.93	0.10	-0.90
	T	-1.12	3.08	0.01	0.51
Cl	B	-1.11	3.10	0.01	0.52
	H	-1.09	3.16	0.01	0.51

The adsorption energy is defined as [5]:

$$\Delta E = E_{total} - E_G - E_a \quad (1)$$

where ΔE, E_{total}, E_G and E_a are the adsorption energy, the energy of the adsorption system, the energy of pristine graphene, and the energy of the single adatom, respectively [5]. Among the parameters, h represents the adsorption height of adatom. Δh_c is represents the maximum deviation distance of the carbon atoms.

For Li adatom, the adsorption system have the minimum energy at H-site. Li adatom almost transfers its whole outmost valence electron to graphene. These results are consistent with previous reports [5-8]. There is negligible difference among the three sites for Cl adsorbed on the surface of graphene. The adsorption height is in excess of 3Å, far beyond the length of the covalent bond that is forming between single Cl and single carbon atoms. The charge that is transferred from graphene to Cl adatom is relatively less, only about 0.5e. All the results are in agreement with previous reports too [9-10].

Table 2 displays the results of Li/Cl adatoms co-adsorbed on opposite sides of graphene. The adsorption energy is:

$$\Delta E = E_{total} - E_G - E_{Li} - E_{Cl} \quad (2)$$

Table 2. Adsorption properties of Li/Cl co-adsorbed on different sides of graphene.

Adatoms	Site	ΔE(eV)	h(Å)	Δq(e)	Δh(Å)
.Li/Cl	H/T	-3.49	1.74/2.81	-0.90/0.69	0.08
	H/B	-3.48	1.75/2.84	-0.90/0.69	0.04
	H/H	-3.44	1.76/2.93	-0.90/0.70	0.01
	B/T	-3.23	1.92/2.83	-0.91/0.68	0.09
	B/B	-3.22	1.95/2.86	-0.91/0.68	0.08
	B/H	-3.15	1.96/2.94	-0.91/0.70	0.06
	T/T	-3.24	1.88/2.80	-0.91/0.65	0.09
	T/B	-3.20	1.93/2.86	-0.91/0.67	0.10
	T/H	-3.12	1.96/2.94	-0.91/0.69	0.10

ΔE, E_{total}, E_{G}, E_{Li} and Ea are the adsorption energy, energy of adsorption system, total energy of pure graphene, the total energy of a single Li adatom and the total energy of single Cl adatom, respectively. The adsorption system has the minimum and maximum energy when Li/Cl adatoms at the H/T and T/H sites. The adsorption height and the charge loss of the Li adatom change very little compared with the situation that only single Li adatom adsorbed; while Cl adatom receives more charge from graphene, it is about 0.17 e and its adsorption height decreases about 0.2 Å.

Li adatom transfer its outmost valence electron to graphene and then Cl gets charge more easily under the co-adsorbed situation. Therefore, the interaction among the adsorbates and substrate is enhanced. The adsorption energy is larger than the sum of adsorption energy of single Li and Cl adsorption systems (as shown in table 1).

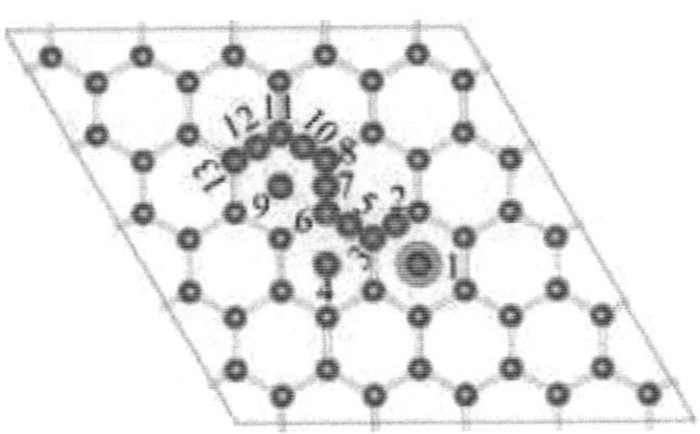

Figure 2. Top view of Cl adatom (red balls) adsrobs on different sites on G-Li adsorption system.

Table 3 summarizes the results when Li adsorbs at H-site and Cl adsorbs on another side of graphene at different sites as figure 2 depicts. Among the parameters, d_{xy} represents the distance between Li and Cl adatoms parallel to the surface of graphene. With the increase of the horizontal distance between the two adatoms, the adsorption energy generally shows a slightly decreasing trend. Among the thirteen sites we considered, the largest changing of the adsorption energy is only about 0.15eV when Cl adatom adsorbed at 3rd and 10th sites respectively. Variations of the adsorption height and the charge transfer of adatoms are very slight too. The insensitive adsorption results with the variation of the horizontal distance between the adatoms imply that graphene can shield the interaction between the adatoms due to the formation of big π bonds.

Table 3. Adsorption properties of Li/Cl co-adsorbed on different sides of graphene with different horizontal distance.

Adatom	No.	Site	ΔE(eV)	h(Å)	Δq(e)	Δh_c(Å)	d_{xy}(Å)
Li/Cl	3	H/T	-3.49	1.74/2.81	-0.90/0.69	0.08	1.42
	6		-3.43	1.74/2.83	-0.90/0.66	0.05	2.85
	8		-3.39	1.74/2.85	-0.90/0.66	0.04	3.77
	11		-3.35	1.74/2.89	-0.90/0.67	0.03	5.14
	13		-3.35	1.74/2.88	-0.90/0.66	0.04	5.71
	2	H/B	-3.48	1.74/2.84	-0.90/0.69	0.04	1.24
	5		-3.46	1.74/2.84	-0.90/0.69	0.05	2.14
	7		-3.40	1.74/2.87	-0.90/0.68	0.02	3.27
	10		-3.36	1.74/2.89	-0.90/0.68	0.02	4.45
	12		-3.34	1.74/2.91	-0.91/0.67	0.02	5.38
	1	H/H	-3.44	1.76/2.93	-0.90/0.70	0.01	0
	4		-3.40	1.74/2.93	-0.90/0.69	0.02	2.47
	9		-3.33	1.74/2.97	-0.90/0.68	0.02	4.28

4. Conclusion

The characteristic of Li/Cl adatoms co-adsorbed on different sides of graphene are studied. The adsorption system become stronger in the case of co-adsorption. However, Cl adatom does not collect all the charge that Li adatom transferred to graphene. And the adsorption energy insensitive to the variation of the horizontal distance between Li and Cl adatoms. It imply that the Li and Cl adatoms bind with graphene independently. This is due to the formation of the big π bond makes graphene has the obvious inertia and the electrostatic screening effect. Graphene can weaken the interaction and hamper the charge transfer between the two adatoms which co-adsorbed on the opposite sides of graphene.

References

[1] Geim A K and Novoselov K S 2007 The rise of graphene *Nat. Mater.* **6** 183-191.

[2] Stankovich S, Dikin D A, Dommett G H B *et al* 2006 Graphene-based composite materials *Nature* **442** 282-286.

[3] Fiori G, Bonaccorso F, Iannaccone G et al 2014 Electronics based on two-dimensional materials Nature 9 768-779.

[4] Matthew J. A, Vincent C T and Richard B K 2010 Honeycomb carbon: a review of graphene Chem. Rev 110 132-145.

[5] Song X, Sun Y, Dong M et al 2015 Effects of Cl adatom on Na-decorated graphene Journal of Physics D: Applied Physics 48. 22 225304.

IOP Conf. Series: Materials Science and Engineering **307** (2018) 012033 doi:10.1088/1757-899X/307/1/012033

[6] Chan T K, Neaton J B and Cohen M L 2008 First principles study of metal adatom adsorption on graphene Phys. Rev. B 77. 23 235430.

[7] Jin K H, Choi S M and Jhi S H 2010 Crossover in the adsorption properties of alkali metals on graphene Phys. Rev. B 82.3033414.

[8] Kiranpreet K, Jasmeet K, Bindiya A, and Sahoo B K 2014 Emending thermal dispersion interactions of Li, Na, K, and Rb alkali metal atoms with graphene in the dirac model Phys. Rev. B 90. 24 245405.

[9] Sahin H and Ciraci S 2012 Chlorine Adsorption on graphene: chlorographene J.Phys. Chem. C 116 24075-24083.

[10] Mei Y M, Lin Z, Ying W J, Fan L Z and Rong L Z 2012 Evolutionary chlorination of graphene: from charge-transfer complex to covalent bonding and nonbonding J. Phys. Chem. C 116 844-850.

Antimicrobial activity of jasmine oil against oral microorganisms

S Thaweboon[1] , B Thaweboon[1] and R Kaypetch[2]

[1] Department of Oral Microbiology
[2] Research Unit
Faculty of Dentistry, Mahidol University, Bangkok 10400, Thailand

E-mail: sroisiri.tha@mahidol.ac.th

Abstract. *Jasmine sambac* is a species of jasmine indigenous to the tropical and warm temperature regions in particular West and Southeast Asia. Essential oil extracted from the flowers of *J. sambac* has been shown to have anti-oxidant activity. However, very little information regarding antimicrobial activity especially oral microorganisms exists. Objective: To investigate antimicrobial effect of essential oil extracted from flowers of *J. sambac* against various oral microorganisms. Materials and Methods: Oral microbial strains used in the study were *Streptococcus mutans* KPSK2, *Staphylococcus aureus* ATCC 5638, *Lactobacillus casei* ATCC 6363, *Klebsiella pneumoniae* (clinical isolate), *Escherichia coli* ATCC 25922, *Candida albicans* ATCC 10231, *Candida krusei* ATCC 6258, *Candida parapsilosis* ATCC 22019, *Candida tropicalis* (clinical isolate), *Candida glabrata* ATCC 90030, *Candida pseudotropicalis* (clinical isolate) and *Candida stellatoidia* (clinical isolate). The potential of microbial growth inhibition of the oil was firstly screened by Kirby-Bauer disk diffusion method and then the minimum inhibitory concentration (MIC) was determined by agar dilution method. Results: Jasmine oil showed antimicrobial activities against *S. mutans, L. casei*, *E. coli* and all strains of Candida species with the zones of inhibition ranging from 9 to 26 mm and MIC values of 0.19-1.56 %v/v. Conclusion: Results from the present study are scientific evidence to demonstrate that jasmine oil could be employed as a natural antimicrobial agent against oral microorganisms.

1. Introduction

Several antimicrobial agents including fluoride, phenol, amoxicillin, erythromycin, tetracycline, vancomycin have been used to a great extent in dentistry. However, over use of these chemicals can result in disturbance of the oral and intestinal microflora and cause side effects such as microbial resistance, vomiting and diarrhea [1-2]. Therefore, the search for natural antimicrobial agents, which are safe for humans, is necessary.

Jasminum sambac L. (Oleaceae) is commonly known as Arabian jasmine or in Thai as Mali. It is a species of jasmine indigenous to regions in Bhutan, India, Nepal and Pakistan. It is grown in many places, especially over much of West and Southeast Asia. *J. sambac* is a small shrub or vine growing up to 1-3 m in height. It is extensively cultivated for its glamorous and gently perfumed flowers. *J. sambac* has been appraised a holy flower in Asian tradition, as it symbolizes purity, spirituality, elegant, clarity and honesty. It also presents the image of beauty of humility; a small and plain white flower that can give such sweet scent. Jasmine flowers are used for the production of perfumes, aromatizing agents, hair decorations and accessories, and ingredient in jasmine tea.

Essential oils from many plants have been used as therapeutic agents since ancient times. The essential oil from *J. sambac* or jasmine oil is used as perfume for skincare products, in anti- irritation and infection, moisturizer, anti-aging, anti-hyperpigmentation and as sun protection [3]. Previous studies have reported several phytochemicals ingredients in this plant including α-amyrin, α-terpineol, daucosterol, dotriacontanoic acid, dotriacontanol, jasminin, kaempferol, linalool, lupeol, oleanolic acid, quercitin, sambacin, sambacoside A, sambacolignoside and glucoside- sambacoside A-G as well as ursolic acid [4-5]. In the case of pharmacological activity, jasmine oil was demonstrated to have antimicrobial properties against many types of bacteria such as *Staphylococcus aureus, Staphylococcus epidermidis, Bacillus cereus, Bacillus subtilis, Enterobacter aerogenes, Escherichia coli, Salmonella typhi, Salmonella typhimurium, Shigella spp.*, and fungus such as *Candida albicans and Saccharomyces cerevisiae* [6], puerperal lactation inhibition [7], anti-oxidant [3], anti-diabetic [8], improve blood cholesterol [9], anti-viral [10], anti-acne [11], heart tonic [12], autonomic nervous system stimulant [13] and anti-tumor activities [14].

Even though ethanolic extract of *J. sambac* has been reported to have antimicrobial against many bacteria, very little information regarding activity on oral microorganisms exists. Therefore, the aim of the present study was to investigate antimicrobial effect of jasmine oil against various oral microorganisms.

2. Materials and method

2.1. *Preparation of essential oil*

Jasmine oil extracted from flowers of *J. sambac* was from Ananda Apothecary, Colorado, USA. The stock solution of the oil was prepared in 10% dimethyl sulfoxide (DMSO) and further diluted in distilled water to give the final concentration of 50% (v/v) for the use in the experiment.

2.2 *Microorganisms*

The oral microbial strains used in this study were *Streptococcus mutans* KPSK2, *Staphylococcus aureus* ATCC 5638, *Lactobacillus casei* ATCC 6363 , *Klebsiella pneumoniae* (clinical isolate), *Escherichia coli* ATCC 25922, *Candida albicans* ATCC 10231, *Candida krusei* ATCC 6258, *Candida parapsilosis* ATCC 22019, *Candida tropicalis* (clinical isolate), *Candida glabrata* ATCC 90030, *Candida pseudotropicalis* (clinical isolate) and *Candida stellatoidia* (clinical isolate). All of them strains from the culture collection of Oral Microbiology Department, Faculty of Dentistry, Mahidol University. Brain heart infusion broth supplemented with 1% yeast extract (Difco Laboratories, Detriot, MI) was used for growing all microbial strains. The suspensions of 10^8 cfu/mL of bacteria and 10^7 cfu/mL of fungi (equivalent to McFarland no. 0.5) were used for antimicrobial determination.

2.3 *Disk diffusion test*

The microbial growth inhibitory potential of the oil was determined using disk diffusion method described by Kirby-Bauer [15]. Freshly prepared microbial suspensions were spread on Mueller Hinton agar (Difco Laboratories, Detriot, MI). The 15 μL of oil (50% v/v) was dropped onto sterile paper disks (6 mm diameter). The same volume of 0.2% chlorhexidine gluconate solution served as positive control for *S. mutans, L. casei* and all strains of *Candida* spp., tetracycline for *S. aureus, K. pneumoniae* and *E. coli*. A 5% DMSO solution was used as negative control. The plates were incubated at 37 °C for 24-48 h. The zones of inhibition were measured after incubation.

2.4 *Minimum inhibitory concentration (MIC)*

The MICs of jasmine oil were determined with the agar dilution method. Mueller Hinton agar plates were prepared with serial two-fold dilutions of oil (ranging from 12.5 to 0.015 % v/v). The microbial suspensions were prepared by adjusting their density to 0.5 McFarland Standard. The 20 μL of suspension were dropped on each plate and incubated at 30 °C for 48 h.

IOP Conf. Series: Materials Science and Engineering **307** (2018) 012034 doi:10.1088/1757-899X/307/1/012034

All tests were performed in triplicate on three separate occasions

3. Results

By using disk diffusion method, jasmine oil showed antimicrobial activities against all strains of Candida species and some bacteria. The zones of inhibition ranged from 9 to 26 mm (Table 1). *S. aureus* and *K. pneumonia* exhibited resistant to the oil.

In the case of MICs, the values for all *Candida* spp. ranged from of 0.19-0.78 %v/v whereas those for susceptible bacteria were from 0.39-1.56 %v/v.

Table 1 Inhibitory effect of jasmine oil against various oral microorganisms

Microorganisms	Inhibition zone (mm)		
	oil	positive control [a]	negative [b] control
Streptococcus mutans KPSK2	23	24	-
Staphylococcus aureus ATCC 5638	-	35	-
Lactobacillus casei ATCC 6363	17	24	-
Klebsiella pneumoniae (clinical isolate)	-	23	-
Escherichia coli ATCC 25922	9	27	-
Candida albicans ATCC 10231	17	18	-
Candida krusei ATCC 6258	26	20	-
Candida parapsilosis ATCC 22019	17	17	-
Candida tropicalis (clinical isolate)	10	18	-
Candida glabrata ATCC 90030	19	20	-
Candida pseudotropicalis (clinical isolate)	13	19	-
Candida stellatoidia (clinical isolate)	20	18	-

[a] positive control for *S. mutans*, *L. casei* and all strains of *Candida* spp. was 0.2% chlorhexidine gluconate and for *S. aureus*, *K. pneumoniae* and *E. coli* was tetracycline

[b] negative control for all the test microorganisms was 5% DMSO

IOP Conf. Series: Materials Science and Engineering **307** (2018) 012034 doi:10.1088/1757-899X/307/1/012034

Table 2 Minimum inhibitory concentration (MIC) values of jasmine oil

Microorganisms	MIC (%v/v)
Streptococcus mutans KPSK2	0.39
Staphylococcus aureus ATCC 5638	>12.5
Lactobacillus casei ATCC 6363	0.39
Klebsiella pneumoniae (clinical isolate)	>12.5
Escherichia coli ATCC 25922	1.56
Candida albicans ATCC 10231	0.39
Candida krusei ATCC 6258	0.19
Candida parapsilosis ATCC 22019	0.39
Candida tropicalis (clinical isolate)	0.78
Candida glabrata ATCC 90030	0.19
Candida pseudotropicalis (clinical isolate)	0.39
Candida stellatoidia (clinical isolate)	0.19

4. Discussion

Jasmine sambac is a plant indigenous to tropical and warm temperature regions particularly West and Southeast Asia. Its flowers are strongly scented. Jasmine essential oil is extracted from flowers by solvent extraction followed by steam distillation. The main constituents found in jasmine oil are benzyl acetate, linalool, phytol, farnesol, benzyl alcohol land cresol.

Our results from screening by using disk diffusion technique revealed that *Candida spp*. had a degree of susceptibility to the oil with maximum activity on *C. krusei* (inhibition zone was 26 mm). For the bacteria, *S. mutans*, *L. casei* and *E. coli* were the three strains of test bacteria that showed susceptibility. When considering the minimum inhibitory concentration of jasmine oil against the test microorganisms, oral yeast seems to be more susceptible than oral bacteria (Table 2). The mechanism of antimicrobial action of jasmine oil is possibly due to the interruption of cell membrane synthesis, specifically membrane proteins interfering [16-17]. The hydrophobic characteristic of the oil enables it to make detachment in the membrane, providing permeability, causing leakage of cell contents and finally resulting in microbial cells death [18].

In conclusion, results from the present study are scientific evidence to demonstrate that jasmine oil could be applied as a natural antimicrobial agent for the prevention or treatment of oral microbial infection.

5. References

[1] Bidault P, Chandad F and Grenier D 2007 *J. Can. Dent. Assoc.***73** 721
[2] Maliz M, Jardim JJ and Alves LS 2010 *Braz. Oral. Res.***1** 18
[3] Abdoul-Latif F, Edou P, Eba F, Mohamed N, Ali A, Djama S, *et al* 2010 *Afr. J. Plant. Sci.* **4** 38
[4] Alrashdi AS, Salama SM, Alkiyumi SS, Abdulla MA, Hadi AH, Abdelwahab ST, *et al* 2012 *Evid. Based. Complement. Alternat. Med.* **5** 786426.
[5] Sabharwal S, Vats M, Sardana S, and Aggarwal S 2011 *I.J.P.P.S.* **1** 237
[6] Chansukh KK 2016 *MFUIC & KTCM.* **2** 38
[7] Shrivastav P, George K, Balasubramaniam N, Jasper MP, Thomas M and Kanagasabhapathy AS 1988 Aust. N. Z. J. Obstet. Gynaecol. **28** 68

IOP Conf. Series: Materials Science and Engineering **307** (2018) 012034 doi:10.1088/1757-899X/307/1/012034

[8] Upaganlawar A, Bhagat A, Tenpe CR and Yeole PG 2009 *Pharmacol.* **1** 1
[9] Kalaiselvi M and Kalaivani K 2011 *Pharmacol.* **1** 38
[10] Chiang LC, Cheng HY, Liu MC, Chiang W and Lin CC 2003 *Biol. Pharm. Bull.* **26** 1600
[11] Tsai T, Wu W, Jonathon TT and Tsai P 2010 *Food. Chem.* **119** 964
[12] Somanadhan B, Varughese G, Palpu P, Sreedharan R, Gudiksen L, Smitt UM, *et al* 1999 *J. Ethnopharmacol.* **65** 103
[13] I Naohiko, Kyoko K, Akiko S, Tagami K and Tohru F 2003 *Biosci. Biotechnol. Biochem.* **67** 1206
[14] Rahman MA, Hasan M, Hossain SMA and, Biswas NN 2011 *Pharmacol.* **1** 124
[15] Bauer AW, Kirby WM and Sherris JC 1966 *Am. J. Clin. Pathol.* **36** 493
[16] Brul B and Coot P 1999 *Int. J. Food. Microbiol.* **50** 1
[17] Sikkema J, de Bont J and Poolman B 1995 *Microbiol. Rev.* **59** 201
[18] Rath CC, Devi S, Dash SK and Mishra RK 2008 *Indian. J. Pharm. Sci.* **70** 238

IOP Conf. Series: Materials Science and Engineering **307** (2018) 012035 doi:10.1088/1757-899X/307/1/012035

The effect of laser process parameters on microstructure and dilution rate of cladding coatings

Liu Bin[1a], **Liu Heping**[12b*], **Jing Xingbin**[1c], **Li Yuxin**[1d], **Bai Peikang**[1e]
1School of Materials Science and Engineering, North University of China, Taiyuan 030051, China
2School of Materials Science and Engineering, Taiyuan University of Technology, Taiyuan 030024, China
E-Mail: peace666@126.com

Abstract: In order to broaden the range of application of Q235 steel, it is necessary to repair the surface of steel. High performance 316L stainless steel coating was successfully obtained on Q235 steel by laser cladding technology. The effect of laser cladding parameters on the geometrical size and appearance of single cladding layer was investigated. The experimental results show that laser current has an important influence on the surface morphology of single channel cladding. When the current is from 155A to 165A, the cladding coating becomes smooth. The laser current has an effect on the geometric cross section size and dilution rate of single cladding. The results revealed that with the rising of laser current, the width, height and depth of layer increase gradually. With the rising of laser current, the dilution rate of cladding layer is gradually increasing.

1. Introduction

Q235 steel is the most common low carbon structural steel. It has the advantages of low carbon content, good strength and toughness, good weldability and processability. It is often rolled into a strip or round bar, flat steel, angle steel, channel steel. It is widely used in petroleum, chemical, mechanical and marine environment[1-5]. The most important place for carbon steel flanges should be in the connection of pipes. Flange corrosion, erosion, temperature, pressure, vibration and other factors will cause corrosion and falling off in the continuous production, and even leakage problems will inevitably occur. If the leakage of flange cannot be dealt with in time, the leakage of the medium will expand rapidly, resulting in loss of material.

In order to prevent corrosion, the traditional solution is to electroplate or brush antirust oil, ion

* Corresponding author.Tel.: +86-0351-3557519
E-mail address: peace666@126.com

implantation and laser surface modification on the flange surface. Among them, the coating and substrate formed by electroplating are mechanical combination, low bonding stress and poor wear resistance. It is difficult to resist water damage, medium erosion and bubble collapse impact. The corrosion resistant layer obtained by ion implantation, laser transformation hardening and alloying is too thin and will soon be lost under the erosion of the medium[6-8]. In recent years, surface treatment of Q235 steel has become a hot topic of research. Wang B et al.[9] reported that the Q235 low-carbon steel can be hardened by plasma electrolytic borocarburizing (PEB/C) method. The microhardness of the prepared layer is up to 1450hv, much higher than that of bare Q235 steel. Hao Y et al.[10] showed that corrosion protection of epoxy coatings on Q235 surfaces was carried out by protecting polyaniline containing phytic acid doped with polyaniline (pani-pa). Gao X et al.[11] suggested phytic acid conversion coating and Zn2 + doped phytic acid conversion coating can be fabricated on Q235 steels by dipping the steel in corresponding coating bath.

In order to promote the surface structure of the parts and improve the surface properties of the surface, it can be laser coated on its surface with a wear-resistant, corrosion-resistant coating, so that the matrix and cladding layer can achieve a good metallurgical combination. In this paper, 316L stainless steel cladding coating was prepared on the surface of low carbon steel substrate by laser cladding. The influence of process parameters on the geometrical section size and macroscopic morphology of single channel cladding was studied.

2. Experiment

In the experiment, Q235 steel was chosen as matrix material and sodium silicate as binder. The powder and binder are modulated into paste and applied to the surface of the matrix. The cladding coating material is made of 316L stainless steel powder with a grain size of 300 mesh, and the main chemical compositions is C 0.03, Mn 1.99, Si 0.80, Cr 17.2, Ni 13.50, Mo 2.30, P 0.022, S 0.019 and Fe in balance. In this experiment, the LWY400P laser device is used and the laser used in the device is YAG pulse laser.

3. Results and Discussion

3.1 The relationship between the change of laser current and the surface morphology

The experimental results show that the coating thickness is 0.5mm, the scanning speed 100mm/s, the pulse width of laser 3.8ms, and the frequency 10Hz. When the spot diameter and defocusing amount are all definite, the forming quality of this single laser coating layer under the laser current of 125A-170A (5A in turn) is studied. **Figure 1** is the surface macroscopic morphology of single clad coatings with different laser currents. **Figure 1** shows that the surface quality of the coating is poor when the laser current is less than 125A-135. The surface is discontinuous and has spheroidization, and the cladding layer is easy to fall off. This is due to the smaller current, resulting in smaller laser energy, poor bonding with the substrate, and a lot of non - molten powder particles. With the rising of the current, the surface of the coating is continuous and spheroidization is less. When the current is from 155A to 165A, the cladding layer becomes smooth. This is because the high laser energy is beneficial to the wetting and spreading of the molten pool[12,13]. With the rising of current, the surface roughness of coating increases. When the current is too large, the power is also increased. Overheating occurs on the surface of the coating layer. The stainless steel powder is oxidized immediately, and the

IOP Conf. Series: Materials Science and Engineering **307** (2018) 012035 doi:10.1088/1757-899X/307/1/012035

surface color and finish gradually decrease.

Figure 1 The surface morphology of single clad coatings with different laser currents

Figure 2 is the cross-sectional morphology of single clad coatings with different laser currents. **Figure 2** shows that when the current is small, the laser energy is smaller. The matrix cannot be melted, and the coating layer and the substrate cannot be combined to influence the quality of coating layer. As the laser current increases, the coating layer and this substrate achieve excellent metallurgical bonding.

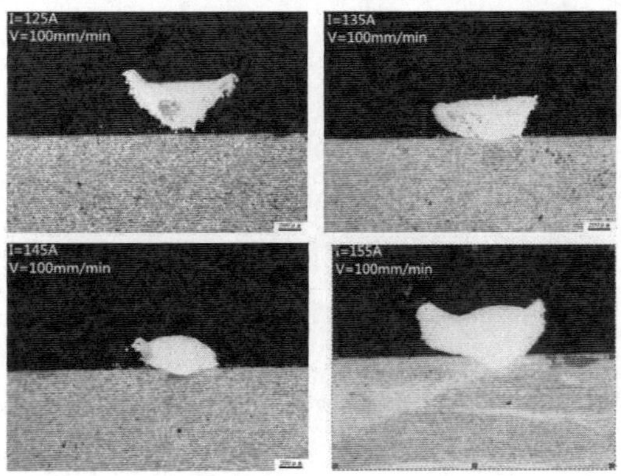

Figure 2 Cross sectional morphology of single channel cladding layer with different laser currents

3.2 Effect of laser current on geometrical cross section size and dilution rate of single cladding coating

Figure 3 is the effect of different laser currents on the height, width, and depth of the single cladding. **Figure 3** notes that with the rise of laser current, the height, width, and depth of cladding layer increase gradually. The depth and width of coating layer change obviously, while the height tends to be even or lower with the increase of laser current. This is because, as the current increases, the laser energy density increases. The amount of molten powder is more, so the width and depth of the bath will increase. Because the thickness of the powder is certain, when the laser energy is larger in unit time, the large impact force of laser beam on the powder can make powder splashing resulting in cladding powder reduced. Finally, the cladding layer height is reduced. On the other hand, because of the deepening of the molten pool, its surface tension of the molten metal is out of balance with gravity, so that the molten metal flows downward, thus increasing the width of the melt and reducing the thickness[14].

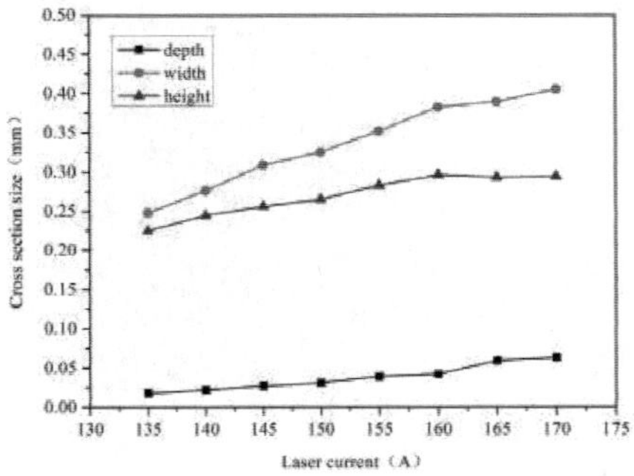

Figure 3 The effect of different laser currents on the width, height, and depth of the single cladding

Figure 4 shows that with the increase of laser current, the dilution rate of this cladding layer increases gradually. Because of the increase of laser energy per unit time, the penetration and melting height both increase, but the melting depth increases faster, so dilution rate of coating layer increases gradually.

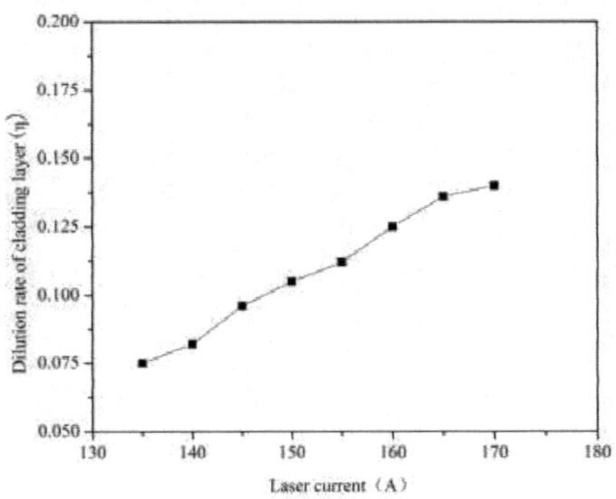

Figure 4 Dilution rate of cladding layer under different laser current

4 Conclusions

The cladding coating of 316L stainless steel has been successfully prepared on Q235 steel by laser cladding technology. The effect of the laser cladding parameters on geometrical size and morphology of single cladding layer was analyzed. The experimental results show that the surface quality of the cladding layer is poor when the laser current is less than 125A-135. The surface is discontinuous and has spheroidization, and the cladding layer is easy to fall off. As the laser current increases, the coating layer and the substrate achieve excellent metallurgical bonding. The results showed that with the increase of laser current, the width, height and depth of cladding layer increase gradually. As the laser current gradually increases, the dilution rate of its cladding layer increases gradually.

Acknowledgments

This research work was financially funded by the Key Research and Development Program of Shanxi Province(No.201603D121002-2), the School Foundation of North University of China, the Science and Technology Innovation Project of Shanxi Province (No.2016156), the Postdoctoral Science Foundation of China (No.2016M590214), and the Natural Science Foundation of Shanxi Province, China (No.2014021019-7, No. 2015011036).

References

[1] Zhou, Y., Zuo, Y., & Lin, B. 2017 The compounded inhibition of sodium molybdate and benzotriazole on pitting corrosion of Q235 steel in NaCl+$NaHCO_3$, solution. Materials Chemistry & Physics, p 86-93.

[2] Lai, C. 2016 Adsorption and corrosion inhibition performance of 2-(p-bromobenzylthio) -1h-benzimidazole for Q235 steel in hcl solution. International Journal of Electrochemical Science,p10462-10472.

[3] Liu, F. G., Du, M., Zhang, J., & Qiu, M. 2009 Electrochemical behavior of Q235 steel in saltwater saturated with carbon dioxide based on new imidazoline derivative inhibitor. Corrosion Science, p102-109.

[4] Jiang, Y., Bao, Y., & Yang, K. 2015 Formation and friction behavior of plasma electrolytically nitrocarburized surface layers on Q235 steel. Surface & Coatings Technology, p 324-328.

[5] Cui, J., Shi, R., & Pei, Y. 2017 Novel inorganic solid controlled-release inhibitor for Q235-b anticorrosion treatment in 1M HCL. Applied Surface Science, p 213-224.

[6] Sakudo, N., Shinohara, T., Amaya, S., Endo, H., Okuji, S., & Ikenaga, N. 2006 Ion implantation into concave polymer surface. Nuclear Inst & Methods in Physics Research B, p 349-352.

[7] Tan, C., Xiao, L., Liu, F., Chen, B., Song, X., & Li, L., et al. 2017 Influence of laser power on the microstructure and mechanical properties of a laser welded-brazed mg alloy/Ni-coated steel dissimilar joint. Journal of Materials Engineering & Performance, p 1-15.

[8] Yang, W., Shi, Y., Huang, J., & Yufen, G. U. 2016 Powder cladding formation with the large spot semiconductor laser. Electric Welding Machine.

[9] Hao, Y., Sani, L. A., Ge, T., & Fang, Q. 2017 Phytic acid doped polyaniline containing epoxy coatings for corrosion protection of q235 carbon steel. Applied Surface Science, p826-837.

[10] Wang, B., Xue, W., Wu, J., Jin, X., Hua, M., & Wu, Z. 2013 Characterization of surface hardened layers on Q235 low-carbon steel treated by plasma electrolytic borocarburizing. Journal of Alloys & Compounds, p162-169.

[11] Gao, X., Li, W., Yan, R., Tian, H., & Ma, H. 2017 Effect of zinc ion on the microstructure and electrochemical behavior of phytic acid based conversion coatings on Q235 steels. Surface & Coatings Technology,p325.

[12] Qiu, C., Adkins, N. J. E., & Attallah, M. M. 2016 Selective laser melting of invar 36: microstructure and properties. Acta Materialia, p382-395.

[13] Yu, J., Lin, X., Wang, J., Chen, J., & Huang, W. 2010 Mechanics and energy analysis on molten pool spreading during laser solid forming.Applied Surface Science, p4612-4620.

[14] Wen, S., Shuai, L., Wei, Q., Yan, C., Sheng, Z., & Shi, Y. 2014 Effect of molten pool boundaries on the mechanical properties of selective laser melting parts. Journal of Materials Processing Tech, p2660-2667.

IOP Conf. Series: Materials Science and Engineering **307** (2018) 012036 doi:10.1088/1757-899X/307/1/012036

Experimental study on deep hole drilling of 17-4PH material

LIUzhanfeng, LIUquantai

School of mechanical engineering, Xi'an Shiyou University,Xian ,China
E-mail:tigerliuquan@foxmail.com

Abstract: This paper uses 17-4PH material as the research object, according to the material characteristics of 17-4PH, designed and carried out deep hole drilling test. The purpose of the experiment is to study and discuss the three major problems of tool wear, chip shape and axial deviation of the hole in the process of deep hole drilling of 17-4PH materials. Through the deep hole drilling test of 17-4PH material, the variation of the chip shape and the deflection of the hole axis was obtained under different wear conditions.

1.Introduction

17-4PH, or 0Cr17Ni4Cu4Nb, is a martensitic precipitation hardened stainless steel, As a result of solid solution strengthening and aging treatment, this kind of stainless steel has better corrosion resistance and better strength, so it is widely used in the manufacture of Petroleum Instrument, chemical equipment, aircraft parts, nuclear reactor structure, etc. In the actual processing, its good mechanical properties cause great difficulties in machining, so the tool is prone to wear and tear and affect surface roughness and dimensional accuracy. In this paper, the wear of cemented carbide cutting tools in deep hole drilling is studied, and the influence of wear on chip morphology and hole straightness is discussed.

2. Analysis of Cutting Performance of 17-4PH Material

The test piece is a 17-4PH bar with a diameter of 850mm long and a hole with a diameter of 50mm is drilled on it. The chemical composition, mechanical and physical properties of 17-4PH are shown in Tables 1, 2, and 3.

Tab. 1 chemical composition of workpiece material(%)

C	Si	Mn	Cr	Ni	Cu	Ni	S	P
≦0.07	≦1.0	≦1.0	15.5～17.5	3.0～5.0	3.0～5.0	0.15～0.45	≦0.030	≦0.035

Tab.2 performance comparison of 17-4PH material and steel 45

material	Tensile strength（MPa）	Yield Strength（MPa）	HBS

17-4ph	≥932	≥726	363
45 steel	≥600	≥336	229

Tab.3 thermal conductivity of 17-4PH and 45 steels(w/mk)

	0	100	200	300	400	500
17-4ph	—	17	19	20	22	23
45 steel	52.34	48.85	44.19	41.87	34.89	—

the comparison of mechanical properties of 17-4PH and 45 steel are shown in Table two, it can be seen that the tensile strength and yield strength of the 17-4PH are much greater than those of the 45 steel, and the cutting force and the torque resistance during machining are also larger, the wear of cutting tool will be very serious, accuracy is difficult to be guaranteed. It can be seen from Table three that the thermal conductivity of 17-4PH is very small and less than half of the 45 steel at the same temperature, so heat dissipation is difficult. In addition, the carbon content of 17-4PH is low, the section stretch ratio reaches 40% - 50%, the toughness is larger, so the chip breaking is difficult during machining. In general, the cutting performance of a material is mainly determined by the cutting resistance, the service life of the tool, the surface roughness after cutting, and the chip breaking ability. Therefore, the cutting performance of this material is poor.

3. Deep hole drilling test of 17-4PH material

3.1. Test method

The testing machine is a deep hole machine converted from an ordinary lathe CW6140, The drilling system uses the system of BTA, The processing method is that the workpiece is rotating while the tool is feeding. the structure of tool is BTA deep hole drill, cutting gear material of drill is ZK30 carbide; the spindle speed is 195-225r/min, the feed is 0.14-0.26mm/r and the cooling fluid pressure is 2.5MPa. In order to ensure the machining accuracy, the vibration of the drill and morphology of chips is closely observed. After the drilling is completed, the cutting teeth should be observed by electron microscope. The thickness of the hole wall should be measured with the ultrasonic thickness gauge every 10cm, and the deflection of hole should be examined.

3.2 *nalysis of test results*

3.2.1. Tool wear analysis

Macroscopic tool wear as shown in Figure one, It can be seen from the figure that the tool's rake face and flank face have different degrees of wear and tear, even the phenomenon of chipping can be seen on the cutter tooth. This is mainly because 17-4PH itself is a kind materials difficultly to machine, its mechanical strength is high and plastic toughness is large, the cutting tool in the machining has a great cutting force and a large torque, there is a sharp friction between the tool and the workpiece surface, which causes wear; In addition, during the cutting process, the workpiece material had plastic deformation, the sliding and dislocation occur between the metal grains, and the cutting heat cause work hardening , all of these intensify the tool wear and make the cutting more difficult. Furthermore, it can seen that the wear occurred mainly in the cutting-edge outside, the blade has a certain degree of burning, the color near the blade was black; and the peeling part is mainly concentrated near the center teeth, which is due to the higher cutting speed of the external teeth, Long time continuous cutting causes high temperature oxidation and wear of cutting tools. The cutting speed of center position of the tool is low, the cutting is mainly done by pressing, the torque of the tool is large, and the high temperature in the cutting reduces the tool strength,these cause the blade to crack.

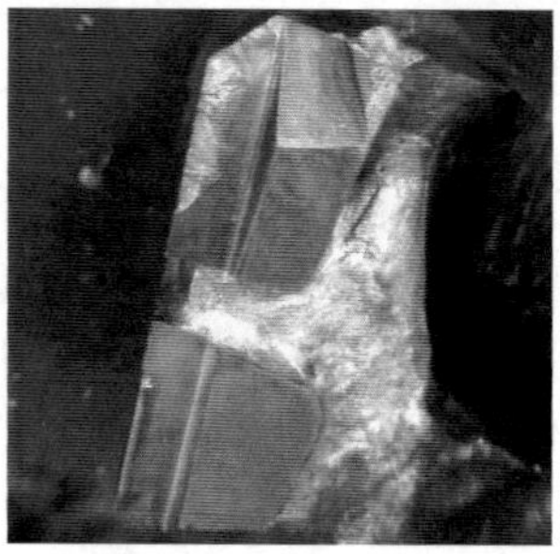

（a）Frontal wear

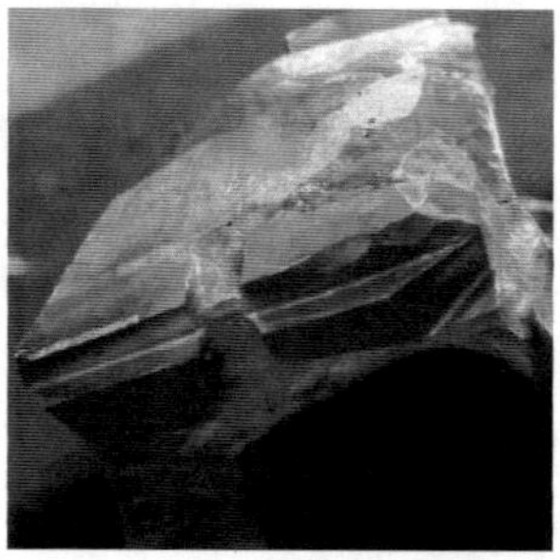

（b）Flank wear

In Figure one (b) the flank has obvious scratches, on one hand the reason for this is the friction between the flank and the rough machined surface. On the other hand because of abrasive grains wear. Carbon and Niobium of 17-4PH material and carbon combine to form hard spots of Cr_3C_2 and NbC, own to the hard spots with high hardness rubbed with the tool, which results in scratches.

Figure two is the tool wear area of the topography, Figure three is the wear area of the EDS element analysis chart, in Figure two it can be seen in the cutting process the tool surface coating has been peeled off. This is due to the continuous cutting process of the tool by the alternating contact stress and thermal stress resulted in stress fatigue, After a period of time, the accumulation of stress make the tool coating produce cracks, which are gradually worn and peeled off under the effect of friction and cutting forces.In addition the blade is blunt and wavy, near the blade there are some crater wear. This is due to adhesive wear and diffusion wear, the tool surface coating due to friction gradually peeling off, the tool surface directly contact with workpiece surface, Mutual adsorption and mutual penetration of atoms diffusion take place under the promotion of high temperature and pressure, The diffusion of atoms between the material to reduce the strength of the tool, so the tool turns blunt, when the tool and the processing of the surface are separated by a part of each other was taken away, and then the tool produces crater wear, Figure 3 Cr, Ni, and Cu elements indicate diffusion wear and adhesion abrasion.

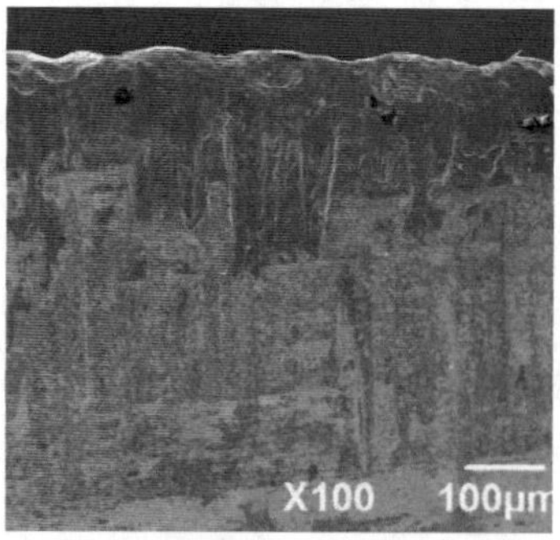

Fig. 2 tool wear under electron microscope

IOP Conf. Series: Materials Science and Engineering **307** (2018) 012036 doi:10.1088/1757-899X/307/1/012036

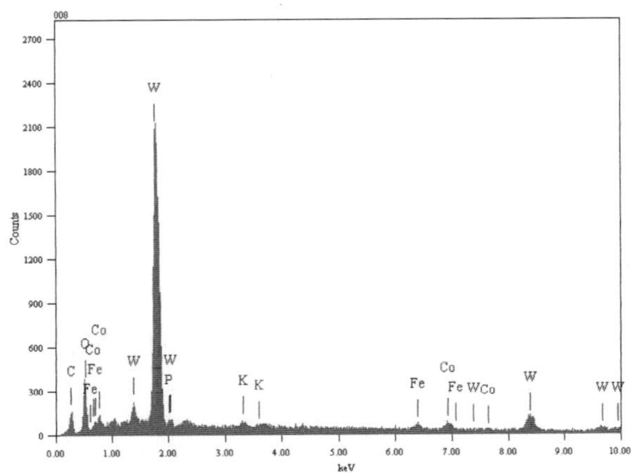

Fig.3 Energy spectrum analysis of wear zone

3.2.2. Analysis of chip morphology

Stainless steel material has a strong toughness, so chip is much difficult to break. Usually in order to facilitate chip breaking, the tool breakers are ground to increase the deformation of the chip and to promote its fracture. During the cutting process, a lot of chip heat will be produced, so the chip has a high temperature. When the hot cutting chips encounter cold cutting fluid, the hardness and brittleness increase, which is beneficial to chip breaking. On the other hand, as the length of the chip increases, the area affected by the impact increases accordingly, the impact of the corresponding increase. When the force of impact exceeds the chip bearing force, the chip breaks.

In the beginning of the cutting depth set 0.14mm, the chip obtained as shown in Figure four (a), the chip is elongated spiral, this is because the tool is not worn, the drilling process is smooth, so the chip surface is also smooth and has little burrs. According to the theory of metal cutting, on the premise of the feed rate and cutting depth remain unchanged, As the cutting speed of external teeth is lager than other, so usually the edge of the external teeth firstly wear and the chip edge has obvious burrs, as shown in Figure four (b). And when the cutting teeth and guide blocks are worn, the wear of tool is more serious, the drill pipe will have a more obvious vibration, The chip is not only burr, but also shorter in length, as shown in Figure four (c) ; Figure four (d) shows the length of the chip is long and full of wrinkles with burrs, the reason for this situation is that when the cutting depth is 0.24mm, the chips are thick and tough, it is difficult to break the chip. In addition, the tool wears and the drill pipe is vibrant violently. Especially in the test, a serious crushing phenomenon occurred, the cutting teeth were significantly cracked, the chip is fine at this time, as shown in Figure four (e), because the tool has completely lost the cutting capacity, drilling must be stopped immediately.

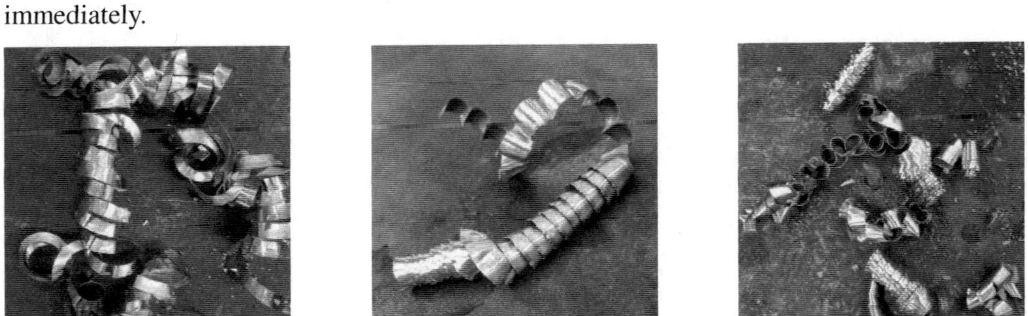

(a) Chip without wear (b) Chip during micro wear (c) Chip during moderate wear

(d) Chip during severe wear (e) Chip during the tool breaks

Fig.4 Chip morphology in different degree of wear of tool

The curve of hole straightness is shown in Figure five, it can be seen that the deviation of the straightness of the hole increases with the increase of the hole depth, and then the deviation decreases. At the beginning, in order to allow the drill bit to enter the guide hole smoothly, the cutting speed is set at 145 m/s and the cutting depth is set to 0.14 mm, At this point the tool is not almost wear, the deviation of the hole is small; After the drill enter the guide hole completely, the cutting speed is raised to 255m/s, and the cutting depth is 0.2m/s, at this time, cutting force become larger, the cutting-edge and guide block began to appear obvious wear and tear, hole deflection is also intensified, the crash of tool causes the hole's deviation to increase rapidly. After the replacement of the bit, the deflection of the hole has been corrected, so the deflection becomes smaller.

3.2.3. analysis of the deflection of the hole axis

In practice, there are many reasons for the deviation of the hole, such as uneven material composition and clamping errors. However, the factor that determines the deflection of the hole is the force in the drilling, One of the factors that affect the drill force is the wear of the drill bit, During the drilling process, the tool will inevitably cause wear and wear, which will cause the change of the force balance, with the continuous development of wear and tear, the tool will continue to change the force, which causes vibration, the axis of the tool machining will deviate from the correct axis of the hole. When the tool is broken, the tool will be violent vibration, which aggravate the deflection of the hole, then we must stop drilling, replace the tool to prevent serious deviation.

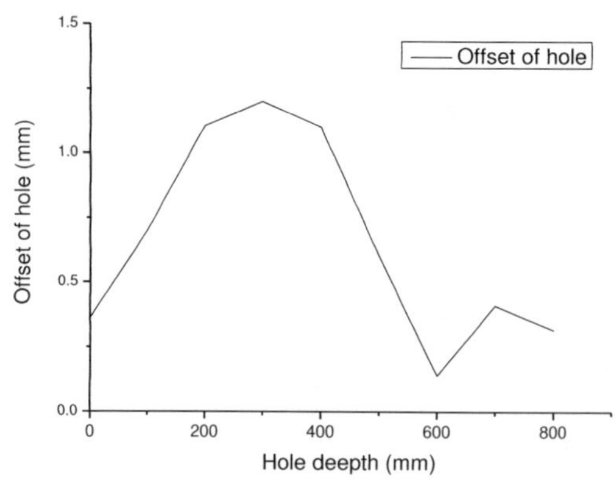

Fig.5 Variation of hole axis deviation with hole depth

3. Conclusions

IOP Conf. Series: Materials Science and Engineering **307** (2018) 012036 doi:10.1088/1757-899X/307/1/012036

The guide block in the drilling will also wear, and the guide block directly guide the direction of the drill bit, if the wear of guide block is very serious, the degree of hole deflection will be particularly serious.

(1) In the drilling process of 17-4PH material, the wear of the cemented carbide tool is mainly manifested as the wear of the rake face, the flank face and the guide block. The wear mechanism includes abrasive wear, bond wear and diffusion wear. Bonding wear is dominant.

(2) Tool wear directly affect the shape of the chip, with the tool wear increases, the chip shape will change accordingly, the actual processing can be observed by cutting the chip shape to determine the tool wear

(3) After the tool wears, the force balance system of the tool is changed and the force of tool is changed. The feed direction deviates from the hole axis, so it has a great influence on the deflection of hole.

4.References

[1] Wang Shiqing, Zhu Lin, Liu Zhaofeng, etc.*Deep hole processing technology* [M]. Northwestern Polytechnical University Press,2003.

[2] Zhang Youzhen. *Difficult cutting material processing problems* [J] .. Jiangsu Machinery, 1979, (1): 12-19

[3] Xu Xin, Yao Jingdong, Li Bing, etc. *Stainless steel roughing tool wear test* [J]. Tool technology, 2016, 50 (10): 36-38

[4] HE Sheng-ming.*Study on Superfine Deep Hole Drilling Technology of Superalloy 718* [D] Xi'an: Xi'an University of Petroleum, 2013.

[5] Wu Xiaodan, He Qiongru, Yan Zhawei et al. *Study on Wear and Damage of Deep-hole Cutter of Titanium Alloy* [J]. Modern Machinery, 1999, (3): 58-60

[6] Wang Han-chen.*Study on deflection mechanism of super-elongated BTA deep hole drilling axis* [D]. Xi'an: Xi'an University of Petroleum, 2012.

[7] Huang Xiaobin, Shen Xingquan, *Miao Hongbin. Research on tool wear characteristics of hollow axle deep hole machining* [J]. Tool Technology, 2014, 48 (8): 35-39

[8] Yang Xiao *.17-4PH stainless steel performance and organization research* [D]. Harbin: Harbin Engineering University, 2007.

[9] Long Zhenhai, Wang Xibin, Liu Zhibing. *Study on wear mechanism and cutting performance of carbide tool rake face in high speed milling of difficult to machine materials* [J]tribology journal, 2005, 25 (1): 83-87

[10] Hu Kuanhui, Shen Xingquan, Ma Teng. *Debris-like study of deep hole drilling steel* [J]. Tool Technology, 2017, 51 (2): 49-51

[11] Luo Zun. *Performance and application of precipitation hardening steel* [J]. Space Technology, 1985, 10 (5): 49-55

Analysis of Fluctuating Friction Version in Sheet Metallic Designing

Jithin Ambarayil Joy[1,a], Dong Won Jung[1,b*]

[1]Department of Mechanical Engineering, Jeju National University, Jeju, Korea
E-mail: [a]jithinjoy30jan@gmail.com, [*b]jdwcheju@jejunu.ac.kr

Abstract. Conservative Coulomb method indicates steady constant of rub in thin metal panel making that appears or feels close to the real thing. It contributes to describing attainable future event too high shear pressure in making ahead in the position of high R-value steel (AHRS). The study is conducted by pretend the making and spring back of a specific panel to understand the characteristic of the stamping procedure. Corresponding of the describe probable future results with (i) physical force-dependent changeable rub method, (ii) perpetual rub method, and the conclusion of exploratory facts point out a significant upgrading of spring back forecast with the prospective method.

1. Introduction

Coulomb rub method has been mainly used to show rub circumstances in the imitation of sheet metal making (SMF) up to the preceding few years [1, 2]. In number-based analysis action of SMF, checking shear pressure (i.e. rub shear pressure) to prevent respective speed up the middle from two interfacial parts frequently changes properly sized related to something else with rub constant and interfacial connecting physical force. Even though checking shear pressure is frequently enforced on the indicated active sliding-sheet growth, its probable usage to create an extremely large quantity of calculable things that change includes an extremely large calculating period. An active decrease of those calculable things that change in the analysis of finite element method is able to be done by presumptuous steady rub constant during the whole number-based repeating procedure. Probably, something is truly what it claims to be such steady constant rule is separating probable once the rise in right connecting part is related to its similar connecting physical force at particular circumstances separating happening under the situation of flexible change the shape with connecting physical force under yield pressure boundary. However, applied SMF procedure frequently includes plastic changes. From the introduction of plastic contortion onwards, additional changes frequently include the higher upward of the enforced usual force than that of the right connecting part. This is because the right connecting part is often held back inside the geometry of its very small quantity of identical part. Below such circumstances of steadily varying connecting physical force the advanced level too, the ideal steady Coulomb rub is probably unacceptable to be enforced because decelerating shear pressure on separating sliding growth is beyond the bounds of possibility to go beyond the shear stamina of not formed sheet. For this reason, the applied/applicable shear pressure is often boundary to the shear yield stamina of decelerating substantially [3]. Once a better shear pressure is enforced by a break, it would susceptibly lead to (i) enduring plastic change the substantial; and/or (ii) the substantial to be cut off in the getting touch with the part. As an outcome, slipping parts are sustained to blended method circumstances of flexible and/or plastic connecting. This type of touch make through the rub circumstances as it differs with: (i) make sure of flexible and/or plastic change the shape of the two sliding substantial; (ii) the character of fabric slashing and shearing; and (iii) the feature of substantial change from soft sheet to the hard tool, etc. Additional outstanding compound difficulty so often decided with the situation of making too many stamina steel sheets for the reason that their better stamina frequently demands much higher making number of tons to change or to shear once they are related to their mild steel identical parts. Such compound situation has the inclination to think supply with also high rub shear pressure price of steady Coulomb rub method is used to analyse rub shear pressure. Penetrating for the advanced version to expecting such shear pressure related to nearer credibly dependable standards extremely important, which after that forms the reason for doing something of the growing investigation.

To be grateful, the benefits of high-stamina and light-weight compounds afterward leads to cumulative the usage of aluminium, magnesium, and ahead in the position of high-stamina steels (AHSS) as substantial for auto-body pieces and assemblies. The use of such mixtures in the vehicle body in a helpful way for better fuel performance and contented the surrounding circumstances openness and security interests [4, 5]. As related with aluminium and magnesium mixtures, related to something else lesser price and advanced stamina of AHSS sheet substantial make it an advanced preference of fabric for related to what holds something together and makes it strong something else along with the front/back bars, A/B important parts, and gate beams, etc. However, AHSS frequently has lesser formability [6], the better level of die attire [7], and the advanced level of spring back [8, 9] once it's far as related to mild steels. Such extremely largely unfavourable making performances are in particular due to the fact AHSS has advanced surface rigidity and advanced yield stamina than modern steels [10]. As a result, the malleable making of AHSS demands too much interfacial physical forces middle from tool and workpiece [11]. In a study of parallel strip drawing, Kirkhorn et al. [12] showed that the rising of frequent load gave to do in a certain way of lesser rub constant. Additional than that, our previous notice [13] also recommended that advanced ordinary load middle from varying the shape of the thin panel and the making die tended to the better level of work- hardening in twin phase (DP) matrix. Such work-hardening dual section matrix eventually led to decreasing the standards of rub constants with cumulative loading. Effects of the research in [12,13] recommend that advanced interfacial physical force contribute to result in lesser rub constant.

Most simple to get the chance to a number-based method of SMF accepts an enduring consistent of rub. Normal illustrations incorporate (i) the theoretical substitution anticipated by Mahdavian and Shao [14] for assessing hydro-dynamic grease of sheet metal making operations; and (ii) the indicative adaptation created by Yang [15], which was joined with adaptable plastic layer limited component code of profound illustration, particularly to try to locate the full film lubricant in profound sketch. In progress of Mahdavian and Shao [14] and Yang [15], surface landscape and harshness interfacing have not been considered because of their acknowledgment of working in thick film foundation. Be that as it may, sensible rub method, as a rule, treat the impact of unpleasantness on the setting free of lubricant and on the touch of awful temper centre from sheet-pass on the container in SMF as external limit oil conditions. In this manner, numerous variable rub techniques in most recent years, been ahead in position to think about the impact of various factors on the rub of activity. Wilson et al. [16] ahead in position strategy by taking into clarification of various oil arrangements in a sheet-tooling associating point. The strategy transformed into furthermore associated with a limit component program for express a result ahead of time of rub way acting in greased up sheet metal making operations. Haluk et al. [17] Accompanying the illustration rules of sheet metal to the neighbourhood grease position amid changed by suitably blending a variable rubbing strategy with a limited component program. The number-based results of the technique for their sheet metal illustration method gave information records precisely worthy of their exploratory indistinguishable part. Through the investigation of the effect of mass plastic strains on genuinely existing interfacing parts, Wiklund et al [18] changed a technique for sheet metal making operations by enhancing the FE examination at first utilizing traditional Coulomb's rub strategy. Despite the fact that, the alterable rub strategy in [16-18] out of a typical and consistent route required with the revelation made ahead of time of various procedure records in sheet metal making, the technique are much of the time not adequate for express result ahead of time the rubbing constants, which are likely to happen changing with the distinction of impacting physical power in the creating of AHSS.

Exact number-based investigation of spring back in SMF of AHSS is an essential issue since it helps: (i) the improvement of polynomial math and parts of a crude clear piece/part; and (ii) the plan of the comparable mechanical assembly for the creation of the segment. There is a wide range of sorts of people groups endeavours made for properly looking at spring back. For instance, Chen et al. [19] built up a kinematic solidifying strategy overall around portrayed the conduct of non-immersed cyclic solidifying to depict a plausible future occasion spring back in DP sheet metals. Rahul and Haldar [20] utilization a method for getting things done of limit component investigation to think about the effect of solidifying conduct and not being indistinguishable every which way on spring back of the Numisheet-2005 twisting test outcome. What's extra, Zang et al. [21] use considerably be equivalent to a way to deal with technique spring back by all around ok contemplating Bauschinger impact and isolating going on for a brief period conduct, everlasting softening, and emptying modulus of the clear. Other than the generous strategy, outstanding number-based computation techniques were likewise inspected nearly so reality can be found by utilizing many research specialists with a specific end goal to express a result ahead of time spring back. Narkeeran and Michael [22] set up the use of an unequivocal way to deal with getting comprehended in a steady progression answer for express a result ahead of time spring back fit as a fiddle of sheet metal. Chatti [23] investigated limit strains by the method for utilizing an elastoplastic (EP) creation rather than the commonly accessible hypo elastoplastic (HEP) manifestations. The utilization of his creation permitted to doing real spring back outcomes with significantly lessened examination period. Be that as it may, the use of alterable rub technique to express a result ahead of time spring back is still scarcely brought into being inside the current writing.

The technique for spring back once the running pass on is taken out basically contemplated weight driven. Generously, the start of spring back is especially a direct result of adaptable strain recuperation of a substance amid the discharging of fluctuating burdens. Accordingly, the significance of spring back can likewise be chosen by the level of additional weight in a shaped sheet-piece. Rub consistent is a number for choosing the significance of rub shear weight in a making technique. Along these lines a lot of impacts the level of additional weight and spring back of shifting workpiece. Subsequently, the correct fuse of a correct or close practical physical power subordinate variable rub technique, into the material number-based investigation of spring back way of acting in the making of sheet metal, for the most part in making AHSS with huge spring back, is important and basic.

Going for extra accurately express a result ahead of time the spring back, a number-based investigation with alterable rub technique for SMF system is directed. This paper is given: (I) the making and spring back impacts for a particular board (Section 2), that is expected with the utilization of (a) three phases of relentless rub steady, and (b) physical power that progressions rub constants, in like manner - it additionally looks at the express a result ahead of time results with their preparatory indistinguishable parts; (ii) exchange of the outcomes (Section 3); and (iii) the diagram of finishing up reaction (Section 4).

2. Making/spring back predicting and validation

2.1 Substantial properties

With the end goal of this paper is explore making and spring back finish up utilizing FEM strategy, a mechanical significant is picked so the outcome will be near the trial as much as likely. Japanese Industrial Standard (JIS) chilly moved steel has been chosen since it is a high R-esteem steel and shows great capacity for making an automotive panel. As acquired uncovered 0.7 mm thick SPRC35E AHSS sheet was utilized as tried specimen and the mechanical properties of this sheet are appeared in Table 1 and the estimation for the flow curve appears in Figure 1.

Table 1. Mechanical properties of SPRC35E sheet

Young's module E (GPa)	Yield Stamina (MPa)	Tensile Stamina (MPa)	Hardening exponent	Hardening coeff. K (MPa)	Poisson's ratio
210	255	720	0.161	1195	0.3

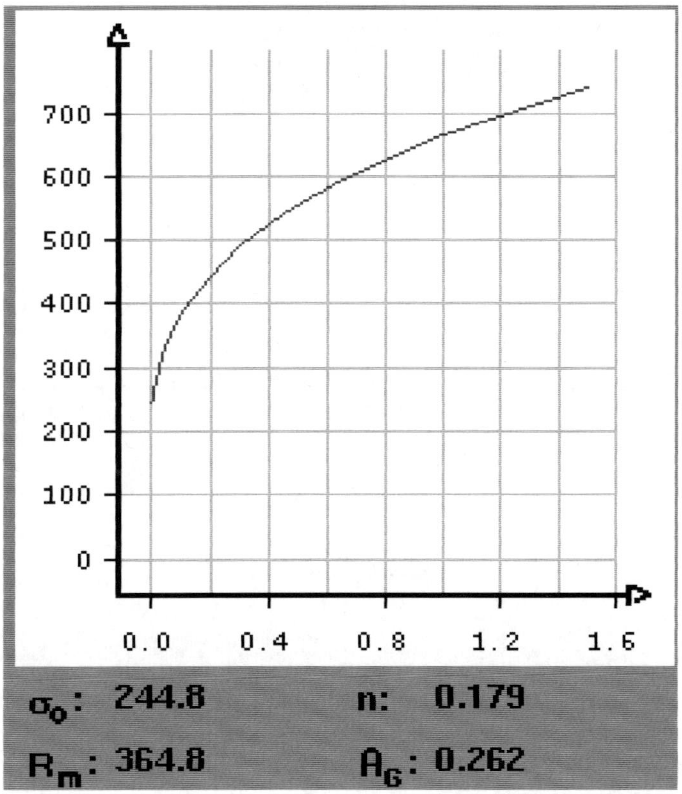

Figure 1. The approximation for the discharge curve of SPRC35E sheet metal

2.2 Forming test

Fig. 2 demonstrates the related dimensional records and logical comprehension of the method. With the course of an action level, the punch was continued unmoving and the fastener kick the bucket was climbed a proper area over the punch. A 250 mm long x 30 mm wide rectangular sheet clear was absolutely put on the highest point of fastener globules. This was then turned of moving the fade away at a velocity of 1 mm/s until the point that the sheet clear transformed into the clasped centre from the bite the dust and the cover so the framework was an altogether shut state. From that point onward, the operation incorporates the movement of the shut bite the dust and cover down towards the settled punch for clasping and stamping the sheet clear into a shape. In the stamping technique, an unfaltering clear holder power of 60 kN was authorized, and the aggregate moving separation of the dust and folio framework permitted a punch stroke drawing crosswise over of 35 mm to have the capacity to be finished.

Figure 2. Stamping procedure set up

2.3 Forming with FEM

2.3.1 Flexible rub modelling

The normal rub strategy is proposed to portray likewise effectively the rub constants under the divergent type of correspondence physical powers in SMF. Thus, the rubbing technique is attempted to relate rub consistent with regular interfacing physical power together graphically or numerically. Since the pattern of the bend in a decaying recognizing, it might be connoted as quick-change physical power subordinate rub administer as sought after.

$$\mu = \mu_o \left(\frac{p}{p_o} \right)^{n-1} \tag{1}$$

Where, PO is the comments/counterbalance physical power which much of the time Po >0 in rub tests and n is a mediator and dependable in the range: 0.0 < n < 1, P is the applying physical power at which rub steady is come to and is the rub consistent at physical power PO. The tirelessness of n was expert by a method for right off the bat setting Po and to 2.05 and 0.125 (Fig. 3), in like manner, after which by the method for controlling n guidelines with the given and its relating P for all the test information focuses (Section 2). The great estimation of n is gotten by utilizing the technique for Minimum Square for all the examined n measures. Through the beforehand said approaches, the physical power subordinate rub lead for the circumstance of a floating panel towards SPRC35E steel can in like manner be composed.

$$\mu = 0.125 \left(\frac{p}{2.05} \right)^{0.793-1} \tag{2}$$

The Curve of the so got numerical definition (Eq. (2)) and the test information purposes of mechanical properties test are charted and recognized in Fig. 3. The assessment indicated the association correct as Eq. (2) nearly to fit well with the greater part of the trying information focuses. Such scientific articulation lets in finish up/count of rub constants for any physical power inside the degree from 2.05 MPa to 18.48 MPa.

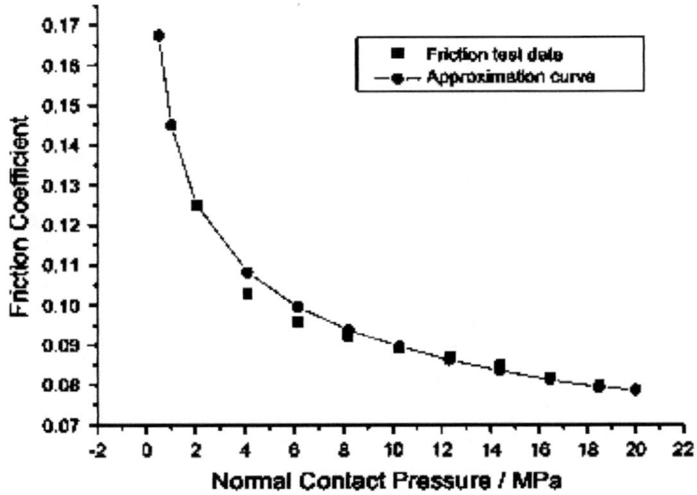

Figure 3. The estimate curve versus rub test data

2.3.2 Forming consequence

To analyse the express an outcome ahead of time spring back outcomes under the unfaltering Coulomb rub technique with the demonstrated utilizing the physical power subordinate variable rub strategy (see: Eq. (2)), approval of the number-based impersonation of SMF can be proficient by enough relating with the trail impacts. Oil circumstance of the sliding rub in making technique might be both dry (> 0.3), or limit (0.1 < > 0.3), blended (0.03 < r 0.1) [3]. For the most part, dry grease strategy implies by no means present of ointment among the sliding part. As this investigation use stamping oil in both PIN-ON-DISK test and stamping, it is like this not marked as dry grease strategy. In stamping, the most extreme extensively as of now met grease technique is external limit oil, in which the strong part is so near one another and their interfacing is ordinarily industrious by the surface transaction centre from mono or multi-sub-atomic movies of oils and the strong cruelty [29]. Be that as it may, consolidated film oil is likewise as often as possible stumbled over in sheet metal making. In blended oil, small-scale tops on the ill tempers of the metal surface might be under limit oil technique, at an indistinguishable period from the harshness miniaturized scale valleys might be pressed with ointment. Hydrodynamic oil technique incredibly exists in metal influencing on the grounds that it to best occurs in the extremely exact condition of speed and temperatures or under the grease conditions of utilizing top-notch fats/waxes.

Therefore, by picking the rub constants estimation of spring back in the frame with the use of persistent Coulomb rub strategy was actualized of (I) 0.05 for joined film oil, and (ii) 0.15 for blended film grease, and (iii) 0.25 for limit oil. Their results were then identified with the showed from the utilizing variable rub strategy. Estimation of unfaltering rub strategy accepted the rub consistent with either = 0.05 or 0.15 or 0.25 to be enduring generally among the limit line of considerable device rub combines in discovering the estimation of the separate circumstance. In the finish up with above strategy, it makes variable rub equation which transfers to physical power, i.e. Eq. (2), for setting up a condition for rub shear weight Tr. In a venture with to condition Coulomb's control expresses, that the rub shear weight Tr centre from the interfaces of generous – instrument of the rub sets ended up noticeably investigated through Eq. (3) under. The condition plainly demonstrates the variety of rub shear weight with the genuine period physical power "p" on every hub.

$$T_R = up = \mu_o \left(\frac{p}{p_o} \right)^{n-1} p = \mu_o p_o \left(\frac{p}{p_o} \right)^{n-1} \quad (3)$$
$$= 0.125 * 2.05 * \left(\frac{p}{2.05} \right)^{0.793}$$

With the estimation of 0.25, of unfaltering Coulomb rub steady, so as to wipe out the up exertion of sheet considerable the aggregate interfacial rub centre from the sheet and bite the dust nose sweep (Fig. 2) was fundamentally ascend to oppose the up development of sheet generous. By doing, it will be going to deliver high extending of sheet metal in the vertical divider. Extending like those is outside the compass of making limit and the low amount of generous at the specific period at extending zone causes in split disappointment. Along these lines, it's important to finish up the right associating physical power and diminishing circulation in the resulting discourse. With respect to the system of examination of the impact of consistent Coulomb rub steady of 0.05 and 0.15, separately, it demonstrates that part can be delivered effectively and no part is watched. Fig. 4 show the formability of particular board under the impact of many rubs consistent

Fig. 5 outlines and relates the uses of recreated physical power for the stamping of sheet metal applying the enduring rub strategy (u = 0.05 and 0.15) and the physical power subordinate variable rub technique. The outcomes in those three circumstances obviously demonstrate that most extreme norms of interfacing physical power occurs right parallel to the area of stamping punch filet corner. Extra finished, the most noteworthy benefit of interfacing physical power is seen for purposes of filet span on both the bite dust corner and draws a dab. Despite the fact that they are worked at a similar clear holder power and drawing speed, diverse level of rub steady changes by fluctuating the character of the significant and a short time later impacts, the genuine touching part and associating physical power. Evidently, a higher unfaltering rub steady can create extra associating physical power with the showed basic parts. In the underlying stage, at shutting stage, the folio is likely to clasp the sheet with an equivalent and enduring interfacing physical power by applying clear holder drive. Once the punch came in associating with the sheet, the generous is subjected to the bite the dust opening. Vast, estimation of rub consistent outcomes in higher controlling power to considerable stream and propelled level of significant solidifying. Therefore, higher making power is required to stamp harder considerable at the same stamping speed, which results in a superior associating physical power. However, it is a compound marvel and is difficult to interface nearby physical power adequacy with steady level quantitatively as the considerable stream additionally recursively influences the neighbourhood associating. As the character of nearby associating and solidifying definitely impacts the arrangement of neighbourhood surface morphology, its then extra varies to the rub consistent. Henceforth, the lead of relentless rub consistent is unquestionably not legitimate to such SMF methodology. It is along these lines of extraordinary significance to utilization physical power subordinate variable rub consistent to characterize such right associating circumstance.

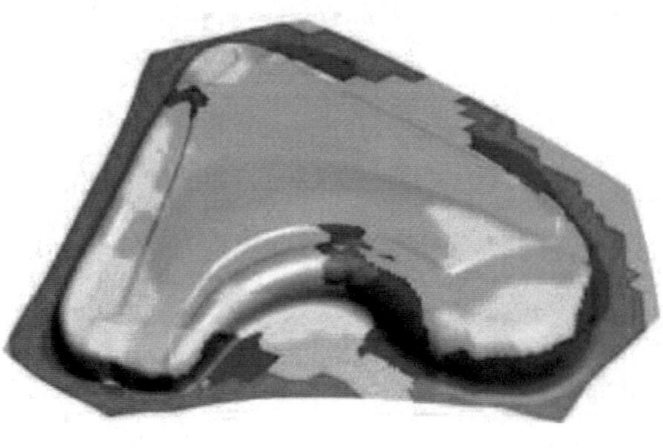

(a)

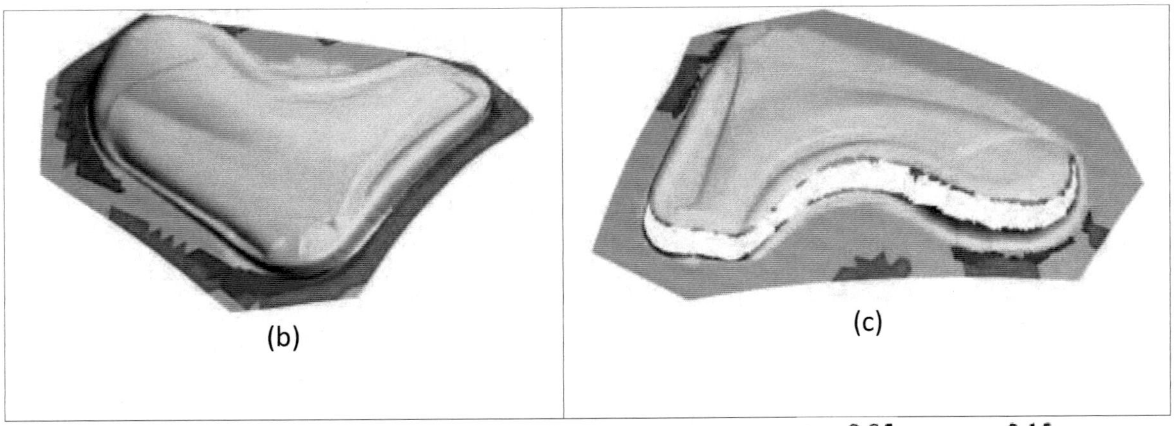

Figure 4. Formability of the panel with steady rub method (a) $\mu = 0.05$ (b) $\mu = 0.15$ (c) Variable rubs method

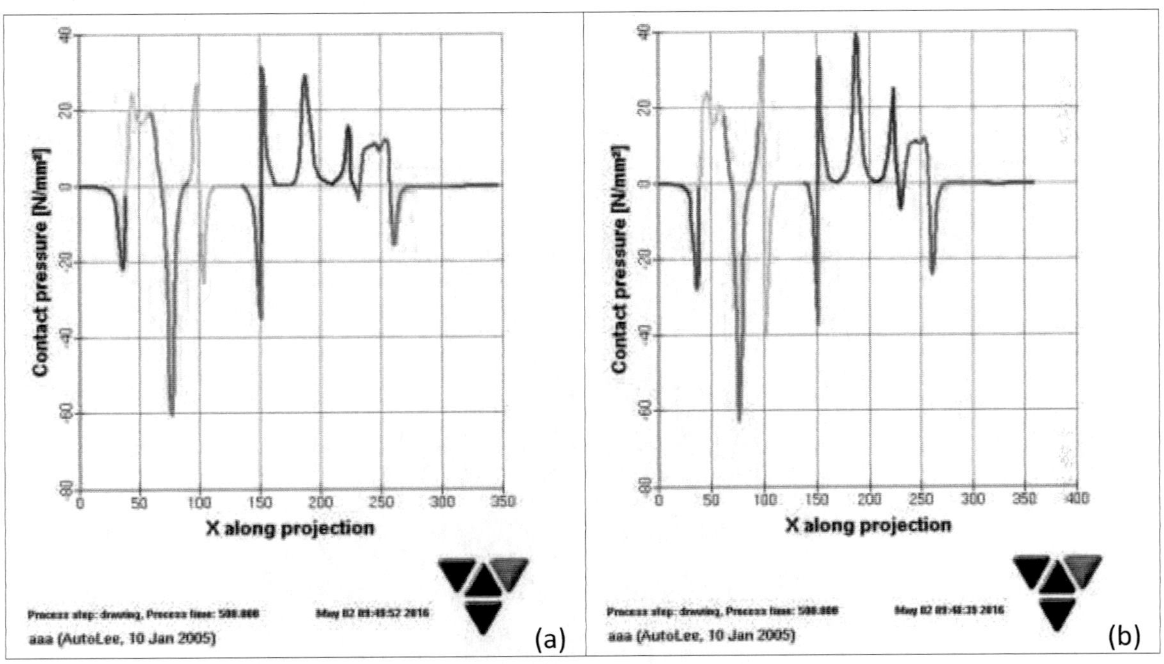

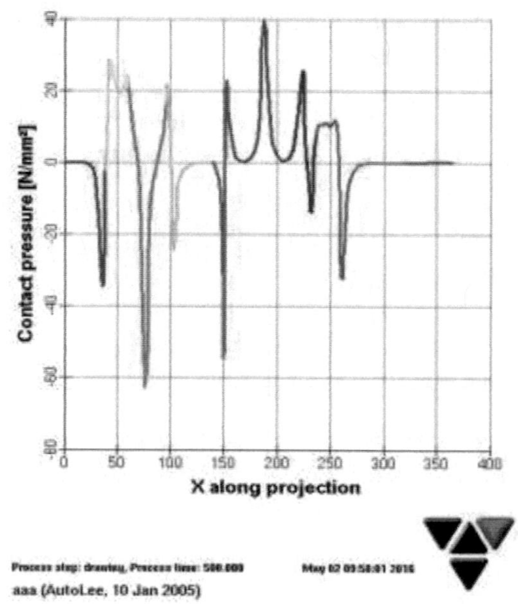

(c)

Figure 5. Steady physical force of sheet metal under different rub constant (a) $\mu = 0.05$ (b) $\mu = 0.15$ (c) Variable rubs method

Under Fig 6 clearly demonstrate the positions for calculating the thickness distribution, alongside the length of the deformed sheet, from the analysis with both (i) the steady Coulomb rub method and (ii) the proposed physical force-dependent variable rub method.

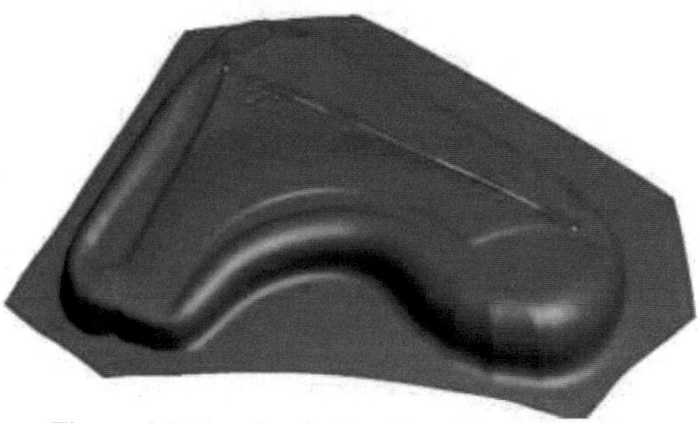

Figure 6. Measuring point at the analysis software

The quantitative and the simulative results of enduring Coulomb rub strategy and physical power subordinate variable technique were appeared in Fig7 under. It is clear from the plot that the real diminishing at the filet range of punch corner, at where the computation of the stamped part plainly demonstrates a greatest of 15% diminishing. By looking at it, the pattern of no diminishing occurring for the enduring bend prescribed that the sheet metal anticipated that would experience unadulterated twisting under blended film oil. The over diminishing to make sheet break with the relentless Coulomb rub consistent of 0.25, and the no diminishing sheet with the unfaltering rub steady of 0.05, are the two uses for clarifying grease state and should be barred from the technique of investigation. The mimicked thickness circulation of stamping sheet greased up (I) with enduring Coulomb rub steady of 0.15, and (ii) with physical power subordinate variable rub looks like to pick up a comparative level of accuracy with the tentatively examined thickness partner Fig. 7. Subsequently, the setting parameters of those two investigations were extra embraced for spring back approval.

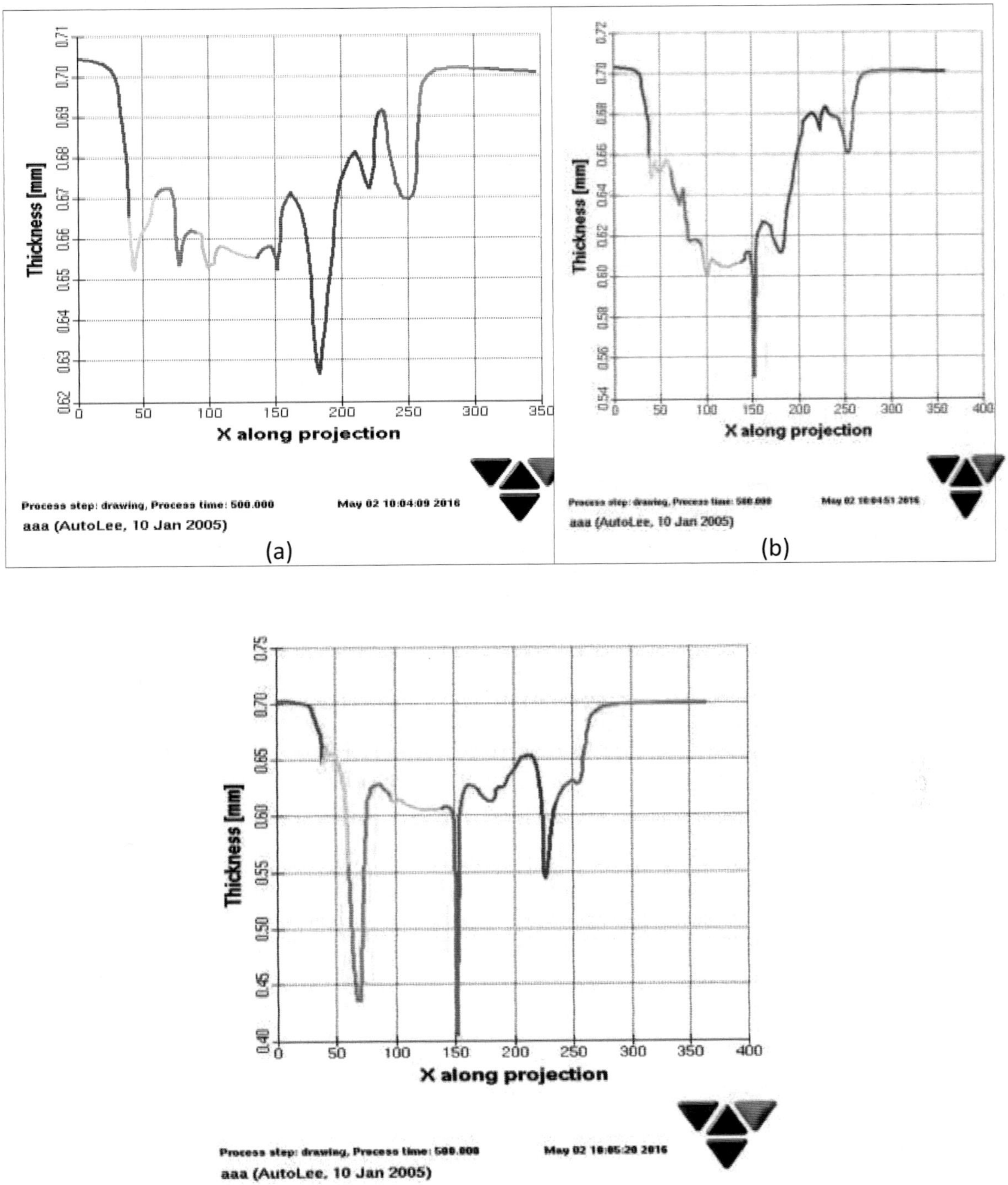

(c)

Figure 7. Thickness corresponding middle from different rubs (a) $\mu = 0.05$ (b) $\mu = 0.15$ (c) Variable rubs method

3. Conclusion

Impact of rub and ampleness of the use of rub technique on spring back close in SMF of AHSS under limit grease conditions were researched in this paper. The exploration was embraced by leading building a physical power subordinate variable rub strategy to portray the variety of constants under various ordinary associating physical powers, went with the guide of assessing the spring back finish up with the use of (I) consistent rub technique and (ii) the proposed physical power subordinate variable rub technique, separately, in making AHSS. The pertinent spring back finishes up were at that point in comparing with outcomes of test data. Results encourage the reaching of following determinations:

1. Approval of making investigation is done through thickness comparing with trial comes about. The last outcome recommends that the close with (I) the unfaltering Coulomb rub technique and rub steady of 0.15, and (ii) the physical power subordinate variable rub performed likewise worthy precision in a real making outcome. Not with standing, examination consequences of that two technique contrast in the dispersion of associating physical power, rub shear weight and primary weight.

2. Unfaltering rub technique adds to weave its impact on the condition of recuperation of inverse strain pressure influencing.

Reference

[1] Hol J, Cid Alfaro MV, Rooij MB, Meinders T. Advanced friction modeling for sheet metal forming. Wear 2012;286-287:66-78.
[2] Dan W, He Y, Heng L. Advance, and trend of friction study in plastic forming. Trans Nonferrous Met Soc China 2014; 24:1263-72.
[3] Schey J. Tribology in metalworking: lubrication, friction, and wear. Ohio: American Society for Metal, Metal Park; 1983.
[4] Novita S, Zahari T, Salwa HA, ARG Raja. Optimal multi-material selection for Lightweight design of automotive body assembly incorporating recyclability. Mater Des 2013; 50:846-57.
[5] Robert AW, Jerome P, Veronique M, Christian L, Jan-Anders EM. Assessing the life cycle costs and environmental performance of lightweight materials in automobile applications. Compos A: Appl Sci Manuf 2011;42:1694-709.
[6] Zhu J, Xia Y, Luo H, Gu G, Zhou Q. Influence of flow rule and calibration approach on plasticity characterization of DP780 steel sheets using Hill48 model. Int J Mech Sci 2014; 89:148-57.
[7] Omer N, Muammer K. Experimental investigations on wear resistance characteristics of alternative die materials for stamping of advanced high-strength steels(AHSS). Int J Mach Tools Manuf 2009;49(12-13):897-905.
[8] Wagoner R, Hojun Lim, Lee M. Advanced issues in springback. Int J Plast2013; 45:3-20.
[9] Chalal H, Racz SG, Balan T. Springback of thick sheet AHSS subject to bending under tension. Int J Mech Sci 2012; 59:104-14.
[10] Hyunok K, Jehu S, Frank EG, Altan T. Investigation of galling in forming galvanized advanced high strength steels (AHSSs) using the twist compression test (TCT). J Mater Process Technol 2013;1-3:459-68.
[11] Kim H, Sung J, Sivakumar R, Altan T. Evaluation of stamping lubricants using the deep drawing test. Int J Mach Tools Manuf 2007;47(14):2120-32.
[12] Kirkhorn L, Frogner K, Andersson M, Stahl JE. Improved tribotesting for sheet metal forming. Procedia CIRP 2012; 3:507-12.
[13] Wang W, Hua M, Wei X. A comparison study of sliding friction behavior between two high-strength DP590 steel sheets against heat treated DC53 punch: Hot-dip galvanized sheet versus cold rolled bare sheet. Tribol Int 2012; 54:114-22.
[14] Mahdavian SM, Shao ZM. Isoviscous hydrodynamic lubrication of deep drawing and its comparison with experiments. ASME J Tribol 1993; 115:111-8.
[15] Yang TS. Full film lubrication of deep drawing. Tribol Int 1999; 32:89-96.

[16] Wilson W, Hsu TC, Huang XB. A realistic friction model for computer simulation of sheet metal forming processes. ASME J Eng Ind 1995; 117:202-9.
[17] Haluk D, Metin A, Can AY. Effect of variable friction coefficient on sheet metal drawing. Tribol Int 2002; 35:97-104.
[18] Wiklund D, Rosen BG, Wihlborg A. A friction model evaluated with results from a bending-under-tension test. Tribol Intl 2009; 42:1448-52.
[19] Chen J, Xiao Y, Ding W, Zhu X. Describing the nonsaturating cyclic hardening behavior with a newly developed kinematic hardening model and its application in spring back prediction of DP sheet metals. J Mater Process Technol 2015; 215:151-8.
[20] Rahul KV, Haldar A. Effect of normal anisotropy on springback. J Mater Process Technol 2007; 90:300-4.
[21] Zang SL, Lee MG, Kim JH. Evaluating the significance of hardening behavior and unloading modulus under strain reversal in sheet spring back prediction. Int J Mech Sci 2013; 77:194-204.
[22] Narkeeran N, Michael L. Predicting spring back in sheet metal forming: an explicit to implicit sequential solution procedure. Finite Elem Anal Des 1999; 33:29-42.

Acknowledgment

This research was supported by the 2018 scientific promotion program funded by Jeju National University.

IOP Conf. Series: Materials Science and Engineering **307** (2018) 012038 doi:10.1088/1757-899X/307/1/012038

Abrasive wear of Hilong BoTN hardfacings

L Fedorova[1], S Fedorov[1], A Sadovnikov[1], Y Ivanova[2] and M Voronina[3]
[1] D.Sc. in Engineering Bauman Moscow State Technical University, Moscow, Russia.
[2] Ph.D. in Engineering , Bauman Moscow State Technical University, Moscow, Russia.
[2] Ph.D. in Engineering , Saint-Petersburg Mining University, Russia.

e-mail: momd@yandex.ru

Abstract. The spread of steels, which are used to produce locks of steel drill pipes, adversely affects their wear resistance, which, in combination with low hardness of HV 2400 ... 2800 MPa as well as of the thread of screw, results in low wear resistance and the need for their reconstruction at the pipe control shop. An efficient way of improving the quality of drill pipe jonts is to hard-face them by the outside diameter with wear-resistant materials (hardbanding). One of the companies engaged in the development of hardfacing materials and hardbanding is Hilong (China) with weld seams of the brand BoTn. According to the results of the studies the following conclusion can be made: hardfacing increases the durability of the hardware, contributing to an increase in wear resistance of locks of DP under the conditions of abrasive action of aggressive geological formations; the usage of DP without wear-resistant weld seams is impermissible, because their further operation, as part of the drill-stem, can lead to emergency consequences; application of the pipes with the hardfacing collars together with the collars without hardfacing, due to varying degree of wear of jonts in the drill-stem, is also impermissible.

1. Introduction

One of the causes of low durability of drill pipes (DP) is an abrasive wear of joints[1]. More than 60 % of pipes are rejected due to the wear of outer joint diameter, while other parameters meet the acceptable norms. The main reason of drill pipe joint wear in an open wellbore is presence of hard layers in aggressive geological bodies. During drilling the lower part of drill stem consists of heavy-duty drill pipe (HDDP), heavyweight drill pipe (HWDP) and steel drill pipe (SDP). HDDP and HWDP are made of single structural alloy steel workpieces 40ХГМА (1.2311), 40ХН2МА (4340). The legnth of DP is 8,5…12,5 metres.

During borehole drilling the lower part of drill stem is in direct contact with abrasive parts of rock-breaking layers. The intensity of wear of DP surface layers during abrasive wear hardly depends on original mechanical properties of joint material. The chemical composition of joint material [5] of steel pipes shows a significant dispersion of steels by carbon content (0,22…0,37), what, in combination with other alloying elements, matches steels 20ХГНМ…40ХГНМ. Hardness of the DP joint material is 28…36 HRC and is achieved by induction heat treatment.

In order to increase the longevity of DP, hardfacings made of wear-resistant materials are welded to the outer diameter of the pipe. In most cases there are 3 of such hardfacings , each of them is 25,4 mm wide and overall width is 76,2±6,35 mm. Hardfacing are jam-welded in argon-shielded environment . Welded material stands above joint body up to 2,5 mm. Such way of increasing durability is widely used by foreign companies [2, 3, 6]. Thus, one of the priorities during production and full repair of drill pipes is reinforcing welding of wear-resistant materials (hardbanding) of outer joint surface[4, 5].

Hardbanding – wear-resistant metal coating, applied by way of electronic welding to the outer surface of drill pipe joint in order to increase the longevity of the joint and decrease the wear of casing pipe.

2. Research methodology

Further use of wear-resistant hardfacings for drill stems is impossible without determining their level of effectiveness . For this it is necessary to carry out a comparative test of different padding materials in relation to DP joints without hardfacings.

Hilong hardfacings BoTn 1000, BoTn 3000, BoTn 5000 (table 1) are welded to joints by “Technomash” of Nevyansk .

Table 1. Parameters of the submitted samples

Collar of drill pipe joint	SDP 89 in the strength group G105
External diameter of the lock, mm	127
Height of coating, mm	2,38 (+0.8)
Width of coating, mm	76,2±6.35
Method of applying hardfacing	Standard

For producing sample parts joints of SDP 89 were mechanically processed: turning to eliminate surface irregularities of hardfacings; inner diameter boring up to 105 mm; cutting a ring with a height of 20±0,1 mm. The ring was cut in segments 35 mm long. Each ring was cut in four segments, three of which were used for wear experiments, and one sample part for metallographic analysis.

The wearing scheme consist of a Sample 1 is fixed on the base of the pendulum arm 2 with lever 3 screwed in from the side. Weight is hanged from the right part of lever 3, which ensures constant pressure in the contact are of sample 1 and turning rubber roller 6. Abrasive material with grit of 16 mm is supplied through the distribution device 5 by the track pan 4 to the contact area of sample piece 1 and rubber roller 6 . The duration of experiments, established by standarts ASTMG65 and GOST 23.208-79, equaled 30 minutes for each sample piece.

Chemical analysis was carried out with laser atomic emission spectrometer «LAES».Cross-sections were made using stand methods with Stuers equipment. Etch processing was done in a 4% solution of nitric acid in ethanol. Etching welded hardfacings was done using the chemical agent Marble (20 g $CuSO_4$, 100 ml HCl, 100 ml ethanol).

Metallographic analysis was carried out using CarlZeiss Axio Observer D1 with zooming in from 200 to 1000 times. Microhardness was measured using a durometer DuraScan 70 with press of 100 grams.

3. Experiment results

Diameters of joint hardfacings for SDP were measured using a micrometer MC 125 with 0,01mm accuracy. Results of measurements of joints and hardfacings are presented in table 2.

Chemical analysis of sample pieces shows that joints were made of medium carbon – low alloyed steel 37XГМА (4041). Alloying elements chrome, manganese, molybdenum, nickel significantly increases hardenability of steels. Hardening effect is amplified when steel is alloyed with several elements.

Table 2. Geometric parameters of the hardfacing collars

The name of the lock	The outer diameter of the lock, mm	The outer diameter of the hardfacing collars, mm (1 plane)			(1 plane)		
		1	2	3	1	2	3
Initial	127,14	-	-	-	-	-	-
BoTn 1000	127,23	133,16	133,18	133,02	133,23	133,18	133,06
BoTn 3000	127,21	132,37	132,13	131,74	132,31	132,36	132,33
BoTn 5000	127,19	132,91	132,56	132,80	132,51	132,61	132,64

All alloying elements decrease critical hardening rate. Therefore, hardness of alloyed steels can be increased by cooling them at lower rates, than for carbon steels. Besides, hardening alloyed steels can be carried out using slower cooling agents, what can decrease the possibility of hardening defects, first of all hardening cracks. This is especially important for DP body and joint, that are assembled by friction welding.

Hardfacings were found to have carbon content up to 0,7% and carbide forming chemical elements: chrome – up to 8%, manganese – up to 0,8%, silicon – up to 0,4%, molybdenum – 0,13% (BoTn 5000). Main alloying elements increase ferrites hardness and durability. Silicon, manganese and nickel increase durability of ferrite the most. Chrome is known to improve mechanical properties of structural steel and special properties of steels, for example, resistance to aggressive environments, that are present during borehole drilling.

Molybdenum is a strong carbide that are forming element. It forms carbides and alloyed ferric carbide, promotes grain size reduction, increases durability and high temperature strength. At the same time, there was no molybdenum found in BoTn 1000 and BoTn 3000 hardfacings. Vanadium and titanium are also used to reduce grain size, but the content of these elements should be limited (up to 0,3…0,5%), as they can form carbides at grain boundaries, which can lead to brittle fracture. Among tested paddings titanium was only found in BoTn 5000, less than 1%. Nickel gives steels corrosion resistance, high durability and ductility, increases hardenability, impact toughness, decreases cold-shortness threshold, thus decreasing the potential of brittle fracture of welded metal. The largest content of nickel (up to 3%) was found in BoTn 3000 hardfacing. Niobium improves acid resistance and promotes the decrease of corrosion in welded constructions. Niobium in BoTn 3000 hardfacing is dispersed irregularly in the cross-section totaling 44 % in spectrum 2.

Microstructure of sample pieces is presented in fig. 1. Thickness of the hardfacing after mechanical processing equaled: about 2 mm for BoTn 1000 padding; up to 3,5 mm for BoTn 3000 and BoTn 5000 hardfacings.

Welded metal has dendrite structure. In contact are of hardfacing and base material dendrite axes are structured perpendicularly to the plate surface. This is especially distinct for BoTn 5000 hardfacing. Hardfacing has an area of thermal influence with changed structure close to it – large-needled martensite near the hardfacing, small-needled martensite further. Structure of the main metal – secondary sorbite. Segregation banding can be visible in the main metal, which is typical for hot forming.

Dispersion of microhardness in welded layers on sample pieces is given in figure 3. Hardfacing hardness decreases with distance from the surface due to dendrite structure enlargement and is within the range: HV7300…8000 MPa for BoTn 1000, HV 8100…9600 MPa for BoTn 3000 и HV 9400…11200 MPa for BoTn 5000.

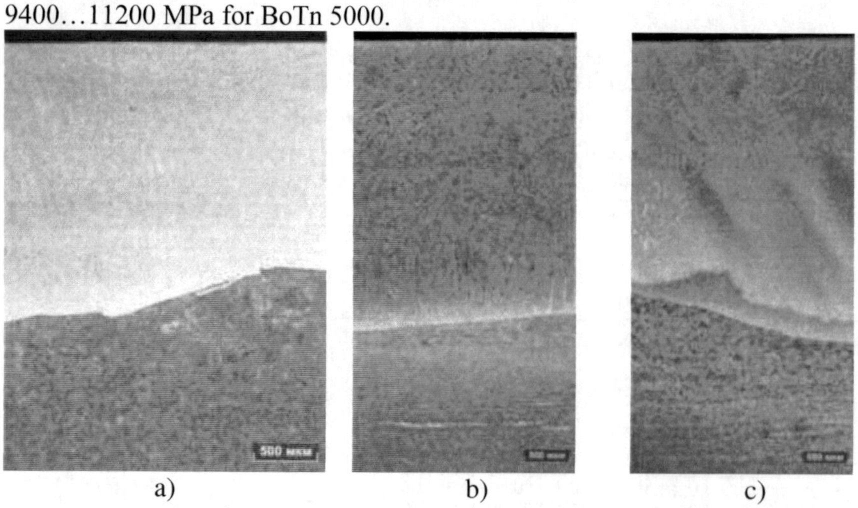

a) b) c)

Figure 1. Fragments of the microstructure of the hardfacing samples, ×50:
a – BoTn 1000;
b – BoTn 3000;
c – BoTn 5000

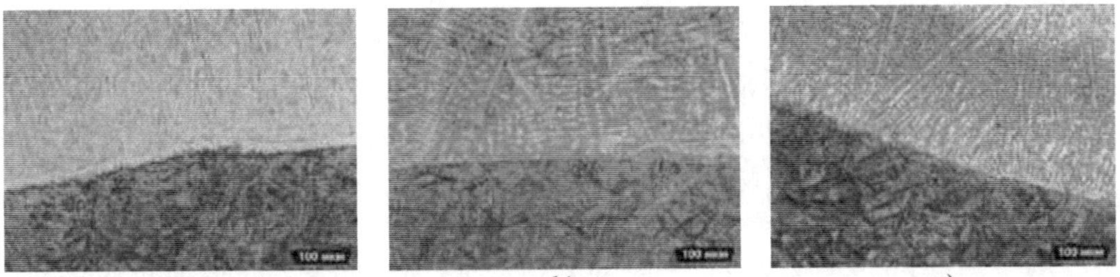

a) b) c)

Figure 2. Contact area of the hardfacing and the base metal, ×200: a – BoTn 1000; b – BoTn 3000; c – BoTn 5000

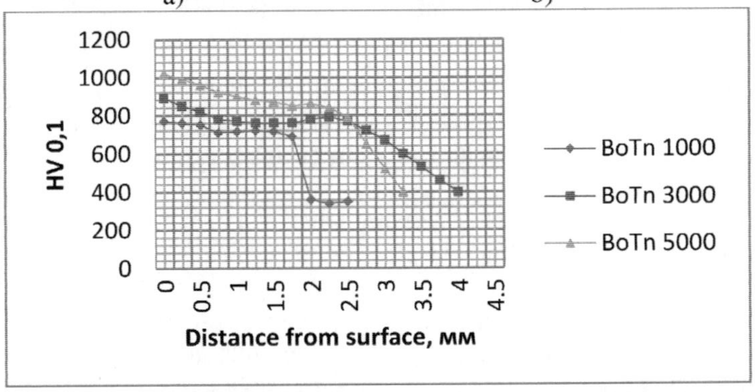

Figure 3. Microhardness of the welded layer in the samples

a) b) c)

Figure 4. Sample pieces after tests: a – BoTn 1000; b – BoTn 3000; c – BoTn 5000

Results of wear test for sample pieces (figure 6) are given in figure 5. Tests of welded materials in relation to joint material used in batch production tell about the effectiveness of each tested sample piece. BoTn 5000 hardfacing provide the best wear resistance. Overall, there is a correspondence between hardfacing hardness and wear resistance.

In order to determine the effectiveness of wear resistant hardfacings for SDP 86, metrological test were carried out on the premises of «BIS-Service» in Nizhnevartovsk, Russia. SDP 86 were supplied for repair due to them having been rejected because of inner and outer joint threads. Measurements of outer joint diameter were carried out using a beam trammel BT 250 with 0,05 mm precision. It was found out that one drill stem contained both SDP 86 with outer welded hardfacings (BoTn 3000), and joints without hardfacing.

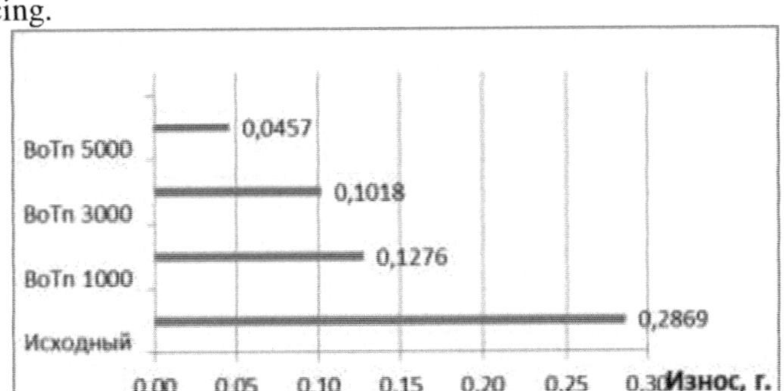

Figure 5. Comparative diagram of wear

The outer diameter of joint without paddings equaled 103,5…104,8 mm, while for joints with paddings – 108±0,1 mm. Measurements of outer diameter for areas with paddings for SDP 86 show a

minor wear of welded material (diameter 110, 8…112,0 mm) and no joint wear. There was also no roughening and detachment of material.
According to metrological analysis of SDP 86, supplied for repair, following conclusions can be made :
– hardfacings increase durability of drill pipe parts, promoting increased abrasive wear resistance for DP joints in the aggressive geological environments ;
– using drill pipes without hardfacings is unacceptable, as their further use as parts of drill stems may lead to emergencies;
- application of the pipes with the hardfacing collars together with the collars without hardfacing, due to varying degree of wear of locks in the drill-stem, is also impermissible.

4. General conclusions and recommendations

1. Dispersion of types of steels, used to make drill pipe joints, has a negative effect on their durability, what in combination with low hardness HV 2400…2800 MPa and strength of thread turns leads to low durability and having to repair them at pipe control shops.
2. Effective way of improving the quality of drill pipe joints is welding wear-resistant hardfacings to them (hardbanding). One of the companies developing hardfacing materials and providing hardbanding services is Hilong (China) with BoTn hardfacings.
3. Chemical composition of hardfacing materials BoTn shows the presence of carbide forming elements (Cr, Ti, Mo,Mn), irregularly dispersed along the welded padding, that increase hardness and form simple and complex carbides, that possess high hardness along with decent viscosity (due to the presence of Ni, Nb).
4. Results of wear test of welded materials, in relation to joint made in batch production (test ASTM G65 degree of resistance to abrasive wear), shows the effectiveness of all the parts made by «Technomash» using the technology from Hilong (China). BoTn 5000 has the best wear resistance and, overall, there is a correlation between hardness of the material and wear resistance.
5. Results of metrological analysis of drill pipes repaired by «BIS-Service» from Nizhnevartovsk by recutting threads of nipple and collar parts, show no wear of the joint body with BoTn 3000 hardfacing. Pipes without hardfacing have outer wear that equals 103,5…104,8 mm and they are subject to demotion to lower quality grade, while their further use as parts of drill stems may lead to emergencies.
6. Due to different wear resistance of original joint materials and welded materials it is highly recommended that the same type of BoTn hardfacing are used as parts of a single drill stem.

References

[1]. Basarygin Y, Bulatov A, Proselkov Y 2000 Complications and accidents when drilling oil and gas wells *Nedra*
[2]. Colin D Duraband N.C. 2016 For more than 7 years successfully protects drill pipes used in Russia *Drilling and oil* **09** pp 72-74
[3]. Colin D 2017 The choice of hardbanding material: why you should never skimp on quality *Drilling and oil* **06**
[4]. Sadovnikov A, Fedorov S, Fedorova L, Lompas A, Fomin O 2017 Hardberding as an effective method of increasing the resource of drill pipes and protecting the casing string. The results of tribological tests of wear-resistant surfacing for boring locks *Drilling and oil* **06** pp 30-35
[5]. Sadovnikov A, Fedorov S, Fedorova L, Lompas A, Fomin O 2017 SSK: Report on hardbanding *ROGTEC. Russian oil and gas technologies* **49** pp 12-25
[6]. Shilovsky A 2016 Hardbanding on the protection of the drill pipe *Drilling and oil* **04** pp 58-59

IOP Conf. Series: Materials Science and Engineering **307** (2018) 012039 doi:10.1088/1757-899X/307/1/012039

Molecular dynamics study on nano-sized wiredrawing: possible atomistic process and application to pearlitic steel wire

K Saitoh[1], K Yoshida[1], K Oda[1], T Sato[1], M Takuma[1] and Y Takahashi[1]

1 Department of Mechanical Engineering, Faculty of Engineering Science, Kansai University
3-3-35 Yamate-cho, Suita, Osaka, 564-8680 JAPAN
E-mail: saitou@kansai-u.ac.jp

Abstract. The process of nano-sized wiredrawing is investigated by using molecular dynamics (MD) simulation in this study. The authors have constructed novel computation models of wiredrawing, in which a single wire of just a several nanometers in diameter is smoothly drawn through a perfectly rigid die together with lubrication mechanism and is forced to be shaped into thinner one. Interatomic potentials used in MD simulation is a conventional pairwise type useable for iron-carbon binary system (for pearlitic steel). For MD model of pearlite steel wire, it is recognized that ferrite-cementite interface effectively offers high-speed diffusion path for carbon atoms from cementite side to ferrite side (elementary mechanism of cementite decomposition). As conclusion, we showed by using atomistic simulation that nano-sized wiredrawing process is theoretically quite possible.

1. Introduction

Wiredrawing is a conventional plastic working technique in which a metallic wire (or bar, for larger diameter) is drawn through hard die and the diameter of material is effectively reduced [1]. In the drawing process, a wire is provided much strengthening as well as finer surface roughness. That is why the wiredrawing process has been used for manufacturing industrial equipment, such as suspension wire for a bridge, steel code for a tire and micro-sized wires for electric device. In recent years, demand for wire with smaller diameter is increasing and the finest wire manufactured in the industry has reached a few micrometers. This trend of wiredrawing technology will be ongoing, so nanometer-sized wire will be produced by drawing technique in near future. Compared with some other processing techniques to make nanowire (e.g. crystal growth techniques such as CVD, nano-metric physical process such as FIB, or nano-sized mechanical machining), drawing will be advantageous because it continuously reduces its diameter from normal size (millimeter or micrometer) to nanometer size. Besides, by drawing technique, microstructure inside of nanowire as well as surface state may be controllable.

However, there remains many problems in miniaturizing drawing process of normal size into nanometer size, such as atomistic friction between wire and die during processing. What is important and needed at this point is to theoretically assess possibility of drawing material in nano-sized scale. In this study, the authors are trying to model nano-sized drawing process by using atomistic simulation, i.e. molecular dynamics (MD) methodology. The MD method has predicted many unknown behaviors of nano-sized material, so it will fit this purpose [2].

On the other hand, from materials science's view point, it is crucial to understand the mechanism of strengthening during drawing. Usual strong wires are made of pearlitic steel, which is a compound between iron (Fe) and carbon (C) atoms. The pearlite phase is a key microstructure, but it has a complicated hierarchical structure in which cementite phase, a quasi-stable phase between Fe and C atoms, and ferrite phase, a pure α-Fe (body-centered cubic(bcc) structure) phase, are combined and aligned in lamellae arrangement [3]. The latest question of microscopic behavior of pearlite phase is how cementite decomposes by releasing carbons into ferrite and how such carbons affect the strength of the wire. This phenomena is often called "cementite decomposition". By using MD model, we will clarify this mechanism from an atomic point of view.

2. Theory and Method

2.1. Potential function for Fe-C system

In this study, as atomistic simulation methodology, molecular dynamics (MD) is used. In MD simulation, only initial atomic configuration and a potential energy function between atoms are specified, then trajectory and velocity of atoms time by time are obtained numerically.

The issue of determination of potential function is important. As of Fe-C system used here, the formulation by Levchenko *et al.* [4] is supposed to be better at this point since it is based on well-used Johnson-type potential for Fe-Fe interaction [5] and is including new *ab initio* data. By using this Fe-C potential function, the authors confirmed that a cementite crystal structure (Fe_3C) is well stably simulated. There have been a lot of elaborative works concerning formulating any precise potential function for Fe-C system or Fe_3C including EAM [6] or MEAM [7]. However, to our present knowledge, any reliable many-body potential function has yet appeared.

2.2. Nano-sized wiredrawing models, pearlite steel model and computation condition

In atomistic modelling of pearlite phase, we should consider microscopic portion of it, where a ferrite (α-Fe) phase having a highly symmetric unit and a cementite phase having the orthorhombic crystal unit are both arranged in narrow filament shape with only a few dozens of nanometres in width (it is often called "lamella" structure) [8][8]. The drawing is performed along the direction of these lamella filament. Due to the limitation of treatable numbers of atoms, the pearlite MD wire structure studied here is composed of one ferrite phase and one cementite phase.

The present MD model is set up as shown in figure 1. The wire has been inserted into a hollow die which has tapered shape in proportion to an industrial larger drawing die. In MD model, the die region comprises discrete atoms but they are totally constrained in space and the die behaves as rigid body block. Geometry of a conical drawing die is almost specified by die semi-angle and length of die land, and a resultant reduction ratio in cross-section area is an important factor.

In drawing, two chucking regions at both ends of the wire are forced to move with constant velocities which are in advance determined so as to maintain constant volume of the wire, supposing almost plastic deformation there [9]. The present wiredrawing is simulated with very high speed but it is not so unrealistic. However, there remains an issue due to miniature scale that friction between die and wire becomes enormously larger [10], regardless of wire speed. Therefore, in order to reduce friction at atomic contact, a kind of artificial lubricating mechanism has to be introduced in the present modeling. In the method, interatomic potential energy is multiplied by an arbitrary factor ω which is less than unity [2]. This affects directly pairwise potential so that cohesive energy, i.e. attractive force, near contact region is reduced by ω. As a result, friction coefficient μ between die and wire is effectively reduced to a smaller value such as μ=0.05 or lower.

In detecting dislocation or other lattice defects, we use a generic software for MD visualization and post-analysis called OVITO [11] along with DXA(dislocation extraction algorithm)[12]. By the DXA, dislocation lines are recognized together with information such as edge/screw type and size and direction of Burgers vector.

The important issue in constructing pearlite model is the relation of crystal orientation at the interface between ferrite (α phase: iron with very low carbon content) and cementite (θ phase: iron carbide) phases. We employ Bagaryatsky relation [13] which is supposed to exhibit relatively lower

IOP Conf. Series: Materials Science and Engineering **307** (2018) 012039 doi:10.1088/1757-899X/307/1/012039

lattice mismatch and lower interfacial energy between two crystal phases. The detailed arrangement is shown in figure 1(a). Several structural crystalline parameters as well as the geometry of drawing and computation conditions are summarized in table 1.

(a) (b)

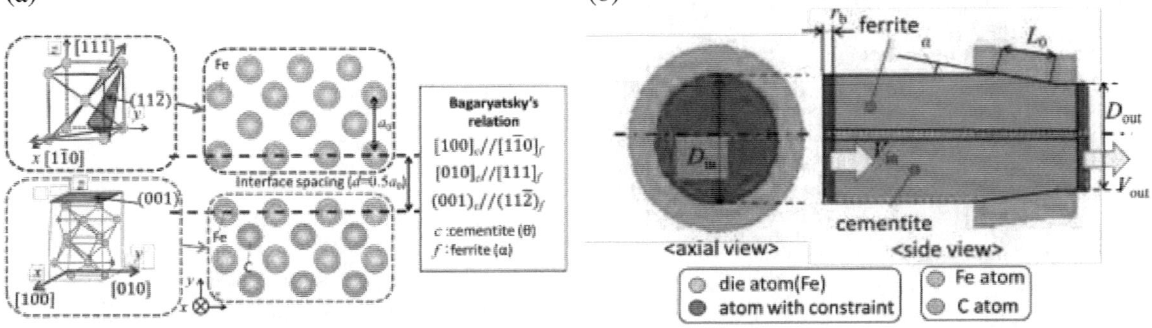

Figure 1. Nano-sized drawing model of pearlitic steel wire used for MD simulation. (a)Arrangement of crystal orientation of ferrite(α phase: α-Fe) and cementite (θ phase: Fe_3C) crystals at the interface (using Bagaryatsky's relation, here). (b)Configuration of two layers and geometry of drawing

Table 1. Geometry of drawing and MD computation conditions

Lattice constant (ferrite (α)) a_0 (nm)			0.28665
Lattice constant (cementite (θ)) a, b, c (nm)			0.4523, 0.5089, 0.6743
The number of atoms	total		77657
	Fe	cementite (θ)	22607
		ferrite (α)	22709
	C	cementite (θ)	7514
	die		24827
Die semi-angle α (deg)			7.0
Inlet diameter D_{in} (nm)			6.8796
Length of die land L_0 (nm)			4.12
Drawing velocity V_{in}, V_{out} (m s^{-1})			36.4, 50.0
Temperature T (K)			10, 600, 1000
Time increment × Steps			2.0 × 40000

3. Results and Discussion

3.1. Behaviour of carbon atoms near ferrite-cementite interface

Figure 2 shows occurrence of lattice defects during drawing by using CNA technique [14], which can make distinction between atoms with usual crystal (ordered) structure and with other (disordered) structure. Atoms in θ-phase are all identified as crystal defect because the CNA does not detect orthorhombic crystal that Fe_3C is supposed to have. The interface position shifts a little bit to easily deformable ferrite side, but it almost retains its original flat shape during drawing. As drawing begins, first the region with phase transition from bcc structure to face-centred cubic(fcc) structure takes place and expands particularly to front portion of the wire, and after then dislocation or grain boundary is left behind around the transition region.

Figure 3 shows time transition of normal stress in drawing direction (i.e. drawing stress) averaged over atoms in the wire. Stress increases rapidly at very beginning of drawing. By the completion of phase transition or formation of defects the drawing stress decreases rapidly, and after that it is kept at almost constant value. That stationary stress is regarded as a flow stress meaning plastic deformation. The smaller the system temperature is, the larger the flow stress is. From this fact we can guess that, in α-θ interface, C atoms are diffusing around θ phase side much more in higher temperature and enter into α phase side that much. When C atoms behave in such manner, they works as resistance of plastic deformation in ferrite phase, then hardening results in. This mechanism is rather similar to strengthening usually explained for the formation of martensite in Fe-C system.

IOP Conf. Series: Materials Science and Engineering **307** (2018) 012039 doi:10.1088/1757-899X/307/1/012039

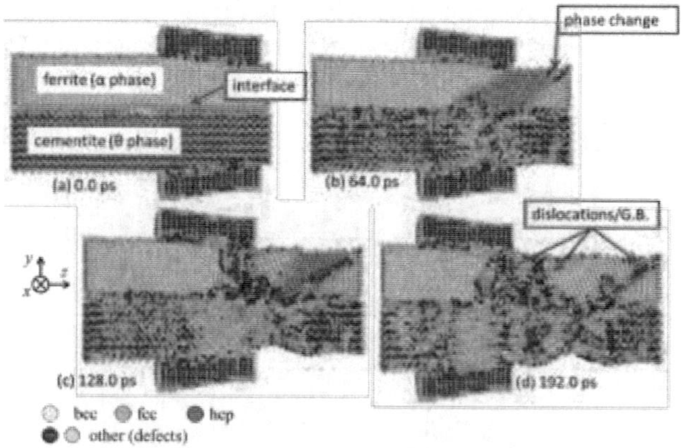

Figure 2. Deformation and lattice defects observed by CNA technique [14] in MD wiredrawing model of pearlite steel (viewing on cross-sections)

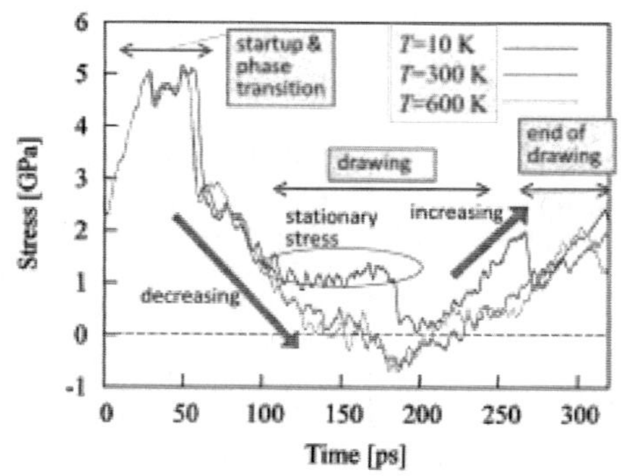

Figure 3. Time transition of averaged stress in drawing direction (drawing stress) during drawing process

3.2. Relation between dislocation motion and carbon diffusion (inspection of "drag mechanism")

Since we understand that C atom diffusion around the interface is key behavior for deformation process in pearlitic wire model, we observe position and trajectory of C atoms during drawing. Figure 4 shows time-evolution of the position of C atoms found in ferrite region at very close to the interface at T=10 K. The reason why observing in such low temperature is that, in high temperature like 1000 K, due to high diffusivity, C atoms move rapidly and are not captured at a stationary position.

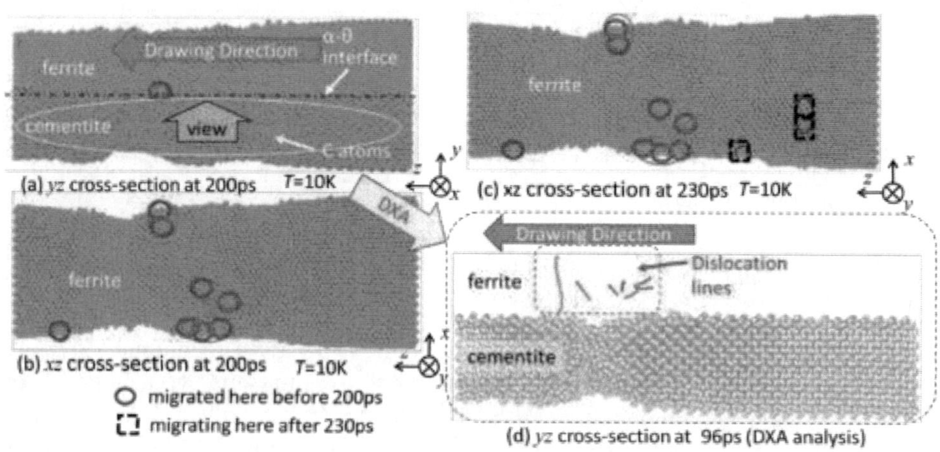

Figure 4. (a),(b),(c)Change of positions of carbon atoms around θ-α interface at which drawing stress rises again, and (d)position of carbon atoms and dislocations lines detected by DXA method [12]

Let us compare results between at times 200 ps (figure 4(a),(b)) and 230ps (figure 4(c)) (time means duration after the start of processing), at which drawing stress rises again from the saturated flow stress. The number of C atoms detected there are increasing. This movement of C atoms is probably being caused by plastic deformation in drawing. In literature, there was a proposal of mechanism that dislocation movement supports the travel of C atoms from θ phase to α phase, which can be called "drag mechanism (by dislocation motion)" [15]. However from the present MD simulation, as shown in figure 4(d), dislocation lines do not observed not so much even in low temperature case (figure 4(d) is obtained as the maximum appearance of dislocations). We can conclude that C atoms are certainly diffusing very much around the interface and such diffusive C atoms can help the wire in resisting to drawing deformation (more generally, plastic deformation), but it is not directly linked to dislocation motion as in "drag mechanism." However, at this point any plastic deformation in θ phase has not captured. We think it needs further discussion and is left to further study.

4. Conclusion

In this study, by using molecular dynamics simulation, possibility of nano-sized wiredrawing of metals was investigated. Although the friction between die and wire atoms tends to become tremendously larger, in MD models using artificial lubrication mechanism drawing is theoretically possible. Pearlitic steel model with two layer of cementite and ferrite were successfully simulated. At the interface carbons diffuse around the interface and work as resistance to drawing stress and plastic deformation. However, "drag mechanism" by dislocation proposed for cementite decomposition mechanism, was not confirmed by the present atomistic scale.

Acknowledgement

This work was was supported by JSPS KAKENHI Grant Number JP16K05994(Grant-in-Aid for Scientific Research (C)) and Nippon Steel & Sumitomo Metal Corporation.

References

[1] Wright R N 2011 *Wire Technology: Process Engineering and Metallurgy* (Burlington: Butterworth-Heinemann(Elsevier)) p1
[2] Saitoh K, Sameshima Y and Daira S, 2014 Nano-scale modelling and simulation of metal wiredrawing by using molecular dynamics method *World J. Nano Sci. Eng.* **4** 70–83
[3] Nematollahi Gh. A, Grabowski B, Raabe D and Neugebauer J 2016 Multiscale description of carbon-supersaturated ferrite in severely drawn pearlitic wires *Acta Mater.* **111** 321–34
[4] Levchenko E V, Evteev A V, Belova I V and Murch G E 2009 Molecular dynamics simulation and theoretical analysis of carbon diffusion in cementite *Acta Mater.* **57** 846-53
[5] Johnson R A. 1964 Interstitials and vacancies in α-iron *Phys Rev* **134** 5A 1329–36
[6] Ruda M, Farkas D and Garcia G 2009 Atomistic simulations in the Fe-C system *Comput. Mater. Sci.* **45** 550–60
[7] Liyanage L S I, Kim S G, Houze J, Kim S, Tschopp M A, Baskes M I and Horstemeyer M F 2014 Structural, elastic, and thermal properties of cementite (Fe_3C) calculated using a modified embedded atom method *Phys. Rev. B* **89** 094102
[8] Borchers C and Kirchheim R 2016 Cold-drawn pearlitic steel wires *Prog. Mater. Sci.* **82** 405–44
[9] Hill R 1950 *The Mathematical Theory of Plasticity* (NewYork: Oxford University Press) p11
[10] Israelachvili J N 2011 *Intermolecular and Surface Forces* (Academic Press (Elsevier))
[11] Stukowski A and Albe K 2010 Dislocation detection algorithm for atomistic simulations *Modelling Simul. Mater. Sci. Eng.* **18** 2 025016
[12] Stukowski A, Bulatov V V and Arsenlis A 2012 Automated identification and indexing of dislocations in crystal interfaces *Modelling Simul. Mater. Sci. Eng.* **20** 085007
[13] Bagaryatski Y A 1950 *Dokl. Akad. Nauk. SSSR* **73** 1161
[14] Honeycutt J D Andersen H C 1987 Molecular dynamics study of melting and freezing of Small Lennard- Jones clusters *J.Phys.Chem.* **91** 4950–63.
[15] Li Y J, Kostka A, Choi P, Goto S, Ponge D, Kirchheim R and Raabe D 2015 Mechanisms of subgrain coarsening and its effect on the mechanical properties of carbon-supersaturated nanocrystalline hypereutectoid steel *Acta Mater.* **84** 110–23.

IOP Conf. Series: Materials Science and Engineering **307** (2018) 012040 doi:10.1088/1757-899X/307/1/012040

Improvement of corrosion resistance of low-alloy steels by resurfacing using multifunction cavitation in water

Masataka Ijiri[1] and Toshihiko Yoshimura[1]

[1] Tokyo University of Science, Yamaguchi
Email: ijiri@rs.tusy.ac.jp

Abstract. Low-alloy steels are based on carbon steel in combination with several percent or less (in many cases, 1 mass%) alloying elements, and they offer improved resistance to corrosion at a cost slightly higher than that of carbon steel. However, these materials do not exhibit the same corrosion resistance as stainless steel. The authors have previously developed a novel multifunction cavitation (MFC) technique, which combines ultrasonic cavitation with water jet cavitation. In this study, MFC was used to modify the surface of Cr-Mo steel (SCM435) and Ni-Cr-Mo steel (SNCM630). MFC was found to improve the residual stress value of the material as the result of surface modification while also imparting high strength and superior corrosion resistance.

1. Introduction

Low-alloy steel is based on carbon steel combined with several alloying elements, such as Ni, Cr, and Mo, in an amount of several mass% or less (in many cases, 1 mass% or less). Different types of low-alloy steels are made for various applications: heat-resistant steel, low-temperature steel, weather-resistant steel, and high tensile-strength steel. This steel costs slightly more than carbon steel but has better corrosion resistance. However, the corrosion resistance of low-alloy steel is inferior to that of stainless steel. In low-alloy steel, rust formation is prevented by two methods. Firstly, the surface can be covered by paint or plating. Secondly, the alloy contains Cu, Cr, P, and Ni, which are corrosion-resistant alloying elements. In appropriate amounts, these alloys encourage the formation of a dense rust on the surface as the steel is exposed to alternating moderate dry and wet conditions in the atmosphere over time. This dense rust protects the underlying steel material, and the formation of deeper rust is gradually suppressed. However, elements effective for improving corrosion resistance are harmful to the excellent mechanical properties and good weldability Low-alloy steel is required to have. As a result, there is a limit to the amount added, and a limit to the corrosion resistance that can be "built in" to the alloy mixture.

The authors developed a new water jet cavitation (WJC) technology [1]. This technique is a mechanical and electrochemical cavitation method because it has electrochemical action by a high-temperature reaction field of ultrasonic cavitation (UC) in addition to mechanical action by high-pressure microjets (MJs). Furthermore, it can be called multifunction cavitation (MFC) because it combines new functions when processing the surface of various materials. In this technology, various material properties have been studied so far. Previously, the authors reported that TiO_2 powder, which is widely used as a photocatalytic material, can be fabricated at the nanoscale when processed with MFC [2]. In addition, in an indium tin oxide (ITO) film, which is applied to the transparent electrode mounted on liquid crystal displays, when the ITO film coating on a transparent electrode is subjected to MFC treatment, indium, which has a low melting point, can be melted and released [1]. This phenomenon demonstrated that MFC heats the specimen surface to a high temperature in water.

In the present study, MFC was applied to the modification of low-alloy Cr-Mo steel (SCM435) and Ni-Cr-Mo steel (SNCM630) surfaces. Then, we investigated the effect of MFC on the alloy microstructure and hardness and compared the results to alloys processed with WJC.

2. Experimental methods

The material used for these tests was Cr-Mo steel (SCM435) and Ni-Cr-Mo steel (SNCM630), both structural steels, the chemical compositions of which are shown in Table 1.

Table 1. Chemical composition of the low-alloy steels used in this work (mass%).

	C	Si	Mn	P	S	Ni	Cr	Mo	Cu	Fe
SCM435	0.37	0.32	0.81	0.014	0.012	0.012	0.95	0.15	0.14	Bal.
SNCM630	0.29	0.24	0.44	0.11	0.018	2.93	3.01	0.56	0.03	Bal.

Figure 1 is a schematic diagram of the equipment used for MFC processing. In conventional WJC apparatus, a nozzle is fixed in water at room temperature and water jets are pumped through it at a discharge pressure of 35 MPa. The nozzle diameter was 0.8 mm. The distance between the nozzle and the specimen was 65 mm. The size of the water tank was 60×45×37 mm^3, and it was filled with tap water. In MFC treatment, an ultrasonic transducer is placed next to the water jet nozzle for conventional WJC, and ultrasonic waves are propagated to the water jet flowing from the nozzle. The ultrasonic wave had an output of 225 W and frequency of 28 kHz.

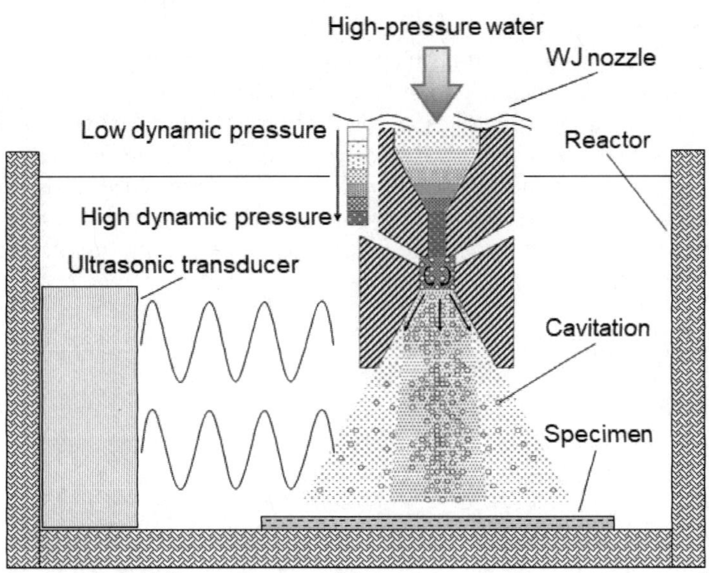

Figure 1. Equipment employed for surface machining by water jet cavitation under ultrasonication.

As shown in Figure 2, bubbles from WJC isothermally expand when the sonic pressure due to ultrasonic irradiation exceeds the Blake threshold, and after the bubbles expand to a certain size, Rayleigh shrinkage occurs rapidly (adiabatic compression). The flow cavitation, including the hot spot, becomes MFC by the repetition of isothermal expansion and adiabatic compression. At the same time that the MFC begins to collapse, it approaches the specimen surface.

IOP Conf. Series: Materials Science and Engineering **307** (2018) 012040 doi:10.1088/1757-899X/307/1/012040

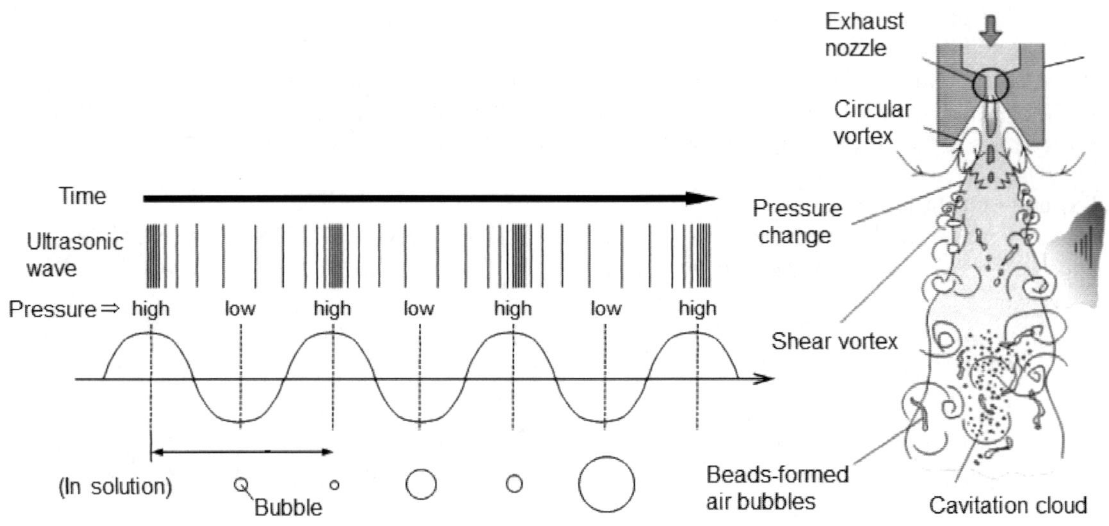

Figure 2. Multifunction cavitation mechanism.

Figure 3 shows the bubble shape change from photographs taken by a high-speed camera during bubble collapse into an aspheric shape [3]. As the volume of the bubble decreases, a MJ with a columnar shape is formed and impinges on the specimen surface. This phenomenon is referred to as micro-forging because the MJ provides high-temperature and high-pressure processing in a microscopic area. Conventional WJC produces large high-pressure bubbles (diameter of several hundred micrometers at approximately 1,000 MPa), whereas bubbles produced with conventional UC are small (several micrometers) with high temperature (several thousand degrees Celsius). In contrast, the MJ that occurs in a bubble during MFC is a deforming liquid-phase body similar to a liquid jet (column) at the terminal stage of bubble collapse, and the interior temperature and pressure of the bubble become high (several thousand degrees Celsius and approximately 1,000 MPa, respectively). The processing time for all three methods was 2 min. Residual stress was measured by the full-width at the half-value breadth method from the (211) lattice spacing strain using Cr Kα X-ray diffraction (MSF-3M, Rigaku Corporation). Micro Vickers hardness tests were performed at room temperature under a load of 1.96 N for 10 s. The hardness values were averaged from eight measurements after the minimum and maximum of ten measurements were discarded.

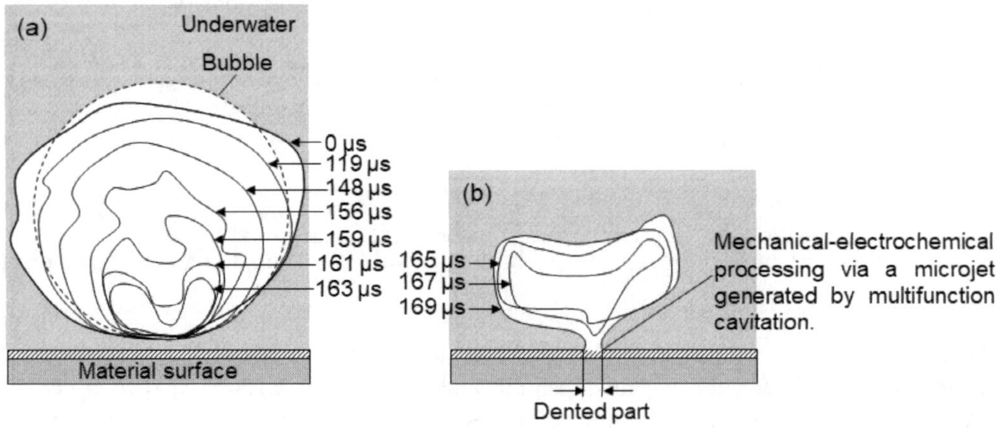

Figure 3. Images showing the aspherical collapse of a bubble at time intervals of (a) 0-163 μs and (b) 165-169 μs.

IOP Conf. Series: Materials Science and Engineering **307** (2018) 012040 doi:10.1088/1757-899X/307/1/012040

In addition, hardness measurements near the surface after processing were performed after slight mechanical polishing to remove the oxide film and peening marks formed during processing. Microstructure observations were conducted using optical microscopy (OM) and scanning electron microscopy (SEM; S-4800, Hitachi). Before observation, the specimens were etched in 5 vol% Nital. The specimens used for all analyses after processing were cut to 1×1 cm^2. For dissolved oxygen (DO) measurement, a portable type DO analyzer (OM–71, Horiba, Ltd.) was used. This measurement was carried out by appropriately collecting treated water in which cavitation occurred within a reactor made of SUS304 for a processing time of 30 min.

3. Results and discussion

Table 2 shows the results of residual stress measurements. To further increase the compressive residual stress generated after WJC and MFC processing, the specimen surface was first subjected to hard grinding, which imparted tensile residual stress by a shakedown effect. These measurements were conducted in a state without rust on the specimen surface just after each processing and were performed in a direction parallel to the grinding direction. It was confirmed that grinding imparted a tensile residual stress in the grinding direction of the unprocessed specimens, while a compressive residual stress was imparted in the vertical direction. In WJC and MFC processing, compressive residual stress is generated when the specimen surface is compressed by the pressure from cavitation because the MJ applies force in the lateral direction, leading to plastic deformation around the region that is elastically constrained by its surroundings.

Table 2. Compressive residual stress of specimens before and after various treatments (MPa).

		SCM435	SNCM630
As received after grinding	Vertical direction	-173	-238
	Parallel direction	+202	+155
UC		-216	-196
WJC		-361	-450
MFC		-293	-481

Figure 4 shows the change in the micro Vickers hardness for each processed surface in the depth direction but does not include the oxide film and peening marks. In the hardness of the near-surface region in the unprocessed specimen, SCM435 was 246 HV and SNCM630 was 256 HV. The hardness distribution of the unprocessed specimen remained unchanged in the depth direction. Thus, the surface grinding did not significantly influence the hardness of the unprocessed specimens. The hardness in the near-surface region of the WJC-processed SCM435 was 271 HV. The hardness decreased at a depth of about 0.75 mm from the surface. On the other hand, the hardness in the near-surface region of the MFC-processed specimen was 276 HV. The hardness decreased at a depth of about 0.5 mm from the surface layer. This hardness change depends on the value (compressive stress field) where compressive residual stress was added in the depth direction. The compressive stress field was deeper in the WJC specimen than in the MFC specimen. The compressive stress field occurred at a depth of about 0.5 mm after WJC processing, and it occurred at a depth of about 0.25 mm after MFC processing. At a depth of about 0.50 mm in the WJC-processed specimen, the hardness decreased compared with the abrasive material due to voids and cracks. The influence of voids and cracks inside the specimen is thought to be the reason why the hardness decreased from a depth of about 0.5 mm in the specimen after WJC processing. Recently, the authors reported that voids and cracks tend to form inside the specimen after WJC processing [4]. The hardness in the near-surface region of the WJC-processed SNCM630 was 270 HV. The hardness decreased at a depth of about 1.00 mm from the surface. On the other hand, the hardness in the near-surface region of the MFC-processed specimen was 264 HV. The hardness decreased at a depth of about 1.00 mm from the surface. The compressive stress field occurred at a depth of about 0.75 mm after WJC processing, and it occurred at a depth of about 0.50 mm after MFC processing.

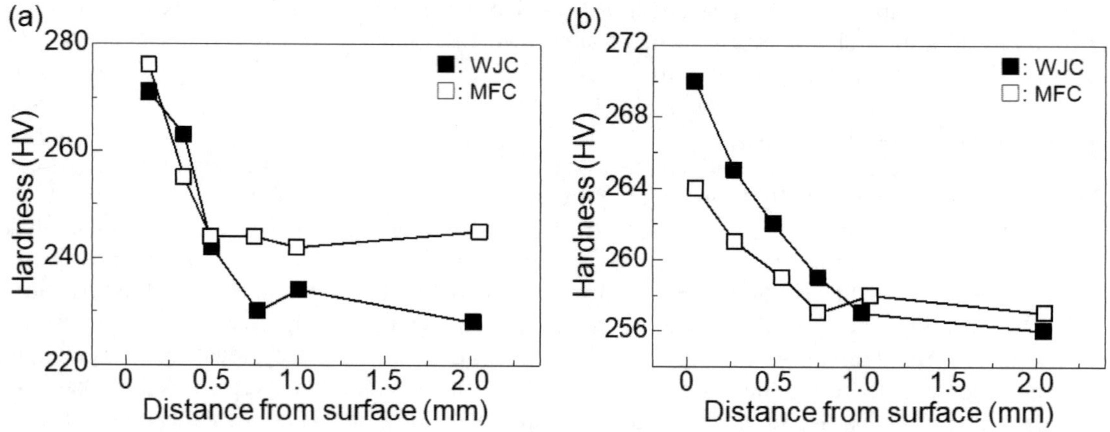

Figure 4. Hardness depth profile of the specimens after WJC and MFC treatment: (a) SCM435 and (b) SNCM630.

Figure 5 shows surface photographs of SCM435 (a, b) and SNCM630 (c, d) specimens after processing. The specimen surface was painted with oil-based ink prior to processing to identify the peening positions. The condition of the ink coating showed that the diameter of the peened portion of SCM435 was 40.8 mm after WJC processing and 42.2 mm after MFC processing. The peened diameter of SNCM630 was 41.3 mm after WJC processing and 42.7 mm after MFC processing. Therefore, MFC produced a slightly larger area than WJC produced. In addition, when the processed specimens were kept at room temperature for several months, corrosion developed after WJC processing, whereas there was almost no corrosion after MFC processing. Perhaps corrosion of the SNCM630 steel was prevented due to the formation of a dense oxide film on the specimen by increased temperature on the surface.

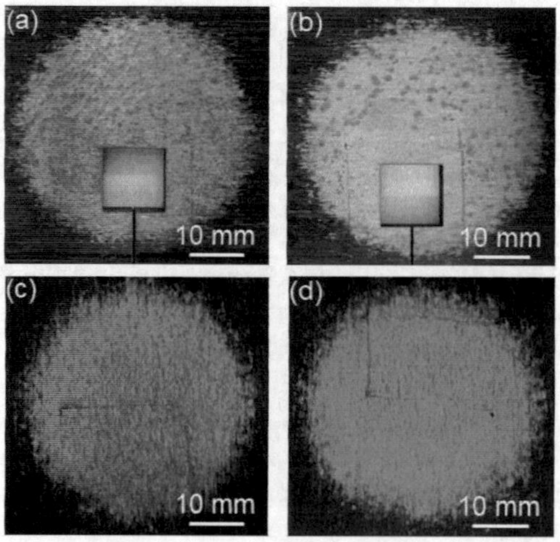

Figure 5. Surface change after treatment: of (a, c) WJC and (b, d) MFC. (a) and (b) are SCM435 and (c) and (d) are SNCM630.

Figure 6 shows OM micrographs of the SCM435 specimens after (a) WJC and (b) MFC. The particle size in the depth direction from the surface after each process was not significantly affected. In the near-surface region, no significant corrosion was observed after WJC processing, although it occurred easily after MFC processing. Corrosion was not observed in the ferrite phase but in the pearlite phase. The rust formed to a significant extent from the surface to a depth of approximately 1 mm.

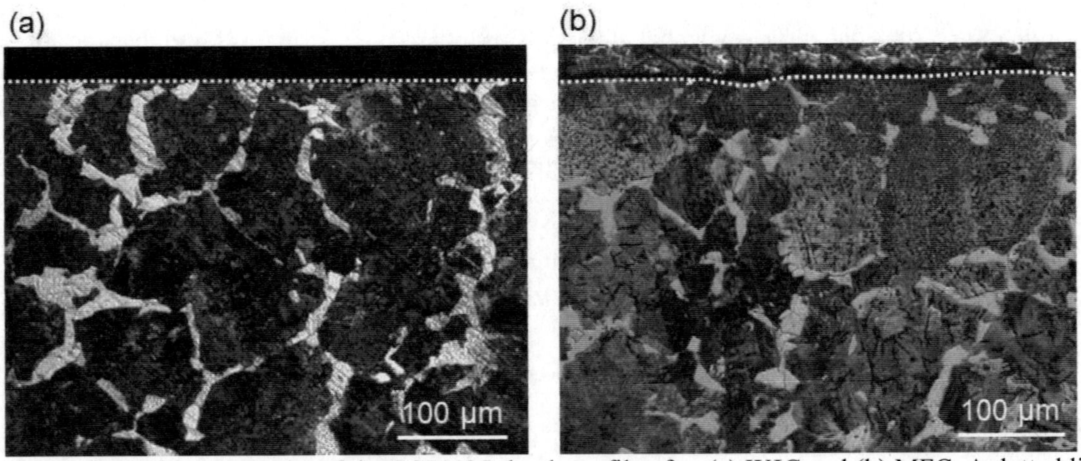

Figure 6. Optical micrographs of the SCM435 depth profile after (a) WJC and (b) MFC. A dotted line indicates a specimen surface.

Figure 7 shows SEM micrographs of the SNCM630 specimen depth profiles after WJC and MFC. The particle size in the depth direction from the surface after each process was not significantly affected. In the near-surface region, no significant corrosion was observed after WJC processing, although it occurred easily after MFC processing. Thus, selective oxidation is considered to have occurred in the bulk of the MFC-processed specimen. The element that is difficult to oxidize in this specimen is Cr, which migrates to the topmost surface and forms a Cr-poor region beneath the surface during heating by MFC processing. As a result, rust can be easily generated because the amount of Cr decreased in the near-surface region.

Figure 7. SEM images of the SNCM630 depth profile after (a) WJC and (b) MFC. A dotted line indicates a specimen surface.

The corrosion resistance of the specimen surfaces was assessed by measuring surface potentials, and the results are summarized in Table 3. The work function corresponding to the surface potential decreases as the surface roughness increases. The cavitation at the material surface during WJC generates impact pressures greater than several thousand megapascals. In addition, because some thermal energy is transferred to the material surface, a passive (corrosion-resistant) layer is formed. Soyama et al. [5] have also reported the formation of a passive layer as part of the mechanism by which the corrosion resistance of carbon steel is improved following the application of a cavitation jet. Regarding SNCM630, the surface potential of the WJC-processed specimen was lower than the abrasive material. It is probably influenced by the increase in surface defects compared with SCM435. During MFC, the bubbles generated by WJC (which are larger than UC bubbles) are irradiated with ultrasonic waves, and thus possess both high temperature (several thousand degrees Celsius) and high pressure (approximately 1,000 MPa). As these bubbles collide with the surface, a more stable passive layer is formed than that obtained with WJC.

Table 3. Surface potential of specimens after various treatments (mV).

	SCM435	SNCM630
As received	234 ± 14	359 ± 25
UC	194 ± 13	342 ± 18
WJC	382 ± 26	329 ± 18
MFC	544 ± 17	445 ± 16

The reason for the selective oxidation inside the specimen during MFC processing was investigated by monitoring the water temperature, and Figure 8 summarizes the results. While UC did not increase the water temperature, both WJC and MFC did raise the temperature in the reactor. This occurred as a result of the pressurization of the water by the high-pressure pump, because the kinetic energy of the water was converted to thermal energy. From the viewpoint of bubble energy, it is thought that both WJC and MFC generate larger bubble collapse energies than UC processing, which also increases the water temperature. The slightly higher temperature obtained during MFC compared to WJC processing is due to the application of ultrasonic energy, reflecting the hot spots generated in the bubbles.

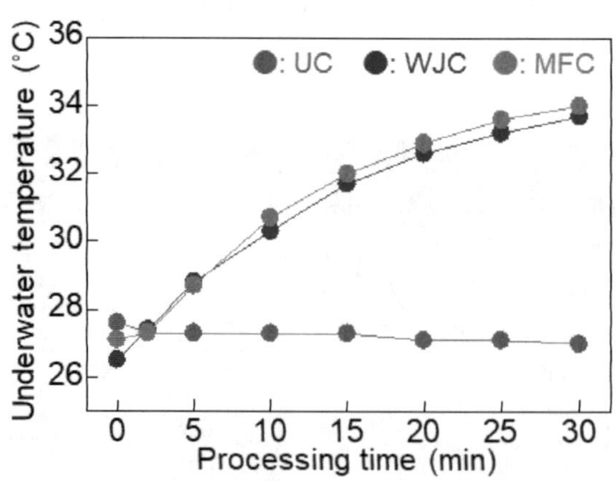

Figure 8. Relationship between processing time and underwater temperature.

Figure 9 plots the DO in the reactor over time for each processing method. The DO concentration was almost unchanged during UC, whereas it decreased during both WJC and MFC processing. In WJC processing, the pressure applied to the specimen surface increases due to the shock wave generated by the MJ. A part of the foam is deformed or bubbles become large again over a very short time span, and so both the water temperature and the temperature of the processed surface increase. As a result, it is believed that DO combines with Cr to form oxides on the processed specimen surface, causing DO in the water to decrease. During MFC, the bubbles (which have a high internal temperature and pressure) increase the sample surface temperature to a greater extent than during WJC processing due to the hot-spot phenomenon inside the bubbles. Thus, the DO concentration is reduced to a greater degree as a passive layer is generated.

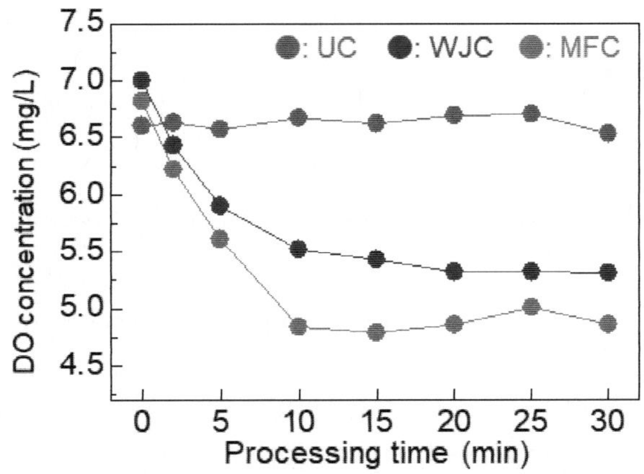

Figure 9. Relationship between processing time and DO concentration.

As described above, in the MFC technique, bubbles having characteristically high temperature and high pressure collide with the specimen surface as the result of irradiating large bubbles generated by WJC with ultrasonic waves. This method effectively improves the surface residual stress via surface modification and also increases the specimen strength and corrosion resistance.

4. Conclusion

Improvements in residual stress and surface modification leading to high strength and corrosion resistance were found to occur at the surfaces of specimens processed with MFC. Corrosion resistance was improved via the formation of an oxide film by selective oxidation, as well as the reduction of surface defects. We concluded that this oxide film was formed by dissolved oxygen in the water reacting with Cr on the metal surface during the MFC treatment.

Acknowledgment

This work was supported by the Innovative Science and Technology Initiative for Security Program of the Acquisition, Technology & Logistics Agency (ATLA) of Japan.

References

[1] T. Yoshimura, K. Tanaka, N. Yoshinaga, Development of mechanical-electrochemical cavitation technology, *J. Jet Flow Eng* **32** (2016) 10-17.

[2] K. Tanaka, T. Yoshimura and N. Yoshinaga, Study on photocatalyst properties improvement of titanium oxide by the mechanical-electrochemical cavitation, *J. Jet Flow Eng* **32** (2016) 10-16.

[3] C. L. Kling and F. G. Hammitt, A photographic study of spark-induced cavitation bubble collapse, *J. Basic Eng* **94** (1972) 825-833.

[4] M. Ijiri, D. Shimonishi, D. Nakagawa, T. Yoshimura, Evolution of microstructure from the surface to the interior of Cr-Mo steel by water jet peening, *Materials Sciences and Applications* **8** (2017) 708-715.

[5] H. Soyama and M. Asahara, Improvement of the corrosion resistance of carbon steel surface by a cavitating jet, *J. Mater. Sci. Lett* **18** (1999) 1953-1955.

IOP Conf. Series: Materials Science and Engineering **307** (2018) 012041 doi:10.1088/1757-899X/307/1/012041

Synthesis and effects on visible light photocatalytic activity of $Bi_2Ti_2O_7$ photocatalyst

Yi-Fan Li[1], Yan Zhong[1,2,*], Jun-Qing Chang[1], Chao-Hao Hu[1,2]

[1] School of Materials Science and Engineering, Guilin University of Electronic Technology, Guilin 541004, China
[2] Guangxi Key Laboratory of Information Materials, Guilin University of Electronic Technology, Guilin 541004, China
E-mail: yanzhong@guet.edu.cn

Abstract. $Bi_2Ti_2O_7$ photocatalysts were successfully synthesized via the co-precipitation method and the solvothermal method and further characterized by XRD, SEM, BET, XPS and UV-vis DRS. Under visible light irradiation, the photocatalytic performance of $Bi_2Ti_2O_7$ photocatalysts were evaluated by the degradation of RhB solution. The results indicated that the as-synthesized $Bi_2Ti_2O_7$ photocatalyst via co-precipitation method has better photocatalytic activity than the photocatalyst prepared by the solvothermal method. The larger surface area and special morphology of the $Bi_2Ti_2O_7$ synthesized by co-precipitation method were responsible to enhanced photocatalytic performance.

1. Introduction

The energy crisis and environmental pollution problem, such as toxic water pollutants and organic pollutants, have become increasingly severe, and the quest to develop and utilize the energy resources of novel and renewable features without harmful effect has been considered as an important test. Photocatalysis employing solar power is friendly looking forward to be a desired "green" technique over the environmental sustainability where the effective photocatalytic materials provide a potential to the treatment of noxious chemicals via its performance and practical applicability [1-2]. TiO_2 is one of the most extensive and in-depth photocatalysts in the application of many photocatalysts while there are still many defects in the application progress such as the limits in the ultraviolet light ($\lambda < 400$ nm) and photo-quantum efficiency [3-5]. In order to use visible light more effectively, developing the effective visible light-respond photocatalyst had been attracting increasing interest. Lately, abundant oxides ($SrTiO_3$, $NaTaO_3$, $BiVO_4$ and Bi_2WO_6) [6-7], due to its high performance, have been investigated widely.

Bi-Ti-O system includes several kinds of phases such as $Bi_2Ti_2O_7$, $Bi_2Ti_4O_{11}$ and $Bi_{20}TiO_{32}$ [8-12], which is formed by the different synthesis condition [13]. $Bi_2Ti_2O_7$, for example, as one of the $A_2B_2O_7$ compounds family, crystalizes with pyrochlore structure. It has plentiful crystalline morphologies, including nanowires, nanoparticles and nanotube, which have been obtained by using the high-temperature quenching method, facile hydrothermal method and wet chemical approach [14]. In previous study, with the visible-light irradiation the structure of pyrochlore ($Bi_2Ti_2O_7$), sillenite ($Bi_{12}TiO_{20}$) and Aurivillius-type ($Bi_4Ti_3O_{12}$) exhibited high performance for decomposing methyl orange [15], but there were few reports on preparing $Bi_2Ti_2O_7$ photocatalyst. The current work reported the synthesis method of $Bi_2Ti_2O_7$ photocatalyst using a chemical solution deposition method and a facile solvotheral method. The influence of the preparation method on the degradation of RhB solution is also investigated and discussed under the visible light irradiation.

2. Experimental

2.1 Synthesis

In this work, all the chemicals reagents were obtained commercially without further purification. The powders of $Bi_2Ti_2O_7$ photocatalyst were synthesized by co-precipitation and solvothermal methods.

Co-precipitation method: First, 0.005 mol $Bi(NO_3)_3 \cdot 5H_2O$ was added into 30 mL CH_3COOH to form solution with 40 minutes continuous stirring at room temperature. Second, 1.7 mL $Ti(OC_4H_9)_4$ was added slowly into above settled solution to form a suspension with white precipitate with 30 minutes constant stirring. Then, 60 mL $NH_3 \cdot H_2O$ was added drop by drop in above solution to form a mass of white precipitate. Next, the white precipitate was filtrated and washed three times with distilled water and ethanol until no impurity was left, dried at 60 ^oC for 12 h. At last calcinated at 500 ^oC for 180 minutes, after grinding, the yellow $Bi_2Ti_2O_7$ was obtained and remarked BT1.

Solvothermal method: First, 50 mL C_2H_5OH and 5 mL $C_3H_8O_3$ were mixed in the 100 mL beaker with magnetic stirring about 1 h. Second, 0.003 mol $Bi(NO_3)_3 \cdot 5H_2O$ was added into this solution with constant stirring about 1.5 h at room temperature until a settled solution was obtained. Third, 2.55 mL $Ti(OC_4H_9)_4$ was added slowly into above settled solution to form a suspension with white precipitate. Before transferred into a 100 mL Teflon-lined autoclave, the obtained mixture in an oven after 15 minutes ultrasonic treatment, then maintained at 120 ^oC for 12 h. Finally, the $Bi_2Ti_2O_7$ precursor was filtered, dried at 60 ^oC for 3 h. At last calcinated at 500 ^oC for 180 minutes, and grind into yellow solid and remarked BT2.

2.2 Characterization

The as-synthesized BT1 and BT2 photocatalysts were characterized by the X-ray power diffraction (Cu Kα radiation, Bruker D8-2-Advanced, Germany), the scanning electron microscopy (Quanta FEG 250, USA), the physical adsorption of N_2 in ASAP2020, UV-vis spectrometer (Puxi TU-1901, China), and X-ray photoelectron spectroscopy (XPS, Axis Ultra DLD, UK)

2.3 Photocatalytic degradation

To explored the photocatalytic activity of the $Bi_2Ti_2O_7$, BT1 and BT2 samples were studied by the degradation of RhB under visible-light irradiation using 300 W xenon lamp with the light of $\lambda > 420$ nm. In the typical photocatalytic degradation activity experiment, 0.1 g BT1 and BT2 photocatalysts were added into 100 mL RhB solution (10 mg/L), respectively. The solution was stirred in the dark for 30 min at room temperature in order to achieve the adsorption-desorption equilibrium. Every 60 minutes, 5 mL of solution was taken out and removed the power by a centrifuge. At last the solution was analyzed by UV-ultraviolet spectrophotometry.

3. Results and discussion

3.1 Crystal structure and morphology

The crystal structure of the $Bi_2Ti_2O_7$ synthesized via two different methods was characterized by XRD, and illustrated in Figure 1. It can be observed that there are two diffraction peaks of XRD patterns, which was consistent with pyrochlore $Bi_2Ti_2O_7$ (JCPDS No.89-4723) perfectly. Additionally, no any secondary phase was detected in the XRD patterns and the sharp diffraction peaks indicated the as-obtained samples were well-crystallized. The main X-ray diffraction peaks of the BT1 and BT2 samples located at 2θ=14.8°, 24.2°, 28.5°, 29.8°, 34.6°, 37.8°, 45.4°, 49.7°, 52.2°, 59.1° and 62.0° corresponded to the facets of (111), (220), (311), (222), (400), (331), (511), (440), (531), (622) and (444), respectively. The grain size was calculated as follows:

$$L = \frac{0.89\lambda}{\beta \cos\theta} \quad (1)$$

where λ is the incident light wavelength, θ is the X-ray diffraction angle of incidence, and β is the full width at half-maximum (FWHM) of the diffraction peak. According to calculation, the calculated average crystallite size of BT1 is around 97 nm, being over two times larger than that of BT2 sample (46 nm).

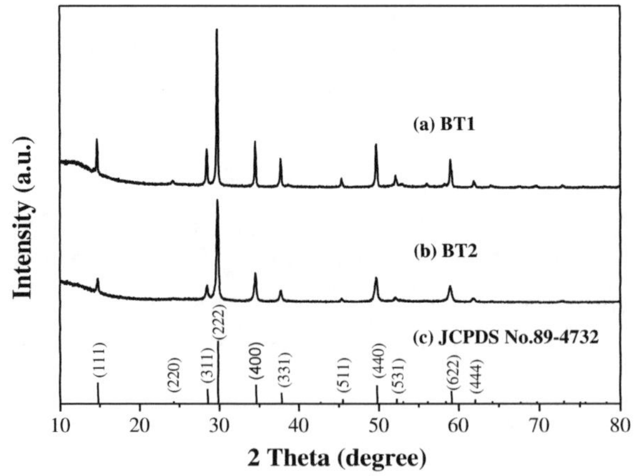

Figure 1. XRD patterns of BT1 and BT2 photocatalysts.

SEM analysis was conducted to observe the morphology of photocatalysts under different preparation conditions. Figure 2 is the SEM images of the BT1 and BT2 samples. It is clearly observed that the morphology is obvious different between the BT1 and BT2 samples. The BT2 photocatalysts have a ball-like morphology with smooth surface (Figure 2(a-c)). The SEM image with higher magnification (Figure 2(d)) indicates that the BT2 sample with diameters about 2-4 μm. The SEM image of BT1 in Figure 2(e) and (f) show that the BT1 is composed of particles with irregular shapes. Besides, the size of particles is chiefly distributed in the range of 1-2 μm.

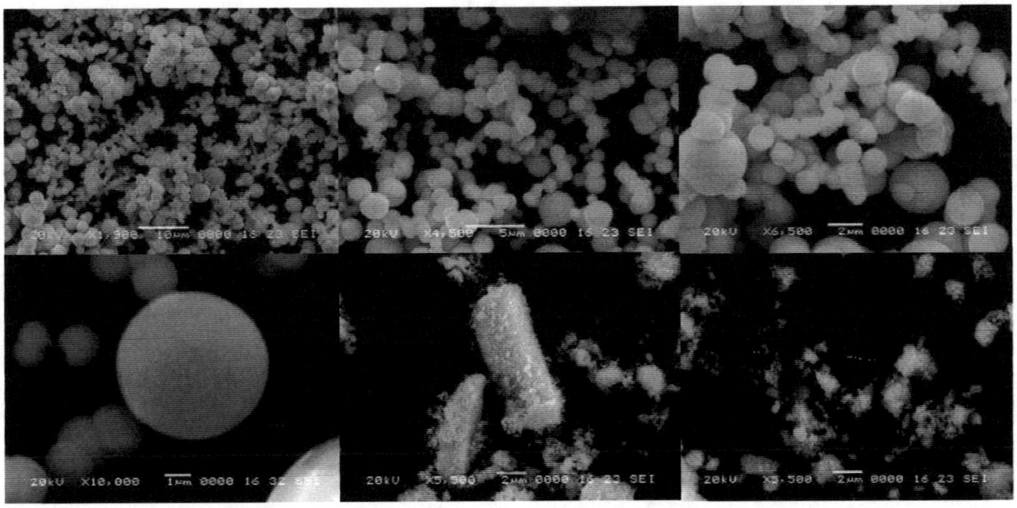

Figure 2. SEM images of pure $B_2T_2O_7$ photocatalytic: (a-d) BT2 sample, and (e-f) BT1 sample

3.2 BET analysis

Figure 3 shows Barret-Joyner-Halenda pore size distributions and nitrogen adsorption-desorption isotherms of the BT1 and BT2 samples. In Figure 3(a), the adsorption-desorption cures of BT2 sample belong to the type V isotherm with hysteresis loops typical of mesoporous materials. The BT1 sample, however, belongs to the type III adsorption-desorption isotherm. From Figure 3(b) and (c), we calculated the average diameters of pore size distribution for the BT2 and BT1 samples were 10.47 nm and 42.15 nm. The BET surface areas of BT2 and BT1 samples were 13.35 and 16.38 cm^2/g, respectively. The BET surface area of BT1 was superior to the BT2 sample. As the previous reports, the increase of average crystallite size could increase the surface areas [16]. This also can be confirmed by the previous results of XRD and SEM analysis.

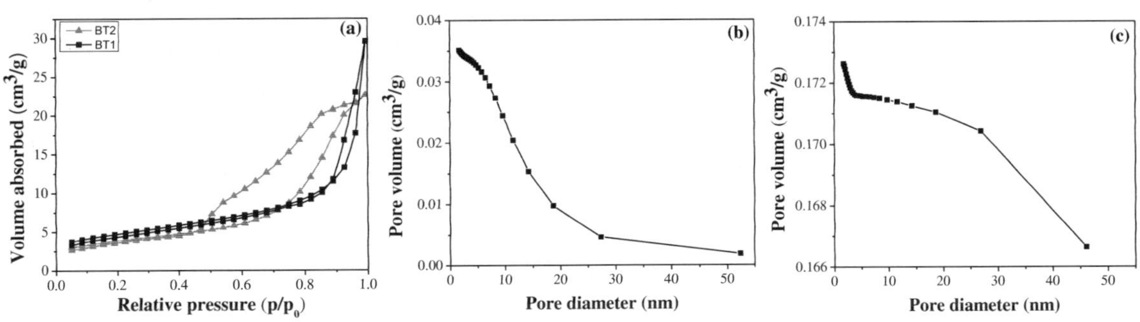

Figure 3. Nitrogen sorption isotherm of (a) BT2 , BT1 sample and pore volume distribution curve of (b) BT2 sample, (c) BT1 sample

3.3 UV-visble adsorption spectra

The optical absorption properties of the semiconductor materials play a major role in the photocatalytic activity. Figure 4 displayed the typical UV-vis absorption spectra of all the as-synthesized samples at the wavelength in the range of 250-850 nm. The band gap absorption edge of BT1 and BT2 samples are determined to be 430 nm and 431 nm, respectively, which shows quite good visible light absorption performance. The band gap can be further calculated from the equation:

$$Eg = \frac{h \cdot c}{\lambda} \tag{2}$$

Where E_g is the optical band gap, h is Planck's constant, λ is the wavelength corresponding to the onset of absorbance, and c is the velocity of light [17]. Thus, the calculated Eg values of BT1 and BT2 samples are 2.88 and 2.87 eV, respectively. These results indicated that the $Bi_2Ti_2O_7$ has potential photocatalytic ability.

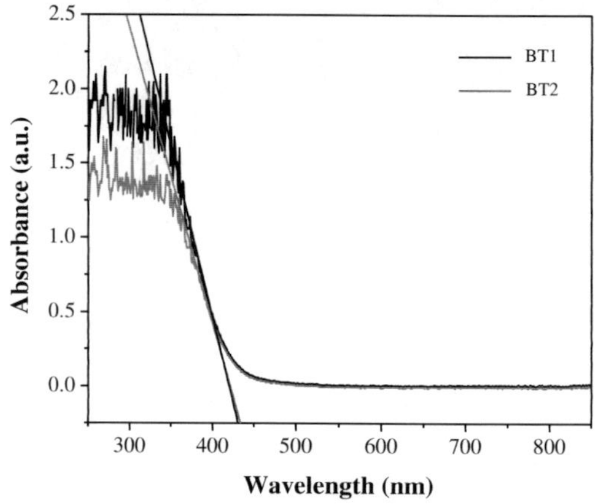

Figure 4. UV-visible diffuse absorption spectra (DRS) of the BT1 and BT2 samples

3.4 XPS analysis

In order to further obtain the insight of the valence state and surface chemical composition of photocatalysts, the survey spectra of BT2 and BT1 are displayed at Figure 5(a), respectively. The high-resolution XPS spectra of the Bi 4*f*, Ti 2*p* and O1*s* for these two photocatalysts were also studied and displayed in Figure 5(b-d). The peak located at 284.68 eV belongs to C 1*s* peak, which is a standard of the peak positions of this element. The XPS spectrum of Bi 4*f* in Fig. 5(b) shows that the higher banding energy peak at 164.58 is Bi $4f_{5/2}$ and the lower banding energy peak at 159.18 eV is characteristic to Bi $4f_{7/2}$. In the Figure 5(c), two peaks are observed around binding energies at 464.88 eV and 458.08 eV, which are ascribed to the Ti $2p_{1/2}$ and Ti $2p_{3/2}$ binding energies, respectively. As displayed in Figure 5(d), the high resolution O 1*s* spectrum of BT2 and BT1shows that two peaks located at 529.58 eV and 531.88 eV. The one of peaks at 529.58 eV attributes to the Bi-O bond at the lower binding energy [18], the other attributes to Ti-O bond at the higher binding energy peaks at 531.88 eV.

3.5 Photocatalytic activity

All the as-synthesized photocatalysts were measured by degradation of RhB (10 mg/L, 100 mL) under visible light irradiation ($\lambda > 420$ nm). According to the formula of $(C_0-C)/C_0 \times 100\%$ (C is the time-dependent concentration of mixture solution, and the C_0 is the initial concentration of mixture solution) and analytical concentration of RhB aqueous solution during the process of photocatalytic experiments, the removal rates of RhB are calculated and summarized in Figure 6. The results revealed that RhB solution can't degrade itself without photocatalyst. This can also demonstrate that under

IOP Conf. Series: Materials Science and Engineering **307** (2018) 012041 doi:10.1088/1757-899X/307/1/012041

visible-light irradiation RhB organic dyestuff is fairly stable. Whereas, the photocatalytic degradation of RhB happened when the BT1 and BT2 photocatalysts were added into solution. The Figure 6(a) revealed that the photo-degradation efficiency of RhB solution for the pure BT1 sample is about 92.8% after 240 minutes under the visible-light irradiation. while that of BT2 sample is only about 74.6%. From the above results, the photocatalytic performance of BT1 was more excellent than BT2, which can be ascribed to the larger specific surface area of BT1. The larger surface area of photocatalyst could provide abundant bonding sites to degradation of organic [19], so that the photocatalytic performance of the photocatalyst has been efficaciously improved.

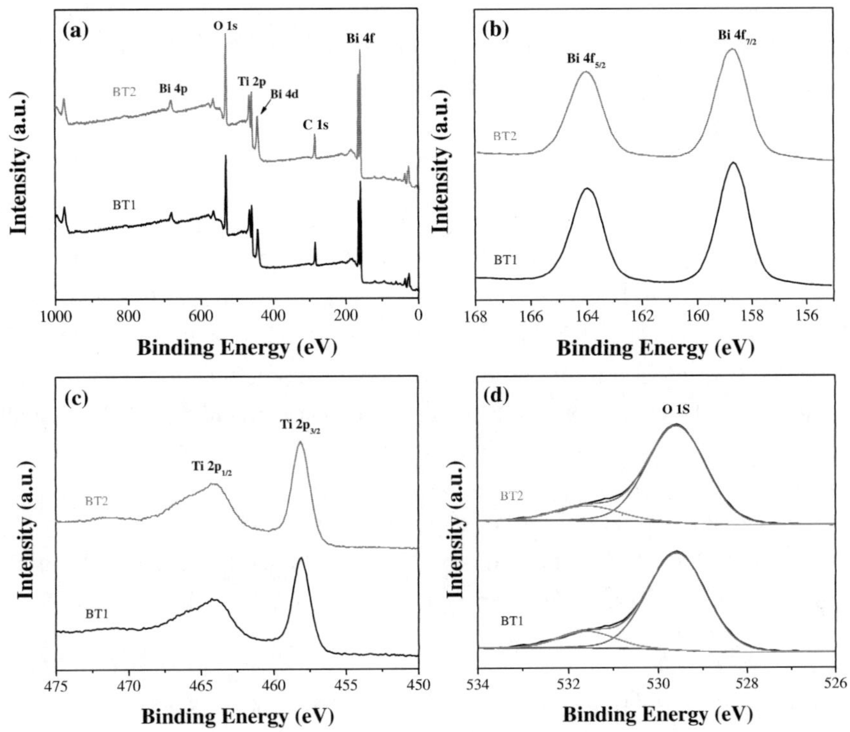

Figure 5. XPS spectra of BT1 and BT2 samples (a) survey spectrum, (b) Bi 4*f*, (c) Ti 2*p*, (d) O 1*s*

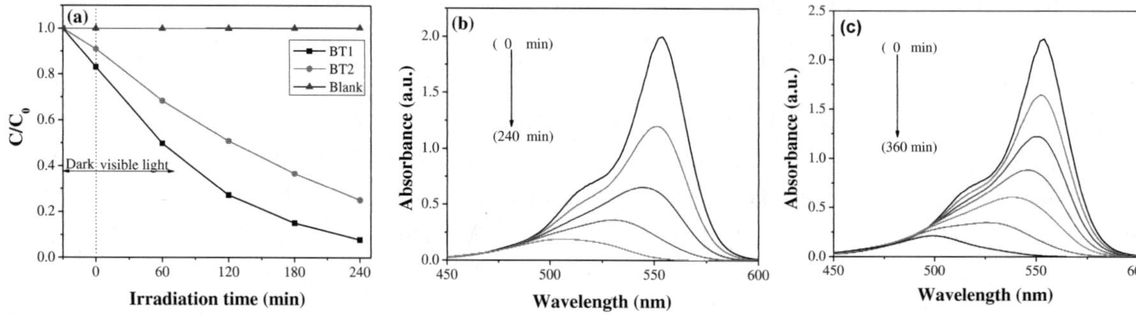

Figure 6. Photodegradation of RhB in the presence of BT1 and BT2 samples (a) the UV-vis absorption of RhB-BT1 (b), the UV-vis absorption of RhB-BT2 (c).

4. Conclusions

$Bi_2Ti_2O_7$ photocatalysts were successfully synthesized via the co-precipitation method and a solvothermal method in this paper. The photocatalytic activity of BT1 synthesized by co-precipitation method was obviously superior to the BT2 synthesized by solvothermal method in degradation of RhB under visible-light irradiation. The degradation efficiency of RhB solution for the BT1 and BT2 samples were about 92.8% and 74% after 240 minutes photocatalytic reactions. The larger surface area and special morphology of the BT1 photocatalyst were responsible to enhanced photocatalytic activity.

Acknowledgements

The financial support came from the Research Foundation of Guangxi Key Laboratory of Information Materials (131022-Z, 1210908-215-Z), the National Natural Science Foundation of China (51401060, 11464008), and Innovation Project of GUET Graduate Education (2017YJCX120).

References

[1] Hoffmann M R, Choi W and Bahnemann D W 1995 *J. Chem. Rev* **95** p 69
[2] Asahi R, Morikawa T, Ohwaki T, Aoki K and Taga Y 2001 *Science* **293** p 269
[3] H Zhu, X Gao, Y Lan and D Song 2004 *J. Am. Chem. Soc* **126** p 8380
[4] Ohko Y, Nakamura Y, Fukuda A, Matsuzawa S and Takeuchi K 2009 *J. Photochem. Photobiol.* A **205** p 28
[5] Karkmaz M, Puzenat E, Guillard C and Herrmann J M 2004 *J. Appl. Catal.* B **51** p 183
[6] Yu J and Kudo A 2006 *J. Adv. Funct. Mater* **16** p 2163
[7] H Fu, S Zhang,T Xu, Y Zhu and J Chen 2008 *J. Environ. Sci. Technol* **42** p 2085
[8] Hou J, Qu Y, Krsmanovic D and Ducati C 2009 J. Chem. Commun 26 p 3937
[9] Hou J, Qu Y, Krsmanovic D and Kumar R V 2010 *J. Nanopart. Res* **12** p 1797
[10] Hou J G, Kumar R V, Qu Y F, Krsmanovic D 2010 *J. Nanopart. Res* **12** p 563
[11] S Xu, W Shangguan, J Yuan, J Shi and M Chen 2007 *Mater. Sci. Eng.* B **137** p 108
[12] Hou J G, Qu Y F, Krsmanovic D and Kumar R V 2009 *J. Scripta Mater* **61** p 664
[13] W F Su and Y T Lu 2003 *Mater. Chem. Phys* **80** p 632
[14] Allured B, DelaCruz S, Darling T, Huda M N and Subramanian V 2014 *Appl. Catal.* B **144** p 261
[15] W F Yao, H Wang, S X Shang and X H Xu 2003 *J. Mol. Catal.* A **198** p 343

[16] F M Meng, H J Li, J F Gong and Z H Fan 2016 *J. Mater. Sci. - Mater. El* **27** p 8434
[17] Y Guo, J H Li, Z Q Gao and X Zhu 2016 *Appl. Catal.* B **192** p 59
[18] Y W Zhao, H Q Fan, K Fu and L T Ma 2016 *Int. J. Hydrogen. Energy* **41** p 16917
[19] W Iqbal, C Y Dong and M Y Xing 2017 *Catal. Sci. Technol* **8** p 1726

IOP Conf. Series: Materials Science and Engineering **307** (2018) 012042 doi:10.1088/1757-899X/307/1/012042

Experimental Study of Laser - enhanced 5A03 Aluminum Alloy and Its Stress Corrosion Resistance

Guicheng Wang [1,2], Jing Chen[1], Tao Pang[1]

1- School of Mechanical Engineering, Jiangsu University(JSU), Zhenjiang, Jiangsu 212013, China
2- Silicon Lake College of Vocational Technology (SLC), Kunshan, Jiangsu China

Abstract. Based on the study of improving the stress corrosion resistance of 5A03 aluminum alloy for ship, this paper mainly studied the tensile test, surface morphology and residual stress under laser shock, high temperature and stress corrosion. It is found that the residual compressive stress and the grain refinement on the surface of the material during the heat strengthening process increase the breaking strength of the sample in the stress corrosion environment. Appropriate high temperature maintenance helps to enhance the effect of deformation strengthening. In the 300°C environment insulation, due to recrystallization of the material, the performance decreased significantly. This study provides an experimental basis for effectively improving the stress corrosion resistance of 5A03 aluminum alloy.

1. Introduction

Aluminum and aluminum alloys have lower density, good mechanical properties, processability, thermal conductivity, electrical conductivity and corrosion resistance, which have been widely used in marine and marine equipment [1,2]. Marine 5A03 aluminum alloy is a typical Al-Mg series ship can welded aluminum alloy, which commonly used in shipbuilding parts, rivets, fuel tanks and pipes and other key components. Studies have shown that corrosion of marine aluminum alloy in the marine environment mainly pitting and stress corrosion, corrosion resistance of aluminum alloy is affecting application in the marine environment, one of the important indicators [3-5]. Therefore, the rolling shot peening [8,9], laser shock [10-12] and other surface strengthening process has been widely used. As one of the main forms of failure, stress corrosion resistance of materials is more and more by scholars both at home and abroad.

So far, no papers and data have been found on the experimental study on the improvement of stress corrosion resistance of marine 5A03 aluminum alloy laser . In this paper, the effects of laser thermal compounding on the stress corrosion resistance of marine 5A03 aluminum alloy are studied, and the mechanism of laser shock on improving metal anti-stress corrosion is discussed in depth.

2. Experimental conditions and testing methods

2.1 Sample preparation and thermal compound strengthening experiment

The experiment uses 5A03 aluminum alloy, which used in commercial marine. Its chemical shown in Tab 1. The specimen ,shown in Figure 1, is obtained by wire cutting.

After the sample processing is completed, it is polished with 400 to 1200 sandpaper and then polished mechanically.Then the sample was cleaned by ultrasonic cleaning in acetone, dried by a blower, which prepared for subsequent thermal compounding and stress corrosion testing.

Tab 1. 5A03-H112 aluminum alloy chemical composition

Element	Mg	Si	Mn	Fe	Cu	Zn	Ti	Al
(Wt%)	3.2-3.8	0.5-0.8	0.3	0.5	0.1	0.2	0.15	Remainder

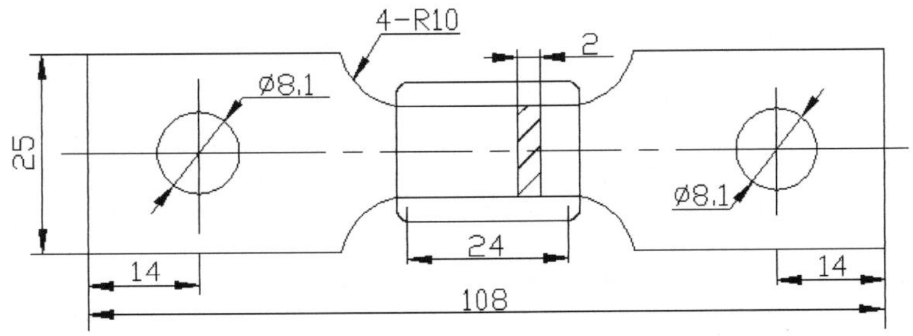

Fig 1. Specimen size after laser shock and stress corrosion

In this paper, the experimental process of the thermal compound to strengthen is to maintain the high temperature after the laser shock. The laser impact experiment was carried out by Shanghai Jiao Tong University laser impact enhanced integrated system. The main process parameters were as follows: the laser pulse width was 10ns, but the pulsed laser energy was 4J, the spot diameter was 2.5mm, the spot overlap rate was 75% ,double-sided impact. High temperature maintenance test is to place the sample in a high temperature box, set the temperature to maintain high temperature and holding time.

2.2 Test of stress corrosion

Stress corrosion tests were performed at room temperature, using a stress corrosion tester as shown. The experiment is divided into 6 groups, the specific process parameters of each group are shown in Table 2. Samples were immersed in 3.5% NaCl solution for stretching. The tensile strain rate was 2 × 10-5 s-1. After the specimen was broken, the fracture time was recorded. The fracture was cleaned by ultrasonic cleaning with acetone. The appearance of the fracture surface was observed with a scanning electron microscope after the blower was blown dry.

Fig 2. Stress corrosion testing machine

Tab 2. Sample grouping and stress corrosion experimental parameters

IOP Conf. Series: Materials Science and Engineering **307** (2018) 012042 doi:10.1088/1757-899X/307/1/012042

Group	Treatment	High temperature to maintain temperature (°C)	High temperature to maintain time (h)
A	Untreated	——	——
B	LSP	——	——
C	LSP+HT	200	2
D	LSP+HT	200	6
E	LSP+HT	300	2
F	LSP+HT	300	6

2.3 Observation and test of residual stress and microstructure

The samples were polished with 400 #, 800 #, 1200 # and 1500 # sandpaper in sequence, then polished and cleaned. The Keller reagent (formula: 1.0% HF + 1.5% HCl + 2.5% HNO3 + 95% H2O) for 30-60s and rinsing with plenty of water. Finally, the morphology was observed with light microscope. Residual stress test uses X350-type X-ray stress meter. During measurement, the impact area by the middle of the point as the measuring point.

3. Experimental results and discussion

3.1 Residual stress induced by laser

Deformation strengthening can induce residual compressive stress on the material surface. As shown in Fig 3, the residual compressive stress on the surface of 5A03 aluminum alloy is up to -100MPa for laser shock treatment. The residual compressive stress on the surface of the material can counteract the tensile stress of the specimen during the stress corrosion stretching. To a certain extent, it can prevent the occurrence of stress corrosion cracking and improve the resistance to the expansion of stress corrosion cracking [15]. However, due to the complexity of marine 5A03 aluminum alloy working conditions, the residual stress will relax during service, which seriously reduces the stress corrosion resistance. Some studies have shown that the heat treatment will not only cause a certain degree of relaxation of the residual stress induced by the deformation strengthening on the one hand, but also enhance the pinning effect by strengthening the relative dislocation due to the dislocation movement during the heat preservation Residual stress in the ensuing service stability.

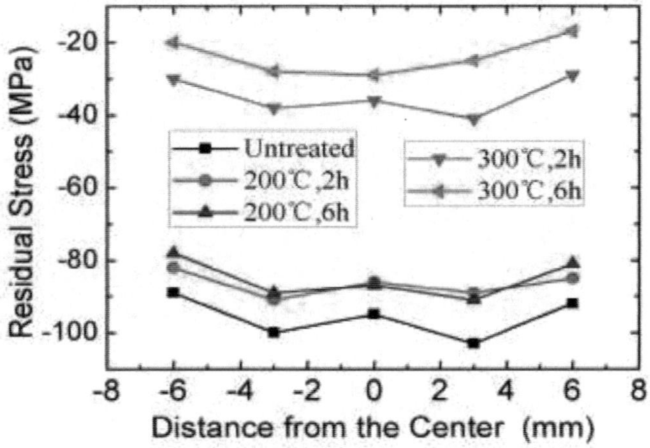

Fig 3. 5A03 aluminum alloy surface residual stress after different treatment

As can be seen from Fig 3, at 200°C for 2h, the sample residual stress relaxation occurred, was about -88MPa. Further incubated to 6h, the residual stress was basically stable. While at 300°C , after incubated 6h, the sample surface Residual stress serious release occurred, which only -30MPa. Obviously, laser shock treatment can produce a certain residual compressive stress on the surface of the material. During high temperature holding stage, due to laser shock generated residual stress relaxation occurs, the specific amount of relaxation and holding temperature and holding time. It is initially judged that the residual stress induced by deformation strengthening can be kept basically stable below the recrystallization temperature of the material.

3.2 Surface morphology and microstructure comparison before and after laser shock

After the laser shock, the macroscopic surface morphology of the sample is shown in Fig 4. It can be found that the laser shock treatment causes macroscopic plastic deformation on the surface of the material. The surface morphology of single point impingement pits is shown in Fig 5. The *diameter* of the pits is about 2500μm, which is basically the same as that of the laser spot used in laser impingement processing.

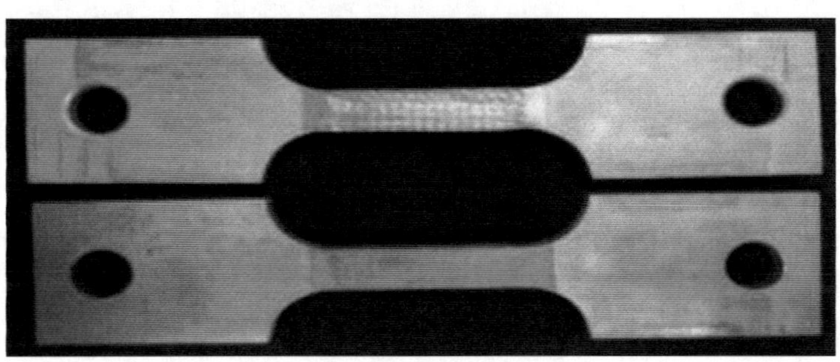

Figure 4. Stress samples after laser shock stress corrosion

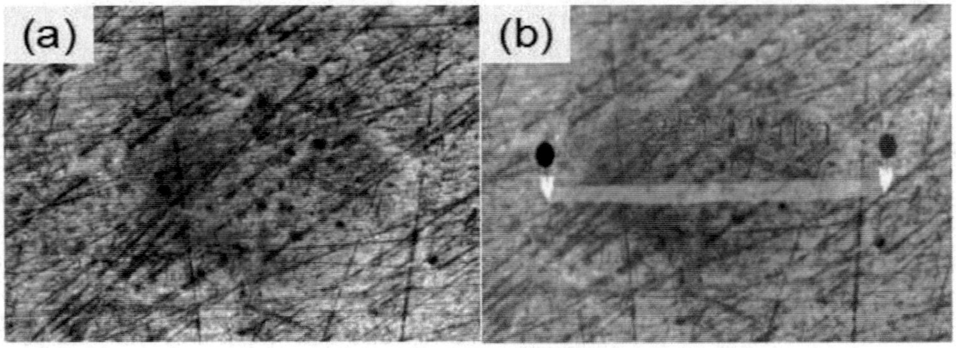

Fig 5. 5A03 aluminum alloy surface after the laser pits

The microstructure of 5A03 aluminum alloy before and after laser shock is shown in Fig 6. It can be seen from the comparison of Fig 6 (a) and Fig 6 (b) that the laser shock caused the surface of the sample to undergo some grain refinement and the grains became more uniform. For the grain refinement behavior, some scholars point out that this is related to the dislocation and agglomeration in the deformation process. Numerous studies have shown that grain refinement induced by deformation strengthening can effectively improve the material fatigue resistance, stress corrosion resistance and other properties.

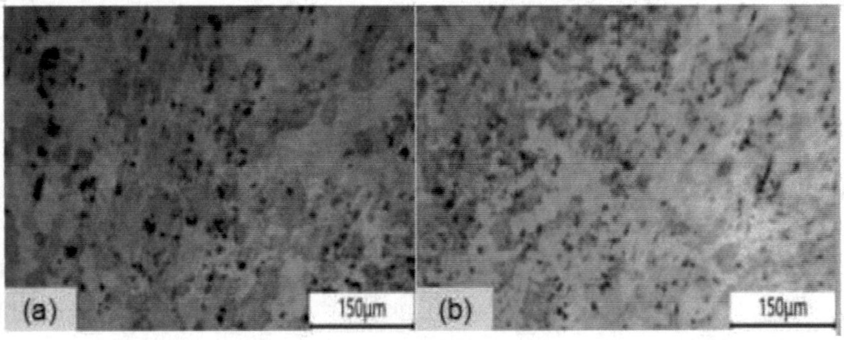

Fig 6. Metallography: (**a**) untreated sample; (**b**) laser shock

3.3 Test of stress corrosion tensile

Tab 3. Test results of 5A03 aluminum alloy after different stress corrosion

Group	Breaking strength (MPa)	Sensitivity coefficient (%)	Elongation (%)
A	210	12.48	12
B	230	10.57	8
C	234	10.28	9
D	237	9.96	15
E	112	27.81	23

F	97	28.94	25

The samples before and after the thermal compound strengthening were placed respectively in a 3.5% NaCl solution in a corrosive medium and subjected to a slow strain rate tensile-stress corrosion test. The test results are shown in Table 3. We can see that ,in the stress corrosion environment, the rupture strength of the marine 5A03 aluminum is about 210MPa, and the rupture strength after laser shock strengthening is increased to 230MPa. After heat preservation at 200°C for 6h, the breaking strength of the sample is 237MPa, reaching the maximum value. At the same time, the corrosion sensitivity index of the sample to 3.5% NaCl solution is 9.96% under the condition of 200°C for 6h after laser shock, which is 20.2% less than that of the untreated sample (12.48%). In view of this, thermal compound strengthening effectively improves the shipboard 5A03 aluminum alloy stress corrosion resistance. The sample was heat-treated at 300°C. The recrystallization of the sample resulted in a significant decrease in the breaking strength and stress corrosion susceptibility of the sample, and the laser shock-strengthening effect basically disappeared.

3.4 Analysis of stress corrosion tensile fracture

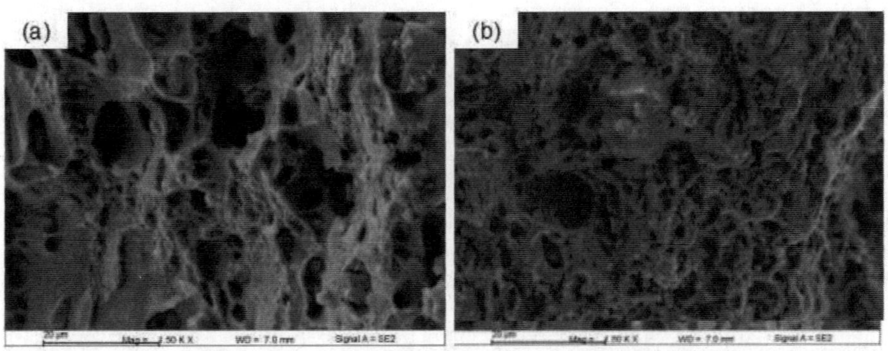

Fig 7. Stress corrosion fracture morphology

(**A**) Untreated; (**b**) Thermal Compound Hardening

In order to further study the effect of thermal compound strengthening on the stress corrosion resistance of marine 5A03 aluminum alloy, SEM was used to observe the stress corrosion fracture. Figure 7 (a) shows the untreated sample in 3.5% NaCl solution stress corrosion tensile fracture morphology. It can be seen from Figure 7 (a) that less dimples on the fracture and more pores Corrosion pits. The fault mode is quasi-cleavage. The stress-corrosion tensile fracture morphology of the samples, shown on Fig 7 (b), exposed to laser shock and incubated at 200°C for 6h in 3.5% NaCl solution. It can be seen from the figure that the corrosion pits and pores are significantly reduced , a considerable number of dimples appear in the fracture, and the dimples are relatively uniform in size. There are a few tearing zones around the dimples. The fault patterns are ductile and locally cleavage faults. Thus, after laser shock, 5A03 aluminum alloy stress corrosion resistance improved significantly, the material in the stress corrosion cracking plasticity is improved.

4. Conclusion

Through the experimental study of laser impact and anti-stress corrosion of 5A03 aluminum alloy, the main conclusions are as follows:

- Laser shock strengthening can introduce residual compressive stress on the surface of marine 5A03 aluminum alloy material with the maximum amplitude of -100MPa. At the same time, the laser shock treatment causes the grain refinement on the sample surface to be refined.
- For marine 5A03 aluminum alloy, heat treatment at 200°C after deformation strengthening helps to enhance the effect of deformation strengthening. However, when the holding temperature is higher than the recrystallization temperature of the material, the strengthening effect will disappear due to recrystallization behavior of the material.
- Compared with the untreated sample, the 5A03 aluminum alloy has a significant improvement in stress corrosion resistance after being laser shocked, which opening up a new path for effectively improving the stress corrosion resistance of 5A03 aluminum alloy.

5. Acknowledgment

This research was supported by the Natural Science Foundation of P. R. China (No.51075192), National key projects of P. R. China（2013ZX04009031）and A Project Funded by the Suzhou City key laboratory of elevator safety technology.

References

[1] P Zeng 2010 Shipbuilding Engineering;39 (1) pp 55-57.
[2] B Liu, Z G Fang , H B Wang and T Wang 2012 Corrosion and Protection; 11 (11) pp 1005-1008.
[3] Y Li, E Daxin and C F She 2014 Precision Forming Engineering; 6 (2) pp 36-41.
[4] J Hou , P F Zhang and W M Guo 2015 Equipment Environmental Engineering; 15 (2) pp 59-63.
[5] G Q Huang 200 Chinese Society of Materials Research, 0.
[6] J W He , S N Ma and D M Ba 2013 Journal of Armored Force Engineering Society; 27 (3) pp 75-81.
[7] H Zhou , X D Zhou and W Zhou 2009 Tool Technology; 43 (2) pp 18-22.
[8] X Dong and X Duan 2004 Mining Machinery; (7) pp 67-70.
[9] Y F Li 2015 Petroleum Storage & Transportation; 34 (1) pp 33-37.
[10] X Q Zhang and Y H Dai 2005 Electroprocessing & Molds; (2) pp 30-32.
[11] G Y Wang and Y B Zhang 2016 Materials Information; 30 (10) pp 46-54.
[12] JZ Lu, et al 2012 Corrosion Science; 60 pp 145-152.
[13] M Z Ge, J Y Xiang and L Yang 2017 Surface & Coatings Technology; 310 pp 157-165.

IOP Conf. Series: Materials Science and Engineering **307** (2018) 012043 doi:10.1088/1757-899X/307/1/012043

Atomic vibrations in alpha-quartz with silicon vacancies

A N Kislov and A F Zatsepin

Ural Federal University, 19 Mira Street, Ekaterinburg, 620002, Russia
E-mail: a.n.kislov@urfu.ru

Abstract. A study of atomic structures and localized symmetrized vibrations in α-quartz with silicon vacancies in different charge states is presented. This study is performed by computer modeling on the basis of the first-principle type potentials. The equilibrium structures are optimized by minimizing the lattice energy. The phonon symmetrized local densities of states are calculated by means of a recursion method. Moreover, frequencies of localized vibrations of A- and B-symmetries induced by silicon vacancies are determined. The results are discussed in comparative manner for the silicon vacancy in three charge states.

1. Introduction

Low-temperature modification of crystalline silica (α-quartz) has been the object of interest of numerous experimental [1-6] and theoretical [7-12] research because it plays key role in various engineering devices. Important physical properties such as structural and vibrational ones were studied in these works. The atomic structure and the physics of vibrations in defect-free α-quartz are well understood.

In the presence of point defects that violate the translational symmetry of the crystal lattice, many extra phenomena are observed: for example, diverse anomalies in spectra of the Raman scattering or infra-red absorption. The microscopic origin of similar phenomena remains unclear and their description refers to a range of difficult solvable problems. In particular, it would be helpful to know the contributions from differently charged defects in such anomalous behaviours.

At the microscopic scale, valuable information about effects of charged defects on the local atomic structure and lattice dynamics is often obtained using atomistic simulation which has become a significant method in research. This simulation can be provided by quantum-mechanical calculations called first-principles or ab initio methods. In the framework of this approach one can distinguish two main currents. There are the Hartree–Fock theory and the density functional theory used with different levels of accuracy: local density approximation, generalized-gradient approximation or hybrid functionals.

When it is necessary to use of very large atomic cluster, for instance, when modelling localized vibrations in solids containing charged defects, ab initio methods are faced with a computational demanding task. In this case, we can apply the computationally less demanding classical mechanics that describe the interactions between the atoms using interatomic potentials known as the force-field approach [13-15].

It has been shown by quantum-mechanical calculations that the local atomic structure of α-quartz is strongly dependent on point defects [16,17] and their charge states [18-23]. Defects in the oxygen sublattice such as oxygen vacancies and interstitials as well as several types of impurities were examined in the above publications. Recently, we were discussing effects of the differently charged oxygen vacancies on the vibrational spectrum of α-quartz [24]. It must however be admitted that the silicon vacancies which can also attend in α-quartz have not received as much attention. Information

IOP Conf. Series: Materials Science and Engineering **307** (2018) 012043 doi:10.1088/1757-899X/307/1/012043

about the influence of charge states of these defects on the atomic structure and vibrational spectrum of α-quartz has not been acquired. A few publications have been devoted to silicon vacancies [21-23]. The purpose of this paper is to investigate and to thoroughly discuss the effects of differently charged silicon vacancies V_{Si}^{-4}, V_{Si}^{-2} and V_{Si}^{0} on the atomic structure and vibrational spectrum of α-quartz.

2. Computational methods

The simulation is based on force-field approach and the numerical methods of the present work are precisely those used in our previous paper on oxygen vacancies [24]. In order to model the structures and the vibrations in quartz with defects in different charged states we used two-body potentials of the classical Born-Mayer model. The potentials with parameters [25] were chosen to describe the interactions between ions. The reliability of this simple model was verified by comparing available experimental data with calculated values for structural, elastic and dynamical properties of pure α-quartz. For example, Figure 1 depicts the dispersion curves in three highly symmetric directions of the Brillouin zone and the density of states (DOS) of phonons. The frequency distribution is determined by solving the Fourier transform of the dynamic matrix for randomly selected wave vectors of the first Brillouin zone and the result is plotted as a histogram.

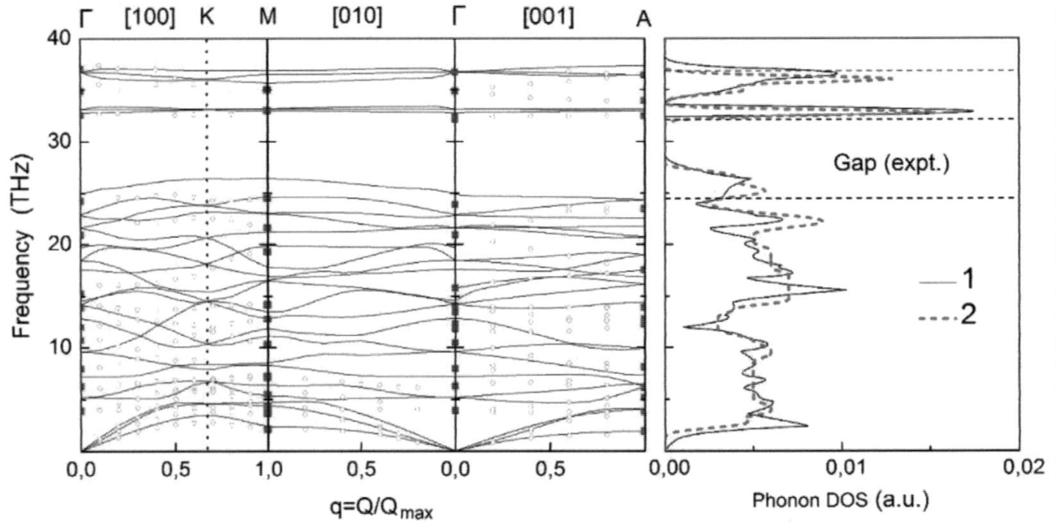

Figure 1. (Left panel) Phonon dispersion in α-quartz along the main symmetry directions. Squares are observed values [3], open circles – [4-6]. (Right panel) Calculated phonon DOS: using chosen point of the Brillouin zone (curve 1), by a recursive method (curve 2).

The formation of silicon vacancies is accompanied by a geometrical distortion of the crystal lattice. The structural optimization was performed using the process of minimizing the lattice energy. All the expressions that are needed for these calculations are presented in detail in [26]. The effect of the silicon vacancies on the vibration spectrum of α-quartz was evaluated by analyzing the local symmetrized densities of states (LSDOS) of phonons in defect-free and defective quartz. The recursion method based on the Lanczos algorithm [27] with the initial vector of a given symmetry is used to calculate the phonon LSDOS. The projection operator method is used to find the symmetry coordinates, which determine this initial vector. The LSDOS is expressed by the imaginary part of the diagonal element of the Green's function matrix, which is written by the easily calculated continued fraction.

A cluster of 4000 ions is sufficient to construct the total phonon DOS and in this case the Lanczos technique is used for 10 levels of recursion. The calculation result is plotted in Figure 1. The comparison of the curves obtained by different methods shows a reasonable agreement.

3. Results and Discussion

We began our investigation with the construct of static equilibrium structures of the α-quartz with Si vacancies in different charge states. The local configurations in the vicinity of V_{Si}^{-4}, V_{Si}^{-2} and V_{Si}^{0} vacancies are shown in Figure 2. The distances between some ions in relaxed configurations are given in Table 1.

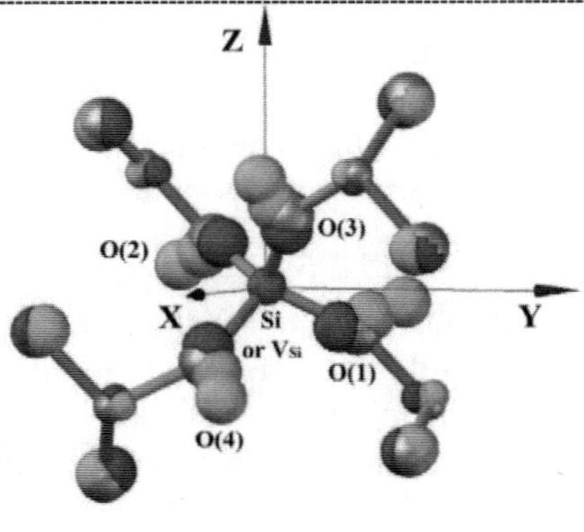

Figure 2. Representation of the local relaxation around silicon vacancy in α-quartz. The Si ion located at the origin of the Cartesian coordinate system is removed to create a Si vacancy. Grey spheres represent the ions in the perfect lattice, blue, green and red ones – around V_{Si}^{-4}, V_{Si}^{-2}, V_{Si}^{0}, respectively.

Table 1. Interionic distances (Å) in α-quartz with Si vacancies.

	Charge state of vacancy		
	–4	–2	0
V_{Si}–[O(1) or O(2)]	2.201	1.713	1.363
V_{Si}–[O(3) or O(4)]	2.080	1.658	1.353
O(1)–O(2)	3.657	2.835	2.266
O(3)–O(4)	3.426	2.740	2.235
Si–[O(1) or O(2)]	1.544	1.549	1.613
Si–[O(3) or O(4)]	1.540	1.551	1.609

A negatively charged vacancy V_{Si}^{-4} is produced by removing a silicon ion. The character of the lattice distortions near this defect correlates with the fact that it has an excess negative charge with respect to the crystal lattice. The distances between vacancy V_{Si}^{-4} and nearest oxygen ions are found to be larger than the Si–O bond lengths in defect-free α-quartz, in which the Si–O distances are determined to be 1.598 (1.605) Å for two oxygen ions and 1.606 (1.614) Å for the other two oxygen ions, where the numbers in parentheses are the experimental data [2].

For a neutral V_{Si}^{0} vacancy, the situation is different. This defect can be formed when extra holes are trapped on nearest-neighbors of the negatively charged Si vacancy. The interaction energy between the neutral silicon vacancy and nearest oxygen ions is increased. This circumstance leads to strengthen an attraction in these pairs and to decrease distances between Si vacancy and O ions.

Then we have calculated the phonon LSDOS projected onto ion displacements of SiO_4 unit. Si ion (Si vacancy) is on the site with point symmetry C_2. The oxygen ions are involved in two vibrations of A-symmetry and one vibration of B-symmetry. The A-type vibrations consist of one symmetric bending mode and one symmetric stretching mode, the B-type vibration is an asymmetric stretching mode. As an example, Figure 3 illustrates the directions of the motions for A-symmetry vibrations.

The phonon LSDOS projected onto A- and B-type displacements of four O ions nearest to Si ion in ideal α-quartz and differently charged Si vacancies in defective α-quartz are shown in Figure 4. According to obtained results, symmetric- and asymmetric-stretch vibrations are mostly present in the high-frequency spectrum band, symmetric bending is responsible for the low-frequency and the medium-frequency bands.

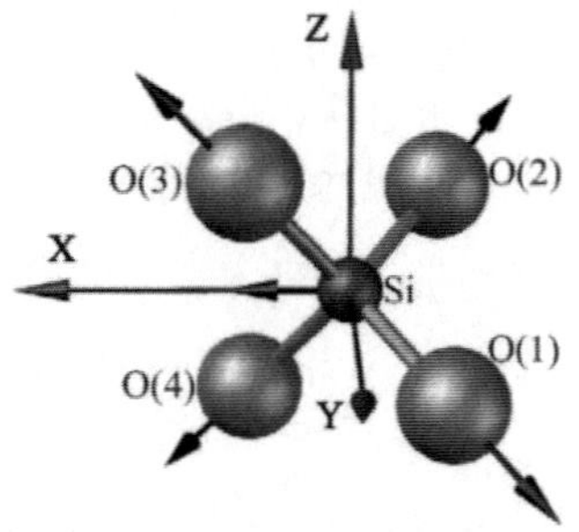

Figure 3. The fully symmetric A-modes. The C_2 axis of rotation through 180° is the *X* axis.

For neutral Si vacancies, the high-band of the phonon LSDOS is shifted to the low-frequency part of the spectrum and the low- and middle bands are shifted to the high-frequency part. There is the large peak in the band gap. These changes may be explained the following. The short-range repulsive interaction is decreased, but the Coulomb interaction remains the same as in the perfect α-quartz. Thus

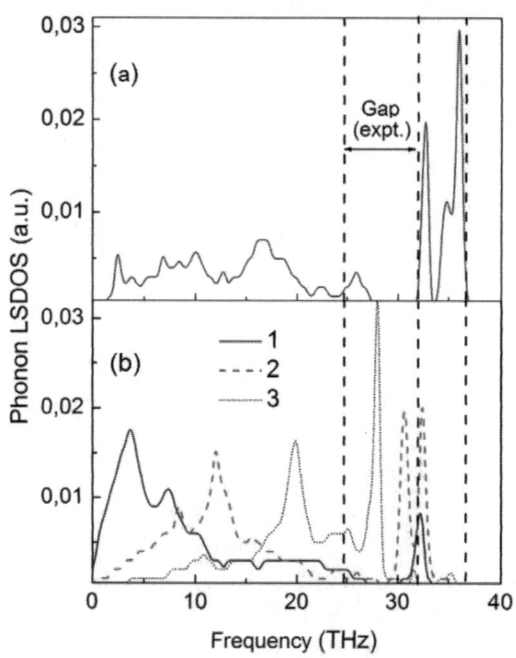

Figure 4. Phonon LSDOS projected onto A- and B-symmetry displacements of four O ions nearest to (a) Si ion in ideal α-quartz and (b) Si vacancies in defective α-quartz: V_{Si}^{-4} (curve 1), V_{Si}^{-2} (curve 2) and V_{Si}^{0} (curve 3). The boundaries of the spectrum bands of pure α-quartz [5] are highlighted by the vertical dashed lines.

Table 2. Frequency (THz) of vacancy-induced defect vibrations in α-quartz with Si vacancies.

Vacancy charge state	Ions	Vibration symmetry	
		A	B
–4	O(1) and O(2)	3.6, 32.4	3.6, 32.4
	O(3) and O(4)	3.6, 7.6, 10.4, 32.4	3.6, 7.6, 10.4, 32.4
–2	O(1) and O(2)	8.4, 12.0, 30.8	30.8
	O(3) and O(4)	8.4, 12.0, 15.6, 30.8	8.4, 12.0, 30.8
0	O(1) and O(2)	20.0, 28.0	25.0
	O(3) and O(4)	10.8, 20.0, 25.0, 28.0	25.0, 28.0

the effective interaction of O ions making the bending movement becomes greater. Consequently, the phonon density located lower than 20 THz shifts to the high-frequency part of the spectrum. The effective interaction of oxygen ions doing stretching movement diminishes and the phonon LSDOS located higher than 30 THz shifts to the low-frequency zone.

The Si vacancies are responsible for resonant vibrations. These defects also induce one gap vibration. The oxygen ions contribute to causing this gap vibration. The frequencies of the vibrations depend on the charge state of Si vacancies. The frequencies of localized symmetrized vibrations induced by Si vacancies are given in Table 2.

4. Conclusion

We investigate the influence of silicon vacancies in three charge states on the structure and the phonon spectrum of α-SiO_2 using first-principle potentials. We predict frequencies of localized vibrations induced by differently charged Si vacancies. Our study indicates that the structural and vibrational properties of α-quartz depend substantially on the charge state of the defect. It demonstrates that the Si vacancies can strongly affect the static lattice deformation and the phonon DOS by changing the charge state. The obtained results can be applied to the interpretation of the vibrational spectra that are measured experimentally by means of IR absorption and Raman spectroscopy.

Acknowledgment

The work was supported by the Ministry of Education and Science of the Russian Federation (Government task № 3.1485.2017/4.6) and Act 211 of RF Government, contract № 02.A03.21.0006.

References

[1] Jorgensen J D 1978 *J. Appl. Phys.* **49** 5473
[2] Levien L, Prewitt C T and Weidner D J 1980 *Am. Mineral.* **65** 920
[3] Gerrais F and Firiou B 1975 *Phys. Rev.* B **11** 3944
[4] Dorner B, Grimm H and Rzany H 1980 *J. Phys.* C **13** 6607
[5] Strauch D and Dorner B 1993 *J. Phys. Condens. Matter* **5** 6149
[6] Burkel E, Seyfert C, Halcoussis C, Sinn H and Simmons R O 1999 *Physica* B **263/264** 412
[7] Nada R, Catlow C R A, Dovesi R and Pisani C 1990 *Phys. Chem. Miner.* **17** 353
[8] Schober H, Strauch D, Nutzel K and Dorner B 1993 *J. Phys. Condens. Matter* **5** 6155
[9] Liu F, Garofalini S H, King-Smith R D and Vanderbilt D 1994 *Phys. Rev.* B **49** 12528
[10] Gonze X, Charlier J C, Allan D C and Teter M P 1994 *Phys. Rev.* B **50** 13035
[11] Boer K, Jansen A P J, van Santen R A, Watson G W and Parker S C 1996 *Phys. Rev.* **54** 826
[12] Sevik C and Bulutay C 2007 *J. Mater. Sci.* **42** 6555
[13] Tsuneyuki S, Tsukada M, Aoki H and Matsui Y 1988 *Phys. Rev. Lett.* **61** 869
[14] Vashishta P, Kalia R K, Rino J P and Ebbsjo I 1990 *Phys. Rev.* B **41** 12197
[15] Combariza A F, Gomez D A and Sastre G 2013 *Chem. Soc. Rev.* **42** 114
[16] Lægsgaard J and Stokbro K 2000 *Phys. Rev.* B **61** 12590
[17] Sulimov V B, Sushko P V, Edwards A H, Shluger A L and Stoneham A M 2002 *Phys. Rev.* B **66** 024108
[18] Jin Y G and Chang K J 2001 *Phys. Rev. Lett.* **86** 1793
[19] Carbonaro C M, Fiorentini V and Bernardini F 2001 *Phys. Rev. Lett.* **86** 3064
[20] Chadi D J 2001 *Phys. Rev.* B **64** 195403
[21] Song J, Corrales L R, Kresse G and Jonsson H 2001 *Phys. Rev.* B **64** 134102
[22] Lægsgaard J 2002 *Phys. Rev.* B **66** 024107
[23] Roma G and Limoge Y 2004 *Phys. Rev.* B **70** 174101
[24] Kislov A N, Toropov M O and Zatsepin A F 2011 *J. Non-Cryst. Solids* **357** 1912
[25] Beest B W H, Kramer G J and Santen R A 1990 *Phys. Rev. Lett.* **64** 1955
[26] Richardson D D 1982 *Comp. Phys. Commun.* **28** 75
[27] Lanczos C 1950 *J. Res. Natl. Bur. Stand.* **45** 225

Influence of the pressure applied to make LATP pellets

Xiaojuan Lu

Department of Environmental Science and Engineering, North China Electric Power University, Mailbox 253, No.689 Huadian Road, Baoding, Hebei, PR China

xiaojuanlv@hotmail.com

Abstract. NASICON structured $Li_{1+x}Al_xTi_{2-x}(PO_4)_3$ ($0.1 \leq x \leq 0.7$) powders were prepared by hydrothermal synthesis method and the pressure applied to press the powders into pellets was investigated in this study. The conductivity was measured by an impedance analyzer and the microstructures were examined by SEM. The variation trend with Al amount was almost identical for LATP pressed at high pressure and low pressure. Both the total conductivities of LATP prepared at high pressures and low pressures peaked at x=0.6. The conductivities of LATP pressed at high pressures were all higher than those at low pressures. The reason was mainly attributed to the denser microstructures achieved at higher pressure.

1. Introduction

Lithium ion batteries have attracted significant interests due to the concerns of limited supply of fossil fuels and the air pollution resulted in burning of fossil fuels. Organic liquid electrolytes which are flammable and inclined to leak, used in commercially available Li ion batteries, hinder the large scale application of Li ion batteries due to the severe safety concerns. Inorganic solid lithium ion conductors, which are chemically stable, safe and easy to miniaturization, have increasingly become popular to replace organic liquid electrolytes in lithium ion batteries. Na-superionic conductor (NASICON) structured lithium ion conductors with a general formula $Li_{1+x}M'_xM''_{2-x}(PO_4)_3$, of which $Li_{1+x}Al_xTi_{2-x}(PO_4)_3$ (LATP) exhibit conductivities in the range of 10^{-3}-10^{-4} S/cm [1–4], show a great prospect. Various methods including solid state reactions, sol-gel [5, 6], co-precipitation [7, 8] and hydrothermal synthesis [4] have been attempted to synthesize LATP. For various synthesis methods, the preparation conditions have played a vital role in affecting the performance of lithium ion conductors made. Synthesis parameters like hydrothermal reaction temperature/time, calcinations temperature and starting materials in the preparation of LATP powders and parameters like sintering temperatures and dwelling time during the subsequent sintering stage have been studied. However, the effect of pressure applied to press the powders into pellets during the preparation of green pellets, which is potentially important, has not been reported yet.

The highest ionic conductivity achieved so far was based on a high preparation temperature over 1400°C, which is detrimental to other components in the batteries. Thus, it is beneficial to lower the synthesis temperatures of LATP. Kim et al. [4] found that hydrothermal method is effective on lowering the sintering temperatures and a decent conductivity has been achieved at 900°C. Therefore, the hydrothermal synthesis was adopted to prepare LATP powers and the effect of the pressure applied to pressing the powders have been investigated in this paper.

IOP Conf. Series: Materials Science and Engineering **307** (2018) 012044 doi:10.1088/1757-899X/307/1/012044

2. Experimental

NASICON structured $Li_{1+x}Al_xTi_{2-x}(PO_4)_3$ ($0.1 \le x \le 0.7$) powders were prepared by hydrothermal synthesis here. Start materials used were $LiOH \bullet H_2O$ (>99%), $Al(NO_3)_3 \bullet 9H_2O$ (>98%) and $Ti(OCH(CH_3)_2)_4$ (>97%) and H_3PO_4(>98%). Firstly, 5mM of $Ti(OCH(CH_3)_2)_4$ in 100mL of deionized water was obtained by stirring at 300rpm for 2h. With the stirring maintained at 300rpm, $LiOH \bullet H_2O$, $Al(NO_3)_3 \bullet 9H_2O$ and H_3PO_4 was added in an interval of 1h. The resultant solution was transferred into a Teflon vessel inside a hermetically sealed stainless steel container and placed in a furnace at 180°C for 6h. And then a powdery product was obtained after drying at 90°C for 24h. The powders were calcined at 700°C for 2h and subsequently ground in a pestle mortar to gain fine powders. Green LATP pellets in 15mm diameter were prepared by pressing the powers with varied pressures and subsequently sintered at 950°C for 4h. The pressure applied was $50kgf/cm^2$ and $100kgf/cm^2$ for 10min. The pellet pressed using a pressure of $100kgf/cm^2$ was designated as LATP-H and the pellet pressed using a pressure of $50kgf/cm^2$ was indexed as LATP-L.

Both the densities of green pellets and sintering pellets were determined by measuring their weights and apparent volumes. The microstructures were examined by using a field emission scanning electron microscope (JSM-7800F). The conductivity measurement was carried out by using a electrochemical impedance spectroscopy [9].

3. Results and Discussions

3.1. Densities evaluations of LATP-H and LATP-L

The densities of sintered LATP-H and LATP-L were determined by weighting the samples and measuring apparent volumes, shown in Fig.1. The densities did not exhibit a clear trend with the doping amount of Al, but the densities of the LATP pressed at high pressures were all higher than those pressed at low pressures.

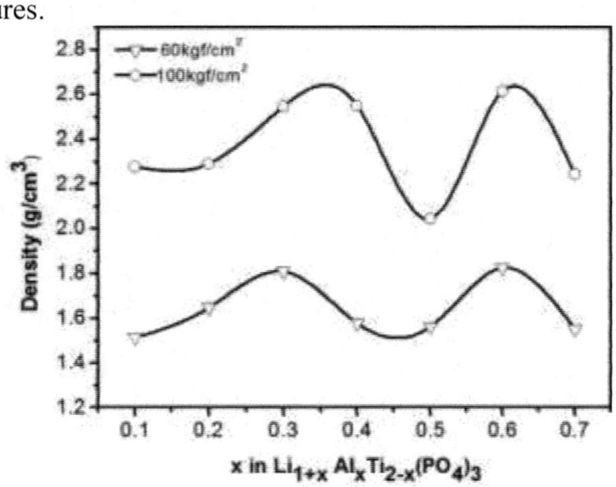

Fig.1 Densities of $Li_{1+x}Al_xTi_{2-x}(PO_4)_3$ pressed at different pressures

3.2. Total conductivities of LATP-H and LATP-L

Total conductivities of sintered LATP-H and LATP-L were calculated based on the impedance obtained from the impedance analyzer. It was found that the variation trend with Al amount was almost identical for LATP pressed at high pressure and low pressure. Both the total conductivities of LATP-H and LATP-L peaked at x=0.6. The value was inconsistent with the value reported previously. The reason is possibly due to the loss of lithium during the sintering stage, since no extra lithium was added to compensate which most of studies did.

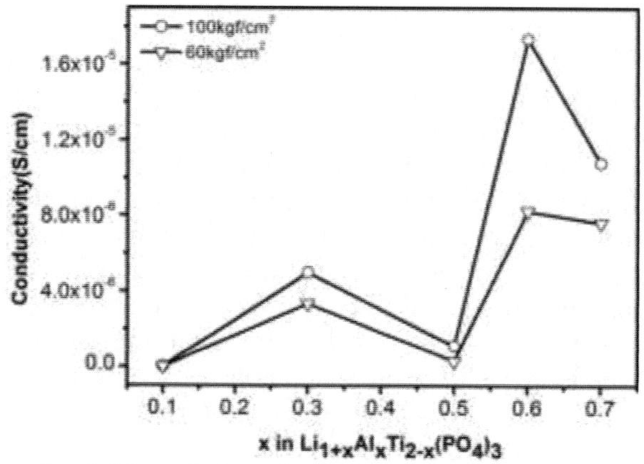

Fig.2 Total conductivities of $Li_{1+x}Al_xTi_{2-x}(PO_4)_3$ pressed at different pressures

The conductivities of LATP-H were all higher than LATP-L. The reason was mainly attributed to the denser microstructures achieved at higher pressure. The higher the density is, the less porosity and defects are in the samples. Therefore, when lithium ions migrated through the grain boundaries, the resistance would be smaller which facilitate the migrations of lithium ions.

3.3. Microstructures of LATP

The conductivity of $Li_{1+x}Al_xTi_{2-x}(PO_4)_3$ with x=0.6 was much higher than the conductivity of $Li_{1+x}Al_xTi_{2-x}(PO_4)_3$ with x=0.1. The microstructures of these two samples were observed by SEM and SEM photos are shown in Fig.3. It can be seen that the microstructure was very rough and the grains were very small when the doping amount of Al was 0.1. When the doping amount of Al increases to 0.6, the morphology became much denser and the grains grew. The change of the morphology may be attributed to the conductivity observed in Fig.2. A denser morphology will be beneficial to the improvement of conductivity which was explained by the steric effect of pores [10]. On the top of that, bigger grains indicated less grain boundaries, which would further improve the conduction of lithium ions.

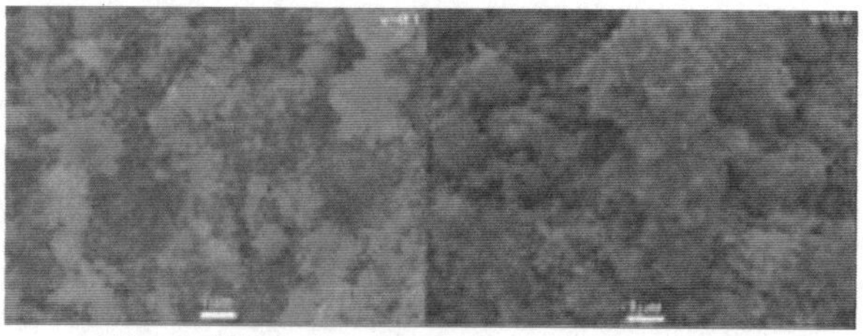

Fig. 3 SEM photos of $Li_{1+x}Al_xTi_{2-x}(PO_4)_3$

4. Conclusions

NASICON structured $Li_{1+x}Al_xTi_{2-x}(PO_4)_3$ ($0.1 \leq x \leq 0.7$) powders were prepared by hydrothermal synthesis method in this study and the pressure applied to press the powders into pellets was investigated. The variation trend with Al amount was almost identical for LATP pressed at high pressure and low pressure. Both the total conductivities of LATP-H and LATP-L peaked at x=0.6. The conductivities of LATP pressed at high pressures were all higher than those at low pressures. The reason was mainly attributed to the denser microstructures achieved at higher pressure.

Acknowledgements

This research was financially supported by the Fundamental Research Funds for the Central Universities (No. 2017MS138) and the Scientific Research Foundation for the Returned Overseas Chinese Scholars, State Education Ministry.

IOP Conf. Series: Materials Science and Engineering **307** (2018) 012044 doi:10.1088/1757-899X/307/1/012044

References

[1] M. Zhang, Z. Huang, J f Cheng, O. Yamamoto, N. Imanishi, B.Chi, J.Pu, J. Li, Solid state lithium ionic conducting thin film $Li_{1.4}Al_{0.4}Ge_{1.6}(PO_4)_3$ prepared by tape casting, J. Alloy. Compd. 590(2014)147–152.

[2] H. Aono, E.Sugimoto,Y. Sadaoka, N. Imanaka, G. Adachi, Ionic conductivity of solid electrolytes based on lithium titanium phosphate, J. Electrochem. Soc. 137 (1990) 1023–1027.

[3] J. L. Narváez-Semanate, A.C.M. Rodrigues, Microstructure and ionic conductivity of $Li_{1+x}Al_xTi_{2-x}(PO_4)_3$ NASICON glass-ceramics, Solid State Ion. 181 (2010) 1197–1204.

[4] K. M. Kim, D. O. Shin, Y.-G. Lee, Effects of preparation conditions on the ionic conductivity of hydrothermally synthesized $Li_{1+x}Al_xTi_{2-x}(PO_4)_3$ solid electrolytes, Electrochimica Acta 176 (2015) 1364–1373.

[5] M. Kotobuki, M. Koishi, Preparation of $Li_{1.5}Al_{0.5}Ti_{1.5}(PO_4)_3$ solid electrolyte via a sol-gel route using various Al sources, Ceram. Intern. 39 (2013) 4645–4649.

[6] G.B. Kunshina, O.G. Gromov, E.P. Lokshin, V.T. Kalinnikov, Sol-gel synthesis of $Li_{1.3}Al_{0.3}Ti_{1.7}(PO_4)_3$ solid electrolyte, Russ. J. Inorg. Chem. 59 (2014) 424–430.

[7] M. Kotobuki, M. Koishi, Y. Kato, Preparation of $Li_{1.5}Al_{0.5}Ti_{1.5}(PO_4)_3$ solid electrolyte via a co-precipitation method, Ionics 19 (2013) 1945–1948.

[8] S. Duluard, A. Paillassa, L. Puech, P. Vinatier, V. Turq, P. Rozier, P. Lenormand, P.- L. Taberna, P. Simon, F. Ansart, Lithium conducting solid electrolyte $Li_{1.3}Al_{0.3}Ti_{1.7}(PO_4)_3$ obtained via solution chemistry, J. Eur. Ceram. Soc. 33 (2013) 1145–1153.

[9] X. Lu, D. Yang, Effect of sintering condictions on perovskite lithium-based ion conductor, Emerging Mater. Res. 6 (2017) 1-19.

[10] J. Tan, Y. Su, H. Tang, T. Hu, Q. Yu, R. Tursun and X. Zhang, Effect of calcined parameters on microstructure and electrical conductivity of 10Sc1CeSZ, J. Alloy Compd. 686 (2016) 394-398.

IOP Conf. Series: Materials Science and Engineering **307** (2018) 012045 doi:10.1088/1757-899X/307/1/012045

Influence of repeated quenching-tempering on spheroidized carbide area in JIS SUJ2 bearing steel

K Egawa [1,2], I Yoshida [1,3], H Yoshida [1,4] K Mizobe[1,5], and K Kida[1,6] *
[1] University of Toyama, Gofuku 3190, Toyama, 930-8555, Japan
[2]m1771203@ ems.u-toyama.ac.jp, [3]m1671239@ems.u-toyama.ac.jp,
[4]s1370485@ems.u-toyama.ac.jp, [5]kmizobe@eng.u-toyama.ac.jp
[6]kida@eng.u-toyama.ac.jp
* The corresponding author

Abstract. High-carbon high-strength JIS-SUJ2 bearing steel is mainly used for rolling contact applications which require high fatigue strength. We had applied repeated quenching which refine the prior austenite grains to this steel. In this work, we prepared JIS SUJ2 bearing steel bar specimens which were quenched three times (Q3T1) and quenched-tempered three times (QTQTQT) in order to investigate the influence of tempering before quenching on the microstructure. The specimens were etched by picral to observe the microstructure. We found that the spheroidized carbide area was important for the prior austenite grain formation.

1. Introduction

High-carbon high-strength JIS-SUJ2 bearing steel is mainly used for rolling contact applications which require high fatigue strength and wear resistance. Quenching is one of the methods which improve the fatigue strength and wear durability.

Repeated quenching is used as one of the microstructure refinement methods[1-3]. In particular, repeated quenching refines the prior austenite grains. Because the prior austenite grain refinement improves the fatigue strength, repeated quenching has been investigated as the method that improves the fatigue strength[4,5]. Ohki et al. reported that additional quenching after nitridizing (FA-treatment) refined the prior austenite grains and which indicated the refinement of martensitic structure. The FA-treatment also increased the retained austenite[6]. Santos et al. reported that the repeated quenching increased the retained austenite and a large amount of retained austenite improved the wear resistance of samples [7]. Mizobe et al. investigated the relation between the repeated quenching and the fatigue strength. They prepared the repeated quenched JIS SUJ2 bearing steel and performed the rotating bending fatigue tests. They reported that the fatigue strength depended on the variation of non-metallic inclusions in the refined the prior austenite grains[4,8,9].

In this work, we prepared JIS SUJ2 bearing steel which were quenched three times (Q3T1) and quenched-tempered three times (QTQTQT). We measured Vickers hardness and the value of the retained austenite, and focused on the microstructure transformation in order to investigate the influence of tempering before quenching.

IOP Conf. Series: Materials Science and Engineering **307** (2018) 012045 doi:10.1088/1757-899X/307/1/012045

2. Test method

2.1. Specimen (JIS SUJ2 steel) and heat treatment

We prepared JIS SUJ2 steel bars whose diameter was 10 mm. Table 1 shows the chemical composition of the specimens. Before the heat treatment, the material was spheroidized annealed.

Figure 1 shows the heat treatment process. Specimens were quenched at 850°C for 40 minutes and tempered at 180°C for 60 minutes. We prepared two type of specimens: specimens which were quenched three times and tempered once (Figure 1(a)), and the specimens which were quenched-tempered three times (Figure 1(b)). The former and later are referred to as Q3T1 and QTQTQT, respectively.

Table 1 Chemical composition of the JIS SUJ2 steel bar (weight %)

C	Si	Mn	P	S	Ni	Cr	Mo	Cu	Ti
0.99	0.26	0.4	0.019	0.007	0.08	1.38	0.02	0.11	0.002

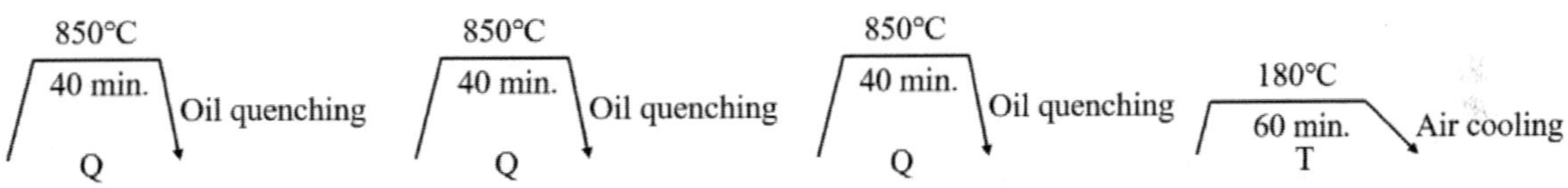

(a) Q3T1

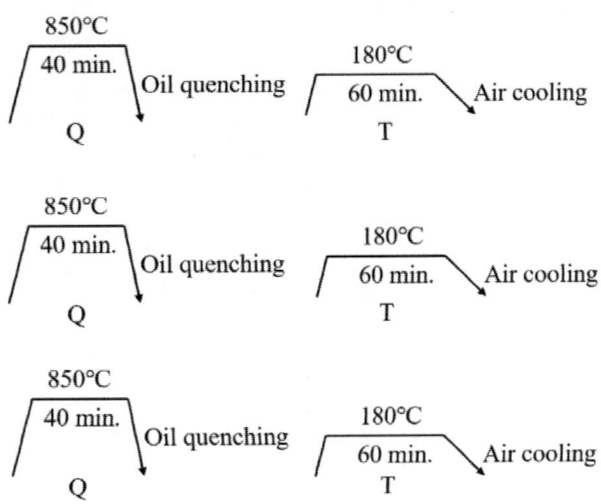

(b) QTQTQT

Figure 1 Heat treatment process.

IOP Conf. Series: Materials Science and Engineering **307** (2018) 012045 doi:10.1088/1757-899X/307/1/012045

2.2. X-Ray measurement and microstructure
We used Bulker D8 DISCOVER with GADDS (2D detector) in order to measure the amount of retained austenite. Table 2 shows the test conditions of the retained austenite measurement. Figure 2 shows the observation points of the retained austenite and the measurement increments were 0.5 mm. Before the measurement, specimens were cut and polished by 1 μm diamond paste.

Vickers hardness distributions of Q3T1 and QTQTQT sumples were measured by a MATSUZAWA VMT-X7s. The test conditions were a load of 10 kgf and a time of 15 seconds. Furthermore, in order to observe the microstructure, the specimens were etched by picral for 360 seconds and observed by laser confocal microscope (LCM).

Table 2 Observation conditions for retained austenite measurement.

X-ray source	Cr Kα
Power	35 kV 40 mA
Diameter of collimator	0.5 mm
Detector distance	150 mm
Measurement angle	60 to 90 (2θ)
Scan time	1200 sec / point

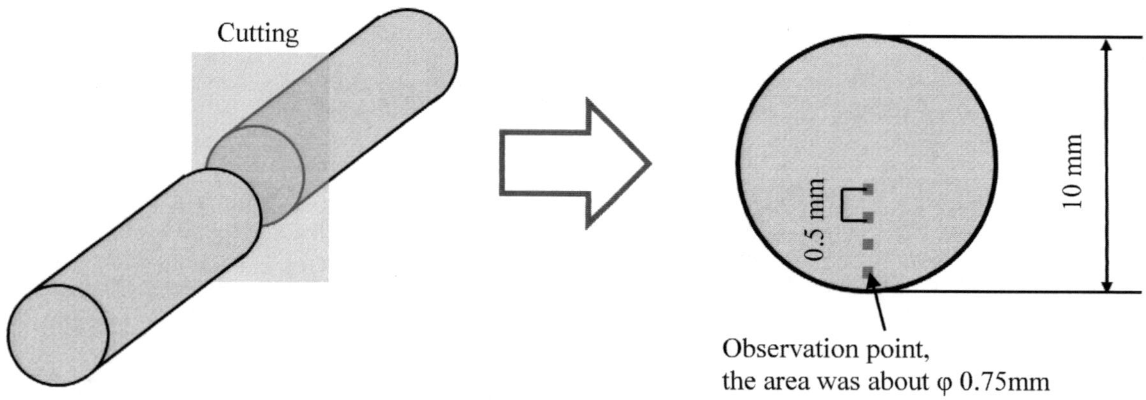

Figure 2 Observation points of the retained austenite.

2.3. Observation of electron probe micro analyser (EPMA)
EPMA(JXA-8230) was used in order to determine the kinds of element. Table 3 shows the test conditions of EPMA observation, and the detection elements were Carbon, Chromium, Sulphur, Phosphorus and Iron. Before the observation, we prepared etched samples in order to emphasize the edge of microstructure.

Table 3 Observation condition of EPMA analysis.

Cathode	W filament
Power	30kV and 500mA
Scan type	Beam scan
Resolution	384×288 pixel
Magnification	×10000
Probe	Electron probe

3. Results and discussion

3.1. Vickers hardness distribution and retained austenite

Figure 3 shows Vickers hardness distribution of Q3T1 and QTQTQT samples. The average of Vickers hardness was 819HV in Q3T1 sample and 840 HV in QTQTQT sample. The average of Vickers hardness of QTQTQT sample was slightly higher than that of Q3T1 sample.

Figure 4 shows the retained austenite distributions of Q3T1 and QTQTQT samples. The average of retained austenite was 24.2% in Q3T1 samples and 22.1% in QTQTQT samples. The average of retained austenite of QTQTQT sample was lower than that of Q3T1 sample. This indicates that tempering before quenching decreased the retained austenite.

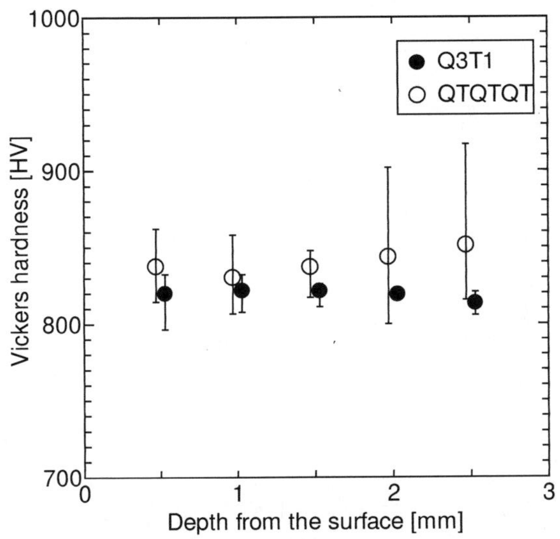

Figure 3 Vickers hardness distributions.

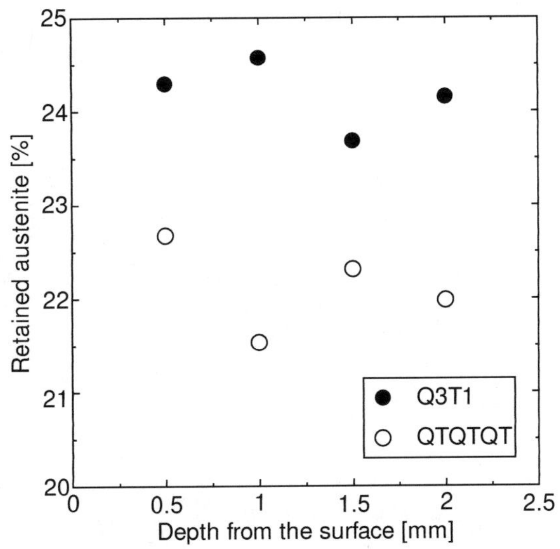

Figure 4 Retained austenite distributions.

3.2. Microstructure

Figure 5 shows the microstructure of the sample etched by picral. In previous work, We applied mathematical method to analyse the patterns of prior austenite grains[10]. In the present work, we observed spheroidized carbides to fur the investigate the effect of quenching times on microstructure. Figure 5(a) shows the microstructure of the Q3T1 sample (×1000). Figure 5(b) shows the microstructure of the QTQTQT sample (×1000). In both pictures, the spheroidized carbide clusters can be seen. We define this area as "spheroidized carbide area". The spheroidized carbide area of QTQTQT sample was larger than that of Q3T1 sample. Figure 5(c) shows the microstructure of the spheroidized carbide area of the Q3T1 sample (×3000). Figure 5(d) shows the microstructure of the area including no spheroidized carbide of the Q3T1 sample (×3000). At the spheroidized carbide area, the prior austenite grains were refined. Figure 5(e) shows the microstructure of the spheroidized carbide area of the QTQTQT sample (×3000). Figure 5(f) shows the microstructure of the area including no spheroidized carbide of the QTQTQT sample (×3000). At the spheroidized carbide area, the prior austenite grains were refined. Figure 5(d) and Figure 5(f) are the area where the spheroidized carbide cluster cannot be seen. Here, the prior austenite grain size of Q3T1 sample was similar to that of QTQTQT sample. This indicates that the spheroidized carbide area is important for the prior austenite grain formation in both samples.

IOP Conf. Series: Materials Science and Engineering **307** (2018) 012045 doi:10.1088/1757-899X/307/1/012045

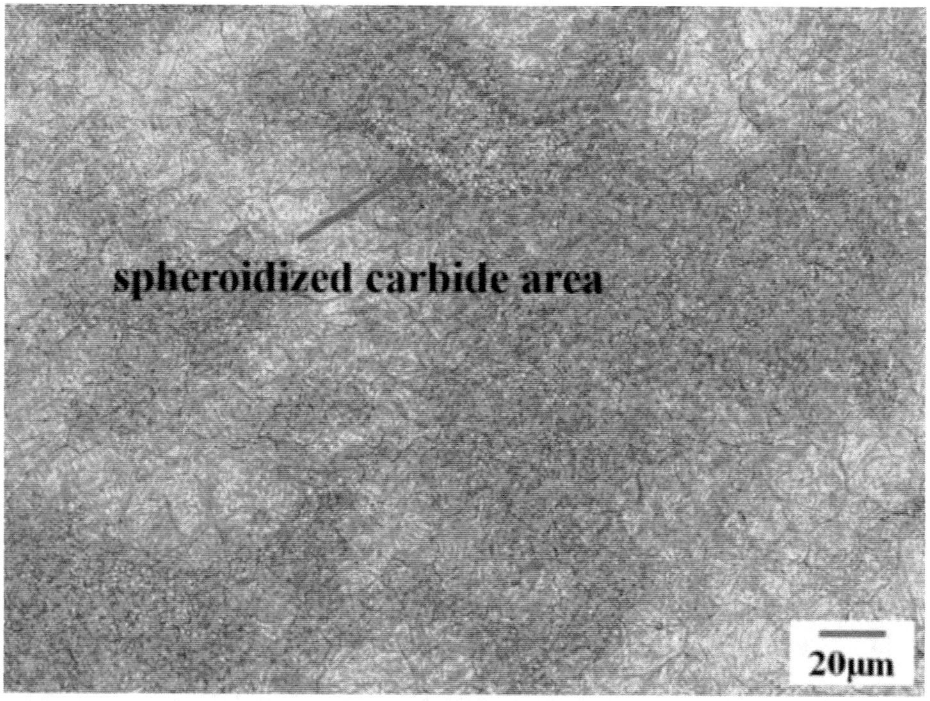

(a) Microstructure of the Q3T1 sample (×1000)

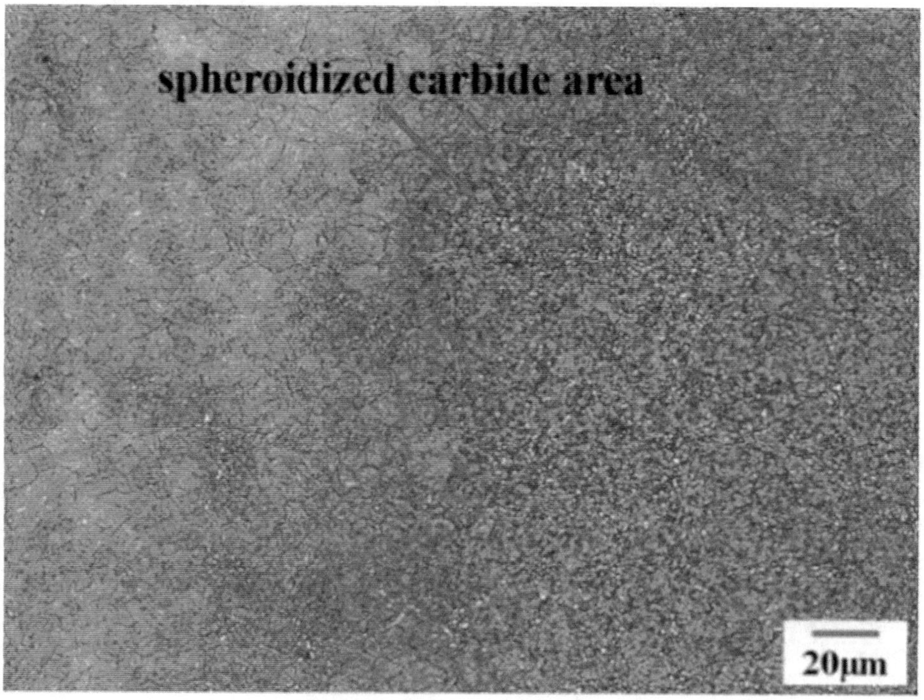

(b) Microstructure of the QTQTQT sample (×1000)

IOP Conf. Series: Materials Science and Engineering **307** (2018) 012045 doi:10.1088/1757-899X/307/1/012045

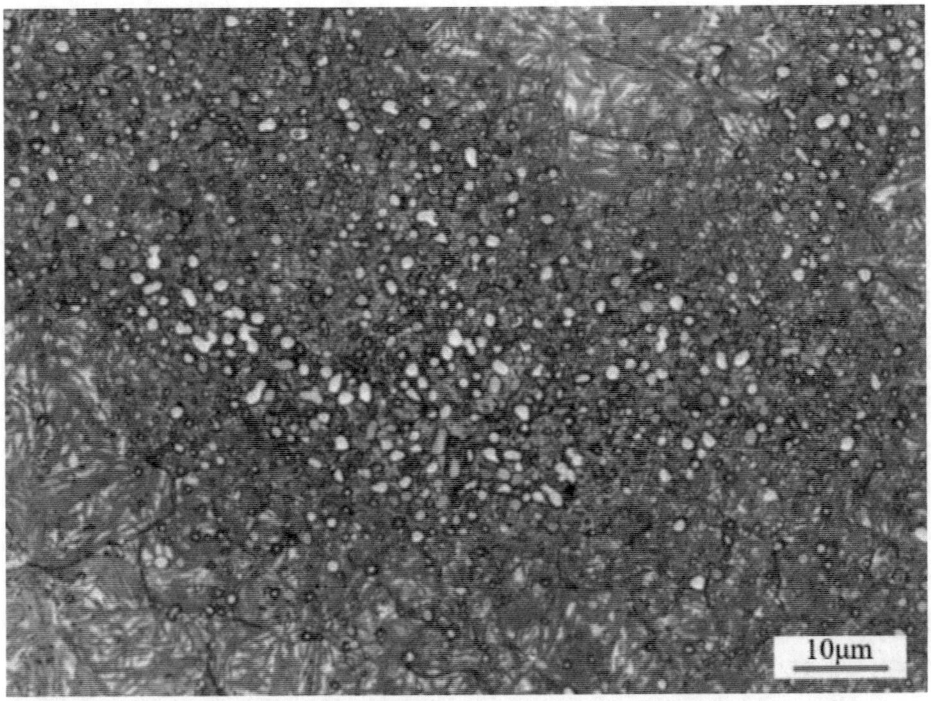

(c) Microstructure of the spheroidized carbide area of the Q3T1 sample (×3000)

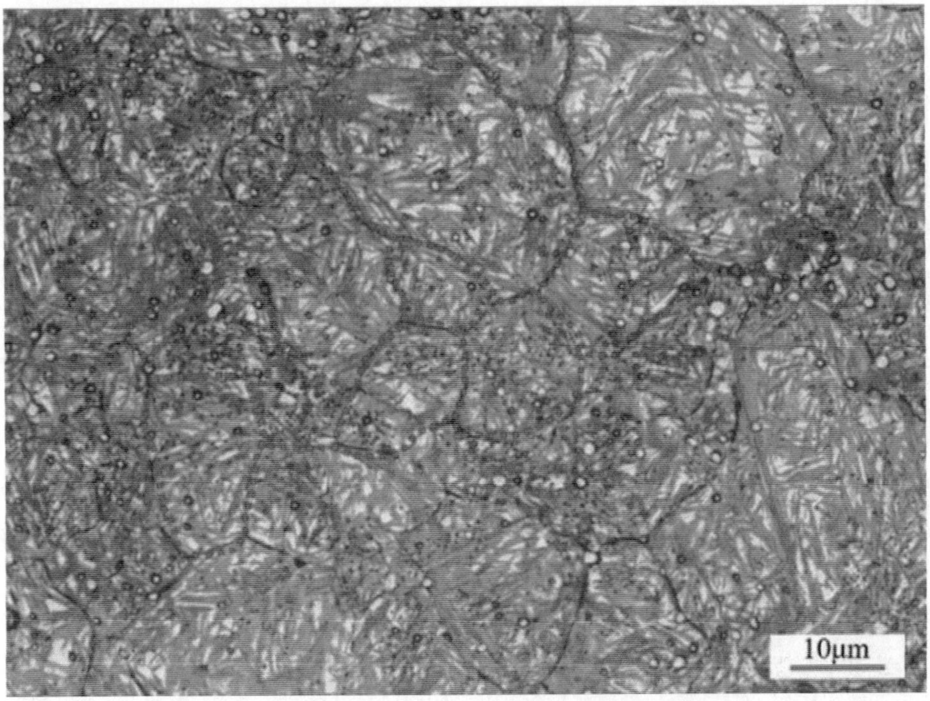

(d) Microstructure of the area including no spheroidized carbide of the Q3T1 sample (×3000)

IOP Conf. Series: Materials Science and Engineering **307** (2018) 012045 doi:10.1088/1757-899X/307/1/012045

(e) Microstructure of the spheroidized carbide area of the QTQTQT sample (×3000)

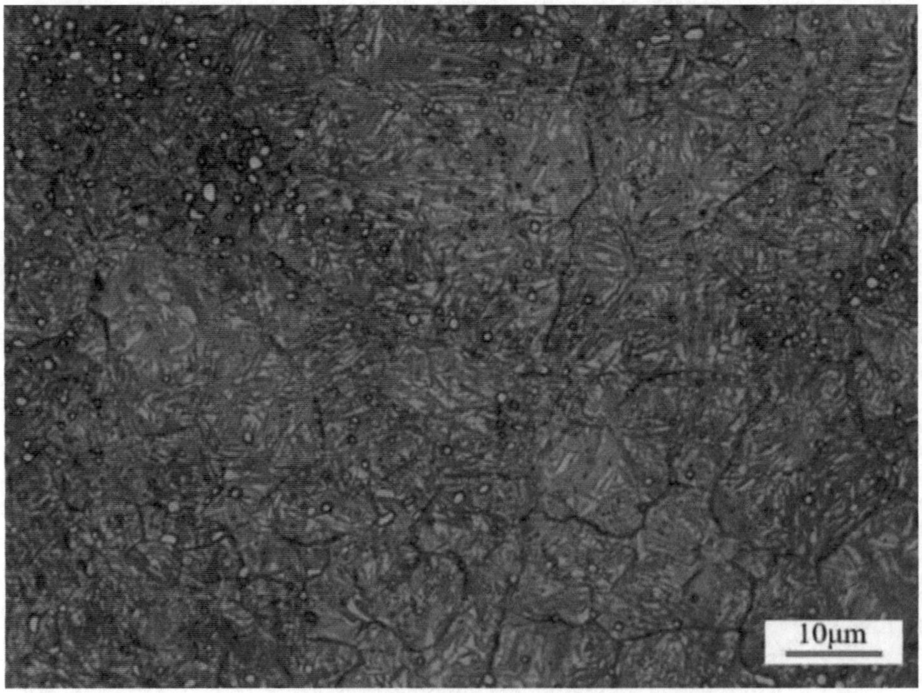

(f) Microstructure of the area including no spheroidized carbide of the QTQTQT sample (×3000)

Figure 5 Microstructure of the samples etched by picral.

3.3. Analysis of EPMA

We used EPMA in order to investigate the characteristic of spheroidized carbide. Figure 6 shows the result of EPMA observation of QTQTQT sample. The spheroidized carbide consists of Carbon and Chromium and the substrate was Iron. Furthermore, a few Iron was detected from the spheroidized carbide.

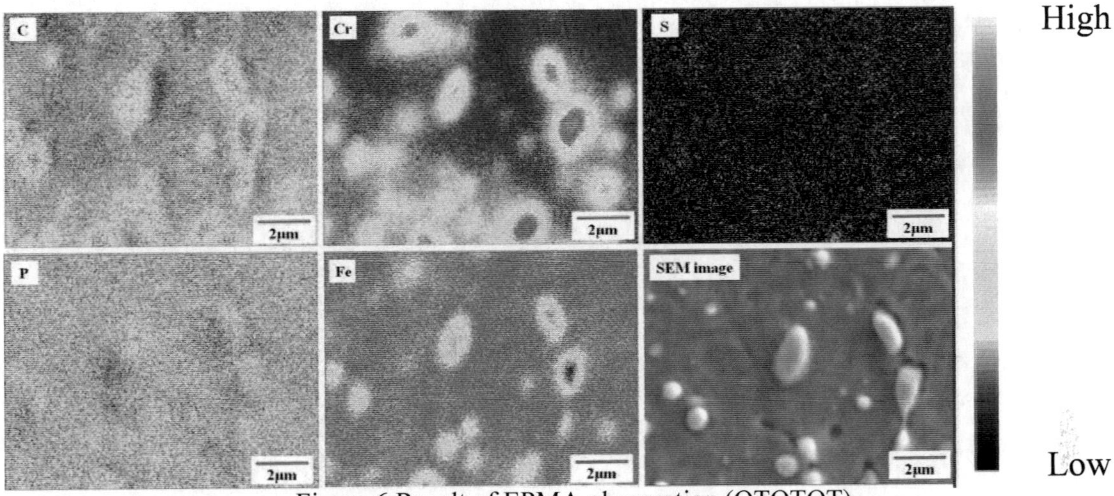

Figure 6 Result of EPMA observation (QTQTQT).

4. Conclusions

We prepared JIS SUJ2 steel bars which were 10 mm diameter. Specimens were quenched three times and tempered once (Q3T1) and quenched-tempered three times (QTQTQT). We measured Vickers hardness and the retained austenite. The specimen was etched by picral in order to observe the microstructure. We used EPMA in order to investigate the characteristic of spheroidized carbide. The conclusions are as follows.

1 The average of Vickers hardness of QTQTQT sample was slightly higher than that of Q3T1 sample.
2 The average of retained austenite of QTQTQT sample was lower than that of Q3T1 sample.
3 Spheroidized carbide area is important for the prior austenite grain formation.
4 The characteristic of spheroidized carbide was chromium carbide.

References

[1] Grange A R 1971 *Metallurgical and Materials Trans. A* **2** pp. 65-78,
[2] Kayali S E, Sunada H, Oyama T, Wadsworth J, Sherby D O 1979 *Jounal of material science* pp. 2688-2692,
[3] Kim J H, Kim H Y and Morris, Jr. W J 1998 *ISIJ International* **38** No. 11 pp. 1277-1285
[4] Mizobe K, Honda T, Koike H, Santos C E, Shibukawa T and Kida K 2014 *Material Research Innovations* **18** S1 pp. 60-65
[5] Yang G Z, Li X S, Zhang M J, Zhang F J, Li Y G, Weng B Z 2004 *Acta Materialia* **52** pp. 5235-5241
[6] Ooki C, Maeda K and Nakashima H 2004 *NTN Technical review* **71**, pp. 2-7

[7] Santos C E, Honda T, Kida K 2010 *Proceedings of SPIE-The International Society for Optical* **7522** pp 1-7
[8] Mizobe K, Santos C E, Honda T, Koike H and Kida K 2012 *Int. J. Materials and Product Technology* **44** Nos. 3/4 pp 227-239
[9] Mizobe K, Honda T, Koike H, Santos C E, Shibukawa T, and Kida K 2013 *Applied Mechanics and Materials* **300-301** pp 1298-1303
[10] Hitoshi Mahara Koshiro Mizobe Katsuyuki Kida and Kazuaki Nakane 2017 *Applied Numerical Mathematics* **114** pp 124-131

IOP Conf. Series: Materials Science and Engineering **307** (2018) 012046 doi:10.1088/1757-899X/307/1/012046

Effect of observation position of SUJ2 bar specimens on inclusions distribution

T Kamiya [1,2], K Mizobe [1,3] and K Kida [1,4*]

[1] University of Toyama, Gofuku 3190, Toyama, 930-8555, Japan
[2] m1671211@ems.u-toyama.ac.jp
[3] kmizobe@eng.u-toyama.ac.jp
[4] kida@eng.u-toyama.ac.jp
[*] corresponding author

Abstract. The size and distribution of non-metallic inclusions are strongly related to rolling contact fatigue (RCF) life. SUJ2 steel is manufactured as a round bar, and is cooled from the surface during casting. This indicates that the inclusions may be distributed non-uniformly in the center direction. The objective of this study is a clarification of the difference of the inclusion distribution in the SUJ2 round bar. It was found that the distribution of inclusions in SUJ2 was distributed uniformly in a transverse direction.

1. Introduction

High carbon chromium steel SUJ2 is used in rolling contact applications when high wear and fatigue resistance are required. It is well known that flaking failures of bearings originate from non-metallic inclusions under Rolling Contact Fatigue (RCF). These non-metallic inclusions are mainly made of oxide, sulfide and nitride. The size of these non-metallic inclusions is strongly related to RCF life.

These inclusions are contaminated in the steel during the steelmaking process. The inclusions are generated in the solidification process by the casting. SUJ2 steel is manufactured as a round bar, and is cooled from the surface during the casting process. This indicates that the inclusions may be distributed non-uniformly in the center direction because of the difference in cooling rate. Therefore, we need to study distribution of inclusions in the round bar.

The statistics of extreme value are one of the major methods to estimate the maximum inclusion size based on the inclusion distributions. Many researchers studied the relation between the inclusions and RCF life. Hashimoto, et. al. [1] studied effect of oxide inclusion size and chemical composition on RCF life of SUJ2 bearing steel. They found that RCF life was improved with decrease in the size of oxide inclusions. Nagao, et. al. [2] studied effect of size and types of inclusions on the RCF life of SUJ2 bearing steel. The RCF life was not related to inclusion type, but to the largest inclusion size.

In this study, we observed inclusion distribution of SUJ2 steel based on the statistics of extreme value method. We measured the maximum size of inclusions in the view field of microscope, and evaluated the inclusion distribution of SUJ2 steel by comparing the inclusions in two areas which are near the core of the bar specimen and below the rolling track.

2. Test method

2.1. Specimens and heat treatment

Figure 1 shows a schematic illustration of plate sample. The outer diameter was 52 mm and the inner diameter was 20 mm. The thickness of the samples was 10 mm. Figure 2 shows a schematic illustration of heat treatment processes. The SUJ2 plate was heated at 830 °C for 40 minutes in a furnace and after that quenched. Subsequently, it was tempered at 180 °C for 120 minutes.

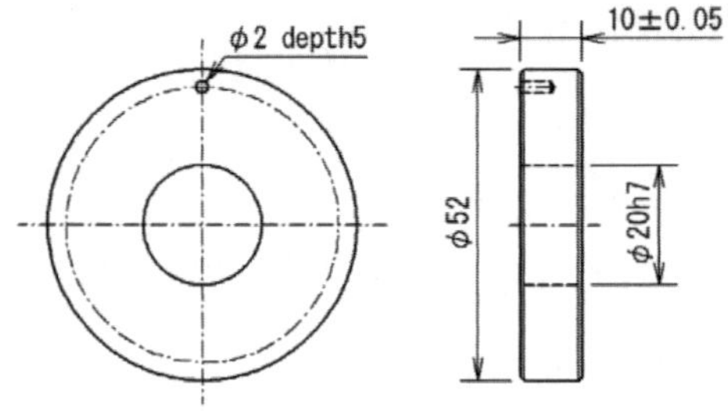

Figure 1 Schematic illustration of sample. (dimension in mm)

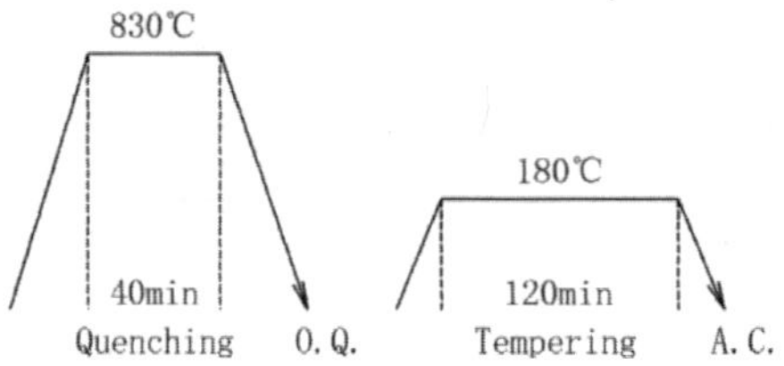

Figure 2 Schematic illustration of heat treatment processes.

2.2. Rolling contact fatigue tests and Hardness tests

We measured Vickers hardness in two regions using a Hardness tester (MATSUZAWA, VMT-X7S). An indentation load was 10 kgf and the loading time was 15 seconds. The measurements were done below the rolling track and at 5 mm depth from the surface. Increment of measuring points was 0.5 mm. Figure 3 shows a schematic illustration of the measurement regions.

We performed RCF tests up to 4.30 × 10^6 cycles using a thrust-type machine [3]. The tests were stopped when we found flaking failures. The rotation speed was 2000 rpm and the maximum contact pressure was 5.3 GPa. The value was calculated by Hertzian contact theory [4]. Young's modulus and Poisson's ratio were 210 GPa and 0.3, respectively.

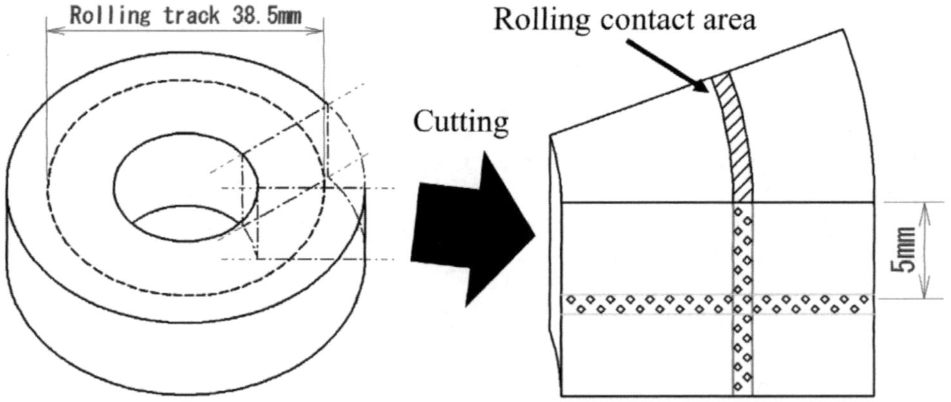

Figure 3 Schematic illustration of the measurement points of Vickers hardness test.

IOP Conf. Series: Materials Science and Engineering **307** (2018) 012046 doi:10.1088/1757-899X/307/1/012046

2.3. Inclusion observation and evaluation methods

After the RCF tests, the specimen was cut and polished to observe the inside. The sectional surfaces were ground by grind papers from #800 to #4000 and polished by 1 μm diamond paste.

Figure 4 shows a schematic illustration of the observation area. The extending direction on the section is shown as ED, and the transverse direction is as TD. We observed two areas as shown in figure 4. They will be referred to as A_{track} and A_{core}, respectively. A_{track} is a red area which is below the rolling track. A_{core} is a green area which is near the core of the bar specimen. The size of both areas was 10 mm^2 (1 mm width × 10 mm length). There is no flaking failure around the A_{track}.

The inclusions were observed by a LCM (laser scanning confocal microscope, Keyence, VK-9700). We measured the size of inclusion which was selected as the largest inclusion in a field of LCM view. The measurement window of the LCM was 90 μm × 67 μm ($S_0 = 6.03 \times 10^{-3}$ mm^2). We took 30 photos from both A_{track} and A_{core}. We measured the maximum inclusion size in each photo and evaluated the distribution of inclusions with the statistics of extreme values of all 30 photos for each area.

Murakami studied the relation between fatigue life and inclusion size [5]. He proposed $\sqrt{area}$ theory to evaluate without considering the shape of inclusion. Based on his theory, we defined the square root of the largest inclusion size as $\sqrt{area}_{max}$ μm in this paper.

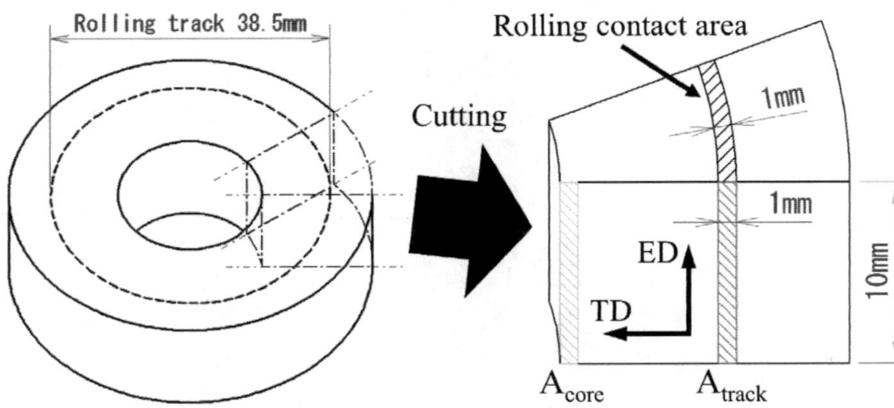

Figure 4 Schematic illustration of the observation area.

3. Results and Discussion

Figures 5 (a) and (b) show the Vickers hardness distributions in the two measurement regions of figure 3. The average values of Vickers hardness below the rolling track and 5 mm depth from the surface are HV730 and HV736. The Vickers hardness values distribute evenly.

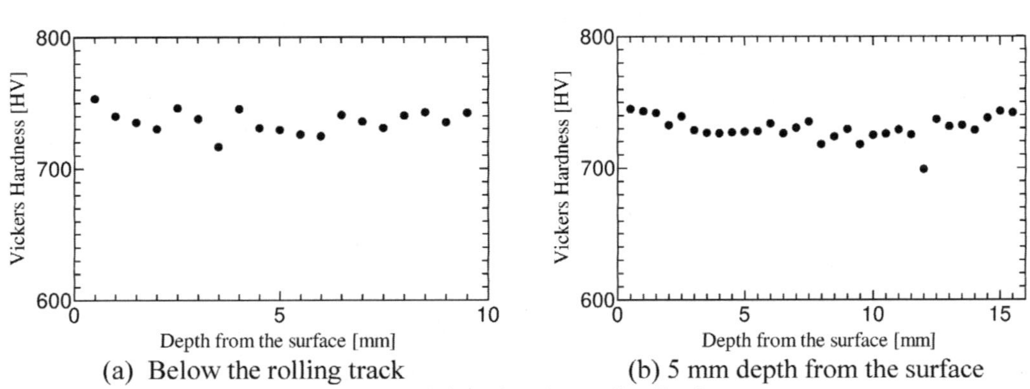

(a) Below the rolling track (b) 5 mm depth from the surface

Figure 5 Vickers hardness distributions.

IOP Conf. Series: Materials Science and Engineering **307** (2018) 012046 doi:10.1088/1757-899X/307/1/012046

Figure 6 shows an example of sulfide inclusions. We found many sulfide inclusions which were extended in the ED direction. The sulfide inclusions have low hardness compared to oxide and nitride inclusions. Therefore, the shape of sulfide inclusions follows the deformation behavior of matrix steel when processing.

Hashimoto et al. [6] reported that the sulfide inclusions which was stretched in a direction parallel to the load movement affected RCF lives. In our observations, we ignored the sulfide inclusions because the ED direction was perpendicular to the load movement. But, the sulfide inclusions which composited with other inclusions were counted. Figure 7 shows an example of the sulfide composition inclusion.

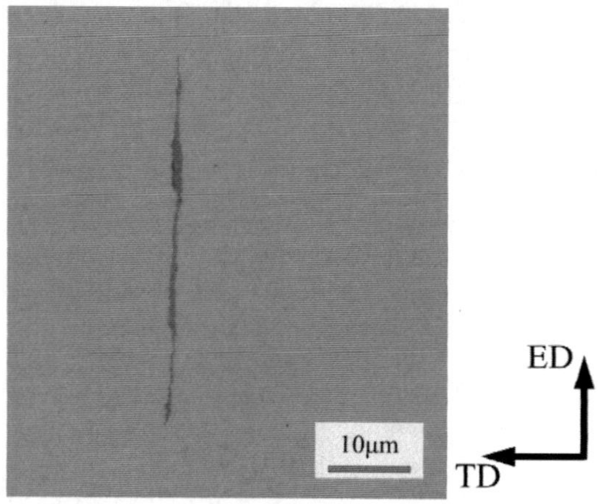

Figure 6 Photograph of sulfide inclusions which were extended in the extending direction.

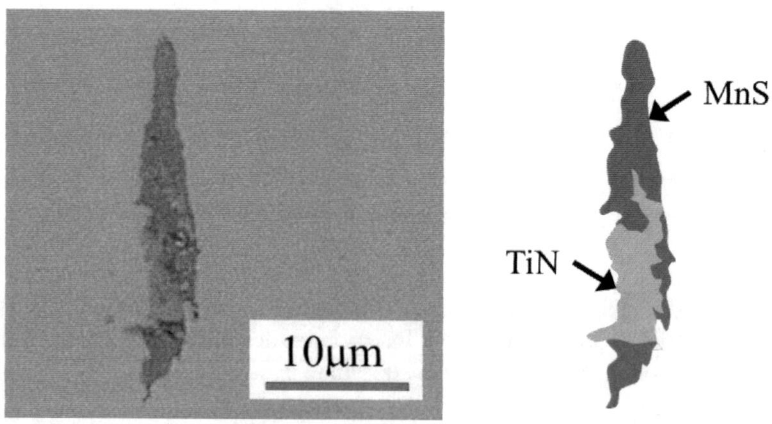

Figure 7 Photograph of sulfide inclusion which composited with other inclusions.

In the observation, we found some types of inclusions, such as oxide, nitride and sulfide. Figure 8 shows the photograph of oxide and nitride inclusions. We counted and evaluated the distribution of these inclusions with the statistics of extreme value method.

IOP Conf. Series: Materials Science and Engineering **307** (2018) 012046 doi:10.1088/1757-899X/307/1/012046

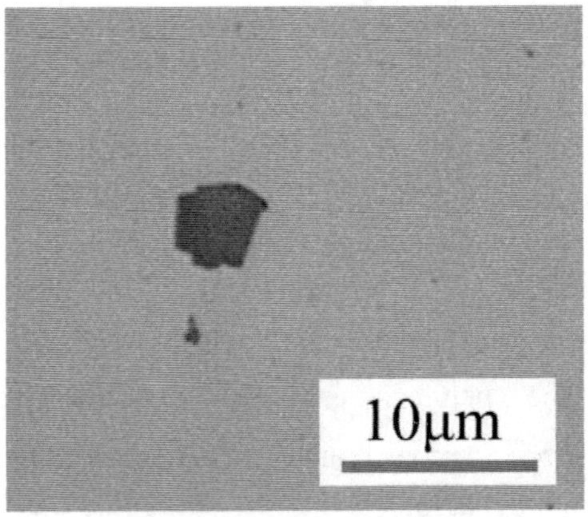

(a) Oxide inclusion (Al_2O_3)

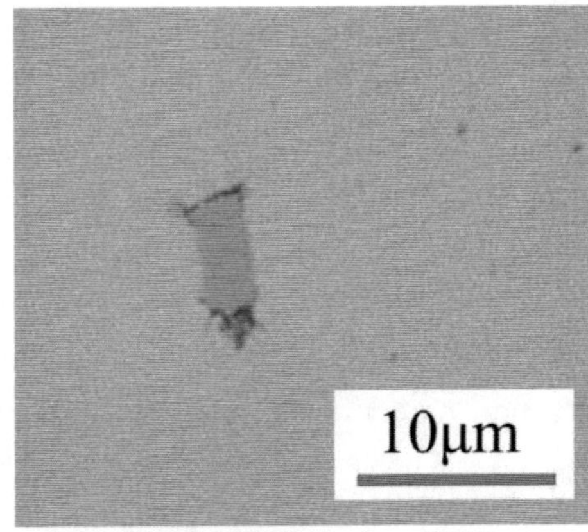

(b) Nitride inclusion (TiN)

Figure 8 Photograph of oxide and nitride inclusions.

Figure 9 shows the extreme values distribution in the A_{track} and A_{core}. The vertical axis of the graph shows the cumulative distribution frequency, reduced variate and return period. The horizontal axis shows the $\sqrt{area}_{max}$ values.

The average values of $\sqrt{area}_{max}$ of A_{track} and A_{core} are 4.68 μm and 4.26 μm. The regression linear equations (y) are described as follows.

$$A_{track} : y = 0.63\sqrt{area}_{max} - 2.41$$

$$A_{core} : y = 0.58\sqrt{area}_{max} - 1.95$$

This indicates that the distribution of the maximum inclusion size in A_{track} is similar to that in A_{core}. Furthermore, the inclusions in SUJ2 bar specimen were uniformly distributed in a direction of TD.

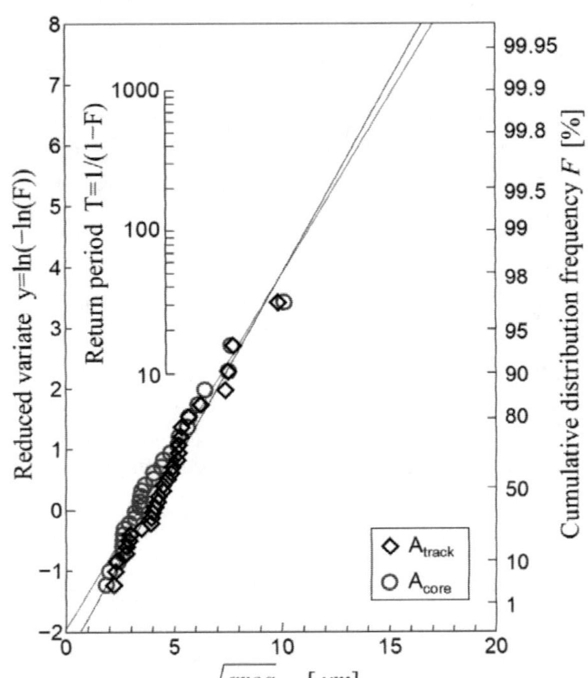

Figure 9 Extreme value distribution in the SUJ2 steel.

We calculate the maximum inclusion size in a hazardous volume from measurement results of the inclusions. In the calculation, we define the maximum inclusion size in the hazardous volume as MCI with the statistics of extreme value distribution of all inclusions. We use a virtual thickness of inclusions to deduce the MCI based on the 2D observation results [7]. The virtual thickness h is calculated from the average value of $\sqrt{area}_{\max}$. We defined hazardous volume V_{RCF} at the rolling contact fatigue tests as equation (1).

$$V_{RCF} = \pi a_c d_s D_p \tag{1}$$

The pitch circle diameter D_p was 38.5 mm. The contact width a_c which was measured after the RCF test was 1 mm. We defined d_s as the depth from the surface when the stress is 80% of maximum shear stress. The maximum shear stress was calculated based on the Hertzian contact theory and Sackfield & Hills formula [4,8]. From the calculation results, it was found that the maximum shear stress was 388.8 MPa and depth from the surface d_s was 0.26 mm. From the results of calculations using Eq. (1), the hazardous volume was 31.4 mm^3. We evaluate the MCI of A_{track} and A_{core} as follows.

$$MCI_{track} = 25.9\ [\mu m]$$

$$MCI_{core} = 27.4\ [\mu m]$$

The difference between both MCI values is negligible. This result shows the distribution of inclusions in SUJ2 is distributed uniformly in a direction of TD and the inclusion size of A_{track} is the same as in A_{core}.

4. Conclusions

We investigated effects of observation position of SUJ2 bar specimens on inclusions distribution. We measured Vickers hardness distribution, and observed the inclusions below the rolling track area and near the core area. After measuring the maximum inclusion size, we evaluated the distribution of inclusions with the statistics of extreme values and calculated the maximum inclusion size in a hazardous volume. The conclusions are as follows.

1) The Vickers hardness values distributed evenly.
2) The distribution of the maximum inclusion size in track area was similar to that in core area. The difference between both the maximum inclusion sizes was negligible. These results show the inclusions in SUJ2 are distributed uniformly and the maximum inclusion size of the track area is the same as in the core area.

References

[1] Hashimoto K, Fujimatsu T, Tsunekage N, Hiraoka K, Kida K and Edson C S 2011 *Materials and Design* **32** pp 1605-11
[2] Nagao M, Hiraoka K and Unigame Y 2005 *Sanyo Technical Report* **12** pp 38-45
[3] Kamiya T, Hashizume Y, Mizobe K and Kida K 2016 *Materials Science Forum* **867** pp 60-65
[4] Hertz H 1881 *Journal für die reine und angewandte Mathematik* **92** pp 156-171
[5] Murakami Y 1994 *Journal of Research of the National Institute of Standards and Technology* **99** pp 345-351
[6] Hashimoto K, Hiraoka K, Kida K and Edson C S 2012 *Materials science and Technology* **28** pp 39-43
[7] Murakami Y 1993 Metal fatigue: effects of small defects and non-metallic inclusions *Yokendo Ltd* pp 321-336
[8] Sackfield A and Hills D A 1983 *The Journal of Strain Analysis for Engineering Design* **18** pp 107-110

IOP Conf. Series: Materials Science and Engineering **307** (2018) 012047 doi:10.1088/1757-899X/307/1/012047

Determination of the boundary conditions of the grinding load in ball mills

Rashid R Sharapov

Moscow State University of Civil Engineering (National Research University), Yaroslavl highway 26, Moscow, 129337, Russia

Corresponding author: ptdm_zavkaf@mail.ru

Abstract. The prospects of application in ball mills for grinding cement clinker with inclined partitions are shown. It is noted that ball mills with inclined partitions are more effective. An algorithm is proposed for calculating the power consumed by a ball mill with inclined inter-chamber partitions in which an axial movement of the ball load takes place. The boundary conditions in which the ball load is located are determined. The equations of bounding the grinding load are determined. The behavior of a grinding load is considered in view of the characteristic cross sections. The coordinates of the centers of gravity of the grinding load with a definite step and the shape of the cross sections are determined. It is theoretically shown that grinding load in some parts of the ball mill not only consumes, but also helps to rotate the ball mill. Methods for calculating complex analytical expressions for determining the coordinates of the centers of gravity of the grinding load under the conditions of its longitudinal motion have developed. The carried out researches allow to approach from the general positions to research of behavior of a grinding load in the ball mills equipped with various in-mill devices.

1. Introduction

Cement production is the basic construction industry. The main aggregates for grinding clinker and additives are ball mills [1-3]. However, calculations have shown that the efficiency of ball mills does not exceed 2% [4]. To increase the efficiency of ball mills, they are equipped in-mill devices [5, 6].
All known theoretical studies are based on the solution of the analytic expression (1).

$$P = \mathrm{A}_A \gamma_b V \sqrt{D} \{S_A\}, \quad (1)$$

where A_A is the constant S. E. Andreev; S_A – high-speed power factor.
Since in all these works conventional ball mills are considered in which the mill load factor does not change during one cycle, the power consumed by the mill drive remains unchanged. However, there are a number of works [5, 6], which show that the function $P(S)$ has a harmonic sinusoidal character.
In our proposed method for calculating the power consumption, we make an attempt, based on experimental data, to take into account the influence of the transverse form of loading along the length of the mill on the amount of power consumed by it, and also the change in the zone of active influence of the partition with a change in the angle of rotation of the mill's drum. In this case, the rotation of the mill drum is considered as a series of quasistatic states at a fixed angle of rotation of the drum ξ.
The algorithm for calculating the power consumed by a ball mill according to the proposed method is as follows:
1. The mill body is divided into sections with a constant cross-sectional shape of the load. 2. Within each section the loading is considered consisting of separate "parts" of constant thickness, normal to

IOP Conf. Series: Materials Science and Engineering **307** (2018) 012047 doi:10.1088/1757-899X/307/1/012047

the axis of rotation of the drum. 3. For a fixed position of the mill drum (ξ_i = const), the coordinates of the center of gravity of each of the "parts" along the length of the mill body are determined. 4. The moment created by the loading within each "part", relative to the axis of rotation of the mill's drum, is calculated. 5. The total load moment relative to the rotation axis for a fixed position of the mill drum (ξ_i = const) is calculated. 6. The power consumed by the ball load is determined at different positions of the mill drum.

2. Determination of mill drive power

The power of the drive mill, spent on moving the load, is determined by the formula, *W*:

$$P_{en} = \frac{P_0}{\eta_t \eta_{en}} \quad (2)$$

where P_0 – output shaft power, W; η_t, η_{en} –the efficiency of the mechanical transmission and the engine, respectively.

$$P_0 = P_b + P_l , \quad (3)$$

where P_b– power used to move grinding media, Вт; P_l – power used to overcome losses in trunnion bearings.

To calculate the power required to move grinding bodies P_b, it is necessary to determine the boundary conditions in which the grinding load is located.

As can be seen from the calculation scheme (Fig. 1), the grinding load in the first mill chamber is limited by: the end cover of the mill, the side surface of the mill housing, the loading level A_1A_2, the surface formed by the displaced volume A_2A_3, the surface of the inclined partition.

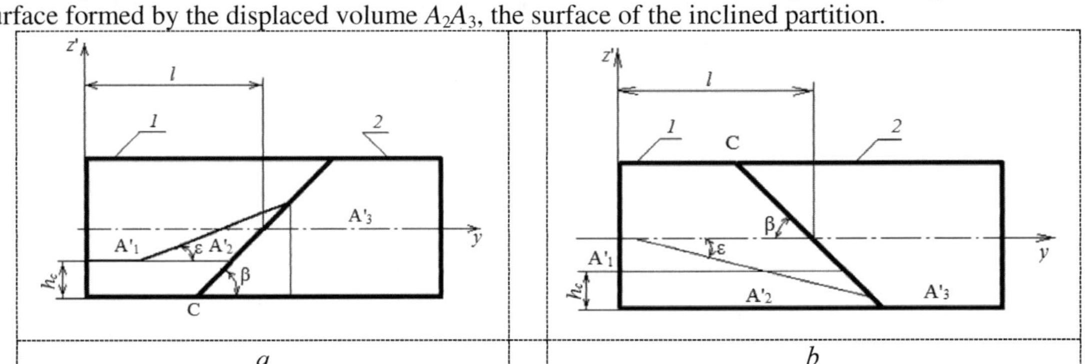

Figure 1. Scheme for calculating the power consumed by a ball mill with an inclined partition $a - \xi = 0°$; $b - \xi = 180°$; *1, 2* – camera mill

3. The definition of the equations of planes, limiting the load

To determine the equation of the surface of a newly formed volume, we consider Fig. 2.

To simplify the calculations, we introduce a mobile coordinate system *x*'0z' rigidly connected with the angle of rotation of the mill's drum ξ.

As is known, the equation of the plane has the form:

$$A(x - x_0) + B(y - y_0) + C(z - z_0) + D = 0 . \quad (4)$$

To determine the coefficients *A*, *B* and *C*, we use the scheme (see Figure 2).

$$\begin{gathered} A = N\cos\varepsilon \cdot \sin\xi = \cos\varepsilon \cdot \sin\xi,\ x_0 = 0; \\ B = -N\sin\varepsilon \cdot \cos\xi = -\sin\varepsilon \cdot \cos\xi,\ y_0 = l - l_{pj}; \\ C = N\cos\varepsilon \cdot \cos\xi = \cos\varepsilon \cdot \cos\xi,\ z_0 = -(R - h_c), \end{gathered} \quad (5)$$

where *N* is the unit normal vector (*N* = 1); ε is the angle of the natural slope of the grinding load, deg; *l* - length of the first chamber of the mill along the axis, *m*; l_{pj} - the size of the zone of active influence of the partition at an arbitrary angle of rotation of the drum (see Figure 1), *m*; *R* is the radius of the mill drum, *m*; h_c is the level of the grinding load, *m*.

It is seen that the normal vector *N* for the coordinates *z*'and *x*' is constant regardless of the rotation angle of the drum ξ, and therefore we obtain the following equation of the plane of the newly formed volume:

$$-[y-(l-l_{pj})]\sin\varepsilon\cdot\cos\xi+[z'+(R-h_c)]\cos\varepsilon=0\,. \tag{6}$$

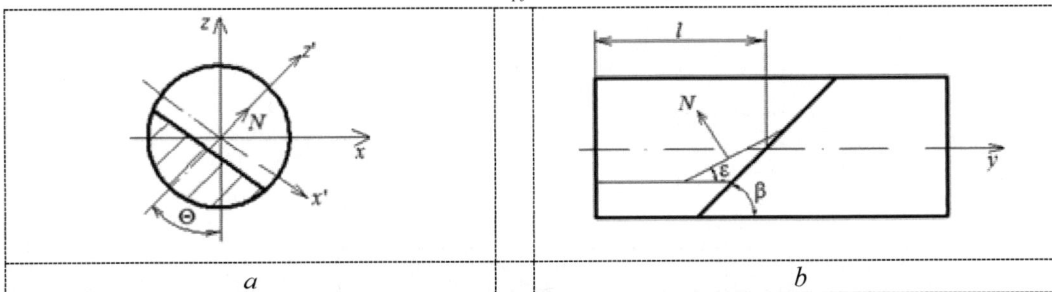

Figure 2. The scheme to determine the coefficients of the equation of the surface of the newly formed volume: *a* - the cross section of the drum; *b* - longitudinal section of the drum

In the formula (6) the parameter l_{pj} the author [6] recommends defining as:

$$l_{pj}=l_{pa}-\frac{1}{2}(l_{pa}-l_{pi})\cdot(1-\cos\,\xi)\,, \tag{7}$$

where l_{pa} is the maximum radius of the zone of active influence of the partition, for $\xi = 0^o$ (see Fig. 1, *a*); l_{pi}- the minimum radius of the zone of active influence of the partition, at $\xi = 180^o$ (see Fig. 1, *b*). Substituting equation (7) into equation (6) we finally obtain the equation of the plane of the newly formed volume:

$$-\left[y-\left(l_{pa}-\frac{1}{2}(l_{pa}-l_{pj})(1-\cos\,\xi)\right)\right]\sin\varepsilon\cdot\cos\,\xi+[z'+(R-h_c)]\cos\varepsilon=0. \tag{8}$$

Similarly, we obtain the equation of the inclined partition, depending on the angle of rotation of the drum. To determine the coefficients *A*, *B* and *C*, we use the calculation scheme (Fig. 3):

$$\begin{aligned}&A=N\,\sin\,\xi\,\cos\,\beta\,;\\&B=-N\,\cos(90-\beta)=N\sin\beta\,;\\&C=N\,\cos\,\xi\,\cos\,\beta\,.\end{aligned} \tag{9}$$

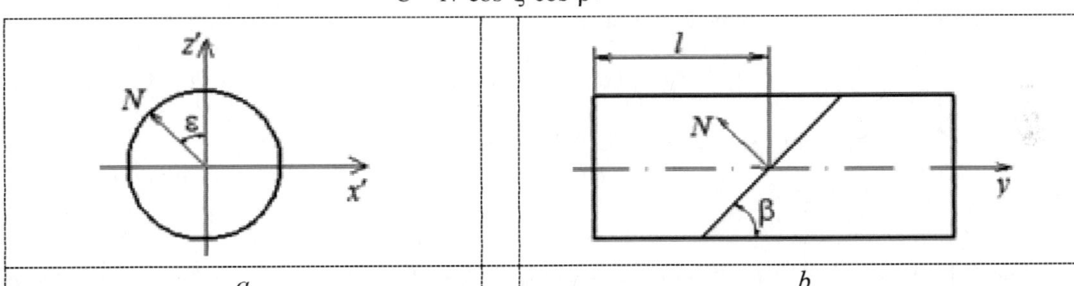

Figure 3. The scheme to determine the coefficients of the equation of the plane of the inclined partition: *a* - the cross section of the mill drum; *b* - longitudinal section of the mill drum

Putting the origin of coordinates in the center of the inclined partition, we get that x_0 and z_0 are equal *l*. Thus, the equation of the partition becomes:

$$-x'\,\sin\,\xi\,\cos\,\beta-(y-l)\sin\,\beta+z'\cos\,\xi\,\cos\,\beta=0\,. \tag{10}$$

Ultimately, the load consuming power in the first chamber is limited to planes:

$$\begin{aligned}&-\left[y-\left(l_{pa}-\frac{1}{2}(l_{pa}-l_{pj})(1-\cos\,\xi)\right)\right]\sin\varepsilon\cdot\cos\,\xi+[z'+(R-h_c)]\cos\varepsilon=0;\\&-x'\cdot\sin\,\xi\cdot\cos\,\beta-(y-l)\cdot\sin\,\beta+z'\cdot\cos\,\xi\cdot\cos\,\beta=0;\\&x'^2+z'^2=R^2\,;\\&z'=-(R-h_c)\,;\\&y=0\,.\end{aligned} \tag{11}$$

In Fig. 4 we can see that in the first chamber of the mill there are three areas with the characteristic behavior of the load:

I area. On this area, the level of loading of grinding bodies does not change and the length of the section is determined from the expression: $0 \le y_1 \le l_1 - l_{pj}$

Here y_1 is the axis coordinate of the area *I*.

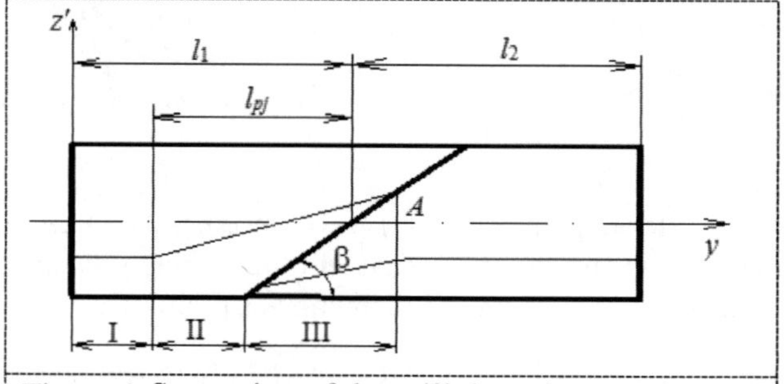

Figure 4. Separation of the mill drum into sections with characteristic load areas

II area. It is characterized by a change in the load by the plane of the newly formed volume from above and is determined from expression:

$l_1 - l_{pj} \le y_2 \le l_1 - R\text{ctg}\,\beta$ under $0 < \xi < 90°$ or $270° < \xi < 360°$;

where y_2 – is the axis coordinate of the area *II*.

III area. It is characterized by a change in the load by the plane of the newly formed volume from above and the plane of the inclined partition from below and is determined from the expression:

$l_1 - R\text{ ctg }\beta \le y_3 \le y_A$ under $0 < \xi < 90°$ or $270° < \xi < 360°$,

$y_A \le y_3 \le l_1 + R\text{ ctg }\beta$ under $90° < \xi < 270°$,

where y_3 – is the axis coordinate of the area *III*; y_A - the maximum value at the intersection of the plane of the newly formed volume, the inclined partition and the drum body. This value is obtained from the solution of a system of three equations:

$$\begin{cases} -\left[y - \left(l_{pa} - \frac{1}{2}(l_{pa} - l_{pj})(1 - \cos\,\xi)\right)\right]\sin\varepsilon \cdot \cos\,\xi + [z' + (R - h_c)]\cos\varepsilon = 0 \\ -x' \cdot \sin\xi \cdot \cos\,\beta - (y - l)\sin\beta + z' \cdot \cos\xi \cdot \cos\beta = 0 \\ x'^2 + z'^2 = R^2. \end{cases}$$

Thus, in each of the three sections where the grinding load consumes power, the equations of the load restricting planes are determined at an arbitrary angle of rotation of the mill drum ξ.

4. Determining the coordinates of the center of gravity of the sections

The coordinates of the center of gravity of a homogeneous planar figure, representing the cross-section of the grinding load along characteristic sections (see Figure 4), are determined by the formulas :

$$x_c = \frac{\int dz \int x dx}{\int dz \int dx};\quad z_c = \frac{\int z dz \int dx}{\int dz \int dx};\quad x_c = \frac{I_z}{S};\quad z_c = \frac{I_x}{S}. \tag{12}$$

We introduce a moving coordinate system rigidly connected with the angle of rotation of the drum of the mill ξ. In this case, the coordinate $x_c' = 0$ (since in the *I* and *II* areas the loading is symmetrical about the $0z'$ axis), and the coordinate z_c' is determined from the expression:

$$z_c = \frac{\int_b^a z dz \int_{\varphi_2(z)}^{\varphi_1(z)} dx}{\int_b^a dz\phi \int_{\varphi_2(z)}^{\varphi_1(z)} dx}. \tag{13}$$

In this case, the limits of integration in expression (13) are equal to:

$$\varphi_1(z') = \sqrt{R^2 - z'^2}\ ; \quad \varphi_2(z') = -\sqrt{R^2 - z'^2}\ ; \quad a = -R\ ; \quad b = -(R - h_c)\ .$$

As is known, the coordinate z_c' for this section is determined by the formula

$$z'_c = -\frac{2}{3} R \sin^3 \frac{\Omega}{2} / \pi\varphi, \tag{14}$$

where Ω – central angle of the load, rad.
According it can be seen that

$$\Omega = 2\arccos\ \frac{R - h_c}{R}\ . \tag{15}$$

Let us now consider the second section, which is characterized by the fact that the level of loading along z' varies along the length of the drum y and with the rotation angle ξ.
According to the equation (8), the change in the loading position z' will occur according to the following law:

$$z' = D = \left[y - \left(l_{pa} - \frac{1}{2}(l_{pa} - l_{pj})(1 - \cos\ \xi) \right) \right] \text{ctg}\ \varepsilon \cdot \cos\ \xi - (R - h_c), \tag{16}$$

where h_c – the load level in the I area.

The central loading angle Ω' will then be equal:

$$\Omega' = 2\arccos \frac{z'}{R}, \tag{17}$$

where z' is determined from equation (16).
The third section is characterized by the fact that the contour of the load section is limited by the plane of the newly formed volume and the plane of the inclined partition.
From the equation of the plane of the partition we have:

$$x' = -\frac{(y - l)\sin\beta - z'\cos\xi\ \cos\beta}{\sin\xi \cdot \cos\beta} = -\frac{(y - l)\text{tg}\beta}{\sin\xi} + z'\text{ctg}\,\xi. \tag{18}$$

Depending on the position of the lower intersection point relative to the loading level and the $0z'$ axis, the integration limits and the form of the formulas for determining the coordinates of the load center of gravity will be different.
Solving equation (10) together with the equation of the lateral surface of the drum, we obtain that the lower point is equal to:

$$\begin{aligned} z = F\ &= \text{tg}\beta \cdot y \cdot \cos\xi - \text{tg}\beta \cdot L \cdot \cos\xi - \\ &- \sqrt{\text{tg}\beta^2 \cdot y^2 \cdot \cos\xi^2 - 2 \cdot \text{tg}\beta^2 \cdot y \cdot \cos\xi^2 \cdot L + \text{tg}\beta^2 \cdot L^2 \cdot \cos\xi^2 - \overline{- \text{tg}\beta^2 \cdot y^2 + 2\text{tg}\beta^2 \cdot y \cdot L - \text{tg}\beta^2 \cdot L^2 + \sin\xi^2 \cdot R^2}}\ . \end{aligned} \tag{19}$$

The lower coordinate x' from equation (10) will be:

$$x' = \frac{(y - l)\sin\beta - z'\cos \cdot \xi \cos\beta}{\sin\xi \cdot \cos\beta} = \frac{(y - l)\text{tg}\beta}{\sin\xi} - F\text{ctg}\,\xi\,. \tag{20}$$

The coordinates of the center of gravity of the cross-sections x_c' and z_c' are determined by the formula (12).
If the lower intersection point of the trace of the plane of the inclined partition with the surface of the drum is greater than the coordinate z' of the intersection of the trace of the plane of the newly formed volume:

$$F \geq= D = \left[y - \left(l_{pa} - \frac{1}{2}(l_{pa} - l_{pj})(1 - \cos\ \xi) \right) \right] \text{tg}\ \varepsilon \cdot \cos\ \xi - (R - h_c), \tag{21}$$

then for all ξ $x' = 0$, and z' is computed by the formulas (14) and (17).
If, in this area, the inclined partition is in contact with the charge,

$$F <= D = \left[y - \left(l_{pa} - \frac{1}{2}(l_{pa} - l_{pj})(1 - \cos\ \xi) \right) \right] \text{tg}\ \varepsilon \cdot \cos\ \xi - (R - h_c),$$

then the coordinates of the center of gravity of the load section are determined by the formulas (12) for each loading position and the inclined partition.

Taking into account the fact that the constantly changing load section in area *III* does not make it possible to derive a general formula for finding the centers of gravity that takes into account the initial parameters, then the area and moments of inertia are determined by calculating the integrals to determine the coordinates of the centers of gravity of the load:

$$I_z = \int_b^a z'\mathrm{d}z' \int_{\varphi_2(z')}^{\varphi_1(z')} \mathrm{d}x' - \int_d^c \mathrm{d}x' \int_{\varphi_4(x')}^{\varphi_3(x')} z'\mathrm{d}z';$$
$$I_x = \int_b^a x'\mathrm{d}x' \int_{\varphi_2(z')}^{\varphi_1(z')} \mathrm{d}z' - \int_d^c \mathrm{d}x' \int_{\varphi_4(x')}^{\varphi_3(x')} x'\mathrm{d}z'; \qquad (22)$$
$$S = \int_b^a \mathrm{d}z' \int_{\gamma_2(z')}^{\varphi_1(z')} \mathrm{d}x' - \int_d^c \mathrm{d}x' \int_{\varphi_4(x')}^{\varphi_3(x')} \mathrm{d}z',$$

where *a*, *b*, *c* и *d* – limits of figures along the axis *z'*; $\varphi_1(z)$, $\varphi_2(z)$, $\varphi_3(z)$ and $\varphi_4(z)$ – equations of surfaces that bound these figures.

the limits of integration in the expressions (22) will be equal to:

$$a = \left[y - \left(l_{pa} - \frac{1}{2}(l_{pa} - l_{pj})(1 - \cos\ \xi) \right) \right] \text{tg}\ \varepsilon \cdot \cos\ \xi - (R - h_c);$$
$$b = -R;$$
$$d = \text{tg}\,\beta \cdot y \cdot \cos\xi - \text{tg}\,\beta \cdot L \cdot \cos\xi -$$
$$- \sqrt{\text{tg}\,\beta^2 \cdot y^2 \cdot \cos\xi^2 - 2\text{tg}\,\beta^2 \cdot y \cdot \cos\xi^2 \cdot L + \text{tg}\,\beta^2 \cdot L^2 \cdot \cos\xi^2 - }$$
$$\overline{- \text{tg}\,\beta^2 \cdot y^2 + 2\text{tg}\,\beta^2 \cdot y \cdot L - \text{tg}\,\beta^2 \cdot L^2 + \sin\xi^2 \cdot R^2};$$
$$c = \left[y - \left(l_{pa} - \frac{1}{2}(l_{pa} - l_{pj})(1 - \cos\ \xi) \right) \right] \text{tg}\ \varepsilon \cdot \cos\ \xi - (R - h_c);$$
$$\varphi_1(z') = \sqrt{R^2 - z'^2}; \qquad \varphi_2(z') = -\sqrt{R^2 - z'^2};$$
$$\varphi_3(x') = \frac{(y - l)\text{tg}\,\beta}{\cos\xi} + x'\text{tg}\,\xi; \qquad \varphi_4(x') = -\sqrt{R^2 - x'^2}.$$

5. References

[1] Duda W H 1985 *Cement-data-book. Band 1* (Internationale Verfahrenstechniken der Zementindustrie, Wiesbaden, Berlin: Bauverlag)

[2] Sharapov R R, Shrubchenko I V and Agarkov A M 2015 Determination of the optimal parameters of the equipment to obtain fine powders Global Projects. *International Journal of Applied Engineering Research* **10**. No 12 pp. 31341–31348

[3] 2007 Global Projects. *World Cement*. December. pp. 87–90.

[4] Telichenko V I, Sharapov R R, Lozovaya S Y and Skel V I 2016 Analysis of the efficiency of the grinding process in closed circuit ball mills. V.I. MATEC Web of Conferences 5. "5th International Scientific Conference "Integration, Partnership and Innovation in Construction Science and Education", IPICSE 2016". p. 04040.

[5] Sevost'yanov V S, Goncharov A A and Potelezhko V P 1993 Improvement of the technology of ball mills. *Glass and Ceramics*. **50**. No 2. pp. 55-59.

[6] Bogdanov V S, Fadin Y M, Lozovaya S Y, Latyshev S S , Bogdanov N E and Vasilenko O S 2016 Ball mill power calculation with inclined partition. *International Journal of Pharmacy and Technology*. **8.** No 3. p. 19031-19041.

IOP Conf. Series: Materials Science and Engineering **307** (2018) 012048 doi:10.1088/1757-899X/307/1/012048

Matrix model of the grinding process of cement clinker in the ball mill

Rashid R Sharapov

Moscow State University of Civil Engineering (National Research University), Yaroslavl highway 26, Moscow, 129337, Russia

Corresponding author's e-mail address: ptdm_zavkaf@mail.ru

Abstract. In the article attention is paid to improving the efficiency of production of fine powders, in particular Portland cement clinker. The questions of Portland cement clinker grinding in closed circuit ball mills. Noted that the main task of modeling the grinding process is predicting the granulometric composition of the finished product taking into account constructive and technological parameters used ball mill and separator. It is shown that the most complete and informative characterization of the grinding process in a ball mill is a grinding matrix taking into account the transformation of grain composition inside the mill drum. Shows how the relative mass fraction of the particles of crushed material, get to corresponding fraction. Noted, that the actual task of reconstruction of the matrix of grinding on the experimental data obtained in the real operating installations. On the basis of experimental data obtained on industrial installations, using matrix method to determine the kinetics of the grinding process in closed circuit ball mills. The calculation method of the conversion of the grain composition of the crushed material along the mill drum developed. Taking into account the proposed approach can be optimized processing methods to improve the manufacturing process of Portland cement clinker.

1. Introduction

The production of the fine powders is a very important task for any sphere of modern industry. One of the most technologically advanced powder, that produce humanity for its use is Portland cement [1, 2]. The production of cement is accompanied by complex processes of transformation of raw materials into finish product. The material passes the stage of grinding, homogenizing, calcining and grinding again the almost finished product in a finely dispersed powder.

A large number of devices proposed by various researchers for grinding the components of the cement and the Portland cement clinker and additives [3-8]. At the same time the requirements for quality of the final product to grinding equipment are very high level. To fulfil these requirements all different types of the grinding equipment works in a closed cycle grinding. Therefore, as practice has shown the most important element of technological systems of grinding are closed circuit ball mills. These type of ball mills in the coming decades will remain the basic units for grinding cement clinker and additives [3].

2. Teoretical

The main task of modeling the grinding process is predicting the granulometric composition of the finished product. This product depends on the granulometric composition of the source material, the constructive and technological parameters of the ball mill and separator. Predicting the granulometric composition of the finished product will allow to choose the optimal configuration of necessary equipment and correct operation modes.

The most complete and informative characterization of the grinding process in a ball mill is a matrix of the grinding process $P = (p_{ij})$. Each element of this matrix p_{ij} equal to the relative share of the j-th fraction of bulk raw materials passing into the i-th fraction of the grinding product ($i, j = 1, 2, ..., m$, where m corresponds to the smallest fraction of the grinding material). All the elements of the matrix of grinding, for which $i > j$ is equal zero, as the particles in the grinding process of their factions to a larger faction can not pass. It can be explain because the reverse process in the mill does not occur. As a result, the grinding matrix has a lower triangular form [9]:

$$P = \begin{bmatrix} p_{11} & 0 & 0 & 0 \\ p_{21} & p_{22} & 0 & 0 \\ p_{31} & p_{32} & p_{33} & 0 \\ - & - & - & - \\ p_{m1} & p_{m2} & & p_{mm} \end{bmatrix} \tag{1}$$

Standing on the main diagonal of the matrix elements p_{ii} show the relative (by mass) of the particles of the j-th fraction, remaining in it because of weak grinding the particles. The elements of the matrix R satisfy the conditions of normalization, reflecting the law of conservation of mass of individual fractions: total probability of transition of particles of the j-th fraction into smaller fractions $i \geq j$ is equal to unit:

$$\begin{aligned} p_{11} + p_{21} + \ \dots + p_{m1} &= 1 \\ p_{22} + p_{32} + \ \dots + p_{m2} &= 1 \\ p_{mm} &= 1. \end{aligned} \tag{2}$$

The transformation of the granulometric composition of the material as a result of grinding process in a ball mill is described by the matrix equation:

$$f' = Pf, \tag{3}$$

where f and f'– are the vectors of the granulometric composition of the disperse material at the inlet to the ball mill and out of it:

$$f = \begin{bmatrix} f_1 \\ f_2 \\ \dots \\ f_m \end{bmatrix}, \quad f' = \begin{bmatrix} f_1' \\ f_2' \\ \dots \\ f_m' \end{bmatrix}. \tag{4}$$

Here $f_1, f_2, ..., f_m$ and $f_1', f_2', ..., f_m'$ are relative the mass share of the particles of material falling into respective fractions. In formulas (1)...(4) adopted the reverse order numbering of the fractions, i.e. in descending order of particle size.

Because of the extreme complexity and variety of processes occurring in ball mills, the theoretical definition of the matrix of grinding currently is not possible. In connection with this very urgent task of reconstruction of the matrix of grinding P on the experimental data obtained on a real ball mill working production.

The matrix equation (3) is equivalent to the system of m linear equations:

$$\begin{cases} p_{11}f_1 = f_1' \\ p_{21}f_1 + p_{22}f_2 = f' \\ \dots\dots\dots\dots\dots\dots \\ p_{m1}f_1 + p_{m2}f_2 + \dots + p_{mm}f_m = f_m' \end{cases} \tag{5}$$

of which independent are only $m - 1$ equations, because the addition of linear equations (5) subject to conditions (2) gives the identity:

$$f_1 + f_2 + \dots + f_m = f_1' + f_2' + \dots + f_m' = 1$$

Thus, at a certain granulometric composition f and f' to determine $m(m+1)/2$ elements of the matrix of grinding in ball mill P we have $2m-1$ equations. It follows that the unique reconstruction of the elements of the matrix of grinding on particle size distribution at the entrance to the mill and exiting is possible only by dividing the material into two fractions – large and small. In this case

$$P=\begin{bmatrix} p_{11} & 0 \\ p_{21} & p_{22} \end{bmatrix}, \quad p_{11}=\frac{f_1'}{f_1}, \; p_{21}=1-p_{11}, \quad p_{22}=1. \tag{6}$$

When $m>2$ the knowledge of only the input or output particle size distribution, the mill charge inside the mill drum material is not enough.

In the article [10] suggested a method of reconstruction of the matrix of the grinding on the results of the m-fold transmission of material through the mill drum. However, the thus obtained experimental data in determining the elements of the initial row of matrix grinding are used only partially, which reduces the reliability of the results. It should also be noted the inability of the proposed experiment in terms of current production. This can be explained by the fact that in the actual production, it is impossible to change either the design features of the equipment or the technological process or the properties of comminuted material.

On the basis of experimental data obtained on industrial installations, using matrix method, we define the kinetics of the grinding process in closed circuit ball mills. For this purpose each chamber of the ball mill are presented in the form of n equal-length parallel working sections, the length of which corresponds to the step of removing the chart of grinding. Extended milling diagram provided in the form of a matrix of size $m\times n$:

$$F=(f_{jk})=\begin{bmatrix} f_{11} & f_{12} & \cdots & f_{1n} \\ f_{21} & f_{22} & \cdots & f_{2n} \\ \cdots & \cdots & \cdots & \cdots \\ f_{m1} & f_{m2} & \cdots & f_{mn} \end{bmatrix}. \tag{7}$$

Here the index $j=1, 2..., m$ corresponds to the number fraction of the grinding material, and the index $k=0,..., n-1$ corresponds to the cross-sections along the length of the mill drum, in which selected material samples for making matrix of the grinding. The section with index k is also an input section $(k+1)$-th section of the mill.

Thinking that the matrix of the grinding in separate sections P_{ck} does not depend of the fractional composition of the material and therefore have the same form $P_{ck}=P_c$, we write the equation of the transformation of the granulometric composition of material:

$$f_{k+1}=P_c f_k, \; k=0, 1, 2, \ldots, n-1. \tag{8}$$

The resulting matrix of grinding for the whole chamber of the mill is determined by the following ratio:

$$P=P_{c1}\cdot P_{c2}\cdots P_{ck}=P_c^n, \tag{9}$$

где P_c – matrix of grinding for each step of sampling in the ball mill.

Matrix of grinding considering the fact that in modern ball mills consist of 2 cameras and for the whole the mill matrix of grinding is:

$$P=P_1\, P_2,$$

where $P_1=P_{c1}^{n1}$, $P_2=P_{c2}^{n2}$ – matrix of grinding along the length of the chambers of ball mill, respectively, the first and second chambers.

Matrix equations (8) are equivalent to n systems of linear equations:

IOP Conf. Series: Materials Science and Engineering **307** (2018) 012048 doi:10.1088/1757-899X/307/1/012048

$$\begin{aligned} & p_{11}f_{1k} = f_{1k+1} \\ & p_{21}f_{1k} + p_{22}f_{2k} + f_{2k+1} \\ & p_{31}f_{1k} + p_{32}f_{2k} + p_{33}f_{3k} = f_{3.k+1} \\ & \ldots\ldots\ldots\ldots\ldots\ldots\ldots\ldots\ldots \\ & p_{m1}f_{1k} + p_{m2}f_{2k} + p_{m3}f_{3k} + \ldots + p_{mm}f_{mk} = f_{mk+1} \\ & k = 0.1,\ldots,n-1, \end{aligned} \tag{10}$$

which must be supplemented by the normalization conditions (2). The total number of independent equations $n\ (m - 1) + m$ greatly exceeds the number to be determined of the elements of the matrix of grinding P_c. These values are obtained by researchers when removing the charts grinding in both chambers of the ball mill. The accuracy is higher, the more often we take the chart of grinding in the drum ball mill. However, taking into account the inevitability of random errors when removing the diagram of grinding (such as the enrichment of sieving large fraction of the products of burnout, large particles in interlining space made bring from outside researchers, etc.), we must do all of the above a specific set of equations (10) using the least squares method [11, 12]. Summing the squares of the residuals, i.e. the difference of the left and right sides of each of equations of system (10) for all partitions, we get:

$$\begin{aligned} & S_1 = \sum_{k=0}^{n-1}\left(p_{11}f_{1k} - f_{1k+1}\right)^2, \\ & S_2 = \sum_{k=0}^{n-1}\left(p_{21}f_{1k} + p_{22}f_{2k} - f_{2k+1}\right)^2, \\ & S_{m-1} = \sum_{k=0}^{n-1}\left(p_{m-11}f_{1k} + p_{m-12}f_{2k} + \ldots + p_{m-1m-1}f_{m-1k} - f_{m-1k+1}\right)^2. \end{aligned} \tag{11}$$

Differentiating the functions S_1, S_2, ..., S_{m-1} within them the elements of the matrix of grinding, obtain the system of normal equations to determine these equations.

In particular, the formula for calculating p_{11} has the form.

$$p_{11} = \sum_{k=0}^{n-1}\left(f_{1k+1}f_{1k}\right) / \sum_{k=0}^{n-1} f_{1k}^2 \ . \tag{12}$$

The system of equations for the determination of p_{21} and p_{22} can be written thus:

$$\begin{aligned} & \left(\sum f_{1k}^2\right)p_{21} + \left(\sum f_{2k}\right)p_{22} = \sum f_{2k+1}f_{1k} \\ & \left(\sum f_{1k}f_{2k}\right)p_{21} + \left(\sum f_{2k}^2\right)p_{22} = \sum f_{2k+1}f_{2k} \end{aligned} \ . \tag{13}$$

The elements of the last row of the matrix of grinding is calculated using the found least-squares method, the elements of the other rows by using the normalization conditions (2).

3. Conclusion

Using found, as described above, the matrix for grinding for the sections of the chambers of ball mill it is possible to investigate the kinetics of the grinding process of the material along each of them, respectively, along the mill as a whole. This will allow us to choose the appropriate size and composition of the grinding media in the chambers of the mills, lining each chamber of the mill, the suction conditions inside the mill drum, to optimize the size of the incoming material and so on.

4. References

[1] Duda W H 1985 *Cement-data-book. Band 1* (Internationale Verfahrenstechniken der Zementindustrie, Wiesbaden, Berlin: Bauverlag)

[2] 2007 Global Projects. *World Cement*. December. pp. 87–90.

[3] Telichenko V I, Sharapov R R, Lozovaya S Y and Skel VI 2016 Analysis of the efficiency of the grinding process in closed circuit ball mills. V.I. MATEC Web of Conferences 5. "5th

IOP Conf. Series: Materials Science and Engineering **307** (2018) 012048 doi:10.1088/1757-899X/307/1/012048

International Scientific Conference "Integration, Partnership and Innovation in Construction Science and Education", IPICSE 2016". P. 04040.

[4] Sevost'yanov V S, Goncharov A A and Potelezhko V P 1993 Improvement of the technology of ball mills. *Glass and Ceramics*. **T. 50**. No 2. pp. 55-59.

[5] Uvarov V A 2006 The use of a countercurrent jet mill in the production of insulating ceramics. *Glass and Ceramics*. **T. 63**. № 7-8. p. 274-276.

[6] 1982 Grinding Mill – Rod, Ball and Autogenous. *Mining magazine*. **147**. No. 9. p. 91.

[7] Nerholm A 1979 *Grinding of cement*. Symposium on cement manufacture. (Moscow: Niicement Research Institute).

[8] Romanovich A A 2013 Definicion of rational conditions of materials griding in energy-saving milling complex. *World Applied Sciences Journal*. **T. 25**. No. 2. p. 214

[9] V. Mizonov V Zhukov V and Bernotat S 1997 *Simulation of Grinding: new approaches*. (Ivanovo: ISPEU Press) 118 p.

[10] Ryadinskaya L V, Lozovaya S Y and Sablin V S 2014Theoretical bases of the calculation of the energy, which it spends for the process of materials grinding in accordance with geometrical and technological characteristics of the chamber of the barrel-shaped form. *International Journal of Applied Engineering Research*. **T. 9**. . No 22. C. 16891-16902.

[11] Kalitkin N N 1998 *Numerical methods*. (Moscow:Nauka) 512 p.

[12] Bogdanov V S, Fadin Y M, Lozovaya S Y, Latyshev S S , Bogdanov N E and Vasilenko O S 2016 Ball mill power calculation with inclined partition. *International Journal of Pharmacy and Technology*. **T. 8.** No 3. C. 19031-19041.

IOP Conf. Series: Materials Science and Engineering **307** (2018) 012049 doi:10.1088/1757-899X/307/1/012049

A Study of Array Direction HDPE Fiber Reinforced Mortar

Trithos Kamsuwan[1]

[1]Civil Engineering Department, Faculty of Engineering, Siam University, Bangkok 10163, Thailand.
Email: skamsuwan@yahoo.com

Abstract. This paper presents the effect of array direction HDPE fiber using as the reinforced material in cement mortar. The experimental data were created reference to the efficiency of using HDPE fiber reinforced on the tensile properties of cement mortar with different high drawn ratio of HDPE fibers. The fiber with the different drawn ratio 25x (d25 with E xx), and 35x (d35 with E xx) fiber volume fraction (0%, 1.0%, 1.5%) and fiber length 20 mm. were used to compare between random direction and array direction of HDPE fibers and the stress – strain displacement relationship behavior of HDPE short fiber reinforced cement mortar were investigated. It was found that the array direction with HDPE fibers show more improved in tensile strength and toughness when reinforced in cement mortar.

1. Introduction

The last four decades have seen a large number of research studies on fiber reinforced cement materials, most of which devoted to the use of metallic (steel), synthetic (polymeric), ceramic (glass), or natural (organic) fibers. In contrast, few studies dealt with the design and development of the fibers themselves. Indeed most polymeric fibers on the market today have been conceived and introduced over thirty years ago. Addressing recent trends and future directions, Fiber Reinforced Cement Composites presents new opportunities for developing innovative and cost-effective materials and techniques in cement and concrete composites manufacturing, testing, and design.

Several types of synthetic polymer fibers such as polyethylene (PE), polypropylene (PP), polyethylene terephthalate (PET), polyvinyl alcohol (PVA) have been incorporated in reinforced cement based matrices. The main problems of the polymer synthetic fiber reinforced cement are the lower tensile strength. The high density polyethylenes (HDPE), a linear chain polymer with a broad molecular weight distribution can be orient into a high strength fiber [1]. There are two processes for producing the high strength fiber including melt-spinning and gel-spinning processes [2].

In this work, the using high strength HDPE fiber prepared from the high draw ratio melt spinning technique. These method can be increase tension strength in fiber depend on the high draw ratio. The properties of ultra draw HDPE fiber are high tensile strength and high elastic modulus. In previous researches, we found that the problem of ball and grouping during the mixing the fiber with cement aggregate can reduce the strength of their composite.

The addition of fibers to cement composites influences its mechanical properties which significantly depend on the type of fiber, the percentage of fiber and fiber orientation. For development of a new type of fiber using in cement composites, ultra draw HDPE fiber can improve more mechanical properties of the fiber reinforced matrix. High tensile strength of HDPE can transfer force and disperse stress through area of the matrix.

This research has the objective to study the effect array direction of HDPE fiber reinforced in cement mortar and investigate the efficiency of the orientation HDPE fiber reinforced cement mortar.

2. Experimental

For experimental design in this research, synthetic fibers, HDPE produced by the technique of pulling fiber high draw ratio to improve the tensile properties called high draw ratio melt spinning technique. High density polyethylenes (HDPE) employed in the present study were 5000S from PTT chemical Public Company Limited. The fiber was prepared by a Randcastle monofilament line as show in figure 1. Fiber was prepared by 2-step method. Extruder temperatures were set at 120, 180, 190 and 215°C for zone 1, 2, 3 and die, respectively. As-spun fiber was spun at a screw speed of 2.2 rpm, or otherwise stated. The fiber was allowed to cool in air. The as-spun fiber had a diameter of about 500 μm. The as-spun fiber was drawn to different draw ratios with two sets of rollers through a glyceral bath set at 115°C. Draw ratio was calculated from the ratio of the speed of the follower to leader roller. Tensile testing was carried out at room temperature with sample gauge length of 50 mm and a crosshead speed of 25 mm/min. The properties of ultra-draw high ratio HDPE fibers as show on Table 1.

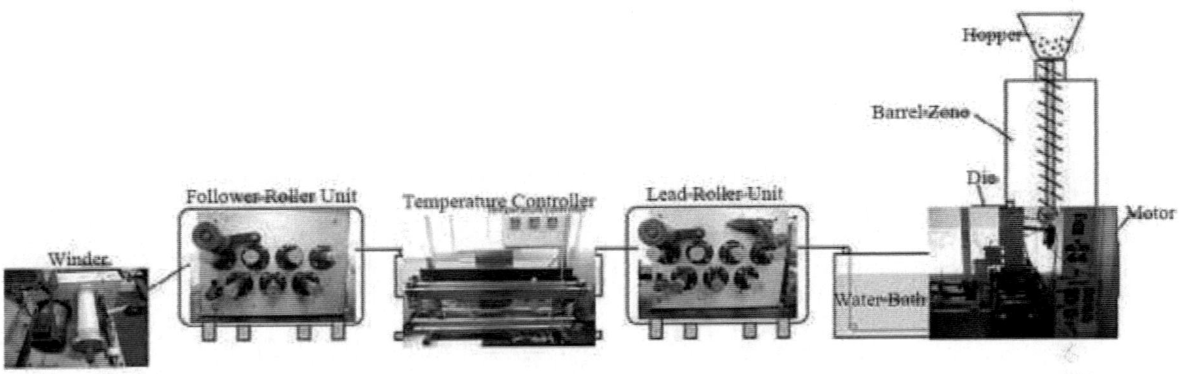

Figure 1. Shcematic diagram of a Randcastle monofilament line employed in the study

Three majors parameters were investigated including (a) tensile strength and the elastic modulus of the fiber where the fiber with different draw ratio at 25(d25) and 35(d35) will be tested (b) fiber content where percent volume fraction of the fibers will be varied from %0, % 0.5, % 1.0, and % 1.5, and (c) the length of the fibers were 20mm. The FRC recipe will be referred to as *d*x*v*x*l*x where *d* follows by the fiber draw ratio, *v* follows by percent volume fraction and *l* follows by fiber length in millimeter, for example *d*25*v*0.5*l*15 is the system with fiber draw ratio 25 having a percent volume fraction of 0.5% and the fiber length of 15 mm. This study was focused on the effect of the fiber orientation between non-oriented and partially oriented system.

In state of art report by ACI committee 544 and the other researches [4-6], we found that the factors influence on the stress and strain curve of the fiber reinforced cement (FRC) mortar comprising of the strength of fiber, volume fraction of the fiber, the length of fiber, and patterns of mixing the fibers. The variable parameters affecting the fiber reinforced cement mortar found that to compare with plain mortar, several variables can effected on the tensile stress and toughness of materials. The critical fiber volume fraction is well below 2% [6]. In this research, the parameter of volume fiber focus on range 0.5%,1.0%, and 1.5%. The parameter of fiber length in term of the aspect ratio (l/d) focus on length 5, 10, 15, and 20 mm. Most of polymer fiber reinforced cement has referenced length 5-20 mm.[7].

The mix proportions used in this study were applicable to cement mortar (the standard mortar shall be one part of cement to 2.75 parts of graded standard sand by weight and water/cement ratio 0.7). In this research, the following mixing procedure was used for all mixes specified in standard specification for mortar cement, ASTM C1329M-12[8] except for the procedure of addition of fibers. First, cement and sand were mixed for approximately 5 minute. Then 50% of the mixing water was added and the mixture was mixed for 1 minute to allow for water absorption. Next, the remaining water was added. The mixing continued for another 3 minutes. All component materials except for fibers were added to ensure proper and uniform mixing. Finally, fibers were added to the mix. Altogether, the additional mixing time took approximately 5 minutes to ensure a uniform fiber distribution and to minimize fiber segregation and balling effects. After completion of the batch, the mix was placed into the appropriate molds, which were then placed on a vibrating table. The vibrating process was continued for approximately 1-2 minutes. Curing of all specimens was carried out in accordance with ASTM C192/C192M-02 (2004): after placement, these specimens covered with a plastic sheet and kept in their plastic for 24 hours. The specimens were then removed from their molds and moved to water bath, which was maintained at 100 percent relative humidity until the time of testing. Continuous curing in fresh water was contained for 28 days.

3. Results and discussion

The HDPE fiber used in this study is the high tensile strength HDPE obtained from melt spinning technique. The tensile properties depend on the draw ratio d25, and d35 as shown in table 1. This superior property of the fiber was originated from the special process where the fiber was drawn to high drawn ratio and the polymeric molecule is highly oriented along the fiber direction. This is resulting in an improvement of the strength of the fiber [9].

Table 1. Properties of Ultra-draw high ratio HDPE fiber

HDPE	D25	D35
Length(mm)	20	20
ϕ (mm)	0.172	0.148
Aspect ratio(l/d)	116	135
σ (GPa)	0.86 ± 0.08	1.05 ± 0.10
ε (%)	5 ± 0.58	4.5 ± 0.60
E (GPa)	24.9 ± 1.80	34.5 ± 2.67
Roughness, R_a(nm)	25.68±0..28	50.31±1.04
τ (MPa)	0.914±0.01	1.015±0.10

In HDPE fibers, the maximum tensile strength of the fiber at d35 and d25 are 1.05 GPa. and 0.86 GPa. respectively. In term of elastic modulus, d35 has the maximum value 34.5 GPa. higher than d25 (24.9 GPa.) 38.5%.

Efficiency of fiber direction

In table 2 showed tensile stress and elastic modulus testing compared between array and random direction HDPE fiber D35 in different volume fraction at length 20 mm. These data were presented the increasing tensile stress of array direction HDPE fiber and improving the elastic modulus of HDPE fiber when volume fraction increased.

Table 2. The tensile stress and elastic modulus of HDPE fiber D35 of array and random direction.

Direction fiber effective	Tensile Stress (MPa.)	Elastic Modulus (GPa.)
arrayD35v1.5l20	0.86	16.40

IOP Conf. Series: Materials Science and Engineering **307** (2018) 012049 doi:10.1088/1757-899X/307/1/012049

arrayD35v1.0l20	0.79	14.51
arrayD35v0.5l20	0.67	12.19
randomd35v1.5l20	0.39	13.26
randomd35v1.0l20	0.30	13.13
randomd35v0.5l20	0.27	12.06

In Figure 2, the relationship between tensile stress and strain in comparing of the patterns of fiber reinforcement and the direction reinforcement are distributed. The factors in direction of the reinforcement and distribution of the fibers effect on the increasing of the tension strength. The pattern of fiber orientation parallel to the line of force could increase the tensile stress as well. In addition, the distribution of the fiber all over cross section composited could increase the properties of the force distribution and improve a lot of the tensile stress. Furthermore, the factor of the distribution fiber would depend on the amount of fiber during mixing. The use of the excessive amount of fiber would effect on the ball and group between the fibers themselves. Of course, the efficiency of the force would be decreased.

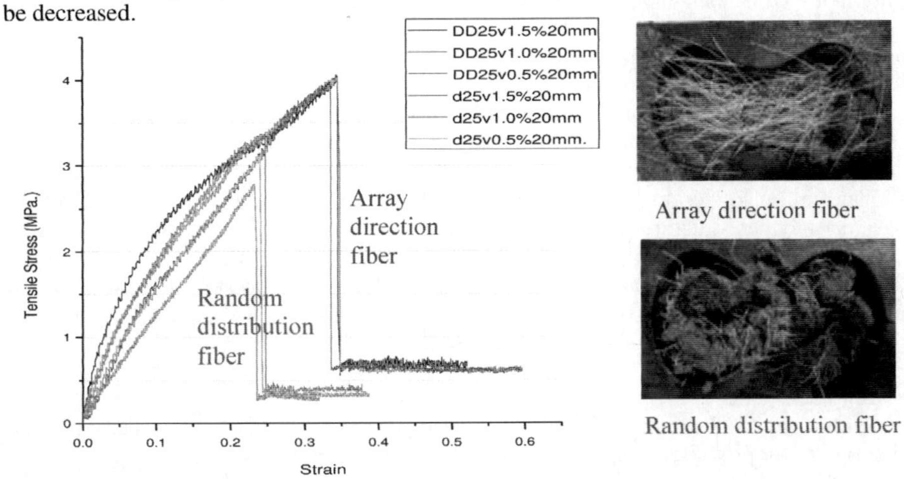

Figure 2. The stress and strain relationship of HDPE fiber D25 of the different distribution.

In figure 3 shown the relationship between stress and strain in case of the different distribution and direction of D23 HDPE fiber reinforced cement mortar. In this relationship scope of elasticity, the sample without fiber presented the efficiency of cement and sand aggregate binding to reinforced. The sample with random direction HDPE fiber reinforced presented the efficiency combine with cement, sand aggregate, and fiber for reinforcing. The value of elastic modulus and tensile stress in the sample with random distribution and direction of HDPE fiber reinforced are higher than the sample without fiber (mortar).

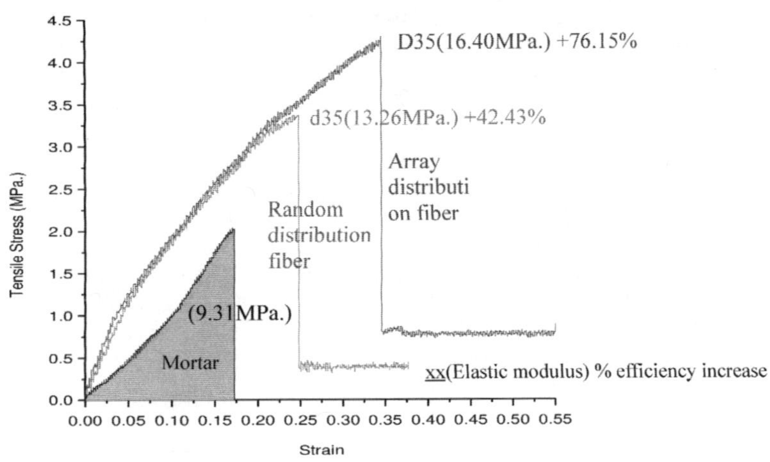

Figure 3. The efficient of array direction HDPE fiber reinforced mortar

The consideration in this study is the efficiency of pattern to arrange the direction and distribution of HDPE fiber reinforced in cement mortar. In figure 8 exhibited the results comparing the random direction and distribution fiber with the arrangement the direction and distribution fiber. They are found that the arrangement direction and distribution HDPE fiber to be parallel with the force axis can increase the tensile stress. Basically, the structure of the sample are prepare to reinforced by align fiber and force as a same direction. They can improve the efficiency for reinforcing. Because the alignment of fiber has a good bonding between cement matrix and fiber length and can reduce the problem of group and ball with fibers. According to the result of the sample with arrange direction and distribute of HDPE fibers is more tensile stress and elastic modulus than the sample with random direction and distribution of HDPE fibers in scope both of elastic region and inelastic region.

4. Conclusion

In this study, we developed the high tensile strength fiber HDPE to be used as reinforcing in mortar by using the technique of melt spinning high draw ratio. The efficiency of HDPE fiber reinforced cement mortar are verified to variable of ultra draw HDPE fiber, volume fraction fiber, effective fiber length, and pattern of mixing fiber. The advantage of this study is understands the efficiency of array direction HDPE fiber reinforced mortar. In term of ultra draw fiber can increase tensile strength HDPE fiber by organize the polymer structure in melt spinning high draw ratio technique. In case of arrangement fiber direction, the pattern of fiber direction and distribution are reform the behavior reinforcing of HDPE fiber in cement mortar, and can increase tensile stress. The pattern for arrangement direction and distribution of fiber helped the fiber resist force throughout the section area of fibers and solved the problem with fibers group and ball together.

5. Acknowledgements

Author would like to thank faculty of science, Mahidol University and faculty of engineering, Siam University for providing supporting machines and materials for this study.

References

[1] Peacock J.A., Handbook of polyethylene: structures, properties, and applications. New York: Marcel Dekke, Inc; 2000.

[2] Chantrasakul S., A study of high density polyethylene/organoclay composite fibers; Mechanical properties and morphology. Mater degree of science. NakhornPathom 2006.

[3] Youjiang W., Stanley B., and Li V.C., "An experimental study of synthetic fiber reinforced cementitious composites" Journal of Materials Science, 22, 1987, pp.4281-4291

[4] Li, V.C. and Leung, C.K.Y., 1992 "Steady-state and multiple cracking of short random fiber composites", J. of Engineering Mechanics, ASCE, 118(11), pp. 2246-263.

[5] Caggiano A., Etse G., and Martinelli E. "Interface model for fracture behavior of fiber reinforced cementitious composites (FRCCs): Theoretical formation and applications". European journal of environmental and civil engineering, 15(9), 2011, pp. 1339-1359.

[6] ACI Committee 544, "Revision of State-of-the-Art Report (ACI 544 TR-73) on Fiber Reinforced Concrete," ACI JOURNAL, Proceedings, Nov. 1973, Vol. 70, No. 11, pp. 727-744.

[7] Amit Rai., Dr. Y.P Joshi. Applications and Properties of Fibre Reinforced Concrete., Issue 5(Version 1), May 2014, pp.123-131

[8] American Society for Testing and Materials. Standard Specification for Mortar Cement. ASTM C 1329/C1329M-120.Book of ASTM Standards. Parte 04.01. ASTM Philadelphia; 2001.

[9] Chaiyut N., Amornsakchai T., Thanawan S., Force volume imaging of defects in highly drawn high-density polyethylene. Polymer testing 2007; 26:396-401.

IOP Conf. Series: Materials Science and Engineering **307** (2018) 012050 doi:10.1088/1757-899X/307/1/012050

Dynamics Modelling of Transmission Gear Rattle and Analysis on Influence Factors

Xiaona He[1], Honghui Zhang[1,2]

[1]*State Key Lab. of Vehicle NVH and Safety Technology, Chongqing 401122, P.R.China*
[2]*Key Lab. for Opto-electronic Technology & Systems of Ministry of Education, Chongqing University, Chongqing 400030, P.R.China*
Email: hhzhang@cqu.edu.cn

Abstract. Based on the vibration dynamics modeling for the single stage gear of transmission system, this paper is to understand the mechanism of transmission rattle. The dynamic model response using MATLAB and Runge-Kutta algorithm is analyzed, and the ways for reducing the rattle noise of the automotive transmission is summarized.

1. Introduction

As the increasing of requirement for automobile quality[1], the transmission rattle noise becomes more and more important for NVH performance. Generally, two factors is significant to the transmission rattle[2]: the external excitation caused by the input speed and torque from the engine, and the internal incentive factors caused by the time-varying stiffness, damping, gear side clearance and mesh error. This paper is trying to analyze the influencing factors for automotive transmission, and summarize the ways for rattle reduction.

2. Dynamics Modeling of Stage Gear

2.1. *Modeling of Single Stage Gear*

Based on the equivalent principle of inertia[3], rattle behavior of the gear system can be equivalent to the superposition of single gear pairs' vibration effect. For convenience, a single straight gear pair is used for the numerical modeling[4]. Dynamic model of single gear pair is shown in Figure 1.

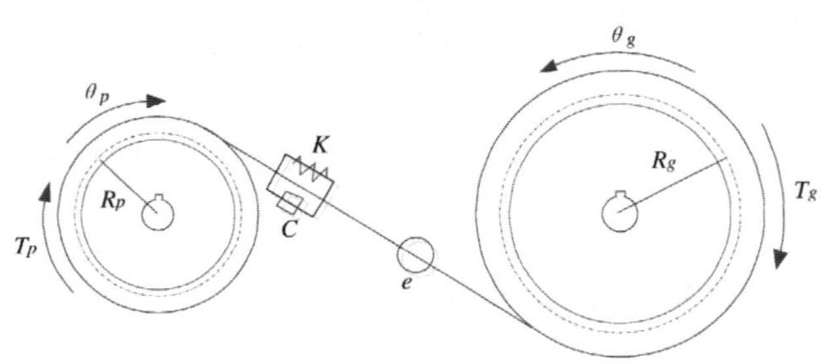

Figure1. Dynamic model of single gear pair

The corresponding mathematical model is built under the second Newton's law. Considering the meshing error, the force analysis on the driving and driven gear is carried out. The equations are as followings:

$$I_p\ddot{\theta}_p + R_p C(R_p\dot{\theta}_p - R_g\dot{\theta}_g - \dot{e}(t)) + R_p K(t)(R_p\theta_p - R_g\theta_g - e(t)) = T_s + T_v(t) \quad (1)$$

$$I_g\ddot{\theta}_g - R_g C(R_p\dot{\theta}_p - R_g\dot{\theta}_g - \dot{e}(t)) - R_g K(t)(R_p\theta_p - R_g\theta_g - \mathrm{e}(t)) = -T_g(t) \quad (2)$$

Where p and g are the subscripts which demonstrating driving and driven gears. I_p, I_g are the rotational inertia; θ_p and θ_g are the torsional vibration angular displacements; R_p and R_g is the base radius; T_p and T_g is the torque; *K, C* is the torsional stiffness and damping coefficient respectively. *e(t)* is the gear dynamic meshing error; T_s , $T_v(t)$ is the stable component and variation component of driving torque.

For the convenience, the relative vibration displacement between the two meshing teeth is [5]:

$$x = R_p\theta_p - R_g\theta_g \quad (3)$$

Considering the clearance, actual deformation and tooth surface vibration displacement are no longer equal [6]. Deformation happens only when the relative displacement is greater than the clearance. Set b as teeth clearance on both sides, the actual deformation can be expressed by function $f_r(t)$:

$$f_r(t) = \begin{cases} x - b, & x > b \\ 0 & , -b \le x \le b \\ x + b, & x < -b \end{cases} \quad (4)$$

Combining the Eqs (1-3) and dynamics equation is as Equation 5:

$$m_e\ddot{x}(t) + C[\dot{x} - \dot{e}(t)] + K(t) f_r(x - e(t)) = F_T - m_e\ddot{e}(t) \quad (5)$$

Where m_e is the equivalent mass of gear pair, F_T is the total load of transferring between teeth. Some text.

2.2. Rattle Criterion of Gear Pair

According to the theory of gear rattle threshold [7], whether rattle happens depends on the drive torque of the driven gear. When the driving torque is positive, meshing between the driving and driven gear is normal, rattle won't occur. When the driving torque is negative, the normal gearing is destroyed and then rattle occurs [8]. The relationship of driving torque, trailing torque, inertia torque is:

$$T_{drive} = T_{drag} - |T_{inertia}| \quad (6)$$

Where T_{drive} is the driving torque of driven gear, T_{drag} is the trailing torque, $T_{inertia}$ is inertia torque.

Contact force, vibration angular velocity and vibration angular acceleration can act as the criterion for rattle occurrence, but for intuitive results this paper takes the actual deformation as the criterion.

In normal gearing condition, the relative displacement between the driving and driven gear is greater than the clearance and deformation exists. Otherwise, the driven gear may work between tooth clearance and knock the driving gear back and forth. The axial relative displacement between two teeth can be described as:

$$\Delta x = R_p\theta_p - R_g\theta_g - e(t) = x(t) - e(t) \quad (7)$$

Taking the deformation as rattle criterion:

$$\begin{cases} \Delta\overline{x} = |\Delta x| - b > 0, & no\ \ rattle \\ \Delta\overline{x} = |\Delta x| - b \le 0, & rattle\ \ occurs \end{cases} \quad (8)$$

3. The Numerical Analysis for Rattle Occurrence

3.1. The Model Pre-processing

In order to get more accurate result, the mesh stiffness data and error data in equations 1 and 2 are both given in the form of harmonic synthesis with 6th order. The equations are as follows:

$$K(t) = k_m + \sum_{i=1}^{6} k_i \cos(i\omega_e t + \phi_i) \quad (9)$$

$$e(t) = e_m + \sum_{j=1}^{6} e_j \cos(j\omega_e t + \phi_j) \quad (10)$$

As the different magnitude of gear parameters such as stiffness and damping in the mode, using the dimensionless processing model is important for comparison. Considering the input torque variation component, the mathematical equation is described as follows:

$$\ddot{y}(t)+2\xi\dot{y}(t)+[1-\sum_{i=1}^{6}\bar{k}_i\cos(i\omega t+\phi_i)]f[y(t)]=F_T{}'+F_V{}'\cos(i\omega t)+\omega^2\sum_{j=1}^{6}m_e{}^2\bar{e}_j\cos(j\omega t+\phi_j) \tag{11}$$

Where $m_e=\dfrac{I_pI_g}{I_pR_g{}^2+I_gR_p{}^2}$, $\xi=\dfrac{C}{2\sqrt{k_m m_e}}$, $\omega=\dfrac{2\pi z_p n_p}{60}\sqrt{\dfrac{m_e}{k_m}}$, $F_T{}'=\dfrac{F_T}{b_0\sqrt{k_m m}}$, $\bar{k}_i=\dfrac{k_i}{k_m}$, $\bar{e}_i=\dfrac{e_i}{b_0}$,

$b_0=10^{-6}$, $f[y(t)]=f_r(x-e(t))\times b_0$

3.2. Analysis of the Result

Using the parameters of a certain type transmission gear, we can carry out the example analysis of the model by Runge-Kutta algorithm. The gear parameters are shown in Table 1.

Table 1. The gear pair parameters

	Number of teeth	Rotary inertia (kg·mm2)	Modulus	Contact ratio	Tooth width (mm)	Clearance 2b (um)
Driving gear(p)	39	1206	2.5	1.7564	85	5
Driven gear(g)	20	681				

3.3. Gear Clearance

Ignoring the influence of the clearance change under the time-varying stiffness, and keeping other parameters unchanged, set clearance 2b as 4, 20, 40 μ m respectively, We can get the dynamic system response respectively as shown in Figure 2.

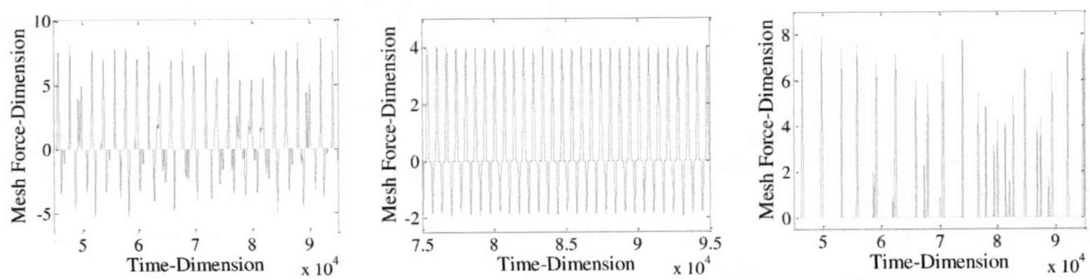

Figure 2a. Meshing force amplitude changes as clearance increases

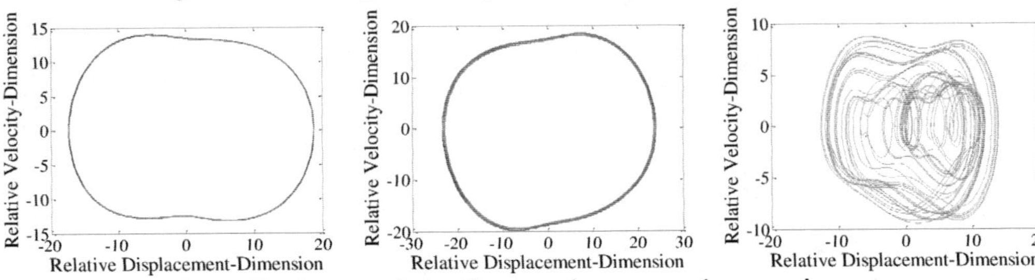

Figure 2b. System phase diagram changes as clearance increases

3.4. Time-varying Stiffness Fluctuation

As shown in the figures, with the increasing of clearance, rattle of gear pair varies from bilateral rap to unilateral tap within a certain range. System phase diagram shows that the original periodic response becoming chaos gradually as the increasing of clearance.

Using the first harmonic component amplitude of time-varying stiffness as the research object, the influence on rattle response is analyzed under stiffness fluctuation. Setting k_1/k_m as 0.1, 0.3, 0.7 respectively, and the dynamic response is shown in Figure 3.

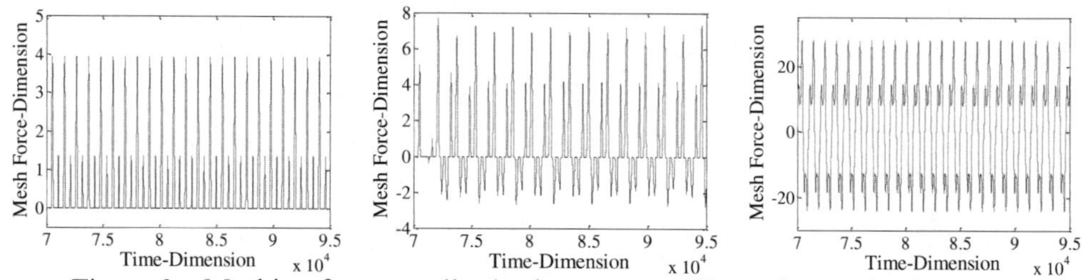

Figure 3a. Meshing force amplitude changes as stiffness fluctuation increases

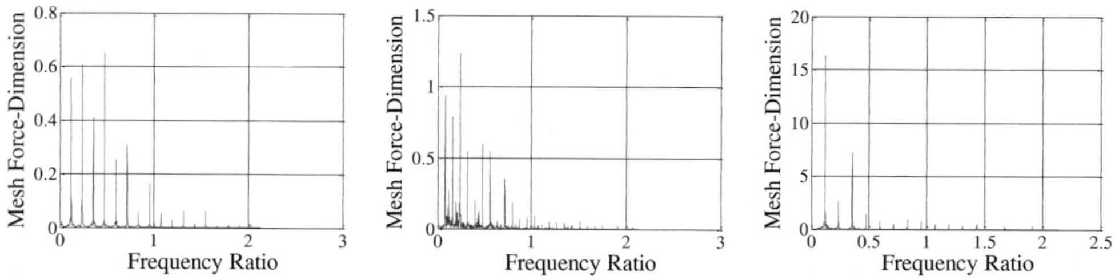

Figure 3b. Meshing force frequency changes as stiffness fluctuation increases

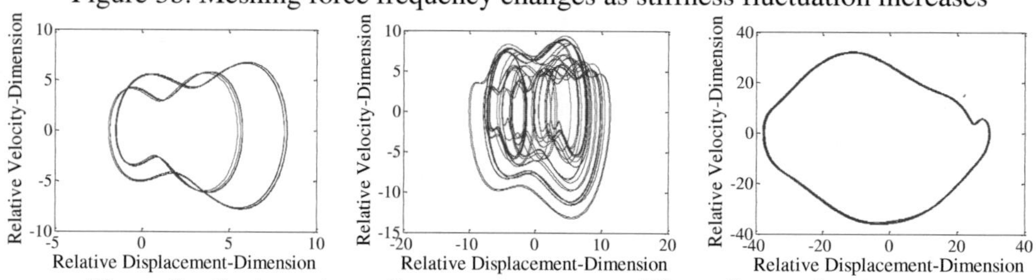

Figure 3c. System phase diagram changes as stiffness fluctuation increases

As shown in the figures, meshing force amplitude increases as the time-varying stiffness fluctuation component increasing. Rattle of gear pair varies from unilateral to bilateral within a certain range, the amplitude of frequency spectrum increases, and the main frequency gradually stands out.

3.5. Meshing Damping
The damping coefficient represents the combination of integration effects on the system. Set ξ as 0.02, 0.05, 0.08 respectively, and the phase diagrams are shown in Figure 4.

As shown in the figures, with the increasing of damping coefficient, the original chaotic response system is gradually becoming harmonic and periodic. It also proves that the appropriately increased damping effect can effectively improve the system stability and reduce the possibility of vibration rattle.

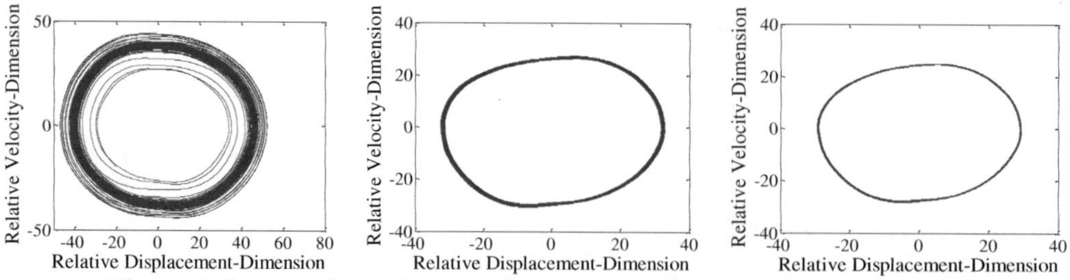

Figure 4. System phase diagram changes as damping coefficient increases

3.6. Meshing Error Fluctuation

IOP Conf. Series: Materials Science and Engineering **307** (2018) 012050 doi:10.1088/1757-899X/307/1/012050

In order to explore the incentive effect of the time-varying meshing error, we use the amplitude of the first harmonic component as the research object. Setting e_1/e_m as 0.15、0.3、0.45 respectively, the dynamic response is shown in Figure 5.

As shown in the figures, the meshing force amplitude increases with the meshing error fluctuation somewhat. Unilateral rattle of gear pair varies to bilateral. The amplitude of frequency spectrum increases, and the main frequency gradually stands out.

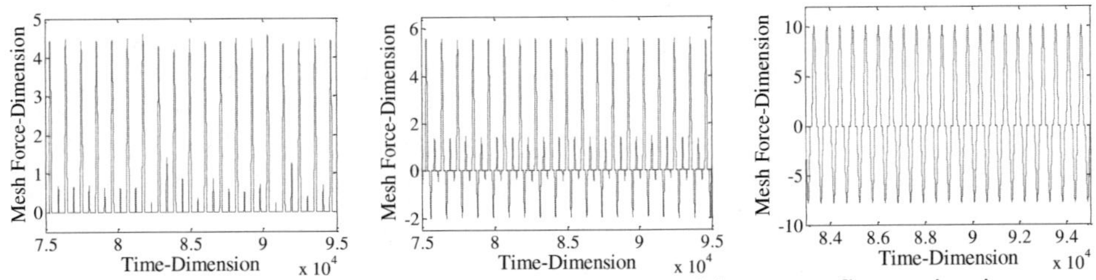

Figure 5a. Meshing force amplitude changes as meshing error fluctuation increases

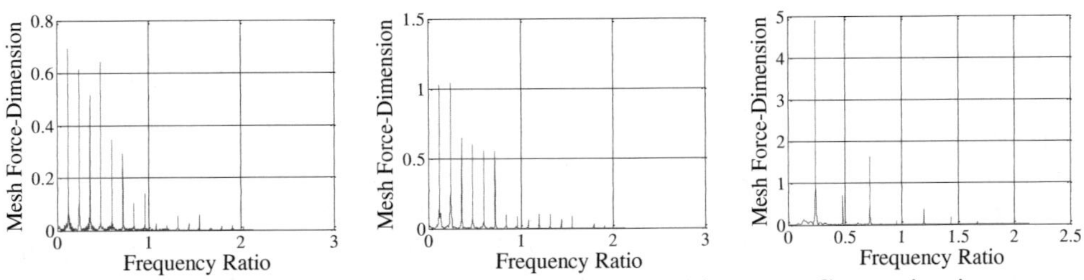

Figure 5b. Meshing force frequency changes as meshing error fluctuation increases

3.7. Input Speed

Input speed determines the meshing frequency of the system. To explore the relationship between the rattle and input speed, we set the wheel input speed n_p as 2000, 3000, 4000 r/min, the response is shown in Figure 6.

The meshing frequency comes close to natural frequency firstly, and then get far away again with the increasing input speed, and the system response is also changed from periodic to chaotic finally.

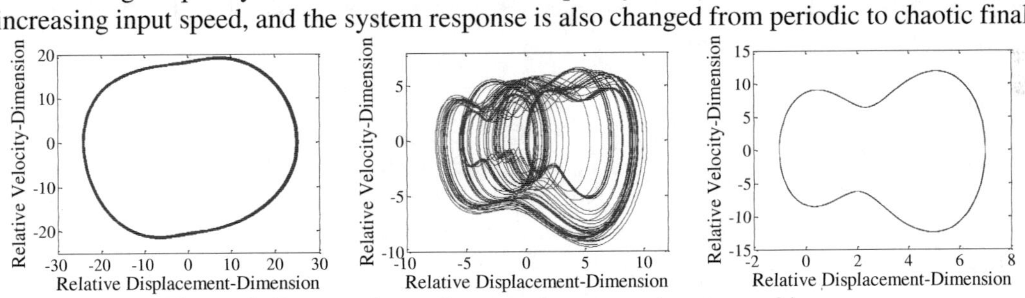

Figure 6. System phase diagram changes as input speed increases

3.8. Input Torque Fluctuation

To explore the effect of torque fluctuation on the gear rattle, keeping other parameters unchanged, set F_V/F_T as 0.05, 0.2, 0.4 respectively, and the response is shown in Figure 7.

Meshing force and frequency amplitude gradually increase with the increment of the torque fluctuation and the rattle of gear pair varies from unilateral to bilateral.

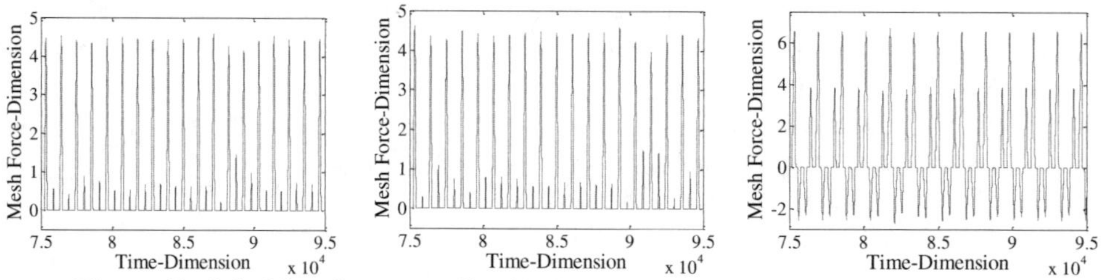

Figure 7a. Meshing force amplitude changes as torque fluctuation increases

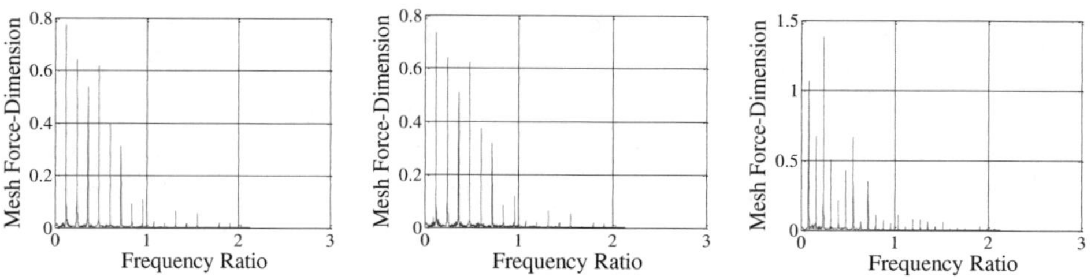

Figure 7b. Meshing force frequency changes as torque fluctuation increases

4. Conclusions

Based on the mechanism of transmission rattle and its influencing factors, a rattle dynamics mode of single gear pair is established and solution is implemented by Runge-Kutta algorithm. It shows that increased amplitude of gear clearance, time-varying stiffness fluctuation, meshing error fluctuation, input torque fluctuation will increase the gear rattle force, make the rattle varies between unilateral and bilateral, and affect the system periodicity. The increased gearing damping will make the dynamic response smooth, and reduce the rattle to some extent. Input speed determines the system meshing frequency and which influences the periodicity of the system obviously.

Acknowledgement

This work is supported by the Opening Foundation of State Key Laboratory of Vehicle NVH and Safety Technology (NVHSKL-201402).

References

[1] R. Brancati, E. Rocca, S. Savino, F. Farroni. Analysis of gear rattle by means of a wavelet-based signal processing procedure[J]. Meccanica, 48(6): 1399-1413, (2013)
[2] O. Tangasawi, S. Theodossiades, H. Rahnejat. Lightly loaded lubricated impacts: Idle gear rattle[J]. Journal of Sound and Vibration, 308(3-5): 418-430 (2007)
[3] M. Barthod, B. Hayne, J. L. Tébec, J. C. Pin. Experimental study of gear rattle excited by a multi-harmonic excitation[J]. Applied Acoustics, 68(9): 1003-1025 (2007)
[4] N. H. Oliver Johnson. Diagnosis and objective evaluation of gear rattle[J]. SAE, 381-396 (1991)
[5] YANG Wen, CHEN Hong-jiang, HUANG Da-xing. Computer Dynamic Simulation of Gears in Gearbox[J]. Tractor & Farm Transporter, 01: 20-21 (2008)
[6] H. Optiz. Dynamic behavior of spur and helical gears[J]. SME Semi-International Symposium Papers, Gears., 9: 89-95 (1967)
[7] Zhang Zhijun, Li Hongcheng, Yang Xianwu, Wang Lanjun, Li Yao, Gear Rattle Analysis of Dual-Clutch Transmission Based on Mulit-Stiffness Damper[C]. SAE-China Congress, Peking, China, 2013. 5 (2013)
[8] WANG Lian-sheng, HAO Zhi-yong, ZHENG Kang, LIU Gong-wen. Simulation and experiment on transmission gear rattle considering drag torque [J]. Journal of Zhejiang University (Engineering Science), 05: 911-916 (2014)

IOP Conf. Series: Materials Science and Engineering **307** (2018) 012051 doi:10.1088/1757-899X/307/1/012051

Graphene based strain sensor with LCP substrate

M Nie, H S Yang and Y H Xia

Key Laboratory of MEMS of Ministry of Education, Southeast University, Nanjing, China
E-mail: mnie1980@163.com

Abstract. A flexible strain sensor constructed by an efficient, low-cost fabrication strategy is presented in this paper. It is assembled by adhering grid-like graphene on LCP substrate. Kinds of measurement setup have been designed to verify that the proposed flexible sensor device is suitable to be used in health monitoring system. From the experiment results, it can be proved that the sensor exhibits the following features: ultra-light, relatively good sensitivity, high reversibility, superior physical robustness, easy fabrication. With the great performance of this flexible strain sensor, it is considered to play an important role in body monitoring, structural health monitoring system, fatigue detection and healthcare systems in the near future.

1. Introduction

Flexible strain sensors are important for various future applications, such as wearable sensors[1-2], touch-on flexible displays[3,4], soft robotics[5,6] and structural health monitoring system[7]. The research concentrates on the characterization of structures, fatigue studies of materials and so on to monitoring damage detection, which can be realized by measuring deformations of the objects with strain sensor embedded inside. Moreover, it has given significant advantages to instead the traditional methods in monitoring human physiological signals to evaluate human health even at the current age of highly developed medical technology. Sensors based on various nano materials such as nanoparticles[8], nanotubes[9], nanowires[10], thin films[11–13] and their assemblages have been attracting interest recently due to their strain sensing characteristics. The strain sensors based on new materials (such as ZnO nanowires, carbon nanotubes etc.) have exhibited distinct advantages and performances, for instance, fast response, excellent strain gauging performances, good conductivity, high-transparency, superior physical robustness, and easy fabrication[10]. For graphene-based sensors, the electrical conductance[14] and the principal vibrational frequencies of graphene strongly depend on its crystal structure which can be controlled by applying strain, such characteristics make graphene film can be utilized in strain sensor to obtain high sensitivity[15].

In addition, to date, the substrate material of the flexible electronic device is popular chosen Polydimethylsiloxane (PDMS), because PDMS has excellent material properties such as high elasticity, biocompatibility, and it is easy to be integrated with two dimension materials. However, the problem is the fabrication process of PDMS is complicated. Especially to fabricate micro patterns inside PDMS, such as cube, line or pyramid to increase the sensitivity of the sensor which always need the patterned Si moulds, the Si moulds are always complicated and expensive. Despite the excellent pressure sensing performance, it still has a lot of difficulties to monitoring human motion signals by the sensors because of the complicated structures, fabrication process, sophisticated equipments.

In this work, a flexible strain sensor with LCP as its substrate was designed, in the meanwhile, the two dimension material graphene has been chosen as the sensitive film of the structure. The proposed device is easily to realize and is readily to measure various strain levels of human motion signals.

Because LCP has it own particular properties such as good dimensional stability, flexibility and biocompatibility, which can offer the proposed sensor has good stability and ultrahigh reversibility. At the same time, the advantages of low-cost and simplicity in device fabrication can be ensured, and the versatility of detecting various signal changes of non-plane objects can be realized as well.

2. Fabrication

The fabrication process is described in the schematic shown in figure 1. For the sensor fabrication, we used LCP(R-F700T, Panasonic Corp provided) shown in figure 1(a), a 125μm thick LCP film with copper cladding on one side, which had the thickness of 18μm. Figure 1(b) shows a 5μm thick photoresist was patterned on the copper cladding. The exposed copper cladding area was etched away by an aqueous solution of $FeCl_3$(0.5mol/L) and HCl(0.5mol/L), the remained copper was used as electrodes of the strain sensor as shown in Fig.1(c). Subsequently, transparent adhesive tape (3M Corp provided) was patterned on the exposed LCP and copper electrodes as shown in figure 1(d). Then graphene oxide (GO, Nanjing XFNANO Materials Tech Co.,Ltd provided) was deposited into exposed LCP, transparent adhesive tape and copper electrodes by simple spin-coating and drying as shown in figure 1(e). The spin-coating was repeated for 45 times and the solution of GO is 4mg/ml. Finally, the transparent adhesive tape was lift off and then the GO was heated in the high-temperature furnace for 2h under 200℃, nitrogen was used as protective gas in the whole heating process. Figure 1(g) is the side view of the sensor structure and figure 1(h) is the top view of the sensor structure.

Figure 2(a) presents a photo image of the strain sensor of about 3cm×3cm, the width of each graphene strip is 0.2cm. As seen in the photo, the mechanical property of LCP substrate was strong enough to prepare a large-scale sheet fabric. The strain sensor can be easily tailored to pieces with appropriate size for sensing applications. Figure 2(b) shows the magnified optical microscope image of the strain sensor. The average resistance of the graphene mesh film was about 45±25 KΩ.

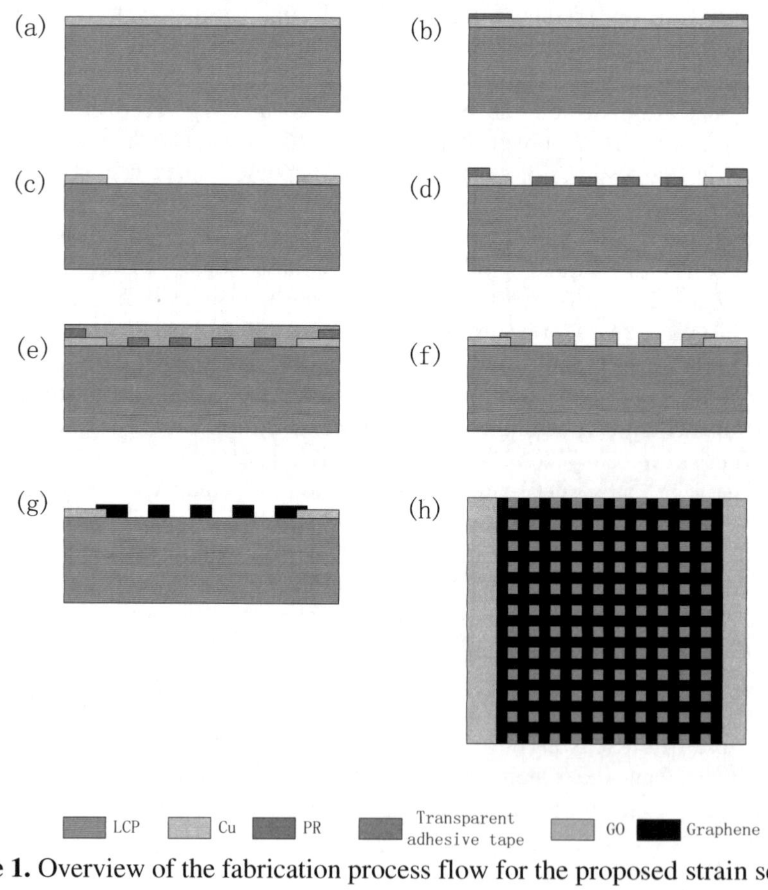

Figure 1. Overview of the fabrication process flow for the proposed strain sensor.

IOP Conf. Series: Materials Science and Engineering **307** (2018) 012051 doi:10.1088/1757-899X/307/1/012051

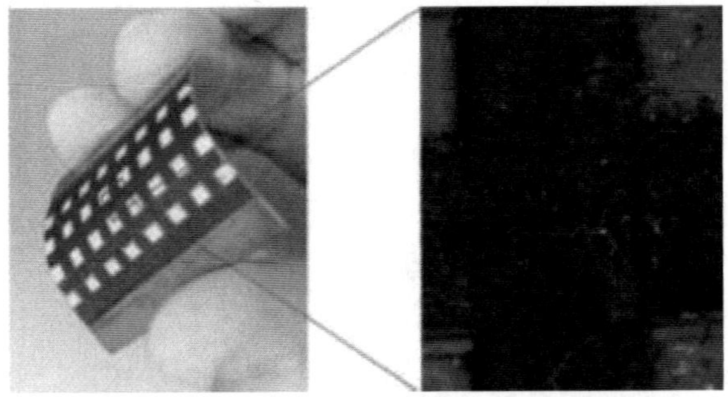

(a) (b)

Figure 2. (a)Photograph showing the bend ability of the strain sensor.
(b)A magnification of the optical microscope image.

3. Experiment and Discussions

As shown in figure 3(a) and 3(b), graphene mesh forms an electrical network with a variable resistor assembled to each current pathway. Partial cracks perpendicular to pathways will lead to an increase of the resistance while the across cracks will result in breaking off the pathways. Figure 3(b) shows a schematic model of the current pathway within the graphene mesh. It was shown that cracks were uniformly distributed in the stretched graphene mesh. As the tensile stresses were applied on the strain sensor, high density cracks generated in the graphene mesh, which causing the current pathways decreased and the resistance increased. Finally, the sensitivity of the strain sensor was improved greatly. On the contrary, when the compressive stress was applied on the sensor, the overlaps of graphene with some micro cracks are created, which lead to decrease the resistance of the strain sensor[7]. And because of the higher Young's moulds of LCP than which of PDMS, strain sensors based on LCP substrate have greater stability and repeatability than that on PDMS substrate.

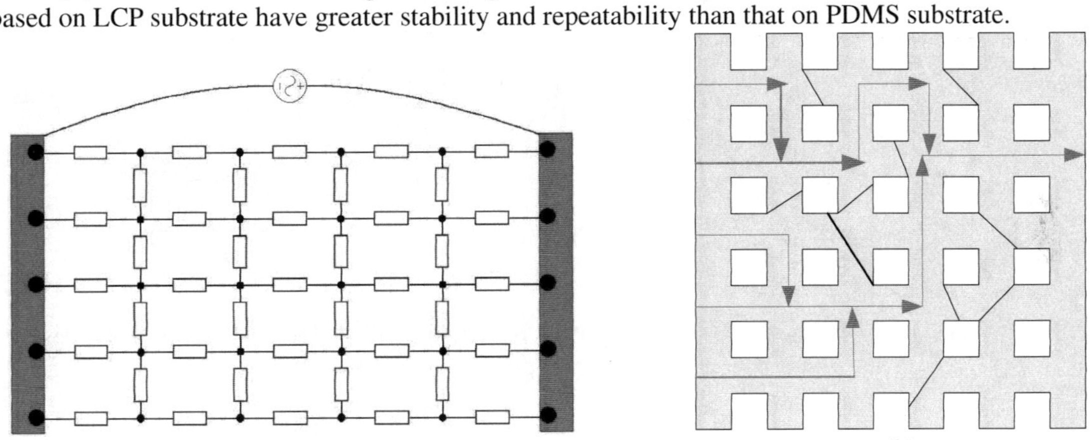

(a) (b)

Figure 3.(a)The equivalent circuit model of the graphene mesh.
(b)Current pathway through the graphene mesh.

The resistance changed significantly with the hand motion from stretch to clench, as shown in figure 4, when the strain sensor was coherent on the back of a hand. As shown in figure 4(b), peaks represent hand motion with fast response, the relative change of the resistance were around 20% for each peak. In addition, the stability and the reversibility of the strain sensor were very good during 200 seconds.

Respiratory is one of the important parameters to be detected in hospitals as a vital sign. The strain sensor can take this sign easily under both stationary state and exercise state. The signals of the breathing rate under stationary state and excise state are shown in figure 5(a) and (b), respectively. The peaks and valleys were assigned to the chest stretch and shrink, respectively, and each cycle represented a breath. Figure 5 demonstrates that the signal had a higher frequency and a larger amplitude after exercise than that of stationary state. The breath numbers in stationary state and after exercise state were 50 and 70, respectively, in 400 seconds. The relative change of resistance of the

IOP Conf. Series: Materials Science and Engineering **307** (2018) 012051 doi:10.1088/1757-899X/307/1/012051

respiratory in still state and after exercise state was 2% and 4%, respectively. Moreover, it can been seen clearly from figure 5(b) that as the testing time went on, the frequency decreased after exercise, the respiratory went towards the stationary state.

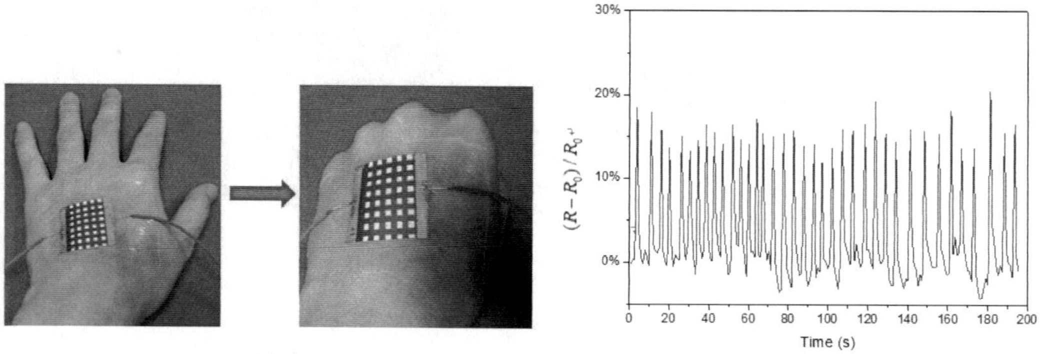

(a) (b)

Figure 4. Relative changes of resistances in hand motion from stretch to clench

The strain sensor also can be used to distinguish out the different signals of the muscle motions when a tester spoke different words or phrases. The results in figure 6 illustrates that the strain sensor exhibited high sensitivity and distinctive patterns when the speaker spoke different words such as "Hi", "Hello" and "SEU (Southeast University)", respectively. The red frame 2 was an amplified view of red frame 1. It can be seen that wave form of single curve was apparently different. All the tests and data were obtained by a healthy adult.

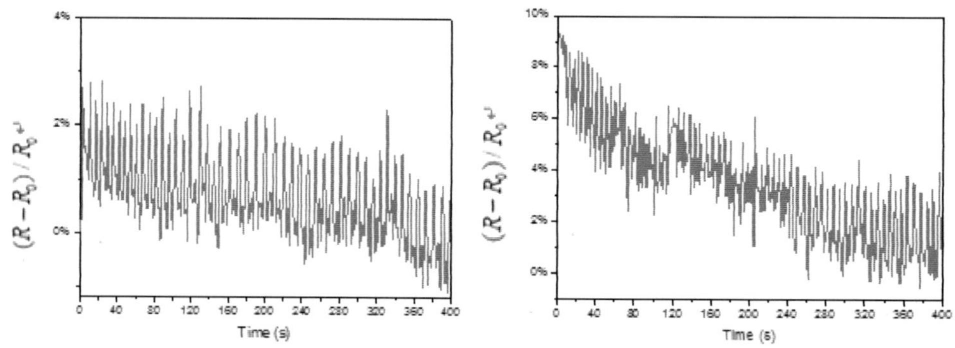

(a) (b)

Figure 5. Relative changes of resistances of the respiratory in (a) stationary state and (b) exercise state.

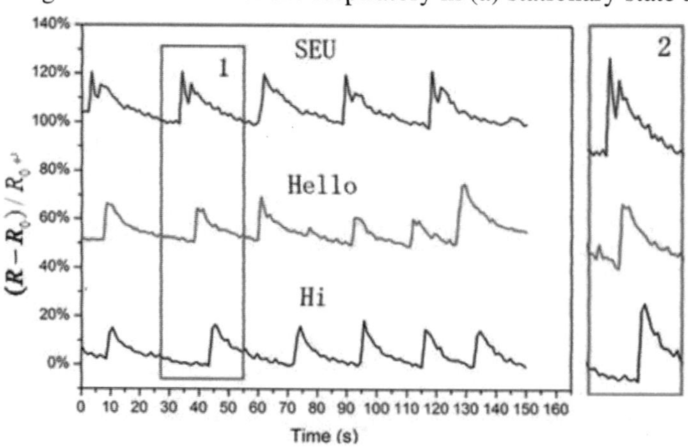

Figure 6. Relative changes of resistances in the muscle motions in speech.

In all the experiments, the double faced adhesive tape (3M Corp provided) was used to stick the strain sensor to the skin or flexible surface. The electrical properties were recorded by a digital sourcemeter (Keithley4200-SCS), and the Source-Drain voltage with 1V DC bias.

4. Conclusion

We have developed a graphene-on-LCP flexible strain sensor to monitoring human physiological signals and microstrain (0.16%) structural variation, and the low-cost feasible fabrication scheme has been designed. The experiment results demonstrated the sensor has a high sensitivity and stability over a wide testing time range. According to the above advantages and facilitative measurement platforms, the proposed flexible sensor can be used in a quite widely applications from the wearable electronic skins to structural health monitoring in bridges, buildings, and other critical infrastructures under severe natural disasters.

5. References

[1] Mannsfeld S C B, Tee B C K, Stoltenberg R M and Bao Z N 2010 *Nat. Mater.* **9** 859
[2] Kaltenbrunner M, Sekitani T, Reeder J and Someya T 2013 *Nature* **499** 458
[3] Lipomi D J, Vosgueritchian M, Tee B C K, Hellstrom S L, Lee J A, Fox C H and Bao Z N 2011 *Nat. Nanotechnol.* **6** 788
[4] Fan F R, Lin L, Zhu G, Wu W Z, Zhang R and Wang Z L 2012 *Nano Lett.* **12** 3109
[5] Someya T, Sekitani T, Iba S, Kato Y, Kawaguchi H and Sakurai T 2004 *Proc. Natl Acad. Sci. USA101* 9966
[6] Cheng M Y, Huang X H, Ma C W and Yang Y J 2009 *J. Micromech. Microeng.***19** 115001
[7] Liao X Q, Liao Q L, Yan X Q, Liang Q J, Si H N, Li M H, Wu H L, Cao S Y and Zhang Y 2015 *Adv. Func. Mat.* **25** 2395
[8] Herrmann J, Muller K H, Reda T, Baxter G R and Raguse B 2007 *Appl. Phys. Lett.* **91** 183105
[9] Obitayo W and Liu T 2012 *J. Sensors* 652438
[10] Ferna´ndez-Regu´lez M Plaza J A, Lora-Tamayo E, and Paulo A S 2010 *Microelectr.Engin.***87** 1270
[11] Loh K J, Kim J H, Lynch J P, Kam N W S and Kotov N A 2007 *Smart Mater. Structures* **16** 429
[12] Song X H, Liu S, Gan Z Y, Lv Q, Cao H and Yan H 2009 *Microelectronic Engin.* **86** 2330
[13] Kumar S B and Guo J 2012 *Nano Lett.* **12** 1362
[14] Sakhaee-Pour A, Ahmadian M T and Vafai A 2008 *Solid State Commun.***147** 336

Acknowledgments

This work was supported by National Natural Science Foundation of China (Grand No. 61474023) and National Key Technology Support Program of China (Grand No.2015BAF16B01).

IOP Conf. Series: Materials Science and Engineering **307** (2018) 012052 doi:10.1088/1757-899X/307/1/012052

On DSS Implementation in the Dynamic Model of the Digital Oil field

Iakov S. Korovin, Maksim V. Khisamutdinov, Anatoly I. Kalyaev

Southern FederalUniversity, Scientific Research Institute of multiprocessor computer systems, 347928, Taganrog, Chekhov Street, 2
korovin_yakov@mail.ru

Abstract. Decision support systems (DSS), especially based on the artificial intelligence (AI) techniques are been widely applied in different domains nowadays. In the paper we depict an approach of implementing DSS in to Digital Oil Field (DOF) dynamic model structure in order to reduce the human factor influence, considering the automation of all production processes to be the DOF model clue element. As the basic tool of data handling we propose the hybrid application on artificial neural networks and evolutional algorithms.

1. The DSS in the DOF framework

The idea to apply definitely decision support systems in the oil production domain – not a full functioning automatic control system (ACS) - was born due the impossibility of the latter's implementation (ACS face with the strict demand of rigid rules which cannot be satisfied because of the fuzzy conditions of an oil field processes).

On the basis on the conducted research results, we suppose that in general, modern DSS, applied in the DOF, provide the following classification as follows [1].

- By the interaction with the user (passive, active, cooperative);
- On conceptual level (communication-driven, data-driven, document-driven, knowledge- driven, model-driven);
- By the application field (task-oriented; hierarchical decision support systems, distributed DSS);
- By way of support (model-oriented, focused on data, focused on documents, focused on knowledge).
- that provide specialized solutions, based on facts.

Speaking on the data processing tools, in the era of Big Data [2] it is fair that software developers for oil industry apply artificial intelligence techniques, and therefore, we can talk about the application the industrial intelligent DSS. This is justified in terms of the efficiency of the results, because in this industry basic requirements to the decision support systems applied are as follows:

- work in real time, under severe time constraints on decision-making, limited computer resources and the uncertainty caused by fuzzy info input;
- continuous operation mode;
- reduction of the operator involvement in the object control;
- real time diagnostics allowing to prevent the equipment failure.

Companies that have ready-made solutions for the industry, represent the possibility of "adaptation" of standard solutions for a specific user, using (including) into the methodology development as their own unique methods and algorithms, as well as the user tested ones.

IOP Conf. Series: Materials Science and Engineering **307** (2018) 012052 doi:10.1088/1757-899X/307/1/012052

Such a situation cannot provide to the researcher "pure" state of art for the proper analysis of methods and algorithms applied.

It can be concluded that the topic of oil industry software methods is almost closed and the materia ls to make reliable and reasoned conclusions on are virtually absent.

In general, the structure of the DSS in the oil industry is not different from the structures of systems used in other domains.

2. The essence of the DOF framework

The basic clue of the digital oil field approach is the real time production management that is intended to somehow enhance the oil production recovery and to significantly assist the human (oil field operator, decision-maker etc).

We claim that the most vital DOF elements are:

- Data management (the level of collecting information from millions of sensors);
- Workflow automation;
- Real time decision support;
- Providing operational efficiency;
- Virtual representation of the production processes;
- Visualisation.

These points include each other either form subsets – and the basic target is to provide the total qualitative control under the oil field. The scheme of the DOF that we propose is represented on the Fig. 1.

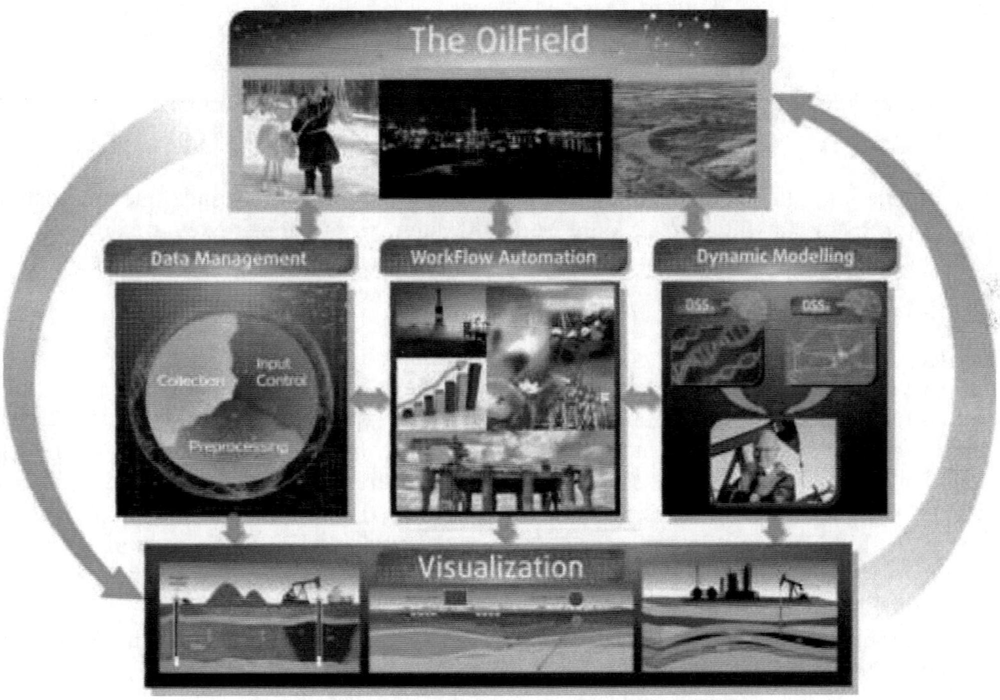

Fig. 1 The proposed DOF structural scheme

3. DOF Dynamic Modelling

3.1. Classical approach of developing dynamic model

Dynamic modelling - is a methodology of analysis, in which the overall process is described in a mathematical model form in which all of the tasks and intermediate processes are presented as a system of interrelated indicators calculated

The method of dynamic modelling is used most often in conditions of inability to automate an entire class of processes, related to decision support systems, and the inability to take into consideration (applying traditional systems) the multi-variance of the problems solutions.

Applying the dynamic modelling the company's business processes are described as a mathematical model, in which all tasks are presented as a system of interrelated calculated indicators.

The particular importance is given to the overall combination of the factors, determining the successfulimplementation of all processes involved. So, the first step in the study the dynamic behaviour is a preliminary analysis of broad industrial aspects that contain a significant number of interactions.

The described method involves the construction of an iterative algorithm in the recursive actions simulation mode. It is peculiar - whatever the first initial decision is, all subsequent decisions must come from the state, resulting from the initial decision.

In general, the object of dynamic modelling is a system in which the processes are described by differential equations, where the system or some of its elements are replaced with a model that possesses the dynamic properties similar to the original ones. As mentioned above, the company is considered as a complicated system. In itself, the modelling provides a scientific basis, around which the controlled objects are grouped. The mathematical models apply method of establishing a causal relationship between any desired number of factors.

Models indicate as the described system will evolve in time. The history of the oil and gas industry is especially important for the dynamic simulation, in particular, that part of it which relates to the specific decisions and the facts explanation, as well as the linear theory of information systems with feedback.

Visualized model creation allows to analyze the possible various processes' development directions, to discover and reflect the interdependence of these processes. For example, applying a dynamic modelling can provide:

- to estimate the effectiveness of space load, the usage of material resources;
- to analyze and model the impact of external factors on the efficiency of the enterprise as a whole, and in certain areas, and so on.

Dynamic modeling, integrated with strategic planning, allows to build standalone applications that solve specific problems, taking into account built-in system simulation models of processes, and to display the modelling results on the specific dashboards, used in the process of decision-making support.

3.2. Building a model, based on AI driven DSS

Taking into account the successful experience of industrial application of the DSS, based on the artificial intelligence (AI) techniques, such as neural network approach in hybrid combination with evolutionary algorithms (genetic algorithm) approach [3]4][5] we propose to withdraw the long and sophisticated process of building a 'classical' dynamic model from the oilfield production process control and to introduce the DSS (subset of DSS's) into the DOF structure (see the Fig. 1).

The idea is as follows. On the basis of the retrospective data, stored in the databases, we train the core of the the system. We apply the methods, depicted in [4][5][6].

After the process of the training is over, the profile of the system is generated and we start to handle the oil production information in real time. Since we apply the approach of implementing DSS, the operator has the opportunity to correct the decisions, made by the core of the system and to initiate the process of post-training (retraining). So, we assume because of the fact of corrections of DDS inference, there is a linear dependence on the terms of DSS industrial exploitation and the quality of solutions, offered to the users.

4. Acknowledgment

This paper is published due the financial support of the Russian Science Foundation (RSF) via the grant No 15-19-00196.

References

[1] Iakov S. Korovin, Igor A. Kalyaev Modern Decision Support Systems in Oil Industry: Types, Approaches and Applications //Advances in Computer Science Research. ISBN 978-94-6252-132-2. ISSN 2352-538x, doi:10.2991/tmcm-15.2015.35. Atlantis Press pp. 141-144.

[2] Iakov S. Korovin The Importance of New Approaches Development and their Implementation in the Oil and Gas Industry in Russian Federation--the Current Situation Analysis // International Conference on Advances in Energy, Environment and Chemical Engineering (AEECE), Changsha, PEOPLES R CHINA; Advances in Engineering Research. ISBN: 978-94-6252-109-4. ISSN: 2352-5401. doi:10.2991/aeece-15.2015.19. Atlantis Press, Volume 23, 2015, p.94-97.

[3] I.S.Korovin, M.V.Khisamutdinov. Neuronetwork decision support system for oilfieldequipment condition online monitoring. Advanced Materials Research (Trans Tech Publications, Switzerland). Volume 902 (2014), Pages 409-415.

[4] Korovin, Ya.S. Decision support system for electrical submersible pumps control on the neural network basis. Neftyanoe khozyaystvo - Oil Industry. Issue 1, January 2007, Pages 80-83.

[5] Korovin, Ya.S., Tkachenko, M.G., Kononov, S.V. Oilfield equipment's state diagnostics on the basis of data mining technologies. Neftyanoe khozyaystvo - Oil Industry. Issue 9, September 2012, Pages 116-118

[6] Korovin. I.S; Khisamutdinov M.V.; Kaliaev A.I. The Application of Evolutionary Algorithms in the Artificial Neural Network Training Process for the Oilfield Equipment Malfunctions' Forecasting. PROCEEDINGS OF THE 2ND INTERNATIONAL SYMPOSIUM ON COMPUTER,COMMUNICATION,CONTROLANDAUTOMATIONBookSeries:Advances in Intelligent Systems Research. Volume: 68. Pages: 253-257. Published: 2013

IOP Conf. Series: Materials Science and Engineering **307** (2018) 012053 doi:10.1088/1757-899X/307/1/012053

Perm State University HPC-hardware and software services: capabilities for aircraft engine aeroacoustics problems solving

A G Demenev[1]

[1]Applied Mathematics and Informatics Department, Faculty of Mechanics and Mathematics, Perm State National Research University, 15 Bukireva street, Perm 614990, Russia

E-mail: a-demenev@mail.ru

Abstract. The present work is devoted to analyze high-performance computing (HPC) infrastructure capabilities for aircraft engine aeroacoustics problems solving at Perm State University. We explore here the ability to develop new computational aeroacoustics methods/solvers for computer-aided engineering (CAE) systems to handle complicated industrial problems of engine noise prediction. Leading aircraft engine engineering company, including "UEC-Aviadvigatel" JSC (our industrial partners in Perm, Russia), require that methods/solvers to optimize geometry of aircraft engine for fan noise reduction. We analysed Perm State University HPC-hardware resources and software services to use efficiently. The performed results demonstrate that Perm State University HPC-infrastructure are mature enough to face out industrial-like problems of development CAE-system with HPC-method and CFD-solvers.

1. Introduction

The present work is devoted to analyzing high-performance computing (HPC) infrastructure capabilities for aircraft engine aeroacoustics problems solving. We explore the ability to develop new structured and unstructured computational aeroacoustics methods/solvers for computer-aided engineering (CAE) systems to handle complicated industrial problems of engine noise prediction. Perm State University's industrial partners, including leading aircraft engine engineering company "UEC-Aviadvigatel" JSC (in Perm, Russia), require that methods/solvers to efficiently optimize geometry of aircraft engine for fan noise reduction.

In 2010 Perm State University (http://www.psu.ru) was awarded the status of National Research University, justifying its reputation as one of the leading Russian universities. The Perm State University Strategic Development Program (2010–2019) has been worked out. The objective of the program is to update and develop Perm State University as a new type research university that integrates multidisciplinary R&D activities and advanced technologies for growing national economic competitiveness. The university participate in national and international activities used parallel and distributed computing (PDC) technologies. For example, Perm State University is:

- a participant of the University Cluster Program (http://unicluster.ru, since 2009),
- a full member of HPC Consortium of Russian Universities (http://hpc-russia.ru, since 2009),
- and a full member National HPC Technology Platform (http://hpc-platform.ru, since 2010).

The university designated by NVIDIA as a CUDA Teaching Center (2012-2015), CUDA Research Center (2014-2015), GPU Education Center (2015-2016), GPU Research Center (2015-2016). The Research-Education Center "Parallel and Distributed Computing (REC P&DC) of Perm State University was founded in 2009. The REC P&DC helps faculties to adopt wide usage of GPU-computing and high-performance computing in general for solving a variety of problems in traditional and interdisciplinary scientific fields. The REC P&DC supports research and educational activities in advanced computing. The project "Development of Joint-Use Center of High-Performance Computing Resources – the REC P&DC" (supervised by this paper author) is a part of the Perm State University Development Program (2010-2019) as a national research university. For the project, Perm State University purchased HPC-systems with hybrid architecture including:

- NVIDIA Tesla S2050-accelerated cluster "PGU-Tesla" with 12 GPU entered the 14th-version (March 2011) of top 50 of all supercomputers in Russia and the CIS and ranked 1st in Perm Krai [1];
- NVIDIA Tesla K20-accelerated cluster "PGNIU-Kepler" with 16 GPU entered the 20th-version (March 2014) of top 50 of all supercomputers in Russia and the CIS and ranked the 1st in Perm Krai [2].

The supercomputers are using actively to solve super-complex computing problems with broad scope; an effective solution is possible only when using HPC-resources. The supercomputers also help to train highly competitive specialists with competences: to solve effectively problems in main areas of science and technology, and to develop critical and breakthrough technologies.

We computed a realistic fan noise emission/radiation problem by using the existing and developing CAE solvers. For example, engineer E.V. Koromyslov (with engineers M.V. Usanin and L.Y. Gomzikov) under supervision of Dr. A.A. Siner continue to develop solver code of GHOST CFD [3] for "UEC-Aviadvigatel" JSC. That code aimed to solve non-stationary Navier-Stokes equations by finite-difference schemes of high order of accuracy on the GPU. We plan to use the solver for task solving in the aerospace industry. This multi-GPU software is essential to perform a precise and robust numerical simulation that requires huge computational resources with Compute Unified Device Architecture (CUDA) [4] to design these eco-friendly engines.

Our vision for furthering the application and technology of GPU computing is fast growth. We believe in the broad opportunities of GPU-computing usage for solving a variety of problems in traditional and interdisciplinary scientific fields. In 2017 at Perm State University start project "Development and implementation of highly efficient computational algorithms for calculating the sound generation of the fan stage of an aircraft engine to reduce the impact of aircraft on the environment and humans".

2. Problem description

We start to use the existing and developing CAE solvers for realistic solving of fan noise emission/radiation problem, which involves both a full-3D exhaust geometry of aircraft engine and a typical fan noise modal content. To date, such modeling, despite its relevance, presents considerable difficulties due to the need for high-order schemes and high resolution (such as, for example, DRP (Dispersion Relation Preserving) schemes [5] for structured mesh and EBR [6] for unstructured ones), as well as detailed calculation grids, which leads to very high computational costs and difficulties in implementation. Previously, different well-known CAE systems (ANSYS CFX and Fluent [7], OpenFOAM[8]) with CFD-solvers already used at Perm State University supercomputers.

First, we considered as computational scheme for structured or unstructured 3D Cartesian meshes to solve the problems of noise reduction at fan stage of an aircraft engine. Second, we considered the problems of developing and effectively implementing of scalable computational algorithms for calculating the noise emission in the aircraft engine stage. We note that the choice of implemented algorithms and development tools determinate with characteristic properties of heterogeneous computing at the target computer architectures. We need to analyze Perm State University HPC-hardware resources and software services to efficiently use.

3. Theory and methods

We analyze modern high-performance architectures used for aeroacoustics task solving with large-scale computation. We consider architectures of high-performance clusters with homogeneous and heterogeneous nodes with multicore vector processors and manycore general-purpose computing for graphics processing units (GPGPUs). We make an emphasis on CUDA capabilities of computational nodes with NVIDIA GPU. We consider software service that are and available at HPC target development platforms. We noted the software sets that are useful for CFD-code development. We use CUDA-enabled development tools, HPC-benchmarks, and CFD-codes interested by our industrial partners. We modernize HPC-infrastructure built on the base of NVIDIA GPU-accelerated hardware and CUDA-enabled software.

4. Hardware and HPC-architectures

Table 1. Hardware description of HPC-systems at Perm State University

Number of single precision (SP) / double precision (DP) floating point (FP) cores			**Hardware of the parallel computing system**		**Performance at benchmark with DP FP, TFlops**	
SP/DP FP vector CPU cores	**SP FP CUDA cores**	**DP FP CUDA cores**	**Node processor numbers (type)**	**Networks types**	**High Performance Linpack[9]**	**Peak**
			"PGU-Tesla" system			
240	5376	2688	20 (2xXeon 5670 2.93 GHz 48 GB RAM) + 3 (Tesla S2050 1.15 GHz 4 GPUs x 3 GB RAM)	Infiniband QDR/ Gigabit Ethernet/ IPMI	4.88	8.99
			"PGNIU-Kepler" system			
128	39936	13312	8 (2xXeon E5-2680 [Acc: 2x Tesla K20] 2.67 GHz 64 GB RAM)	Infiniband FDR-10/ Gigabit Ethernet/ IPMI	17.48	21.49

5. Software services

The REC P&DC staff under supervision Assoc. Prof. A.G. Demenev deployed on the supercomputers and tested more actual versions OS Linux, MPI Libraries, GCC Compiler set, Intel Cluster Software Suite, NVIDIA GPU Drivers, CUDA Toolkit, PGI Accelerator Server.

To carry out calculations on "PGNIU-Kepler", the following software is installed:

- Linux operating system (RedHat Enterprise Linux 7.3 64-bit);
- cluster file system (GPFS);
- set for managing tasks and resources (IBM Platform HPC 3.2, including IBM Platform LSF Task Scheduler);
- scripting languages (Perl, Python, bash);
- MPI execution environments (OpenMPI 1.5.4, MVAPICH2 1.6.3, IntelMPI 13.1, IBM Platform MPI 8.3.0);
- compiler sets (Intel 12, GNU 4.7, PGI Accelerator Server 14) for different programming languages (C / C ++, Fortran 77/90/95);
- HPC math libraries from NVIDIA (CUDA Toolkit v.6.0, 7.5, 8.0) and Intel (MKL 2017);
- tools for debugging and performance analysis (grpof 4, Intel® vTune 4, Intel® Thread Checker, IDB);
- version control systems (CVS, GIT, SVN) and data archiving tools (TAR, BZ2, GZ);
- SSH Terminal Services and remote desktop software (X2Go);
- file transfer software (SCP, SFTP).

The general difference of "PGU-Tesla" system software system:

- Linux operating system (CentOS 7.2 64-bit);
- cluster file system (PanFS);
- set for managing tasks and resources (OpenHPC 1.3, including PBS Pro 13).

We can install the existing and developing CAE solvers to compute a realistic fan noise emission/radiation problem. For example, Perm State University have academic licenses of well-known ANSYS CFX and ANSYS Fluent software.

6. Results and analysis

The "PGU-Tesla" system can simultaneously perform:

- up to 240 tasks on the Intel Xeon 5670 processor cores with SSE (Streaming SIMD Extensions) 4.2;
- up to 2688 subtasks with DP FP instructions or 5376 subtasks with SP FP instructions on the Nvidia Tesla S2050 CUDA cores with Compute Capability 2.0;
- up to 40 times faster data exchange between tasks over high-speed network QDR Infiniband than over conventional Gigabit Ethernet network.

Theoretical peak performance the "PGU-Tesla" system is up to 9.0 Teraflops as peak (in operations with DP FP). But real performance is up to 54.3 % of the peak performance on the High Performance Linpack. The tasks can use up to 960 GB DDR3 + 36 GB GDDR5 of the total amount of random access memory of computing nodes and up to 20 terabytes of data storage cluster with Panasas File System (PanFS).

The "PGNIU-Kepler" system can simultaneously perform:

- up to 128 tasks on the Intel Xeon E5-2680 processor cores with AVX (Advanced Vector Extensions) 1.0;
- up to 13312 subtasks with DP FP instructions or 39936 subtasks with SP FP instructions on the NVIDIA Tesla K20 CUDA cores with Compute Capability 3.5;
- up to 40 times faster data exchange between tasks over high-speed network FDR-10 Infiniband than over conventional Gigabit Ethernet network.

Theoretical peak performance of the "PGNIU-Kepler" system is about 21.5 Teraflops (with DP FP). But real top performance the "PGNIU-Kepler" is about 81.4% of the peak performance on the High Performance Linpack (with DP FP). The tasks can use up to 512 GB DDR3 + 80 GB GDDR5 of the total amount of random access memory of all computing nodes and up to 36 terabytes of data storage with IBM GPFS.

7. Conclusions

We analyzed Perm State University HPC-hardware resources and software services to use efficiently. We determined the characteristics of the architecture and the software of target computing platforms. We evaluated the upper limits of the actual performance expected for the CFD-solvers being developed. The expected performance of CFD-problems solving at the supercomputers is many times smaller for solving linear algebra problems with sparse matrix than for the HPL-benchmark. It may to verify on a suitable set of benchmarks (for example, HPCG benchmark [10]).

We ready to compute a realistic fan noise emission/radiation problem by using the existing and developing CAE solvers. The performed results demonstrate that Perm State University HPC-infrastructure are mature enough to face out industrial-like problems of development CAE-system with HPC-method and CFD-solvers.

Acknowledgments

The research is carried out using the equipment of the shared research facilities of HPC computing resources of the Research-Development Center "Parallel and Distributed Computing" at Perm State National Research University. This work was financially supported by the Ministry of Education and Science of the Russian Federation (unique project identifier RFMEFI57417X0145, the Agreement 14.607.21.0041 of September 26, 2017) under framework of the Federal Targeted Programme for Research and Development in Priority Areas of Development of the Russian Scientific and Technological Complex for 2014-2020. First and foremost, we would like to thank to supervisor of the above-mentioned project, Dr. Alexander A. Siner for the technical discussions on the project aims and details. I thank Professor Sergey V. Rusakov for the valuable guidance and advice.

IOP Conf. Series: Materials Science and Engineering **307** (2018) 012053 doi:10.1088/1757-899X/307/1/012053

References

[1] The Top 50 ranking of the most powerful supercomputers in Russia and CIS: the 14th edition of March 29, 2011 [webpage]: URL: http://top50.supercomputers.ru/?page=archive&rating=14
[2] The Top 50 ranking of the most powerful supercomputers in Russia and CIS: the 20th edition of April 1, 2014 [webpage]: URL: http://top50.supercomputers.ru/?page=archive&rating=20
[3] Gomzikov L U, Koromyslov E V, Usanin M V and Siner A A 2014 *Proc.Third International Workshop "Computational Experiment in Aeroacoustics"* (Svetlogorsk, Russia) pp 136-138
[4] CUDA Zone [website]: https://developer.nvidia.com/cuda-zone
[5] Bogey C and Bailly C 2004 *J. Comp. Phys* **194** pp. 194-214.
[6] Abalakin I, Bakhvalov P, Kozubskaya T K 2014 *International Journal of Aeroacoustics* **13** no.3 pp 207-234
[7] ANSYS [website]: URL: http://www.ansys.com/
[8] OpenFOAM [website]: URL: https://openfoam.com/
[9] HPL - A Portable Implementation of the High-Performance Linpack Benchmark for Distributed-Memory Computers [website]: http://www.netlib.org/benchmark/hpl/
[10] HPCG Benchmark [website]: URL: http://hpcg-benchmark.org/

IOP Conf. Series: Materials Science and Engineering **307** (2018) 012054 doi:10.1088/1757-899X/307/1/012054

Similarity study on chloride corrosion of prestressed concrete in marine atmosphere

Congqi Li [1,2], Ruojun Wang [3] and Ronggui Liu [1]

1 College of Civil Engineering and Mechanics, Jiangsu University, Zhenjiang , China
2 College of Civil Science and Engineering, Yangzhou University, Yangzhou, China
3 College of Civil and Construction Engineering, National Taiwan University of Science and Technology, Taipei, Taiwan

Abstract. All articles must contain an abstract. The abstract text should be formatted using 10 point Times or Times New Roman and indented 25 mm from the left margin. Leave 10 mm space after the abstract before you begin the main text of your article, starting on the same page as the abstract. The abstract should give readers concise information about the content of the article and indicate the main results obtained and conclusions drawn. The abstract is not part of the text and should be complete in itself; no table numbers, figure numbers, references or displayed mathematical expressions should be included. It should be suitable for direct inclusion in abstracting services and should not normally exceed 200 words in a single paragraph. Since contemporary information-retrieval systems rely heavily on the content of titles and abstracts to identify relevant articles in literature searches, great care should be taken in constructing both.

1. Introduction

There are many influencing factors on the mechanism of concrete erosion in the marine environment. For different research objectives, its corresponding durability evaluation index, test methods are different. The concrete exposure test shows the real degradation process and regularity of the research object, but it usually takes 2 to 3 years or longer to obtain meaningful test data. At the same time, there are many unpredictable factors that interfere with the test. For the corrosion degradation mechanism and regularity of concrete structures in marine environment, most of the research is based on the method of indoor rapid erosion test, which can simulate a single indoor erosion factor or multi-factor interaction to make the research object closer to its real working state. Although the indoor experiment has many merits such as strong pertinence and short test period, the accelerated test time is short. And the relationship between them is fuzzy and complex compared with the degeneration law of the object in the actual erosive environment. Therefore, the correlation between artificial simulation environment and corrosion law of concrete structure in real marine environment has aroused the concern of relevant scholars. Zhejiang University[1] proposed the "Multi-Environmental Time Similarity Theory" (METS) based on the similarity theory of the traditional theory by introducing a third-party reference with similar environmental conditions and a certain service life. Under natural and artificial simulated environments, the differences of chloride ingress into concrete are in terms of surface chloride content, convective zone depth, and diffusion coefficient[2]. Clearing the similarity between the indoor rapid test and the long-term natural exposure test and the intrinsic quantitative link between the two test results is the prerequisite for the effective application of the research results of the indoor acceleration test to the durability design of the new concrete structure and the durability evaluation of the in-service concrete structure. Therefore, the study of experimental methods based on indoor accelerated simulation of durability degradation similar to that of long-term exposure test is a long-term concern for researchers.

In this paper, we take the port environment of east coast of China as the background, and the material and component stress state of site prestressed concrete port structure as reference. The free chloride ion content at different depths of prestressed concrete structures under the environment of marine atmospheric environment and artificial climate simulation is measured by experiment. This paper also analyzes the influence of the ratio of water-cement ratio, prestressed tendon tension and fly ash content on the chloride ion transport mechanism in concrete.It establishes the time similarity between the field test and the indoor chloride ion transport in the indoor accelerated test concrete, which is of great significance to the actual life prediction.

2.Time-Similarity Relation of Chloride Transport in Prestressed Concrete

2.1 Simulating Process and Similarity Analysis Based on Similarity Theory

Artificial simulation system and the actual system should have some form and a certain degree of similarity, so the establishment of artificial simulation system, the similar principles and similar methods should be throughout the entire artificial simulation process, so as to make artificial simulation system more accurate, efficient and Credible[3].

The study of chloride ion transport similarity in prestressed concrete structures is based on the similarity theory. Based on the similarity between the laboratory accelerated test and the on-site environmental erosion effect, the service life prediction and the durability design of the proposed structure are realized. For the service structure, based on similar environmental effects, similar load history, the same material composition, the establishment of artificial climate simulation accelerated erosion system, to carry out artificial climate simulation acceleration test. The apparent chloride ion diffusion coefficient is calculated by measuring, comparing and analyzing the chloride ion content in the field service structure and the artificial climate simulation acceleration test. The influence of water-cement ratio, prestressing tension and fly ash content on the diffusion of chloride ions is analyzed with the time attenuation coefficient.The similarity relation between concrete parameters and durability parameters is established by using the time similarity constants as the link, and the life prediction of the objects is carried out.

2.2 Time Similarity Constant

Chloride ion diffusion coefficient is an important indicator of the durability of concrete. Generally, the diffusion coefficient is often called the apparent diffusion coefficient of chloride ions, which is used to describe the diffusion rate of chloride ions under unsteady conditions. Its value in addition to the material properties of concrete itself, but also with the erosion medium solution chloride ion concentration, binding effect, age, etc. are also closely linked. The apparent chloride diffusion coefficient was adopted as equation (1). Where, $D_{RCM,,0}$ is the effective chlorine ion diffusion coefficient (m^2/s) measured by RCM at time t_0; k_r is the temperature influence coefficient; k_w is the influence coefficient of concrete water saturation w; n is the time decay coefficient.

$$D_a(t)=\frac{\int_{t_0}^{t}\left(D_{\mathrm{RCM,0}}(t)k_T k_w\left(\frac{t_0}{t}\right)^n\right)\mathrm{d}t}{t-t_0} \tag{1}$$

From Eq. (1), the apparent diffusion coefficient of the concrete structure decreases with the increase of the exposure time, and even decreases in the order of magnitude. The relationship can be described by Eq. (2). Where, $D_{a,i}$ (i=1,2) is the apparent diffusion coefficient (m^2/s) for the corresponding exposure time t_i (s), and n is the time decay coefficient.

IOP Conf. Series: Materials Science and Engineering **307** (2018) 012054 doi:10.1088/1757-899X/307/1/012054

$$D_{a,2} = D_{a,1}\left(\frac{t_1}{t_2}\right)^n \tag{2}$$

When the apparent chloride diffusivities of the concrete and the artificial climate simulation environment are all $D_{a,0}$, the erosion time is t ', t, the time similar constants is:

$$\lambda_t = \frac{t'}{t} \tag{3}$$

If erosion under the artificial climate simulation environment is known as 28d, the apparent chloride diffusion coefficient of the concrete is $D_{a,28}$. Then according to Eq. (2),we have:

$$t = 28 \cdot \sqrt[n]{D_{a,28}/D_{a,0}} \tag{4}$$

So, the time constant of chloride ion transport λ_t can be expressed by Eq. (5):

$$\lambda_t = \frac{t'}{28 \cdot \sqrt[n]{D_{a,28}/D_{a,0}}} \tag{5}$$

2.3 Time Decay Coefficient n

As shown in equation (1), n is the time decay coefficient. Professor Mangat[4] proposed the following empirical formula for the relationship of ordinary concrete between the time attenuation coefficient and water-cement ratio W/C:

$$n = 2.5(W/C) - 0.6 \tag{6}$$

Equation (7) was recommended as the empirical formula for the time attenuation coefficient by the American Society of Concrete Life-365, where, %FA is the percentage of fly ash in the cementitious material.

$$n = 0.2 + 0.4 \cdot \%FA/50 \tag{7}$$

Considering the effect of prestressed tension on the prestressed concrete members under the erosion of chloride salt, the influence function f and the fitting coefficient P are introduced to correct the time attenuation coefficient:

$$n = f(\eta) \cdot (2.5W/B - 0.6 + P \cdot \%FA) \tag{8}$$

Where η is the ratio of the effective tensile stress to the tensile strength of the prestressing tendon; $f(\eta)$ is the influence function of the prestressing level on the time decay coefficient; *W/B* is the water cement ratio; P is the fitting coefficient, dimensionless.

3.Test Scheme

3.1 Determination of Chloride Contentin for Service Structures

The concrete structure of a port in the east coast of China is selected in the field service structure. The port was built in 1974, for the prestressed reinforced concrete high-piled wharf.

The part of the service structure in Marine atmospheric zone was selected as the research object. The depth and content of chloride intrusion in structural concrete are detected by using the method of drilling and taking powder. Take the powder point distribution diagram in the beams, in a random manner, in the region are taken ten points are recorded as A0-A9. In this paper, the RCT method is used to detect the chloride ion content in the concrete powder.

3.2 Artificial Climate Simulation Accelerated Test

Prestressed concrete specimen size is 150 × 200 × 1500mm; specimen uses prestressed steel side of the configuration, using post-tensioning method. The material composition and structural features of the specimen are basically the same as those of the on-site service structural members. The specimen number is shown in Table 1.

The experiments are conducted in an artificial climate simulation laboratory. The concentration of salt solution used in the salt fog is 5% according to the environmental conditions of the port. The temperature is based on the monthly average temperature of the port environment. The time scale of the salt spray test procedure is 1: 1 for the spray time and drying time. Every 24 hours for a cycle process.After the samples are put into the artificial climate simulation laboratory, the samples are collected at the top of the specimen after 30 days, 90 days, 180 days, 270 days, and 360 days after several times of salt spray - drying and wetting cycles. The method of powder sampling is consistent with the field test method.

Table 1. Concrete member number for artificial climate simulation test

number	water cement ratio (W/B)	prestress level	amount of fly ash (%)	number	water cement ratio (W/B)	prestress level	amount of fly ash (%)
A-a- I	0.349	$0.5f_{ptk}$	0	B-a- I	0.412	$0.5f_{ptk}$	0
A-a- II	0.349	$0.5f_{ptk}$	15	B-a- II	0.412	$0.5f_{ptk}$	15
A-b- I	0.349	$0.7f_{ptk}$	0	B-b- I	0.412	$0.7f_{ptk}$	0
				B-b- II	0.412	$0.7f_{ptk}$	15

4. Test Results and Analysis

4.1 Result and Analysis of Chloride Ion Content in Service Structure Concrete

The results of the chloride ion content of marine concrete structures in the marine atmosphere are shown in Fig. 1, in which the thick solid lines are multi-curve averages. The test results show that, at the depth of 50mm from the concrete surface, the chloride ion content in the range of 0.13% -0.30%. It is assumed that the concrete material is an isotropic homogeneous material, regardless of the chemical bonding and physical adsorption between the chloride ion and the concrete material, and the pore of the concrete is always saturated and non-flowing, the chloride diffusion behavior Can be expressed in Fick's second law. Fick's Second Law The analytical solution of a constant boundary condition can be expressed as[5]:

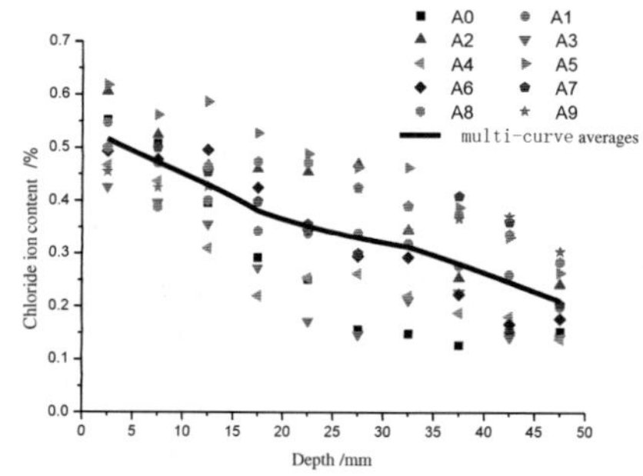

Fig. 1 Chloride ion content in the marine atmosphere

$$C(x,t) = C_0 + (C_s - C_0)\left[1 - erf\left(\frac{x}{2\sqrt{Dt}}\right)\right] \quad (9)$$

Where, t is the exposure time(s); x is the distance from the concrete surface (mm); C is the content of chloride ion in concrete (%); C_s is the chloride ion mass fraction (%) of the surface concrete in contact with the environment (%);C_0 is the initial chloride ion content in concrete (%);D is the apparent diffusion coefficient of chloride ion (mm²/s).

Based on Eq. (9), according to the measured values of chloride ion content in each point, the apparent point of diffusion coefficient of chloride ion in concrete can be calculated as shown in Table 2, where R^2 is the coefficient of determination. The average chloride ion diffusion coefficient in the region D_0=2.3720×10^{-6}mm²/s.

Table 2. Chloride ion diffusion coefficient in the marine atmosphere

number	D(×10^{-6}mm²/s)	R^2	number	D(×10^{-6}mm²/s)	R^2
A0	0.7468	0.9260	A5	3.5950	0.9307
A1	2.5880	0.9199	A6	1.7110	0.9470

A2	2.2910	0.9516	A7	4.4190	0.7858
A3	1.5580	0.9781	A8	4.9090	0.9086
A4	1.5770	0.9404	A9	5.3400	0.6150

4.2 Experimental Result and Analysis of Accelerated Test Component by Artificial Simulation

The results of chlorine ion content in the accelerated test component show that: (1) The erosion of chloride ion is more serious with time; (2) For the same erosion time, the chloride content at the same depth of the prestressing tendon is $0.7f_{ptk}$ higher than that of the beam with tension of $0.5f_{ptk}$; (3) When the size of prestressing tendon is the same, the incorporation of fly ash can effectively improve the resistance of concrete to chloride ion erosion .The apparent chloride diffusion coefficients D (t) of 12 prestressed concrete beams are calculated as shown in Fig. 2, according to the measured chloride ion content.

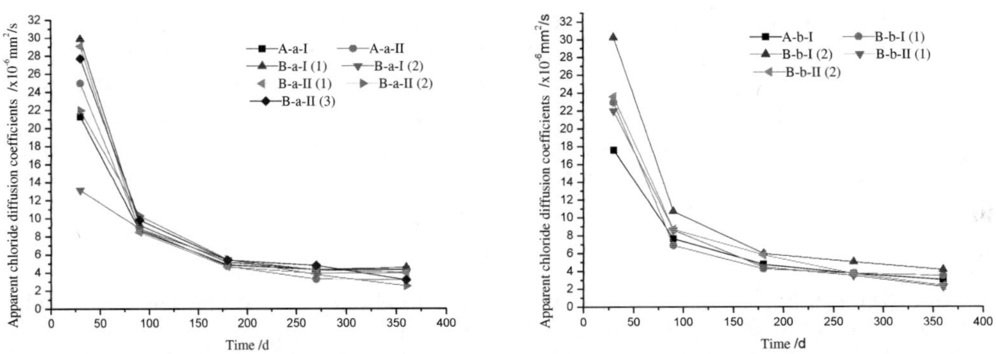

(a) prestressing tendon $0.7f_{ptk}$ (b) prestressing tendon $0.5f_{ptk}$

Fig. 2 Apparent chloride diffusion coefficients D (t) of 12 prestressed concrete beams

According to the formula (2), D_{28} and time attenuation coefficient n of concrete samples with different mix proportions and prestressing levels are curve fitted. The results show that the time attenuation coefficient n should be related to three parts: (1) the decay coefficient increases with the increase of water-binder ratio; (2) the addition of fly ash has physical water-reducing effect; (3) the level of pre-stress increases, the chloride ion diffusion slows down, and the decay cycle increase. By fitting the test data multiple times, the error is minimized when the time attenuation coefficient is linearly related to the square root of the prestress level. After the simplified treatment, the time attenuation coefficient of the prestressed concrete member in the atmosphere area can be expressed as:

$$n = 3\sqrt{\eta}\cdot\left(2.5W/B - 0.6 + 0.45\cdot\%FA\right) \quad (10)$$

According to Eq.(5)、(10) , we can have chloride ion transport time similarity:

$$\lambda_t = \frac{t'}{28\cdot\sqrt[3\sqrt{\eta}\cdot(2.5W/B-0.6+0.45\cdot\%FA)]{D_{a,28}/D_{a,0}}} \quad (11)$$

4.3 Time Similarity Relation Verification

In order to verify the similarity of chloride ion transport time, three trabeculae,B-a-Ⅱ(1)、(2)、(3), which are similar to the material and stress conditions, are taken as the members of the artificial climate simulation accelerated test.

From the test results of the chloride content of the field components, we have D_0=2.3720×10^{-6}mm^2/s, T=14400d. According to Eq. (2), when the apparent chloride diffusion coefficient is D_0, the erosion time is 559d, 414d and 532d respectively. Using the formula (3) and formula (11) we have the experimental value $\lambda_{t,1}$, the theoretical value$\lambda_{t,2}$ and the error shown in Table 3.

Table 3. Chloride ion transport time similarity

number	experimental value $\lambda_{t,1}$	Theoretical value $\lambda_{t,2}$	error（%）
B-a- II (1)	25.76	28.93	12.31
B-a- II (2)	34.78	34.60	-0.52
B-a- II (3)	27.07	28.48	5.21

5.Conclusions

In this paper, a pre-stressed concrete structure in the oceanic atmosphere is taken as a reference, and the artificial climate simulation acceleration test is carried out on the basis of its environment, material and prestress level. The chloride ion content of the concrete specimen in two environments is tested, The influence of prestressing level on the diffusion of chloride ions in concrete is studied. The influence of prestressing level on the chloride diffusion in concrete is also discussed. The experimental results show that the time similarity coefficient is less when the time attenuation coefficient is linear with the square root of the prestressing force. The time similarity relation of the chloride ion transport of the prestressed concrete in the marine atmosphere is analyzed by the measured value of the service structure The results are in good agreement with the experimental results, indicating that the expression is of practical significance for the life prediction of the existing marine prestressed concrete structures.

References

[1] Jin Libin. 2008. Multi-environmental time similarity(METS) theory and its application in coastal concrete struetural durability(in Chinese, dissertation). Hangzhou: Zhejiang University.

[2] Zhiwu Yu, Ying Chen, Peng Liu & Weilun Wang, 2015. Accelerated simulation of chloride ingress into concrete under drying-wetting alternation condition chloride environment. *Construction and Building Materials*. 93 (2015): 205-213

[3] Zhou Meili. 2004.*Similarity science*. Beijing: Science Press.

[4] Mangat P S, Molly B T. 1994. Prediction of long term chloride concentration in concrete. *Materials and Structures*, 1994, 27(6): 338-346

[5] JIN Weiliang，ZHANG Yi，LU Zhenyong. 2008. Mechanism and mathematic modeling of chloride permeation in concrete under unsaturated state. *Journal of the Chinese Ceramic Society*, 2008,36(10):1362-1369

Acknowledgements

The authors appreciate the support of the National Science Foundation of China (51778272), Research and innovation plan of postgraduate research of ordinary university in jiangsu province (CXZZ12-0655), and science and technology project of Jiangsu province construction system (2015ZD47).

IOP Conf. Series: Materials Science and Engineering **307** (2018) 012055 doi:10.1088/1757-899X/307/1/012055

Determination on Damage Mechanism of the Planet Gear of Heavy Vehicle Final Drive

RD Ramdan[1,*], R Setiawan[1], F Sasmita[1], R Suratman[1], Taufiqulloh[2]

T[1]Faculty of Mechanical and Aerospace Engineering, InstitutTeknologi Bandung, Bandung, Indonesia
[2]Komatsu Indonesia, Jakarta,Indonesia

*dadan@material.itb.ac.id

Abstract. The works focus on the investigation of damage mechanism of fractured in the form of spalling of the planet gears from the final drive assembly of 160-ton heavy vehicles. The objective of this work is to clearly understand the mechanism of damage. The work is the first stage of the on-going research on the remaining life estimation of such gears. The understanding of the damage mechanism is critical in order to provide accurate estimate of the gear's remaining life with observed initial damage. The analysis was performed based on the metallurgy laboratory works, including visual observation, macro-micro fractography by optical stereo and optical microscope and micro-vickers hardness test. From visual observation it was observed pitting that form lining defect at common position, which is at gear flank position. From spalling sample it was observed ratchet mark at the boundary between macro pitting and the edge of fractured parts. Further observation on the cross-section of the samples by optical microscope confirm that initial micro pitting occur without spalling of the case hardened surface. Spalling occur when pitting achieve certain critical size, and occur at multiple initiation site of crack propagation. From the present research it was concluded that pitting was resulted due to repeated contact fatigue. In addition, development of micro to macro pitting as well as spalling occur at certain direction towards the top of the gear teeth.

1. Introduction

Spalling is a type of damage, which involving detachment of small to medium parts of component. This type of damage is predicted related with common damage mechanism of gear component that works as mating parts and susceptible to various fatigue failure [1-3]. This type of failure can be the extension of other premature surface defect, namely pitting [4], and can also be accompanied and become worst with the presence of corrosion [5]. Gear component commonly manufactured in the form of case-hardened part [6-9], that offer high hardness at the surface and toughness at the inner parts. The first condition is required since gear is commonly used as mating parts where contact fatigue cannot be avoided. In addition the inner parts of the gear should has sufficient toughness in order to avoid brittle fractures once crack present in the component. Regarding this requirement gear component is preferably made from low carbon steel that having sufficient toughness. On the other hand, carburizing, which is thermo-chemical treatment is normally applied on the low carbon steel gear, in order to diffuse carbon element into the steel surface and forming harder phase in the surface.

Investigation on this type of damage is important, and at least there are three points can be the objective of such investigation. The first is to understand the mechanism of the damage, second is to avoid the similar failure in the future and to estimate the remaining life of gears with pitting defect before finally spalling occur on the component. By understanding the mechanism of damage it can be followed with appropriate treatment on the manufacturing of the component therefore similar failure can be avoided in the future. On the other hand, by the ability to estimate the life of gear component with pitting defect, it will significantly help in making appropriate preventive maintenance. All of these benefits will also directly contribute in the cost saving of vehicle maintenance and services.

One method of works that has been widely implemented as well as developed is failure analysis works that commonly used to investigate various failures in the various industries. Different with root cause analysis that considers the human error as the probable cause of failure, failure analysis is merely count on the physical aspect of failure. It can be due to the quality of material [10], inappropriate services or manufacturing. The investigation was carried out based on metallurgical failure analysis works, as will be explained subsequently.

2. Results and Discussion

There are several works have been done in order to obtain the objective of the present research. The first step is macrofractography by using DLSR camera with macro lense, in order to obtain the area where pitting as well as spalling defects occur. Further detail on macrofractography was performed by optical stereo microscope. In addition, in order to obtain cross-section data especially the case hardening profile of the gears, macro-structure characterization were performed by optical stereo microscope characterization. This characterization was performed both on the spalling section and the normal section. In order to support this characterization, micro-hardness test was also performed on the cros-section parts of the gears samples and the description of the data is supported with microstructure data of the corresponding section by optical microscope characaterization. Cross-section observation was also performed by optical stereo microscope in order to obtain the size and morphology of the pitting. By evaluating all of these data, damage mechanism that causing spalling on the gears samples can be determined.

3. Research Methodology

The Figure 1 shows two gear samples with initial pitting and spalling defect respectively. It can be seen from Figure 1(a) that commonly initial pitting exists at certain position on the gears and forming line pattern. It should be noted that pitting terminology here is not representing pitting defect resulted from corrosion process. However it is resulted from mechanical works between two mating parts. From observation of various gears sample with pitting defects, it was observed that typical position of the defect mainly occur at 1/3 distance from the lower section of the tooth (gear flank position). On the other hand Figure 1(b) shows uncommon failure of the gear that experience spalling due to uncommon incident as well.

Figure 2 shows one closer view of figure 1(b) which showing the boundary between the spalling parts and the pitting parts. It can be observed that ratchet marks appear at the boundary of the spalling parts. These marks appears near big pitting (resulted from coalescence of several smaller pitting) and indicates that spalling initiate from multiple site close to big pitting. The edge of pitting is having high stress concentration due to sharp edges and it is predicted that there is critical size of pitting that finally induce in the bigger fracture of the parts (spalling case). Other interesting features are fine and coarse ridges on the crack surface. It can be seen from Fig. 3 that rupture crack propagated by forming fine ridges before coarse ridges occurs at the final stage of fracture. On the further observation, the

IOP Conf. Series: Materials Science and Engineering **307** (2018) 012055 doi:10.1088/1757-899X/307/1/012055

ridges texture is found to be appeared on both macro pitting and rupture surface which is the characteristic of fatigue fracture (Fig.4).

(a) (b)

Figure 1. Two gear samples with initial pitting (a) andfractured failed (b)

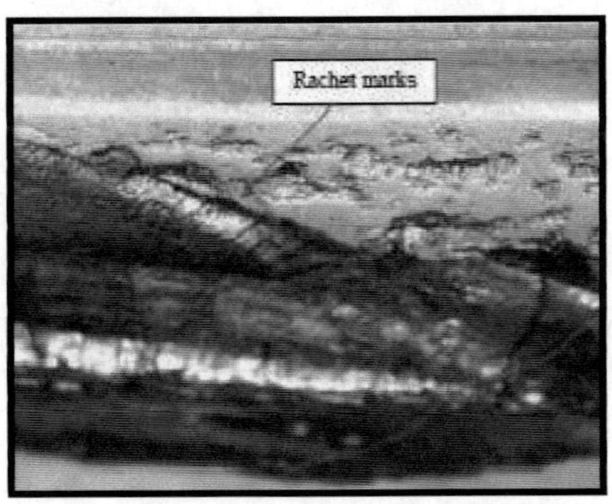

Figure 2.Rachet marks on fracture surface

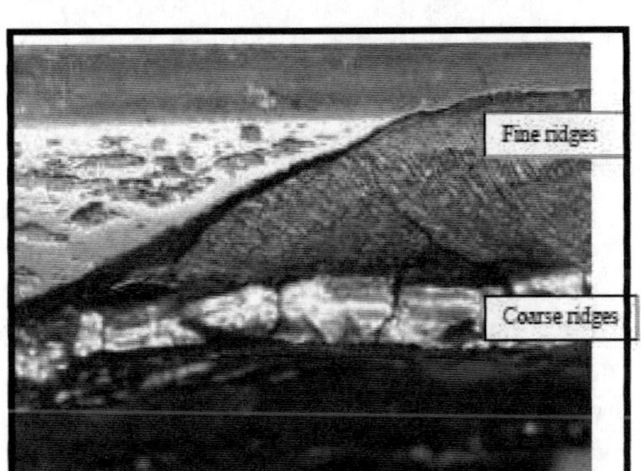

Figure3. Fine and coarse ridges on fracture surface

IOP Conf. Series: Materials Science and Engineering **307** (2018) 012055 doi:10.1088/1757-899X/307/1/012055

Figure 4. Ridges pattern on macro pitting and rupture crack surface

In addition, samples inspection at cross section area were conducted in several processes which are macro case-hardened profile, micro-structure characterization and microhardness test. Case-hardened profile inspection was conducted in order to identify and compare the cross section between gear tooth which has micro pitting and another gear tooth that has been fractured. Figure 5 shows gear teeth that having micro pitting (left figure) and the fractured gear (right figure). From the left figure it can be seen that the gear teeth has homogeneous and compact case-hardened surface that indicate pitting does not preceded with the breaking of case-hardened surface. On the other hand, cross section figure of the fractured gear tooth, shows that the case-hardened surface has been degraded with rupture tend to proceed toward the top of the gear. Refer to the position of the initial pitting which is at the bottom of the gear (flank position), it is predicted that the crack propagation starting from this position up to the top of the gear teeth.

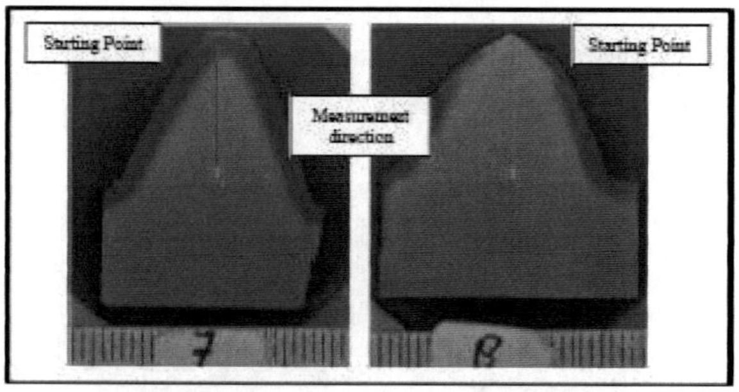

Figure 5. Cross section figure of initial pitting sample (left) and fractured sample (right)

Figure 6 shows cross section microstructure of initial pitting and fractured gear teeth shown in the Figure 5. This characterization was performed in order to observe any microstructure degradation after operation. From these figures it can be seen that both of gear tooth contains the same phase which is TemperedMartensite. However there is a distinct phase morphology between them. The left figure contains surface structure that finer than its core structure, while the right figure for the fractured gear, the case-hardened surface has been diminished and the surface structure become similar with its core.

In addition, micro-hardness profiles were evaluated in order to emphasize the distinction between two different cross section profile especially on its mechanical properties. The inspection was conducted by micro-vickers hardness test and 7 hardness (HV) values points starting from gear teeth's top edge into its core with distance 0.5 mm between every measurement point were performed. Table 1 shows micro-hardness test data, and it can be seen that surface hardness of initial pitting gear teeth is significantly higher than its core value. On the other hand, hardness value of fractured gear tooth is relatively similar between both surface and core area. This phenomenon confirms microstructure data in the Figure 6, and signify the differences between gear tooth that only contain micro pitting and spalling gear tooth. One point to be noted from the above data is the initiation of pitting is formed without the breaking of case hardened surface. Spalling or fracture of the gear surface initiate when pitting rearch certain critical size, and this crack propagation initiate from the multiple site of critical pitting.

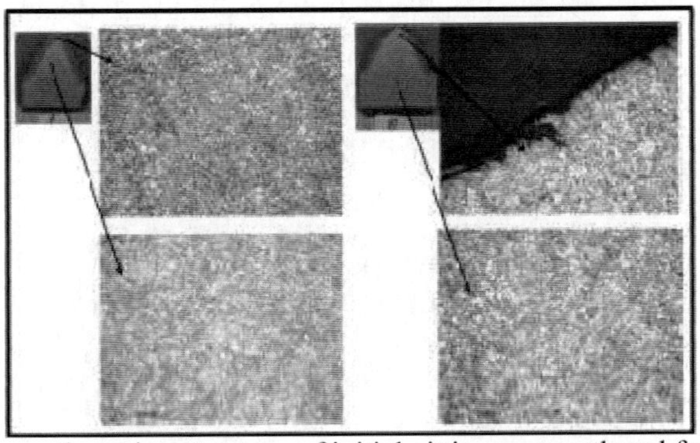

Figure 6. Cros section microstructure of initial pitting gear teeth and fractured gear teeth

Table 1 Micro-Hardness Value of initial pitting and Fractured Gear Tooth

Sample	Hardness (HV)						
	1	2	3	4	5	6	7
Pitting sample	727	673	524	475	475	460	407
Spalling sample	439	439	446	453	432	439	378

Figure 7 shows cross-section surface texture of initial pitting gears by optical microscope. It can be observed uneven texture with the appearance of sharp high peaks and valley texture. According to Moorthy [11] this peak and valley texture can be atrigger for the the initiation of micro-crack and micro-pitting. Therefore up to this point, we have obtain two condition for the initiation of pitting, occur at the gear plank position and preferably on the coarse surface with sharp peak and valley texture.

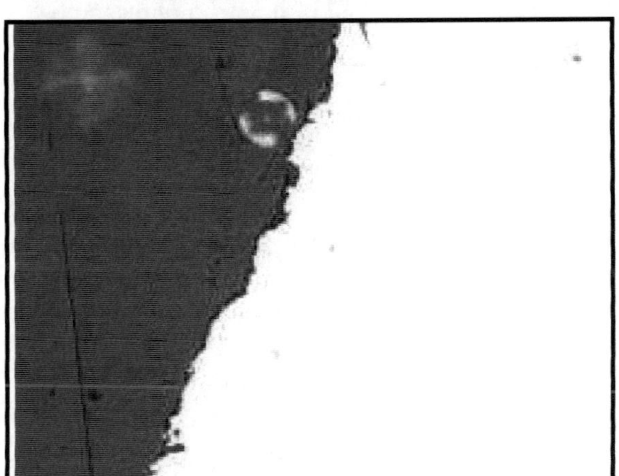

Figure 7. Surface Texture of Gear Teeth with initial pitting

IOP Conf. Series: Materials Science and Engineering **307** (2018) 012055 doi:10.1088/1757-899X/307/1/012055

In addition, Figure 8 shows the appearance of deep micro pitting and coarse crater defects. Both of these defects are predicted resulted from the different modes of pitting growth. The deep micro pitting is predicted resulted from single micro pitting that grows into sharp deep pitting. The other one is predicted resulted from the coalescence of more than one pitting and creating coarse crater. Furhermore, micro pitting will growinto its preferential direction which is against sliding direction (Moorthy, 2013) and for this case is toward the top of gear tooth.

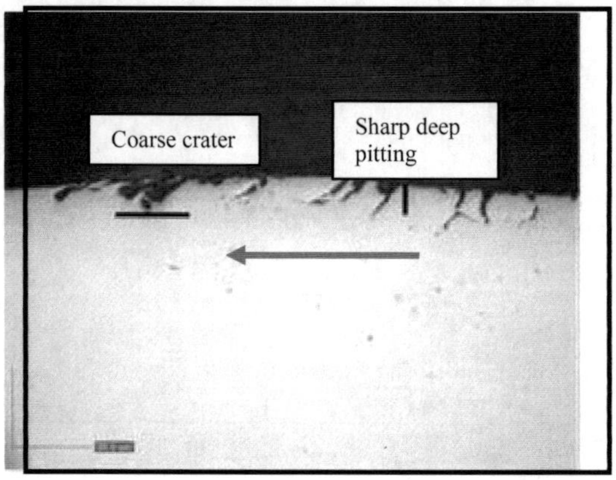

Fig. 8 Prediction of pitting growth direction

3.1. Damage Mechanism

As previously said that the damage initiated due to peak and valley texture of gear tooth's surface which then become micro-crack or micro-pitting, and the mechanism of defect is repeated contact fatigue, as can be predicted from the position and morphology of the defects. Micro-pitting then will grow within two modes whether sharp deep pitting (single) or coarse crater. Furthermore, micro-pitting will continue grow by contact fatigue mechanism. The coalescence of multiple micro-pitting will produce macro-pitting. Macro-pitting then became initiation site of fatigue cracks. The cracks on the gears itself generated from multiple initiation site, based on observation of ratchet marks at fracture surface. Cracks then propagated towards gear's top section by forming fine ridges until certain cross-section area before cracks propagates into coarse ridges at the final stage of fracture. Figure 9 show the scheme of damage mechanism observed in the present research.

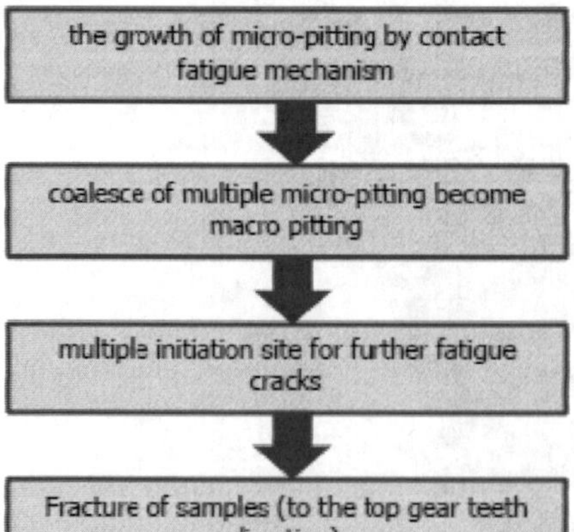

Fig. 9 Scheme of Damage Mechanism from Micro-Pitting to Rupture

4. Conclusion

From the investigation, the following damage mechanism has been concluded.

a. It was observed gears samples with pitting and fractured/ spalling defects condition
b. Pitting was resulted from repeated contact fatigue of the mating gear
c. Progressive macropitting causing spalling (pits coalesce and form irregular craters over a large area)
d. Coalesce macro pittings can be the sites for multiple initiation sites for further fatigues

The result of this investigation is used to develop the in-situ methodology of the remaining life estimate of the gear, which will be reported elsewhere.

Acknowledgment

Faculty of mechanical and aerospace engineering ITB and Komatsu Ltd, for the full support of the present research.

References

[1] P. Kumar, H. Hirani, A.Agrawal, "Fatigue failure prediction in spur gear pair using AGMA approach", Mater. Today: Proceed., 4(2) A, 2470-2477 (2017)
[2] E. Conrado, C. Gorla, P. Davoli, M. Boniardi, "A comparison of bending fatigue strength of carburized and nitrided gears for industrial applications", Engineering Failure Analysis, 78, 41-54 (2017).
[3] A. Terrin, C. Dengo, G. Meneghetti, "Experimental analysis of contact fatigue damage in case hardened gears for off-highway axles", Engineering Failure Analysis, 76,10-26 (2017)
[4] S. Glodez, Z. Ren, J. Flasker, "Surface fatigue of gear teeth flanks", Comput. Struct. 73, 475–483(1999)
[5] A.K. Jha, V. Diwakar, "Metallurgical analysis of failed gear", Eng. Fail. Anal. 9, 359–365 (2002)
[6] A. Terrin,C. Dengo, G. Meneghetti, "Experimental analysis of contact fatigue damage in case hardened gears for off-highway axles", Engineering Failure Analysis, 76, 10-26 (2017)
[7] I. Boiadjiev,J. Witzig,T. Tobie, K., "Tooth flank fracture – basic principles and calculation model for a sub surface initiated fatigue failure mode of case hardened gears", International Gear Conference, 670-680 (2014)
[8] Selçuk. C.Y.l, Levent. Ö.T., Gençol. E.A., "Flywheel starter ring gear failures and hardness variation reduction in surface hardening process", Case Studies in Engineering Failure Analysis, 4, 8-19 (2015)
[9] Bonglae J., Shahriar S., Yongbo S., Ali .F., "Cyclic deformation and fatigue behavior of carburized automotive gear steel and predictions including multiaxial stress states", International Journal of Fatigue, 100(2), 454-465 (2017)
[10] A. Lanzuttia, A. Gagliardia,b, A. Raffaellia, M. Simonatob, R. Furlanettob, M. Magnana, F. Andreattaa, L. Fedrizzia, "Failure analysis of gears, shafts and keys of centrifugal washers failed during life tes", Engineering Failure Analysis, 79, 634–641(2017)
[11] V. Moorthy, B.A. Shaw, "An observation on the initiation of micro-pitting damage in as-ground and coated gears during contact fatigue", Wear, 297 (1), 878-884(2013)

IOP Conf. Series: Materials Science and Engineering **307** (2018) 012056 doi:10.1088/1757-899X/307/1/012056

Metallurgy and mechanical properties variation with heat input,during dissimilar metal welding between stainless and carbon steel

RD Ramdan[1,*], AL Koswara[2], Surasno[3], R Wirawan[4], FFaturohman[1], B Widyanto[1],& R Suratman[1]

[1]Faculty of Mechanical and Aerospace Engineering, InstitutTeknologi Bandung, Jl. Ganesa no 10, Bandung, 40132, Indonesia
[2]CMechanical Engineering Department, UniversitasSanggaBuana, Jl. PHH.Mustofa no 68, Bandung, West Javaa 40124
[3]Center for Material and Technical Products, Jl. Sangkuriang 14 Bandung, West Java 40135
[4]Faculty of Engineering, UniversitasNegeri Jakarta, Jl. RawamangunMuka, Jakarta Timur, 13220, Indonesia

dadan@material.itb.ac.id

Abstract. The present research focus on the metallurgy and mechanical aspect of dissimilar metal welding.One of the common parameters that significantly contribute to the metallurgical aspect on the metal during welding is heat input. Regarding this point, in the present research, voltage, current and the welding speed has been varied in order to observe the effect of heat input on the metallurgical and mechanical aspect of both welded metals. Welding was conducted by Gas Metal Arc Welding (GMAW) on stainless and carbon steel with filler metal of ER 309. After welding, hardness test (micro-Vickers), tensile test, macro and micro-structure characterization and Energy Dispersive Spectroscopy (EDS) characterization were performed. It was observed no brittle martensite observed at HAZ of carbon steel, whereas sensitization was observed at the HAZ of stainless steel for all heat input variation at the present research. Generally, both HAZ at carbon steel and stainless steel did not affect tensile test result, however the formation of chromium carbide at the grain boundary of HAZ structure (sensitization) of stainless steel, indicate that better process and control of welding is required for dissimilar metal welding, especially to overcome this issue.

1. Introduction

Dissimilar metal welding can be found in various industries, such as automotive industry, air craft, electric generator, oil and gas, etc. Practically, dissimilar metal welding involving various metals such as between carbon steel and stainless steel [1], TWIP and TRIP steel [3], steel with high strength to weight ratio [6], aluminum and stainless steel [7], duplex and austenitic stainless steel [11], etc.

Apart of its wide applications, there are several remaining problems that need to be solved. Among these problems including unmixed composition at the fusion boundary [1], different microstructure characteristic at the weld interface of different metal [2], brittleness [3], macro segregation [8], corrosion [11], etc. Therefore several works need to be performed such as study on fluid flow of metal melting and element transport process during dissimilar metal welding [1,10]. Other matters that

IOP Conf. Series: Materials Science and Engineering **307** (2018) 012056 doi:10.1088/1757-899X/307/1/012056

become concern are mixing characteristic of two different metals during welding including microstructure, composition, hardness at the fusion [1,2,3,10], tensile strength [3], corrosion [11], effect of heat treatment before and after welding [6], and heat input [2,3,4,12].

Regarding heat input, welding parameter that related with this value is current, voltage, welding speed [3] and applied current mode [4]. For the case of welding on stainless steel, variation on heat input might affect the thickness of carbide layer on the weld interface [2]. Heat input also can influence microstructure and the size of welding nugget and therefore influence the mechanical properties of the weld. Based on the above facts the present paper describes several characteristics on dissimilar metal welding between carbon steel and stainless steel using gas metal arc welding (GMAW) method, with heat input as the variable parameter.

2. Experimental Method

Carbon steel and stainless steel used as the based metals in the present research are ASTM A53 and AISI 304 respectively, whereas weld/filler metal used is AISI 308. Before the process, based metals were prepared by cleaning, grinding and polishing, therefore homogeneous surface can be obtained before the process.

Welding processes were performed by gas metal arc welding (GMAW) method. In order to establish variation on heat input, welding current and voltage were varied at 105-190A and 18.5-25V respectively. The welding speed was also varied at around 0.4mm/s up to around 1mm/s. This variation resulted in the variation of heat input from around 0.4 kj/mm up to 1kj/mm.

After welding, evaluation on the microstructure, macrostructure, composition profile at HAZ and fusion area, hardness and tensile strength were performed. Comparison of microstructure resulted at HAZ and fusion area at both different metals were performed based on these data.

3. Results and Discussion

Figure 1 shows typical microstructure of base metal carbon steel and its HAZ. It can be seen both figure showing pearlite structure, a structure in steel containing ferite (white feature) and cementite (dark feature) phases. However for the case of HAZ structure shows finer structure than the base metals. This fact also shows that the present welding processes did not create martensite phase, which normally induce in the brittleness at this area as well as corrosion problem. The HAZ area for carbon steel in the present research reaching more than 1.5 mm in width as can be seen in the Figure 2. Based on these data, it can be expected that HAZ of carbon steel were not negatively affected by the welding process.

On the other hand Figure 3 (a) shows typical microstructure obtained at the base of stainless steel and its HAZ area. It can be seen the base area showing austenite phases with twinning structures, typical microstructure for austenitic stainless steel. In addition Figure 3(b) shows typical HAZ microstructrure of stainless steel obtained in the present research and it was observed that sensitization occur on this steel. The appearance of this microstructure is due to the formation of chromium carbide at the grain boundary of austenite at temperature between 500-850°C, leaving the area around the grain boundary poor with chromium and therefore susceptible to corrosion attack. This microstructure is appeared as darker grain boundary as compared with common austenite phase as can be seen in the figure. The area of sensitization at HAZ area of stainless steel reaching within 100-150□m. Therefore the HAZ area is significantly narrower than carbon steel, however for the carbon steel with no appearance of harmfull phase, whereas for the case of stainless steel with the appearance of sensitization.

IOP Conf. Series: Materials Science and Engineering **307** (2018) 012056 doi:10.1088/1757-899X/307/1/012056

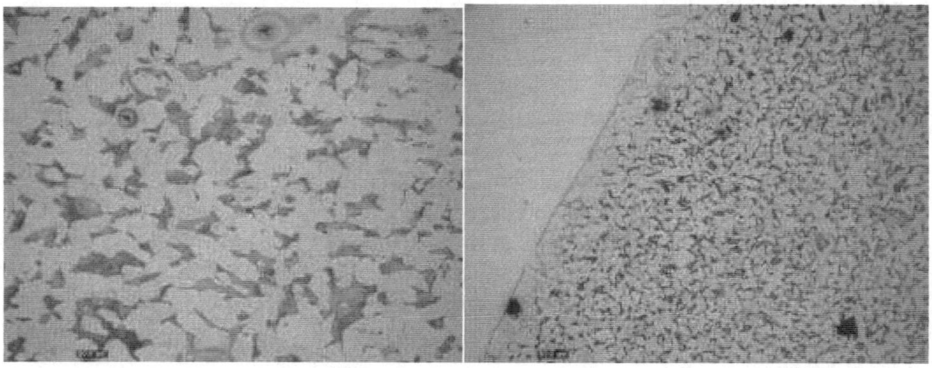

(a) (b)

Figure 1 Microstructure of Based metal carbon steel (a) HAZ of carbon steel (b)

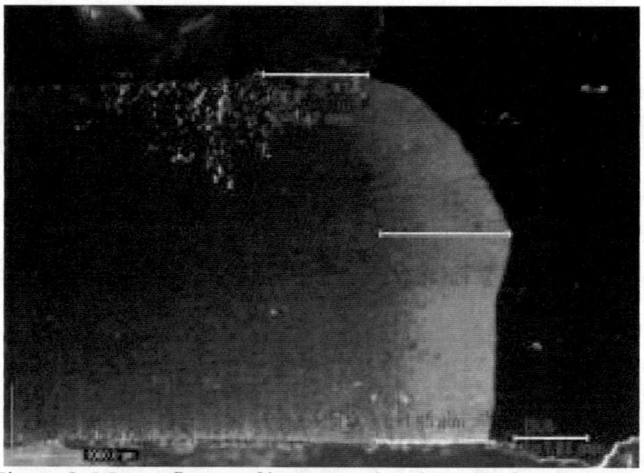

Figure 2. Macro figure of base metal and HAZ of carbon steel

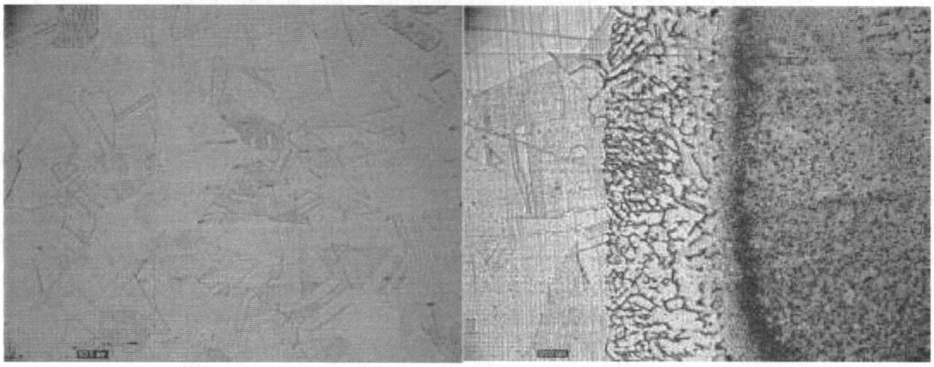

(a) (b)

Figure 3 Microstructure of based metal (a) stainless steel (b) and HAZ area

Figure 4 shows the effect of heat input on the tensile strength of welded samples. It can be seen that initially there is an increasing of tensile strength with increasing of heat input, however after heat input reaching around 1kJ/mm it tends to be stable. Another point to be noted is, none of the sample fracture at the weld or HAZ area, indicating that the samples have been well welded and microstructure resulted at the HAZ area did not cause in early or catasthropic failure during the test. From the

IOP Conf. Series: Materials Science and Engineering **307** (2018) 012056 doi:10.1088/1757-899X/307/1/012056

microstructure of HAZ of stainless steel it was observed the appearance of sensitized structure. However eventhough this structure might create brittleness as well as susceptible to corrosion, data from tensile test show that it did not significantly affect on the strength of the welded sample.

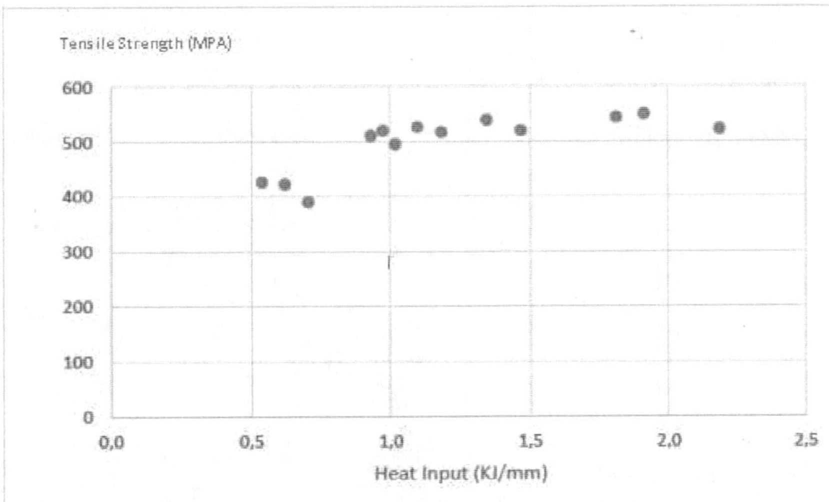

Figure 4 Effect of heat input on the tensile strength

Figure 5 shows hardness of various samples with different heat input. For each samples it can be seen that welded area showing the highest hardness, which is considered more due to different steel grade. Higher hardnesses were found at both HAZ of carbon steel and stainless steel as compared with their based area for all samples. For the case of carbon steel, eventhoughmartensite structures were not resulted at the HAZ area, however finer pearlite structure as can be seen in the Figure 1 causing higher hardness as compared with the based area. On the other hand, for stainless steel, higher hardness at the HAZ area are resulted from the finer structure as well as due to the formation chromium carbide at this area.

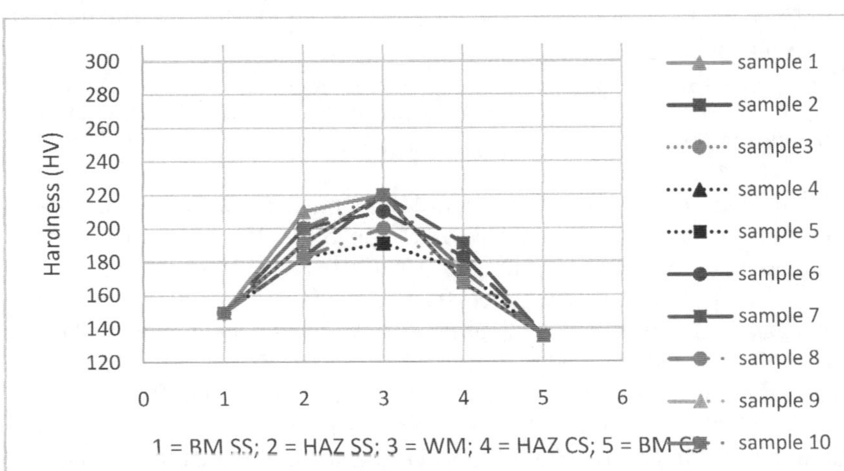

Figure 5 Hardness at based metal (BM), HAZ of stainless steel(SS) and carbon steel (CS), and weld metal (WM)

In addition Figure 6 shows common composition line data by energy dispersive spectroscopy (EDS) characterization method, obtained in the present research. It can be seen from Figure 6(a) gradation of elemental composition of Fe, Ni and Cr from the weld to the based metal of carbon steel. Since Ni and Cr were not observed in the single based metal characterization, therefore it can be concluded that there is significant diffusion of Ni and Cr from the weld to the carbon steel based metal. On the other hand, no significant diffusion was observed from based metal of stainless steel to the weld area and vice versa.

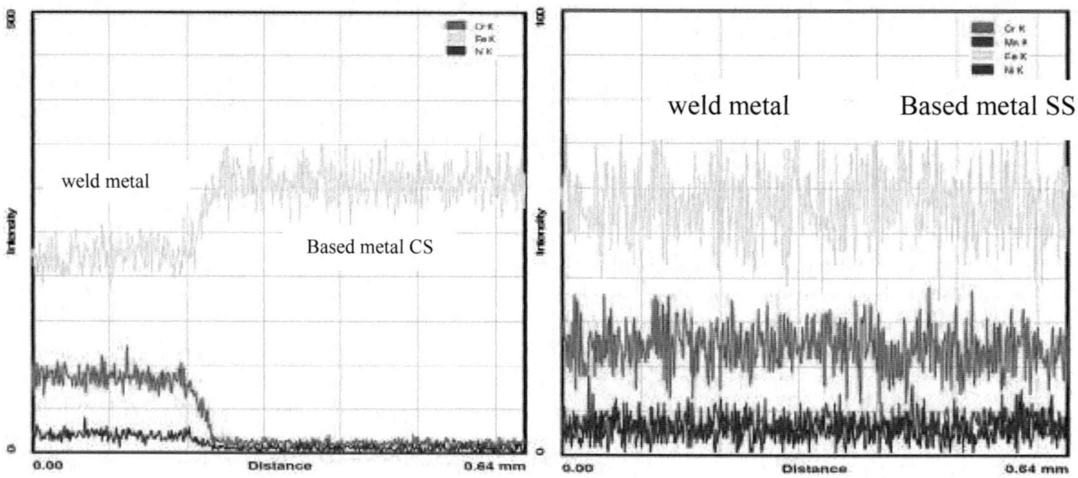

(a) (b)

Figure 6 Line EDS data of CS and weld area (a) ; SS and weld area (b)

4. Conclusions

Several characterizations have been performed on the dissimilar welding sample between carbon steel and stainless steel. Generally finer pearlite structure was obtained at the HAZ of carbon steel with no appearance of martensite phases, whereas sensitization was observed at the HAZ of stainless steel. Due to these structure both HAZ of carbon steel and stainless steel show higher hardness than their based metals. From tensile test data it was obtained that no negative effect from the HAZ and weld structure on the tensile test results. From the present resulted, it can be concluded that controlling welding process to avoid sensitization in the HAZ of stainless steel is more difficult than martensitic transformation in the HAZ of carbon steel. Therefore this condition still be remaining challenge to be solved for further research.

Acknowledgment

To the Faculty of Mechanical and Aerospace Engineering, InstitutTeknologi Bandung Indonesia, for supporting the present research.

References

[1] A. Bahrami et al.: Int. Journal of Heat and Mass Transfer. 93 (2016) pp. 729-741
[2] H. Ma et al.: Materials and Design. 86 (2015) pp. 587-597
[3] K. Májlinger et al.: Materials and Design. 109 (2016) pp. 615-621
[4] Y. Su et al.: Journal of Materials Processing Technology. 214 (2014) pp. 81-86
[5] K. Nandagopal, C. Kailasanathan: Journal of Alloys and Compounds. 682 (2016)
[6] P. Naveen Kumar et al.: Procedia Materials Science. 5 (2014) pp. 2382-2391
[7] V.I. Isaev et al.: International Journal of Heat and Mass Transfer. 99 (2016) pp. 711-720
[8] T. Soysal et al.: ActaMaterialia. 110 (2016) pp. 149-160
[9] A. Bahrami et al.: International Journal of Heat and Mass Transfer. 85 (2015) pp. 41-53
[10] W. Wang et al.: Materials Characterization. 107 (2015) pp. 255-261
[11] J. Verma, R.V. Taiwade: Journal of Manufacturing Processes. 24 (2016) pp. 1-10
[12] Y. Zhang et al.: Materials and Design. 106 (2016) pp. 235–246
[13] H. Ming et al.: Materials Science & Engineering. A 669 (2016) pp.279–290

IOP Conf. Series: Materials Science and Engineering **307** (2018) 012057 doi:10.1088/1757-899X/307/1/012057

The Influence of Shaping Air Pressure of Pneumatic Spray Gun

Wenzhuo Chen[1], Yan Chen[1], Haiwei Pan[1], Weiming Zhang[1] and Bo Li[1]

1 Petroleum Department, Army Logistics University, Chongqing, China
E-mail: yansohucom@sohu.com

Abstract. The shaping air pressure is a very important parameter in the application of pneumatic spray gun, and studying its influence on spray flow field and film thickness distribution has practical values. In this paper, Euler-Lagrangian method is adopted to describe the two-phase spray flow of pneumatic painting process, and the air flow fields, spray patterns and dynamic film thickness distributions were obtained with the help of the computational fluid dynamics code—ANSYS Fluent. Results show that with the increase of the shaping air pressure, the air phase flow field spreads in the plane perpendicular to the shaping air hole plane, the spray pattern becomes narrower and flatter, and the width of the dynamic film increases with the reduced maximum value of the film thickness. But the film thickness distribution seems to change little with the shaping air pressure decreasing from 0.6bar to 0.9bar.

1. Introduction

Robotic painting technology is widely applied in the manufacture of automobile, plane, furniture and many other products [1-3]. In order to optimize paint process, ideal painting parameters are needed, such as paint flow rate, atomization air pressure, shaping air pressure, painting distance, which are crucial to painting quality. Among these, shaping air jetting out from the shaping air holes on the horn-like structures of pneumatic paint guns affects the shape of spray pattern and the paint thickness distribution on workpiece: the cross section of the spray cone would turn elliptical with the opening of the shaping air hole during painting process, and major and minor axes of the elliptical cross section would be enlarged and reduced, respectively, with the increase of shaping air pressure. Therefore, the shaping air pressure is an important parameter that should be taken into consideration when planning the path of painting robot.

With the development of computer and computational fluid dynamics(CFD), a large number of researchers have studied the painting process through CFD methods. Hicks *et al.* [4], Fogliati *et al.* [5], Ye *et al.* [6-7], Toljic *et al.* [8] and Osman *et al.* [9] have researched the spray flow field of painting the flat surface using the painting mode established by the Euler-Lagrangian method. This method intends to describe the two-phase paint spray flow in two frames, with the air flow described in Euler frame and the motion of paint droplets described in Lagrangian frame. Chen Yan *et al.* [10] researched spray flow field of painting process on flat surface using the Euler-Euler method, which is a more general method of describing two-phase flows, both the phases described in Eulerian frame and with similar governing equations. However, few of them investigate the effects of shaping air pressure through CFD simulation.

Therefore, the paper aims to study the effects of shaping air pressure based on CFD simulation. A paint deposition model is established based on Euler—Lagrangian method in CFD theory. Simulations

IOP Conf. Series: Materials Science and Engineering **307** (2018) 012057 doi:10.1088/1757-899X/307/1/012057

for the cases with different shaping air pressures are carried out. The effects of shaping air pressure on the air flow field and paint thickness distribution are obtained and analysed.

2. Governing equations for paint deposition process

The flow field of paint deposition process can be regarded as a two-phase flow where the atomized liquids are transported under the action of high-speed air phase to the workpiece. Euler-Lagrangian method is adopted to describe the two-phase flow of painting process.

2.1. Basic governing equations for air phase

The description of the gas phase is obtained by solving the continuity and momentum equations coupled with a suitable closure model for turbulence.
Continuity equation for the air phase is

$$\frac{\partial \rho}{\partial t}+\nabla\cdot(\rho \boldsymbol{u})=0 \tag{1}$$

where, ρ is the gas density and $\boldsymbol{u}$ the gas velocity.
Momentum equation for the air phase is

$$\frac{\partial}{\partial t}(\rho \boldsymbol{u})+\nabla\cdot(\rho \boldsymbol{u}\boldsymbol{u})=-\nabla p+\nabla\cdot\left(\overline{\overline{\tau}}\right)+\rho \boldsymbol{g} \tag{2}$$

in which, p the pressure, $\overline{\overline{\tau}}$ viscous stress tensor, $\boldsymbol{g}$ the gravitational acceleration.
The equations (1) and (2) is closed by standard *k-ε* turbulence model:

$$\frac{\partial}{\partial t}(\rho k)+\nabla\cdot(\rho k\boldsymbol{u})=\nabla\left(\left(\mu+\frac{\mu_t}{\sigma_k}\right)\nabla k\right)+G_k-\rho\varepsilon \tag{3}$$

$$\frac{\partial}{\partial t}(\rho\varepsilon)+\nabla\cdot(\rho\varepsilon\boldsymbol{u})=\nabla\cdot\left(\left(\mu+\frac{\mu_t}{\sigma_\varepsilon}\right)\nabla\varepsilon\right)+C_{1\varepsilon}\frac{\varepsilon}{k}G_k-C_{2\varepsilon}\rho\frac{\varepsilon^2}{k} \tag{4}$$

2.2. Governing equations for the liquid phase

The liquid phase is treated by the Lagrangian approach, where the trajectory of the droplet can be calculated by solving the equation of motion for a single droplet according to equation (5).

$$\frac{d\boldsymbol{u}_p}{dt}=f_D(\boldsymbol{u}-\boldsymbol{u}_p)+\frac{\boldsymbol{g}(\rho_p-\rho)}{\rho_p} \tag{5}$$

in which, $f_D(\boldsymbol{u}-\boldsymbol{u}_p)$ is the unit mass drag force with a coefficient $f_D=\frac{18\mu}{\rho_p d_p^2}\frac{C_D \mathrm{Re}}{24}$, $\boldsymbol{u}_p$ the droplet velocity, $\boldsymbol{u}$ the instantaneous air velocity that can be calculated as the sum of the local mean velocity and a fluctuating velocity corresponding to the local turbulence degree, by using a stochastic tracking model.

2.3. Paint impingement and sticking model

When the paint liquid phase from the spray flow field impinge and stick onto the workpiece, its mass and momentum are removed from the spray flow, and added as source terms to the governing equations for the film on workpiece.
The conservation of mass for the film is:

$$\frac{\partial h}{\partial t}+\nabla\cdot(h\cdot\boldsymbol{V})=\frac{\dot{m}_s}{\rho_l} \tag{6}$$

$$\dot{m}_s=\alpha_d\rho_d V_{dn}A \tag{7}$$

where, ρ_l is the liquid density, *h* the film height, ∇_s the surface gradient operator, $\boldsymbol{V}$ the mean film velocity, $\dot{m}_s$ the mass source, where α_d is the volume fraction of the paint, ρ_d is the paint phase density, V_{dn} is the paint phase velocity normal to the wall surface, and *A* is the wall surface area.

The conservation of momentum for the film is:

$$\frac{\partial h\boldsymbol{V}}{\partial t}+\nabla_s\cdot(h\boldsymbol{V}\boldsymbol{V})=-\frac{h\nabla_s P_L}{\rho_l}+\frac{3}{2\rho_l}\boldsymbol{\tau}_{fs}-\frac{3v_l}{h}\boldsymbol{V}+\frac{\dot{q}_s}{\rho_l} \tag{8}$$

$$\dot{\boldsymbol{q}}_s=\dot{m}_s\boldsymbol{V}_d \tag{9}$$

The terms on the left-hand side of equation (7) represent transient and convection effects, respectively. On the right hand side, the first term includes the effects of gas-flow pressure, the gravity component normal to the wall surface (known as spreading), and surface tension; the second and third terms represent the net viscous shear force on the gas-film and film-wall interfaces, based on the quadratic film velocity profile representation; and the last term is the effect of the momentum source. In equation (8), the $\dot{\boldsymbol{q}}_s$ is the momentum source and $\boldsymbol{V}_d$ is the velocity vector of the paint liquid.

3. Simulation setup

The geometry of the pneumatic atomizer adopted in this study is shown as Figure 1. The paint hole with diameter 1.1 mm is coaxially surrounded with an annular air orifice which is an annular ring with outer diameter 2 mm and inner diameter 1.6 mm. On each side of the annular air orifice are two assisting air holes with diameter 0.6 mm. On each of the horn-like structures, there are two shaping air holes, which can jet shaping air so that spray cone with elliptic cross-section can be formed for painting larger workpieces.

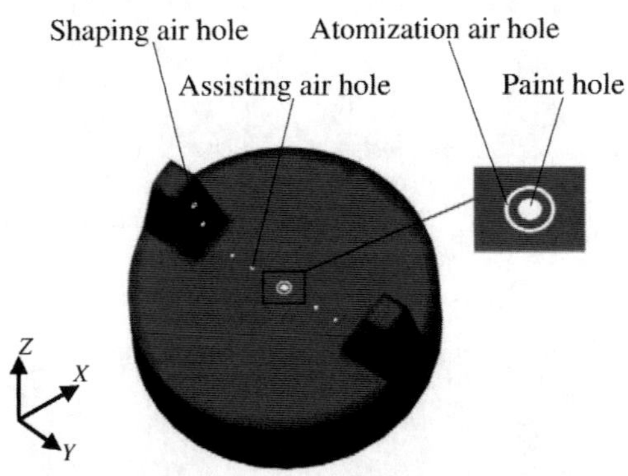

Figure 1. 3-D model of the spray gun

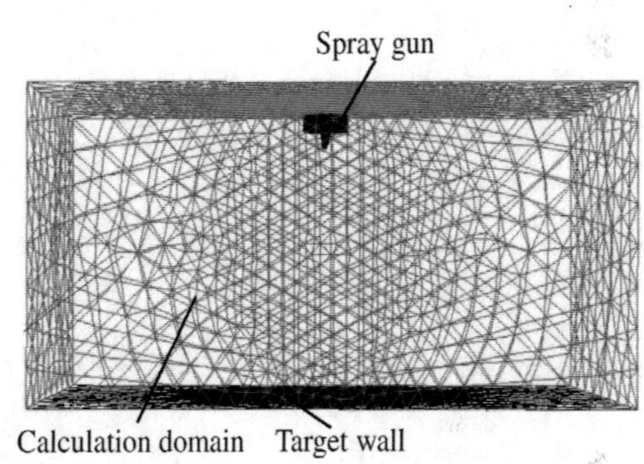

Figure 2. Mesh of the computational domain

The simulation cases in this study were carried out for painting a 300 mm×500 mm flat plate. For all the simulation cases, the axis of the atomizer was perpendicular to the target wall, and the paint hole was set at a distance of 180 mm from the wall. The calculation domain was built up as shown in Figure 2 and was meshed with 500,000 unstructured grids. 71 droplet injection sources were set uniformly in a 1.1 mm diameter-circular plane 2 mm downstream the paint hole. Atomization hole was set as the pressure inlet hole, the pressure of which was 1.2 bar. The shaping air pressures corresponding to the four cases in this study were set to 0 bar, 0.3 bar, 0.6 bar, 0.9 bar respectively. The simulation was carried out on the platform of ANSYS Fluent, where the model was discretized by the 2nd order upwind scheme and the discretized equations were solved by the SIMPLE algorithm. The time step was $\Delta t = 1\times10^{-4}$ s.

4. Simulation results and analysis

Figure 3 shows the contours of the air velocities of the four cases with different values of shaping air pressures. Only the flow fields in *XZ*-plane (i.e. the plane perpendicular to the shaping air hole plane) are presented, because they are larger and more obvious to measure than that in *YZ*-plane due to the effect of shaping air. The highest value of the air phase is around 512 m/s at the exit of the air nozzle, but the air velocity contours are depicted in the range of 0–50 m/s in order to provide reasonable shapes of the entire flow fields.

The spray patterns of the paint films on the flat plate obtained under different shaping air pressures are shown in Figure 4; these can also be called static film growth rate (μm/s), for the spray gun did not move in the simulation.

As is shown, with the increase of the shaping air pressure, the air phase flow field spreads in XZ-plane, and the spray pattern becomes narrower being flattened in *X* direction and extended in *Y* direction.

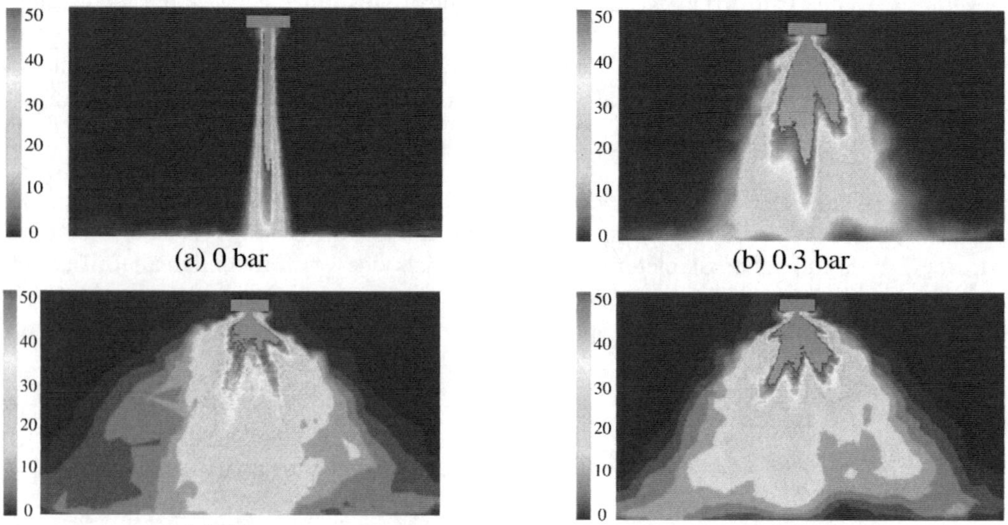

(a) 0 bar (b) 0.3 bar

(c) 0.6 bar (d) 0.9 bar

Figure 3. Air velocity contours (m/s) under different shaping air pressures

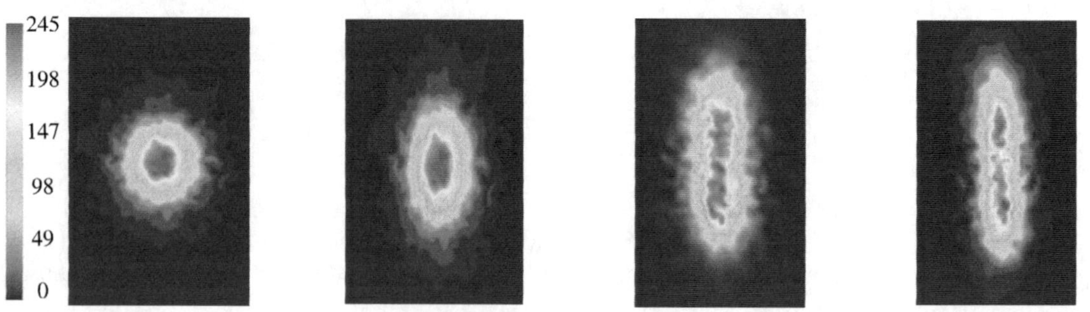

(a) 0 bar (b) 0.3 bar (c) 0.6 bar (d) 0.9 bar

Figure 4. Spray patterns or static paint deposition rates (μm/s) obtained by painting the flat plate with different shaping air pressures

In practical painting process, the spray gun is usually moved along the direction of the minor axis of spray pattern (i.e. *Y*-axis direction) to achieve high painting efficiency. Therefore, in order to better illustrate the influence of shaping air pressure, the dynamic film distributions are acquired as shown as Figure 5 by integrating the fitted static film growth rate along the *Y* axis for the movement of the spray gun at a linear speed of 180 mm/s.

It can be seen that as the shaping air increases from 0 bar to 0.9 bar, the width of the film increases from 100 mm to about 264 mm, and the maximum value of the film thickness decreases from around 50 μm to around 18 μm. It is also worth noting that the film thickness distribution seems to change little with the shaping air pressure decreasing from 0.6 bar to 0.9 bar.

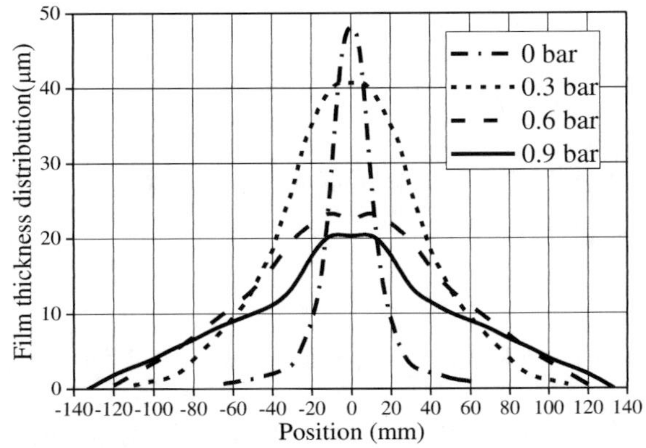

Figure 5. Dynamic film thickness distributions obtained under three different shaping air pressures

5. Conclusions

In order to analyse the influence of shaping air pressure, Euler-Lagrangian method has been used to build a model to describe the paint spray flow field based on CFD theory. The cases of painting a flat plate with four different shaping air pressures were simulated based on the model. Results were achieved on the commercial codes—ANSYS Fluent.

With the increase of the shaping air pressure, the air phase flow field spreads in *XZ*-plane, and the spray pattern becomes narrower being flattened in *X* direction and extended in *Y* direction. As the shaping air increases from 0 bar to 0.9 bar, the width of the dynamic film increases, and the maximum value of the film thickness decreases. But the film thickness distribution seems to change little with the shaping air pressure decreasing from 0.6 bar to 0.9 bar.

Acknowledgments

This project is supported by National Natural Science Foundation of China (Grant No. 51475469 and No. 51505494).

References

[1] Chen W, Chen Y, Chen K, Li B and Zhang W, Chen K 2016 *Industrial Robot: an International Journal* **43** pp 58-64.
[2] Chen Y, Chen W, Chen K and Li B 2016 *Journal of Robotics* **2016**
[3] Chen Y, Chen W, Li B, Zhang G and Zhang W 2017 *Industrial Robot: an International Journal* **44** pp 629-38
[4] Hicks P G and Senser D W 1995 *Journal of fluids engineering* **117** pp 713-19
[5] Fogliati M, Fontana D, Garbero M and Vanni M 2006 *Journal of Coating Technology and Research*, **3** pp 117-125
[6] Ye Q, Shen B, Tiedje O, Bauernhansl T and Domnick J 2015 *Atomization & Sprays* **25** pp 643-56
[7] Ye Q and Pulli K 2017 *Coatings* **7** 13
[8] Toljic N, Adamiak K, Castle G S P, Kuo H and Fan H 2013 *Journal of Electrostatics* **71** pp 299-304
[9] Osman H, Adamiak K, Castle P G S, Fan H and Simmer J 2015 *Proceedings of the 2015 International Mechanical Engineering Congress & Exposition (Houston)*
[10] Chen Y, He S, Zhang G, Chen W and Jiang J 2016 *Proceedings of the 4th International Conference on Machinery, Materials and Computing Technology (Hangzhou)*

IOP Conf. Series: Materials Science and Engineering **307** (2018) 012058 doi:10.1088/1757-899X/307/1/012058

Differential quadrature method of nonlinear bending of functionally graded beam

Xu Gangnian[1], Ma Liansheng[2], Wang Youzhi[1], Yuan Quan[1] and You Weijie[1]

1 School of Civil Engineering, Shandong University, Ji'nan 250061, China
2 School of Science, Lanzhou University of Technology, Lanzhou 730050, China
E-mail: wangyouzhi@sdu.edu.cn

Abstract. Using the third-order shear deflection beam theory (TBT), nonlinear bending of functionally graded (FG) beams composed with various amounts of ceramic and metal is analyzed utilizing the differential quadrature method (DQM). The properties of beam material are supposed to accord with the power law index along to thickness. First, according to the principle of stationary potential energy, the partial differential control formulae of the FG beams subjected to a distributed lateral force are derived. To obtain numerical results of the nonlinear bending, non-dimensional boundary conditions and control formulae are dispersed by applying the DQM. To verify the present solution, several examples are analyzed for nonlinear bending of homogeneous beams with various edges. A minute parametric research is in progress about the effect of the law index, transverse shear deformation, distributed lateral force and boundary conditions.

1. Introduction

It is very clear that analysis of the nonlinear bending for FGM beams under various shear deflection theories consists of two accented phases, for example, building the mathematical pattern. These problems are represented by a crowd of the partial differential equations, either nonlinear or linear, it is difficult to establish the analytical solutions. Therefore, the approximate numerical solution is adopted to meet the requirements of the project. At present, the focus of a large amount of literature is on FGM beams under the mechanical load.

For exact solutions, Sankar [1] obtained an exact elasticity solution for FG beams that considers the static transverse loading; this was later adopted for getting analytical solutions regarding stresses about a sandwich beam. Yu and Zhong [2] obtained a general solution for a beam with the uniform loading on the upper surface by using a semi-inverse method. Huang [3] et al. studied the exact solutions for anisotropic FG material beams with the different boundary conditions. Based on the shear deformation theory, Ma and Lee [4] gave a analytical solution about nonlinear responses of a FG beam. Lü et al. [5] obtained a exact solution for the bending of FGM beams to adopt the differential quadrature method.

For numerical solutions, using TBT, the shear, axial stresses and the transverse deflection in the thick FGM bar under an uniform loading for the various boundary conditions are studied in detailed [6]. By using the mathematical similarities among the governing equations, Li et al. [7] derived the analytical relationship of bending solutions. Li [8] obtained a new way to study the static responses of classic and first-order beams. Application of TBT and physical neutral surface, Zhang and Zhou [9] built a mathematical model of beams to obtain the numerical results of bending by the Multi-term Ritz method. Under mechanical loading, Niknam et al. [10] studied nonlinear static responses of the tapered beam under the various conditions. For the static and dynamic analysis of beams subjected to thermal load, Ma and Lee [11]carried out the related research by a shooting method. In this paper,

IOP Conf. Series: Materials Science and Engineering **307** (2018) 012058 doi:10.1088/1757-899X/307/1/012058

based on TBT, a numerical result by DQ method is obtained about nonlinear bending of FG beams subjected to mechanical loading under C-C and S-S boundary conditions.

2. Mathematical problem

2.1. Model material properties

Take into account a FGM beam (width b ,length l , *and* height h) with a rectangular section (Figure 1), which is mainly made up of ceramics and metals. The material properties vary continuously along the height direction from the pure ceramic surfaces to the pure metal surfaces. Young's modulus E and the passion ratio v are expressed as following:

$$P(z) = P_C + \left(P_M - P_C\right)V_M \tag{1}$$

where M and C denote the components of metal and ceramic; and V_M is the percentage of the volume of metal components. This formula changes according to below function:

$$V_M = \left(\frac{1}{2} - \frac{z}{h}\right)^n \tag{2}$$

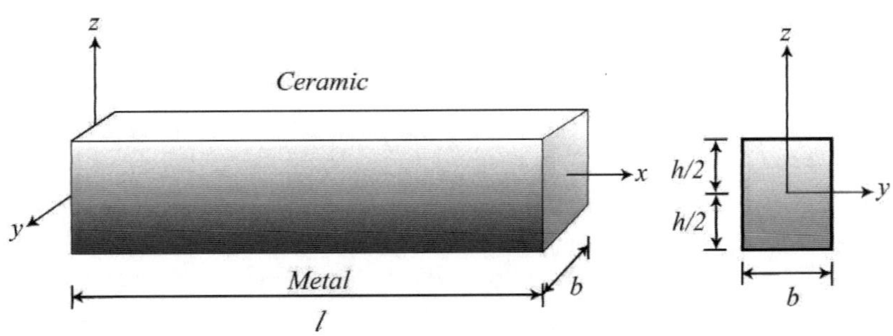

Figure 1. Geometry model of the studied FG beam.

2.2. The equilibrium equations

Displacement field as following:

$$\begin{cases} U_x(x, z) = \phi(x) + u(x) - \beta z^3\phi(x) - \beta z^3 w_{,x}(x) \\ U_z(x, z) = w(x) \end{cases} \tag{3}$$

where, x is the horizontal ordinate along the longitudinal direction of model beam; $U_x(x,z)$ and $U_z(x,z)$ are the deformation of x, z axial at arbitrary points of the studied beam; $w(x)$ denotes the vertical deflection; $\phi(x)$ expresses as the rotational angle.

Formula expression of displacement-strain are quoted by Heyliger and Reddy[12]:

$$\begin{cases} \varepsilon_x = z\phi_{,x} + \dfrac{w_{,x}^2}{2} + u_{,x} - \beta z^3(\phi_{,x} + w_{,xx}) \\ \gamma_{xz} = w_{,x} + \phi - 3\beta z^2(\phi + w_{,x}) \end{cases} \tag{4}$$

where $\beta = \dfrac{4}{3h^2}$

As a result of the materials of the FG beam obey Hooke's law, we find the following stress-displacement relations:

$$\sigma_x = E(z)\varepsilon_x \tag{5}$$

$$\tau_{xz} = \frac{E}{2(1+v)}\gamma_{xz} \tag{6}$$

IOP Conf. Series: Materials Science and Engineering **307** (2018) 012058 doi:10.1088/1757-899X/307/1/012058

2.3. Third-order shear deformation theory (TBT)

Based on TBT, The equilibrium equations can be expressed by:

$$\begin{cases} \dfrac{d^2u}{dx^2}+\dfrac{dw}{dx}\dfrac{d^2w}{dx^2}-\dfrac{\beta E_{11}\Omega_2}{\Omega_1}\dfrac{d^3w}{dx^3}-\dfrac{\beta E_{11}\Omega_3}{A_{11}}\varphi=0 \\ \dfrac{d^2\varphi}{dx^2}-\dfrac{A_{11}\Omega_2}{\Omega_1}\dfrac{d^3w}{dx^3}-\Omega_3\varphi=0 \\ (\dfrac{\beta^2 E_{11}{}^2\Omega_2{}^2}{\Omega_1}-\beta F_{11}+\beta\bar{F}_{11}\Omega_2)\dfrac{d^4w}{dx^4}+\Omega_2\Omega_3\tilde{D}_{11}\dfrac{d\varphi}{dx}+\left\{A_{11}\left[\dfrac{du}{dx}+\dfrac{1}{2}(\dfrac{dw}{dx})^2\right]-\beta E_{11}\dfrac{d\varphi}{dx}\right\}\dfrac{d^2w}{dx^2}+q=0 \end{cases} \tag{7}$$

Where

$$\Omega_1=A_{11}-\frac{\beta^2 E_{11}{}^2}{\tilde{D}_{11}},\Omega_2=1+\frac{\beta\bar{F}_{11}}{\tilde{D}_{11}},\Omega_3=\frac{A_{11}\tilde{A}_{44}}{\Omega_1\tilde{D}_{11}},\varphi=\phi+\frac{dw}{dx},\tilde{A}_{44}=\bar{A}_{44}-3\beta\bar{D}_{44},\bar{A}_{44}=A_{44}-3\beta D_{44},$$

$$\tilde{D}_{11}=\bar{D}_{11}-\beta\bar{F}_{11},\bar{D}_{11}=D_{11}-\beta F_{11}.$$

According to the above equations, the non-dimension quantities are defined by:

$$\xi=\frac{x}{l},W=\frac{w}{h},\ \bar{\varphi}=\frac{\varphi l}{h},U=\frac{ul}{h^2},\beta_0=\frac{l}{h},p=\frac{Pl^2}{D_x},S_{T0}=\frac{\beta^2 E_{11}{}^2\Omega_2{}^2}{\Omega_1}-\beta F_{11}+\beta\bar{F}_{11}\Omega_2,S_{T1}=\frac{\beta E_{11}\Omega_2}{\Omega_1 h},$$

$$S_{T2}=\frac{\beta E_{11}\Omega_3 l^2}{A_{11}h},S_{T3}=\frac{A_{11}\Omega_2}{\Omega_1},S_{T4}=\Omega_3 l^2,S_{T5}=\frac{\Omega_2\Omega_3\tilde{D}_{11}l^2}{S_{T0}},S_{T6}=\frac{A_{11}h^2}{S_{T0}},q_{T0}=\frac{ql^4}{hS_{T0}}.$$

From Eq. (7), we obtain the non-dimension equilibrium equations as following:

$$\begin{cases} \dfrac{d^2U}{d\xi^2}+\dfrac{dW}{d\xi}\dfrac{d^2W}{d\xi^2}-S_{T1}\dfrac{d^3W}{d\xi^3}-S_{T2}\bar{\varphi}=0 \\ \dfrac{d^2\bar{\varphi}}{d\xi^2}-S_{T3}\dfrac{d^3W}{d\xi^3}-S_{T4}\bar{\varphi}=0 \\ \dfrac{d^4W}{d\xi^4}+S_{T5}\dfrac{d\bar{\varphi}}{d\xi}+S_{T6}\left\{\left[\dfrac{dU}{d\xi}+\dfrac{1}{2}(\dfrac{dW}{d\xi})^2\right]-\dfrac{S_{T2}}{S_{T4}}\dfrac{d\bar{\varphi}}{d\xi}\right\}\dfrac{d^2W}{d\xi^2}+q_{T0}=0 \end{cases} \tag{8}$$

3. Mathematical formulation

3.1. The weighting coefficients

The DQM must discretize the definition domain into m points. Along the definition domain, a weighted nonlinear summation of all of the functional values in m points approximately expresses the derivatives of any point [13-14].

$$\frac{d^k f(x_i)}{dx^k}=f_i^{(k)}=\sum_{j=1}^{m}A_{ij}{}^{(k)}f_j,\quad i=1,2,\cdots,m \tag{9}$$

where m denotes the number of grid points, $A_{ij}^{(k)}$ is the weighting coefficient for the kth derivative.

3.2. Selection of the sampling nodes

The sampling points of differential quadrature representations are selected as shifted Chebyshev-Gauss-Lobatto points, which are given as follows:

$$x_i=\frac{1}{2}\left\{1-\cos\left[\frac{\pi(i-1)}{m-1}\right]\right\},\ i=1,2,\cdots,m \tag{10}$$

4. Numerical results and discussions

For numerical analysis, we calculate an Si3N4-SUS304 FG beam under a uniformly distributed loading $q(\text{KNcm}^{-1})$, with a length of $l = 100\text{cm}$, a height of $h = 1\text{cm}$, a width of $b = 1\text{cm}$, and the number of nodes is $m = 11$. The material properties: $E_{\text{Si3N4}} = 348.43\text{GPa}$, $E_{\text{SUS304}} = 201.04\text{GPa}$, $\upsilon_{\text{SUS304}} = 0.24$, $\upsilon_{\text{SUS304}} = 0.33$.

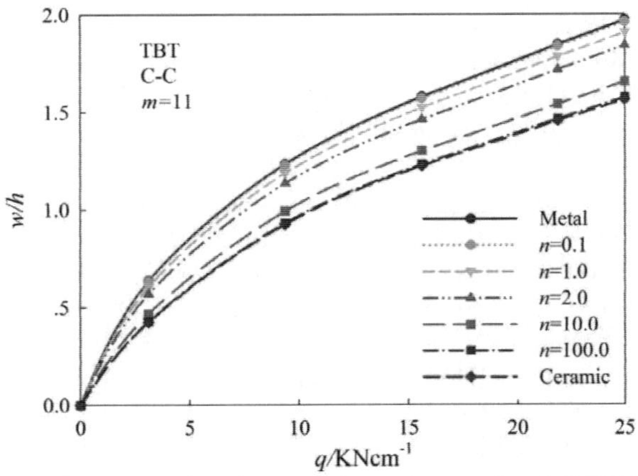

Figure 2. Central deflection-loading curves of a C-C FG beam.

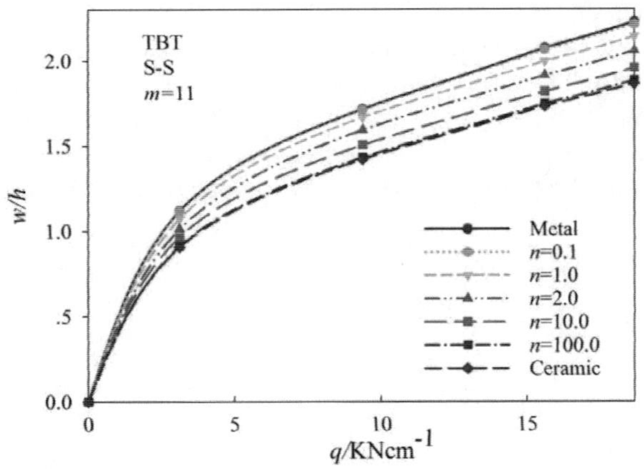

Figure 3. Central deflection-loading curves of a S-S FG beam.

From figures 2 to 3, it can be observed that the shear deformation effects is the critical factor to cause an increase of the central deflection. For an Si3N4-SUS304 FG beam with the C-C and S-S boundary conditions, using a shear deformation beam theory, dropping the power law index will cut down the stiffness of FG beam, in succession, results in a decrease of the deflections.

5. Conclusions

In this article, based on TBT, the nonlinear bending of FG beams exerted under two different edges is analytically investigated. Nonlinear curves are gradually obvious with the increasing of load. Due to consider shear deformation, the central deflection of an FG beam using TBT is greater compared to homogeneous beam. The central deflection-load curves obtained by exerting an simply supported edge is faster, and gradually becomes smoothly with load increase. Using TBT, the maximum central deflections of Si_3N_4-SUS304 beams is not significantly different from the center deflections found using FBT. In addition, dropping the power law index will cut down the stiffness of FG beam, in succession, results in a decrease of the deflections.

References

[1]Sankar B V 2001 *Compos. Sci. Technol.* An elasticity solution for functionally graded beams **61** 689-96

[2]Yu T and Zhong Z 2006 Acta Mech. Solida Sin. A general solution of a clamped functionally graded cantilever-beam under uniform loading **27** 15-20

[3]Huang D J, Ding H J and Chen W Q 2007 Int. J. Solids Struct. Analytical solution for functionally graded magneto-electro-elastic plane beams **45** 467-85

[4]Ma L S and Lee D W 2012 Eur. J. Mech. A/Solids Exact solutions for nonlinear static responses of a shear deformable FGM beam under an in-plane thermal loading **31** 13-20

[5]Lü C F, Chen W Q, Xu R Q and Lim C W 2008 Int. J. Solids Struct. Semi-analytical elasticity solutions for bi-directional functionally graded beams **45** 258-75

[6]Kadoli R, Akhtar K and Ganesan N 2003 Appl. Math. Model. Static analysis of functionally graded beams using higher order shear deformation theory **32** 2509-25

[7]Li S R, Cao D F and Wan Z Q 2013 Appl. Math. Model. Bending solutions of FGM Timoshenko beams from those of the homogenous Euler-Bernoulli beam **37** 7077-85

[8]Li X F 2008 J. Sound Vib. A unified approach for analyzing static and dynamic behaviors of functionally graded Timoshenko and Euler-Bernoulli beams **318** 1210-29

[9]Zhang D G 2013 Compos. Struct. Nonlinear bending analysis of FGM beams based on physical neutral surface and higher-order shear deformation theory **100** 121-6

[10]Niknam H, Fallah A and Aghdam M M 2014 Int. J. Nonlin. Mech. Nonlinear bendingof functionally graded tapered beams subjected to thermal and mechanical loading **65** 141-7

[11]Ma L S and Lee D W 2011 Compos. Struct. A further discussion of nonlinear mechanical behavior for FGM beams under in plane thermal loading **93** 831-42

[12]Heyliger P R and Reddy J N 1998 J. Sound Vib. A higher order beam finite element for bending and vibration problem **126** 309-26

[13]Pu J P 2005 J. Zhejiang Univ. Technol. The differential quadrature method and its application 33 429-33

[14]Wang L 1983 J. Liaoning Arch. Eng. Large deflection calculation of beams **1** 17-36

IOP Conf. Series: Materials Science and Engineering **307** (2018) 012059 doi:10.1088/1757-899X/307/1/012059

Analytical Calculation for slope stability based on the Combination of Horizontal Slice Method and Oblique Slice Method

Chao Qiu and Ping-an Wang

College of Civil Engineering, Xi'an University of Architecture and Technology, Xi'an, Shaanxi, China
E-mail:m15667087215@163.com

Abstract. An analytical method for analyzing the stability of the homogeneous slope based on horizontal slice method is presented. A combination model of horizontal slice and oblique slice are used for analysis the inter-slice forces. The analytical expression of factor of safety is obtained with the assumption that the shear strength of soil meets the Mhor-Coublum strength criterion. The position of critical slip surface and its corresponding factor of safety computed by the development of the optimization technique are in good agreement with Bishop's method. The advantages of presented method are straightforward in analyzing the stability of slope.

1. Introduction

The equilibrium methods for analysis of slope stability are mostly vertical slice methods [1-3].For homogeneous rock and soil, the variation of soil mass is less affected by the internal variation of the vertical slice. However, when the slope is layered soil, it will bring a great inconvenience to the calculation, and horizontal slice method is more applicable. The idea of horizontal slice method is proposed for stability analysis of reinforced slope by Lo and Xu [4]. Shangholi et al.[5] applied the horizontal slice method in analyzing the reinforced slopes seismic stability based on pseudo-static method. Nouri et al.[6-7] extended the latter work of reinforced slope stability by developing a number of formulations and studying the effect of earthquake. Chandaluri et al.[8] used the horizontal slice method suggested by Shangholi et al.[5]to Study the influence of various slope parameters on slope stability. These approaches for analysis of reinforced soil slope are based on the assumption that the failure surface pass through the toe of slope. Therefore, it is necessary to carry out stability analysis of slope with general sliding surface based on horizontal slice method. Furthermore, when the slope sliding surface pass below the slope toe, the use of horizontal slice method for analysis of slope stability will bring a problem that two sliding bottom surfaces will appear, which will bring a trouble to analysis of the inter-slice forces. Chen et al.[9] proposed a method for analysis of soil nailing that the failure wedge of soil is divided into a number of oblique slices. Ge et al.[10] analyzed the slope stability using the oblique slice method .The presented paper proposed analytical method for analysis of slope stability based on the combination of horizontal slice method and oblique slice method. And related optimization technique is developed for obtaining the slip sliding surface and its corresponding factor of safety.

2. Basic assumptions

In the study, the following assumptions are made:

(1)The vertical stress is equal to the overburden pressure for both horizontal soil slice and oblique soil slice.

(2)The slip sliding surface is circular, and the factor of safety for all slices along the critical slip surface is equal.

(3)The shear strength of soil meets the Mhor-Coublum strength criterion, that is

$$\tau_f = \sigma \tan \varphi + c \tag{1}$$

Where τ_f and σ is the shear stress and normal stress along the failure surface at failure, respectively.

Figure.1 shows a critical slip surface with centre O and radius r_0 for a homogeneous slope with the height H , inclination angle θ ,unit weight γ , internal friction angle φ and cohesion c . x_0 , y_0 are the centre of the abscissa and ordinate coordinates of the circular sliding surface, respectively, *D*, *F* are the intersections of the vertical line at *C,A* and the failure surface, respectively. μ is the abscissa of point *G*. α_1 , α_2 , η_1 , η , β are the vertical centre angles of the points *G,F, E,D,B* , respectively, with respect to the centre *O,* ϕ is the central angle of any point on the circular sliding surface, and the vertical centre angle is positive counter clockwise.

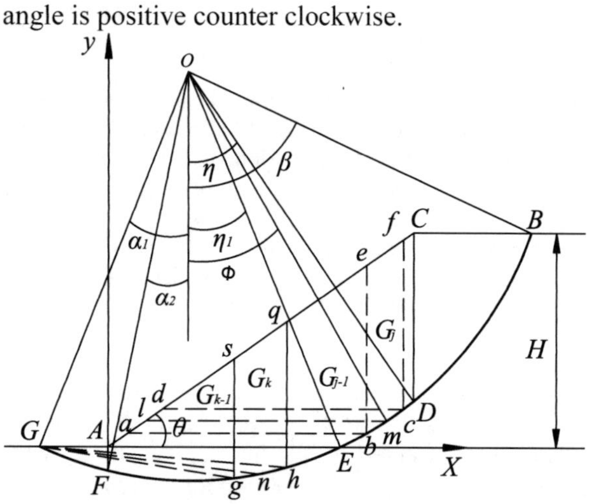

Figure.1 Parameters diagram of the critical slip surface

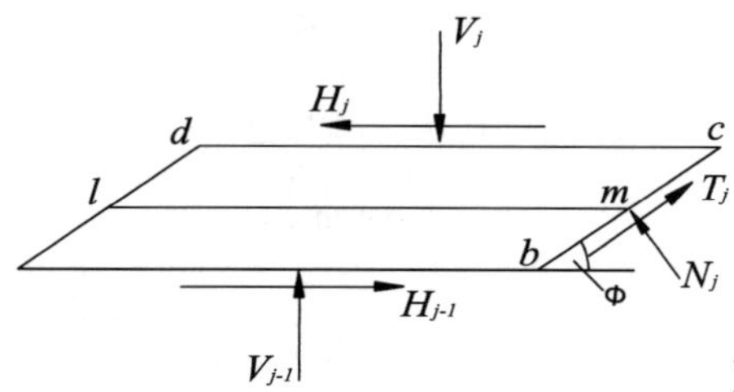

Figure. 2a Forces diagram of the horizontal slice

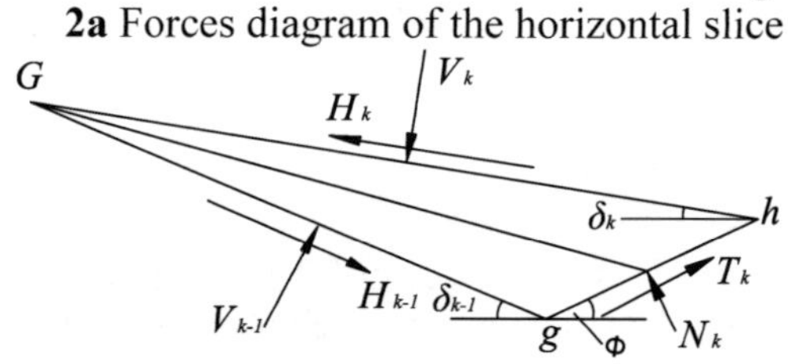

Figure.2b Forces diagram of the oblique slice

Horizontal slice method and oblique slice method are combined to analysis of slope stability. The soil above the toe of slope is divided in horizontal slices, and the soil below the toe of slope is divided into oblique slices. Forces diagram of the horizontal slice are shown in Figure.2a.According to the assumption used by Shangholi et al.[5].The inter-slice forces have the following relationship:

$$\begin{cases} V_j - V_{j-1} = G_j - G_{j-1} \\ H_j - H_{j-1} = 0 \end{cases} \tag{2}$$

Where V_j and V_{j-1} and H_j , H_{j-1} are the vertical inter-slice and the horizontal inter-slice forces on the horizontal slice, respectively. G_j , G_{j-1} are the gravity of block *def* and *abe*, respectively.

Forces diagram of the oblique slice are shown in Figure.2b.In order to obtain the relationship of inter-slice forces, the block *GhqA* and *sghq* are analyzed shown in Figure.3a and 3b. X_k , X_{k-1} and E_k , E_{k-1} are the horizontal inter-slice forces and the vertical inter-slice forces on the vertical slice, respectively. According to the equilibrium conditions, the inter-slice forces have the following relationship:

$$\begin{cases} -V_k = (G_k + E_k)\cos\delta_k + X_k \sin\delta_k \\ -H_k = -(G_k + E_k)\sin\delta_k + X_k \cos\delta_k \end{cases} \tag{3}$$

After the conversion of equation (3), X_k , E_k can be represented by:

$$\begin{cases} X_k = -V_k \sin\delta_k - H_k \cos\delta_k \\ E_k = H_k \sin\delta_k - V_k \cos\delta_k - G_k \end{cases} \tag{4}$$

For the traditional method, the relationship between the inter-slice forces has different assumptions. Bishop[2] ignored the inter-slice forces, and Morgenstern NR[1] considered the inter-slice forces in great detail. According to relationship between the inter-slice forces assumed by Bishop, and it has the following relationship:

$$\begin{cases} E_k - E_{k-1} = 0 \\ X_k - X_{k-1} = 0 \end{cases} \tag{5}$$

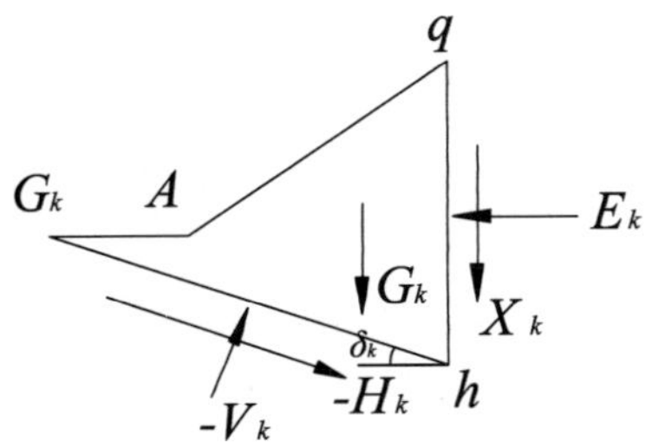

Figure.3a Forces acting on the overburden soil

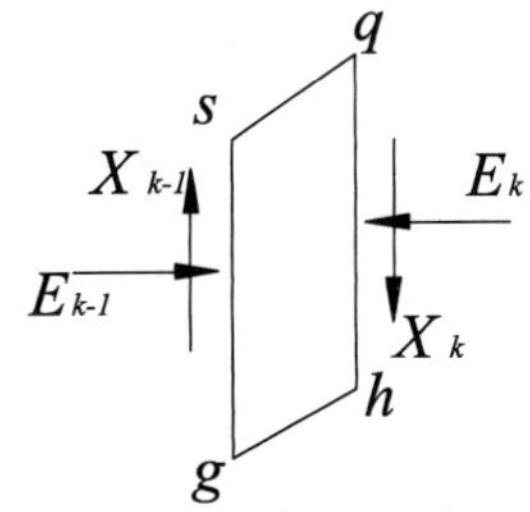

Figure.3b Forces acting on the vertical slice

Substituting equation (5) into equation (4):

$$\begin{cases} V_k \cos\delta_k - H_k \sin\delta_k - V_{k-1}\cos\delta_{k-1} + H_{k-1}\sin\delta_{k-1} = G_k - G_{k-1} \\ V_k \sin\delta_k + H_k \cos\delta_k - V_{k-1}\sin\delta_{k-1} - H_{k-1}\cos\delta_{k-1} = 0 \end{cases} \tag{6}$$

Where V_k and V_{k-1} and H_k , H_{k-1} are the vertical inter-slice and the horizontal inter-slice forces on the oblique slice, respectively. G_k , G_{k-1} are the gravity of block *GhqA* and *GgsA*, respectively. δ_k is the angle between the upper side of oblique slice and the horizontal direction. δ_{k-1} is the angle between the lateral side of oblique slice and the horizontal direction, and clockwise is positive. The following conclusions that the resultant force of the vertical forces is equal to the overburden pressure, and the resultant force of the horizontal force is equal to zero obtained are the same as the assumption presented by Shangholi et al.[5] .

3. Analytical method for safety factor

Horizontal and vertical equilibrium are considered for individual slice, no account being taken of moment equilibrium. The following equilibrium equations can be obtained from Figure.2a and 2b:

$$\begin{cases} N_j = (W_j + V_j - V_{j-1})\cos\phi \\ T_j = (W_j + V_j - V_{j-1})\sin\phi \end{cases} \tag{7}$$

$$\begin{cases} N_k = H_{k-1}\sin(\delta_{k-1}+\phi) - V_{k-1}\cos(\delta_{k-1}+\phi) - H_k \sin(\delta_k+\phi) + V_k\cos(\delta_k+\phi) + W_k\cos\phi \\ T_k = -H_{k-1}\cos(\delta_{k-1}+\phi) - V_{k-1}\sin(\delta_{k-1}+\phi) + H_k\cos(\delta_k+\phi) + V_k\sin(\delta_k+\phi) + W_k\sin\phi \end{cases} \tag{8}$$

Where W_j and W_k are the gravity of horizontal and oblique slice, respectively. N_j , T_j are the tangential and normal forces on the slip surface of horizontal slice. N_k , N_{k-1} are the tangential and normal forces on the slip surface of oblique slice. The expressions of x , y can be expressed as:

$$\begin{cases} x = x_0 + r_0 \sin\phi \\ y = y_0 - r_0 \cos\phi \end{cases} \tag{9}$$

And the expressions of x_0 , y_0 and r_0 can be represented by:

$$\begin{cases} x_0 = -\dfrac{H\sin\alpha_2}{\cos\alpha_1 - \cos\beta} \\ y_0 = \dfrac{H\cos\alpha_1}{\cos\alpha_1 - \cos\beta} \\ r_0 = \dfrac{H}{\cos\alpha_1 - \cos\beta} \end{cases} \tag{10}$$

Furthermore, equations (2) and (6) are introduced into the equations (7) and (8), respectively. The following relationship can be obtained, that is

$$\begin{cases} N_i = (W_i + G_i - G_{i-1})\cos\phi \\ T_i = (W_i + G_i - G_{i-1})\sin\phi \end{cases} \quad i = j,k \tag{11}$$

The slope soils meet the Mohr-Coublum strength criterion, and the safety factor can be represented by:

$$F_s = \frac{\sum(cl_i + N_i \tan\varphi)}{\sum T_i} = \frac{\sum_{i=1}^{n} cl_i + \tan\varphi \sum_{i=1}^{n}(W_i + G_i - G_{i-1})\cos\phi}{\sum_{i=1}^{n}(W_i + G_i - G_{i-1})\sin\phi} \tag{12}$$

Where, l_i is the length of the bottom surface of the slice, n is the number of slices. In order to analyze the slope stability more accurately, the slope is presented in the terms of integral rather than the sums of terms for all slices. The following expressions can be derived from Figure.1:

$$G_i - G_{i-1} = \begin{cases} \gamma(H-y)dx - \gamma(x - y\cot\theta)dy(\eta \le \phi \le \beta) \\ \gamma(x\tan\theta - y)(dx - \cot\theta dy)(\eta_1 \le \phi < \eta) \\ \gamma x\tan\theta dx - 0.5\gamma(x-\mu)dy - 0.5\gamma y dx(\alpha_2 \le \phi < \eta_1) \\ -0.5\gamma(x-\mu)dy - 0.5\gamma y dx(\alpha_1 \le \phi < \alpha_2) \end{cases} \tag{13}$$

$$W_i = \begin{cases} \gamma(x - y\cot\theta)dy(\eta_1 \le \phi \le \beta) \\ 0.5\gamma[-ydx + (x-\mu)dy](\alpha_1 \le \phi < \eta_1) \end{cases} \tag{14}$$

After the conversion and simplification of equations (9), (10), (12),(13)and(14), the analytical expression of F_s is derived, that is

$$F_s = \frac{\dfrac{c}{\gamma H}(\beta - \alpha_1)(\cos\alpha_1 - \cos\beta) + \tan\varphi \cdot I_r}{I_c} \tag{15}$$

Where :

$$\begin{aligned} I_r = {} & (\cos\alpha_1 - \cos\beta)[\frac{1}{2}(\beta-\eta) + \frac{1}{4}(\sin 2\beta - \sin 2\eta)] - \tan\theta\sin\alpha_2[\frac{1}{2}(\eta - \alpha_2) + \frac{1}{4}(\sin 2\eta - \sin 2\alpha_2\ )] \\ & -\cos\alpha_1[\frac{1}{2}(\beta - \alpha_1) + \frac{1}{4}(\sin 2\beta - \sin 2\alpha_1)] + \frac{1}{3}\tan\theta(\cos^3\alpha_2 - \cos^3\eta) + \frac{1}{3}(\sin^3\alpha_1 - \sin^3\beta) + \sin\beta \\ & -\sin\alpha_1 \end{aligned} \tag{16}$$

$$\begin{aligned} I_c = {} & \frac{1}{2}(\cos\alpha_1 - \cos\beta)(\sin^2\beta - \sin^2\eta) - \frac{1}{2}\tan\theta\sin\alpha_2(\sin^2\eta - \sin^2\alpha_2) - \frac{1}{2}\cos\alpha_1(\sin^2\beta - \sin^2\alpha_1) \\ & + \frac{1}{3}\tan\theta(\sin^3\eta - \sin^3\alpha_2) + \frac{1}{3}(\cos^3\alpha_1 - \cos^3\beta) \end{aligned} \tag{17}$$

It is a commonly used method to determine the critical slip surface with the minimum factor of safety for analysis of slope stability. In this paper, the safety factor is a function of four variables. MABLAB software is used to optimize the function (15), which is a nonlinear constrained optimization problem, to obtain the minimum value of F_s .The minimum factor of safety can be

IOP Conf. Series: Materials Science and Engineering **307** (2018) 012059 doi:10.1088/1757-899X/307/1/012059

quickly obtained with the related optimization procedures(presented in Appendix), and the constraint conditions are:

$$\begin{cases} \cos\alpha_1 - \cos\alpha_2 \le 0 \\ (\cos\alpha_1 - \cos\beta)\cot\theta - \sin\eta + \sin\alpha_2 = 0 \end{cases} \quad (18)$$

4. Verification

The cases are analyzed in order to verify the reasonableness of the proposed analytical method in this study for a homogeneous slope with $\gamma = 18.5kN/\text{m}^3$ .Slope height is 8m with the configurations of 1:1 slope and 2:1 slope. In order to analyze the influence of slope soil strength on slope stability more comprehensively, cohesions ranging from 5 to 30kPa and angles of internal friction ranging from 10° to 20° .The rationality of proposed analytical method is verified with two methods of slopes stability analysis, Bishop's method and the local minimum factor of safety method. The results are represented in Table 1.

Table 1 Comparison of safety factor

Slope (radio)	Cohesion $c(kPa)$	Friction angle φ (deg.)	Safety factor by proposed method	Safety factor by Bishop's method	Safety factor by Local minimum factor of safety method(1993) [11]
1:1	25	20	1.68	1.74	1.87
1:1	20	20	1.46	1.50	1.68
1:1	15	20	1.24	1.29	1.46
1:1	10	20	1.00	1.05	1.00
1:1	30	15	1.73	1.75	1.85
1:1	25	15	1.52	1.53	1.65
1:1	20	15	1.30	1.32	1.45
1:1	15	15	1.09	1.11	1.24
1:1	25	10	1.35	1.35	1.42
1:1	20	10	1.15	1.15	1.23
2:1	20	20	1.96	2.09	2.05
2:1	15	20	1.71	1.82	1.85
2:1	10	20	1.44	1.54	1.60
2:1	5	20	1.15	1.21	1.23
2:1	25	15	1.78	2.05	1.87
2:1	20	15	1.55	1.78	1.72
2:1	15	15	1.31	1.53	1.54
2:1	10	15	1.06	1.29	1.29
2:1	15	10	1.21	1.27	1.19

It is found that the safety factors obtained by the proposed analytical method are in good agreement with those determined by the Bishop's method and the local minimum factor of safety. From the above analysis, it is reasonable for assumption of the inter-slice forces on the oblique slices, and it is safer for analysis of slope stability. The presented method solves the problem that the failure surface goes down the toe of slope by using the combination of horizontal method and oblique method, and it is a development of the horizontal slice method for analysis of slope stability.

5. Examples

Two cases[12] are analyzed to verify the reasonableness of the proposed analytical method. The material properties are shown in Table 2.The position of the critical slip surface and its corresponding factors of safety calculated by proposed analytical method and Bishop's method for a circle slip surface are shown in Figure.4a and 4b.

In addition, the position of critical slip surface and its corresponding factor of safety are in good agreement with each other for 1:1 slope, and they all pass thought the toe of slopes as expected. Although the proposed analytical method defined the critical slip surface passing the toe of slope that is different from the critical slip surface passing below the toe of slope that determined by Bishop's method, the position of slip surface and its corresponding factor of safety can be considered to be substantially the same for 1:2 slope. Moreover, the safety factor calculated by the analytical method is relatively safer compared with the safety factor calculated by the Bishop's method. It is note that critical slip surfaces calculated by presented method are deeper than the computed by Bishop's method.

Table 2 Material properties

examples	H / m	Slope(radio)	γ(kN/ m^3)	$c(kPa)$	φ(deg.)
1	20	1:1.5	18.82	41.65	15.0
2	5	1:2.0	17.64	9.80	10.0

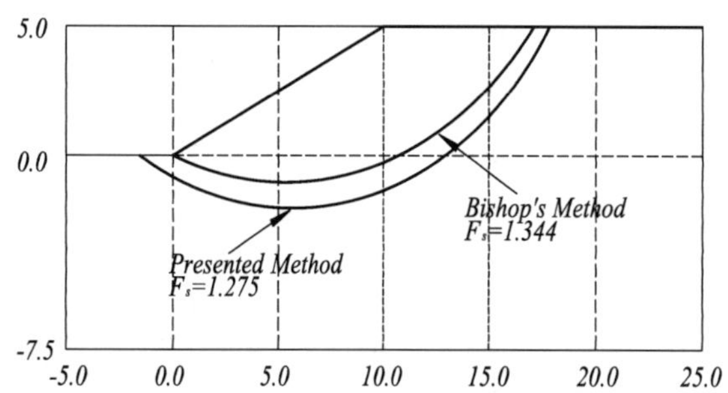

Figure.4a The critical slip surface in example 1

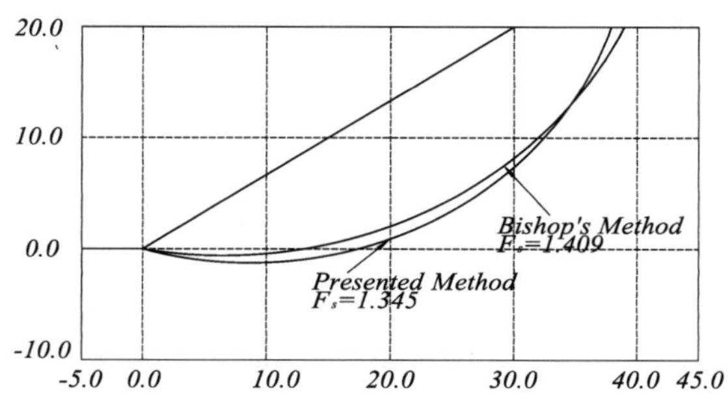

Figure.4b The critical slip surface in example 2

6. Conclusions

The situation that the critical slip surface does pass below the toe of slope was considered in the study of slope stability. The analytical method is proposed based on the combination model of horizontal slice and oblique slice. The location of critical slip surface and its corresponding factor of safety are in good agreement for both proposed analytical method and Bishop's method.The advantages of proposed analytical method are straightforward, ease to use, less time-consuming in analyzing the stability of slope. The proposed analytical method is for homogeneous slopes, this is a limitation.

7. References

[1] Morgenstern NR, Price VE. The analysis of the stability of general slip surfaces. Geotechnique 1965;15(1): 79–93.
[2] Bishop AW.The use of slip circle in the stability analysis of slopes. Computers and Geotechnics 1955; 5(1): 7-17.
[3] Sarma SK . Stability of analysis of embankments and slopes. Geotechnique 1973; 23(3): 423–433.
[4] Lo SCR, Xu DW. A strain-based design method for the collapse limit state of reinforced soil walls orslopes. Can.Geotech.J 1992; 29(8): 832-842.

[5] Shangholi M, Fakher A, Jones CJFP. Horizontal slice method of analysis. Geotechnique 2001; 51(10):881-885.
[6] Nouri H, Fakher A, Jones CJFP. Development of Horizontal Slice Method for seismic stability analysis of reinforced slopes and walls. Geotext.Geomembr. 2006; 24: 175-187.
[7] Nouri H, Fakher A, Jones CJFP. The effect of earthquake on the seismic stability of reinforced slopes using horizontal slice method. In:4th international conference on earthquake geotechnical engineering 2007.
[8] Chandaluri VK, Sawant VA, Shukla SK. Seismic Stability Analysis of Reinforced Soil Wall Using Horizontal Slice Method. Int. J.l of Geosynth. and Ground Eng. 2015; 1.
[9] Chen CF, Yang Y. The parallel slice method for analysis of soil nailing based on genetic algorithm.Advanced in Earth Structures: Research to Practice 2006; GSP151: 188-194.
[10] Ge HY, Tu JS, Wang JG. With the oblique slice method for slope stability safety factor. Chinese Quarterly of Mechanics,2010,31(3):430-435.
[11] Huang SL, Yamsaki K. Slope failure analysis using local minimum factor-of-safety approach. J. Geotech.Eng. ASCE 1993; 1974-1987.
[12] Rocscience Inc.2D limit equilibrium slope stability for soil and rock slopes.Slope Stability Verification Manual Part1 2016.

APPENDIX

Optimization procedure for equation (15):

For example: $\theta = 45°$, $c/\gamma H = 0.20$, $\varphi = 20°$

$x(1) = \alpha_1$, $x(2) = \alpha_2$, $x(3) = \eta$, $x(4) = \beta$

```
------------------------------------------------------------------------------------------------------------
fun='abs((0.2*(x(4)-x(1))*(cos(x(1))-cos(x(4)))+tan(pi/9)*((cos(x(1))-cos(x(4)))*(0.5*(x(4)-
x(3))+0.25*(sin(2*x(4))-sin(2*x(3))))-cos(x(1))*(0.5*(x(4)-x(1))+0.25*(sin(2*(x(4)))-sin(2*x(1))))-
tan(pi/4)*sin(x(2))*(0.5*(x(3)-x(2))+0.25*(sin(2*x(3))-sin(2*x(2))))+(1/3)*tan(pi/4)*((cos(x(2)))^3-
(cos(x(3)))^3)+(1/3)*((sin(x(1)))^3-(sin(x(4)))^3)+sin(x(4))-sin(x(1))))/(0.5*(cos(x(1))-
cos(x(4)))*((sin(x(4)))^2-(sin(x(3)))^2)-0.5*cos(x(1))*((sin(x(4)))^2-(sin(x(1)))^2)-
0.5*tan(pi/4)*sin(x(2))*((sin(x(3)))^2-(sin(x(2)))^2)+(1/3)*tan(pi/4)*((sin(x(3)))^3-
(sin(x(2)))^3)+(1/3)*((cos(x(1)))^3-(cos(x(4)))^3)))';
x0=[0 0.1 0.5 1.3];
A=[1 -1 0 0;0 1 -1 0;0 0 1 -1;0 0 0 1];
b=[0;0;0;pi/2];
options=optimset('TolX',1e-12);
[x,fval]=fmincon(fun,x0,A,b,[],[],[],[],@comfun,options)
-------------------------------------------------------------------------------
Comfun:
function [c,ceq]=comfun(x)
c=(tan(x(2)))^2-(tan(x(1)))^2;
ceq=(cos(x(1))-cos(x(4)))*1-sin(x(3))+sin(x(2));
```

Dynamic performance of a suspended reinforced concrete footbridge under pedestrian movements

I Drygala [1], J Dulinska[2] and K Kondrat[3]

1 Assistant Professor, Cracow University Of Technology, Cracow, PL
2 Professor, Cracow University Of Technology, Cracow, PL
3 Student, Cracow University Of Technology, Cracow, PL
E-mail: idrygala@pk.edu.pl

Abstract. In the paper the dynamic analysis of a suspended reinforced concrete footbridge over a national road located in South Poland was carried out. Firstly, modes and values of natural frequencies of vibration of the structure were calculated. The results of the numerical modal investigation shown that the natural frequencies of the structure coincided with the frequency of human beings during motion steps (walking fast or running). Hence, to consider the comfort standards, the dynamic response of the footbridge to a runner dynamic motion should be calculated. Secondly, the dynamic response of the footbridge was calculated taking into consideration two models of dynamic forces produced by a single running pedestrian: a 'sine' and 'half-sine' model. It occurred that the values of accelerations and displacements obtained for the 'half-sine' model of dynamic forces were greater than those obtained for the 'sine' model up 20%. The 'sine' model is appropriate only for walking users of the walkways, because the nature of their motion has continues characteristic. In the case of running users of walkways this theory is unfitting, since the forces produced by a running pedestrian has a discontinuous nature. In this scenario of calculations, a 'half-sine' model seemed to be more effective. Finally, the comfort conditions for the footbridge were evaluated. The analysis proved that the vertical comfort criteria were not exceeded for a single user of footbridge running or walking fast.

1. Introduction

Many pedestrian bridges erected nowadays have modal properties (natural frequencies) that are close to the critical frequencies of dynamic excitations produced by pedestrians. Potentially, users of this structures walking or running on a footbridge may cause a resonance phenomenon. There are several models of forces produced by a single human being running or walking along a footbridge [1, 2, 3].
In this work the dynamic investigation of a suspended footbridge that carry people over the S7 national roadway in Gaj, South Poland was performed. The central purpose of the research was to compare the dynamic behaviour of the structure obtained for two different models of dynamic forces generated by a single running user of the footbridge. On the basis of the achieved effects of the study, the vibration comfort standards were also assessed.

2. Vibration comfort criteria

The leading aim of dynamic study on footbridges is the evaluation of their vibration comfort standards. There are some standards which are dedicated to the assessment of the influence of oscillations on human beaings who are the users of walkways, such as British Standards [4] or ISO [5]. The rules for design practice are also delivered by Eurocodes [6]. The critical allowed value of vertical acceleration, given by Eurocode (EN 1990:2002/Al:2005), is 0.7 m/s^2. The rules are appropriate for pedestrian bridges with natural frequencies less than 5 Hz. Other useful rules for the evaluation of the vibrational comfort of pedestrian bridges are proposed by SÉTRA [7]. The acceleration ranges in horizontal and vertical course with comfort stages assigned, provided by this document, are summarized in Table 1. The classification in the frequency domain of the human being motion in the Tab. 2 is presented.

Table 1. Comfort of use levels in the acceleration domain (SÉTRA 2006) [7].

Comfort level	*Ranges of comfort [m/s^2]*	
	Vertical	*Horizontal*
Maximum comfort	0.0-0.5	0.00-0.15
Mean comfort	0.5-1.0	0.15-0.30
Minimum comfort	1.0-2.5	0.30-0.80
Uncomfortable	> 2.5	> 0.8

Table 2. The classification in the frequency domain of the human being motion [1, 2].

Type of movement	*Frequency ranges [Hz]*			
	Total range	*Slow*	*Normal*	*Fast*
Walking	1.40-2.40	1.40-1.70	1.70-2.20	2.20-2.40
Running	1.90-3.30	1.90-2.20	2.20-2.70	2.70-3.30

3. Models of dynamic forces produced by a runner

Vibrant loading normally produced by people on walkways results from walking or running. The dynamic behavior of a footbridge under pedestrians movements is usually obtained by the time history analysis in which the dynamic loading has to be modelled. This type of forces are periodic as well as change in time and space [8]. There are number of mathematical models of forces caused by a single user walking or running. In this study two different models are introduced for the dynamic analysis.
Firstly, the force generated by a single walker or runner was used as a sum of static and dynamic parts by the equation [3]:

$$F(t) = G\left[1 + \sum_{i=1}^{n} A_i \sin(2i\pi f_s t - \varphi_i)\right] \qquad i = 1, 2, \ldots, n \tag{1}$$

where: G – weight of the human body; f_s – major frequency of the motion; φ_i and A_i– the phase angel of the i-th harmonic and amplitude, respectively. The Fourier constants in formula (1) for different forms of people motion are presented in [3].
The 'sine' model provided by formula (1) is adequate in the case of forces produced by walking people because each step overlaps with the previous one thus making the function of loading force continuous. However, in the situation of running this postulation seems to be unfitting, since the loading caused by a running user of footbridge has a discontinuous nature. In this situation, a 'half-sine' mathematical model [1] appears to be more adequate. Forces produced by a runner are calculated from the formula:

$$F_z(t) = \begin{cases} k_z G \sin\left(\pi \dfrac{t}{t_c}\right) & for \quad t \le t_c \\ 0 & for \quad t_c < t < t_u \end{cases} \tag{2}$$

where: k_z – impact factor ($k_z = F_{max}/G$); F_{max} – maximum load produced by runner; G – weight of the human body; t_c – period of contact; t_u – period of load produced by runner; f_u – frequency of running.

4. Geometry and material data of the footbridge

The span of the analysed structure is 50.0m, the whole width is equal to 4.50m and the dimension (height) of the both pylons are equal to 15.5m and 17.5m (Fig. 1). The structural system of the investigated footbridge consists of two reinforced concrete girders (85.00 x 40.00cm). The girders are linked by ten crossbars (thickness 0.16cm, diameter 2.44cm, material: steel) along the span and two crossbars at both ends (90.00 x 50.00cm, material: concrete). The thickness of the footbridge deck is equal to 22.00cm. The deck is supported by twenty trusses (diameter 0.56cm) from steel material pylons.

Elastomeric bearings (height 20.00cm, cross-section 30.00 x 30.00cm) support the footbridge deck. The values of material parameters are summarized in Table 3.

Figure 1. The footbridge located in Gaj (South Poland) along the S7 roadway

Table 3. Parameters of materials of the structure of the footbridge.

Material	*Elasticity modulus [GPa]*	*Poisson's ratio [-]*	*Mass density [kg/m³]*
Concrete	32	0.2	2500
Structural steel	210	0.3	7800
Steel trusses	200	0.3	7800

5. The numerical model and the dynamic characteristics of the structure

The finite element model (FEM) of the investigated footbridge with control points of the walkway selected for dynamic study is presented in Fig. 2. The model was assembling in the ABAQUS software program [9].

In the FEM of the footbridge beam elements were used for crossbars and girders. Deck was solved with shell elements. Solid finite elements were applied for the foundations. Fixed boundary conditions were applied in all cases. This numerical approach reflect the high rigidity of subsoils. The steel hangers were solved as truss finite elements with the 'no compression' material option [9] in order to assurance that compressive effects would not be produced during dynamic time history analysis (THA).

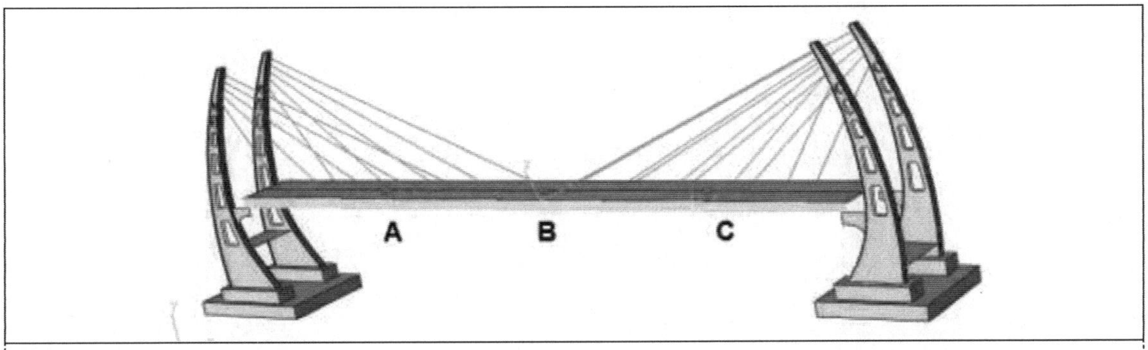

Figure 2. FEM of the investigated footbridge with the selected control points

The evaluation of modal characteristics of pedestrian bridges has been extensively studied [1, 2, 10, 11]. On the basis of the other authors results, it was recognised that the basic natural frequencies of these structures are located generally in the range of 1.4-3.4 Hz. The typical frequency of vertical oscillations produced by running or walking user is also 1.4-3.4 Hz (see Table 2). The evaluated lowest natural frequency of vibrations in vertical direction of the studied footbridge is equal to 1.92 Hz. This value is adequate for the walking as well as running user of the walkways. Taken into consideration the maximum values for the forces generated by walkers and runners, the dynamic evaluation for runner was conducted in the next stage of the study.

6. Two models of forces generated by a runner

In the second step, the response of the footbridge to a single running user were evaluated. The THA was prepared with ABAQUS. The calculations were carried out, during which, the twenty lowest modes were considered with a damping ratio equal to 2% for each mode. Two mathematical models of forces generated by a runner were introduced and the results obtained for both variants were compared. Firstly, the function of forces produced by a single running user, provided by equation (1), was discussed [3]. For the analysis purposes, the values were assumed as follow: pedestrian weight – $G = 800\ N$; frequency of running – $f_u = 1.92\ Hz$; period of running –$t_u = 0.52\ s$; period of contact – $t_c = 0.26\ s$. In the Fig. 3 the function of the force produced by a runner is presented.

In the next stage of the investigation, the 'half-sine' mathematical model of loading generated by a single running user, provided by equation (2), was discussed [1]. For this model, the values were assumed as follow: pedestrian weight – $G = 800\ N$; impact factor – $k_z = 3$ (according to [2]); frequency of running – $f_u = 1.92\ Hz$; period of running – $(t_u = 1/f_u)\ t_u = 0.52$ sec; period of contact with the deck – $t_c = 0.26\ s$. The curve of the function of forces produced by a runner is presented in Fig. 4.

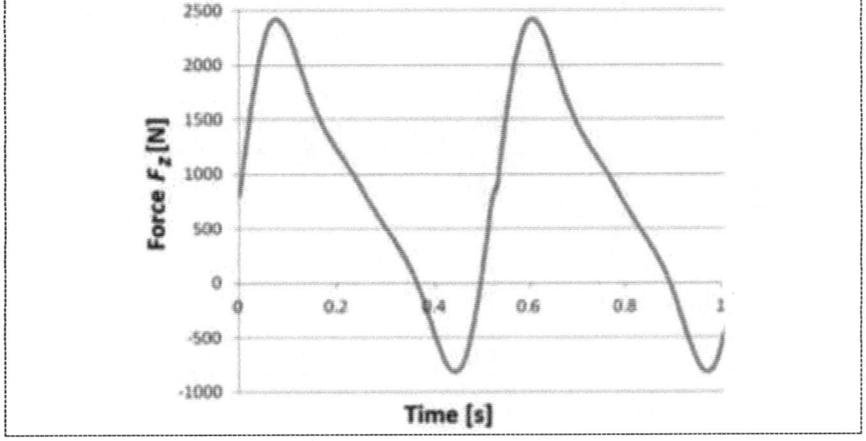

IOP Conf. Series: Materials Science and Engineering **307** (2018) 012060 doi:10.1088/1757-899X/307/1/012060

Figure 3. Periodic forces F_z generated by a runner – 'sine' model [3]

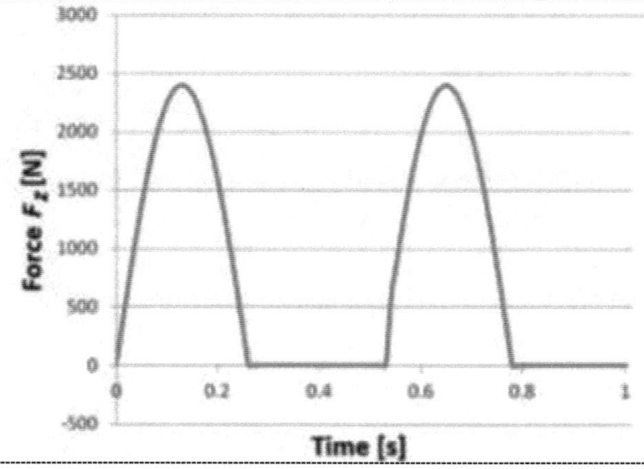

Figure 4. Periodic forces F_z generated by a runner – *'half-sine' model [1]*

7. Comparison of the results obtained for the different models of dynamic forces

The time histories of accelerations and displacements obtained at point B (midspan), for the first 'sine' model of dynamic forces [3] (see Fig. 3) are presented in Fig. 5 and 6. The accelerations and displacements obtained for the second 'half-sine' model [1] (see Fig. 4) are presented in Fig. 7 and 8.

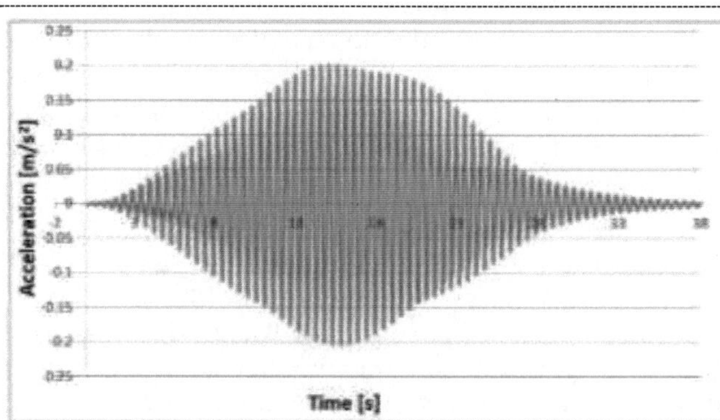

Figure 5. Vertical accelerations occurring at point B for the 'sine' model of dynamic forces adopted from [3]

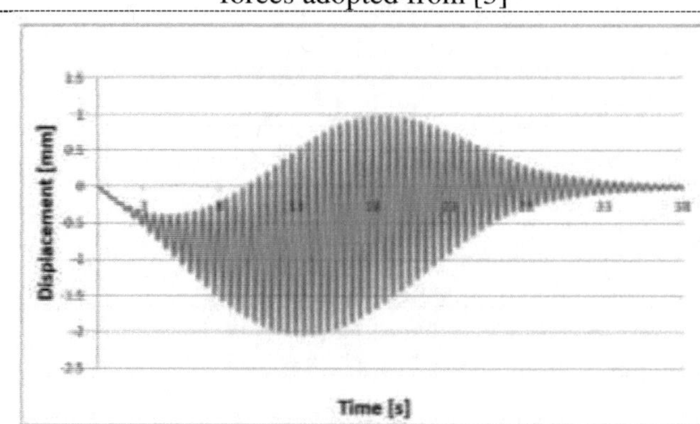

Figure 6. Vertical accelerations occurring at point B for the 'sine' model of dynamic forces adopted from [3]

IOP Conf. Series: Materials Science and Engineering **307** (2018) 012060 doi:10.1088/1757-899X/307/1/012060

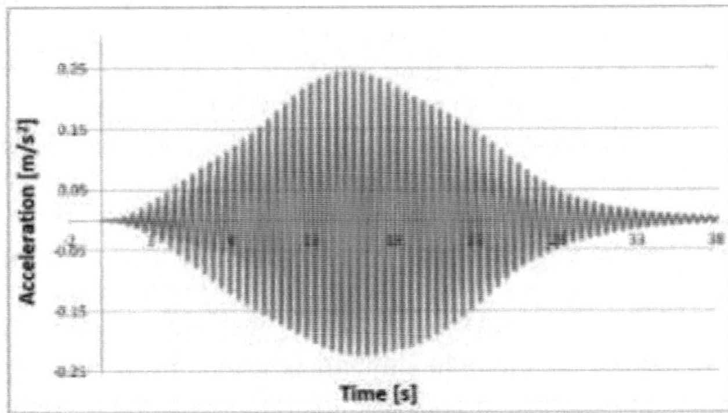

Figure 7. Vertical accelerations occurring at point B for the 'half-sine' model of dynamic forces adopted from [1]

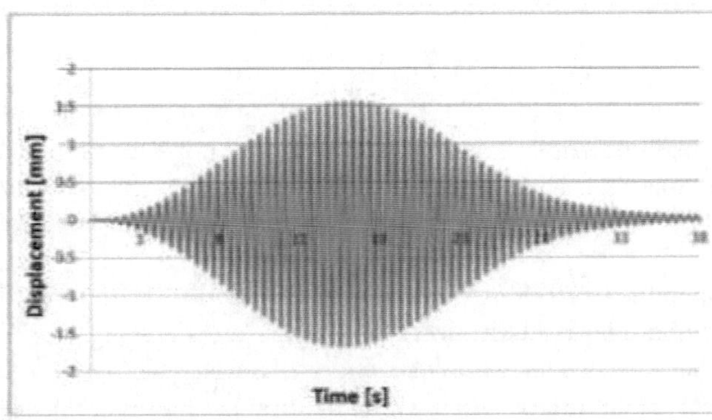

Figure 8. Vertical accelerations occurring at point B for the 'half-sine' model of dynamic forces adopted from [1]

The maximum values of accelerations and displacements that occurred at points A, B and C during the pedestrian passage along the footbridge, for both models of dynamic forces are summarised in Table 4.

Table 4. Comparison of maximum accelerations and displacements at points A, B and C obtained for two models of dynamic forces.

Point	*Maximum accelerations [m/s²]*		*Maximum displacements [mm]*	
	'Sine' model of forces [3]	*'Half-sine' model of forces [1]*	*'Sine' model of forces [3]*	*'Half-sine' model of forces [1]*
A	0.18	0.23	1.98	1.58
B	0.20	0.24	2.03	1.67
C	0.18	0.17	1.42	1.23

It can be observed from Table 4 that for both models of forces generated by a runner, the requirements of vibration comfort standards recommended by Eurocode for vibrations in vertical direction, are fulfilled. The level of oscillation does not exceed 0.7 m/s^2. Taken into consideration, the recommendation taken from of SÉTRA document (see Table 1), the comfort of use the investigated footbridge is on the maximum level.

IOP Conf. Series: Materials Science and Engineering **307** (2018) 012060 doi:10.1088/1757-899X/307/1/012060

8. Conclusions

Firstly, modal properties, i.e. modes and natural frequencies of the suspended footbridge were calculated. Secondly, the dynamic response of the footbridge was assessed taking into consideration two models of dynamic forces produced by a single running user of the structure. Finally, the comfort of footbridge use was checked. The following concluding remarks can be expressed:

- The results of the modal analysis shown that the first, vertical natural frequencies of the structure coincided with the frequency of pedestrian generated loading (walking or running) which could result in the resonance phenomenon.
- The values of accelerations and displacements obtained for the second half-sine model of dynamic forces [1] were greater than those obtained for the first model [3] up 20%. The mathematical model recommended by [3] is adequate only in the case of walking, because each step overlaps with the previous one thus making the force continuous. In the example of running this postulation is inappropriate. The loading generated by runner has a discontinuous nature. For the loading produced by a runner the 'half-sine' model is more complete. The dynamic analysis of a footbridge, perform with first 'sine' model of dynamic force, may underestimate the dynamic response of the pedestrian bridge.
- The vertical comfort criteria are not exceeded for a single pedestrian walking fast or running.

9. References

[1] Flaga A 2011 *Footbridges* (Warsaw: WKŁ)

[2] Zivanovic S Pavic A and Reynolds P 2005 *Vibration serviceability of footbridges under human-induced excitation: a literature review* Journal of Sound and Vibration (vol 279) pp 1-74

[3] Occhiuzzi A Spizzuoco M Ricciardelli F 2008 *Loading models and response control of footbridges excited by running pedestrians* Structural control and health monitoring (vol 15) pp 349-368

[4] BS5400 Part 2 Appendix C 1978 – Vibration Serviceability Requirements for Foot and Cycle Track 1985 (Bridges Standards Institution)

[5] ISO 10137 2007 *Bases for Design of Structures. Serviceability of buildings and walkways against vibrations* (Geneva: International Organization for Standarization)

[6] EN 1990:2002/Al 2005 *Eurocode - Basis of structural design* (Brussels: CEN)

[7] SÉTRA 2006 *Assessment of vibrational behaviour of footbridges under pedestrian loading. Technical guide* (Paris: Technical Department for Transport, Roads and Bridges Engineering and Road Safety)

[8] Bachmann H and Ammann H W 1987 *Vibrations in Structures Induced by Man and Machines* (Zürich: IABSE - AIPC – IVBH)

[9] ABAQUS Users' Manual v. 6.13. Dassault Systemes Simulia Corp. 2013 (Providence, RI)

[10] Bachmann H 2002 *Lively footbridges – real challenge* Proc. Int. Con. on the Design and Dynamic Behaviour of Footbridges – Footbridge (Paris)

[11] Dulinska J M Murzyn I J and Pluta K 2016 *Evaluation of dynamic response of a footbridge to human movement and traffic-induced vibrations* Proc. 6th Int. Con. on Structural Engineering, Mechanics and Computation (Cape Town: CRC Press, Taylor&Francis Group) pp 183-188

IOP Conf. Series: Materials Science and Engineering **307** (2018) 012061 doi:10.1088/1757-899X/307/1/012061

Assembly force modelling method on fixtures of automotive body compliant sheet metal parts

Yu Kuigang[1,2], Gao Xu[1], Wang Yanan[1,2] and Yang Zhihong[1,2]

1 Shandong University, School of Mechanical Engineering, Jinan, 250061, China
2 Key Laboratory of High-efficiency and Clean Mechanical Manufacturing (Shandong University), Ministry of Education, Jinan, 250061, China
E-mail:yukuigang@sdu.edu.cn

Abstract. Some assembly forces will be generated on fixtures because of some existing variation sources during the welding process of automotive body compliant metal parts. This article proposes an assembly force modelling method on fixtures during the welding process of compliant sheet metal parts to decrease the assembly forces. The assembly force models on fixtures are established to show the influence relationship between assembly forces and all variation sources, and the finite element method is used to obtain all influence coefficients of assembly forces. The obtained assembly force models on fixtures describe the propagating principle of assembly forces in all welding subprocesses. Cases of different fixture layouts show that the fixture layout can greatly influence assembly forces on fixtures. So assembly forces on fixtures can be decreased through the optimization of variation sources and the fixture layout. This method can also be referred for other assembly processes, such as adhesive connection and fastener connection of compliant sheet metal parts.

1. Introduction

Some assembly forces will be generated on fixtures because of some existing variation sources during these assembly processes. These forces will cause some assembly variation and some residual stresses for the assembly after fixtures are released. So decreasing these assembly forces will greatly decrease the residual stresses risk, and improve the assembly quality. This research will focus on the generating principles of assembly forces on fixtures during the welding process of automotive body compliant sheet metal parts.

It is very important to analyze the generating principles of assembly forces on fixtures to decrease assembly forces of automotive body compliant metal parts. Some relevant research has been done in recent years. Dong and Kang developed an approach based on the response surface methodology to analyze the assembly variation and residual stress of compliant structures [1]. The finite element method and regression models were used to obtain the virtual experiment data and fit them. The illustrating result using a composite-metal assembly showed that the method was a practical and reliable solution to the analysis of compliant structures. Yi et al. used a 3D thermo-mechanical finite element analysis method to predict the welding distortion and residual stress of cylinder-shaped multi-pass layer weldments. A continuous welding procedure was adopted to investigate the influence of deposition sequence and welding heat input on the welding distortion and residual stress [2]. Zia and Qiao studied the influence of the clearances in mating parts on different stresses and deformation

IOP Conf. Series: Materials Science and Engineering **307** (2018) 012061 doi:10.1088/1757-899X/307/1/012061

adopting a structure from automotive spot welding equipments. The effect of these clearances was determined using finite element analysis simulations for the process model of the product [3]. Faraz et al. used Monte Carlo simulations to investigate the expected size and location of the working regions of parts in a sheet metal family subjected to random resistance spot welding forces [4]. Astarita et al. finished the study and the experimental characterization of the influence of the process parameters on forces acting on the tool during the friction-stir welding process, and studied the correlation between the forces and the grain size [5].

To investigate the variation propagating principle of automotive compliant sheet metal parts during the welding process, some finite element modeling method had bee proposed to effectively model the assembly variation of compliant sheet metal parts. Liu and Hu firstly proposed the Method of Influence Coefficients (MIC) using the Finite Element Method (FEM) to set variation relationship between assembly variation and variation sources when deformable sheet metal parts were welded [6]. Camelio et al proposed a multi-station assembly variation model of compliant parts considering part geometric variation, fixture variation and welding gun variation [7]. Liao and Wang applied wavelets analysis and the FEM in the variation modelling of compliant assembly [8]. Yu and Yang proposed a system assembly variation modelling method of multi assembly subprocesses using the FEM [9].

In above research, though some modelling methods of assembly forces or residual stresses were given, the generating principle of assembly forces or residual stresses and how to decrease the assembly forces or residual stresses by optimizing the assembly process were not clear. So in this research the assembly variation finite element modeling method of compliant parts is referred to set the generating principle of assembly forces on fixtures during the welding process of compliant sheet metal parts. And assembly force modelling methods in every assembly subprocess on fixtures and the influencing relationship between assembly forces and variation sources and the fixture locating layout will be given.

2. Assembly process and assembly force propagating process of compliant sheet metal parts

The assembly process is shown in Figure 1. To analyze the assembly process it can be divided into four subprocesses, i.e. the fixture locating subprocess, the welding gun clamping subprocess, the welding operation subprocess and the releasing subprocess. Assembly forces on all fixtures of parts must be generated during these assembly subprocesses because of the variation sources, i.e. manufacturing variation of parts and fixture locating variation.

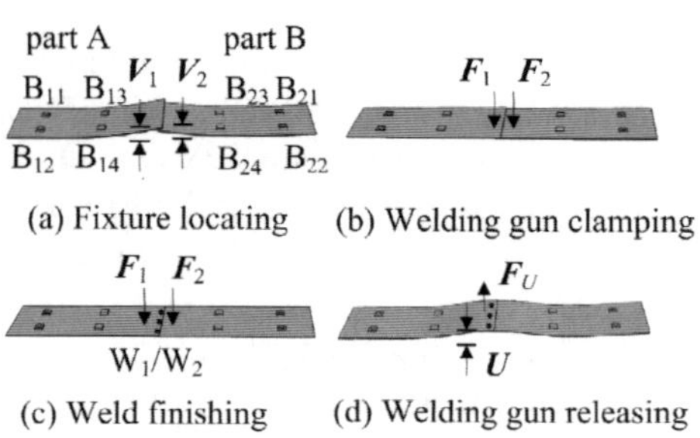

Figure 1. Assembly process of compliant sheet metal parts.

Figure 2 demonstrates the assembly force propagating process during these assembly subprocesses caused by variation sources. An over located fixture layout is usually adopted for compliant sheet metal parts because of the compliant characteristic of parts. Usually there are some manufacturing variation in the additional locating points and welding points of parts and fixture locating variation. These variation sources will cause assembly forces on fixtures to overcome themselves during the assembly process.

In Liu and Hu [6], to set up the assembly spring-back model after the weld, some forces relationship caused during the welding process as shown in Figure 1 were described. In this research the referred assembly forces are focused on fixtures during the assembly process because the assembly forces on fixtures greatly influence the assembly variation and the residual stresses in the next assembly process. In some assembly subprocesses of Figure 2, the existing variation sources will cause new assembly forces on fixtures. So the assembly forces on fixtures are generated and propagated during the whole assembly process. The detail assembly force models in every assembly subprocess will be described in the next section.

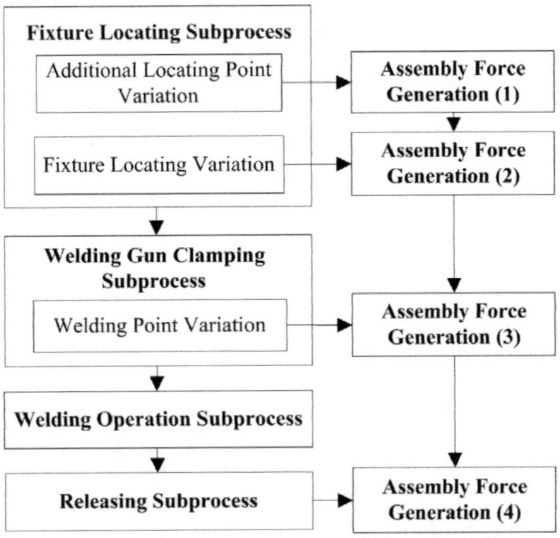

Figure 2. Assembly force propagation during assembly subprocesses.

3. Assembly force modelling method of compliant sheet metal parts

To introduce the assembly force modelling method, Figure 3 gives an assembly of compliant sheet metal parts. It shows that the part 1 and part 2 adopted a 5-2-1 and a 4-2-1 fixture layouts, respectively. All fixture locating points on parts are divided into determinate locating points and additional locating points. The determinate locating points are consisted of six locating points composing a 3-2-1 locating layout to constrain six motion freedoms of a part, and the additional locating points are added to constrain more compliant deformations. In Figure 3 the locating points on B_{14}, B_{15} and B_{24} are all additional locating points.

In the part 1, the locating pins P_{11} and P_{12} constrain the movements in the x and y directions and the rotating motion by the z direction, and the locating blocks B_{11}~B_{15} constrain the z direction movement and rotating motions by the x and y directions. Similarly, in the part 2 the locating pins P_{21} and P_{22} constrain the movements in the x and y directions and the rotating motion by the z direction, and the locating blocks B_{21}~B_{24} constrain the z direction movement and rotating motions by the x and y directions. W_{11}~W_{13} and W_{21}~W_{23} are the welding points of these parts.

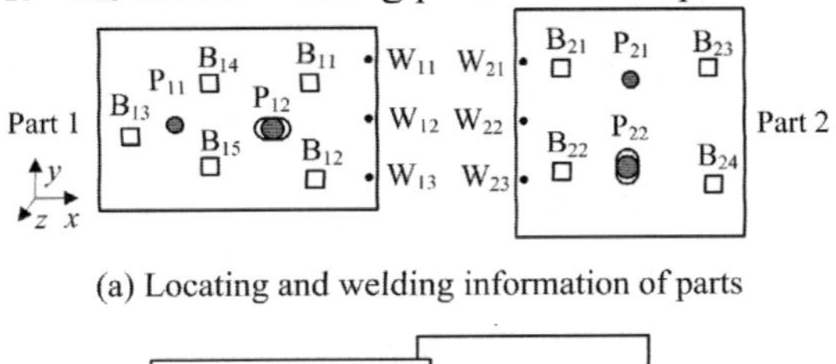

(a) Locating and welding information of parts

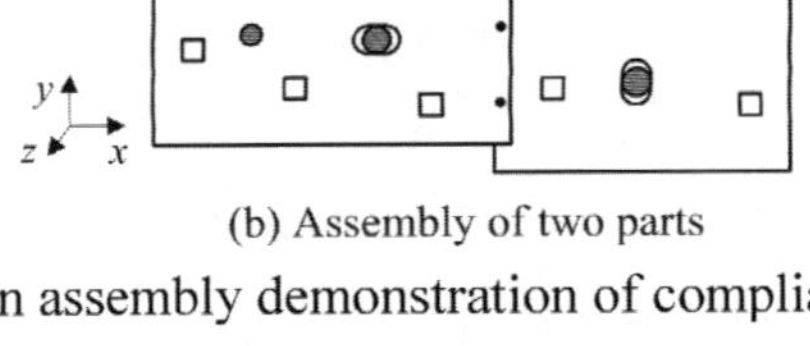

(b) Assembly of two parts

Figure 3. An assembly demonstration of compliant sheet metal parts.

Only assembly forces on locating blocks are considered in this research, so the fixture variation in locating pins are ignored. Then the existing variation sources are given as vector denotations.

The additional point variation of the part 1 is $\left\{V_{\mathrm{B}_1}^{(0)}\right\}=\left\{v_{\mathrm{B}_{14}}^{(0)} \quad v_{\mathrm{B}_{15}}^{(0)}\right\}^T$.

The additional point variation of the part 2 is $\left\{V_{\mathrm{B}_2}^{(0)}\right\}=\left\{v_{\mathrm{B}_{24}}^{(0)}\right\}$.

The welding point variation of the part 1 is $\left\{V_{\mathrm{W}_1}^{(0)}\right\}=\left\{v_{\mathrm{W}_{11}}^{(0)} \quad v_{\mathrm{W}_{12}}^{(0)} \quad v_{\mathrm{W}_{13}}^{(0)}\right\}^T$.

The welding point variation of the part 2 is $\left\{V_{\mathrm{W}_2}^{(0)}\right\}=\left\{v_{\mathrm{W}_{21}}^{(0)} \quad v_{\mathrm{W}_{22}}^{(0)} \quad v_{\mathrm{W}_{23}}^{(0)}\right\}^T$.

The fixture locating variation of the part 1 is $\{V_{f1}\}=\{V_{f11}\ V_{f12}\ V_{f13}\ V_{f14}\ V_{f15}\}^T$.

The fixture locating variation of the part 2 is $\{V_{f2}\}=\{V_{f21}\ V_{f22}\ V_{f23}\ V_{f24}\}^T$.

The superscript (0) denotes the initial status before the additional location of parts.

3.1. Assembly force models on fixtures in the fixture locating subprocess

In this subprocess, two types of variation sources can cause assembly forces on fixtures of one part, i.e. the additional locating point variation and fixture locating variation.

(1) Assembly force modelling caused by the additional locating point variation

If there exists some manufacturing variation on the additional locating points of one part, some forces will be produced to overcome these variation when these additional locating points are clamped to the nominal. Then these additional locating point forces will cause some assembly reaction forces on the other locating points, i.e. the determinate locating points.

Assume that the additional locating forces on B_{14} and B_{15} are $f_{14}^{(1)}$ and $f_{15}^{(1)}$, and that on B_{24} is $f_{24}^{(1)}$. The solving process using the finite element method is shown as follows.

The finite element model of the part 1 is created in which the boundary conditions on the locating points B_{11}, B_{12}, B_{13}, P_{11} and P_{12} are shown in Figure 3. The movements on these locating points are all constrained to be zero. Apply one unit force on B_{14} and B_{15} which is in accordance with the locating direction on B_{14} and B_{15}, respectively, and the displacements on B_{14} and B_{15} can be obtained. They are denoted as vectors $\begin{Bmatrix} c_{11} \\ c_{21} \end{Bmatrix}$ and $\begin{Bmatrix} c_{12} \\ c_{22} \end{Bmatrix}$. Then the relationship between the caused displacements and the applied forces on the additional locating points is

$$\{V\}=\begin{bmatrix} c_{11} & c_{12} \\ c_{21} & c_{22} \end{bmatrix}\begin{Bmatrix} f_{14}^{(1)} \\ f_{15}^{(1)} \end{Bmatrix}=[C]\begin{Bmatrix} f_{14}^{(1)} \\ f_{15}^{(1)} \end{Bmatrix} \tag{1}$$

So the caused forces on the additional locating points of the part 1 to overcome the additional locating point variation are

$$\begin{aligned}\left\{F_{1(N-3)}^{(1)}\right\}&=\begin{Bmatrix} f_{14}^{(1)} \\ f_{15}^{(1)} \end{Bmatrix}=[C_1]^{-1}\{V\}=[K_1]\begin{Bmatrix} -v_{14}^{(0)} \\ -v_{15}^{(0)} \end{Bmatrix} \\ &=[K_1]\left\{-V_{B_1}^{(0)}\right\}\end{aligned} \tag{2}$$

where the force vector $\left\{F_{1(N-3)}^{(1)}\right\}$ is the additional locating point forces of the part 1, $[K_1]$ is the rigidness matrix of the part 1 on the additional locating points B_{14} and B_{15}, which is the inversion of the matrix $[C]$.

Also, the additional locating force of the part 2 on B_{24} can be obtained as follows.

$$\left\{F_{2(N-3)}^{(1)}\right\}=\left\{f_{24}^{(1)}\right\}=[K_2]\left\{-v_{14}^{(0)}\right\}=[K_2]\left\{-V_{B_2}^{(0)}\right\} \tag{3}$$

where the force vector $\left\{F_{2(N-3)}^{(1)}\right\}$ is the additional locating point force of the part 2, $[K_2]$ is the rigidness matrix of the part 2 on the additional locating point B_{24}.

After the caused forces on the additional locating points are obtained, assembly forces on the determinate locating points can be solved through the propagating relationship analysis between the additional locating points and the determinate locating points using the finite element method. Then all locating point force vector containing the additional locating points and the determinate locating points is

$$\left\{F_1^{(1)}\right\} = \left[G_{1B}\right]\left\{F_{1(N-3)}^{(1)}\right\} \tag{4}$$

$$\left\{F_2^{(1)}\right\} = \left[G_{2B}\right]\left\{F_{2(N-3)}^{(1)}\right\} \tag{5}$$

where $\left\{F_1^{(1)}\right\}$ and $\left\{F_2^{(1)}\right\}$ are the assembly force vectors of the part 1 and part 2, respectively, $\left[G_{1B}\right]$ and $\left[G_{2B}\right]$ are the reaction coefficients between all locating points and the additional locating points of the part 1 and part 2, respectively.

The expressions of $\left[G_{1B}\right]$ and $\left[G_{2B}\right]$ are

$$\left[G_{1B}\right] = \begin{bmatrix} g_{1B(11)} & g_{1B(12)} \\ g_{1B(21)} & g_{1B(22)} \\ g_{1B(31)} & g_{1B(32)} \\ 1 & 0 \\ 0 & 1 \end{bmatrix} \tag{6}$$

$$\left[G_{2B}\right] = \begin{bmatrix} g_{2B(11)} \\ g_{2B(21)} \\ g_{2B(31)} \\ 1 \end{bmatrix} \tag{7}$$

where the parameters $\left\{g_{1B(11)} \quad g_{1B(21)} \quad g_{1B(31)}\right\}$ and $\left\{g_{1B(12)} \quad g_{1B(22)} \quad g_{1B(32)}\right\}$ are the caused forces on B_{11}, B_{12} and B_{13} when one unit force is applied on B_{14} and B_{15}, respectively.

Similarly, after the additional locating points B_{14} and B_{15} are located, the welding point variation of the part 1 will be changed to

$$\{V_{W_1}^{(1)}\} = [S_{WB_1}]\{-V_{B_1}^{(0)}\} + \{V_{W_1}^{(0)}\} \tag{8}$$

where the matrix $[S_{WB_1}]$ is the sensitivity matrix of the welding point variation of the part 1 to its additional locating variation. The following gives the solving method using the finite element method.

Apply one unit displacement separately on the additional locating points and compute the displacement variation of all welding points of the part 1. The displacement variation vectors composed of all the welding point variation constitute the matrix $[S_{WB_1}]$.

Similarly, the welding point variation of the part 2 is also obtained as follows.

$$\{V_{W_2}^{(1)}\} = [S_{WB_2}]\{-V_{B_2}^{(0)}\} + \{V_{W_2}^{(0)}\} \tag{9}$$

where the matrix $[S_{WB_2}]$ is the sensitivity matrix of the welding point variation of the part 2 to its additional locating variation.

(2) Assembly force modelling caused by fixture locating variation

When the fixtures of one part are located and clamped, some new assembly forces will be added on these fixtures and some new assembly variation will be added on welding points if there exists some fixture locating variation.

Assume that the fixture locating variation for the part 1 is $\{V_{f1}\}$, then the welding point variation of the part 1 is changed to

$$\{V_{W_1}^{(2)}\} = [S_{fW_1}]\{V_{f1}\} + \{V_{W_1}^{(1)}\} \tag{10}$$

IOP Conf. Series: Materials Science and Engineering **307** (2018) 012061 doi:10.1088/1757-899X/307/1/012061

where the matrix $[S_{fW_1}]$ denotes the sensitivity matrix of the part 1's welding point variation to its fixture locating variation. Its solving process using the finite element method is as follows.

Separately apply one unit displacement on all fixtures of the part 1, and compute the variation vectors of its all welding points. All variation vectors compose the matrix $[S_{fW_1}]$.

Similarly, the welding point variation of the part 2 is obtained as follows.

$$\{V_{W_2}^{(2)}\} = [S_{fW_2}]\{V_{f2}\} + \{V_{W_2}^{(1)}\} \tag{11}$$

where the matrix $[S_{fW_2}]$ denotes the sensitivity matrix of the part 2's welding point variation to its fixture locating variation.

For the fixtures on the part 1 and part 2, assembly forces are changed to

$$\{F_1^{(2)}\} = [G_{1f}]\{V_{f1}\} + \{F_1^{(1)}\} \tag{12}$$

$$\{F_2^{(2)}\} = [G_{2f}]\{V_{f2}\} + \{F_2^{(1)}\} \tag{13}$$

where the matrices $[G_{1f}]$ and $[G_{2f}]$ are the influence coefficients of assembly forces on fixture locating points of the part 1 and part 2 to their own fixture locating variation, respectively. The following gives the solving process using the finite element.

Apply one unit displacement on the *j*-th fixture, and compute the force vector $\{G_{1fj}\}$ on all fixtures of the part. After one unit displacement is separately applied on all fixtures, all obtained force vectors compose the coefficient matrix $[G_{1f}]$. Similarly, the matrix $[G_{2f}]$ can also be obtained.

3.2. Assembly force models on fixtures in the welding gun clamping subprocess

In this subprocess, welding areas of parts will be clamped by welding guns to the nominal. So when there exists some variation on these welding points, some forces $\{F_W\}$ are needed to be applied on these welding points to overcome these variation. These forces will cause new reaction forces on fixtures.

The solving process of $\{F_W\}$ is shown as follows.

Apply one unit force on the *j*-th welding point of one part, and the displacements of all welding points of this part $\{C_{Wj}\}$ (*j*=1,…,*N*) can be obtained. *N* is the welding point number of this part. After one unit force is separately applied on all welding points of this part, the obtained displacement vectors on all welding points compose one displacement matrix. The relationship between the displacements and forces of all welding points is

$$\{V_W\} = \begin{bmatrix} c_{W11} & c_{W12} & \cdots & c_{W1N} \\ c_{W21} & c_{W22} & \cdots & c_{W2N} \\ \vdots & \vdots & \vdots & \vdots \\ c_{WN1} & c_{WN2} & \cdots & c_{WNN} \end{bmatrix} \begin{Bmatrix} f_{W1} \\ f_{W2} \\ \vdots \\ f_{WN} \end{Bmatrix} = [C_W]\{F_W\} \tag{14}$$

Then the needed forces applied on all welding points of the part 1 and the part 2 can be obtained as follows, respectively.

$$\{F_{W_1}\} = [C_{W_1}]^{-1}\{-V_{W_1}^{(2)}\} \tag{15}$$

$$\{F_{W_2}\} = [C_{W_2}]^{-1}\{-V_{W_2}^{(2)}\} \tag{16}$$

The caused reaction forces on fixtures of the part 1 and the part 2 can be obtained as follows, respectively.

$$\{F_1^{(3)}\} = [G_{1W}]\{F_{W_1}\} + \{F_1^{(2)}\} \tag{17}$$

$$\{F_2^{(3)}\} = [G_{2W}]\{F_{W_2}\} + \{F_2^{(2)}\} \tag{18}$$

where the matrices $[G_{1W}]$ and $[G_{2W}]$ are the influence coefficients between the fixture forces and the welding point forces of the part 1 and the part 2, respectively. The solving process is shown as follows.

Separately apply one unit force on all welding points of one part, and compute the force vector on all fixtures of this part. All force vectors caused by all the welding point forces of this part compose the matrices $[G_{1W}]$ or $[G_{2W}]$.

3.3. Assembly force models on fixtures in the welding gun releasing subprocess

After the welding operation is finished between parts, welding guns will be removed. Because there exist the welding point forces $\{F_{W_1}\}$ and $\{F_{W_2}\}$, the assembly will cause the spring-back after the welding guns are released. The spring-back forces are equal to the forces $\{F_{W_1}\}$ and $\{F_{W_2}\}$, but their directions are reverse.

Solving the influence relationship between the spring-back forces and the reaction forces on all fixtures of the part 1 and the part 2, then assembly forces on all fixtures of the part 1 and the part 2 are obtained after the spring-back as follows.

$$\{F\} = \begin{Bmatrix} F_1^{(4)} \\ F_2^{(4)} \end{Bmatrix} = [G_S]\begin{Bmatrix} -F_{W_1} \\ -F_{W_2} \end{Bmatrix} + \begin{Bmatrix} F_1^{(3)} \\ F_2^{(3)} \end{Bmatrix} \tag{19}$$

where the matrix $[G_S]$ is the influence coefficient matrix between all fixture forces and all welding point forces of the assembly.

The solving process using the finite element method is shown as follows.

Apply one unit force on every welding point of the assembly, and compute the reaction forces on all fixtures of the assembly which compose the matrix $[G_S]$.

In the given case in Figure 3, there are nine fixtures in which five is on the part 1 and four is on the part 2, and six welding points in which the part 1 and the part 2 all have three. So the matrix $[G_S]$ is a 9×6 matrix, it can be changed to

$$[G_S] = \begin{bmatrix} G_{S11} & G_{S12} \\ G_{S21} & G_{S22} \end{bmatrix} \tag{20}$$

where the matrix $[G_{S11}]$ is a 5×3 matrix which denotes the influence relationship between the five fixture forces of the part 1 and its three welding point forces under the assembly status, the matrix $[G_{S12}]$ is also a 5×3 matrix which denotes the influence relationship between the five fixture forces of the part 1 and three welding point forces of the part 2 under the assembly status, the matrix $[G_{S21}]$ is a 4×3 matrix which denotes the influence relationship between the four fixture forces of the part 2 and three welding point forces of the part 1 under the assembly status, and the matrix $[G_{S22}]$ is a 4×3 matrix which denotes the influence relationship between the four fixture forces of the part 2 and its three welding point forces under the assembly status.

Then the equation (19) is changed to

$$\begin{cases} F_1^{(4)} = -G_{S11}F_{W_1} - G_{S12}F_{W_2} + F_1^{(3)} \\ F_2^{(4)} = -G_{S21}F_{W_1} - G_{S22}F_{W_2} + F_2^{(3)} \end{cases} \tag{21}$$

3.4. Combined assembly force models

Combine the above assembly force models of fixtures in every assembly subprocess, and the finial combined assembly force models of fixtures are shown as follows.

$$F_1^{(4)}=[S_{B11}]\{V_{B_1}^{(0)}\}+[S_{B12}]\{V_{B_2}^{(0)}\}+[S_{W11}]\{V_{W_1}^{(0)}\}+[S_{W12}]\{V_{W_2}^{(0)}\}+[S_{f11}]\{V_{f1}\}+[S_{f12}]\{V_{f2}\} \quad (22)$$

$$F_2^{(4)}=[S_{B21}]\{V_{B_1}^{(0)}\}+[S_{B22}]\{V_{B_2}^{(0)}\}+[S_{W21}]\{V_{W_1}^{(0)}\}+[S_{W22}]\{V_{W_2}^{(0)}\}+[S_{f21}]\{V_{f1}\}+[S_{f22}]\{V_{f2}\} \quad (23)$$

where $F_1^{(4)}$ and $F_2^{(4)}$ are the assembly force vectors caused on all fixtures of the part 1 and the part 2, respectively, $[S_{B11}]$ and $[S_{B21}]$ are the sensitivity matrices of the fixture forces of the part 1 and the part 2 to the additional locating point variation of the part 1, respectively, $[S_{B12}]$ and $[S_{B22}]$ are the sensitivity matrices of the fixture forces of the part 1 and the part 2 to the additional locating point variation of the part 2, respectively, $[S_{W11}]$ and $[S_{W21}]$ are the sensitivity matrices of the fixture forces of the part 1 and the part 2 to the welding point variation of the part 1, $[S_{W12}]$ and $[S_{W22}]$ are the sensitivity matrices of the fixture forces of the part 1 and the part 2 to the welding point variation of the part 2, $[S_{f11}]$ and $[S_{f21}]$ are the sensitivity matrices of the fixture forces of the part 1 and the part 2 to the fixture locating point variation of the part 1, $[S_{f12}]$ and $[S_{f22}]$ are the sensitivity matrices of the fixture forces of the part 1 and the part 2 to the fixture locating point variation of the part 2.

These sensitivity matrix expressions are listed as follows.

$$\begin{cases} [S_{B11}]=[G_{1W}C_{W_1}^{-1}S_{WB_1}-G_{1B}C_1^{-1}-G_{S11}C_{W_1}^{-1}S_{WB_1}] \\ [S_{B12}]=[-G_{S12}C_{W_2}^{-1}S_{WB_2}] \\ [S_{W11}]=[G_{S11}C_{W_1}^{-1}-G_{1W}C_{W_1}^{-1}] \\ [S_{W12}]=[G_{S12}C_{W_2}^{-1}] \\ [S_{f11}]=[G_{1f}-G_{1W}C_{W_1}^{-1}S_{fW_1}+G_{S11}C_{W_1}^{-1}S_{fW_1}] \\ [S_{f12}]=[G_{S12}C_{W_2}^{-1}S_{fW_2}] \end{cases} \quad (24)$$

$$\begin{cases} [S_{B21}]=[-G_{S21}C_{W_1}^{-1}S_{WB_1}] \\ [S_{B22}]=[G_{2W}C_{W_2}^{-1}S_{WB_2}-G_{2B}C_2^{-1}-G_{S22}C_{W_2}^{-1}S_{WB_2}] \\ [S_{W21}]=[G_{S21}C_{W_1}^{-1}] \\ [S_{W22}]=[G_{S22}C_{W_2}^{-1}-G_{2W}C_{W_2}^{-1}] \\ [S_{f21}]=[G_{S21}C_{W_1}^{-1}S_{fW_1}] \\ [S_{f22}]=[G_{2f}-G_{2W}C_{W_2}^{-1}S_{fW_2}+G_{S22}C_{W_2}^{-1}S_{fW_2}] \end{cases} \quad (25)$$

Using these equations, assembly forces on fixtures after the weld can be predicted according to variation sources, and the sensitivity analysis of assembly forces to variation sources can be made using these obtained assembly force models to optimize variation source design to decrease assembly forces.

The above proposed modelling method of assembly forces on fixtures can also be referred for other assembly processes to help the optimization of assembly processes, such as adhesive connection and fastener connection of compliant sheet metal parts.

4. Assembly case analysis

4.1. Case description

The assembly case in Figure 3 is adopted to demonstrate the proposed assembly force modelling method. The parameters of the part 1 are $160\times240\times1mm^3$ and those of the part 2 are $200\times200\times1mm^3$. The modulus of elasticity and poisson's ratio are adopted as $207{,}000N/mm^2$ and 0.3, respectively. The software of ABAQUS is used to analyze the assembly process. The finite element models are shown in Figure 4.

In Figure 4 the symbol 'Δ' denotes the fixture constraint position corresponding to one node of the finite element model below which the letter means the constrained direction of the part. The symbol '●' denotes the welding point position between these two parts. In this analysis the tie constraint is adopted for welding points of the two parts during the welding operation. The information of fixtures and welding points are listed in Table 1.

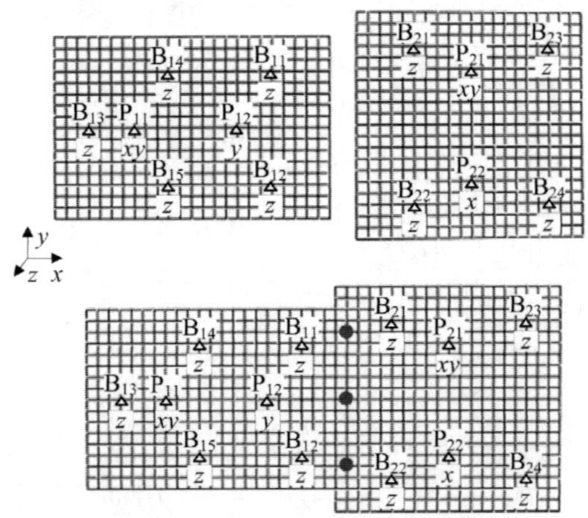

Figure 4. Finite element models.

Table 1. Information of fixtures and welding points.

Fixtures		Coordinates	Welding points
Part 1	B_{11}	(-30,50,0)	W_{11}
	B_{12}	(-30,-50,0)	W_{12}
	B_{13}	(-190,0,0)	W_{13}
	B_{14}	(-120,50,0)	
	B_{15}	(-120,-50,0)	
	P_{11}	(-150,0,0)	
	P_{12}	(-60,0,0)	
Part 2	B_{21}	(50,70,-1)	W_{21}
	B_{22}	(50,-70,-1)	W_{22}
	B_{23}	(170,70,-1)	W_{23}
	B_{24}	(170,-70,-1)	
	P_{21}	(100,50,-1)	
	P_{22}	(100,-50,-1)	

4.2. Assembly force computation

According to the solving method of process parameters in former sections, the sensitivity matrices in Equations (24) and (25) can be obtained using the finite element models in Figure 4 as follows. They are

IOP Conf. Series: Materials Science and Engineering **307** (2018) 012061 doi:10.1088/1757-899X/307/1/012061

$$S_{B11}=\begin{bmatrix} 209.49 & -209.3 \\ -209.29 & 209.49 \\ 0.1896 & 0.1095 \\ -210.27 & 210.04 \\ 210.04 & -210.28 \end{bmatrix},\ S_{B12}=\begin{bmatrix} -0.0053 \\ -0.0549 \\ -0.0063 \\ 0.0163 \\ 0.0114 \end{bmatrix},\ S_{W11}=\begin{bmatrix} 6.1384 & 0.0365 & -6.0072 \\ -6.0131 & 0.0605 & 6.1233 \\ -0.0085 & 0.0476 & -0.0129 \\ 0.6421 & -0.0611 & -0.6572 \\ -0.6548 & -0.0746 & 0.6514 \end{bmatrix},$$

$$S_{W12}=\begin{bmatrix} -6.25 & -0.0018 & 6.0721 \\ 6.0663 & 0.0168 & -6.2647 \\ -0.0022 & -0.0290 & -0.0064 \\ -0.5976 & 0.0140 & 0.6661 \\ 0.6686 & 0.0036 & -0.5885 \end{bmatrix},\ S_{f11}=\begin{bmatrix} 217.2 & -216.9 & 0.0683 & -209.86 & 209.67 \\ -216.9 & 217.2 & 0.0703 & 209.67 & -209.87 \\ 0.0666 & 0.0618 & 0.1163 & -0.1090 & -0.1095 \\ -209.85 & 209.65 & -0.1120 & 210.65 & -210.41 \\ 209.65 & -209.85 & -0.1131 & -210.41 & 210.65 \end{bmatrix},$$

$$S_{f12}=\begin{bmatrix} -5.4253 & 5.1858 & 0.0546 & 0.0053 \\ 5.1883 & -5.4305 & 0.0056 & 0.0549 \\ -0.0233 & -0.0269 & 0.0063 & 0.0063 \\ -0.4891 & 0.5992 & -0.0112 & -0.0163 \\ 0.5970 & -0.4855 & -0.0165 & -0.0114 \end{bmatrix},\ S_{B21}=\begin{bmatrix} 0.4968 & -0.5835 \\ -0.5939 & 0.4853 \\ 0.0114 & 0.0083 \\ 0.0129 & 0.0162 \end{bmatrix},\ S_{B22}=\begin{bmatrix} -112.5 \\ 112.5 \\ 112.5 \\ -112.5 \end{bmatrix},$$

$$S_{W21}=\begin{bmatrix} -4.8388 & 0.0476 & 4.6578 \\ 4.6999 & -0.0593 & -4.7877 \\ 0.0136 & -0.0215 & 0.0401 \\ 0.0211 & 0.0245 & -0.0076 \end{bmatrix},\ S_{W22}=\begin{bmatrix} 5.0042 & 0.0390 & -4.8923 \\ -4.851 & -0.0439 & 5.0542 \\ -0.0670 & -0.0172 & 0.0473 \\ 0.0287 & 0.0184 & -0.0878 \end{bmatrix},$$

$$S_{f21}=\begin{bmatrix} -5.2601 & 5.0555 & -0.0154 & -0.4968 & 0.5835 \\ 5.0372 & -5.2686 & -0.0243 & 0.5939 & -0.4853 \\ 0.0099 & 0.0388 & 0.0031 & -0.0114 & -0.0083 \\ 0.0457 & 0.0145 & 0.0069 & -0.0129 & -0.0162 \end{bmatrix} \text{ and } S_{f22}=\begin{bmatrix} 116.85 & -116.65 & -112.5 & 112.5 \\ -116.65 & 116.86 & 112.5 & -112.5 \\ -112.5 & 112.5 & 112.5 & -112.5 \\ 112.5 & -112.5 & -112.5 & 112.5 \end{bmatrix}.$$

The variation measurement data are usually followed by the normal distribution. Then assume that manufacturing variation data of the part 1 and the part 2 follow N (0, 0.33332) and N (0, 0.52), and fixture locating variation data of the part 1 and the part 2 follow all N (0, 0.12). The assembly variation results are obtained by the Monte Carlo Simulation Method. Assembly force results are listed in Table 2.

Table 2. Assembly force statistical results.

Fixtures	Assembly force (N)	
	μ	6σ
B_{11}	2.1316	647.46
B_{12}	-2.1283	647.46
B_{13}	0.0003	0.3589
B_{14}	-2.045	647.22
B_{15}	2.0435	647.22
B_{21}	-0.1606	366.62
B_{22}	0.1577	366.77
B_{23}	0.0981	364.75
B_{24}	-0.0974	364.79

From the results of the part 1, the assembly forces on all fixtures are greatly unequal. The assembly force on B_{13} is much less than those on others. The forces on fixtures of the part 2 are nearly equal.

To compare the different fixture layouts, another two fixture layouts are given for the part 1 which are shown in Figure 5. Comparing to the first fixture layout in Figure 4, the second fixture layout in Figure 5(a) removes the fixture B_{13}, and the third fixture layout in Figure 5(b) removes the fixtures B_{14} and B_{15}. Assembly forces on fixtures are computed in these two fixture layouts and the results are shown in Table 3.

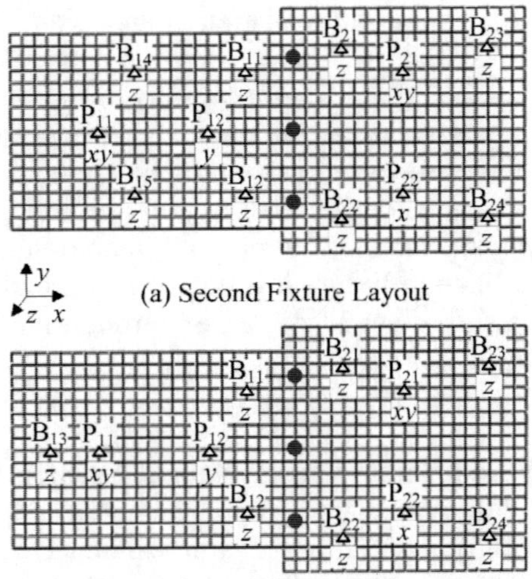

Figure 5. Another two fixture layouts.

Table 3. Assembly force statistical results.

Fixtures	Second Layout		Third Layout	
	μ	6σ	μ	6σ
B_{11}	0.13	532.81	0	35.3
B_{12}	-0.13	532.9	0	35.29
B_{13}			0	0.18
B_{14}	0.18	491.47		
B_{15}	-0.19	491.49		
B_{21}	0.73	391.76	0.27	367.83
B_{22}	-0.72	391.72	-0.27	367.9
B_{23}	-0.94	366.79	-0.27	365.92
B_{24}	0.94	366.8	0.27	365.94

Combing the results in Table 2 and 3, two conclusions can be made.

On the one hand, it can be seen that the fixture layout will greatly influence assembly forces on fixtures. The assembly forces on fixtures of the part 1 in the second fixture layout are less than those in the first fixture layout, and those in the third fixture layout are much less than those in the other two layouts while they are unequal on all fixtures of the part 1. Generally, the third fixture layout is best in these three fixture locating layouts. And the most optimization fixture locating layout can be obtained to minimize the assembly forces on fixtures.

On the other hand, in these three fixture layouts the assembly forces on fixtures of the part 2 are nearly equal. It shows that the fixture layout of one part only influences the assembly forces on its own fixtures. So the fixture locating layouts of all assembled parts can be separately optimized.

5. Conclusion

This article proposes an assembly force modelling method on fixtures during the welding process of compliant sheet metal parts to control the assembly force. This method can also be referred for other assembly processes, such as adhesive connection and fastener connection.

In this method the welding process is divided into four subprocesses. The assembly force propagating model in every subprocess is given. And the combined assembly force model is built after the weld is finished. The assembly force models on fixtures are established to show the influence relationship between assembly forces and all variation sources, and the FEM is used to obtain all influence coefficients of assembly forces. These variation sources contain the additional point variation, the welding point variation and the fixture locating variation. Using this modelling method, assembly forces on fixtures after the weld can be predicted according to variation sources, and the sensitivity analysis of assembly forces to variation sources can be made using these obtained assembly force models to optimize variation source design to decrease assembly forces.

A compliant assembly is adopted to analyze assembly forces on fixtures considering different fixture layouts. The result shows that the fixture layout can also greatly influence assembly forces on fixtures. So assembly forces on fixtures can be decreased through the separate fixture locating layout optimization of all assembled parts.

Acknowledgments

This work was supported by Shandong Provincial Natural Science Foundation, China (Grant No. ZR2016EEM31), Excellent Middle-Aged and Youth Scientist Award Foundation of Shandong Province (Grant No. BS2014ZZ003) and National Natural Science Foundation of China (Grant No. 51405269 and 51375277).

References

[1] Dong C S and Kang L 2012 Deformation and stress of a composite-metal assembly *Int. J. Adv. Manuf. Technol.* vol 61 p 1035-1042
[2] Yi H J, Kim J Y, Yoon J H and Kang S S 2011 Investigation on welding residual stress and distortion in a cylinder assembly by means of a 3D finite element method and experiments *Journal of Mechanical Science and Technology* vol 25 p 3185-3193
[3] Zia A and Qiao L 2012 The influence of assembly clearance on the deformation and stress of an assembly structure through FEA simulation *J. Mech. Sci. Technol.* vol 26 p 823-1827
[4] Faraz Z, Uihaq S W, Baqai A A, Tarar W A and Ali L 2014 Reconfigurable fixture locating layout for compliant sheet metal welded assemblies subjected to welding force variations *Proc. Inst. Mech. Eng. Part B-J. Eng. Manuf.* vol 228 p 740-750
[5] Astarita A, Squillace A and Carrino L 2014 Experimental Study of the Forces Acting on the Tool in the Friction-Stir Welding of AA 2024 T3 Sheets *J. Mater. Eng. Perform.* vol 23 p 3754-3761
[6] Liu S C and Hu S J 1997 Variation simulation for deformable sheet metal assemblies using finite element methods *J. Manuf. Sci. Eng.-Trans. ASME* vol 119 p 368-373
[7] Camelio J, Hu S J and Ceglarek D 2003 Modelling variation propagation of multi-station assembly systems with compliant parts *J. Mech. Des.* vol 125 p 673-681
[8] Liao X and Wang G G 2005 Wavelets-based method for variation analysis of non-rigid assemblies *Int. J. Mach. Tools Manuf.* vol 45 p 1551-1559
[9] Yu K G and Yang Z H 2015 Assembly variation modeling method research of compliant automobile body sheet metal parts using the finite element method *Int. J. Automot. Technol.* vol 16 p 51-56

Development of Lateral Prestress in High-Strength Concrete-Filled FRP Tubes

T Vincent[1] and T Ozbakkaloglu[2]

[1] College of Science and Engineering, Flinders University, Adelaide, South Australia, Australia 5042
[2] School of Civil, Environmental and Mining Engineering, The University of Adelaide, Adelaide, South Australia, Australia 5005

E-mail: thomas.vincent@flinders.edu.au

Abstract. This paper reports on an experimental investigation into the axial and lateral strain development of fiber reinforced polymer (FRP) confined high-strength concrete (HSC) with prestressed FRP shells. A total of 24 aramid FRP (AFRP)-confined concrete specimens were manufactured as concrete-filled FRP tubes (CFFTs) with instrumentation to measure the strain variations during application of prestress, removal of end constraints and progressive prestress losses. Prestressed CFFT specimens were prepared with three different dose rates of expansive mineral admixture to create a range of lateral prestress applied to AFRP tubes manufactured with sheet thicknesses of 0.2 or 0.3 mm/ply and referred to as lightly- or well-confined, respectively. In addition to these three levels of prestress, non-prestressed companion specimens were manufactured and tested to determine baseline performance. The experimental results from this study indicate that lateral prestressing of CFFTs manufactured with HSC can be achieved by varying the expansive mineral admixture dose rate with a lateral prestress of up to 7.3 MPa recorded in this study. Significant strain variations were measured during removal of the end constraints with up to 700 microstrain recorded in the axial direction. Finally, the measurement of prestress losses for the month following prestress application revealed minimal progressive losses, with only 250 and 100 με recorded for the axial and hoop strains, respectively.

1. Introduction

The use of fiber reinforced polymer (FRP) composites as a wrap or jacket for the confinement of concrete columns has gained significant research attention over the last two decades. Recent reviews [1-5] have revealed that many studies have examined both FRP-wrapped concrete (e.g., [6-9]), intended for strengthening existing concrete columns and concrete-filled FRP tubes (CFFTs) (e.g., [10-24]), intended for construction of new columns, which has resulted in a large number of proposed models (e.g., [25-30]).

Recently the construction industry has experienced a progressive increase in the use of higher strength concretes (HSCs) over normal strength concrete (NSC) due to the improved performance and reduced dead loads offered. However, as was recently demonstrated in Vincent and Ozbakkaloglu [31], FRP-confined HSC columns have a tendency to experience a more rapid and uncontrolled expansion during activation of the confinement mechanism compared to the more progressive expansion of NSC

IOP Conf. Series: Materials Science and Engineering **307** (2018) 012062 doi:10.1088/1757-899X/307/1/012062

due to the higher brittleness of HSC [3, 14]. The addition of prestrain to the FRP fibers is a potential technique to overcome the uncontrolled expansion, however limited studies exist on the application of prestress to FRP confined concrete.

To address this research gap, this paper reports on an experimental program to monitor prestress development for FRP-confined HSC specimens manufactured with four different dose rates of expansive mineral admixture. Variations in axial and lateral strains in the FRP shell are reported for the 3 key prestress stages: application of prestress; removal of end constraints; and progressive prestress losses. Initially, the details of the test program are presented followed by a comparison of the axial and hoop strain development during these 3 key stages. Finally, discussions on the effects of the key parameters are provided.

2. Test Database

2.1. Details of manufactured specimens

A total of 24 test specimens were manufactured and laboratory tested for this experimental program. All specimens were manufactured as cylindrical aramid FRP (AFRP)-confined concrete specimens with a diameter of 152 mm and height of 305 mm. Four different concrete mixes were used in the experimental program with average compressive strengths (f'_{co}) of 100.2 and 110.3 MPa for the expansive and non-expansive mixes, respectively. The FRP tubes for all specimens were constructed using a manual wet layup technique from a single continuous sheet to create four layers. The tubes were allowed 24 hours to cure before concrete was cast inside to create concrete-filled FRP tubes (CFFTs). Nominal fiber thicknesses of either 0.2 or 0.3 mm/layer were used to manufacture specimens with a total fiber thickness (t_f) of either 0.8 or 1.2 mm, referred to as lightly- or well-confined, respectively. Table 1 provides a summary of the test specimens.

Table 1. Test specimen details

Total fiber thickness, t_f (mm)	Concrete Mix	Concrete strength, f'_{co} (MPa)	f_{lu}/f'_{co}	Number of specimens
0.8	Mix 1	110.3	0.25	3
	Mix 2	100.2	0.27	3
	Mix 3	100.2	0.27	3
	Mix 4	100.2	0.27	3
1.2	Mix 1	110.3	0.37	3
	Mix 2	100.2	0.41	3
	Mix 3	100.2	0.41	3
	Mix 4	100.2	0.41	3

To laterally prestrain the fibers in the AFRP shell, the CFFT specimens were placed in steel prestressing rigs which consisted of 16-mm thick steel end caps connected by 16-mm diameter threaded rod. The specimens were placed in these rigs immediately after casting of concrete into the FRP tubes to restrain the extension of the concrete in the axial direction and limit the expansion to the lateral direction. For consistency, control specimens from all four concrete mixes were cast in steel tubes to replicate curing conditions experienced in the FRP tubes. This configuration also limited the tendency for loss of integrity for the expansive concrete mixes as they cured. These steel tubes were carefully cut and removed from the control cylinders immediately prior to testing f'_{co}. An example of the prestressing rigs can be seen in Figure 1.

a) *b)*

Figure 1. Steel end caps and threaded rod used to create the prestressing rig: a) steel tube specimen prior to casting of control specimen; b) CFFT specimen during concrete curing

To determine dose rates of expansive admixture numerous trial mixes were prepared to target fiber prestrain levels at 0, 10, 20 and 30% of their ultimate tensile strain (ε_f) for the lightly-confined specimens. Final mix designs and material properties of the fiber and FRP composite are provided in Tables 2 and 3 respectively, with a more detailed explanation of the experimental program provided elsewhere [32].

Table 2. Concrete mix proportions

Concrete constituent	Mix 1	Mix 2	Mix 3	Mix 4
Cement (kg/m^3)	462	462	462	462
Silica fume (kg/m^3)	88	88	88	88
Sand (kg/m^3)	700	647	629	618
Coarse Aggregate (kg/m^3)	1049	970	942	927
Water (kg/m^3)	136	163	166	172
Superplasticiser (kg/m^3)	15	14	14	13
Expansive admixture (kg/m^3)	0	105	143	165
Water to binder ratio*	0.269	0.266	0.255	0.255

* *Binder content includes cement, silica fume and expansive admixture*

Table 3. Fiber and FRP composite material properties

Type	Nominal thickness t_f (mm/ply)	Material properties					
		From manufacturer			Determined by coupon tests		
		Ultimate Tensile stress f_f (MPa)	Ultimate Tensile strain ε_f (%)	Elastic modulus E_f (GPa)	Ultimate tensile stress f (MPa)	Ultimate tensile strain ε (%)	Elastic modulus E (GPa)
Aramid fiber	0.2 or 0.3	2600	2.20	118.2	2390*	1.86	128.5*

* *Values determined by using nominal thickness of fibers*

2.2. Instrumentation

Each of the 24 test specimens was instrumented with two 5-mm unidirectional strain gauges attached to the external FRP tube surface at specimen mid-height. These strain gauges were aligned horizontally on opposing sides of the AFRP shell outside the overlap region to measure hoop strain development. These strain gauges recorded the prestrain development in the FRP tubes for over 50 days as the concrete cured, with measurements commencing immediately after casting of the concrete.

To monitor the influence of end constraints removal on prestress loss, the six specimens with the highest dose rate of expansive admixture were instrumented to measure axial and hoop strain development during this stage. These specimens were allocated two additional 5-mm strain gauges located at mid-height on opposing sides of the FRP tube mounted in the axial direction.

A single specimen was instrumented to monitor the amount of progressive prestress relaxation over a four week period after removal of the prestressing rig. This specimen had four 5-mm strain gauges attached to the surface of the FRP tube to monitor axial strain development and an additional four 5-mm gauges to monitor lateral strain development. These strain gauges were located at specimen mid-height outside the overlap region and were mounted with equal spacing around specimen perimeter.

3. Test Results

3.1. Lateral prestrain development

Lateral prestress development due to expansion of the curing concrete was recorded for over 50 days. Figure 2 presents the average lateral prestrain development and corresponding lateral prestress for the four concrete mixes presented in Table 2.

Figure 2(a) presents the lateral prestrain development for the lightly-confined test specimens, where it can be seen that a prestrain of approximately 4000 to 7000 με developed. Figure 2(b) presents a similar strain development trend for the well-confined test specimens with a prestrain of approximately 3000 to 5000 με recorded. It can be seen in these figures that, similar to concrete compressive strength development, the majority of the lateral prestress was observed to develop within 28 days from concrete casting. It is also evident in this comparison that as expected, higher lateral strains developed in the lightly-confined specimens compared to their well-confined counterparts and an increase in lateral strain development resulted from an increase in expansive mineral admixture dose rate. A closer inspection of these figures reveals that both series of specimens without expansive admixtures exhibited a slight compressive strain of approximately 50 με. This outcome can be attributed to a slight amount of concrete shrinkage overcoming the initial tensile strains caused by the hydrostatic pressure of the fresh concrete mix.

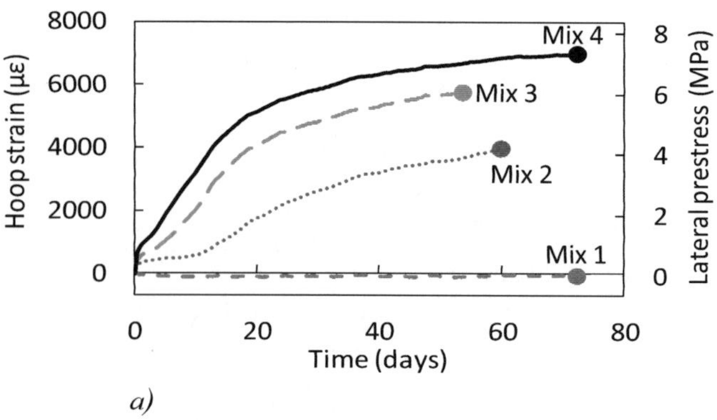

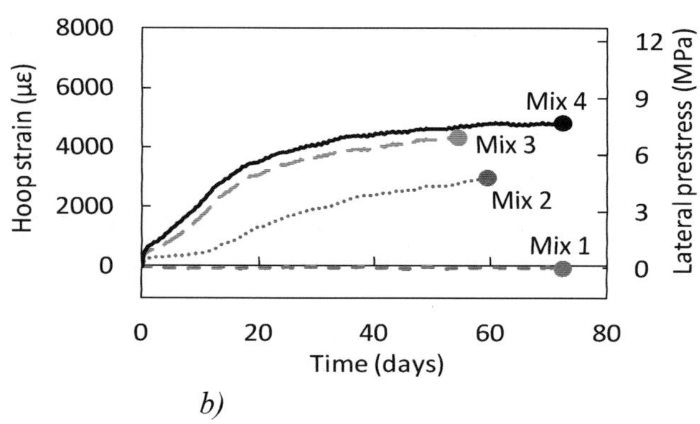

Figure 2. Hoop strain and corresponding lateral prestress development in the FRP tube: a) specimens with $t_f = 0.8$ mm; b) specimens with $t_f = 1.2$ mm.

3.2. Removal of end constraints

As shown in Figure 3, removal of the prestressing rig lead to the specimens experiencing a reduction in the tensile hoop strain of up to 60 με. In addition to this, up to 700 με tensile strain in the axial direction was developed due to the concrete expansion as the constraints were released. This outcome indicates that axial strains were significantly more influenced by removal of end constraints compared to variations in lateral prestress. It also indicates that tensile strain variations were sufficient for the internal concrete to experience tensile cracks. It is worth noting that in practice these axial tensile strains are expected to be overcome by the axial loads present in the column during construction.

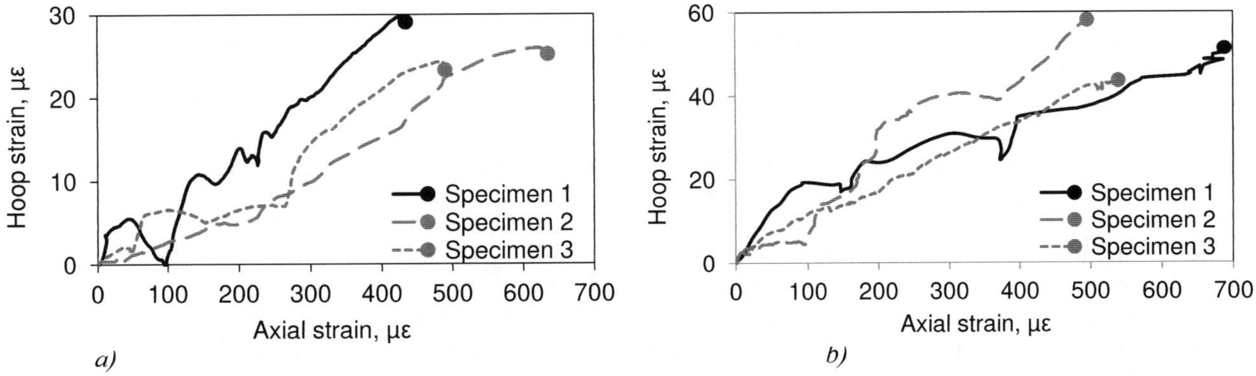

Figure 3. Hoop strain variations during removal of specimens from steel end constraints: a) specimens with $t_f = 0.8$ mm; b) specimens with $t_f = 1.2$ mm.

3.3. Prestress losses

In addition to the prestress development and recorded prestrain reductions, further time dependent prestress losses were examined on a single well confined specimen with the highest level of prestress from Mix 4. Figures 4(a) and 4(b) present the average development of the prestress losses in the axial and hoop directly, respectively. This outcome indicates minimal progressive prestrain losses with only approximately 250 and 100 με recorded over a month for the axial and hoop strains, respectively.

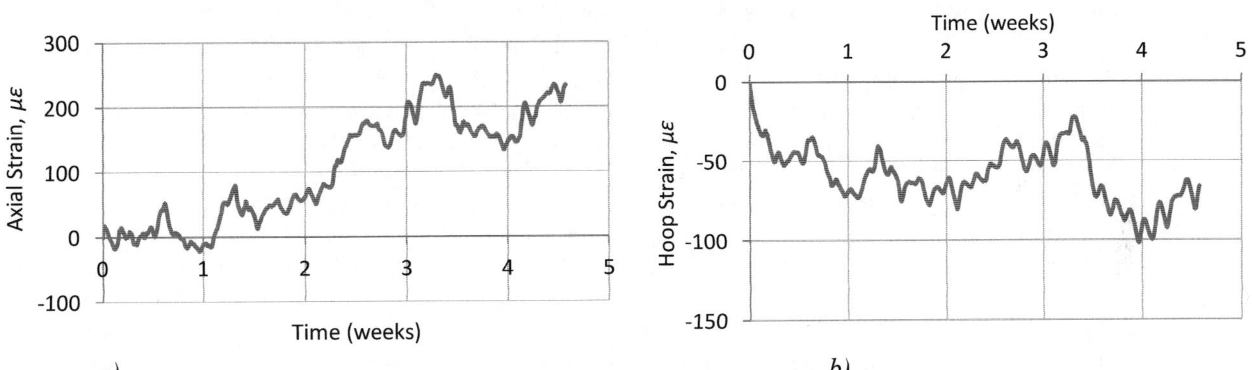

Figure 4. Time dependent variations in prestress losses after removal of end constraints: a) axial strain losses; b) hoop strain losses.

4. Conclusions

This paper has reported the results of an investigation into the axial and lateral strain development of FRP confined high-strength concrete with prestressed FRP shells and the following conclusions are presented:

1. Development of lateral prestress in CFFTs manufactured with HSC is achievable by adding expansive mineral admixture to the concrete mix. Similar to concrete compressive strength development, the majority of the lateral prestress was observed to develop within 28 days from concrete casting.
2. The amount of lateral prestress in the FRP fibers can be controlled by varying the expansive mineral admixture dose rate in the concrete mix design. A dose rate of up to 30% addition to

the total cementitious content resulted in a lateral prestress of up to 7.3 MPa for lightly-confined specimens.

3. Removal of end constraints for prestressed CFFTs has only a minor effect on the amount of prestress developed in the specimen. However, this procedure influences the axial strains significantly and the tensile strains recorded are sufficient to cause tensile cracks in the internal concrete.
4. CFFT columns with prestressed fibers experience minimal progressive prestrain losses within the first four weeks following prestress application.

References

[1] Ozbakkaloglu, T., J.C. Lim, and T. Vincent, 2013 FRP-confined concrete in circular sections: Review and assessment of the stress-strain models. *Eng. Struct.* **49**: p. 1068-1088.
[2] Ozbakkaloglu, T. and J.C. Lim, 2013 Axial compressive behavior of FRP-confined concrete: Experimental test database and a new design-oriented model. *Compos. Part B*. **55**: p. 607 - 634.
[3] Lim, J.C. and T. Ozbakkaloglu, 2014 Confinement model for FRP-confined high-strength concrete. *ASCE, J. Compos. Constr.* **18**(4): p. 04013058.
[4] Lim, J.C. and T. Ozbakkaloglu, 2014 Design model for FRP-confined normal- and high-strength concrete square and rectangular columns. *Mag. Conc. Res.* **66**(20): p. 1020-1035.
[5] Ozbakkaloglu, T., 2013 Compressive behavior of concrete-filled FRP tube columns: Assessment of critical column parameters. *Eng. Struct.* **51**: p. 151-161.
[6] Ilki, A., et al., 2008 FRP Retrofit of Low and Medium Strength Circular and Rectangular Reinforced Concrete Columns. *ASCE J. Mater. Civ. Eng.* **20**(2): p. 169 - 188.
[7] Rousakis, T. and A. Karabinis, 2008 Substandard reinforced concrete members subjected to compression: FRP confining effects. *Mater. Struct.* **41**(9): p. 1595 - 1611.
[8] Wang, Z., et al., 2012 CFRP-confined square RC columns. I: Experimental investigation. *ASCE, J. Compos. Constr.* **16**(2): p. 150 - 160.
[9] Ozbakkaloglu, T. and E. Akin, 2012 Behavior of FRP-confined normal- and high-strength concrete under cyclic axial compression. *ASCE, J. Compos. Constr.* **16**(4): p. 451-463.
[10] Xie, T. and T. Ozbakkaloglu, 2015 Behavior of Steel Fiber-Reinforced High-Strength Concrete-Filled FRP Tube Columns under Axial Compression. *Engineering Structures*. **90**: p. 158-171.
[11] Ozbakkaloglu, T. and T. Xie, 2016 Geopolymer Concrete-Filled FRP Tubes: Behavior of Circular and Square Columns under Axial Compression. *Compos. Part B*. **96**: p. 215-230.
[12] Xie, T. and T. Ozbakkaloglu, 2016 Behavior of Recycled Aggregate Concrete-Filled Basalt and Carbon FRP Tubes. *Constr. Build. Mater.* **105**: p. 132-143.
[13] Fam, A. and S. Rizkalla, 2001 Behavior of axially loaded concrete-filled circular fiber-reinforced polymer tubes. *ACI Struct. J.* **98**(3).
[14] Ozbakkaloglu, T. and T. Vincent, 2014 Axial compressive behavior of circular high-strength concrete-filled FRP tubes. *ASCE, J. Compos. Constr.* **18**(2): p. 04013037.
[15] Vincent, T. and T. Ozbakkaloglu, 2014 Influence of slenderness on stress-strain behavior of concrete-filled FRP tubes: an experimental study. *ASCE, J. Compos. Constr.* **19**(1): p. 04014029.
[16] Lim, J.C. and T. Ozbakkaloglu, 2015 Influence of Concrete Age on Stress-Strain Behavior of FRP-Confined Normal- and High-Strength Concrete. *Constr. Build. Mater.* **82**: p. 61-70.
[17] Vincent, T. and T. Ozbakkaloglu, 2016 Influence of overlap configuration on compressive behavior of CFRP-confined normal- and high-strength concrete. *Mater. Struct.* **49**(4): p. 1245-1268.
[18] Ozbakkaloglu, T., 2013 Axial compressive behavior of square and rectangular high-strength concrete-filled FRP tubes. *ASCE, J. Compos. Constr.* **17**(1): p. 151-161.
[19] Ozbakkaloglu, T. and Y. Idris, 2014 Seismic behavior of FRP-high-strength concrete-steel double skin tubular columns. *ASCE, J. Struct. Eng.* **140**(6): p. 04014019.
[20] Idris, Y. and T. Ozbakkaloglu, 2013 Seismic behavior of high-strength concrete-filled FRP tube columns. *ASCE, J. Compos. Constr.* **17**(6): p. 04013013.

[21] Idris, Y. and T. Ozbakkaloglu, 2016 Behavior of square fiber reinforced polymer-high-strength concrete-steel double-skin tubular columns under combined axial compression and reversed-cyclic lateral loading. *Eng. Struct.* **118**: p. 307 - 319.
[22] Lim, J.C. and T. Ozbakkaloglu, 2015 Hoop strains in FRP-confined concrete columns: experimental observations. *Mater. Struct.* **48**(9): p. 2839-2854.
[23] Chen, L. and T. Ozbakkaloglu, 2016 Corner strengthening of square and rectangular concrete-filled FRP tubes. *Eng. Struct.* **117**: p. 486 - 495.
[24] Ozbakkaloglu, T., 2015 A novel FRP-dual-grade concrete-steel composite column system. *Thin-Walled Structures*. **96**: p. 295-306.
[25] Vincent, T. and T. Ozbakkaloglu, 2017 Lateral strain-to-axial strain model for concrete-filled FRP tube columns incorporating interface gap and prestressed confinement. *ASCE, J. Compos. Constr.* **21**(5): p. 04017021.
[26] Lim, J.C. and T. Ozbakkaloglu, 2016 Influence of Size and Slenderness on Compressive Strain Softening of Confined and Unconfined Concrete. *Journal of Materials in Civil Engineering, ASCE*. **28**(2): p. 06015010.
[27] Ozbakkaloglu, T., A. Gholampour, and J.C. Lim, 2016 Damage-Plasticity Model for FRP-Confined Normal- and High-Strength Concrete. *Journal of Composites for Construction, ASCE*. **20**(6): p. 04016053.
[28] Lim, J.C. and T. Ozbakkaloglu, 2015 Lateral strain-to-axial strain relationship of confined concrete. *ASCE, J. Struct. Eng.* **141**(5): p. 04014141.
[29] Mansouri, I., et al., 2016 Predicting Behavior of FRP-Confined Concrete using Neuro Fuzzy, Neural Network, Multivariate Adaptive Regression Splines and M5 Model Tree Techniques. *Materials and Structures*: p. DOI:10.1617/s11527-015-0790-4.
[30] Lim, J.C. and T. Ozbakkaloglu, 2014 Unified Stress-strain model for FRP and actively confined normal-strength and high-strength concrete. *ASCE, J. Compos. Constr.* **19**(4): p. 04014072.
[31] Vincent, T. and T. Ozbakkaloglu, 2013 Influence of concrete strength and confinement method on axial compressive behavior of FRP-confined high- and ultra high-strength concrete. *Compos. Part B*. **50**: p. 413-428.
[32] Vincent, T. and T. Ozbakkaloglu, 2015 Compressive behavior of prestressed high-strength concrete-filled aramid FRP tube columns: Experimental observations. *ASCE, J. Compos. Constr.* **19**(6): p. 04015003.

IOP Conf. Series: Materials Science and Engineering **307** (2018) 012063 doi:10.1088/1757-899X/307/1/012063

Simulation Modeling of the 4DL Reinforced Imperfect Composite Elastic Behavior

A E Dvoretskiy, I V Magnitsky, F R Odinabekov, E S Sergeeva, S V Tashchilov

Non-Metal Materials Complex, JSC "Kompozit", 4 Pionerskaya Street, Korolyov, Moscow Region, 141070, Russia
Email: info@kompozit-mv.ru

Abstract. Two models simulating the 4DL-reinforced imperfect material elastic behavior are analyzed. The model that involves direct introduction of an imperfect fiber-matrix contact is found to predict higher values for the compressive modulus of elasticity, rather than for the tensile one. This model also gives non-linear stress-strain diagram along some directions. Both models demonstrate a significant rigidity scale effect. Given the effects observed in actual practice with the materials of this class, conclusion was made about second model's usefulness in engineering practice.

1. Introduction

Currently, two basic simulation modeling concepts exist for simulating the behavior of composite material structures, namely, the effective modulus method and various approaches considering the material microstructure. The effective modulus method involves composite material replacement with a uniform, generally isotropic, medium, its properties being determined experimentally using representative samples of the material. As this method does not involve solving of any special mechanics problems, it is not resource-intensive and has been widely implemented. However, it fails to account for a number of effects observed in actual practice such as various scaling effects. Due consideration for the material microstructure provides closer reproduction of the actual material behavior, while requiring development of special simulation models and means.

In terms of the microstructure approach, different concepts of material irregularity representation have been considered. The best understood and most widely used in engineering practice are methods based on analytic macro-/microstress-strain dependencies [1-4]. Such dependencies, while being relatively simple and easy-to-implement as a software program, provide limited accuracy due to the necessity of multiple arbitrary assumptions. Among numerical methods, those based on asymptotic approximation can be distinguished, as they provide more accurate solutions, while requiring an assumption about a small scale of irregularities (as related to the actual size of the structure in question) and their periodic arrangement [5, 6]. However, the method based on direct simulation of the material irregularity, followed by solving the problem using the finite element method or its equivalent, should be regarded as the most generic approach in terms of both the material nature and the simulated structure behavior. At the same time, its implementation consumes considerable computing resources.

This paper addresses the 4DL-reinforced composite material elastic behavior with due account for the defect that is most frequently encountered in practice, while being ignored generally in the simulation models in use, namely, separation of the reinforcing elements (fibers) both from the matrix and from each other. The computations were based on direct simulation of irregular volumes of the material using the finite element method in the ANSYS software environment.

2. Task definition

IOP Conf. Series: Materials Science and Engineering **307** (2018) 012063 doi:10.1088/1757-899X/307/1/012063

For the ease of material simulation, a single structural cell of the material was taken (see Figure 1). The computational volume of the material was treated as composed of those structural cells. The volume length in the a direction – as l_a. The component elastic properties used in the computations are in accordance with the state-of-the-are composite material (see Table 1).

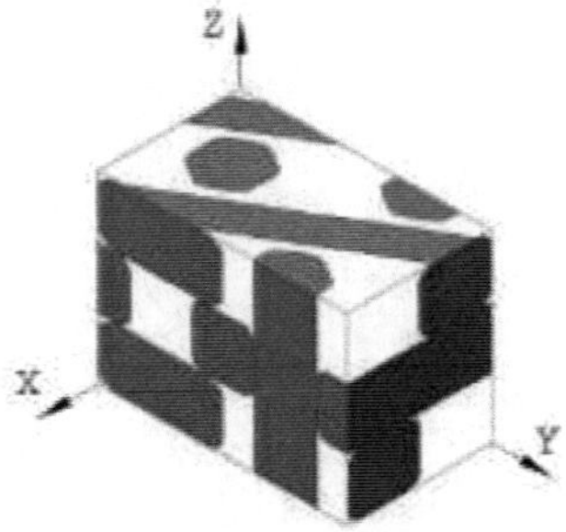

Figure 1. The single structural cell of the material and chosen coordinate system.

Table 1. The components elastic properties.

Characteristic	Value
Longitudinal elastic modulus of the fiber	165 GPa
Transverse elastic modulus of the fiber	3.59 GPa
Longitudinal shear modulus of the fiber	3.84 GPa
Longitudinal and transverse Poisson's ratios of the fiber	0.25
Elastic modulus of the matrix	2.60 GPa
Poisson's ratio of the matrix	0.25

The composite material was deemed as composed of two components, a matrix and a reinforcing element – a core. The components were considered to be solid and uniform, the matrix – isotropic, and the fibers – transversally isotropic.

The boundary conditions set for the material volume simulated the uniaxial loading of a test sample during experimental investigation of the material properties. Thus, under tension along the x axis, they were defined in accordance with [7]

$$\begin{gathered} x=0 \Rightarrow \begin{cases} u_x = 0, \\ \sigma_{xy} = \sigma_{xz} = 0; \end{cases} \\ x=l_x \Rightarrow \begin{cases} u_x = s > 0, \\ \sigma_{xy} = \sigma_{xz} = 0; \end{cases} \\ y \in \{0, l_y\} \Rightarrow \sigma_{yy} = \sigma_{xy} = \sigma_{yz} = 0; \\ z \in \{0, l_z\} \Rightarrow \sigma_{zz} = \sigma_{xz} = \sigma_{yz} = 0. \end{gathered} \tag{1}$$

Here σ_{ij} – stress tensor components, u_i – displacement vectors. The s value was chosen so as to provide deformation equal to 0.001, unless otherwise stated. For other load patterns, the boundary conditions were defined in similar fashion.

The component separation in the material was simulated in two ways. In the first case, a dry friction contact was defined at the components interface. Such problem definition, while providing closer reproduction of the actual material, results in physical nonlinearity, thus adding complexity to the solution and failing to adjust the degree of material imperfection. In the second case, a special isotropic medium featuring small thickness and extremely low rigidness was introduced at the components interface, simulating the air gap. In this case, the problem remained linear, while by adjusting the rigidness of the medium it was possible to simulate partial destruction of the interface.

However, the operational convenience of this model was achieved at the expense of its adequacy to the actual conditions.

As the value characterizing the material elastic behavior, the elasticity modulus was taken, being determined as follows:

$$E_a = \frac{2Wl_a^2}{s^2 l_x l_y l_z}. \quad (2)$$

Here W – total elastic energy of the volume. The choice of the calculation method was based on the stability of the results that depend only weakly on the special features of the finite element grid and the contact problem settings.

In the computations, a finite element grid composed of mainly SOLID186 second-order 20-node solid elements was used.

As a comparison object, the results obtained using the method described above for the separation-free material were taken.

3. Main results

In the first case (with the separation model), the elasticity module values obtained do not depend on the frictional coefficient between the material components. Moreover, with the frictional coefficient ranging from 0 to 0.5, the elasticity modulus data scattering makes up about 0.5%, meaning that the force transfer between the material components due to the frictional force is negligibly small.

The elasticity moduli along the x and y axes show presence of the scaling effect. For the y axis elastic modulus dependence of the cell number is shown in figure 2. As can be seen in Fig. 2, magnitude of the scaling effect in each certain case being dependent not so much on the absolute volume of the material block under consideration as on the presence or absence of the reinforcing cores passing through the volume from one load-bearing facet to the other. The absolute value of the scaling effect, when loading along the x axis, is lower than that along the y axis, due to the presence of the cores oriented along the x axis.

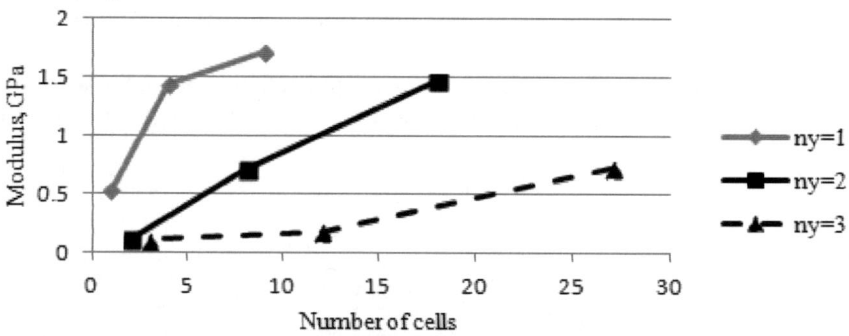

Figure 2. The elastic modulus dependence of the cell number for the y axis.

Both effects described above take place both under tension and under compression; however, the elasticity modulus value itself depends on the load direction. That said, it is essential that the elasticity modulus value obtained under tension is smaller than that obtained under compression. This effect is at its greatest when applying loads along the y axis, in which case there may be a difference of tens of percent. This can be attributed to the fact that, when compressing a material volume, any separations contained within the material tend to close, therefore, the cores bear the loads almost similar to those in case of a solid material, while when under tension, any cracks tend to open wider, and the presence of the cores transverse to the load direction in the material has almost no effect on the material rigidness. On the other hand, this effect contradicts to the available experimental data, according to which, materials of the type under consideration usually show better rigidness under tension than under compression, thereby demonstrating that the modulus difference revealed in the material can be attributed primarily to the special nature of the material components rather than the presence of component separations within the material volume.

IOP Conf. Series: Materials Science and Engineering **307** (2018) 012063 doi:10.1088/1757-899X/307/1/012063

The compressive stress-strain diagram is of non-linear nature, with the material rigidness growing as the deformation increases. Figure 3 shows an example of the compressive stress-strain diagram along the *y* axis for a single cell. The tensile stress-strain diagram is approximately linear. Such a difference between material behaviors under tension and compression loads may explain the corresponding elastic moduli values inequality.

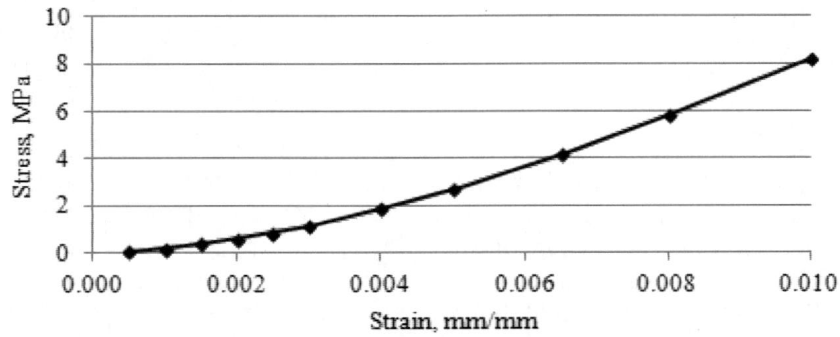

Figure 3. The compressive stress-strain diagram along the *y* axis for a single cell.

As should be expected, the general level of the material rigidness is lower than that in case of a perfect material, which is well illustrated by the data given in Table 2.

Table 2. Elastic modulus of the perfect and imperfect composite under the tensile stress.

Volume structure	Elastic modulus of the perfect material, GPa			Elastic modulus of the imperfect material, GPa		
	x axis	*y* axis	*z* axis	*x* axis	*y* axis	*z* axis
1×1×1	29.8	7.1	37.3	26.45	0.53	33.20
2×2×2	29.9	10.9	37.4	26.87	0.72	34.84
3×3×3	30.0	14.3	37.4	26.86	0.73	34.85

Since no complete core/matrix separation has been observed in reality, to substantiate the aforesaid effects, computation was performed for a material volume composed of 3×3×3 cells, where the separations were localized in the central cell. The results of the calculations show that the delamination effect on the average elasticity modulus does not exceed 3%. Also average value of elasticity moduli in tension and compression coincide with high accuracy. Thus, there is influence of local delamination on the average stiffness is virtually nonexistent in the real material.

Computation results obtained using the second model showed that in this case, the material behave as a linear one both under tension and under compression. Moreover, no material modulus difference is shown by the results. Comparing to first model results, obtained *y* axis modulus is significantly lower because of there is no direct contact between cores and matrix in the second model. In respect that the level of imperfect material *y* axis modulus is extremely low and the local delamination influence at large volume elastic moduli is rather negligible, we may disregard such a difference. It should be noted, however, that within the framework of such a problem definition it is incorrect to perform computations for large deformations, since starting from a certain deformation point, the separation-simulating gaps close, thus necessitating relevant changes in the finite element grid.

Also it was shown that the elastic moduli values depend significantly on the material structural cell choice. As can be seen in Table 3, the elasticity modulus values obtained with different cell assignment differ highly. This effect should be investigated in more detail further.

IOP Conf. Series: Materials Science and Engineering **307** (2018) 012063 doi:10.1088/1757-899X/307/1/012063

Table 3. Elastic modulus of the imperfect composite under the tensile stress.

Volume structure	Elastic modulus of the general cell, GPa			Elastic modulus of the modified cell, GPa		
	x axis	*y* axis	*z* axis	*x* axis	*y* axis	*z* axis
1×1×1	26.78	0.53	34.82	26.45	2.59	33.20
2×2×2	26.87	0.72	34.84	28.53	8.45	34.49
3×3×3	26.86	0.73	34.85	28.49	13.22	35.18

4. Conclusion

Analysis of the computation results shows that all the effects observed agree qualitatively with the existing understanding of the peculiarities of the imperfect material behaviour. The results obtained using different models show sufficient agreement.

The first model allows for observing the difference in the material stress-strain state that develops under tensile and compressive loads. With this model, it is also possible to predict successfully the material behavior, including non-linear one, at large deformations; however, the effects predicted therewith do not manifest themselves to a large extent in real materials. This fact makes it possible to conclude that the material defect such as component separation has a very small impact on the observed values of the actual 4DL reinforced material elasticity modulus.

That being said, given the negligibility of these effects in the actual practice, the second model seems to be more useful due to its simplicity, smaller computational resource requirement and the ability of simulating, to some extent, partial destructions in the material, reducing the whole of the local separations to a single value very similar to the degree of destruction in some kinetic criteria of strength. At the same time, when it is necessary to simulate exactly the local state of the material in the fiber/matrix separation area, the first model seems to be quite applicable as well.

The results obtained attest to long-term benefits of this line of research aimed at the development of structural simulation models for 4DL reinforced materials. As an interesting subject for further study, the local stress state of material components, either with or without components separation, can be investigated.

List of Reference

[1] Malmeister A.K., Tamuj V.P., Teters G.A. Resistance of Polymer and Composite Materials. — Riga, Zinatne, 1980. — 572 pp.

[2] Voigt W. Lehrbuch der Kristallphysik. Berlin—Leipzig, Teubner-Verlag, 1910. 964 c.

[3] Hill R. Theory of mechanical properties of fibre-strengthened materials. III. Selfconsistent model. — J. Mech. Phys. Sol., 1965, vol. 13, N 4, p. 189 — 198.

[4] Z. Hashin, B.W. Rozen. Elasticity Models of Fiber Reinforced Materials. Transl. from English. — Applied Mechanics. Series E. 1964, No. 2, pp. 223 — 232.

[5] I.M. Lifshits, L.N. Rosenzweig. On the Theory of Elastic Properties of Polycrystals. — JETF, 1946, 16, 11.

[6] I.M. Lifshits, L.N. Rosenzweig. On the Construction of Green's Tensor for the Fundamental Equation of the Theory of Elasticity in the Case of an Unbounded Anisotropic Elastic Medium. — JETF, 1947, 17, Release 9.

[7] B.S. Sarbayev, I.V. Magnitsky. Method for Estimating the Effective Elasticity Characteristics of 3D Reinforced Composite Materials // Composite Material Structures, 2014, No. 2. pp. 3 — 9.

IOP Conf. Series: Materials Science and Engineering **307** (2018) 012064 doi:10.1088/1757-899X/307/1/012064

Design and Evaluation of Glass/epoxy Composite Blade and Composite Tower Applied to Wind Turbine

Hyunbum Park[1]

1 Department of DefenceScience &Technology - Aeronautics, Howon University, Rep. of Korea
E-mail: swordship@daum.net

Abstract. In the study, the analysis and manufacturing of small class wind turbine blade was performed. In the structural design, firstly the loading conditions are defined through the load case analysis.The proposed structural configuration of blade has a sandwich type composite structure with the E-glass/Epoxy face sheets and the Urethane foam core for lightness, structural stability, low manufacturing cost and easy manufacturing process. And also, this work proposes a design procedure and results of tower for the small scale wind turbine systems.Structural analysis of blade including load cases, stress, deformation, buckling, vibration and fatigue life was performed using the finite element method, the load spectrum analysis and the Miner rule. Moreover, investigation on structural safety of tower was verified through structural analysis by FEM. The manufacturing of blade and tower was performed based on structural design. In order to investigate the designed structure, the structural testswereconducted and its results were compared with the calculated results. It is confirmed that the final proposed blade and tower meet the design requirements.

1. Introduction

The technology for wind energy was being studied and researched in manymethods in the midst of the recentstatewhere researcheswas being widely performed on alternative wind powerresearch and development results from the exhaustion of fossil fuel. Recently, the supply of electric power was being substituted through wind turbine technologyin Europe and that is said that over 10 percent is being covered in areasof Germany and Denmark[1]. In the studied tendency of electric energies from wind, wind turbine technologies were being studied in manysizes from thesmall scale to large electric powers. Since the small scale wind turbine can be operated individually in a small size, involvedresearches are being carried out lately[2]. Moreover, a utilization of composites has led to the result of important betterment of structural configuration. But, many researched blades of small scale wind turbine technologywerestudied in Europecountries. Accordingly, the rated wind velocityis not suitable for the domestic condition.Therefore, there is the necessity to research and develop wind turbine blade suitable for the local climate conditions.And also, tower design is very important for wind turbine system. Even though the towers are simple structural type to other structures, development of a cost effective tower is important because the tower has 25% of the whole wind turbinc tcchnology.

In this work, design, analysis and manufacturing of the blade and composite tower of wind turbine was performed. It can be worked at domestic region and manufactured to display good performance in low wind velocityregion. Manufacturing about prototype and structure tests were performed to investigate whether real motion of designed bladesare consistent with the theoretical calculated results. And also, investigations on safety of structurewere carried out through the FEM analysis method of the tower for developed small wind turbine.

2. Composite bladedesign and manufacturing

For blade design, the specification of blade system was investigated. The type of wind turbine system is 1kW class in this study. And also, the horizontal axis system for blade was adopted. The rated wind speed is 12m/s. The used airfoil is NACA 632-615[3]. The diameterof blade is 2.54m. The torsion angle is 22°. Wind turbine blade system to be adopted is a direct-driven axial flux permanent magnetic equipment. This magnet equipment has high energy conversion technology at low velocityspeed. Specific aerodynamic designed results are specified in Table 1. Figure 1 shows designed aerodynamic configuration.

The design and analysis blade structure was carried out after investigation on aerodynamic configuration to investigation the structural configuration safety and stability of composite blade through stress analysis using FEM method. Structural designed load includes the aerodynamic calculatedloadsof blade and rotational load as major load. For aerodynamic loads to be calculated, load conditions specified in Table 2 were considered. The load analysis result confirmed that load case 2 causes the biggest moment to blade. Accordingly, the structural design of the blade in this paper was performed based on the load in load case 2.

Figure 1. Designed aerodynamic results of composite blade

Table 1. Designed results of blade

Power	1,000 W
Diameter	2.54 m
Root chord	143.42 mm
Tip chord	64.84 mm
Twist	22°
Airfoil	NACA 63_2-615

The skin and foam core sandwich composite structural type were applied as the structural configuration based on the design load. In structural designed configuration, spar and skin were laminated number of layers through longitudinal direction from the root area of blade for weight reduction by calculated sectional loads.

IOP Conf. Series: Materials Science and Engineering **307** (2018) 012064 doi:10.1088/1757-899X/307/1/012064

Table 2. Design loads

Load cases	1	2	3
Wind velocity	12.8m/s	30.0m/s	55.0m/s
Gust (±20m/s,±40°)	No gust	With gust	Storm
Rotational velocity	500rpm	800rpm	Stop

For the analysis of blade structureperformed to investigate the safety of structural designedconfiguration, finite analysis method, MSC. NASTRAN software was used, and failure criteria index theory [4] was adopted as the failure criterion for the safety. The structural analysis results for each load conditions defined that it was designed with a safe structural configuration ensuring sufficient safety rate. Table 3 shows the result of load condition-specific finite element linear static analysis result and the review of according failure. Eigenvalue analyses were analyzed to examine any resonance problem of blade and a result showed that resonance problemdidn't happen around workedconditions. The evaluation of buckling stability also confirmed that it is a structure ensuring a sufficient safety and stability during work with over the maximum loads. The maximum loads of fatigue were investigated based on equations of Spera for fatigue life estimation result was 577Nm and the fatigue estimation result showed 35.4MPa. This value is maximum compressive stress. The stress of tensile is 28MPa. Therefore, the designed blade was safe as allowable fatigue strength (1445MPa)[5]. A result of structural designed blade wasconfirmed that it is the blade that ensured required fatigue life.

The resultsof structural designed blade were compared with structural tested result for verification.The stress, deformation and fatigue life analysis result of designed blade were investigated respectively. Figure 2 shows first flap mode configuration and frequency and Fig. 3 shows prototype blade.

Table 3. Structural analysis results

Results		1	2	3
Maximum Stress[MPa]	Tension	27.2	73.0	28.6
	Compression	19.8	69.8	20.2
Max. Disp. [mm]		27.3	82.8	28.3
Tsai-wu failure criterion		0.071	0.278	0.126

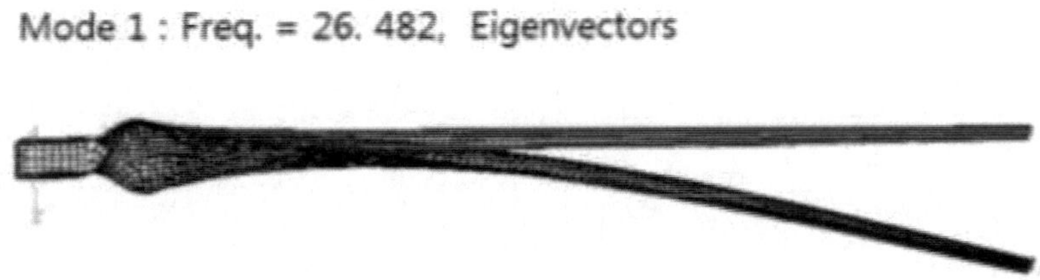

Figure 2. First flap mode shape and frequency

IOP Conf. Series: Materials Science and Engineering **307** (2018) 012064 doi:10.1088/1757-899X/307/1/012064

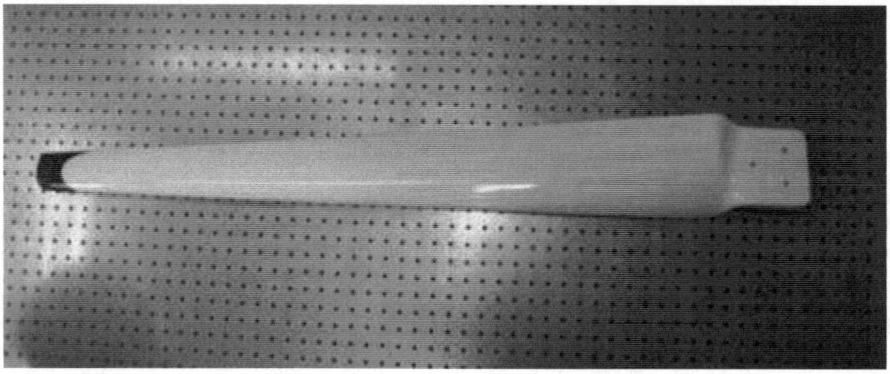

Figure 3. Manufactured prototype blade

3. Evaluation of designed tower

Safety evaluation of tower was carried out.In this study, it was evaluated structural safety of supported system for the blade[6,7]. Firstly, the investigation of tower structure werecarried out by determining design loads by considering each loading conditions and analyzing deformation and stress analysis after verifying the load. As for various loading condition according to each cases, the tower loads can be defined.

In the study, two conditions were investigated with the loads placed on tower during normal operation condition and the loads placed on tower when it was stopped during the storm condition. Critical loading that need to be investigated during structural analysis of tower are load from thrust occurredfrom blade rotating and distributed load placed on tower from wind and load placed from blade weight, generator weight and tower weight. They can be considered as non-uniform load with concentrated load placed in the tower upper section and distributed loadingplaced on the wholestructure. Figure4 shows load configuration of tower. W is weight of blade and nacelle. P is load of wind turbine and q is distributed load of wind. The analysis results of tower showed that stress placed on tower during normal operation is 23MPa and the stress analysis result when it stopped during a storm is 154MPa, thereby identifying that it is a tower that ensured a sufficient state of operational safety. Figure 5 and Fig. 6 show the results of natural frequency analysis.

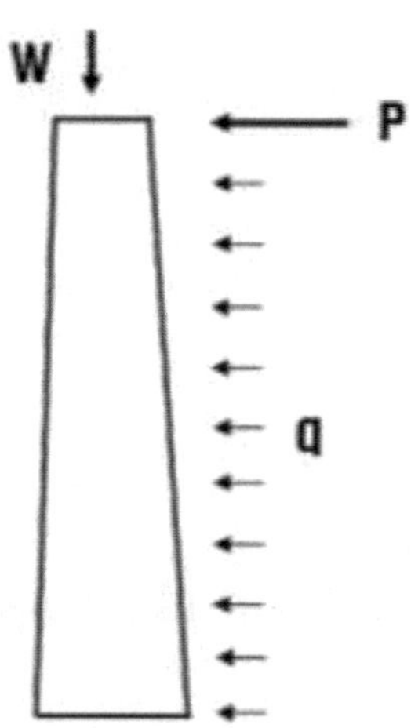

Figure 4. Load of tower

Figure 5. Natural frequency analysis: first mode

Figure 6. Natural frequency analysis: second mode

4. Conclusions

In this work, structural design and manufacturing of blade system was carried out. Firstly blade design was performed and then tower design was conducted.It can be worked in domestic region and manufactured to display good performances in low wind speed regions such as Asia. Manufacturing of prototype and experimental testswere performed to verify whether the real motion of the designed blade and tower is consistent with the theoretical calculated result. In addition, investigations on structural safety were carried out through the FEM of the blade and tower for installing the developed small wind turbine.

Acknowledgments

The work was performed by Howon University fund.

References

[1] Desire Le Gourieres 1982 *Wind Power Plants Theory and Design* (Pergamon Press: Oxford)
[2] G. Park, C. Kong and H. Park 2015 Design and analysis ofsmall wind turbine blade using natural fiber composite *Proc. of the 8th Asian-Pacific Conference on Aerospace Technology and Science*p 761-765
[3] FrankB. 2001 *Wind turbine airfoil catalogue* (Riso national laboratory)
[4] M. J. Robert 1999*Mechanics of Composite Materials* (Taylor and Francis: London)
[5] Spera D. A. 1994 *Wind Turbine Technology* (ASME Press) p488-589
[6] N.Bazeos2002 Static, seismic and stability analyses of a prototype wind turbine steel tower (*Engineering Structure* vol 24) pp 1015-1025
[7] J. P. Dimos, G. R. Ioannis and U. Nibong 2009 Static and dynamic characteristics of multi-cell jointed GFRP wind turbine towers (*Composite Structure* vol 90) pp 34-42

IOP Conf. Series: Materials Science and Engineering **307** (2018) 012065 doi:10.1088/1757-899X/307/1/012065

Design and Manufacturing of Composite Tower Structure for Wind Turbine Equipment

Hyunbum Park[1]

1 Department of Defence Science & Technology - Aeronautics, Howon University, Rep. of Korea
E-mail: swordship@daum.net

Abstract. This study proposes the composite tower design process for large wind turbine equipment. In this work, structural design of tower and analysis using finite element method was performed. After structural design, prototype blade manufacturing and test was performed. The used material is a glass fiber and epoxy resin composite. And also, sand was used in the middle part. The optimized structural design and analysis was performed. The parameter for optimized structural design is weight reduction and safety of structure. Finally, structure of tower will be confirmed by structural test.

1. Introduction

In recent years, the wind generation technologies have been ameliorated by intent investment and researched work on improved technique based on former developed technique due to fossil fuel depletion, and become big size wind turbine technologies more than megawatt blade. Thus, a supporting tower structure of large class wind power plant equipment was became big size. The composite material is multiply used to reduce the blade's weight in the structural design for wind power plant blade. However, various large scale towers for wind turbine systems were made by steel. If a tower applied to composite is designed, the market of wind power plant will be increased.

According to literature survey results, M. Feyzollahzadeh studied investigation on wind load response of offshore wind turbine towers with fixed monopile platform [1]. N. Bazeos studied on stability analysis about tower manufactured by steel structure [2]. Dimos J. P. was performed the dynamic and static characteristics about glass fiber reinforced plastic towers [3]. From the literature reviewed results, many studies of tower made by steel were performed. In the research work, study on manufacturing procedure of glass and epoxy material tower was carried out.

2. Design requirement of tower

It is necessary to specify the specification of system for the purpose of wind turbine tower design. It is horizontal axis wind turbine system with rate output of MW class as for the specification of the wind turbine tower. A normal wind speed is 12.8m/s. The tower structure specification to be designed was shown Table 1. From the literature review of 2 megawatt class wind power plant blade, the most of towers are height of 80m [4]. But, the study chooses the height of 100m tower for application of 4 megawatt scale wind power plant blade system as well as 2 megawatt. Designed tower was partitioned as four parts for easy transportation. The first section is diameter of 3m. The diameter of second part is 3.4m. The diameter of third part structure is 3.7m. The section diameter of bottom structure is 4m. Structural analysis of load cases were performed for the tower design and manufacturing. The three

IOP Conf. Series: Materials Science and Engineering **307** (2018) 012065 doi:10.1088/1757-899X/307/1/012065

wind load cases were investigated. The one case is rated wind velocity condition. The two cases is cut-out wind velocity condition. The three cases is storm wind velocity condition. In the wind load cases, a thrusted force condition occurred by wind turbine rotor system. And also, various equipment and tower weight will be reflected as design load cases [5,6]. Table 2 shows loads for tower analysis and manufacturing. A filament winding process was adopted for tower manufacturing.

Table 1. Design requirement of tower

Design requirement	
Rated power	2MW
Diameter	3.0~4.0m
Height	100m

Table 2. Load case of tower

Load cases	Case 1	Case 2	Case 3
Wind velocity	8.0m/s	20.0m/s	55.0m/s

3. Structural design and analysis

Structural design was performed after tower load case analysis. The design of tower was previously performed using designed metal tower specifications, the structural shape and configuration, a manufactured process. After investigation on metal wind power plant structure, composite tower structure was designed. Fig. 1 shows a design method in this work. A section of tower was partitioned into four structures such as top section, second section, the third section and the bottom section. The jointed parts between each section part have a truncated shape. Fig.2 shows the designed tower configuration.

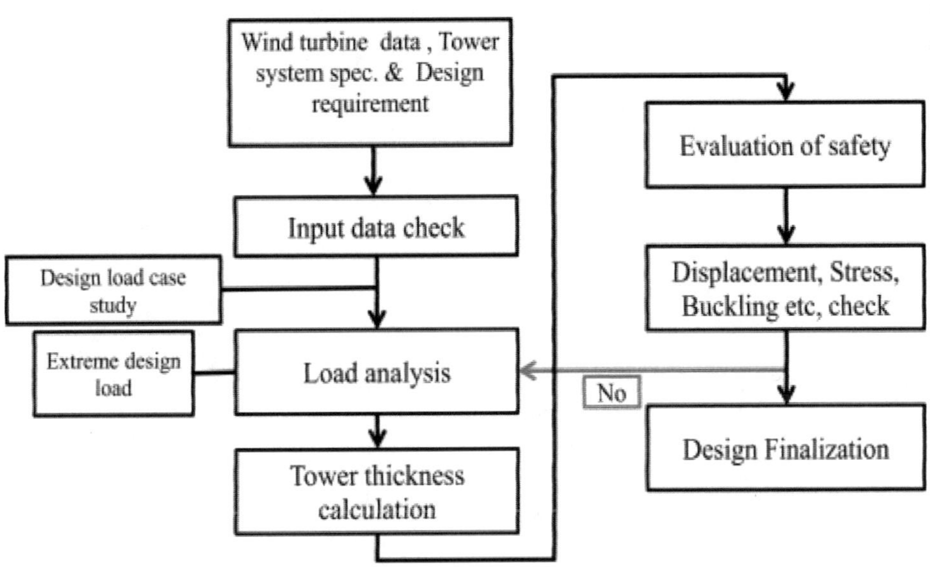

Figure 1. Design procedure of tower for wind turbine system

The structural analysis was performed to investigate structural safety of designed structural result. In this study, finite element method was adopted for structural analysis. The failure criterion analysis [4] was applied for the safety check. The result of structural analysis for each load cases confirmed that it has been designed with a safe tower structural configuration ensuring sufficient safety factor. Table 3 shows the result of load condition-specific finite element linear static analysis result. In this study, the optimized thickness and winding pattern is defined considering proper structural safety, deformation at the top of tower. Among the various stacking patterns, the pattern $[0/90]_{20s}$ has the lowest displacement and stress. Fig. 4-7 show stress contour of each lay-up patterns. It is confirmed that the best stacking sequence condition is $[0/90]_{20s}$.

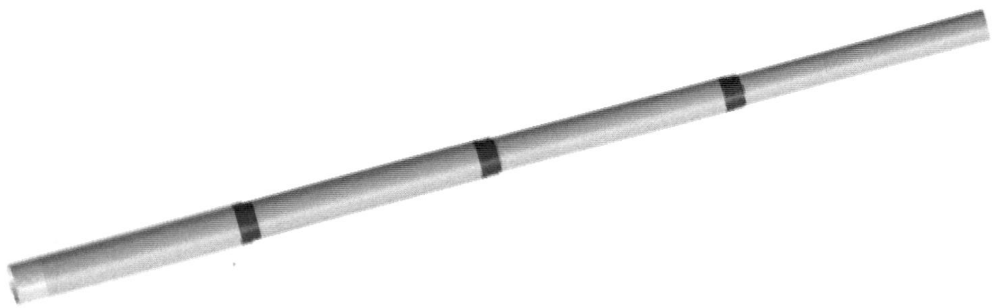

Figure 2. Configuration of designed tower

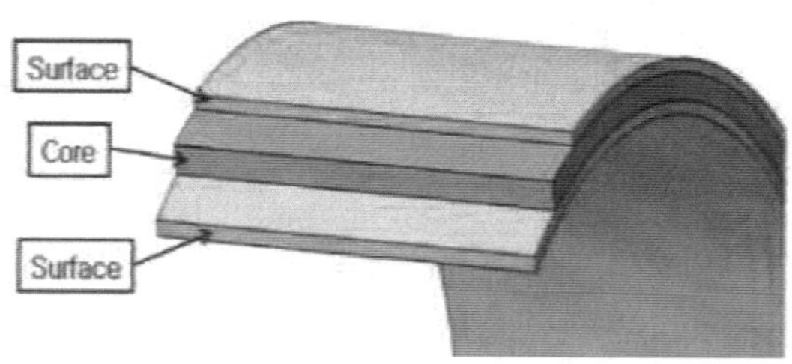

Figure 3. Section view and structural shapeof designed tower

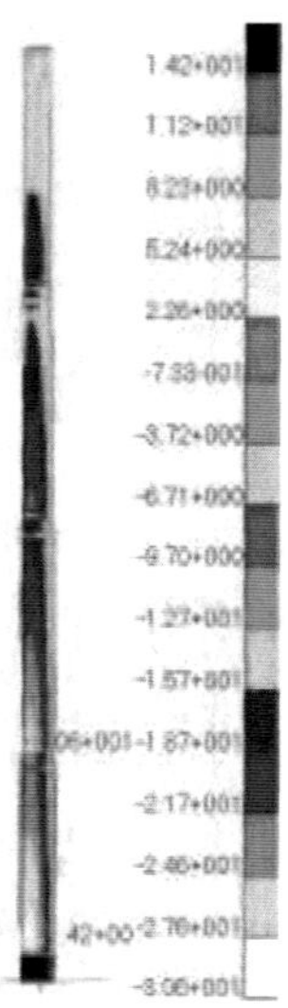

Figure 4. Stress analysis result of pattern 1

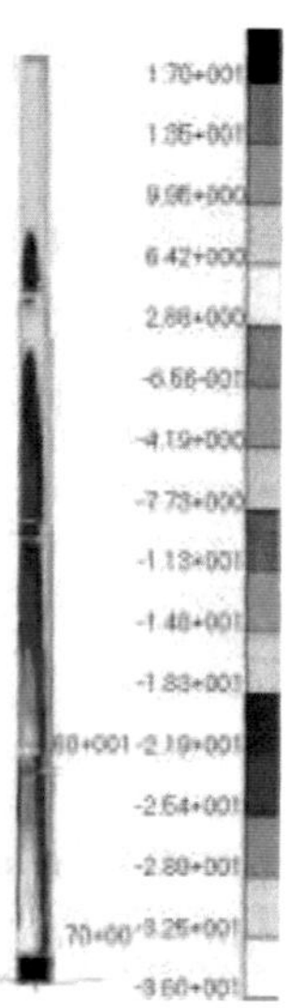

Figure 5. Stress analysis result of pattern 2

IOP Conf. Series: Materials Science and Engineering **307** (2018) 012065 doi:10.1088/1757-899X/307/1/012065

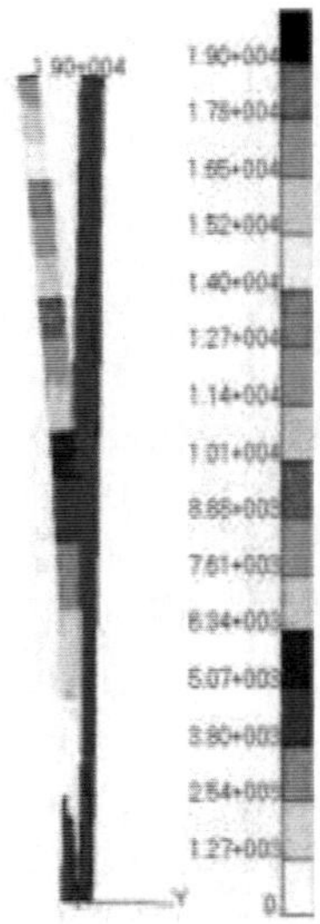

Figure 6. Deformation of pattern 1

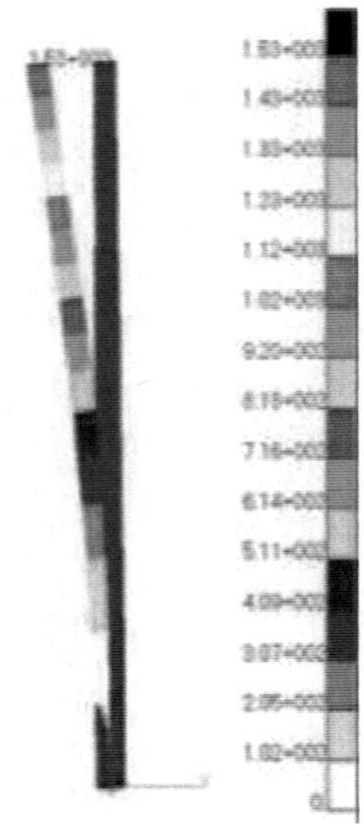

Figure 7. Deformation of pattern 2

Table 3. Structural analysis results

Lay-up pattern	$[0/90]_{20s}$	$[45/90]_{20s}$	$[0/90/45]_{20s}$
Compressive stress	31 MPa	36 MPa	43 MPa
Tensile stress	14 MPa	17 MPa	21 MPa

4. Structural optimization, manufacturing and structural test

According to the deformation and stress analysis, tower deformation is large deformation. Thus, both applying composite structural type applied to glass, epoxy and sand core and increasing thickness is adopted to decrease the deformation as well as to get the safety of tower. The optimal design of tower thickness was conducted by the increasing through confirming both the structural deformation and the stability. Among ten design results, the 10th case is selected as final tower designed results that satisfy both the safety and structural stability and the top deformation. Finally, the weight of proposed 100m height tower applied to composite is 270ton. That is lighter than the 80m metal tower having from 450 to 700ton. In order to analyze the cost of the composite material tower, the cost of production is compared with the metal tower's manufactured cost. A filament winding manufacturing process was adopted for cost effective manufacturing. After manufacturing tower, it was performed the structural test. The compressive strength test of prototype tower was performed by UTM equipment. The compressive analysis result is 3.41MPa and the test result is 3.42MPa. Therefore, it is investigated that the composite structure is safe through stress analysis and test result. Fig. 8 shows the manufacturing process of composite tower using filament winding manufacturing method. Fig. 9 shows compressive strength test of manufactured tower.

Figure 8. Manufacturing process of composite tower

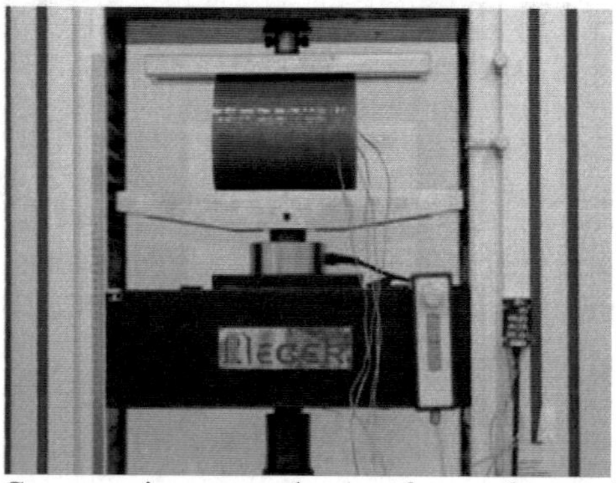

Figure 9. Compressive strength test of manufactured tower

5. Conclusions

This study performed the design and analysis of composite tower about two megawatt scale wind power plant system equipment. A glass and epoxy skin part and sand core composite structural shape was applied to considering on manufactured expense. And also, optimal tower designed reducing the weight with structural safety, structural stability. It was investigated that the composite material wind power plant tower structure is safe through analysis results of stress contour, deflection and eigen value, and lighter than metal structures.

Acknowledgments

This study was supported by Howon University research funds.

References

[1] M. Feyzollahzadeh, M. J. Mahmoodi, S. M. Yadavar-Nikravesh and J. Jamali 2016 Wind load response of offshore wind turbine towers with fixed monopile platform (*Journal of Wind Engineering and Industrial Aerodynamics* vol 158) pp 122-138

[2] N.Bazeos2002 Static, seismic and stability analyses of a prototype wind turbine steel tower (*Engineering Structure* vol 24) pp 1015-1025

[3] Dimos J. Polyzois, Ioannis G. Raftoyiannis and Nibong Ungkurapinan 2009 Static and dynamic characteristics of multi-cell jointed GFRP wind turbine towers(*Composite Structure* vol 90) pp 34-42

[4] Spera D. A. 1994*Wind turbine technology*(ASME Press)

[5] Desire Le Gourieres 1982*Wind Power Plants Theory and Design*(Prefamon Press: Oxford)

[6] Robert M. Jones 1999 *Mechanics of Composite Materials* (Taylor and Francis: London)

IOP Conf. Series: Materials Science and Engineering **307** (2018) 012066 doi:10.1088/1757-899X/307/1/012066

The fibre composite macroscale response to the impact force and its microscale structure

Z Murčinková, A Šmeringaiová, I Vojtko and M Šebestová

Faculty of Manufacturing Technologies, Bayerova 1, 080 01 Prešov, Slovak Republic,
zuzana.murcinkova@tuke.sk

Abstract. The paper provides the study of macroscale response of composites reinforced by long and short fibres to the impact force. The measured and analysed quantities are natural frequency and damping time. As the inner composite microscale structure can be of various modified configurations affected by production, we analysed influence to macroscale response.

1. Introduction

Economic reasons support the use of predictive methods for the study of dynamic behaviour of machines from the earliest stages of a design [1]. Dynamic response of material is in form of mechanical waves (vibrations) that propagate in continuum. Vibration is the oscillation of elastic body or environment where individual particles make mechanical oscillation [2,3]. In present, there is a great variety of the developing advanced materials and material combinations. Selection of the appropriate material for a specific machine structure is a crucial but challenging task [4]. Fibre reinforced and composite materials are particularly interesting for applications in machine tool structures due to their ratio of mechanical strength to density [4].

Moreover, the mentioned ratio influences the natural frequency and thus the dynamic properties that are so important for machine tools. Our research is focused on fibre composites and their reinforcing effect (more in [5]). As the composites consist of two or more distinct constituent phases (in our case fibres and matrix), they provide possibility to control their microscale structure in process of production and thus to control their directional mechanical (static and dynamic), also thermal, properties according to requirements, i.e. higher stiffness in required direction, high natural frequency, higher material damping in required direction, thermal isolator in required direction etc.

2. Measuring description

Figure 1 shows the measuring chain and samples of the fibre composites. To measure dynamic response of composites the sample is fixed into jaw of clamp of measuring stand on the one side and it is bumped by impact ball on the opposite side (figure 2). The vibrations (velocity of vibrations) caused by impact are measured by vibrometer Polytec. The impact ball rolls on the inclined groove of the measuring stand, then it impacts the sample (figure 2) by specific velocity applying the specific impact force on the sample.

The precise measuring of vibration velocity is based on a phenomenon of Doppler Shift in laser beam (laser-Doppler vibrometry). The light beam is reflected by moving surface and thus the frequency of light is shifted proportionally to its velocity. The accuracy of measurement is independent on reflected light intensity.

To be able to make repeated measuring (accuracy of measuring) and impact the different samples with the same values of forces, the measuring stand was designed. It allows to change the angle φ, length of trajectory l and weight of impact body m (figure 2, equation (4)). Thus the value of force is affected.

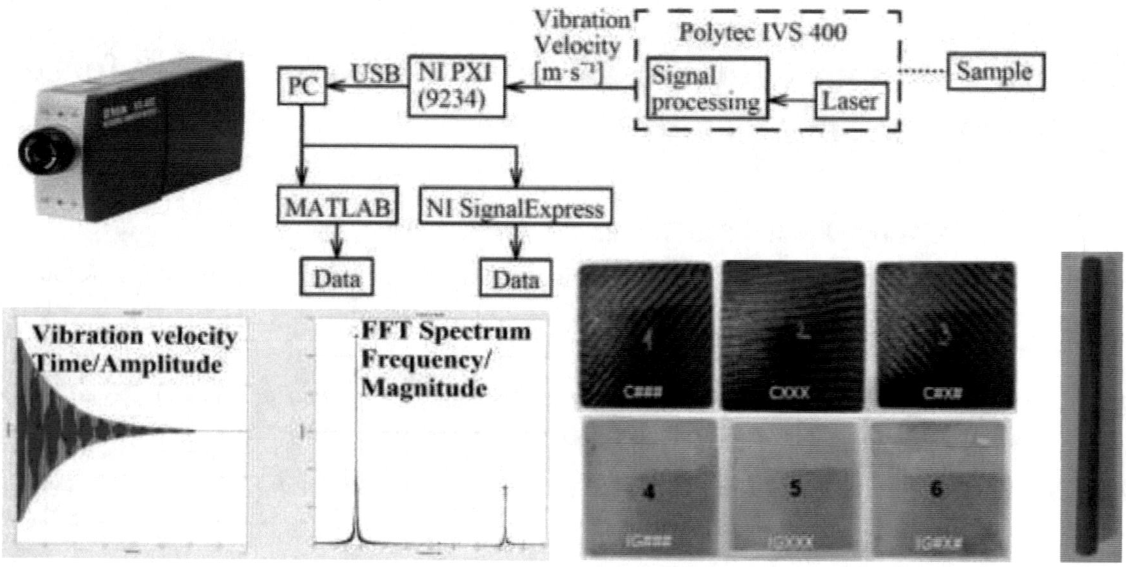

Figure 1. Measuring chain and samples

2.1Materials and dimensions of samples

1. Long-fibre composite (laminate): 3 layers, twill fabric, carbon fibres C200g/m^2, epoxy resin and fixative of mutual ratio 5:2;
   a. C###, fibres orientation parallel to the edges in all layers of the sample,
   b. CXXX, diagonal fibre orientation (45°) in all layers of the sample,
   c. C#X#, combination.
2. Long-fibre composite (laminate): 3 layers, twill fabric, glass fibres IG280g/m^2, epoxy resin and fixative of mutual ratio 5:2.
   a. IG###, fibres orientation parallel to the edges in all layers of the sample,
   b. IGXXX, diagonal fibre orientation (45°) in all layers of the sample,
   c. IG#X#, combination.
3. Short-fibre composite: ceramic composite, matrix Sic, carbon fibres in rovings 12k (1k=1000 filaments) of length 3-6 mm.
   a. SiC/C random, carbon fibres are randomly distributed,
4. Standard construction steel.

Dimensions of samples: Long-fibres composite: 115x115x0.9 mm; short-fibre composite: tube, Ø 8 mm, wall thickness 1 mm, length 110 mm; standard construction steel: tube, Ø 8 mm, wall thickness 2 mm, length 110 mm.

The fibres are usually oriented either randomly or preferably with unidirectional orientation or woven. The fibre composites can be reinforced by long continuous (continuous) and short (discontinuous) fibres. The criteria of classification are very coarse and needs to be clarified and completed. According to [6], if the aspect ratio, i.e. the ratio of the fibre length L to the diameter D, L/D is less than 100 (L/D <100) then the fibre is short. However, the diameter of nanotubes can be 5 nm and length even more than 1 mm, i.e. length to diameter ratio L/D is more than 200 000. Such fibres should be named relatively short fibres [5,7].

2.2 Impact force

Impact force for measurement was estimated according to theoretical method using several assumptions based on Newton's elementary theory of impact (according to [2]). In generally, the impact can be divided into two periods: 1st period starts with contact of bodies (collision) and finishes in moment of their maximum deformation when the bodies are moving by common velocity. 2nd period starts in moment of maximum deformation and finishes by separation of bodies. The large surface load appears in contact surface. The resultant force of surface load is impact force lying on the normal line of impact (figure 2).

The impact body (ball) of mass m moves downward the inclined groove with the angle of inclination φ (figure 2). The initial velocity at point A is $v_0 = 0$. After passing the trajectory l, the ball impacts a barrier (test sample) at the velocity v_1. The ball performs an inelastic (do not conserve kinetic energy) inclined impact on a fixed barrier.

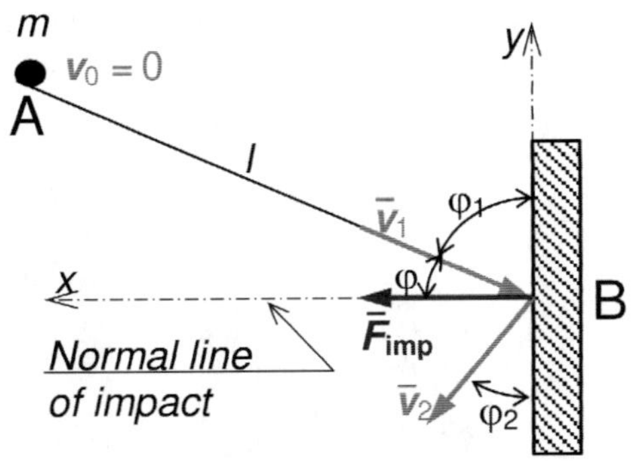

Figure 2. Schema of impact and experiment

The magnitude of the dynamic effect of the impact force on the body is expressed by the impulse of the impact force $\vec{I}_{imp}$:

$$\vec{I}_{\text{imp}} = \int_0^{\tau} \vec{F}_{\text{imp}}\,\mathrm{d}t = \int_0^{t_1} \vec{F}_{\text{imp}}\,\mathrm{d}t + \int_0^{t_2} \vec{F}_{\text{imp}}\,\mathrm{d}t = \vec{I}_{\text{imp1}} + \vec{I}_{\text{imp2}} = \vec{F}_{\text{imp}avg}\,\tau \tag{1}$$

where $\vec{F}_{\text{imp}}$ is impact force acting on body, t_1 time of 1st period of impact, t_2 time of 2nd period of impact, $\tau = t_1 + t_2$ is time of impact, $\vec{I}_{\text{imp1}}$ is impulse of impact force in 1st period of impact, $\vec{I}_{\text{imp2}}$ is impulse of impact force in 2nd period of impact, $\vec{F}_{\text{imp}avg}$ is mean impact force.

Line of action of impact force is perpendicular to surface of object. The change of the momentum of the impact ball is only in direction of impact force. In direction of x-axis, the ball makes the inelastic direct central impact. Since the location of the test sample does not change, the common (centroid) velocity at the end of two impact phases is $v_s = 0$. For our case, based on (1) and the momentum change theorem, the impulse of the impact force in the first stage of the collision is in the x-axis direction (with respect to coordinate system in figure 2):

$$I_{\text{imp1}} = \int_{v_1}^{v_s} m\mathrm{d}v = m\left(v_s - v_{1x}\right) \quad \Rightarrow \quad I_{\text{imp1}} = -m\,v_{1x} \tag{2}$$

Based on the kinetic energy change theorem, we determine the velocity of the impact body on the AB trajectory. The velocity of the impact body at the moment of impact is the following:

$$v_1 = \sqrt{2\,g\,l\sin\varphi} \tag{3}$$

Taking into account and adjusting the previous equations, we get a formula for the calculation of the impact force magnitude at the end of the first period of the impact:

$$F_{\text{imp}}\, t_1 = \left| m\, v_1 \cos\varphi \right| \quad \Rightarrow \quad F_{\text{imp}} = \frac{m \cos\varphi \sqrt{2\, g\, l \sin\varphi}}{t_1} \tag{4}$$

3. Results

The samples of long fibre composites are of different material configuration (described in 2.1) which cause the modification of natural frequency values and material damping time. The graph in figure 3, left, evaluates the measured first and second natural frequencies of samples based on the FFT (Fast Fourier Transformation) analysis of the measured vibration velocity values. The measuring was repeated by exciting the samples of four impact forces values (in direction of velocity v_1) 332 N, 680 N, 2917 N and 5978 N in order to obtain more reliable results. We found out some modification of natural frequency by material configuration, but the natural freqency is not influenced significantly.

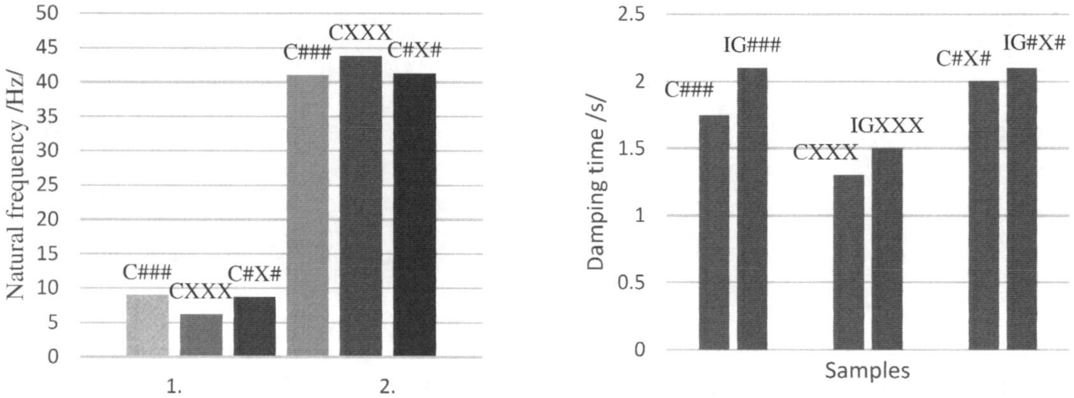

Figure 3. Natural frequency (left) and damping time

The samples of different material configuration are of different material damping time (figure 3, right). We confirm that the damping properties of carbon fibre composite are better comparing to glass fibre composite. The shortest damping time is for configuration XXX and even the configuration #X# make the damping time longest in that boundary conditions. We can make statement that the material configuration significantly change the damping time.

Figure 4 presents vibration velocity plot. The circles marks the jumps in plot. These jumps are of various size and in different time for individual samples. The detail of such jump involve the phase change, i.e. the distortion of waviness. We assume the reason of the phase change is interface layer-layer that shortens the damping.

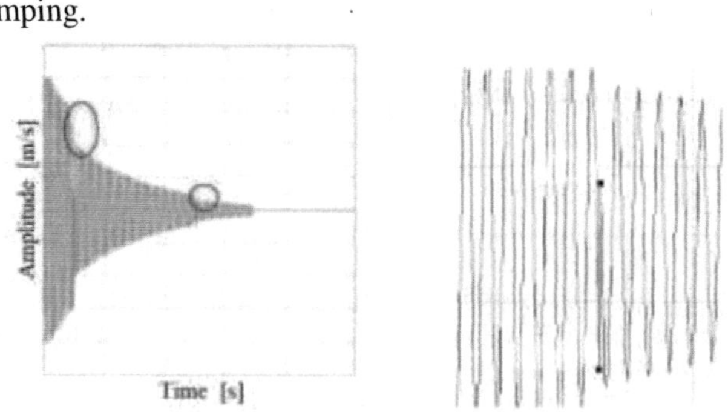

Figure 4. Amplitude of vibration velocity behaviour and detail

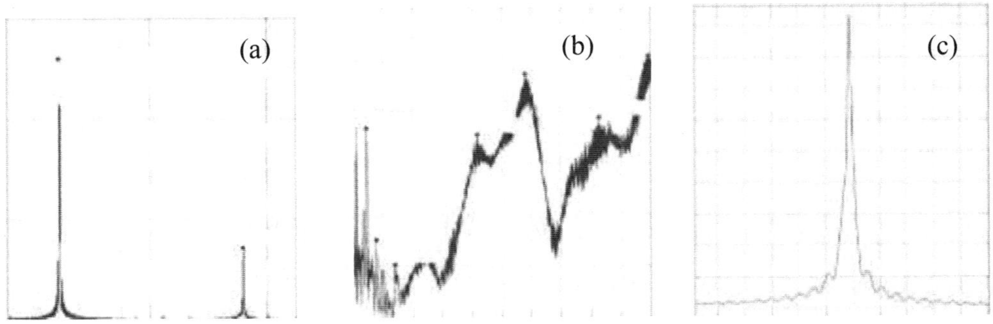

Figure 5. Typical courses of FFT spectrums (natural frequency/magnitude) of layered long fibres laminate, (a), short fibre composite (b), steel (c)

FFT (Fast Fourier Transformation) spectrums (figure 5) of layered long fibre laminate and steel is characterized by one sharp resonance peak comparing to several peaks of short fibre composite. These several peaks form the wide resonance range which cause the desired weak resonance phenomenon. Comparing FFT spectrums of both composites (figure 5, (a), (b)), the difference is visible. Mainly, the short fibre composites provide appropriate dynamic response because of high energy dissipation capability due to larger interface surface.

4. Conclusions

The study of various fibre (long and short) composites focused on impact load showed the specific and useful properties of fibre composites for dynamic load. Fibre reinforcement is beneficial for dynamic performance since the short fibre interfaces contributes energy dissipation.

We confirm that macroscale dynamic properties (mainly material damping) of fibre composites can be modified by microscale inner structure. It provide options to control macroscale dynamic properties through the microscale material configuration.

Acknowledgement
Authors thank for supporting this research by grant VEGA 1/0910/17 of Agency of Ministry of education of Slovak Republic.

References

[1] Genta G 2012 Vibration of Structures and Machines: Practical Aspects, 2nd ed., (New York: Springer-Verlag)
[2] Záhorec O and Caban S 2002 Dynamika (Dynamics) (Košice: Faculty of Mechanical Engineering, Technical University of Košice)
[3] Norton M P and Karczub D 2003 Fundamentals of Noise and Vibration Analysis for Engineer, 2nd ed., (Cambridge: Cambridge University Press)
[4] Möhring H C, Brecher C, Abele E, Fleischer, J and Bleicher F 2015 Materials in machine tool structures. *CIRP Annals-Manufacturing Technology*, *64*(2), pp 725-748
[5] Kompiš V, Murčinková Z and Žmindák M 2015 Toughening mechanisms for the fibre of middle-large-aspect-ratio-reinforced composites *Toughening mechanisms in composite materials*, ed Qin, Q H and Ye J (Cambridge: Elsevier; Woodhead Publishing) chapter 5, pp 137-159
[6] Jancár J 2003 Úvod do materiálového inženýrství polymerních kompozitu (Introduction to Materials Engineering of polymer composites) (Brno. VUT in Brno, Chemical faculty)
[7] Murčinková Z, Kompiš V, Adamčík P, Dobrovič S and Murčinko J 2017 Design of Fibrous Composite Materials for Saving Energy. *Handbook of Composites from Renewable Materials, Design and Manufacturing*, Vol. 2, ed Thakur V K, Thakur M K and Kessler M R (Scrivener Publishing Wiley), chapter 3, pp 49-93

IOP Conf. Series: Materials Science and Engineering **307** (2018) 012067 doi:10.1088/1757-899X/307/1/012067

Structure and Properties of Melt-spun Bio-based Polyamide/Eu(TTA)$_3$Phen Composite fibers

Yunye Li[1], Pengfei Lou[2] and Qingxiu Jia[1, 2]

1 School of Material Science and Engineering, Beijing Institute of Fashion Technology, Beijing, 100029, China
2 Beijing Engineering Research Center of Textile Nanofiber, Beijing, 100029, China
E-mail: clyjqx@bift.edu.cn

Abstract. In this paper, the bio-based polyamide (PA) was melt polymerized from four bio-based monomers. Composites of the bio-based PA and europium complex Eu(TTA)$_3$Phen were prepared through solution mixing using N, N-Dimethylformamide (DMF) and formic acid as the mixed solvent, and then composite fibers were obtained by melt spinning method. The structure and properties of the melt-spun composite fibers were characterized by FTIR and SEM. The results indicated that the Eu(TTA)$_3$Phen complex, with the average diameter below 300 nm, was homogeneously dispersed in the PA matrix. FTIR spectra indicated that the coordination bond between carbonyl of BDIS and Eu(TTA)3Phen complex formed, which was also confirmed by the mechanical properties. The initial modulus and breaking strength of these fibers can arrived at 2.5GPa and 0.3GPa, respectively.

1. Introduction

During the past decades, much attention was paid on the trivalent rare earth ions due to their unique emission properties from the intra-4f transitions, which are shielded by outer 5s5p orbital [1]. The incorporation of rare earth organic complexes into polymer matrixes can combine the characteristics of both the complexes and polymers, can display high lighting properties, good processing abilities, chemical stabilities, and good mechanical strengths, making them applicable in the areas of optical, electrical, magnetic materials [2, 3]. To date, the potential applications of the polymer-capped rare earth complexes in display, probes, and laser materials have been conducting extensive research [4-6].

Polymer-capped rare earth complexes can be prepared by physical blending and chemical incorporation. Physical blending can be easily achieved through melt blending, mechanical blending, or solution mixing, which is simple and convenient. With the increasing content of complexes added into the polymer matrixes, it becomes more difficult to prepare a uniformly distributed polymer-capped rare earth complexes because of the intermolecular aggregation of complexes and the poor solubility of rare earth complexes in the polymer matrices [7, 8].

In this paper, we used N, N- Dimethylformamide (DMF) and formic acid as the mixed solvent during the preparation of the composites to ameliorate this situation. Furthermore, homogeneous mixing of the complexes in the bio-based PA matrix could be achieved by virtue of the large solubility of the Eu(TTA)$_3$Phen into DMF. The aim of this study was to search a simple and feasible method for increasing the concentration of rare earth complexes into the bio-based PA, and at the same time, to improve the mechanical properties of materials.

2. Experimental

2.1. Materials

All the materials used in this paper are listed in Table 1.

Table 1. materials used in this paper

samper	purity	provider
Eu_2O_3 powder	99.99%	Haimen Material Technology Co., Ltd
Itaconic acid (IA)	98%	Alfa-Aesa
Sebacic acid (SA)	98%	Alfa-Aesa
1,4-diaminobutane (BD)	98%	Alfa-Aesa
1,10-diaminodecane (DD)	98%	TCI-EP (Tokyo)
others	AR	Commercial available

Four monomers (IA, SA, BD, and DD) are melt polymerized to produce the bio-based PA. In the polymerization, mol ratios of four monomers are same. Here the bio-base PA has $\overline{M}_n$ of 70000, and a polydispersity index of 1.80.

2.2. Preparation of Europium Complex, Eu(TTA)$_3$Phen

Eu_2O_3 (0.0125mol) was dissolved in HCl followed by successively evaporating the excessive acid, and then dissolved in the ethanol, and moved to a 250 ml sidearm flask. 2-Thenoyltrifluoroacetone (0.075mol) and 1,10-phenanthrolinec (0.025mol) ethanol solution were dripped to the flask, respectively. The mixture was under a continue stirring at 60 °C with a time of 5h. Then the solutions were cooled at a room temperature, and its pH value was adjusted to 6~7 by ammonia. The system was allowed to stand overnight. After the filtering and washing with ethanol, the precipitate was obtained and dried standby. Its chemical formula as Eu(TTA)3phen was proved by the elementary analysis.

2.3. Melt-spinning of PA/Eu(TTA)$_3$phen fiber

The composite with 1wt% Eu(TTA)$_3$Phen content in bio-based PA was prepared as follows. First, 20g of the bio-based PA was dissolved into 20ml formic acid and 0.202g Eu(TTA)$_3$Phen was dissolved into 30mL DMF, after completely dissolving, mixed the two solutions together, stirred for 6h under 60 °C, and then evaporated to dryness. In this paper, the weight content of Eu(TTA)$_3$Phen in the composite is variable, 5 %, 8 %, and 12 %, respectively. Then the fibers were prepared by the melt-spinning method. Those fibers were stretched to a optimum ratios at 130°C, and heat-set at 140°C.

2.4. Characterization

For all sample, Fourier Transform Infrared spectra in the range of 2000-500 cm^{-1} was completed on a NEXUS FT-IR spectrophotometer. Scanning electron micrographs (Cambridge S-250-III) was used to character the dispersion of rare earth in the PA matrix. The mechanical properties of PA/Eu(TTA)$_3$phen fibers were achieved by Instron 5969 according the ASTM standard.

3. Results and discussion

3.1. Morphology of the composite fibers

SEM images in Fig. 1 show the morphology of cross-section of composite fibers with various mass ratios 1%, 5%, 8% and 12% of Eu(TTA)$_3$Phen. It can be observed that the Eu(TTA)$_3$Phen complex particles were well dispersed in the polymeric matrix with dimension about 100-300 nm, which was smaller than their original sizes. one possible reason for this obvious decrease in size of

IOP Conf. Series: Materials Science and Engineering **307** (2018) 012067 doi:10.1088/1757-899X/307/1/012067

Eu(TTA)$_3$Phen particles is that due to the complexation effect between Eu(TTA)$_3$Phen complex particles and PA matrix, moreover, the mechanical agitation and shearing effects during the processes of melt spinning improved the disperse state of Eu(TTA)$_3$Phen in PA.

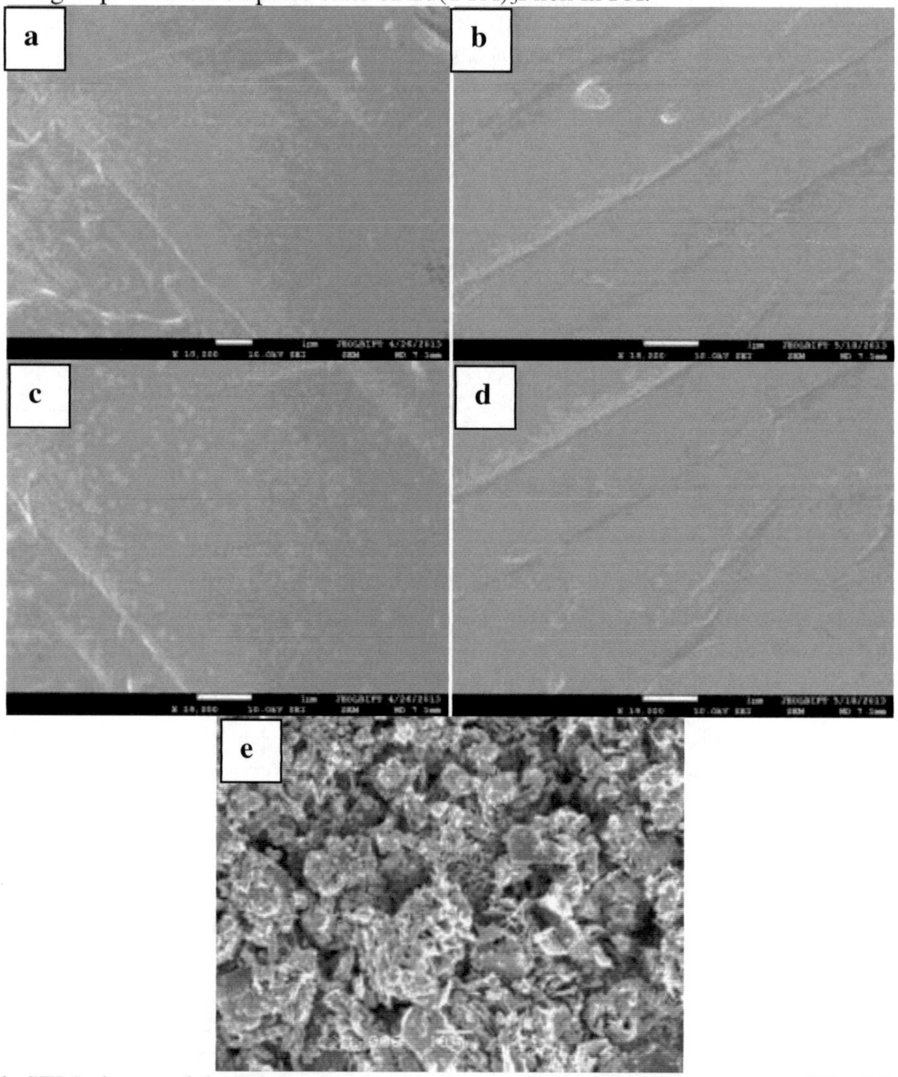

Figure 1. SEM photos of the PA composite fibers containing different contents of Eu(TTA)$_3$phen complex(a)1wt%, (b)5wt%, (c)8wt%, (d)12wt%, and (e) Eu(TTA)$_3$phen complex.

3.2. Interaction between RE and PA

The IR spectra (1000~2000cm^{-1}) of the bio-based PA/Eu(TTA)$_3$phen composites and pure bio-based PA are shown in Fig.2. In the spectrum of pure PA, the band at 1635cm^{-1} (Amide I, C=O stretch) and 1542 cm^{-1} (Amide II, C-N stretch + CO-N-H bend) are assigned to the characteristic vibrational modes of PA. The characteristic peaks of composite fiber shift to higher wavenumber, which may be caused by the coordination bond formed between the carbonyl of the PA and Eu^{3+} ions. These indicate that PA, acting as a lewis base, its carbonyl oxygen atoms and the nitrogen atoms of the amide bond in the polymer chain tend to share a free-electron pair with the f electron orbits available from Eu^{3+} ions and form coordination bond.

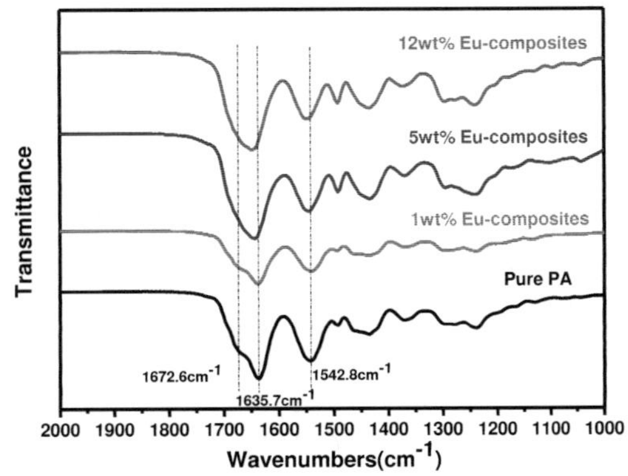

Figure 2. FTIR spectra of pure PA and PA/Eu(TTA)$_3$phen composite fibers.

3.3. Mechanical properties

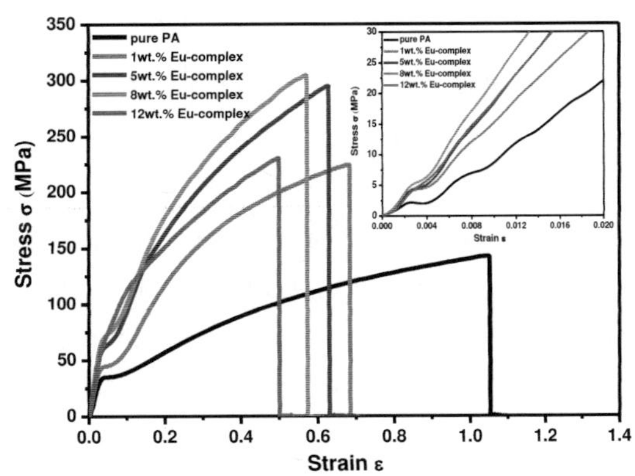

Figure 3. Stress-strain curves of PA and the composite fibers with various contents of Eu(TTA)$_3$Phen.

Fig.3 shows the typical stress-strain curves of PA and the composite fibers doped with different contents of Eu(TTA)$_3$phen complex. Inset shows the enlarged initial portion of all these curves. Detailed mechanical property results of PA with different weight ratios of Eu(TTA)$_3$Phen are listed in table 3. From the shape of these curves and the specific data we can deduce that the materials exhibit a trend of being much more unyielding and stronger progressive, and transform from ductile fracture to brittle rupture (when weight ratio of Eu(TTA)$_3$Phen comes to 12%). Meanwhile, as is shown in table 3, the initial modulus of composite fibers increases from low Eu(TTA)$_3$Phen complex content, and reached a maximum. Similar trend is found in the change of breaking strength values. When Eu(TTA)$_3$Phen complex content reaches 8 wt%, the maximum modulus and breaking strength of the composite fibers were 2504.0 and 305.8MPa, respectively.

There are plenty of explanations for the reinforcement mechanism. According to the above characterizations, the reasons may be listed as follows: firstly, good compatibility between PA and Eu(TTA)$_3$Phen leads to the nice dispersion of the dopant in the matrix, in addition, within the interpenetrating polymeric network structure of the composite fibers, the coordination bond can transfer and distribute the applied stress of molecular chains. However, when the Eu(TTA)$_3$Phen content comes to 12%, much aggregations of the complex result in stress concentration of composite fibers, the mechanical properties deteriorated accordingly.

4. Conclusions

In this work, the bio-based PA/Eu(TTA)$_3$Phen fibers were successfully obtained through melt-spinning. Within the composite materials, a considerable amount of the coordinate bond PA-Eu^{3+} between carbonyl of PA and Eu(TTA)$_3$Phen complex was formed, the rest of complexes dispersed in the matrix with a dimension smaller than 300 nm. The maximum initial modulus and breaking strength of composite fibers obtained were 2504.0 and 305.8MPa, respectively, when Eu(TTA)$_3$Phen content was 8 wt%, which improved significantly compared with pure PA.

Acknowledgments

The authors sincerely acknowledge the financial support from the National Natural Science Foundation of China (51273005), and the Project of Higher Innovation Ability Enhancement of Beijing Municipal Institutions (TJSHG201510012014). The authors declare that they have no conflict of interest.

References

[1] Melby L, Rose N, Abramson E, and Caris J 1964 *J. Am. Chem. Soc.* **86** 5117

[2] Lin S, Feuerstein RJ, and Mickelson AR 1996 *J. Appl. Phys.* **79** 2868.

[3] Slooff LH, Polman A, Klink SI, Hebbink GA, Grave L, Van Veggel FCJM, Reinhoudt DN, and Hofstraat JW 2000 *Opt. Mater.* **14** 101

[4] Liao YC, Lin CH, and Wang SL 2005 *J. Am. Chem. Soc.* **127** 9986

[5] Hassey R, Swain EJ, Hammer NI, Venkataraman D, and Barnes MD 2006 Science **314** 1437

[6] Robinson MR, Ostrowski JC, Bazan GC, and McGehee MD 2003 *Adv. Mater.* **15** 1547

[7] Sun Z, Bai F, Wu H, Boye DM, and Fan H 2012 *Chem. Mater.* **24** 3415.

[8] Niyama E, Brito H, Cremona M, Teotonio EES, Reyes R, Brito GES, and Felinto MCFC 2005 Spectrochim. Acta, Part A **61** 2643

IOP Conf. Series: Materials Science and Engineering **307** (2018) 012068 doi:10.1088/1757-899X/307/1/012068

Polyaniline/Carbon nanotube Electrochromic Films: Electrochemical Polymerization and characterization

Xiao-Xia LI[1, 2, a*], Liang ZHAO[3, b], De-Yue MA[1, c] and Yu-Run ZENG[1]

[1] State Key Laboratory of Pulsed Power Laser Technology, Hefei 230037, China
[2] Key Laboratory of IR and Low Temperature Plasma of Anhui Province, Hefei 230037,China
[3] Beijing Institute of Remote Sensing, Beijing 100192, China
[a*]lxxhong@163.com

Abstract: Polyaniline/Carbon nanotube (PANI/CNT) composite films doped with dodecylbenzene sulfonic acid were synthesized by cyclic voltammetry on an ITO-coated glass substrate. FTIR, XRD and electrochemical analyzer were used to characterize the micro-morphology, chemical structure, crystallinity and electrochromic behavior of the films, respectively. The effect of CNT content on the properties of the films was investigated. Results show that the introducing CNTs make aniline polymerize easier than before. Within a range, the conductivity and crystallinity of PANI/CNT composites improves with CNT content increasing. The electrochromic device made from the PAN/CNT film with a CNT content of 2.5wt% presents a reflectance contrast of 38.8%, a mean response time of 2.3s and a coloration efficiency of 386.4cm^2/C at 540nm. The PAN/CNT film shows better electrochromic behaviors due to some interaction between CNTs and the PANI backbones than PANI film.

1. Introduction

Polyaniline (PANI) with metal-like conductivity and electrochromic behavior can be obtained from inexpensive aniline and some doping agents by electrochemical polymerization or chemical oxidation. It has diverse potential applications such as electrochromic devices (ECDs), super capacitors and electromagnetic shielding, etc. However, as a versatile conducting polymer, PANI has some unavoidable shortcomings, such as lower charge transferring capacity and shorter life cycles of duty-cycle operation, which need to be improved by acid-doping or compounding with other materials [1].

Carbon nanotubes (CNTs) have recently attracted much attention owing to their unique spiral structure, quantum effect, surface action, electronic properties and so on [2]. Some studies indicate that the electric conductivity and mechanical properties of polymers can improve by introducing CNTs into a polymer matrix [3]. Therefore, nano-composites that combine CNTs and conducting polymers have recently attracted considerable interest. As PANI has high affinity for the surface of CNTs, PANI-CNT composites have been synthesized in situ by chemical or electrochemical oxidation of aniline in the presence of CNTs. The composites show enhanced electrical, thermal or mechanical properties by comparison to PANI or CNTs alone, and exhibit potential applications including of cells, sensor, micro/nanoelectronic components and optoelectronic devices [4].

At present, PANI/CNT composites are mainly prepared in situ by chemical oxidation, and the powdered product limits their further application. In this paper, PANI/CNT composite films were synthesized by electrochemical polymerization, and their behaviors were characterized.

IOP Conf. Series: Materials Science and Engineering **307** (2018) 012068 doi:10.1088/1757-899X/307/1/012068

2. Material and Methods

2.1. Materials
Aniline (≥99.5%) was distilled twice before used for polymerization. Carbon nanotubes (96%) with a diameter of 20-40nm and a length of 10-30μm were received from Beijing Boyu Technology Co. Ltd. Dodecylbenzene sulfonic acid (DBSA, ≥96%); alcohol (C_2H_5OH, ≥99.7%) and deionized water were all needed. A 1.0cm×3.0cm ITO-coated glass with a resistivity of 6-8Ω/sq was used as a substrate.

2.2. Electrochemical polymerization of PANI/CNT film
PANI/CNT films were synthesized in a three-component cell, and the potential was scanned between -0.2 and 1.0V at a rate of 50mV/s, where an ITO-coated glass slide, platinum strip and saturated calomel electrode were used as working electrode, counter and reference electrodes.

The electrolyte for polymerization was prepared as follows: The mixture of 0.25mol·L^{-1} aniline, 0.8mol·L^{-1} DBSA and 5mol·L^{-1} C_2H_5OH was ready, and then, the suspension of various weight percent of CNTs by ultrasonic dispersion was added to the mixture, respectively. Finally, PANI/CNT composite films with various CNT content were prepared by electrochemical polymerization as mentioned above.

2.3. Measurements
An ECD in which the PAN/CNT film with 2.5wt% CNTs acted as electrochromic layers was made, and its electrochemical and electrochromic behaviors were performed by a CHI760 electrochemical analyzer. IR spectra were recorded by VER TEX 80+HYPERION 2000 FTIR spectrometer. XRD patterns were obtained on an X′Pert Pro X-ray diffractometer (Cu Kα radiation). The conductivity was measured by four-point probe technique with a 1934 Conductivity Meter of China. HATCHI U-4100 Spectrophotometer was used to measure the reflectivity of films.

3. Results and Discussion

3.1. Cyclic voltammetric analysis
The electrochemical polymerization of PANI/CNT films can be described by cyclic voltammograms (CVs). The CVs for sample CP-3 are showed in Fig 1. Compared with pristine PANI [5], a pair of redox peaks appears and the current density rises obviously at 0.2V and 0.9V at the first cycle, which implies higher electroactivity. The interaction between CNTs and PANI through π-stacking can enhance the delocalization of electrons [6]. And meanwhile, we can see the film color change cyclically among light yellow, yellow green, green and dark green.

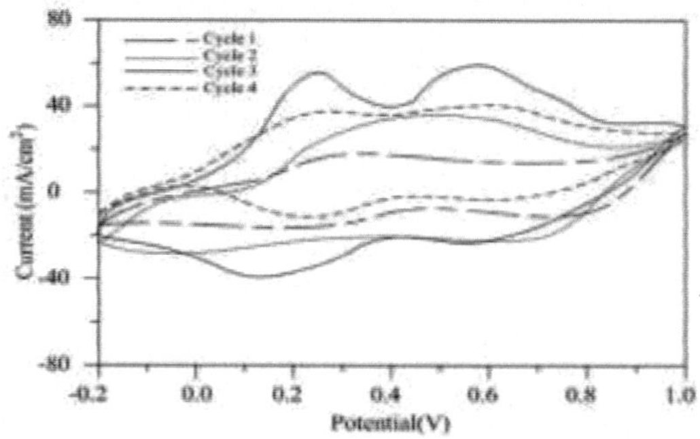

Fig 1 CVs of PANI/CNT films

3.2. FTIR spectra

FTIR spectra of PANI/CNT composites and PANI are showed in Fig 2. All samples present similar peaks, such as aromatic ring stretching at 1555cm^{-1}, quinoid ring stretching at 1470 cm^{-1}, C-N stretching of aromatic amines at 1300cm^{-1}, C-N stretching of benzenoid ring at 1238cm^{-1}, a vibration mode of N=Q=N at 1108 cm^{-1}, an out-of plane stretching vibration of C-H of aromatic ring at 790 cm^{-1}, etc. Remarkably, the wide and strong band at 1108 cm^{-1} is described as "electronic-like band" and is considered a measure of electron delocalization degree [7]. Thus, this peak can be used to characterize the conductivity of PANI. It is clear that increasing CNT content results in a distinct increase of the band intensity at 1108 cm^{-1}, which indicates that PANI/CNT composites have a better conductivity than PANI alone. This result is in good agreement with the conductivity test data (Table 1) and indicates that CNTs are well mixed with PANI in the composite films.

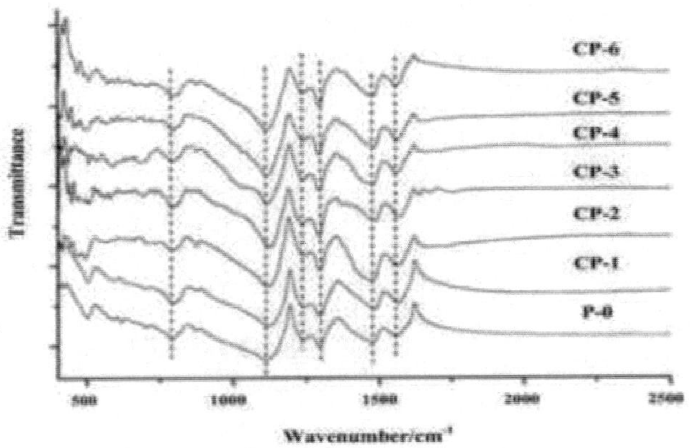

Fig 2 FT-IR spectra of PANI/CNT films with various CNT contents

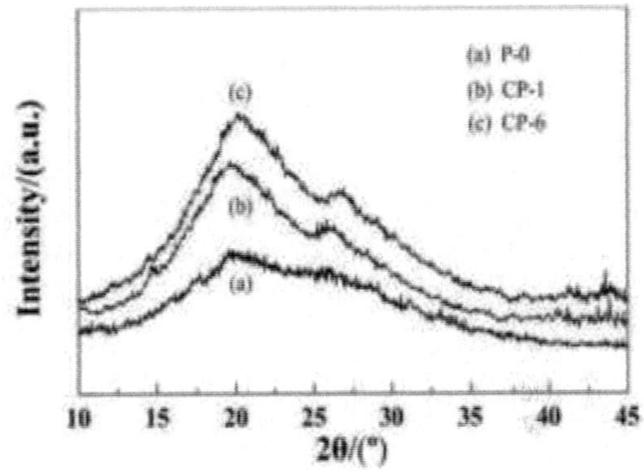

Fig 3 XRD patterns of PANI/CNT composites with various CNT contents

3.3. XRD patterns

Fig 3 shows XRD patterns of PANI/CNT composites. The peak at 2θ=20.7° is attributed to the period structure of PANI chains, and that at 25.2° is for the amorphous state of PANI. As for CP-1 and CP-6, the peaks at 2θ=20.7°, 25.2° are stronger than those of sample P-0, which implies that PANI crystallinity in PANI/CNT composites improves obviously while the formation of PANI side chains is inhibited by introducing CNTs. Meanwhile, the introduction of CNTs leads to two peaks at 2θ=25.9° (002) and 43° (100) corresponding to a graphite-like CNT structure [8].

3.4. Electrical conductivity

Table 1 shows the electrical conductivity of PANI/CNT films with various CNT content. The conductivity rises sharply as the CNT content increases from 0.5 to 2.5wt%, and then does slowly and approaches to a certain value. As CNTs are good electron acceptor owing to their large length-diameter ratio and specific surface, they can serve as conductive bridges to connect conducting domains of PANI and improve charge transfer. A proper content CNT can offer enough conductive bridges to form some conductive paths and result in a high conductivity.

Table 1. Conductivity of PANI/CNT films with various CNT contents

Samples	CNT content (wt%)	Conductivity(S.cm^{-1})
P-0	0	1.123
CP 1	0.5	1.927
CP-2	1.5	9.224
CP-3	2.5	19.465
CP-4	3.5	24.653

CP-5	4.5	25.127
CP-6	5.5	25.437

3.5. Electrochromic behaviors

Fig 4 shows Vis-NIR reflectance spectra for CP-3 film when applied potentials step between –0.4V and 0.8V with a step of 0.1V. With the potential increasing, the reflectivity at 500-600nm reduces gradually until the peak disappears, and a maximal color contrast of 38.8% appears at 540nm. Meanwhile, we can see the film color change from dark green to yellow green.

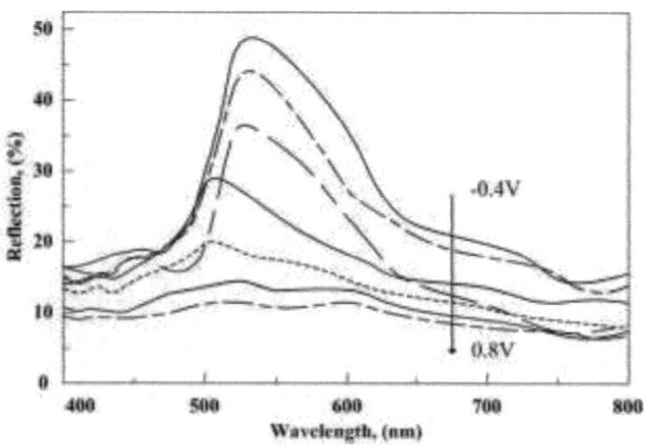

Fig 4 Vis-NIR reflectance spectra of PANI/CNT films at -0.4~0.8V

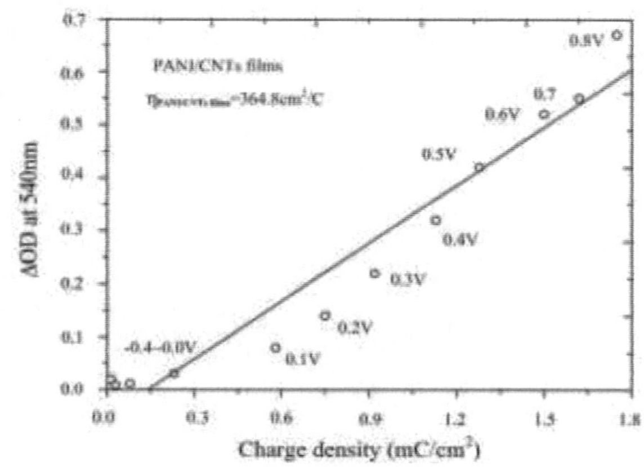

Fig 5 Optical density and charge density of PANI/CNT films at 540nm

Coloration efficiency (CE) is a key parameter to characterize the electrochromic behavior of materials and devices [9], which can be derived from the slope of optical density~current density curve as showed in Fig 5. Therefore, the CE at 540nm for CP-3 film is 368.4 cm^2/C. The film color becomes darker at 0.8V with the drop of reflectivity at 540nm, and lighter at -0.4V with reflectivity rising.

As showed in Fig 6(a), applied potentials step between 0.8V and -0.4V with a residence time of 5s. Fig 6(b) shows the reflectivity changes at 540nm for CP-3 film after several cycles. It reveals that the average response time is about 2.3s both for coloring state and bleaching one, which means a fast response of PANI/CNT electrochromic film. The reflectivity contrast nearly keeps unchangeable after four cycles, which means a good memory effect and stability of the film.

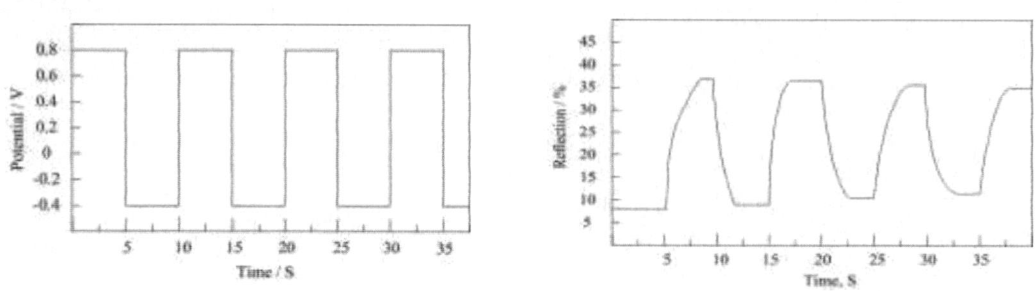

(a) (b)

Fig 6 Charge time (a) and response time (b) profiles of PANI/CNT composites at 540nm

IOP Conf. Series: Materials Science and Engineering **307** (2018) 012068 doi:10.1088/1757-899X/307/1/012068

4. Conclusion

DBSA-doped PANI/CNT composite films were synthesized by cyclic voltammetry. The introduction of CNTs promotes the polymerization of aniline. A reflectance contrast of 38.8% and a coloration efficiency of 386.4cm^2/C at 540nm occurs in PANI/CNT film with CNT content of 2.5 wt%. Compared with the pristine PANI film, the PAN/CNT composite film shows a higher conductivity, a shorter response time, a higher electrochromic efficiency and better stability due to some interaction between CNTs and the PANI backbones.

Reference

[1] Li-Ming Huang, Cheng-Hou Chen, Ten-Chin Wen, A Gopalan. Effect of secondary dopants on electrochemical and spectroelectrochemical properties of polyaniline. Electrochimica Acta. 51 (2006) 2756-2764

[2] Ken P. Chong. Nano science and Engineering in solid mechanics. Acta Mechanica Solida Sinica, 20 (2008) 95-103

[3] P.C. Ramamurthy, W.R. Harrell, R.V. Gregory, B. Sadanadan and A.M. Rao. Electronic Properties of Polyanline/Carbon Nanotube Composites. Synthetic Metals. 137 (2003)1497-1498

[4] Cynthia Oueiny, Sophie Berlioz, Francois-Xavier Perrin. Carbon nanotube-polyaniline composites. Progress in Polymer Science. 39 (2014) 707-748

[5] Sun Tong, Li Xiaoxia, Guo Yuxiang, Zhao Jijin, etc. Electrochemical polymerization of polyaniline nanofiber doped with neopelex by cyclic voltammetry. Polymer Materials Science and Engineering, 30(2014) 118-121.

[6] J. E. Huang, X.H. Li, J.C. Xu,H.L. Li. Well-dispersed single-walled carbon nanotube/ polyaniline composite films. Carbon. 41 (2003) 2731-2736

[7] E. JKulszeohny Jelmy, S. Ramakrishnan, Murali Rangarajan, Nikhil K. Kothurkar. Effect of different carbon fillers and dopant acids on electrical properties of polyaniline nanocomposites. Bull. Mater. Sci. 36 (2013) 37-44.

[8] T.H. Ting, Y.N. Jau, R.P. Yu. Microwave absorbing properties of polyaniline/multi-walled carbon nanotube composites with various polyaniline contents. Applied Surface Science. 258 (2012) 3184-3190.

[9] Kuan-Yu Shen, Chih-Wei Hu, Li-Chi Chang, Kuo-Chuan Ho. A complementary electrochromic device based on carbon nanotubes/conducting polymers. Solar Energy Materials & Solar Cells, 98 (2012) 294-299.

IOP Conf. Series: Materials Science and Engineering **307** (2018) 012069 doi:10.1088/1757-899X/307/1/012069

The fibre orientation influence in cementitious composite against extreme load resistance

R Lovichova[1], J Fornusek[1], M Mara[1], M Kocova[1] and Z Rihova[1]

1 CTU in Prague, Czech Republic, Faculty of Civil Engineering, Experimental Centre
E-mail: romana.lovichova@fsv.cvut.cz

Abstract. This paper is focused on resistance of steel fibre-reinforced cement composite against impact of the deformable projectile shot from the 7.62 × 39 caliber. Different values of resistance against impact of the projectile are caused by different orientation of the fibres. The influence of formwork position, which is the main cause of the different orientation of the fibres, is investigated. The resistance was examined on thirty slabs made of ultra-high performance fibre-reinforced concrete (UHPFRC). Fifteen specimens with vertical orientation of formwork and fifteen with horizontal orientation of formwork was made. The resistance is classified according to the visual evaluation and local damage measurement on the front side and the rear side of the examined specimens. The experiment shown positive influence of vertically oriented formwork on the slabs according to their resistance against impact of the projectile.

1. Introduction

Currently, main protection against shooting from handguns at military and police posts in areas of armed conflicts are used a sandbags and sacks filled with other loose materials. Construction of "sack walls" is overall technically and physically complicated. Furthermore, sacks are mostly made of jute because of its low price. But they degrade relatively quickly under weather conditions and damaged bags spill out filing material which reduce the protection in time. Resistance against extreme loads is increased by the use of ultra-high-grade reinforced steel fibers to reduce brittleness and increase material ductility. The slabs of UHPFRC provide good ballistic protection and their relatively low weight allows their quick assembly and disassembly of protective constructions. Moreover high quality products of UHPFRC enable a long service life even when they are exposed to heavy weather conditions. It is necessary to deal with this topic because positive results could lead to the introduction and putting of these materials into the construction procedures and practice. Especially in structures requiring protection against shooting or explosions.

2. Material Description

This chapter deals with the production of tested samples and description of their mechanical properties. Ultra-high-performance fiber-reinforced concrete containing fine-grained aggregate up to fraction 1 mm was used for our experiments. The composition of the mixture was taken according to the utility model by the originators Kolář et al. [1]. The weight ratio of the individual components of the mixture is shown in Table 1. The amount of steel fiber content was stated as 2 % of the mixture volume. The content of fibers and their orientation affects the mechanical properties of fiber concrete. Therefore, the different positioning of the formwork during the laying of the concrete mixture can be a way of achieving different orientation of the fibers on one flat element.

IOP Conf. Series: Materials Science and Engineering **307** (2018) 012069 doi:10.1088/1757-899X/307/1/012069

Table 1. UHPC mixture expressed in weight ratio [1].

Components of UHPC	
Cement CEM I 42.5R	1.00
Water	0.30
Microsilica	0.10
Silica fume	0.25
Silica sand 0.1 - 1.2	1.60
Plasticizer 1	0.01
Plasticizer 2	0.001

Thirty slabs were made for the planned experimental program on the shooting range. These concrete slabs with reinforced steel fibers had dimensions of 300 × 400 mm and thickness of 50 mm. Fifteen of them were made in a vertical formwork (see Figure 1a) and fifteen in a horizontally stored formwork (see Figure 1b). The samples were de-molded after 24 hours. Also testing samples with dimensions of 40 × 40 × 160 mm were produced with each series of slabs (see Figure 1c) to obtain the average results of mechanical. Prisms were weighed and measured to determine the volume density of each testing mixture. Three-point bend test was carried out after 28 days and then a compressive strength was also tested on the one part of specimen left after the three point bending test - dimensions of contact area 40×40 mm. Amount of samples was set on three for both tests. The average volume density was measured as 2373 kg/m^3. Average results of obtained mechanical properties such as the average of compressive strength (f_c) was 157.1 MPa and the average of flexural strength (f_b) was 26.6 MPa [2].

a) b) c)

Figure 1. Formworks for slabs and testing samples
a) vertical slabs 300×400×50 mm; b) vertical slabs 300×400×50 mm;
c) testing samples 40×40×160 mm

3. Experimental Program

Experimental program was realized on the shooting range. Testing conditions were adjusted as close as possible according to a standard EN-1522 and EN-1523 [3, 4] and to the shooting range options. Concrete penetration resistance was performed on the thirty slabs made from UHPFRC with dimensions 300 × 400 mm and thickness of 50 mm. The half of the slabs was made in a horizontally placed formwork and the other half in a formwork placed vertically. The aim was to verify if the formwork position in the production of the specimens had an effect on the resistance against an extreme loads, such as projectile hit. Each specimen was placed into the special mount with simulation of point supports by two screws in each corner. Projectile was fired from the distance 20 m to the specimen (Figure 2). Each slab was hit only once and to the center. Then, condition of the slab after the hit was captured by digital camera and analyzed. Proposed types of projectiles were 7.62 × 39 FMJ with lead core in all cases (Figure 3). The appointed projectile has ogive nose, diameter of 7.92 mm, mass of 8.04 g and initial energy 2030 J [5, 6].

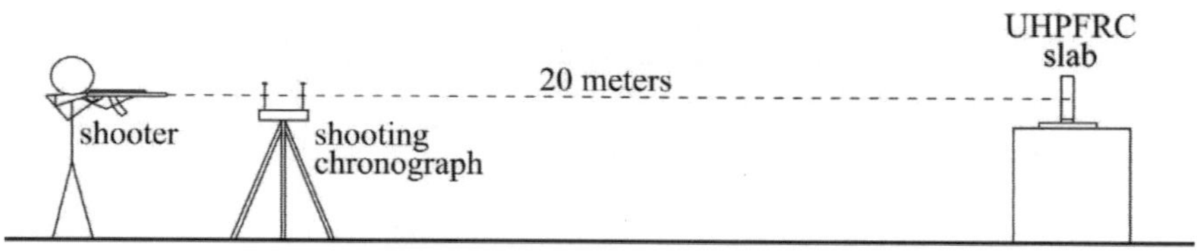

Figure 2. Schema of the applied testing system shooting range.

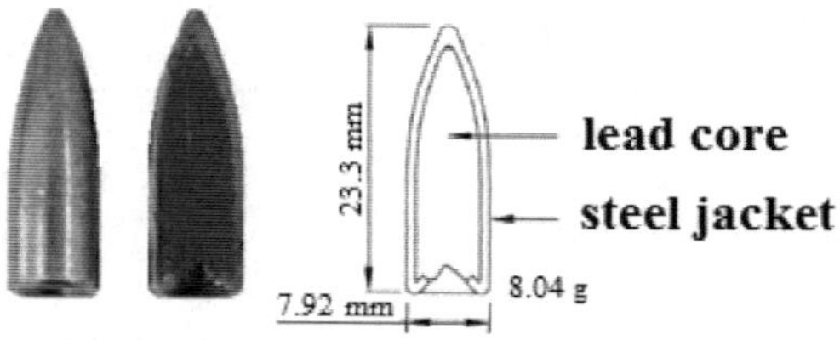

Figure 3. Deformable projectile 7.62 × 39 FMJ with lead core.

4. Results and Discussion

The impact assessment of the projectile is based on the criteria used in the work of Vossoughi et al. [7]. The effects are divided into the following four categories: P - perforated (the projectile passed through the specimen entirely), PL - perforation limit (projectile was stuck), UP - unperforated (specimen was punched but projectile bounced), PB - perforated and then bounced. The observed effects on the individual slabs are in Table 2 - the formwork laid horizontally is described on left and the formwork stored vertically is on the right.

Table 2 Effects of deformable projectile on the slabs made of UHPFRC.

Horizontally stored formwork				Vertically stored formwork			
Impact response of concrete	Excavation of material on the back side	Avg. crater width - front side [mm]	Avg. crater width - back side [mm]	Impact response of concrete	Excavation of material on the back side	Avg. crater width - front side [mm]	Avg. crater width - back side [mm]
UP	NO	45	120	UP	NO	60	105
UP	YES	55	120	UP	YES	60	80
UP	YES	50	80	UP	NO	55	95
UP	NO	75	75	UP	NO	55	80
UP	YES	55	90	UP	NO	60	100
UP	NO	60	100	UP	NO	55	115
UP	YES	65	100	UP	NO	75	80
UP	YES	55	100	UP	NO	60	115
UP	YES	50	125	UP	NO	70	115
UP	YES	55	105	PL	NO	55	100
UP	NO	60	120	UP	NO	55	75
UP	YES	60	125	UP	NO	55	75
P	YES	75	100	UP	NO	55	115
UP	NO	70	100	UP	NO	55	60
PB	YES	55	100	UP	NO	60	75
	Average [mm]	59	104		Average [mm]	59	92

The left part of the Tab. 2 shows that a complete shoot through the slab occurred in only one case (Fig. 4) and in another only case a hole was formed, with the projectile bouncing off. Since both of these cases are from one series, it can be assumed as an error during the production which subsequently manifested this behaviour. In other cases, the projectile was deflected but it created a crater at the front (Fig. 5) and back side of the slab. There was a scabbing on the back of the slab in some cases and the material was broken out by the tensile stress (Fig. 6), it was occurred in 66 % cases. In the remaining cases, there was also a breakdown due to tensile stress, but the structure and the greater tensile strength of the material allowed only a material deformation and cavity formation within the investigated slab (Fig. 7). The average width of crater on the front and back side was 59 and 104 mm.

Fig. 4 UHPFRC slab with hole passing through overall thickness

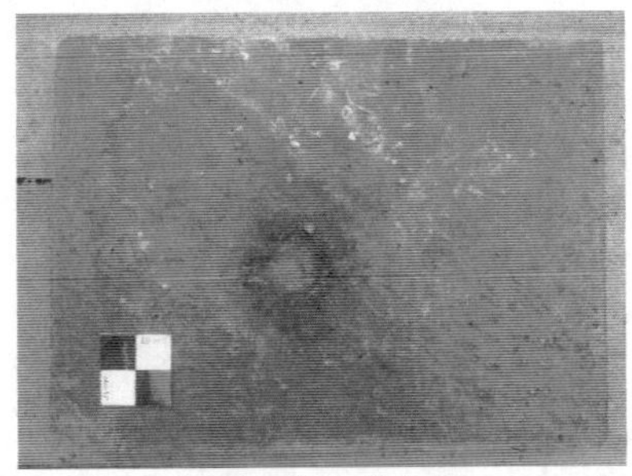

Fig. 5 Crater after projectil impact on the front side of the slab

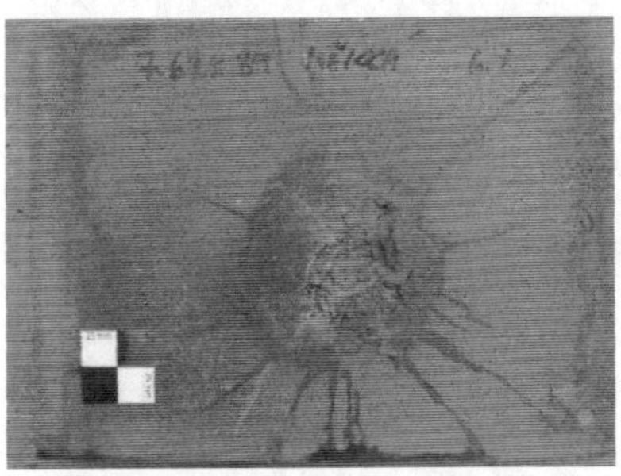

Fig. 6 Tearing of material from the back side of the board due to tensile stress

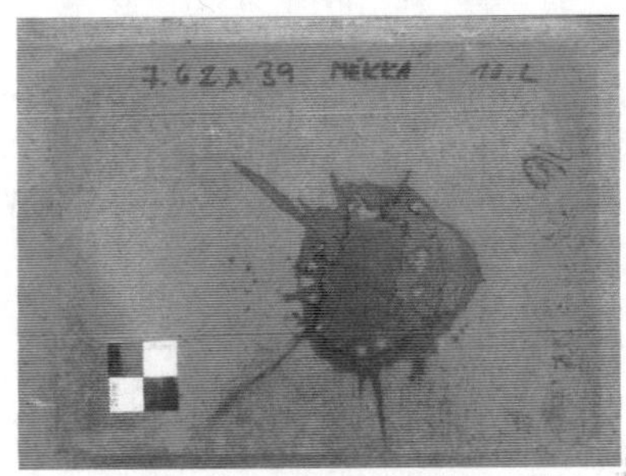

Fig. 7 The back of the board where the material has been stuck

Between samples made in a vertically stored formwork (right side of Table 2), there was only one slab where the projectile stayed stuck inside the slab (Fig. 8). In other cases, a projectile created a crater and then bounced off after impact (Fig. 9). There were also signs of tensile stress on the back side which was caused by the spread of the pressure wave after the impact of the projectile. It can be seen from the table 2 that material tear-out on the back side of these slabs has occurred only in one case out of the fifteen samples, i.e. for 93 % of the vertical slabs, the material still held together in one piece (Fig. 10). The created craters on the front and the back of the slabs formed in the horizontal formwork had an average width of 59 mm and 92 mm.

Fig. 8 Projectile stuck in the UHPFRC

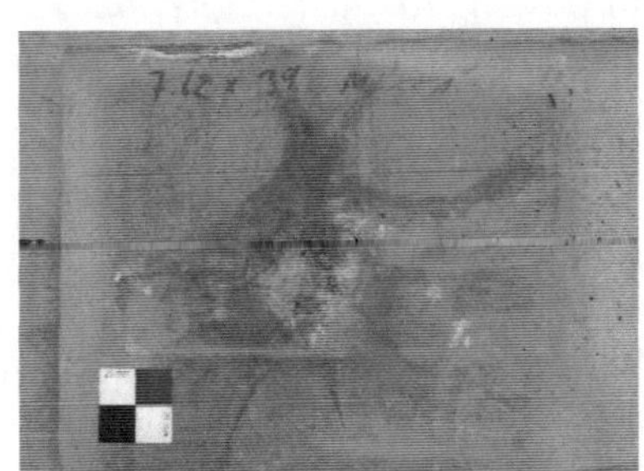

Fig. 9 Crater on the front side after impact

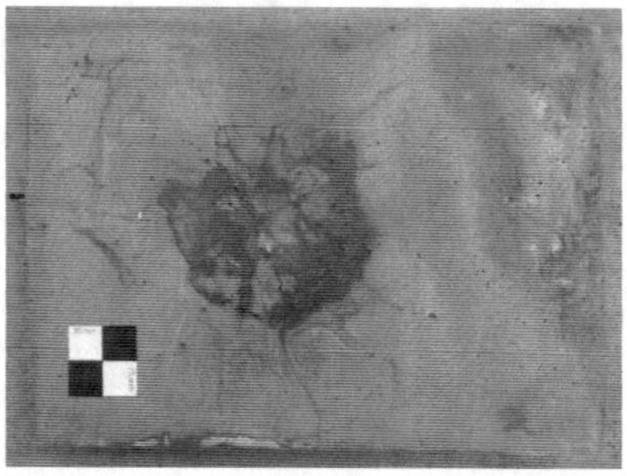

Fig. 10 The back side of the slab where the material has not been riped out

5. Conclusions

The test evaluations obtained the following results. There was two-thirds of horizontal UHPFRC slabs where the material was torn-out from the rear side and only one case out of fifteen vertically poured UHPFRC slabs. The average crater width at the front side of the UHPFRC slabs is the same for both types of formwork orientation. Its value is 59 mm. The average width of the crater at the back side of the vertically oriented UHPFRC slabs is by 12 mm smaller. The results show a considerable positive effect of the formwork orientation on the extreme load resistance, which is achieved in this work by the projectile impact of the 7.62 × 39 caliber rifle. The formwork orientation has almost no effect on the endurance of the front side of the UHPFRC board. The stress strengths of the load tests ranged from 130 MPa to 199 MPa and a flexural tensile strength of 15 MPa to 41 MPa. However, it is clear from the results of the UHPFRC projectile impact resistance testing that the strengths within these ranges do not have a visible effect.

The separation of material on the rare side is decisive - slabs with a vertical formwork does not occur rapture of the material on the back side. This is due to a better transfer of tensile stress induced by the pressure wave after the impact of the projectile. Based on this work, it is recommended to use slabs made of ultra-high performance concrete reinforced with steel fibers, oriented vertically.

6. References

[1] Czech Technical University in Prague, Faculty of Civil Engineering, Prague 6. Dry prefabricated mixture of multifunctional silicate composite (in Czech: Suchá prefabrikovaná směs multifunkčního silikátového kompozitu). Czech Republic. Patent No. CZ 306663.

[2] Hurtig K 2017 The influence of fibre orientation in cementitious composite on extreme load resistance. *Bachelor thesis*. CTU in Prague. Faculty of Civil Engineering, Experimental Centre.

[3] European Comittee for Standardization, EN 1522, Windows, doors, shutters and blinds – Bullet resistance – Requirements and classification, 1998, Brussel.

[4] European Comittee for Standardization, EN 1523, Windows, doors, shutters and blinds – Bullet resistance – Test method, 1998, Brussel.

[5] Vavřiník T 2013 Experimental and Numerical Analysis of Projectile Impact on Cementitious Composite. *Diploma thesis*. CTU in Prague. Faculty of Civil Engineering, Experimental Centre.

[6] T. Vavřiník, J. Zatloukal, J. Fornůsek, P. Konvalinka. 2013 Numerical Analysis of Projectile Impact on Cementitious Composite Slabs, In: *11th International Conference of Numerical Analysis and Applied Mathematics 2013*, Rhodos. p. 1012-1015.

[7] Vossoughi F, Ostertag C P, Monteiro P J M, Johnson G C, Resistance of concrete protected by fabric to projectile impact, In: *Cement and Concrete Research*, Volume 37, Issue 1, 2007, p. 96-106, ISSN 0008-8846, https://doi.org/10.1016/j.cemconres.2006.09.003.

Acknowledgments

Authors gratefully acknowledge the financial support from CTU foundation number SGS17/069/OHK1/1T/11 and the Ministry of Industry and Trade of the Czech Republic, the grant No. MPO FV10547.

IOP Conf. Series: Materials Science and Engineering **307** (2018) 012070 doi:10.1088/1757-899X/307/1/012070

A new local thickening reverse spiral origami thin-wall construction for improving of energy absorption

C H Kong[1], X L Zhao[2], I R Hagiwara[3]

[1,2]Department of Mechanical Engineering, Saitama Institute of Technology, 1690 Fusaiji, Fukaya, Saitama, Japan
[3]Meiji Institute for Advanced Study of Mathematical Sciences, Meiji University, 4-21-1, Nakano, Tokyo, Japan

[1]kongchenghai@gmail.com [2]Zhaoxilu@sit.ac.jp [3]ihagi@meiji.ac.jp

Abstract. As an effective and representative origami structure, reverse spiral origami structure can be capable to effectively take up energy in a crash test. The origami structure has origami creases thus this can guide the deformation of structure and avoid of Euler buckling. Even so the origami creases also weaken the support force and this may cut the absorption of crash energy. In order to increase the supporting capacity of the reverse spiral origami structure, we projected a new local thickening reverse spiral origami thin-wall construction. The reverse spiral origami thin-wall structure with thickening areas distributed along the longitudinal origami crease has a higher energy absorption capacity than the ordinary reverse spiral origami thin-wall structure.

1. Introduction

In the recent years, in order to realize the sustainable development strategy, especially in the field of energy conservation and environment protection, many lightweight and safety structures have been proposed.

Side members are the primary energy-absorbing components that are set up in the front right and left sides of a vehicle. To ensure adequate survival space for passengers in the case of a vehicle crash, the deformation length of the front side members is required to be long enough and the energy absorption should be as much as much as possible. Crash energy absorption is always accompanied by Euler buckling, and once the structure bends to one side, it can no longer absorb energy. Hagiwara[1] (1991) worked out the buckling wavelength of a side member: the structure underwent an accordion crease during the vehicle crash process but the deformation length was only set to 70%. Nojima[2] (2002) first defined the concept of origami engineering and proposed many origami structures. Zhong[3] (2014) introduced folded structures constructed from flat materials, thereby broadening and enhancing the origami theory. Hagiwara et al. [4,5](2003, 2004), proposed the most representative tubular origami structure named reverse spiral origami thin-wall structure that is able to achieve a deformation length of 90%.Zhao[6](2010) utilized optimization technology on the reverse spiral origami structure. The literature profiled indicates that energy absorption of this structure is about 1.91 times that of the collapsed structure of a traditional rectangular section, and1.37 times that of an uncollapsed traditional rectangular section of the same mass.

Although the structure receives a full performance in deforming process, because of the origami creases, the support force along the longitudinal direction is decreased. In order to increase the support strength of the reverse spiral origami structure, we proposed a new local thickening reverse spiral origami thin-wall structure for better energy absorption.

2. Reverse spiral origami thin-wall structure

Figure.1 shows the side member of the vehicle. In order to design high energy absorption and avoiding of Euler bucking side member, the reverse spiral origami thin-wall tube is proposed. As is shown in Figure.2, a reverse spiral origami structure is designed from a twisted polygon section pipe with a vertical cylindrical shell along the axial direction. When there is an impact load on the end surface of the structure, crash deformation can be guided by the folding edges. Because of its energy absorption character, the structure is applied to the side member of the vehicle.

Figure.1 Side member of the vehicle

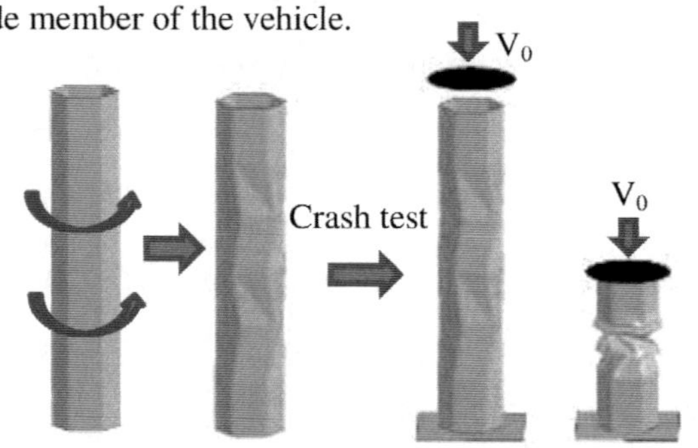

Figure.2 Reverse spiral origami thin-wall structure

In the deformation process, many wrinkles are generalized and folded like a spring, which can effectively avoid collapse in the transverse direction. However, the origami creases also weaken the support force of origami structure and decrease its energy absorption.

3. Reverse spiral origami thin-wall structure of local strengthening

When the reverse spiral origami structure is processed by hydroforming, the thickness distribution can be effected by the original dimeter of pipe. In the hydroforming process, the separated mould moved to form a sealed space, then the liquid is loaded inside the tube and the tube deformed to be the shape of mould.

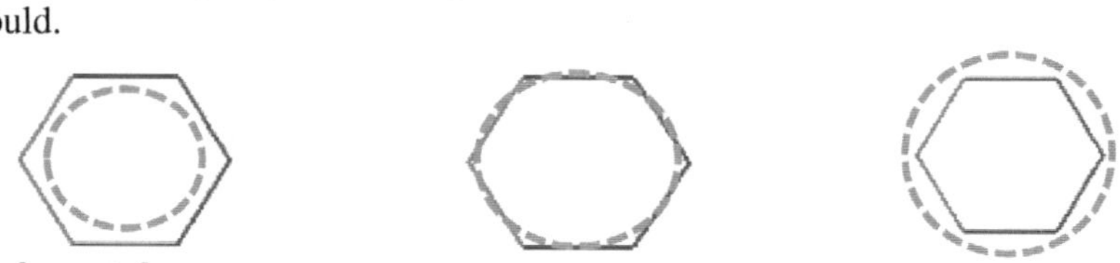

(a) $d_{org} < d_{in}$ (b) $d_{in} < d_{org} < d_{cir}$ (c) $d_{cir} < d_{org}$

Figure.3 Relationship of pipe and mould（d_{org} is diameter of pipe、d_{in} is inscribed diameter of mould, d_{cir} is circumradius of mould）

As is shown in Figure.3, when $d_{org} < d_{in}$, before hydroforming, the pipe doesn't contact with mould. When the hydraulic pressure is loaded inside the pipe, because of the tension force, the pipe is becoming thinning in the whole hydroforming process. And the pipe may fracture at last. When $d_{in} < d_{org} < d_{cir}$, before hydroforming, the pipe contact with mould, thickness of longitudinal creases area of the pipe is thickening while thickness of transverse and diagonal creases area of the pipe is thinning. Under this condition, the thickness distribution of pipe is easily controlled. When $d_{cir} < d_{org}$, the material is accumulated along the longitudinal direction, the pipe may wrinkle at last.

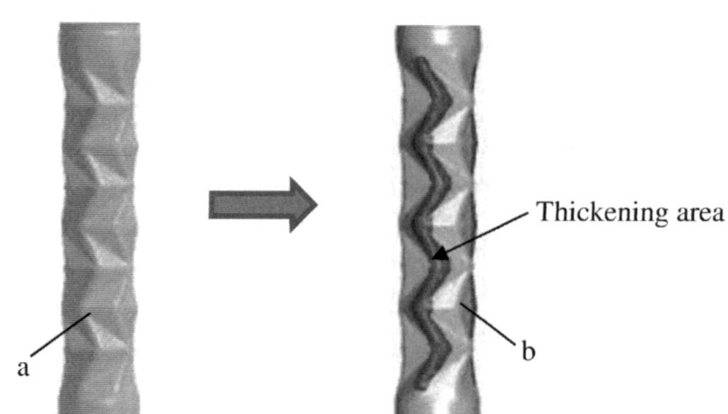

IOP Conf. Series: Materials Science and Engineering **307** (2018) 012070 doi:10.1088/1757-899X/307/1/012070

Figure.4 A local strengthening reverse spiral origami structure

When $d_{in} < d_{org} < d_{cir}$, it is easy to adjust the thickness distribution without rupture and wrinkle. In order to enhance the support force along the longitudinal direction, we designed a local strengthening reverse spiral origami structure as shown in Figure.4. The area along longitudinal direction of the tube is thickened and the support force is enhanced and the enhanced structure can absorb more energy than ordinary reverse spiral origami structure.

4. Reverse spiral origami thin-wall structure of local strengthening

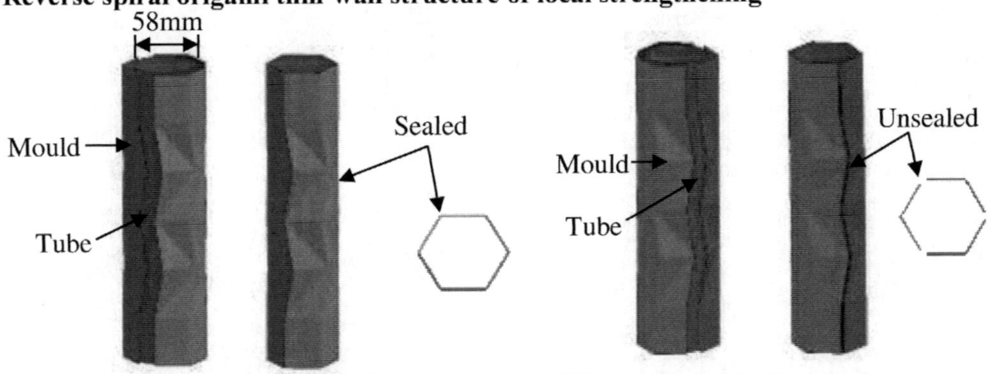

(a)Ordinary model of origami structure (b)Unsealed model of origami structure

Figure.5 Simulation model of hydroforming

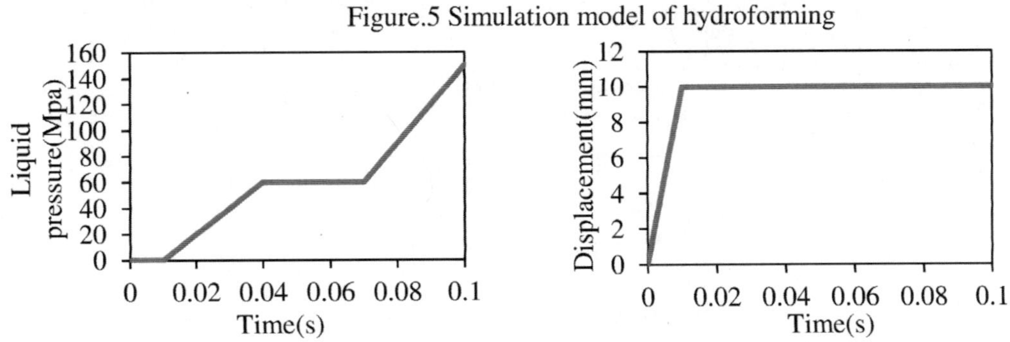

Figure.6 Loading curve of hydraulic pressure and displacement of mould

In order to research the energy absorption character of local strengthening reverse spiral origami structure, we made use of unsealed mould for hydroforming. The width of separated mould is 3mm smaller than the ordinary mould, and after the separated moulds assembly a whole mould, there is 6mm space left between the separated moulds as shown in Figure.5. A 3D model of a reverse spiral origami structure was built and the deforming process was calculated by Ls-Dyna software. The pipe has a length of 270 mm, diameter of 58 mm, thickness of 1 mm, torsion angle of 15°, and segment length of 45 mm. The element sizes for the die, punch, and tube were 1 mm. The moulds are assumed to be rigid bodies.Figure.6 shows the loading curve of hydraulic pressure and displacement of mould.

5. Simulation results and discussion

The Figure.7 showed the thickness distribution of reverse spiral origami structure. The maximum thickness of structure processed by ordinary sealed mould and unsealed mould is 1.201mm and 1.211mm. The minimum thickness of origami structure is reduced from 0.977mm to 0.968mm. The thickness difference has been improved by 8.48%.

In order to research the energy absorption character of designing origami structure, the simulation of crash test as is shown in Figure.8 has been carried out. The simulation model includes the model of origami structure processed by ordinary sealed mould, unsealed and the model with mean distribution of 1.028mm.

Figure.9 shows the absorption energy in the crash test. The absorption energy of origami structure processed by unsealed mould is greater than the structure processed by ordinary sealed mould. The energy absorption of origami structure by sealed and unsealed mould is 33.3% and 43.8% more than an origami structure with mean thickness.

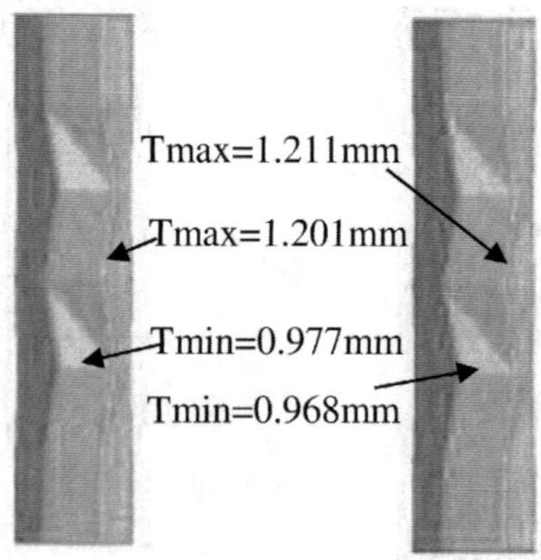

(a) By sealed mould (b) By unsealed mould

Figure.7 Thickness distribution after hydroforming

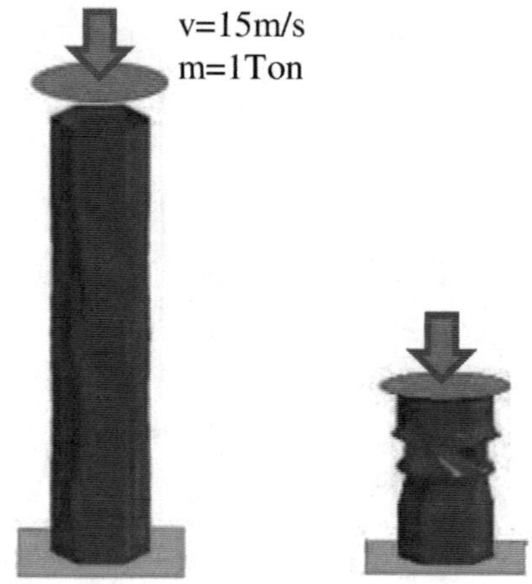

Figure.8 Crash test model

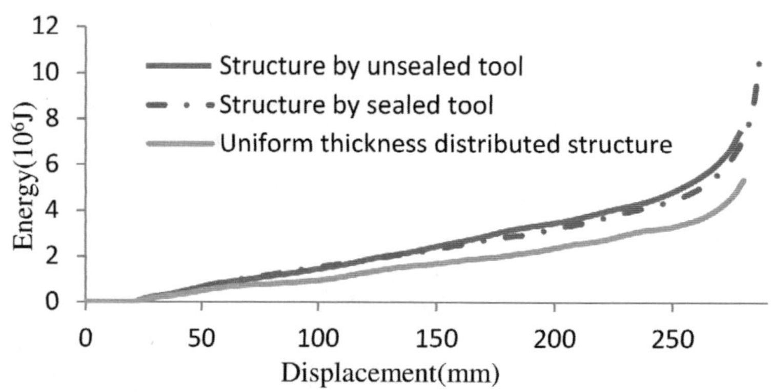

Figure.9 Crash model different reverse spiral origami

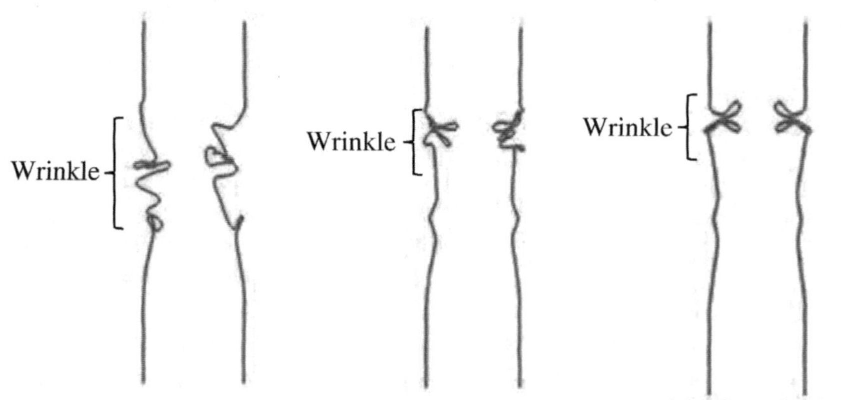

(a) By sealed mould (b) By unsealed mould (c) Mean thickness

Figure.10 longitudinal section of different origami structure in crash test

Figure.10 shows the longitudinal section of the different origami structures in crash test. The origami structure processed by ordinary and unsealed mould has more wrinkle than the structure with mean thickness distribution. Thus, these origami structures can absorb more energy than the structure with mean thickness distribution. But because the winkles density of origami structure processed by sealed mould is sparser than the winkles density of origami structure processed by unsealed mould, the structure processed by sealed mould absorb less energy than the structure processed by unsealed mould in crash test.

6. Conclusion

In this paper, we simulated the hydroforming process of reverse spiral origami structure and discussed the formability and energy absorption ability. On the base of thickness distribution, we proposed a local strengthening reverse spiral origami structure and compared the energy absorption of origami structure processed by ordinary mould and unsealed mould. The conclusions are written as follows.

(1)The thickness distribution of reverse spiral origami structure has a relationship with the diameter of original tubes. When the diameter of pipe is smaller than inscribed diameter of mould, the pipe may be rupture in hydroforming process. When the diameter of pipe is bigger than circumradius of mould, there may be wrinkle in the hydroforming process. When the diameter of pipe is bigger than inscribed diameter of mould and smaller than circumradius of mould, thickness of structure can be adjustable without rupture and wrinkle.

(2) The local strengthening reverse spiral origami structure processed by unsealed mould has a thicker thickness distribution along the longitudinal origami crease. This kind of thickness distribution can improve support force of origami structure. The energy absorption of this structure is 43.8% more than origami structure with mean thickness and 7.87% more than origami structure processed by ordinary sealed mould.

(3) The local strengthening reverse spiral origami structure has a relationship to the diameter of pipe and the space of separated mould. It is necessary to discuss their relationships and get the best design of this structure.

References

[1] Hagiwara I, Tsuda M, Kitagawa Y and Futamata T 1991 Method of determining positions of beads *United States Patent*. Patent Number 5045345.

[2] Taketoshi N 2002 Foldable tubes and conical shells consisting of differently shaped elements(Origami-modelling of no-circular tubes and conical shells) *Transactions of the Japan society of Mechanical Engineers (C)*.**68** 303 (In Japanese).

[3] You Z 2014 Folding structures out of flat materials *Science*.**345** 623.

[4] Hagiwara I and Nadsyoshi S 2003 Consideration on crush characteristics of cylindrical structures using origami engineering *Transactions Society of Automotive Engineers of Japan*. **34** 145 (In Japanese).

[5] Hagiwara I, Yamamoto C, Tao K and Nojima T 2004 Consideration on crush characters of cylindrical structures using origami engineering *Transactions of the Japan society of Mechanical Engineers (A)*. **70** 36 (In Japanese).

[6] Zhao X L, Hu Y B and Hagiwara I R 2010 Optimization design of impact crushing characteristics for origami thin wall tube *The Japan Society of Mechanical Engineers*.**76** 10 (In Japanese).

IOP Conf. Series: Materials Science and Engineering **307** (2018) 012071 doi:10.1088/1757-899X/307/1/012071

Fracture Analysis of Cast Steel Sling

Xinghui Li

Department of Architecturl Engineering,Jiangsu Prvince Xuzhou Technician Institute Xuzhou, Jiangsu, China

Abstract: The fracture reasons of ZG270-500 cast steel sling are analyzed through such means as macroscopic morphology analysis, chemical composition analysis, and microscopic metallography analysis. Results: coarse Widmanstatten structure and casting defects occurring in casting and subsequent heat treatment process reduce the strength, plasticity and toughness of the steel, which is the main reason of brittle fracture of the sling during work, and corresponding improvement suggestions are proposed herein.

Brittle fracture is one of the main failure modes of metal structures; in particular, materials with relatively large brittleness are more likely to produce brittle fracture. ZG270-500 cast steel has a good comprehensive mechanical properties, widely used in a variety of important structural parts, especially those working under alternating load connecting rods, hooks, bolts, gears and shafts and so on. One of the rings is a kind of spreader commonly used in hoisting machinery. However, if the manufacturing process is improper or the processing quality is not good, it may cause brittle fracture of the rings to cause the hanging material to fall and bring serious consequences.

A sudden fracture occurred when the clamping ring made of ZG270-500 cast steel by a mechanical company during use, which caused hoisting workpiece damages, impacted the safety of personnel and equipment, and caused loss of production, Broken rings is shown in Fig. 1. Coarse morphology, chemical composition, and microscopic metallography are analyzed for the fracture site in order to find out the cause of the sling fracture and prevent fracture.

1. Macroscopic Morphology Analysis

The rings are broken into two parts, some of which remain on the base, The fracture site of clamping sling ring is shown in Fig 2. The component integrally fractures; the fracture surface is relatively large; the macroscopic morphology shows that the fracture is a brittle fracture, the grains of the fracture surface are obviously coarse, and the crack originates from the interior of the material.

Fig 1. Broken rings

Fig 2. Macroscopic Morphology of Fracture

2. Chemical Composition Analysis

The chemical composition of the failure sample is analyzed with a spectrometer and the results are shown in Table 1.Table 1 shows that the chemical composition of the material conforms to the provisions of ZG270-500 in the Carbon Steel Castings for General Engineering Purposes (GB11352-89) and the analysis of chemical composition also shows that the design requirements are met.

Table 1. Chemical Composition of Cast Steel (Mass Fraction)

元素 **Element**	C	Si	Mn	P	S
含量 **Content**	0.38	0.32	0.65	0.030	0.028

3. Metallography Analysis

In the hanging ring on the back of the longitudinal section of the specimen, polished after observation under the optical microscope,as shown in Fig. 3. Fig. 3 shows that the structure of the material is the typical Widmannstatten structure, namely free ferrites precipitate along grain boundaries, some ferrites extend to the grain interior from grain boundaries, and some ferrites precipitate independently in the grain interior. The metallography result also shows that obviously coarse grains appear in the material interior but it is hard to find a complete grain under 50x optical microscope. According to GB13299-1991, the level of Widmannstatten structure is Level 3-4.

The Widmannstatten structure has coarse grains and reduces all the strength, plasticity and toughness of the steel, especially toughness. Therefore, Widmanstatten structure is generally not allowed in important products.

Fig 3. Metallurgical Structure of Failure Sample

4. SEM Analysis

4.1 Microstructure

Will ring broken with acetone repeated ultrasonic cleaning,the microstructure of sample fracture is examined with a SEM and the sample microstructure is shown in Fig.4. Fig. 4 shows that massive pearlite exist among ferrite chips in the Widmanstatten structure.

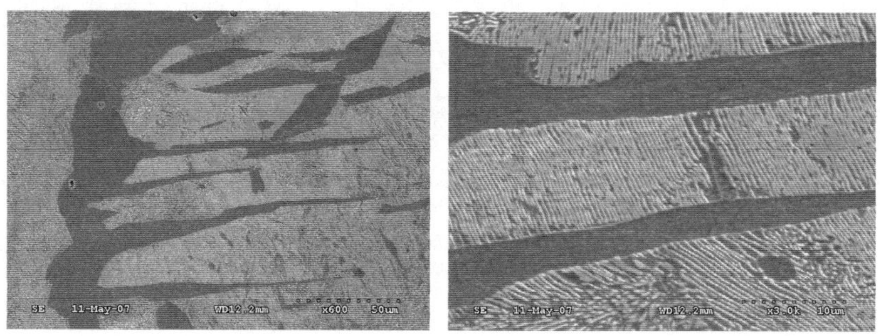

a b

Fig 4. Microstructure of Sample Fracture under SEM

4.2 Morphology of Sample Fracture

The morphology of the sample fracture is shown in Fig. 5. Fig. 5 shows that the fracture is an obvious brittle transgranular fracture and the characteristics of radial stripes can be observed. There are a lot of secondary cracks on the fracture surface and there are a large number of casting defects (pore) in the fracture surface material.

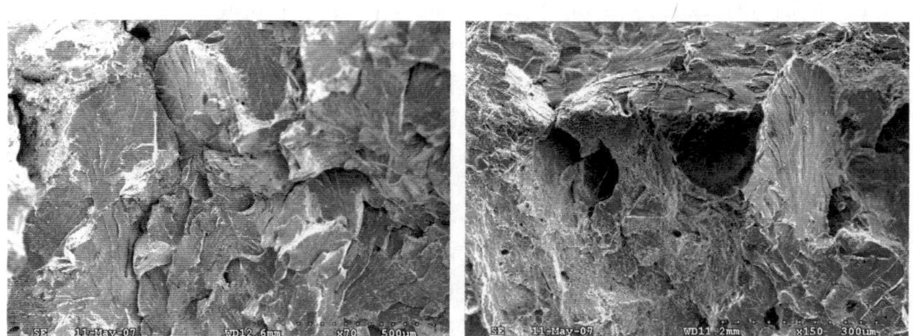

a b

Fig 5. Fracture of Failure Sample under SEM

IOP Conf. Series: Materials Science and Engineering **307** (2018) 012071 doi:10.1088/1757-899X/307/1/012071

5. Conclusions and Suggestions

From the above analysis we can see that the chemical composition of the ring meet the standard requirements, but the microstructure shows the material was overheated.

The main reasons of sling fracture are improper casting and subsequent heat treatment process which caused obvious Widmanstatten structure and coarse grains in the material.

As the service status of the rings for the dynamic load, often by the tensile load and the impact of the repeated impact of the load,The Widmanstatten structure reduces the strength, plasticity and toughness of the steel; in particular, the reduction of toughness reduces the capacity of the cast steel sling to withstand the impact load and brittle fracture occurs during work. In addition, many micropore defects during sling casting are also a factor causing its brittle fracture.

The method for preventing brittle fracture is mainly to prevent the generation of Widmanstatten structure and normal normalizing process can be utilized to eliminate the Widmanstatten structure. As for this product, it can be treated with two normalizing processes; relatively high normalizing temperature is adopted in the first time and relatively low normalizing temperature in the second time, which can not only eliminate the Widmanstatten structure but also can refine grains.

References

[1] Fan Dongli M 2009 Heat Treatment Technical Manual Chemical Industry Press

[2] Ye Weiping Practical Steel Metallographic Test M 2012 Beijing: China Machine Press

[3] Wang Guangsheng, Shi Kangcai, Zhou Jing'en, et. al J 2007 *Metal Heat Treatment Defect Analysis and Cases* 13-51

[4] Yao Zhengyao M 2007 Casting Manual - Cast Steel Version 2

IOP Conf. Series: Materials Science and Engineering **307** (2018) 012072 doi:10.1088/1757-899X/307/1/012072

Effect of water-cooling treatment times on properties of friction stir welded joints of 7N01-T4 aluminum alloy

T H Zhang[1], Y Wang[2], X F Fang[3], P Liang[4], Y Zhao[5], Y H Li[6], and X M Liu[*]

[1] CRRC QINGDAO SIFANG CO., LTD., Qingdao, China
[2] CRRC QINGDAO SIFANG CO., LTD., Qingdao, China
[3] CRRC QINGDAO SIFANG CO., LTD., Qingdao, China
[4] School of materials science and engineering, Shandong University, Jinan, China
[5] School of materials science and engineering, Shandong University, Jinan, China
[6] School of materials science and engineering, Shandong University, Jinan, China
[7] School of materials science and engineering, Shandong University, Jinan, China

*E-mail: xuemei_buaa@sdu.edu.cn 981424318@qq.com

Abstract. Due to the deformation caused by residual stress in the welding process, welded components need treatment to reduce welding distortion. In this paper, several different times of flame-heating and water-cooling treatment were subjected to the friction stir welding joints of 15mm thick 7N01P-T4 aluminum alloy sheets to study the microstructure variation of friction stir welding joints of 7N01P-T4 aluminum alloy, and to analyze the effect on micro-hardness, tensile and fracture mechanical properties. This investigation will be helpful to optimize treatment methods and provide instruction on industrial production.

1. Introduction

Nowadays, friction stir welding technology has been widely used in welding of high-speed EMU, intercity EMU, urban rail, and aluminum alloy car-body parts. Friction stir welding joints have the advantages of good mechanical properties and small welding distortion. As for welding products, the distortion can be well controlled. However, the aluminum alloy car-body structure is not completely welded by friction stir welding, some welds are arc welded. In this case, distortion still remains in the structure after welding process and the weld area needs to be adjusted to meet the requirements of geometric tolerance. Flame-heating followed by water-cooling treatment on weld joints is the common adjustment method. However, references on the influence of flame straightening on the performance of friction stir welding joint are still rare both at home and abroad. Therefore, it is necessary to start a research on the influence of water-cooling treatment on the performance of friction stir welded joints.

This paper is concerned with the influence of different water-cooling treatment times on the properties, macro and micro microstructure of FSW joints of 7N01-T4 aluminum alloy. The influence and regularity of mechanical properties of FSW joints were also studied.

2. Materials, methods and equipment

1100mm ×310mm 7N01P-T4 aluminum alloy sheets with 15mm thickness were used in this research. The sheets were welded by MTI static LONGMEN numerical control friction stir welding machine in butt joints.

Figure 1. Welded wrokpiece

Figure 2. Water-cooling treatment

The weld joints were heated to 350℃ by oxy propane flame after welding, and then the water-cooling treatment was applied in different times. Specimens with once, twice and thrice water-cooling treatment and specimens with no water-cooling treatment were made for corresponding testing. The microstructures of the joint area were observed by VHX-500FE ultra-high resolution digital microscopy system, and the micro-hardness of the joint area was measured by DHV-1000 digital micro-hardness tester, following the standard of GB/T 2654-2008 (ISO 9015-1:2000, IDT) and GB/T 27552-2011 (ISO 9015-2:2003, MOD). The hardness curve was drawn according to the hardness tests on BM, HAZ, TMAZ and WN, including near surface, middle part and near bottom 3 layers. According to GB/T 2651-2008(ISO 4136:2001, IDT), tensile test specimens were made to measure the tensile strength, yield strength, extensibility of the joints and record the fracture position. Bending tests were conducted in accordance with GB/T 2653-2008(ISO 5173: 2000, IDT). The bending angle was 180° . Cracks existing on the workpiece longer than 3mm would be regarded as non-conformity. Charpy impact test was conducted in room temperature in accordance with ISO9016:2001 to study the effect of FSW on the impact performance of welding joints. The notch was taken from the weld and the notch surface was parallel to the surface of the specimen.

3. Results and discussion

3.1. Effect of water-cooling treatment times on microstructure

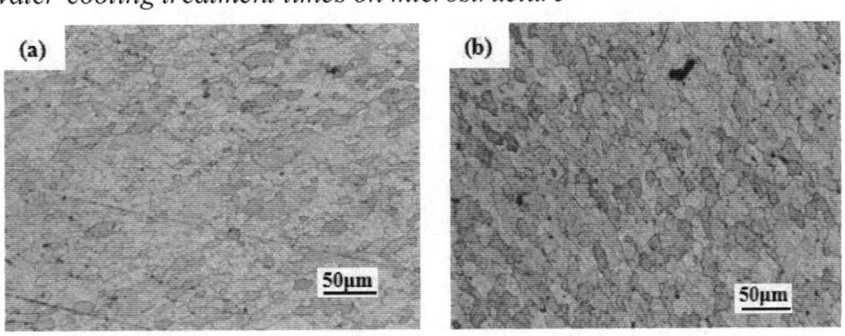

Figure 3. Comparison of microstructure of welding nuclear area with different times of water-cooling treatment. (a) One time. (b) Two times. (c) Three times. (d) None.

The microstructure photos from same section of welding nuclear area are shown in the Figure 3. Different water-cooling treatment time have some influences on the grain size. The grain size of joints with no water-cooling treatment is about 20-50 μ m, and it increased with once water-cooling treatment. But with the increasing treatment times, the grain size starts to reduce, which is close to the one with no water-cooling treatment.

3.2. Tensile mechanical properties

As is shown in Table 1 and Figure 4, four specimens in each group were subjected to tensile test, and the average data was considered to reflect the mechanical properties of the joints.

Table 1. Mechanical properties of joints under different times of water-cooling treatment.

	Tensile strength (MPa)	Yield strength (MPa)	Elongation (%)
Base metal	436.6	308	20.336
No treatment	350.2	220.4	10.828
One time of treatment	345.8	217.4	11.14
Two times of treatment	351.4	191.2	14.754
Three times of treatment	354.6	215.6	16.558

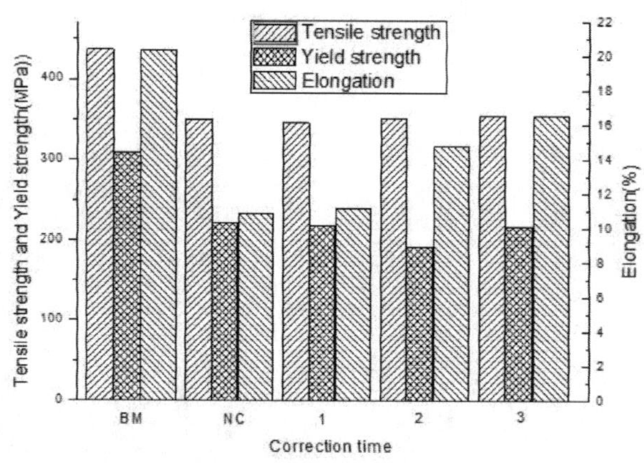

Figure 4. Influence of different water-cooling treatment times on tensile mechanical properties

The study shows that after once 350 ℃ flame conditioning, the tensile strength of the joint decreases with a small amplitude. However, with the increasing times of treatment, the tensile strength has minor increase to about 350MPa. The tensile strength of joints is about 80% of the measured strength of the base metal, and about 111% of the base metal standard strength (315MPa). The yield strength is about 200MPa. The elongation of joints increases compared with joints without treatment after 350 ℃ flame conditioning. With the increase in the number of water-cooling treatment, the elongation shows an upward trend. After three times of flame conditioning, the elongation can reach 81.4% of the measured value of the base metal, and 150% of the base metal standard value. After once flame conditioning, the tensile strength declines due to the growth of grains and the decreasing density of precipitated phase. But after two and three times of flame conditioning, with the increase of precipitated phase which has strengthening effect, the tensile strength of the joint rises again.

3.3. Micro-hardness

The average micro-hardness value of 7WN1,7WN2,7WN3 and 7WN4 in upper parts of the joints is 113.2Hv, 111.2Hv, 109.9Hv and 108.2Hv. As is shown in Figure 5, for the upper part, joints with once treatment have the highest micro-hardness in weld nuclear area, then joints with twice treatment, thrice treatment and no treatment. After once treatment, the hardness value of the back side becomes significantly higher than the forward side.

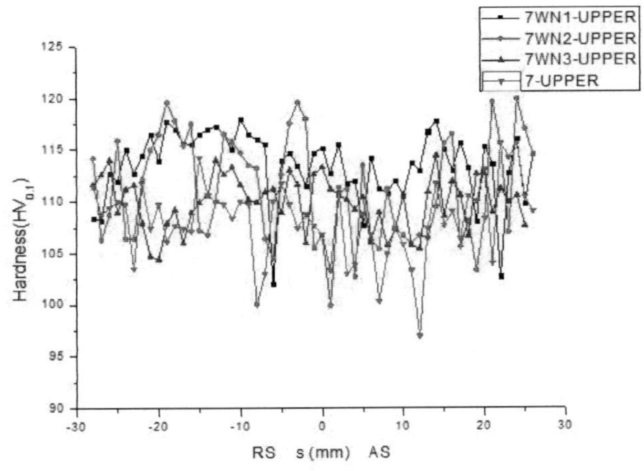

Figure 5. Hardness distribution of joints under different times of water-cooling treatment in upper part

The average micro-hardness value of 7WN1,7WN2,7WN3 and 7WN4 in weld central layer of the joints is 109.9Hv, 107.7Hv, 106.2Hv and 109.4Hv. As is shown in Figure 6, for the central layer, joints with once water-cooling treatment have the highest hardness value and the weld nuclear area has the lowest hardness value. The study also shows that joints with water-cooling treatment have higher hardness value than those with no treatment in weld nuclear area.

The average micro-hardness value of 7WN1,7WN2,7WN3 and 7WN4 in lower part of the joints is 114Hv, 113.3Hv, 110.2Hv and 107.4Hv. According to Figure 7, the joints hardness value is significantly improved after treatment. Joints with once water-cooling treatment have the highest hardness except the weld nuclear area, and twice treatment can lower the hardness of weld nuclear area. The weld nuclear area has the lowest hardness value.

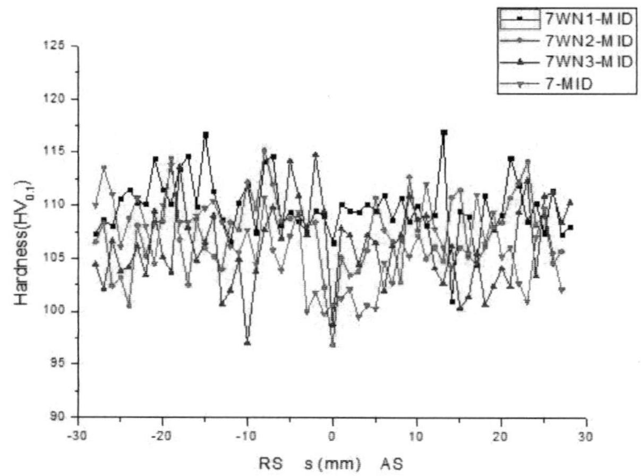

Figure 6. Hardness distribution of joints under different times of water-cooling treatment in middle part

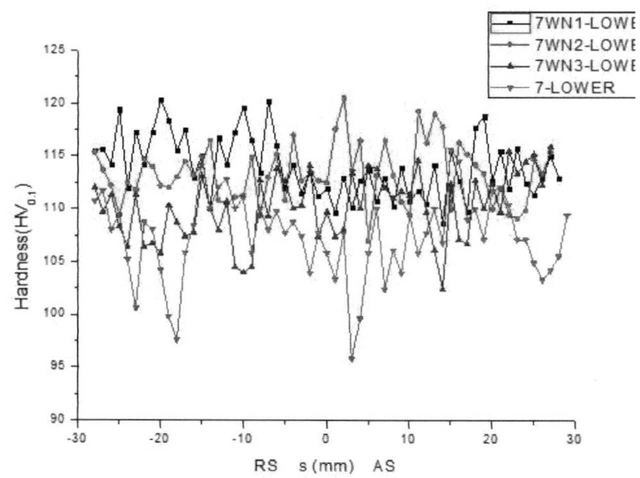

Figure 7. Hardness distribution of joints under different times of water-cooling treatment in lower part

4. Results and discussion

According to the experiment about flame-heating and water-cooling treatment on 7N01-T4 aluminium alloy FSW joints, conclusions can be drawn that the grain size in weld nuclear area of FSW joints starts to be smaller with the increasing times of water-cooling treatment. Water-cooling treatment barely has influence on micro-hardness and tensile strength of the joints. But the elongation increases with the increasing times of water-cooling treatment.

References

[1] Lv C J and Li L L 2015 *J. Informatization Construction* **09** 59-60+62
[2] Ma K X 2013 *J. Electric Locomotives & Mass Transit Vehicles* **36** 1-6
[3] He G Z, Wang F, Li J and Hu Z C 2015 *J. Hot Working Technology* **05** 206-8+211
[4] Han X H, Zhou C Q, Zhang T H and Wang X 2016 *J. Modern Welding Technology* **06** 16-20
[5] Zhang T H, Han D C and Liu S L 2013 *J. Welding & Joining* **10** 25-6
[6] Song X C, Zhu Z Q and Chen Y F 2013 *J. Hot Working Technology* **13** 5-7+12
[7] Zhao Q Y, Wu Z S, Liu C R, Zheng H H and Wu C X 2016 *J. Welding Technology* **01** 1-5

IOP Conf. Series: Materials Science and Engineering **307** (2018) 012073 doi:10.1088/1757-899X/307/1/012073

Improving the Quality of Welding Seam of Automatic Welding of Buckets Based on TCP

Min Hu
Jiangsu province xuzhou technician institute,Xuzhou,Jiangsu 221151,China
331947080@qq.com

Abstract: Since February 2014, the welding defects of the automatic welding line of buckets have been frequently appeared. The average repair time of each bucket is 26min, which seriously affects the production efficiency and welding quality. We conducted troubleshooting, and found the main reasons for the welding defects of the buckets were the deviations of the center points of the robot tools and the poor quality of the locating welding. We corrected the gripper, welding torch, and accuracy of repeat positioning of robots to control the quality of positioning welding. The welding defect rate of buckets was reduced greatly, ensuring the production efficiency and welding quality.

1. Introduction

The Tool Center Point (TCP) of robots is the relative position between the center of the six-axis flange plate of the robots and the end of welding torch. The robots record the position information of each axis of the robots through a Resolver Digital Converter (RDC). Because of the welding deformation of the small assembly or the dimension error of the single plate, The deviation of workpiece is unavoidable in the process of assembly and these deviated robots conduct automatic compensation by the locating and arc tracing, while TCP is the precondition of locating compensation error. When the TCP is deviated, the corrected arc-start point and the arc-end point will also be deviated, which leads to the deviated welding of the arc-start point and the arc-end point. For welding seam that is suitable for arc tracing, even if there is a small deviation in the arc-start point, the welding seam can automatically make alignment through arc tracing. But when welding is not suitable for the welding seam with arc tracing (such as the butt welding seam, the welding seam with too large swing, the welding seam with too small welding angle, etc.), if the arc-start point or the arc-end point is deviated, the whole welding seam will have deviated welding; the structures of buckets are complex; there are many types of the butt welding seam or lap welding seam; the plates are thin, which is not conducive to locating of robots. When TCP is deviated, welding seam defects are continuously appeared. When welding arc, the robots automatically establish a theoretical arc through the arc-start point, the arc-auxiliary point, and the arc-end point recorded by RDC, and the theoretical arc establishes the

IOP Conf. Series: Materials Science and Engineering **307** (2018) 012073 doi:10.1088/1757-899X/307/1/012073

actual arc through the TCP relationship. When TCP is deviated, the established actual arc will be deviated, which will lead to deviated welding and collided welding torch.

2. Summary Questions

Since February 2014, the welding defects of the welding line of buckets have been frequently appeared. The time of repair welding was long with high labor intensity. At the same time, 2-3 technicians in each shift were required to carry on-site tracking service for the robot groups, seriously affecting the production efficiency. Table 1 The quality feedback of automatic welding of 5T bucket on March 2, 2014.

Table 1. The quality feedback of automatic welding of 5T bucket

Buckets No.	Welding defect positions and defect types	Defect length	Repair time
11402465	1. Deviated weldings in 6 reinforcing bar plates. 2. Bended reinforcing bar plates-deviated weldings in wall plate of buckets. 3. Plates of large blade-incompletely filled welding in the bottom protection plates of bucket.	1160mm	24 min
11402468	1. Deviated weldings of U-type cover plates. 2. Deviated weldings in 6 reinforcing bar plates. 3. Wall plate of bucket-deviated weldings in the side plate of bucket. 4. Plates of large blade-Air hole and deviated weldings in wall plate of bucket. 5. Incompletely filled welding in the bottom protection plates of bucket of plates of large blade.	1370mm	30 min
11402480	1. Wall plate of bucket-deviated weldings in the side plate of bucket. 2. Incompletely filled welding in the filled plug welding seam of T-type bar plates. 3. Plates of large blade-Deviated weldings in wall plate of bucket.	1160mm	26 min
11402201	1. Deviated weldings in 6 reinforcing bar plates. 2. Wall plate of bucket-deviated weldings in the side plate of bucket. 3. Bended reinforcing bar plates-deviated weldings in wall plate of buckets.	1000mm	18 min
11402483	1. Deviated weldings in 6 reinforcing bar plates. 2. Bended reinforcing bar plates-deviated weldings in wall plate of buckets.	700mm	14 min
11402507	1. Deviated weldings in 6 reinforcing bar plates. 2. Wall plate of bucket-deviated weldings in the side plate of bucket. 3. Bended reinforcing bar plates-deviated weldings in wall plate of buckets. 4. Incompletely filled	1800mm	38 min

	welding in the filled plug welding seam of T-type bar plates.		
11402505	1. Deviated weldings in 6 reinforcing bar plates. 2. Plates of large blade-incompletely filled welding in the bottom protection plates of bucket. 3. Bended reinforcing bar plates-deviated weldings in wall plate of buckets.	950mm	16 min
11402503	1. Deviated weldings in 6 reinforcing bar plates. 2. Wall plate of bucket-deviated weldings in the side plate of bucket. 3. Plates of large blade-incompletely filled welding in the bottom protection plates of bucket. 4. Bended reinforcing bar plates-deviated weldings in wall plate of buckets.	1750mm	34 min

From table 1, it can be seen that the deviated weldings of buckets, air holes, undercutting, incompletely filled welding, and other defects are serious. The average repair time is 26.4min, becoming the bottleneck of loader production.

Problem Analysis and SolutionWe arranged and analyzed the welding defects, and found that the root reasons for the welding defects of the buckets were the TCP deviations and the poor quality of the locating welding table 2.

Table 2. Causes and solutions for the welding defects of the buckets

Defect Types	Defect Reasons	Solutions
Deviated weldings in 6 reinforcing bar plates.	1. The reinforcing bar plates in positioning welding were not straight (tooling deformation of reinforcing bar plates; operation of positioning welding was not standard). 2. TCP deviation.	1. Adjust tooling of reinforcing bar plates. 2. Control the quality of positioning welding. 3. Correct TCP.
Deviated weldings in U-type plates	1. TCP deviation. 2. The greater error of the assembly (operation of positioning welding was not standard; some operators directly conducted assembly without make lines).	1. Correct the TCP. 2. Control the quality of the assembly, and strengthen the inspection of incoming materials before complete welding.
Plates of large blade-Air hole and deviated weldings in wall plate of bucket.	1. TCP deviation. 2. The incoming materials of knife plates were unqualified.	1. Correct TCP. 2. Control the pass rate of incoming materials; prohibit assembly of unqualified single board pieces.

3.Quality Improvement of Positioning Welding

When the buckets were under assembly, there were many single board pieces; the types of welding seam were complex; the cumulative deformation was relatively large; some operators did not have strong sense of responsibility, and lacked of work experience. All these resulted into poor quality of assembly point of the buckets. Combined with the structure forms of the loader products, the welding programming office of the technical department now develops the general norms for positioning welding as follows.

We strictly control the pass rate of incoming materials, prohibit the use of expired the single board pieces not meeting the requirements of the process, and timely respond to the leaders in related process section. 2) We require the assembly to be conducted in strict accordance with drawings. For the welding seam required to leave a gap by process, we strictly control the gap of assembly within the tolerance range. 3) We try our best to use the tooling or the underlined template to make the assembly, in order to ensure the consistency of assembly of workpiece. 4) We forbid personal increase or decrease of spare parts. If the structure is changed, this must be notified to the programming office in advance. 5) If the inner opening welding seam requires the robots to conduct welding, the process tie bar must be underlined for assembly. The length of the tie bar should be equal to or slightly larger than the opening size, and the height of the tie bar should be controlled15mm below as far as possible. 6) If the gap of assembly caused by accumulated error of the assembly is greater than 1.2mm, the small welding parameters must be used to fill the gap. The height of filled welding seam is advisable to the height of filling the gap of assembly. It is strictly prohibited to appear soybean welding and residual welding wire. 7) It is strictly prohibited to conduct assembly in the single board pieces with label paper. We must fix the label paper position in the structure of the finished spot welding, and the label paper position shall be away from the welding seam and shall not fall off easily. 8) Within the 30mm range from the arc-start point, it is prohibited to conduct spot welding in the arc-start point and the arc-end point, so as to prevent that the welding points affect the robots' data acquisition on arc tracking and debugging of welding posture. 9) For the T-joints whose height of vertical plate are 5mm higher than those of the theoretical welding, the spot welding requirements are shown in the requirements 8. 10) For the T-joints whose height of vertical plate are 5mm lower than those of the theoretical welding (such as the pasting board and the joints requiring the end of the full welding to connect the T-joints), within the 60mm range from the arc-start point or the arc-end point, it is prohibited to conduct spot welding, but we require the spot welding with multiple welding points, small and high welding (the welding height of spot welding is not 0.5 times greater than that of the theoretical welding height), so as not to affect the robot locating and welding. 11) For the butt welding seam in the end of single groove, we should avoid the spot welding within the groove. If the spot welding within the groove is unavoidable, the welding height of spot welding should not be 0.5 times greater than that of the theoretical welding height. 12) For the welding seam with ladder-like transition, we should conduct spot welding at the ladder, so that the welding of robots can be smoothly transited during the transition (such as the welding seam composed by bucket hinge plate-wall plate of bucket-bended reinforcing bar plates). 13) Within the 30mm range from the straight line segment, the tangent point of straight line and the arc, and the midpoint of the arc, it is prohibited to conduct welding, so as not to affect the robot locating and welding posture debugging. 14) In the positions of 0°, 90°, 180°, and 270° of the circle, it is prohibited to conduct spot welding, so as not to affect the robot locating. The length direction of the workpiece is X-axis) 15.When conducting spot welding on the first new product, the technicians of the programming office must confirm it, and also determine the unfinished spot welding requirements on-site.

Through training the operators on site, we control the quality of spot welding of buckets, paving the

IOP Conf. Series: Materials Science and Engineering **307** (2018) 012073 doi:10.1088/1757-899X/307/1/012073

ground for subsequent automatic welding of robot.

4. Improvement of TCP Precision

The reasons of the robot's TCP deviation mainly include four aspects: welding torch deformation, gripper deformation, anti-collision deformation, and motor reducer failure. The failure of the motor reducer directly affects the repeated positioning accuracy of the robots. We used the inside micrometer to detect the repeated positioning accuracy of the various axes of 24 robots, got the results, and we found that the repeated positioning accuracy of the 24 robots were all within the error range, thus eliminating the influence of the fault of the motor reducer on the TCP deviation.

As a safety protection device for the robot system, anti-collision Figure 1 is installed in the center of 6-axis flange of robots and is connected with the upper part of the gripper of welding torch. Anti-collision is a constant closed loop under work. When the welding torches are collided, the anti-collision built-in spring Figure 2 occurs elastic deformation, and the constant closed loop disconnects. The anti-collision effects, and the robot control system is outage shutdown. After the long-term use of anti-collision, the built-in spring occurs plastic deformation, and TCP is deviated. We conducted troubleshooting on the collision prevention of the 24 bucket robots, and found that the anti-collision springs of No. 2 machine in 5T bucket line and No. 12 machine in 3T bucket line were deviated, and we replaced the anti-collision springs in time.

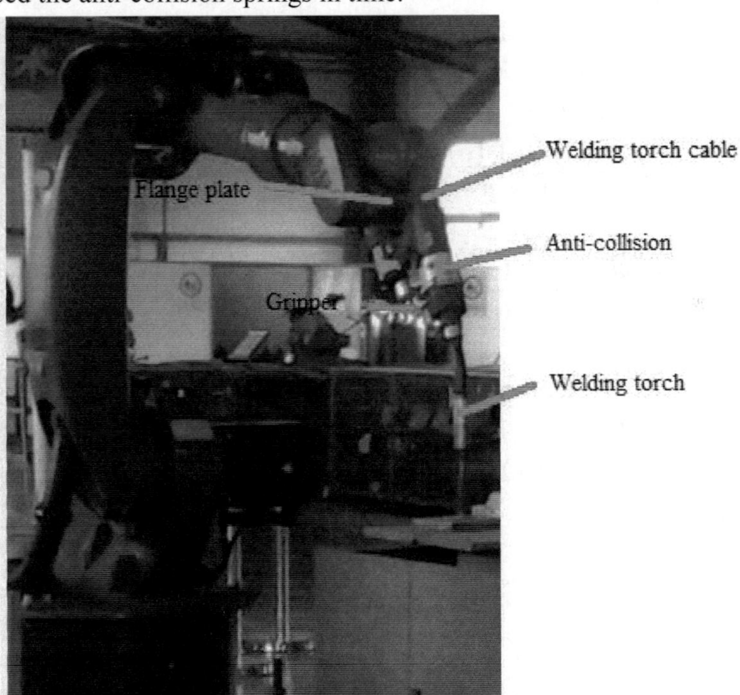

Figure1. Assembly diagram of accessory parts of robots

IOP Conf. Series: Materials Science and Engineering **307** (2018) 012073 doi:10.1088/1757-899X/307/1/012073

Figure 2 .Anti-collision structure diagram

5. Conclusion

The product quality is related to the survival and development of enterprises. With the aging of equipment and product transformation and upgrading, the market has put forward higher requirements on product quality, and the bucket welding production line also have a long way to go. However, no matter how the market changes, as long as we adhere to the quality concept, we adopt scientific analysis means coupled with effective coordination among departments, and we believe that we can solve any encountered difficulties in the future.

6.Reference

[1] Liu Jie, Yang Daojin, Zhang Xiaocui, et al. Application of welding automation technology in the manufacture of belt conveyor directional roller [J]. China Science and Technology Expo, 2015 (32): 211-212.

[2] Huang Jiqiang, Li Mingzhe, Zou Yong, et al. Pipeline welding automation in engineering construction [J]. electric welding machine, 2010, 40 (10): 73-75

[3] Zhang Tao, Wang Chuanyang. Optimization of welding sequence of grader bucket based on deformation simulation [J]. welding technology, 2010, 39 (6): 56-57.

IOP Conf. Series: Materials Science and Engineering **307** (2018) 012074 doi:10.1088/1757-899X/307/1/012074

Examination of thermophotovoltaic GaSb cell technology in low and medium temperatures waste heat

Z Utlu[1] and B S Önal[2]

[1] İstanbul Gedik University, Engineering Faculty,
Department of Mechanical Engineering, 34876, Kartal, İstanbul,Turkey
[2]İstanbul Aydin University, Institute of Science and Technology, Turkey
E-mail: zafer.utlu@gedik.edu.tr ; selenay.onal@hotmail.com

Abstract. In this study, waste heat was evaluated and examined by means of thermo-photovoltaic systems with the application of energy production potential GaSb cells. The aim of our study is to examine GaSb cell technology at low and medium temperature waste heat. The evaluation of the waste heat to be used in the system is designed to be used in the electricity, industry and iron and steel industry. Our work is research. Graphic analysis is done with Matlab program. The low and medium temperature waste heat graphs applied on the GaSb cell are in the results section. Our study aims to provide a source for future studies.
Keywords: Thermophotovoltaic, Electricity generation, GaSb cell, Waste Heat

1.Introduction

Waste heat is the low-energy heat generated by the work done in the system. Systems such as machines, ovens and stoves emit heat for the duration of their work [1]. Waste heat can be produced by evaluating in thermophotovoltaic systems. The working temperature of thermophotovoltaic systems is in the range of 1000-1500 ° C. The waste heat at this temperature value is converted to electricity energy by passing through the system components heat source, selective emitter, filter and photovoltaic (PV) cell. The discovery of TPV dates back to about 1956. Most literature references, MIT in 1956, also made the concept suggests a series of conferences during the Aigra TPV. [2, 3]. Nelso Un, TPV system and by Kolmar 'Solar cells power supply,' 'about a post that has been given [4, 5]. Until the mid-1970s, research conducted in the US, uses fossil fuels as a heat source independent military focused on the low-noise electrical generator. During this period three main heat source (solar, nuclear and combustion) and spectral control options (eluent radiator, filter, PV cell front and back surface reflector) is defined [4, 6]. Industrial waste heat recovery using TPV conversion was proposed by Coutts at the end of 1990s [7-9]. These studies show that TPV applications are increasingly increasing, contributing greatly to energy conversion and productivity. The applications of TPV are residential, automotive, and industrial and so on. It is seen that it has entered the sectors and brought an alternative to electricity generation.
In this study, firstly waste heat is classified. In the second stage, electricity production by thermophotovoltaic methods is investigated by using GaSb cell from low and medium temperatures waste heat.

2.Classification of Waste Heat

Waste heat; Waste heat from low-temperature heat sources, waste heat from medium-temperature heat sources, and waste heat from high-temperature heat sources. Low temperature waste heat can be useful as a complementary way to low vapour pressure needs and preheating purpose [1]. Waste gas temperature values in the medium temperature range of industrial process equipment are shown in Table 1. Most of these mid-temperature values are obtained from combustion processes [1,14]. Waste gas temperature values at high temperature ranges of industrial process equipment are also shown in Table 1. All these results are obtained directly from the combustion processes [1].

Table 1. Waste heat values obtained from different sources at low, medium and high temperatures [1].

Low Temperature		Medium Temperature		High Temperature	
Heat Source	**Temperature (°C)**	**Heat Source**	**Temperature (°C)**	**Heat Source**	**Temperature (°C)**
Steam Condensation Processes	55-88	Steam Exhaust Exhausts	230-480	Nickel Refine Ovens	1370-1650
Cooling Water	-	Gas Turbine Exhausts	370-540	Aluminium Refined Ovens	650-760
Welding Machines	32-88	Piston Engine Exhausts	315-600	Zinc Refined Ovens	760-1100
Injection Machines	32-88	Piston Engine Exhausts (Turbo Charged)	230-370	Copper Refined Ovens	760-815
Annealing Furnaces	66-230	Heat Treatment Furnaces	425-650	Steel Heating Furnaces	925-1050
Internal Combustion Engines	66-120	Drying and Cooking Ovens	230-600	Copper Reverber Oven	900-1100
Mold Forming	27-88	Catalytic Crackers	425-650	Open hearth furnaces	650-700

3. Termophotovoltaic System and Structure

Thermophotovoltaic systems are systems that generate heat energy and electric energy from high temperature waste heat and solar radiation [7-9]. Thermophotovoltaic system includes selective emitter, heat source, filter and a photovoltaic cell. The heat source is the source of photons. Heat sources with operating temperatures between 1000° C and 1500°C can be used in TPV systems [1, 10-15]. The heat energy from the heat source passes through the selective emitter, filter and cells by radiation. The heat source comes from photovoltaic cells and allows photons to be obtained. Selective spreader is used to increase system efficiency. The selective emitter translates the emission spectrum from the heat source to the emission spectrum by providing the appropriate receiver cell sensitivity before transferring the filters. Because the receiving cells can only use an energy absorber above the band spacing. This leads to less electricity generation [8-11]. The photons from the selective emitter reach the filter before they reach the cells. The filters have the same characteristics as the selective emitter. Reflects non-energized radiation and sends the selective emitter back. Photovoltaic cells; Absorb photons from the emitter and convert them into thermal energy electrical energy. The necessity of absorbing as many photons as possible obligates the use of materials with low band gap [1,8-11].

4. Analyses

Analyses TPV low and medium temperature waste heat values were obtained using GaSb cells. The analyses were done in MatLab program. The graphs obtained according to this analysis result are in our study. In the analysis made, TPV low and medium temperature graphs were obtained using GaSb cell. The parameters used are; the temperature of the cell is the source temperature and the radiation temperature. With these graphs, energy efficiency, filling factor, effect of open circuit voltage and short circuit current values are determined. In Figure 1. the energy efficiency was examined by varying the cell temperature at different source temperatures of 450-950 K. For example, when the cell temperature is 300 K at a 650 K source temperature, the energy efficiency is 11% while the cell temperature is 350 K, resulting in an energy efficiency of 3%. In this case, as cell temperature increases, energy efficiency decreases.

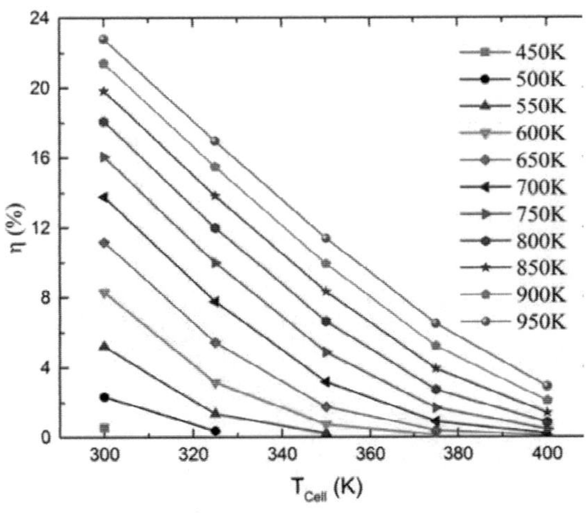

Figure 1. Effect of change in cell temperature on energy efficiency

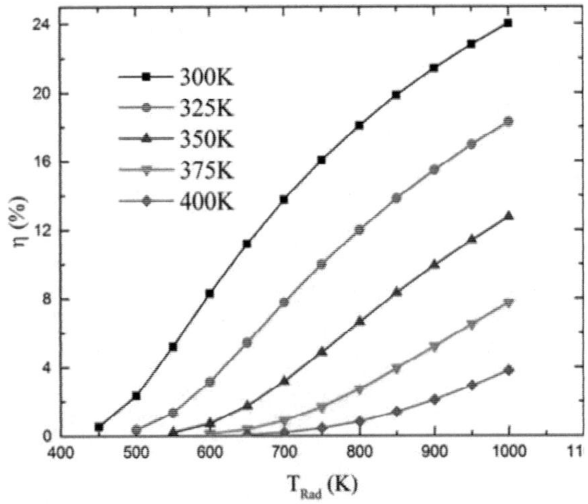

Figure 2. Effect of radiation temperature change on energy efficiency

In Figure 2. the energy efficiency was investigated by changing the radiation temperature at different sources temperatures of 300-400 K. For example, at a radiation temperature of 325 K at a radiation temperature of 800 K, the energy efficiency increased to 16% while the energy efficiency was around 12% and the radiation temperature was 1000 K. In this case, the higher the radiation temperature, the higher the energy efficiency.

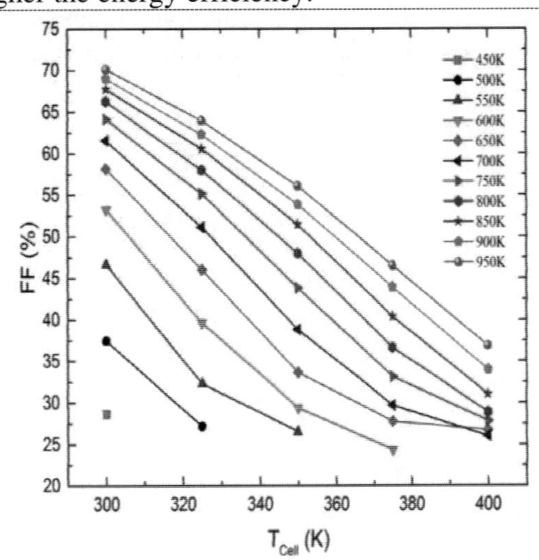

Figure 3. Effect of cell temperature change on filling factor

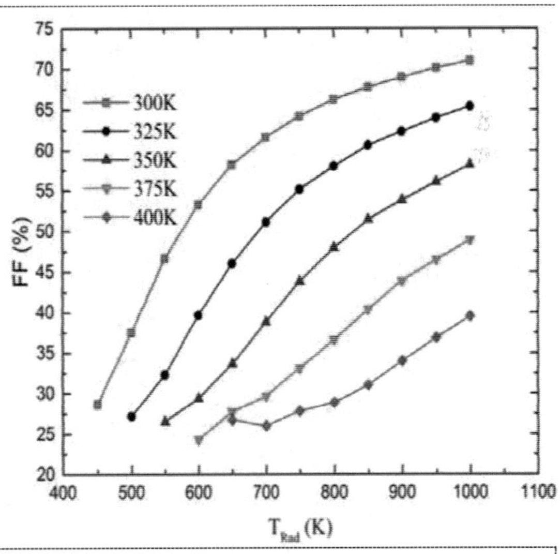

Figure 4. Influence of radiation temperature change on filling factor

In Figure 3. shows the variation of cell temperature and filling factor at different source temperatures of 450-950 K. For example, when the cell temperature is 300 K at 750 K source temperature, the filling factor is 65% while the cell temperature is 380 K, the filling factor has decreased to about 33%. In this case, as the cell temperature increases, the filling factor appears to decrease. In Figure 4, the fill factor was investigated by varying the radiation temperature at different source temperatures of 300-

400 K. For example, when the radiation temperature is 600 K at 300 K sources temperature, the filling factor has increased to about 67% when the filling factor is about 53% and the radiation temperature is 900 K. In this case, as the radiation temperature increases, the filling factor seems to increase.

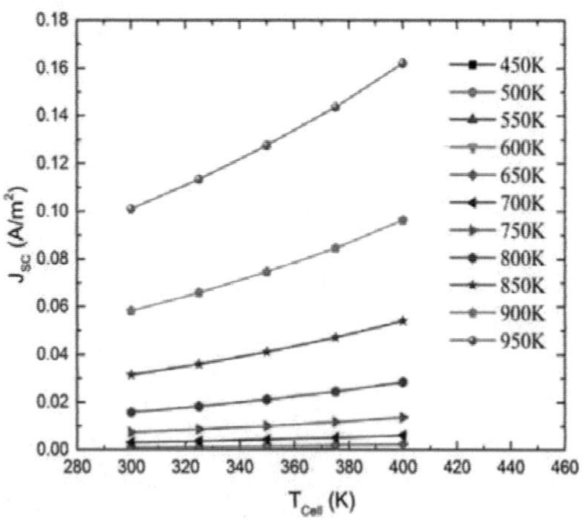

Figure 5. Effect of cell temperature change on short circuit current

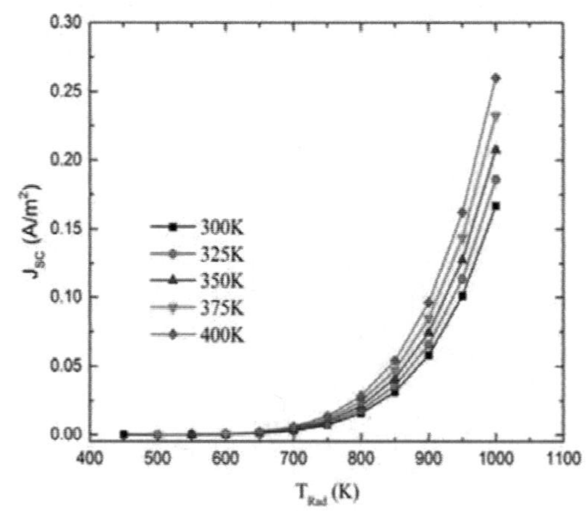

Figure 6. Effect of radiation temperature change on short circuit current

In Figure 5. short-circuit currents were investigated by varying the cell temperature at different source temperatures of 450-950 K. For example, when the cell temperature is 350 K at 800 K source temperature, the short circuit current is about 0.02 A / m² while when the cell temperature is 400 K, the short circuit current is about 0.03 A / m². In this case, short circuit current seems to increase as cell temperature increases.Figure.6. shows the variation of radiation temperature and the short circuit current at different source temperatures of 300-400 K. For example, when the radiation temperature at 350 K is 950 K, the short circuit current is around 0.13 A / m², and when the radiation temperature is 1000 K, the short circuit current is increased by about 0.20 A / m².

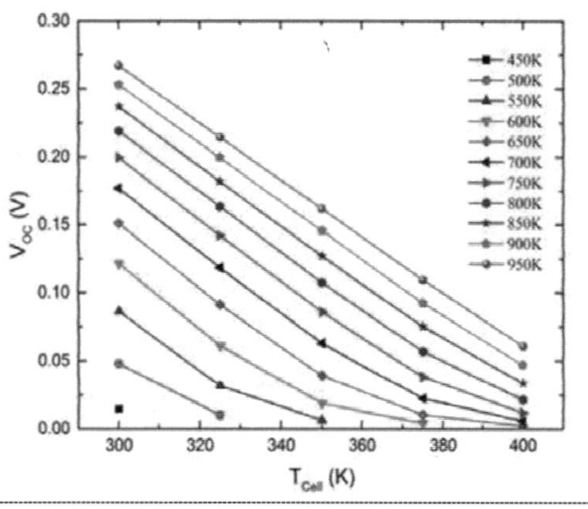

Figure 7. Effect of cell temperature change on open circuit voltage

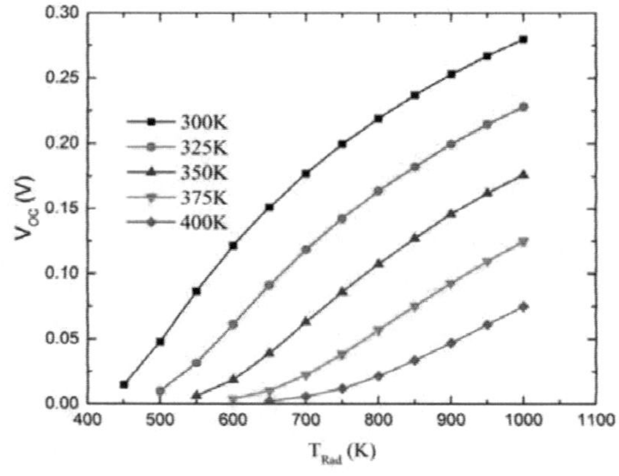

Figure 8. Effect of radiation temperature change on open circuit voltage

Figure 7. examines the open circuit voltage by varying the cell temperature at different source temperatures of 450-950 K. For example, when the cell temperature is 320 K at 700 K source temperature, the open circuit voltage is around 0.14 V, while when the cell temperature is 380 K, the open circuit voltage is reduced to 0.03 V. In this case, as the cell temperature increases, the open circuit voltage decreases. In Figure 8, the open circuit voltage was investigated by varying the radiation temperature at different source temperatures of 300-400 K. For example, at a radiation temperature of 325 K at a radiation temperature of 800 K, when the open circuit voltage is around 0.17 V and the radiation temperature is 1000 K, the open circuit voltage has increased to about 0.23 V. In this case, as the radiation temperature increases, the open circuit voltage increases.

5. Conclusion

The main conclusions drawn from present study may summarize as follows;

- Cell temperature at different sources temperatures that the effect of the change in energy efficiency decreases when the cell temperature increases energy efficiency examined.
- The effects of radiation temperature changes in the energy efficiency of different sources that the temperature increases of the investigation when the radiation temperature increases energy efficiency.

References

[1] Utlu Z 2015 Investigation of the potential for heat recovery at low, medium and high stages in the Turkish industrial sector (TIS) *Energy* **81**: 394-405.

[2] Mustafaa K F, Abdullahb S, Abdullaha M Z and Sopianb K 2017 A review of combustion-driven thermoelectric (TE) and thermophotovoltaic (TPV) power Systems *Renew. Sustainable Energy Rev.* **71**:572–584.

[3] Steinhuser A, Hille G, Kugele R, Roth W and Schulz W 1999 Photovoltaic-hybrid power supply for radio network components *In Telecommunication Energy Conf, INTELEC'99. The 21st International* (p. 142).

[4] Lodhi M A.K , Vijayaraghavan P and Daloglu A 2001 An overview of advanced space/terrestrial power generation device: AMTEC *J. Power Sources* **103(1)**, 25-33.

[5] Bauer T 2011 Berlin Heidelberg. Applications of tpv generators in thermophotovoltaics *Springer* (pp. 147-196).

[6] Horne E 2002 Hybrid thermophotovoltaic power systems *EDTEK. Inc., US, Consultant Report* P500-02-048F.

[7] Qiu K and Hayden ACS 2012 Development of a novel cascading TPV and TE power generation system *Applied Energy* **91**, 1: 304-308,

[8] De Pascale A, Ferrari C, Melino F, Morini M and Pinelli, M 2012 Integration between Thermo-Voltaic generator and an Organic Rankin Cycle *Applied Energy* **97**:695–703.

[9] Barbieri E, De Pascale A, Ferrari C, Melino F, Morini M, Peretto A and Pinelli M 2012 Performance evaluation of the integration between a Thermo-Photo-Voltaic generator and an organic Rankin cycle *J Eng Gas Turbine Power* **134**, 10:1-11.

[10] Kittl E and Guazzoni G 1972 Design analysis of TPV generator system Proc. *25th Power Sources Symp.* **106**–9.

[11] Guazzoni G 1972 High temperature spectral emittance of oxides of erbium, samarium, neodymium and ytterbium *Appl. Spectrosc* **26**: 60–5.

[12] Gee JM 2002 Selective emitters using photonic crystals for thermophotovoltaic energy conversion *Conference twenty-ninth IEEE photovoltaic spec conf.* 896- 99.

[13] Shukla K, Doyle E and Becket F 1998 Thermophotovoltaic energy development program conversion *NASA/CR*- 208512 TR7020-003-98.

[14] Utlu Z and Parali U 2013 Investigation of the potential of thermophotovoltaic heat recovery for Turkish industrial sector *Energy Convers.Manag*. **74**:308–322.

[15] Utlu Z, Parali U and Gültekin Ç. Investigation of applicability of thermophotovoltaic technologies in iron and steel sector Energy Technology DOI: 10.1002/ente.201700607

Acknowledgments

This study was produced from the project 114R088 which is carried out within the scope of Scientific and Technological Research Council of Turkey (TUBITAK) 1001 from the project titled "Development of Electricity Energy Production Technology by Thermophotovoltaic Methods Utilizing Existing Waste Heat Potential in Thermal Systems". We would like to thank TUBITAK for their support to the Project.

IOP Conf. Series: Materials Science and Engineering **307** (2018) 012075 doi:10.1088/1757-899X/307/1/012075

Performance evaluation of thermophotovoltaic GaSb cell technology in high temperature waste heat

Z Utlu[1] and B S Önal[2]

[1] İstanbul Gedik University, Engineering Faculty,
Department of Mechanical Engineering, 34876, Kartal, İstanbul,Turkey
[2]İstanbul Aydin University, Institute of Science and Technology,Turkey
E-mail: zafer.utlu@gedik.edu.tr ; selenay.onal@hotmail.com

Abstract. In this study, waste heat was evaluated and examined by means of thermo-photovoltaic systems with the application of energy production potential GaSb cells. The aim of our study is to examine GaSb cell technology at high temperature waste heat. The evaluation of the waste heat to be used in the system is designed to be used in the electricity, industry and iron and steel industry. Our work is research. Graphic analysis is done with Matlab program. The high temperature waste heat graphs applied on the GaSb cell are in the results section. Our study aims to provide a source for future studies. **Keywords:** Thermophotovoltaic, Electricity generation, GaSb cell, High Temperature, Waste Heat

1.Introduction

Waste heat is the low-energy heat generated by the work done in the system. Systems such as machines, ovens and stoves emit heat for the duration of their work [1,2]. Waste heat can be produced by evaluating in thermophotovoltaic (TPV) systems. The working temperature of thermophotovoltaic systems is in the range of 1000-1500 ° C. The waste heat at this temperature value is converted to electricity energy by passing through the system components heat source, selective emitter, filter and photovoltaic (PV) cell.
The discovery of TPV dates back to about 1956. Most literature references, MIT in 1956, also made the concept suggests a series of conferences during the Aigra TPV. [3, 4]. Industrial waste heat recovery using TPV conversion was proposed by Coutts at the end of 1990s [5-8]. In addition, at the end of the 1990s, basic research on the Near Field TPV (NF-TPV) started. Electric power since the early 2000s has accelerated the development of the bottom 10 W miniature TPV generator. These studies show that TPV applications are increasingly increasing, contributing greatly to energy conversion and productivity.
In this study, electricity production by thermophotovoltaic methods is investigated by using GaSb cell in the high waste heat temperatures.

2.Description of Termophotovoltaic System

Thermophotovoltaic systems are systems that generate heat energy and electric energy from high temperature waste heat and solar radiation [8-13]. The solar rays on the photovoltaic cell are absorbed by the cell and turn the heat energy into electrical energy. They operate in the same way as photovoltaic devices convert visible light (solar energy) into electricity. Unlike solar photovoltaics [12]. TPV systems can provide higher efficiency and higher output energy density due to the lower energy band gap and closer distance between the active area of the TPV diode and the emitter. Thermophotovoltaic system includes selective emitter, heat source, filter and a photovoltaic cell as indicated Figure 1. The heat source in the system conveys the heat energy to the selective emitter and the photovoltaic cell module converts it to the thermal energy electrical energy. This transformation is considered as an alternative to existing electricity generation. At the same time, the electric energy to be obtained is obtained from the waste heat from the production stage in industrial systems. This saves energy and cost. In addition, waste heat released to the environment is evaluated.

IOP Conf. Series: Materials Science and Engineering **307** (2018) 012075 doi:10.1088/1757-899X/307/1/012075

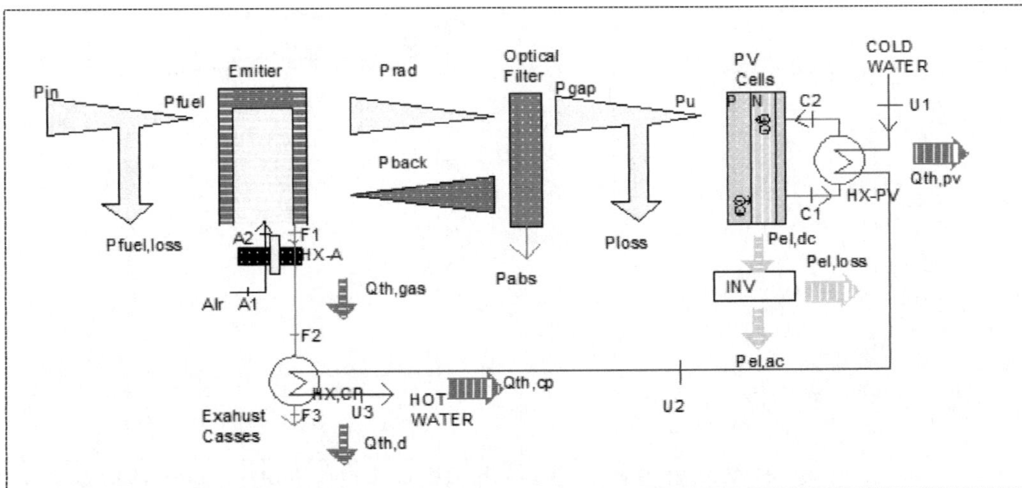

Figure 1. Schematic representation of the TPV system

3.Material and Methods

Thermophotovoltaic systems are systems that generate electricity from waste heat energy. Photovoltaic cells turn into thermal energy electrical energy. Analyses TPV high temperature waste heat values were obtained using GaSb cells. This analyses were done in MatLab program. The graphs obtained according to this analysis result are in our study.In the analysis made, TPV high temperature graphs were obtained using GaSb cell. The parameters used are; the temperature of the cell is the source temperature and the radiation temperature. With these graphs, energy efficiency, filling factor, effect of open circuit voltage and short circuit current values are determined.

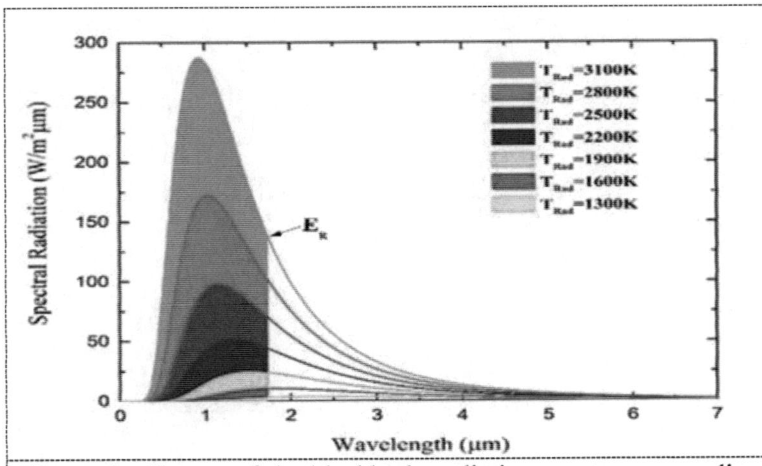

Figure 2. Change of the blackbody radiation spectrum according to the radiation temperature.

Figure 2 shows the change of the black body radiation spectrum at different wavelengths depending on the changing radiation temperature. For example, at a radiation temperature of 3100 K, the spectral radiation at 1 μ m wavelength is approximately 280 W / m^2 μ m, while it drops to approximately 100 W / m^2 μ m at 2 μ m wavelength. In this case, as the wavelength increases, the spectral radiation decreases.

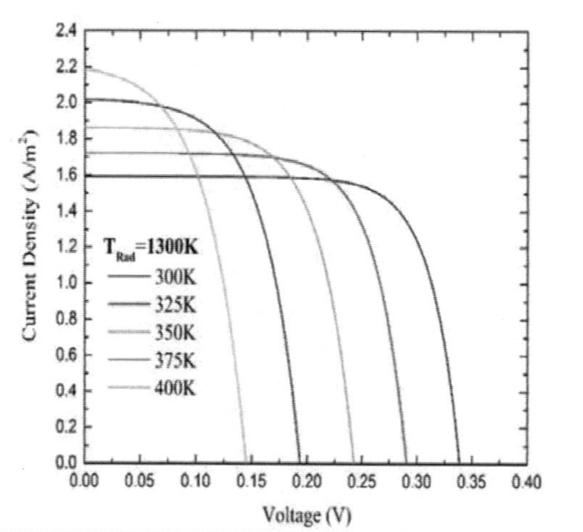

Figure 3. . Dependent on cell temperature for different current density - Voltage graph for Tradiation = 1300K.

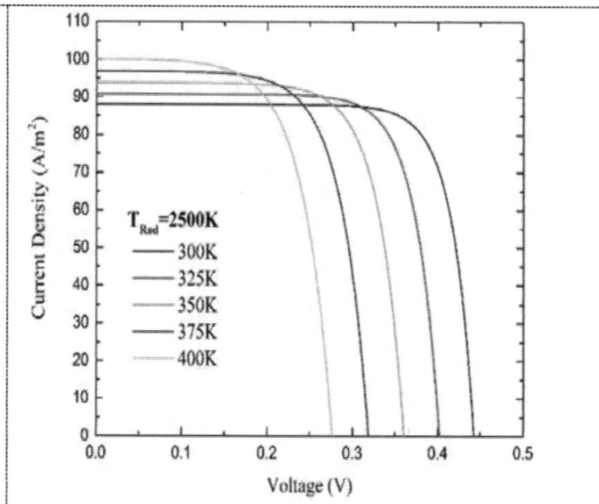

Figure 4. Dependent on cell temperature for different current density - Voltage graph for Tradiation = 2500K.

Figure 3 shows current densities at different cell temperatures at 1300 K radiation temperature versus varying voltage values. For example, at 400 K cell temperature, when the voltage value is 0.05 V, the current density is 2.2 A / m² whereas when the voltage value is 0.10 V, the current density drops to 1.7 A / m². In this case, as the voltage increases, the current density decreases. Figure 4 shows current densities at 2500 K radiation temperatures versus varying voltage values at different cell temperatures. For example, at 375 K cell temperature, when the voltage value is 0.1 V, the current density is about 97 A / m² whereas when the voltage value is 0.3 V, the current density drops to about 45 A / m². In this case, as the voltage increases, the current density decreases.

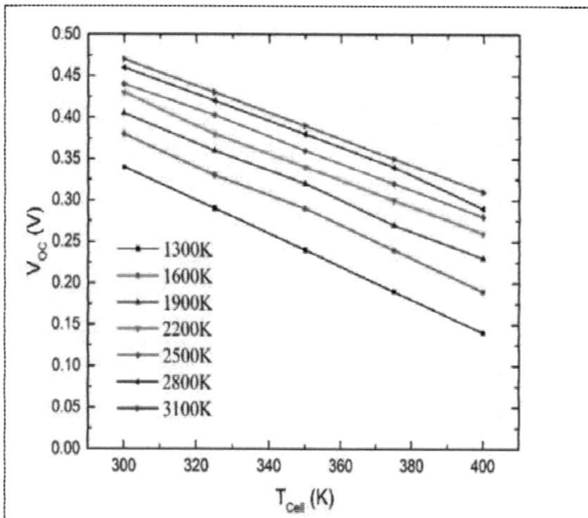

Figure 5. Tcell-Voc graph depending on the changing radiation source

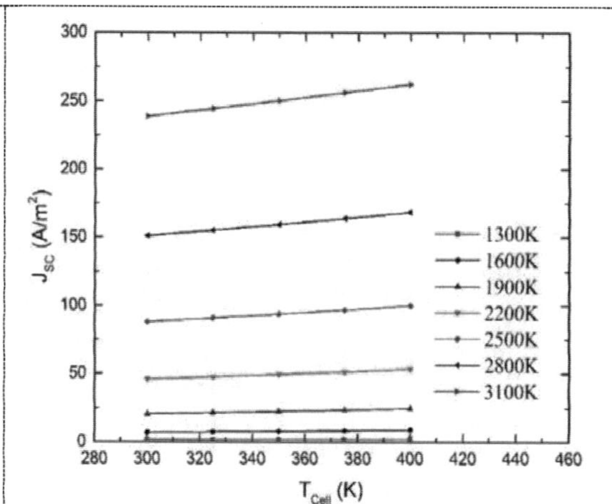

Figure 6. Tcell-Jsc graph depending on the changing radiation source temperatures

Figure 5 shows the relationship between cell temperature and open-circuit voltage, depending on the changing source temperature. For example, when the cell temperature is 300 K at 1900 K source temperature, the open circuit voltage is 0.4 V while the cell temperature is 400 K, the open circuit voltage decreases to 0.25 V.In this case, as the cell temperature increases, the open circuit voltage decreases. Figure 6 shows the relationship between cell temperature and short-circuit current depending on the changing source temperature. For example, when the cell temperature is 300 K at a source temperature of 2500 K, the short circuit current is about 80 A / m² while when the cell temperature is 400 K, the short circuit current is about 100 A / m². In this case, it is observed that the short circuit current rises as the cell temperature increases.

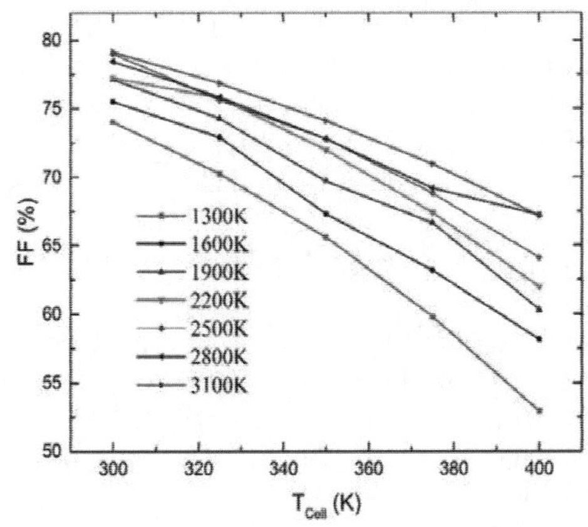

Figure 7. Cell-to-FF graph depending on varying radiation source temperatures

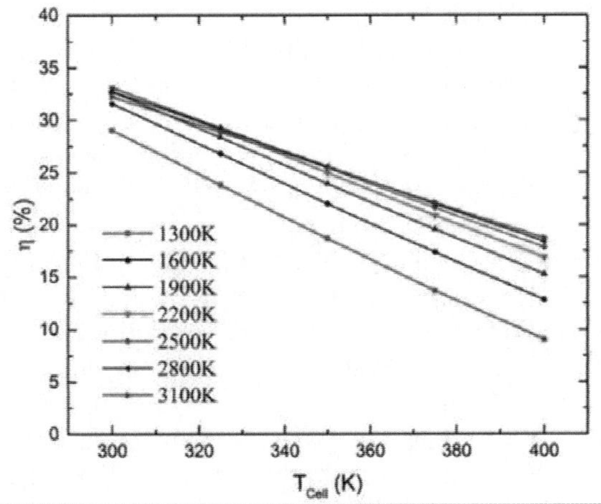

Figure 8. Cell- η graph based on varying radiation source temperatures.

Figure 7 shows the relation between cell temperature and filling factor depending on the changing source temperature. For example, when the cell temperature is 300 K at 1300 K source temperature, the filling factor is about 75% while when the cell temperature is 400 K, the filling factor is about 53%. Figure 8 shows the relation between cell temperature and energy consumption depending on the changing source temperature. For example, at a source temperature of 1300 K, when the cell temperature is 300 K, the energy yield is about 30% while when the cell temperature is 400 K, the energy yield drops to about 10%. In this case, as the cell temperature increases, the energy efficiency decreases.

4. Conclusion

The main conclusions drawn from present study may summarize as follows;

• The change of the black body radiation spectrum at different wavelengths depending on the changing radiation temperature were examined. The highest spectral radiation was obtained at 3100 K source temperature.

• The current densities at different cell temperatures at 1300 K radiation temperature versus varying voltage values were analyzed.

• The highest starting current density was 2.2 A / m² at 400 K source temperature, while the lowest starting current density was 1.6 A / m² at 300 K source temperature.

• It has been determined that GaSb TPV cell applications can be applied to industrial systems in the direction of the obtained data, provide energy efficiency and provide an alternative to electricity generation.

References

[1] Utlu Z 2015 Investigation of the potential for heat recovery at low, medium and high stages in the Turkish industrial sector (TIS) *Energy* **81**: 394-405.

[2] Utlu Z and Hepbaşlı A 2007 A Review and assessment of the energy utilization efficiency in the turkish industrial sector using energy and exergy analysis method *Renew. Sustainable Energy Rev.* **11(7)** 1438-1459.

[3] Mustafaa K F, Abdullahb S, Abdullaha M Z and Sopianb K 2017 A review of combustion-driven thermoelectric (TE) and thermophotovoltaic (TPV) power Systems Renew. *Sustainable Energy Rev.* **71**:572–584.

[4] Steinhuser A, Hille G, Kugele R, Roth W and Schulz W 1999 Photovoltaic-hybrid power supply for radio network components *In Telecommunication Energy Conf, INTELEC'99. The 21st International* (p. 142).
[5] Bianchi M, Ferrari C, Melino F and Peretto, A 2012 Feasibility study of a Thermo-Photo–Voltaic system for CHP application in residential buildings *Applied Energy* **97**:704-13.
[6] Stone KW, Chubb DL, Wilt DM and Wanlass MW. 1996 *2nd NREL Conf. on Thermophotovoltaic generation of electricity* (AIP Conf. Proc. vol 358) p 198.
[7] Durisch W, Grob B, Mayor JC, Panitz JC and Rosselet A 1999 *4th NREL Conf. on Thermophotovoltaic generation of electricity* (AIP Conf. Proc. vol 460) p 403.
[8] Nelson RE 1996 *2nd NREL Conf. on Thermophotovoltaic generation of electricity* (AIP Conf. Proc. vol 358) p 221.
[9] Qiu K and Hayden ACS 2012 Development of a novel cascading TPV and TE power generation system *Applied Energy* **91**, 1: 304-308.
[10] De Pascale A, Ferrari C, Melino F, Morini M and Pinelli, M 2012 Integration between Thermo-Voltaic generator and an Organic Rankin Cycle *Applied Energy* **97**:695–703.
[11] Barbieri E, De Pascale A, Ferrari C, Melino F, Morini M, Peretto A and Pinelli M 2012 Performance evaluation of the integration between a Thermo-Photo-Voltaic generator and an organic Rankin cycle *J Eng Gas Turbine Power* **134**, 10:1-11.
[12] Utlu Z and Parali U 2013 Investigation of the potential of thermophotovoltaic heat recovery for Turkish industrial sector *Energy Convers.Manag.* **74**:308–322.
[13] Utlu Z, Parali U and Gültekin Ç. Investigation of applicability of thermophotovoltaic technologies in the iron and steel sector *Energy Technology* DOI: 10.1002/ente.201700607

Acknowledgments
This study was produced from the project 114R088 which is carried out within the scope of Scientific and Technological Research Council of Turkey (TUBITAK) 1001 from the project titled "Development of Electricity Energy Production Technology by Thermophotovoltaic Methods Utilizing Existing Waste Heat Potential in Thermal Systems". We would like to thank TUBITAK for their support to the Project.

IOP Conf. Series: Materials Science and Engineering **307** (2018) 012076 doi:10.1088/1757-899X/307/1/012076

Study on heat transfer performance of water-borne and oily graphene coatings using anti-/de-icing component

Long Chen[1], **Yidu Zhang**[1], **Qiong Wu**[1*], **Zhang Jie**[2]

[1] School of Mechanical Engineering and Automation, Beihang University, Beijing, China

[2] Department of Science and Technology, AVIC Changhe Aircraft Industry Co., Ltd. Jingdezhen, China

E-mail: wuqiong@buaa.edu.cn

Abstract. A graphene coating anti-/de-icing experiment was proposed by employing water-borne and oily graphene coatings on the composite material anti-/de-icing component. Considering the characteristics of helicopter rotor sensitivity to icing, a new graphene coating enhancing thermal conductivity of anti-/de-icing component was proposed. The anti-/de-icing experiment was conducted to validate the effectiveness of graphene coating. The results of the experiment show that the graphene coatings play a prominent role in controlling the heat transfer of anti-/de-icing component. The anti-/de-icing effect of oily graphene coating is superior to water-borne graphene.

1. Introduction

The helicopter will encounter cold water droplets in the air when it is in the cold area or passing through the cloud, and the rotor surface is easy to freeze [1-3]. The icing on the rotor surface will lead to decreasing in flight lift and imbalance of rotor weight. At this stage, the main mean of protection is to prevent ice and deicing by installing and controlling the anti-/de-icing components on the rotor [4-5]. However, this method is more expensive for airborne energy, so it is necessary to improve this method.

Graphene, as a single layer graphite material, has been widely used in aviation and aerospace because of its excellent thermal conductivity and electrical conductivity [6-9]. A large number of studies have shown that graphene coating has good thermal conductivity and mechanical stability. Grzegorz [6], Prolongo [7] and Raji [8] studied the thermal conductivity of graphite coating respectively. The researches focus on the Joule heat produced by the graphene coating. Wang [9] proposed a graphene nanoribbon film that take advantage of both the low polarizability and conductive.

In this paper, we mainly focus on the thermal conductivity of graphene itself. Due to the heat conduction structure of graphene, the generated heat of the heating pad in the anti-/de-icing component is transferred to the surface by graphene coating, which accelerating the rotor surface ice melting, and improving the anti-icing and de-icing efficiency of helicopter rotor. The thermal conductivity of graphene coating was studied by preparing anti-/de-icing component and surface sprayed graphene coating, and conducting the anti-/de-icing experiment.

2. Heat transfer mathematical model

The heat transfer mathematical model mainly consists of the internal heat transfer model and surface heat transfer model. The internal heat transfer model can be written as follows [10]:

IOP Conf. Series: Materials Science and Engineering **307** (2018) 012076 doi:10.1088/1757-899X/307/1/012076

$$T_{wi} = T_{w1} - \phi\sum_{j=1}^{i-1}\frac{\delta_j}{A\lambda_j} = T_{w1} - \phi\sum_{j=1}^{i-1}\frac{\delta_j}{A(\lambda_{j1}\sin^2\beta + \lambda_{j2}\cos^2\beta)} \quad (1)$$

At ambient temperature, Balandin [11] measured thermal conductivity of single layer graphene prepared by mechanical peeling. The test result range from 4840 to 5300 $W \cdot mK^{-1}$, and the deduced thermal conductivity formula can be written as follows:

$$k = \chi_G\left(L/2hW\right)\left(\delta w/\delta p\right)^{-1} \quad (2)$$

Where L denotes the length of Graphene film; W stands for graphene width; h denotes the thickness of graphene; χ_G is temperature coefficient: $\chi_G = \delta w/\Delta T$.

3. Experiment setup

The anti-/de-icing experiment was set up in the refrigeration environment under -15 ℃. As shown in Figure 1, the graphene coating samples were prepared by using the standard spray process and the composite material were paved and cured under high temperature of 120 ℃.

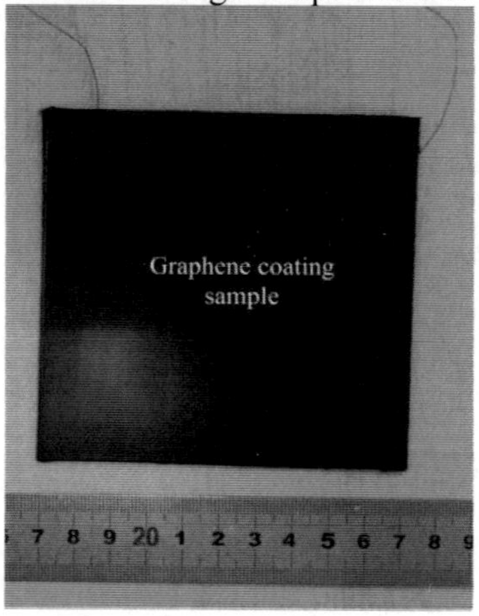

Figure 1.Schematic of the graphene sample.

Figure 2 shows the anti-/de-icing device in de-icing experiment, the pulling force is used, which is equivalent to centrifugal force generated by rotor rotating at high speed. The surface temperature was measured by thermocouples. The accuracy of the thermocouple is 0.5K, and the measuring range is 243.15K-353.15K.

In this paper, water-soluble graphene coating and oily graphene coating tests were carried out. By comparing the heat transfer effect of these two kinds of graphene coating, different characters of graphene were analyzed. Two kinds of graphene have the same content of carbon, and are provided by Jiangsu Graphene Company.

IOP Conf. Series: Materials Science and Engineering **307** (2018) 012076 doi:10.1088/1757-899X/307/1/012076

Figure 2. Anti-/de-icing experiment device.

4. Result and discuss

Through conducting the de-icing experiments, the de-icing performance of the graphene coating on the anti-/de-icing component was verified by recording the de-icing time in each de-icing experiment.

Firstly, the effect of thermal conductivity of graphene coating on the surface temperature distribution of anti-/de-icing components was tested under the same experimental conditions. Plot the maximum temperature value as shown in Figure 3.

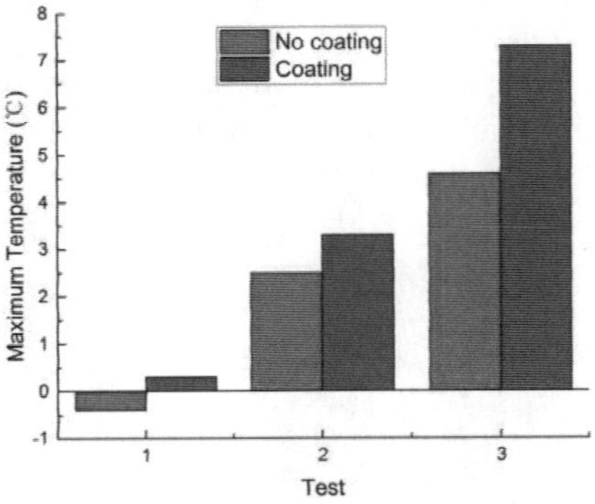

Figure 3. Comparison of maximum temperature of coating sample and uncoated sample.

By comparing the experimental data, the graphene coating surface temperature controlling ice components increased significantly, indicating that the graphene sheet coating can change the thermal conductivity of ice on the surface of the rotor assembly, improve the anti icing and de-icing efficiency.

The de-icing performance of graphene coating with different properties was compared by de-icing experiments. The water-soluble graphene and the oily graphene were coated on the surface of the anti-/de-icing component by the same spraying process. Under the same heating scheme, the rotor surface temperature is measured and the de-icing test is carried out under the same icing condition.

Figure 4 shows the temperature distribution of different kinds of graphene. From the figure we can know that water-soluble graphene has better heat transfer performance, which indicating the de-icing time is shorter than that of oily graphene.

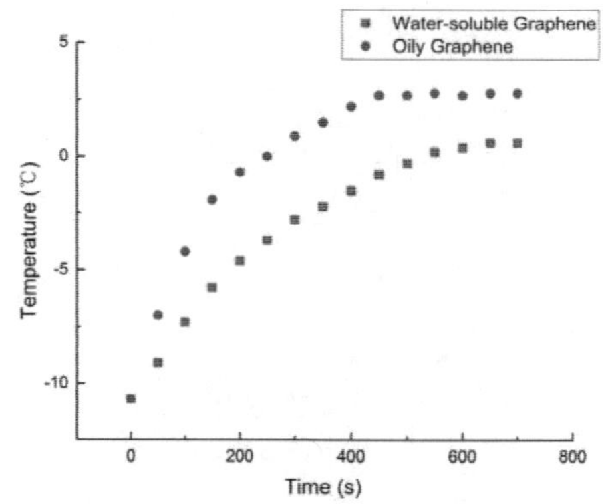

Figure 4. Temperature distribution of different kinds of graphene.

At the same time, the de-icing tests were taken three times to measuring the de-icing time. The experiment results verify that the heat transfer performance of the oily graphene coating is better than that of the water-soluble graphene coating. The data listed in Table 1 are experimental results. Analyzing experiment results show that with the change of water graphene coating to oily graphene coating, the de-icing time is obviously shortened, and the de-icing efficiency is improved obviously.

Table 1. De-icing time of two kinds of graphene.

De-icing time (s)	Water-soluble Graphene	Oily Graphene
1	125	83
2	137	92
3	119	78

The different spraying process was conducted in the de-icing experiment. By changing the spraying process and changing the spraying thickness of the graphene coating, 4 tests were carried out, and the coating thickness of the four experiments was 2 um, 4.3 um, 5.5 um, and 6.7um respectively. The maximum temperature of the anti-/de-icing surface was shown in Figure 5.

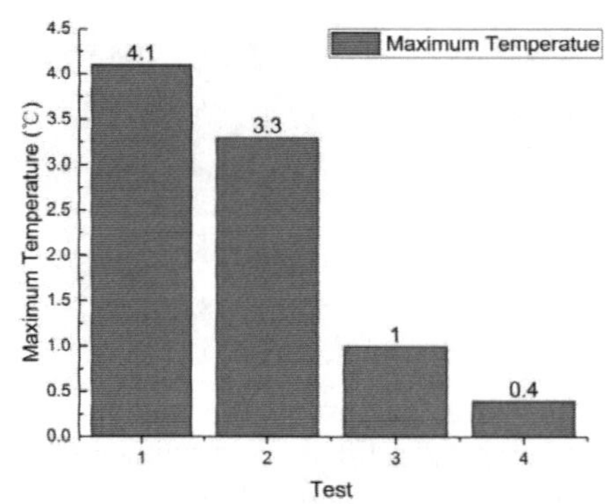

Figure 5. Maximum temperature of different spray layers.

From the data, it is obvious that the thickness can change the heat transfer performance of the graphene. With the increase of graphene coating thickness, the thermal conductivity of coating decreases gradually. The results also verify the thermal transfer mathematic model referred in Section 2.

5. Conclusions

Through the experiments of de-icing on the graphene coated on the anti-/de-icing component, the improvement of graphene coatings on the heat transfer performance is studied, which providing a basis for the use of graphene coating in the anti-/de-icing for helicopter rotor.

The experimental results show that the graphene coating can improve the thermal conductivity of the helicopter rotor anti-/de-icing component to prevent icing. When the graphene coating is employed, the de-icing time of the rotor anti-/de-icing component is shortened obviously. Moreover, the results show that the thermal conductivity of the oily graphene coating is better than the thermal conductivity of the water-soluble graphene coating. When increasing the thickness of graphene coating, the heat transfer capacity of anti-/de-icing component decreased.

Acknowledgement

This work was supported by Defense Industrial Technology Development Program (No. A0520110009), State Key Laboratory of Virtual Reality Technology Independent Subject (No. BUAA-VR-16ZZ-07). The authors thank the referees of this paper for their valuable and very helpful comments.

References

[1] Yang C, Heping H U, Yanling M A, et al. An New Idea on Anti-icing and De-icing of Helicopter Rotor Blade. *Helicopter Tech.* 2009

[2] Jose L, Placios, Yun Z, et al. Ultrasonic Shear and Lamb Wave Interface Stress for Helicopter Rotor De-Icing Purposes . *Aiaa J.* 2006

[3] Overmeyer A, Palacios J, Smith E. Ultrasonic De-Icing Bondline Design and Rotor Ice Testing [J]. *Aiaa J.* **51** 2965

[4] Thomas S K, Cassoni R P, Macarthur C D. Aircraft Anti-icing and De-icing Techniques and Modeling. *J. Aircraft* **33** 841

[5] Lin R, Yan Z G, Liu T, et al. Anti-icing/de-icing Fluid Anti-icing/de-icing and Lubricating fluid Brief Article. *Chin. J. Process Eng.* **10** 655

[6] Wroblewski G, Kielbasinski K, Swatowska B, et al. Carbon Nanomaterials Dedicated to Heating Systems. *Circuit World* **41** 102

[7] Prolongo S G, Moriche R, Rosario G D, et al. Joule Effect Self-heating of Epoxy Composites Reinforced with Graphitic Nanofillers. *J. Polym. Res.* **23** 189

[8] Raji A R O, Varadhachary T, Nan K, et al. Composites of Graphene Nanoribbon Stacks and Epoxy for Joule Heating and Deicing of Surfaces. *Acs Appl. Mater. Inter* **8** 3551

[9] Wang T, Zheng Y, Raji A R O, et al. Passive Anti-Icing and Active Deicing Films. *Acs Appl. Mater. Inter.* **8** 151

[10] Chen L, Zhang Y, Wu Q, et al. Numerical Simulation and Optimization Analysis of Anti-/De-Icing Component of Helicopter Rotor Based on Big Data Analytics. *Theory, Methodology, Tools and Applications for Modeling and Simulation of Complex Systems*. Springer Singapore, **11** 1107

[11] Balandin A A, Ghosh S, Bao W, et al. Superior Thermal Conductivity of Single-Layer Graphene. *Nano Lett.* **8** 902

IOP Conf. Series: Materials Science and Engineering **307** (2018) 012077 doi:10.1088/1757-899X/307/1/012077

Methods to include the influence of thermal bonds on the calculation of the energy performance of buildings and their influence on the heat demand for building heating

D Valachova[1] and N Zdrazilova[2] and B Chudikova[3]

1, 2, 3 VSB – Technical University of Ostrava, Faculty of Civil Engineering, Czech Republic
E-mail: blanka.chudikova.st@vsb.cz

Abstract. The paper deals with the effect of thermal bonds on heat transmission of a building envelope. Then it deals with ways to include of thermal bonds in the calculation of heat loss through the building envelope and the calculation of energy efficiency of buildings. A solution of thermal bonds is very important, because it fundamentally influences the energy efficiency of the buildings. It is important to realize that building envelope comprises not only the peripheral surface structures but also thermal bonds in areas where building structures join.

1. Introduction

Nowadays the emphasis is put on reducing the energy performance of buildings [1], which is influenced by the so-called the envelope of the building, which means the perimeter structures at the boundary of the heated zone and outside or unheated space. These building structures are strictly evaluated in terms of thermo-technical qualities. When designing buildings, care must be taken to ensure that these qualities are as good as possible. This means that they try to achieve the lowest values of the coefficient of heat transfer in the U constructions [2]. However, it is necessary to solve the constructions in a complex way, including the right solution of the details [1]. Although the peripheral structures will have excellent thermal insulation qualities, this does not mean that the envelope of the building is designed properly. The quality of the building details solution significantly influences the resulting energy demand of the building, especially for low energy and passive objects [1, 3].

2. Thermal bonds

Thermal bonds are defined according to ČSN 73 0540-1 as boundary between two or more structures, which can be of a point or linear character (point and linear thermal bonds). In these places, the heat flux is significantly altered due to their interaction. Thermal bonds affect thermal flux through the building jacket, especially in buildings with low energy performance. The building envelope can not be perceived only as a summary of the perimeter structures, but as a system of perimeter structures with two-way systemic thermal bonds. However, the design of structural details - thermal bonds is not a regular part of the project documentation, although it is an important part of it. Due to the improper design or construction of the incorrect construction details, the thermal flow is increased by thermal bonding. Examples of wall-ceiling thermal bonding are shown in Figure 1.

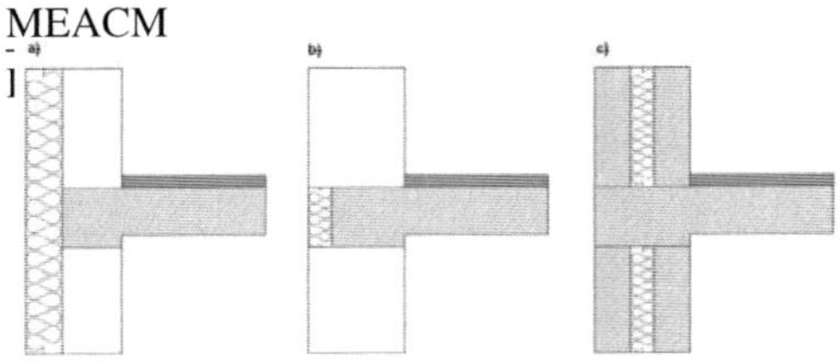

Fig. 1 : Example of thermal bonding: a) Dragging of thermal insulator on the outside of the structure, b) Insertion of heat insulator at the place of thermal bonding, c) Neglected thermal bonding solution in panel sandwich construction

A panel house can serve as an example of poorly designed and neglected building details. The part of sandwich peripheral structures, were thermal insulators and the perimeter structures had good thermal-technical qualities, but the thermal bonding were very neglected. This fact led to a significant increase in the heat flow at those parts of the constructions and to the increase of the energy performance of the building. At the same time, the increased heat flux was manifested by troubles - mildew growth, which had a negative effect on the indoor environment. Fig. 2 shows an increased heat flux through thermal bonding.

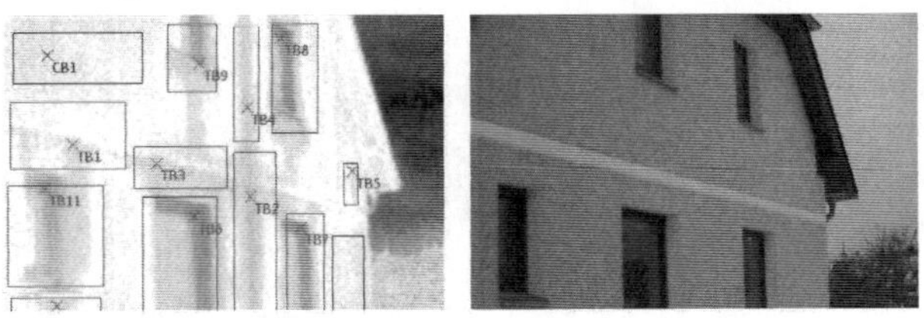

Fig. 2: Thermal imaging of a panel house (increased thermal flow in the place of thermal bonds - contact of panels, ceiling and perimeter walls)

3. Including the influence of thermal bonds on calculation of heat transfer through the building envelope

When calculating the thermal flow through the envelope of the building, which is further applied in the calculations of the heat losses, the energy performance of the building and the average coefficient of heat transfer, it is possible to include the thermal bonds in two ways [5]. The influence of thermal bonds in these calculations includes so-called average influence by thermal bonds.

3.1. Simplified possibility to include the influence of thermal bonds

The easiest way is to roughly include thermal bonds in the calculations with an approximate value that increases the heat flow rate by building envelope by type of building. This is the most commonly used method, however, the results of experimental calculations point to the crucial effect of such a method of including thermal bonds on inaccuracies in heat penetration calculations through the building envelope. The calculation of the specific heat loss in this case is determined by the relationship (1).

$$H_T = \sum \left(U_j \cdot A_j \cdot b_j \right) + A \cdot \Delta U_{tbm} \tag{1}$$

, where H_T is specific loss of heat transfer (W/K)

U_j heat transfer coefficient of construction "j" (W/(m^{2}K))

A_j the surface of the cooled structure "j" at the system boundary of the building envelope (m^2)

b_j thermal reduction factor of construction "j" (-)

ΔU_{tbm} the average influence of thermal bonds between cooled constructions on the system boundary of a building (W/(m^{2}K))

A the area of all cooled structures at the system boundary of the building envelope (m^2)

In the case of this simplified calculation, the values are considered as follows:

- ΔU_{tb} = 0,02W/(m^{2}K)) for buildings with consistently optimized thermal bonding, this value is commonly used for new buildings.
- ΔU_{tb} = 0,05W/(m^{2}K)) for buildings with mild thermal bonds, this value is commonly used for insulated buildings.
- ΔU_{tb} = 0,10W/(m^{2}K)) for buildings with normal thermal bonding.
- ΔU_{tb} = 0,20W/(m^{2}K)) and more for neglected solutions.

IOP Conf. Series: Materials Science and Engineering **307** (2018) 012077 doi:10.1088/1757-899X/307/1/012077

3.2. Exact inclusion of the influence of thermal bonds

For a more accurate calculation, however, it is possible to make a calculation by precise input of thermal bonds [4], where the thermal flux through thermal bonding and its length (in the case of point thermal bonding his count) is taken into account. Calculation of specific heat loss through heat transfer is then performed according to the relationship (2).

$$H_T = \sum(U_j \cdot A_j \cdot b_j) + \sum(\psi_j \cdot l_j \cdot b_j) + \sum(\chi_j \cdot b_j) \quad (2)$$

, where H_T is specific loss of heat transfer (W/K)

U_j heat transfer coefficient of construction "j" ($W/(m^2K)$)

A_j the surface of the cooled structure "j" at the system boundary of the building envelope (m^2)

b_j thermal reduction factor of construction "j" (-)

ψ_j linear factor thermal transmittance linear thermal bond "j" (W/(m·K))

l_j the length of the linear thermal bond "j" between the structures within the building (m)

χ_j point factor thermal transmittance of the j-point thermal bond between buildings within the building ($W/(m^2K)$)

All thermal bonds that are part of the building envelope considered must be included in the calculation. This can be complicated in buildings of non-compact shape. The linear and point heat transfer factors are input into the calculation. Their quantities are determined by calculation based on thermal bonding. At thermal bonding sites multidimensional heat dissipation takes place and calculation of thermal permeability can only be done by solving the so-called thermal field. The calculation is done using finites elements software and is time consuming.

4. Heat flow through the envelope of the building

Thus, the choice of the method of incorporating the influence of thermal bonds can significantly influence the calculation of heat flow through the envelope of the building envelope. This is especially important for the evaluation of the thermal-technical properties of the whole envelope of the building using the so-called average coefficient of heat transfer U_{em}, but also for calculating the heat losses and energy performance of buildings. The difference in the calculation will not be noticeable in the case of buildings that do not have thermal-technical properties at the low energy standard level. However, in the case of low-energy and passive houses [1], the specific heat flux through the thermal bonds forms a significant share of the total heat flow through the building envelope. In the case of these buildings, it is important to emphasize on right solution the thermal-technical of thermal bonds to avoid increased heat fluxes. Experimental calculations have shown that in passive buildings the choice of the method of incorporation of thermal bonds is essential and will affect the specific heat demand for heating E_a [6].

4.1. Heat demand for heating

Thermal loss characterizes the thermal flow of the building under extreme conditions. The heat demand for heating is determined by the amount of energy required to maintain a pleasant indoor temperature. The specific heat demand for heating is rated. This variable is not dependent only on the properties of the building envelope, but also on other parameters, such as the way of ventilation and thermal gains of the building. In the case of passive houses, it has been shown that due to the precise inclusion of thermal bonds in the calculation of the energy intensity, there is usually a difference of about 5 kWh/(m^2.year) in case of a specific annual heat demand for heating E_a. The specific heat demand for heating a passive house is max to 20 kWh/(m^2.year). Thus, the difference between the results according to the way of incorporating the influence of the thermal bonds can thus account for up to 25% of the total heat demand for heating [3], assuming a high quality thermal bonding. In the passive houses, we solve the details so that there is no increase in thermal bonds [1]. The use of a flat rate then significantly increases the heat flux that is not actually occurring. In the case of the energy performance assessment of the passive building, it is therefore advantageous to make a precise calculation of the influence of the thermal bonds. Figure 3 shows a graphical representation of the temperature field, the calculation of which was made to accurately include the influence of thermal bonds in the calculation, several building details of the passive house.

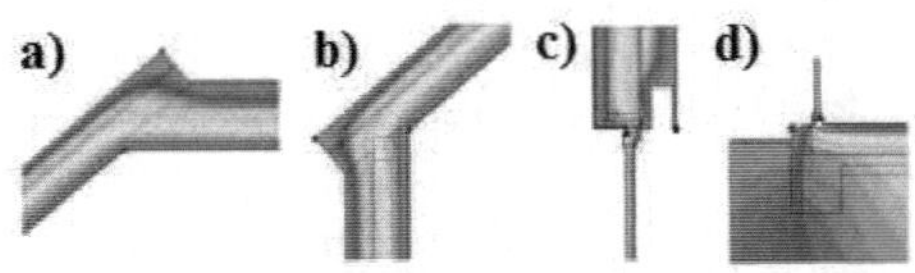

Fig. 3: Graphical representation of the thermal field of several building details of the passive house (thermal bond: a) oblique roof and ceiling under unheated soil, b) perimeter wall and oblique roof in the place of the wall plate, c) the perimeter wall and the window top, d) the floor and door threshold)

5. Experimental analysis

For three types of family houses, the calculation of the energy performance was always performed with the inclusion of thermal bonds in a simplified and detailed manner. The calculation was made for a typical non-insulated existing house and for the new building of brick and lightweight perimeter constructions.

5.1. Computational model

The energy performance of buildings is also influenced by the so-called volume factor of the A / V building shape. Four models of family houses were was created, see Fig. 4. The building factor volume factor determines the ratio between the total area of the building envelope and the outside volume of the heated zone. The lower the volume factor, the more energy-efficient the building is. The less compact buildings have a larger amount of thermal bonds.

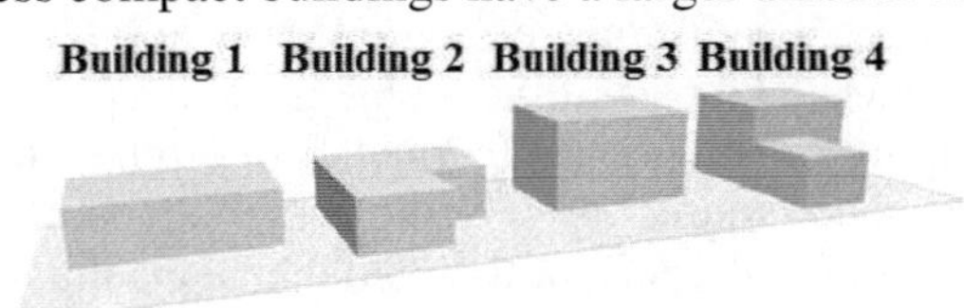

Fig. 4: Computational building models

Table 1: Influence of the compactness of the house on the volume factor

Model	Volume factor *A/V* [-]	Description
1	0,97	Single-storey building of compact shape (bungalow type)
2	1,00	Single-storey L-shaped building (bungalow type)
3	0,79	Two-storey building of compact shape
4	0,88	Two-storey building of non-compact shape

The calculation was made for three houses. The first is a brick family house, a perimeter construction typical of houses built around the 1950s. The other two models are a typical new building whose perimeter construction is at the level of passive buildings. The model was designed for the construction of masonry with ETICS and for a light wooden structure. For the simplified method of incorporating the influence of thermal bonds, the values ΔU_{tb} = 0,10 W/(m^2.K) for the existing insulated building and ΔU_{tb} = 0,02 W/(m^2.K) for new buildings. In case of accurate inclusion of the influence of thermal bonds, the values of linear heat transfer factors ψ, which were calculated based on the results of two-dimensional temperature field solution for typical solution of building details for the given building type, were considered.

5.2. Calculation results

The results of the calculations are shown in Tab. 2 As an evaluation criterion, the annual specific heat demand for heating E_a v kWh/(m^2.year) was chosen. In the case of new buildings with well-designed details, the method of determining the influence of thermal bonds can significantly influence the results of the calculation. At the same time, of non-isolated existing buildings, the differences in the calculation results are not so severe. In this case, the calculation result may even deteriorate due to poorly solved thermal bonds.

Table 2: Specific heat demand for heating by specific and exact calculations

Type of building	Method of calculating the influence of thermal bonds	Specific heat demand for heating E_a (kWh/(m^2·year))			
		1 A/V = 0,97	2 A/V = 1,00	3 A/V = 0,79	4 A/V = 0,88
Existing bricked	Simplified $\Delta U_{tb} = 0{,}10$ W/(m^2·K)	349	362	263	302
	Exact by ψ	337	350	268	304
New bricked	Simplified $\Delta U_{tb} = 0{,}02$ W/(m^2·K)	45	45	31	35
	Exact by ψ	38	38	28	30
New (light const.)	Simplified $\Delta U_{tb} = 0{,}02$ W/(m^2·K)	52	53	37	41
	Exact by ψ	43	47	31	35

6. Conclusion

The article deals with the influence of thermal bonds on two-way heat penetration through the envelope of the building. It emphasizes that the envelope of the building is not only created by its flat construction, but also by thermal bonds. Percentage increase depends only on the general solution of thermal bonds. The article shows that the flat increase can be very inaccurate and it is only an indicative determination of the influence of thermal bonds. More precise results can be obtained by specifying the influence of thermal bonds by means of linear and point heat transfer factors. This procedure is time consuming and is almost unused in practice. The part of the article is an analysis of the influence of the method of assigning thermal bonds to calculation of thermal flux through the building envelope. The results have shown that in the case of thoroughly solved thermal bonds, it is appropriate to perform the calculation by accurately determined influence of thermal bonds. Quality of the solution of the building details has a crucial influence on the energy performance of the building and it shows how important it is to think of the correct implementation of the building details.

7. References

[1]PALKO, M. House in Passive Standard - Thermal Bridges. *In: Advanced Materials Research. 2014; vol. 899. DOI: 10.4028/www.scientific.net/AMR.899.42.*

[2]ZACH, J., HROUDOVA, J. Study of the Properties of the Environmentally Friendly Insulation Materials. *Advanced Materials Research. 2014; vol. 899, pp. 435-439. DOI: 10.4028/www.scientific.net/AMR.899.435*

[3]INGELI, R., VAVROVIC, B., CEKON, M. Thermal Bridges Minimizing through Typical Details in Low Energy Designing. *In: Advanced Materials Research. 2014; vol. 899. DOI: 10.4028/www.scientific.net/AMR.899.62.*

[4]CAPOZZOLI, A., GORRINO, A., CORRADO, V.,A building thermal bridges sensitivity analysis. *In: Applied Energy. 2013; vol. 107. pp 229-243.*

[5]VIOT, H., SEMPEY, A., PAULY, M., MORA, L.Comparison of different methods for calculating thermal bridges: Application to wood-frame buildings. *In: Building Environment. 2015; vol. 93 (P). pp 339-348.*

[6]KAMENSKY, M., VASKOVA, A., VERTAL, M. Analysis of Energy and Internal Environment Improvements in a Single Family House. *In: Advanced Materials Research. 2014; vol. 899, pp. 3-6. DOI: 10.4028/www.scientific.net/AMR.899.3.*

Acknowledgement

The paper is supported from founds of the Conceptual development of science, research and innovation for 2017 allocated to the VŠB-TUO by the Ministry of Education, Youth and Sports of the Czech Republic.

Effect of an additional accumulation wall on course of interior temperatures in a light modular construction

M Černíková [1], V Panovec[2]

[1,2] VSB – Technical University of Ostrava, Faculty of Civil Engineering, Ostrava, Czech Republic

marcela.cernikova@vsb.cz

Abstract. This paper is focused on influence of an additional concrete wall on indoor temperature in the light modular construction. Typically, there are no accumulation materials used in these constructions. The study was based on measurements obtained from a full-scale experimental set-up built for this purpose. Two identical modular constructions were used; one of them was equipped with an additional concrete wall. Indoor temperatures were measured in the course of time and compared.

1. Introduction

Evaluation of heat accumulation is very important in case of modular constructions that are permanently used as residential buildings. Light modular constructions with low heat accumulation started to be popular after the First World War [1] but the heating mode remained the same as for the brick houses, i.g. with night declines. As a consequence, the light constructions quickly cool down.
This was an impulse that started evaluation of heat accumulation of perimeter walls. It is important to bear in mind that the temperatures in the buildings and walls are changing with time. The real temperatures of the constructions are always variable, unstable.

2. Description of the light modular constructions

The study was based on measurements obtained from a full-scale experimental set-up built for this purposes. Two identical light modular construction, on Figure 1, were compared; one with a concrete accumulation wall and the other without any additional wall.

The modular construction has standard dimensions (length is 6,058m, width is 2,990m, height is 3,010m) and it can be carried by a truck. There are two triple glass insulated windows and one entrance door. Inside is an electric heater which is controlled manually.

Figure 1. Two light modular constructions used for the experiment

A concrete wall from from dry-concrete paving stones was built in one of the modular constructions, Figure 2. Thanks to two different set-up, it was possible to compare the course of interior temperatures in both constructions.

Figure 2. Construction of the concrete accumulation wall

3. Thermal resistance of the perimeter constructions

When choosing the material of perimeter walls, it is desirable to select a material with the highest value of thermal resistance R [m^2K/W] so that the heat loss would be the smallest possible. The value of thermal resistance depends on thermal conduction and on the thickness of individual layers of the wall construction, but it is not depending on accumulation abilities of the used materials [2].

Nevertheless, accumulation ability of the material is one of the crucial factors that influence thermal comfort in the building, both in summer and winter. While assessing thermal resistance of a material, there is an assumption that the temperature on both sides of the construction is constant in time [3]. However, this scenario is rare since the indoor and outdoor temperatures are variable in time.

The temperature of air outside of the construction is affected by many attributes; changing of day and night, changing of seasons or changing of intensity of solar radiation. The temperature inside of the construction is affected by discontinuous heating, ventilation or by immediate changes in the environment.

Evaluation of heat accumulation of perimeter constructions is important when deciding if the heating in the building is going to be continuous or discontinuous [4]. If discontinuous mode is chosen, the temperature of indoor air as well as the surface temperature of the constructions are decreasing while the heating is off. The value of the temperature decrease during the heating break is assessed by criteria called thermal stability of a room during winter [5].

4. Properties of the accumulation wall

The accumulation wall was built in one of the two light modular constructions from dry-concrete paving stones. The thickness of one paving stone was 100mm. While considering the accumulation properties of the accumulation wall the maximum width of would be 150mm; thicker wall will not show significantly better results [6].

Table 1. Properties of the accumulation wall.

	Units	
material	-	**concrete**
thickness	(mm)	**100**
density	kg/m^3	**2100**
specific heat capacity	J/kgK	**1020**
thermal conductivity	W/mK	**1,23**

5. Course of temperature

Data from this experiment were collected in 2015 during winter. Each light modular construction was equipped with an electric heating that was turned on an off controllably. The daily cycle, on Figure 3, started in the morning when the heater was turned on. Specific amount of time is needed to warm up the interior air on desired temperature. During the day, the heating is maintaining the temperature on a stable level. During the night, the heating is turned off and the temperature is dropping. The cycle starts again in the morning next day.

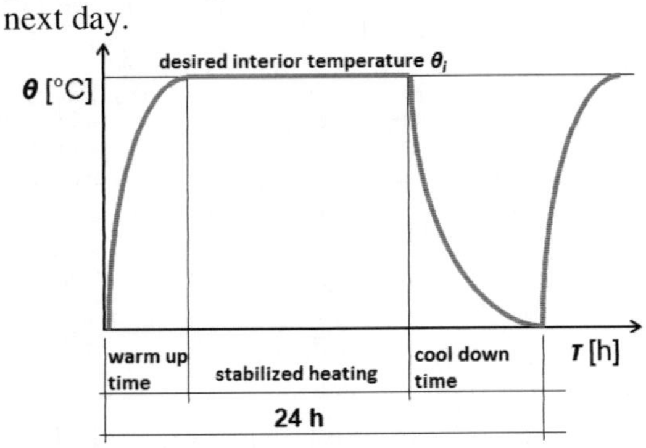

Figure 3. Daily cycle of course of temperature in a room with discontinuous heating

6. Results

6.1. Time needed for warming up interior temperature

The heating in the two light modular constructions was turned off for several days so the initial indoor temperature was the same in both objects. After the heating was turned on, temperature increase was monitored. As compared on Figure 4,5,6, the experiment was repeated three times always with different course of exterior temperature. It is obvious, that the light modular construction with the accumulation wall needs longer time to warm up the interior on a desired temperature. Depending on exterior temperature, the needed time to warm up the interior was between four to six hours. The desired interior temperature in the construction without any accumulation wall was usually reached in two to four hours.

IOP Conf. Series: Materials Science and Engineering **307** (2018) 012078 doi:10.1088/1757-899X/307/1/012078

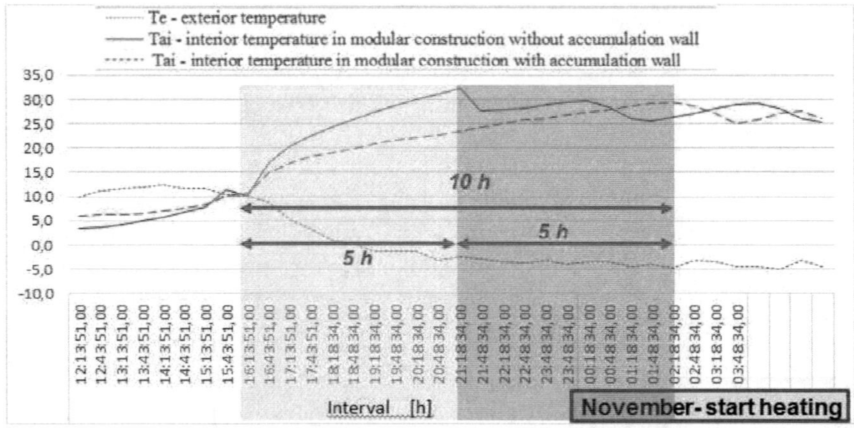

Figure 4. Time needed to warm up indoor temperatures in the constructions and the course of outdoor temperature in November 2015

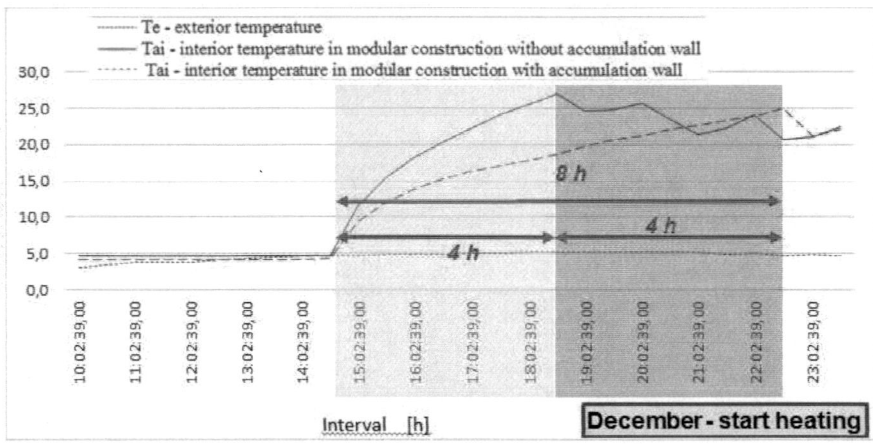

Figure 5. Time needed to warm up indoor temperatures in the constructions and the course of outdoor temperature in December 2015

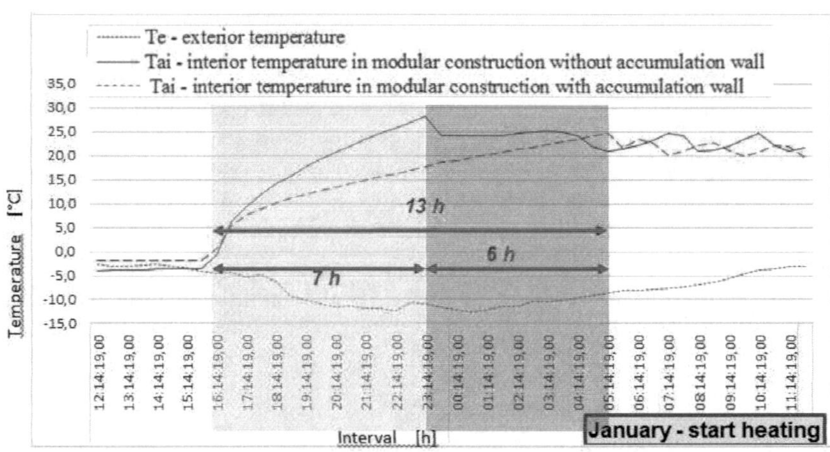

Figure 6. Time needed to warm up indoor temperatures in the constructions and the course of outdoor temperature in January 2016

IOP Conf. Series: Materials Science and Engineering **307** (2018) 012078 doi:10.1088/1757-899X/307/1/012078

6.2. Time needed to cool down the interior temperatures

Time needed to cool down the interior temperatures after the heating is turned off depends on the course of exterior temperatures and on the amount of accumulated heat in accumulation constructions. As it was observed, Figure 7,8, on the set-up, the interior temperature was equal to exterior temperature in shorter time in the light modular construction without any accumulation wall.

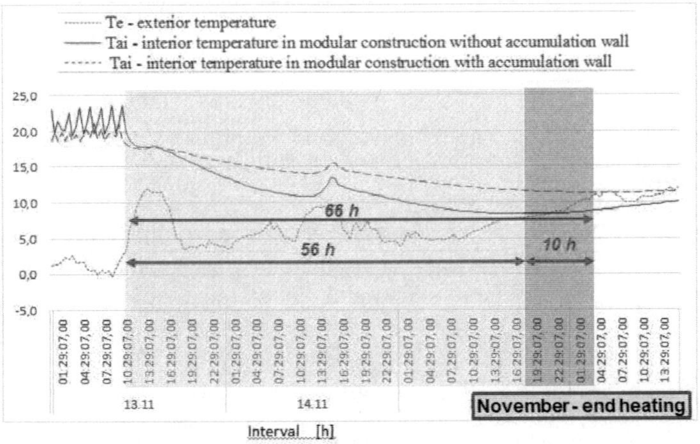

Figure 7. Time needed to cool down indoor temperatures in the constructions and the course of outdoor temperature in November 2015

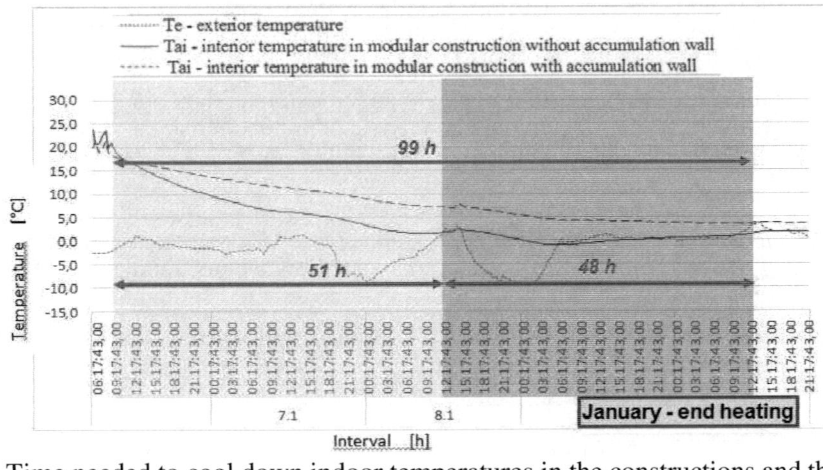

Figure 8. Time needed to cool down indoor temperatures in the constructions and the course of outdoor temperature in January 2016

7. Conslusion

The comparison of the course of interior temperatures in two light modular constructions showed that the concrete accumulation wall has a significant impact on the cooling time. The light modular construction without any accumulation wall was warmed up/cooled down faster than the light modular construction with the accumulation wall. Buildings without any accumulation constructions are suitable for short-time use, when the heating is running continuously and when quick cool down of the indoor temperature is of small importance.

IOP Conf. Series: Materials Science and Engineering **307** (2018) 012078 doi:10.1088/1757-899X/307/1/012078

Nevertheless, if the light modular construction is meant for long term residential use it is more than desirable to integrate an accumulation construction. Such construction will raise thermal comfort in the building if the heating is discontinuous. The accumulation wall is balancing the indoor temperature more smoothly and if the heating discontinuity is controlled, it can lead to energy savings and therefore money savings. This aspect is going to be even more noticeable in so called zero buildings, where the heating is depending on direct demand of power from a local (renewable) source, e.g. solar power.

8. References

[1] Ingeli, R., Vavrovič, B., Čekon, M., *Thermal bridges minimizing through typical details in low energy designing,* In: International Conference on Buildings and Environment, EnviBUILD 2013, Code 103212. Vol. 899, 2014, pp. 62-65. ISBN 978-303835040-8. ISSN: 10226680. DOI: 10.4028/www.scientific.net/AMR.899.62

[2] Kalánek, J., Šteffek, L., Ostrý, M., *Calculation and experimental evaluation of the thermal resistance of a structure with reflective insulation,* In: International Conference on Buildings and Environment, EnviBUILD 2013, Code 103212. Vol. 899, 2014, pp. 374-380. ISBN: 978-303835040-8. ISSN: 10226680. DOI: 10.4028/www.scientific.net/AMR.899.37

[3] Charvat, P., Ostry, M., Klimes, L., Stetina, J., *A validated TRNSYS model of thermally activated layer with phase change material.* In: ASME 2015 International Mechanical Engineering Congress and Exposition, IMECE 2015, Houston; United States; 13 November 2015 through 19 November 2015; Code 121052. Vol. 8A-2015, 2015, ISBN: 978-079185749-6, DOI: 10.1115/IMECE2015-51705

[4] DONOVA, D., KUCERIKOVA, V., *The Relative Humidity inside the Wall Structure According to the Type of Façade*, In: CEEE 2014, Hong Kong, pp. 1-5. ISBN 978-1-908074-18-8.´

[5] BILANIČ, M., KUBEČKA, K., VLČEK, P., *Application of risk analysis for building evaluation,* In: 6th International Conference on Contemporary Problems of Architecture and Construction; Ostrava; Czech Republic; 24 June 2014 through 27 June 2014; Code 108709. Vol. 1020, 2014, pp 879-882.ISBN: 978-303835237-2.ISSN: 10226680.
DOI: 10.4028/www.scientific.net/AMR.1020.879

[6] LABUDEK, Jiri, TESLIK, Jiri, VODICKOVA, Martina, WALDSTEIN, Petr. *Settlement of cellulose insulation.* In: EESD (3rd International Conference on Energy, Environment and Sustainable Development)2013:Shanghai,China, vol. 860-863, 2014, pp 1204-1207. ISBN 978-303785972-8.ISSN10226680.DOI: 10.4028/www.scientific.net/AMR.860-863.1204.

Acknowledgment

The work is supported from the funds of conceptual development of science, research and innovation for 2017 allocated to the VŠB-Technical University of Ostrava by the Ministry of Education, Youth and Sports of the Czech Republic.

IOP Conf. Series: Materials Science and Engineering **307** (2018) 012079 doi:10.1088/1757-899X/307/1/012079

Effect of aluminum contents on sputter deposited CrAlN thin films

A Vyas[1, *], **Z F Zhou**[2] and **Y G Shen**[3]

[1]Department of Mechanical Engineering, The Hong Kong Polytechnic University, Hung Hom, Kowloon, Hong Kong, China
[2,3]Department of Mechanical and Biomedical Engineering, City University of Hong Kong, Kowloon, Hong Kong, China
[*]Email: mmavyas@polyu.edu.hk

Abstract. Pure CrN and CrAlN films with varied Al concentrations were prepared onto Si(100) substrates by an unbalanced reactive dc-magnetron sputtering system. The crystal structure, chemical states, and microstructure of the films were characterized by X-ray diffraction, X-ray photoelectron microscopy, transmission electron microscopy whereas mechanical properties were determined by nano-indentation measurements. XRD results showed a prominent (200) reflection in both CrN and CrAlN films. Results demonstrate that CrAlN films formed a solid solution and doping of Al atoms replace the Cr atoms affecting the lattice parameter and crystallization of the films. All Al doped films were of B1 NaCl-type structure, demonstrating that CrAlN films primarily crystallized in cubic structure. Microstructural investigation by TEM for a CrAlN film containing Al content of 24.1 at.%, revealed that there exists an amorphous/nanocrystalline domains (grains of about ~ 11 nm) and hardness increases 22% when compared with pure CrN film.

1. Introduction

Binary and ternary transition metal based nitride materials are applied in various industrial as well as engineering applications owing to their significant physical, chemical and mechanical properties including enhanced oxidation and corrosion resistance [1-3]. Nowadays, most metal formed or molded components are coated with chromium nitride (CrN) to improve the corrosion and oxidation resistance under severe environmental conditions [4-5]. However, due to limited hardness of CrN and its poor wear resistance, restricts its applications in the field of engineering such as machine parts, drills and cutting tools [6]. On the other hand, addition of Al, into CrN leads to the formation of a nanocomposite structure comprising of two phases, CrN/AlN, which helps in overcoming the issue of high temperature instability and gives higherhardness (>30 GPa) and enhanced anti-corrosion resistance, making chromium-aluminium-nitride (Cr-Al-N) films superior to CrN especially enhancing oxidation resistance at high temperatures. Microstructurally, formation of AlN phase at the CrN grain boundaries, helps suppress the oxidation in these high diffusive paths, and shields the enclosed two-phase CrN/AlN nanocomposite structure [7-9].

Up to now, nanocrystalline CrAlN composite films have been deposited by various techniques which include vacuum arc reactive deposition process, [10, 12], chemical vapor deposition (CVD) [13], and dc/rf reactive magnetron sputtering [14,15]. The crystal structure, microstructure, concentration, and corrosion resistance of these nanocomposite films have been studied in comparison with conventional binary nitrides of CrN and AlN films. However, due to lack of in-depth study on chemical bonding status and microstructure for sputter-deposited CrAlN films generates curiosity to investigate. In our

IOP Conf. Series: Materials Science and Engineering **307** (2018) 012079 doi:10.1088/1757-899X/307/1/012079

investigation, the effect of the small amount of Al added to CrN matrix on bonding status, microstructure and mechanical properties of CrAlN films is reported, in which new features which were not covered previously are presented here. X-ray photoelectron spectroscopy (XPS) was employed to investigate the chemical bonding status in the films. Phases in the films were revealed by X-ray diffraction (XRD). Nano-structural evolution was studied by employing high-resolution transmission electron microscopy (HRTEM) and transmission electron diffraction (TED). The mechanical properties were obtained by nanoindentation technique.

2. Experimental

CrN and CrAlN films with various Al contents were deposited onto Si(100) substrates using a close-field unbalanced dc-magnetron sputtering system (UDP650, Teer Coating Limited) with four high-purity targets (three Ti and one Al) in a mixture of Ar and N_2 gases. The substrates were ultrasonically cleaned in acetone then dried andplaced into the vacuum chamber. The base pressure in the chamber was 2×10^{-6} Torr, and the working pressure, was set at 1.5 mTorr. The Cr target current was set at 5 A and Al target current ranged from 0 to 10 A in order to acquire films with varied Al concentrations. The films deposited are namely, A0, A1, A2, A3, A4, A5, and A6. Film A0 is pure CrN and A6 with Al current of 10 A has the highest Al content of 24.1%. With target current ratio (I_{Al}/I_{Cr}) ranging from 0 to 2.0, the atomic ratio of Al/(Cr+Al) could be increased from 0 to 0.48 (in Table 1 in Section 3). The substrate rotation speed was 10 rpm. Films nominal thickness was 600 nm at a bias voltage of -80 V (250 kHz).

The crystal structure of the films was characterized by XRD using a Bragg-Brentano diffractometer (Simens D500). The elemental concentration and bonding status were investigated by XPS (PHI Quantum 2000) having monochromatic Al Kα X-ray source (hν=1486.6 eV) induced by 13 kV, 25 mA Al kα radiation. The nanostructure was analyzed by HRTEM (field emission JEOL 2010F) at 200 kV. Nanoindentation tests were repeated 8 times and depth was at <10% of the film thickness.

3. Results and discussion

The measured results for the films are shown in Table 1.

Table 1. The I_{Al}/I_{Cr} target current, elemental concentration, grain size, hardness, Young's Modulus and H/E ratio of sputter-deposited CrAlN films.

Sample	Target Current ratio (I_{Al}/I_{Cr})	Al content (at. %)	Al/(Cr+Al) ratio	N content (at. %)	Grain size *D* (nm)	H(GPa)	E(GPa)	H/E ratio
A0	0	0	0	47.5	11.8 ± 0.2	18.8 ± 0.7	225.0 ± 6.2	0.083
A1	0.2	2.3	0.04	48.4	11.5 ± 0.3	21.8 ± 1.0	233.2 ± 8.6	0.093
A2	0.4	6.3	0.12	49.8	11.3 ± 0.2	22.5 ± 1.3	242.1 ± 11.4	0.097
A3	0.8	12.8	0.25	48.7	11.4 ± 0.2	21.4 ± 1.3	233.0 ± 9.6	0.092
A4	1.2	17.1	0.33	49.2	11.1 ± 0.2	22.0 ± 1.2	233.5 ± 10.6	0.094
A5	1.6	21.5	0.43	50.3	10.7 ± 0.2	22.8 ± 1.2	238.0 ± 8.9	0.095
A6	2.0	24.1	0.48	49.2	10.8 ± 0.2	24.0 ± 0.9	255.0 ± 8.5	0.096

3.1 Crystal structure by XRD

The influence of Al doping on the lattice parameter of CrAlN films can be seen in Fig. 1. The inset of Fig. 1 shows the XRD patterns of CrAlN films (A0 to A6).

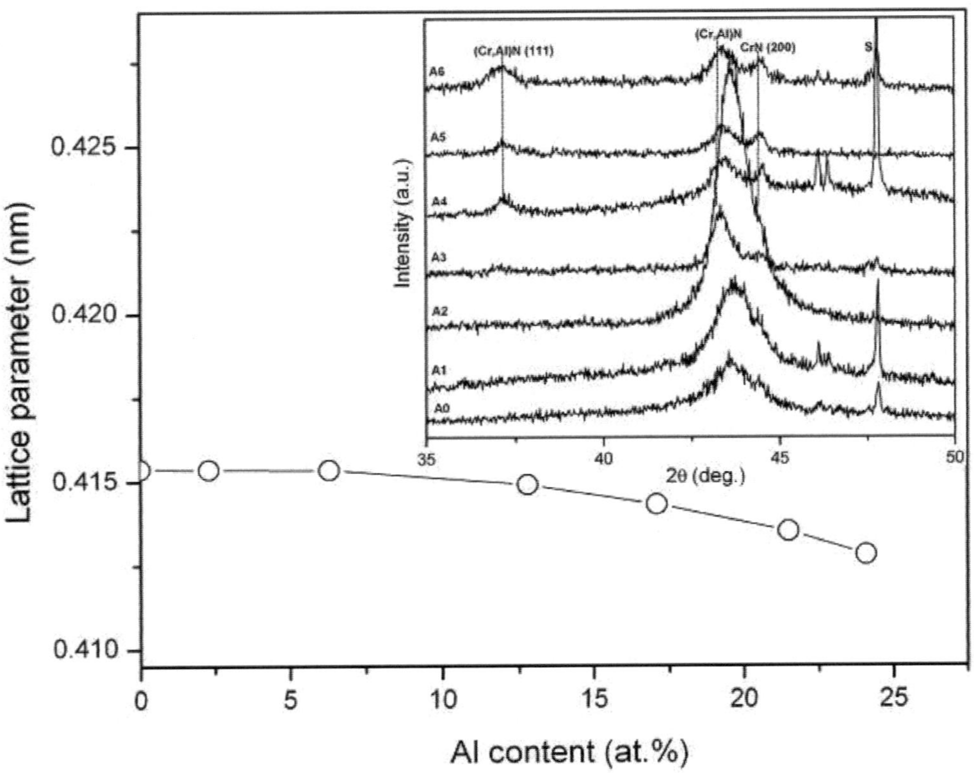

Figure 1. Dependence of lattice parameter in CrAlN films against Al content. Inset shows XRD patterns of pure CrN (A0) and CrAlN (A6) nanocomposite film with Al content of 24.1 at.% (S represents the substrate peak).

All deposited films exhibit polycrystalline structure. A B1-NaCl type cubic structure is observed in all the films [16-18]. Pure CrN film (A0), like all other films, prepared at substrate bias voltage -80V shows orientation of CrN (200) plane, which is a low intensity reflection. However, the broad peak indicates that the deposited film is not well crystallized at all and mostly amorphous at low substrate bias voltage, which is consistent with the previous report [19]. No other crystalline phase is identified in the XRD pattern of this film. In all our films, the orientation of the CrN (200) plane is due to the lowest surface energy of the crystal plane, similar to the results previously reported [20]. When the energy of incident ion bombardment increased during deposition, the CrN (200) plane was preferentially grown in order to reduce the surface energy [21]. The orientation (200) for A0 film, is identified at ~43.5°, which is lower than the standard powder diffraction peak according to PCPDFWIN Version 2.3, JSPDF-ICDD (2002), suggesting occurrence of some stresses in the film. The diffracted (200) peaks obtained from film A3 with 12.8 at.% of Al content and film A6 with 24.1 at.% of Al are identified at ~43.6° and ~43.7°, respectively. It is noted that addition of Al results in enhanced crystallization, which is evidenced by the increase in CrN (200) peak intensities. It is also noted that a peak split occurs from film A3 onwards, resulting in a mixed peak consisting of (Cr,Al)N and CrN(200). Increasing Al content slightly shifts (200) peak towards a higher diffraction angle. Such a deviation towards a higher diffraction angle between the standard CrN (200) peak and that of CrAlN films is attributed to decrease of a lattice parameter in CrAlN films. The obvious gradual decrease in lattice parameter from 0.415 nm (A0 film) to 0.412 (A6 film) can be seen in Fig. 1, confirming the integration of Al atoms into cubic CrN lattice. The reason being,the covalent radius of Al (0.121 nm) is smaller as compared to Cr (0.139 nm) and makes possible to initiate the crystal lattice distortion (lattice contraction) which is in good agreement with previous studies [22, 23]. Also, when Al content is ≥ 17.1 at.%, the CrN (200) peak intensities are significantly decreased implying that crystallites in the film became smaller and moreover, the CrN (111) peak with a weak intensity appears at ~37.2°. At Al content of 24.1 at.%, the peak intensity of CrN (111) is found to increase. It has been reported [22] that since increasing Al substitution causes lattice distortion resulting in an increase in the film's strain energy, thus to combat the increase in the strain energy, a tendency of change of preferred orientation from (200) to (111) is seen. Moreover, for the film with Al content of 24.1 at.% (A6), there is no evidence of the presence of *w*-AlN (B4-Wurtzite type) structure [18].The reason is that there is a maximum solubility of cubic Al ~ 77 at.% in cubic CrN in order to see the crystal structure change from B1-NaCl to B4-wurtzite (which is AlN phase) [16-18]. Al contents in our films are much less than the maximum solubility limit. But such a change in crystal structure has been also reported [17, 18] at much lower Al content (58 at.%). In our case, the fact that the observed peak (200) broadening as well as splitting in the films is very prominent, suggesting probable phase formation of either *c*-AlN and *c*-CrN or combination of other (Cr,Al)N phases or sub-stoichiometric grain formation.

3.2 Chemical analysis by XPS and Microstructure evaluation by HRTEM analysis

Figure 2 shows Cr 2p, N 1s and Al 2p XPS core-level spectra of pure CrN (A0) and CrAlN (A6) films to illustrate the development of chemical states in the films due to Al addition. The metal Cr peak consists of two peaks originating from Cr 2p3/2 and Cr 2p1/2 bonding states. Fig. 3(a) is the deconvolution of the Cr 2p3/2 peak of A0 film, indicating the existence of two peaks centered at 575.0 eV representing a CrN phase whereas peak at 576.6 eV occurs because of the existence of Cr_2O_3. Peaks related to pure Cr at 574.3 eV and Cr_2N at 574.5 eV were not present [24]. The N 1s spectra of A0 film in Fig. 3(b), shows the existence of CrN peak at 396.8 eV and a weaker peak at 398.3 eV which can be related to surface oxidation [25]. The XPS core-level spectra of CrAlN film (A6, with al content of 24.1 at.%) are shown in Fig. 3(c). The peak related with Cr was centered at 575.0 eV originating from Cr 2p3/2. The deconvolution of Cr 2p3/2 peak showed that it contained three peaks positioned at 575.0 eV, representing a CrAlN phase, peak at 575.7 eV being a CrN phase and 577.5 eV representing a Cr_2O_3 phase, respectively. Metallic Cr and Cr_2N were once again not observed [24]. The Al 2p spectra in Fig. 3(d) showed characteristic peaks at binding energy 73.5 eV originating from CrAlN, whereas peak at 74.3 eV is from AlN and the peak at 77.6 eV represents Al_2O_3 phase. In Fig. 3(e) the deconvoluted spectra of N 1s showed a typical peak of CrN at 396.21 eV along with weaker peak at 397.4 eV that could be related to AlN [23, 26].

TEM observations were performed to determine the microstructural difference and phase formation of the films due to addition of Al. Figure 3 depicts the plane-view TEM images and the insets are corresponding selected area diffraction patterns (SAED) of pure CrN (A0) and CrAlN (A6) films. The inset in Fig. 3(a) shows diffused scattering, producing a continuous ring pattern indicating lack of significant crystallinity, which is in agreement with the XRD results. Basically, can be understood as, a microstructure that comprises of crystallinity and amorphous region that co-exists. A continuous ring is a consequence of the fine and small size of grain distribution whereas diffuse scattering is a consequence of mostly amorphous AlN_x region in the film. Fig. 3(b) shows a HRTEM image of CrAlN (A6) film with Al content of 24.1 at.%. The corresponding electron diffraction rings are shown in the inset. These diffraction rings show sharp diffraction spots in a circle typical for polycrystalline material. Moreover, the Cr(Al)N solid solution which is a nanocrystalline phase has already been seen in our XRD and XPS results. The amorphous phase AlN is marked by white circles in the TEM image. On comparing the two SAED inset images in Fig. 3, we can see that there is a marked difference between diffraction ring patterns of both CrN (A0) and CrAlN (A6) films.

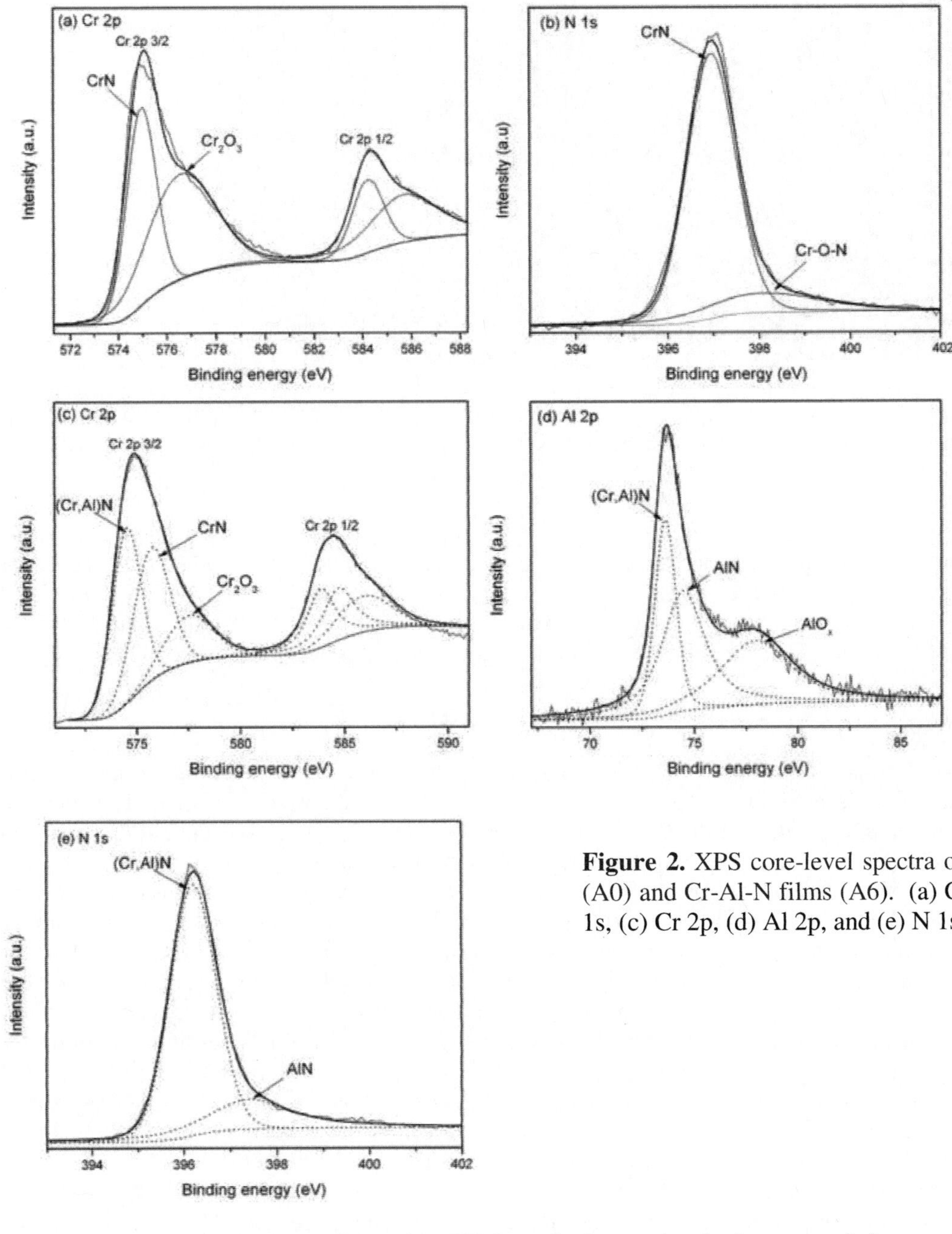

Figure 2. XPS core-level spectra of pure CrN (A0) and Cr-Al-N films (A6). (a) Cr 2p, (b) N 1s, (c) Cr 2p, (d) Al 2p, and (e) N 1s.

Such an observation was possible with TEM method as a local character of the material can be achieved when compared with XRD or XPS results. The SAED pattern in the inset of Fig. 3(b) clearly shows the diffraction rings for cubic phase nano-crystalline structure, which can be understood as solid solution of Cr(Al)N matrix embedded with few amorphous region (AlN). The film with Al content of 24.1 at.% in Fig. 3(b) shows a grown crystalline phase with grain sizes of ~11 nm, in good agreement with our XRD peak width results (calculated average grain size of ~10.8 nm).

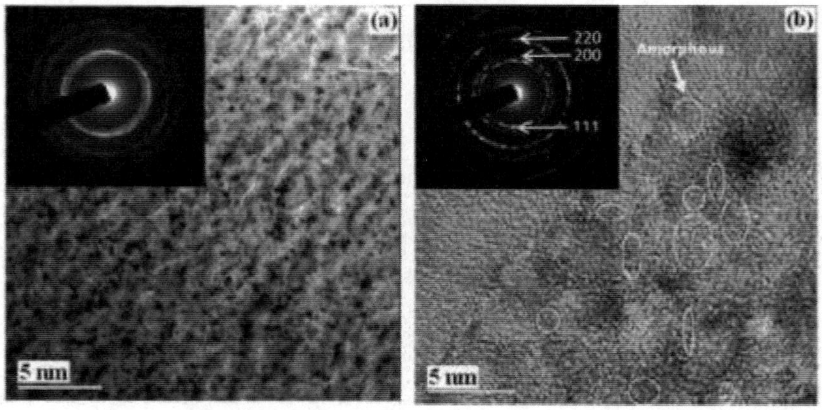

Figure 3. (a) Plane-view TEM image of pure CrN film A0 with SAED pattern in the inset; (b) HRTEM image of CrAlN film A6 with SAED pattern in the inset depicting amorphous/crystalline microstructure.

It is noted that although the Al content is only 24.1 at.%, but still promotes amorphous/crystalline microstructure. Also, to quite an extent, the amorphous/crystalline microstructure shows distinction of crystalline and amorphous phases that occur in this film. Low concentration of Al atoms will either substitute Cr atoms forming solid solution or having an interstitial substitution due to small atomic radii of Al. In addition, distinctive phase separation might be also possible with Al content of 24.1 at.% in the film.

3.3 Nanohardness

The mechanical properties of the CrAlN films were significantly affected by Al concentration. Nano-hardness and Young's modulus of the films as a function of Al content are shown in Table 1. The hardness and Young's modulus of the CrN (A0) film are 18.8 and 225 GPa, respectively. For the film A2 with 6.3 at.%, both hardness and Young's modulus values increased to 22.5 and 242 GPa. Evidently, the solid solution hardening effect contributed to the increase in the hardness of the film [27]. When the Al content was increased to 24.1 at.%, the hardness and modulus of the film further increased to maximum value of 24 GPa and 255 GPa, respectively. It is reported that the effects of solid solution hardening as well as nanocomposite interface strain strengthening mechanisms [1,2] are considered together for the hardness enhancement in the film. Increase in Al content has resulted in the decreases of the grain size too from 11.8 to 10.8 nm (based on our XRD results), consequently increasing the hardness which also suggests that high Al concentration in the film supported crystalline grain formation with less volume fraction of AlN amorphous regions. The ratio of hardness to Young's modulus *H/E* is considered as an important parameter that reflects the fracture toughness and wear properties of the films [28, 29]. The *H/E* ratio against Al content shows an analogous progression with films hardness. The film with Al content of 6.3 at.% has the highest *H/E* ratio 0.097. The hardness of A6 film is low as compared to the previously published result [30] reporting ~36 GPa (with 30 at.% of Al). In some regions, interdiffusion between the CrN and CrAlN phases could have led to comparatively a lower hardness. Previous report [26] explained that the presence of an interface energy difference between the crystallite/amorphous phase and grain boundary leads to the formation of amorphous phase which is settled at the grain boundary and can be seen in Fig. 3(a). For that reason, the mechanical properties of CrAlN films are superior than pure CrN film (A0) film, respectively.

4. Conclusions

At substrate bias voltage of -80 V and by varying target current I_{Al}/I_{Cr} ratio, a set of CrAlN films consisting ofvarious Al concentrations were successfully deposited by dc magnetron sputtering technique. All films exhibited B1 NaCl structure. The changes in the film chemistry due to the addition of Al in cubic CrN were investigated by XRD, XPS, TEM and nanoindentation tests. The results showed that increasing target current I_{Al}/I_{Cr} ratio during deposition produced films with different Al contents, which affect the structure, phase, microstructure and hardness of the films. XRD data of CrN film showed orientation along (200) and also revealed the transition from amorphous to promotion of crystallinity in the films due to Al addition. Cr-Al-N films crystallize in cubic structure, with a preferred (200) orientation for all. The XRD results showed that fcc CrAlN films form a solid solution whereby Al atoms substitute Cr atoms. Results of XPS and TEM show clear incorporation of Al into CrN matrix resulting innanocomposite microstructure comprising of nanocrystallites and amorphous regions. The nano-hardness increased to 24 GPa with the increase in Al concentration. Maximum hardness obtained for CrAlN film is 22% superior to pure CrN film.

Acknowledgement
This work was supported by the Research Grant Council of the Hong Kong Special Administra-tive Region, China [Project No. CityU 120611 (9041679)]. YGS and ZFZ are very grateful to Dr. Y.F. Han for the contributions made in the early stages of this investigation.

References
[1] Chang YY, Chiu WT, Hung JP 2016 Surf. Coat. Technol. **303** 18.
[2] Park I W, Kang DS, Moore JJ 2007 Surf. Coat. Technol. **201** 5223.
[3] Zhang YJ, Yan PX, Wu ZG, Xu JW, Zhang WW, Li X, Liu WM, Xue QJ 2017, Tribol. Int. **115** 591.
[4] Creus J, Indriss H, Mazille H, Sanchette F, Jacquot P 1998 Surf. Coat. Technol. **107** 183-190.
[5] Gwang KS, Sang YL 2002 Surf. Coat. Technol. **171** 91.
[6] Yu C, Wang S, Tian L 2009 J Mater Sci **44** 300.
[7] Li W, Zheng KP, Liu P, Zhu PW, Zhang K, Ma FC, Liu XK, Chen XH, He DH **2016** Mat. Character. **118** 79.
[8] Martinez E, Sanjines R, Karimi A, Esteve J, Lévy F 2004 Surf. Coat. Technol. **180** 570.
[9] Sabitzer C, Paulitsch J, Kolozsvari S, 2016 Thin Solid Films, **610** 26.
[10] Ding XZ, Tan ALK, Zeng XT, Wang C, Yue T, Sun CQ 2008 Thin Solid Films **516** 5716.
[11] Tillmann W**,** Stangier D**,** Schroder P 2016 Surf. Coat. Technol. **308** 147.
[12] Long Y, Zeng J, Donghai Y, Wu S 2014 Ceramics International **40** 9889.
[13] Heim D, Holler F and Mitterer C 1999 Surf. Coat. Technol. **116** 530.
[14] Beliardouh NE, Bouzid K, Nouveau C, Tlili B, Walock MJ 2014 Tribol. Inter. **82** 443.
[15] Chen X, Yeting X, Jie M, Pang X, Yang H 2016 Journal of Alloys and Compounds, **665** 210.
[16] Lin J, Mishra B, Moore JJ, Sproul WD 2006 Surf. Coat. Technol. **201** 4329.
[17] Makino Y, Nogi K 1998 Surf. Coat. Technol. **98** 1008.
[18] Sugishima A, Kajioka H, Makino Y 1997 Surf. Coat. Technol. **97** 590.
[19] Yanhong LV, Li J, Liu X, Li H, Zhou H, Chen J 2012 App. Surf. Sci. **258** 3864.
[20] Lin J, Moore JJ, Mishra B, Pinkas M, Sproul WD, Rees JA 2008 Surf. Coat. Technol. **202** 1418.
[21] Abadias G 2008 Surf. Coat. Technol. **202** 2223.
[22] Wang L, Zhang G, Wood RJK, Wang SC, Xue Q 2010 Surf. Coat. Technol. **204** 3517.
[23] Li Z, Munroe P, Jiang ZT, Zhao X, Xu J, Zhou ZF, Jiang JQ, Fang F, Xie ZH 2012 Acta Materialia **60** 5735.
[24] Emery C, Chourasia AR, Yashar P 1999 J. Electron Spectrosc. Relat. Phenom. **104** 91.
[25] Lippitz A, Hubert T 2005 Surf. Coat. Technol. **200** 250.
[26] Barshilla HC, Selvakumar N 2006 Surf. Coat. Technol. **201** 2193.
[27] Liu ZJ, Shen YG 2004 Acta Mater. **52** 729.
[28] Musil J 2000 Surf. Coat. Technol. **125** 322.
[29] Khlifi K, Dhiflaoui H, Zoghlami L 2015 J. Mater. Eng. Perfor. **24** 4077.
[30] Tlili B, Mustapha N, Nouveau C, Benlatreche Y, Guillemot G, Lambertin M 2010 Vacuum **84** 1067.

IOP Conf. Series: Materials Science and Engineering **307** (2018) 012080 doi:10.1088/1757-899X/307/1/012080

Influence of Cooling Condition on the Performance of Grinding Hardened Layer in Grind-hardening

G C Wang[1,2], J Chen[1], G Y Xu[1] and X Li[1]

[1]School of Mechanical Engineering, Jiangsu University, Zhenjiang, Jiangsu 212013, China

[2]Silicon Lake College of Vocational Technology, Kunshan, Jiangsu 215332, China

Abstract: 45# steel was grinded and hardened on a surface grinding machine to study the effect of three different cooling media, including emulsion, dry air and liquid nitrogen, on the microstructure and properties of the hardened layer. The results show that the microstructure of material surface hardened with emulsion is pearlite and no hardened layer. The surface roughness is small and the residual stress is compressive stress. With cooling condition of liquid nitrogen and dry air, the specimen surface are hardened, the organization is martensite, the surface roughness is also not changed, but high hardness of hardened layer and surface compressive stress were obtained when grinding using liquid nitrogen. The deeper hardened layer grinded with dry air was obtained and surface residual stress is tensile stress. This study provides an experimental basis for choosing the appropriate cooling mode to effectively control the performance of grinding hardened layer.

Keywords: Grind-hardening; Cooling condidtion; Microstructure; Hardness; Residual stress

1. Introduction

In the process of grind-hardenimg, workpiece surface temperature increases to austenitizing temperature in short time due to violent extrusion and sliding motion between grinding wheel and workpiece and then were quickly cooled down,so that workpiece surface layer transform to martensite[1].Cooling is an essential part of the grind-hardening process, and cooling mode has an important impact on the formation and performance of grinding hardened layer. Therefore, based on the grind-hardening process, experiments were conducted in this research, in order to study the influences of three different cooling media, such as emulsion, dry air and liquid nitrogen, on the microstructure and properties of the hardened layer.This study provides theoretical and experimental

IOP Conf. Series: Materials Science and Engineering **307** (2018) 012080 doi:10.1088/1757-899X/307/1/012080

guidence for selecting appropriate cooling method and effectively controlling the quality of grind-hardening process.

2. Cooling methods and features

The purposes of cooling in grinding surfece between grind-hardening process and ordinary grinding are different .Ordinary grinding process aims to reduce the grinding temperature for avoiding grinding burn, so the jet cooling area is set in front of wheel-workpiece contact surface, as shown in Figure 1a, to facilitate the cooling medium into the arc zone [2] .In grinding hardened process, grinding energy is used to heat the workpiece, and then cooling, therefore the machined surface behind the wheel-workpiece contact surface is regared as the cooling area, as shown in Figure 1b. Providing cooling medium for the machined surface of the workpiece aim to obtain hardened surface and reduce temperature of the grinding wheel for longer service life [3].

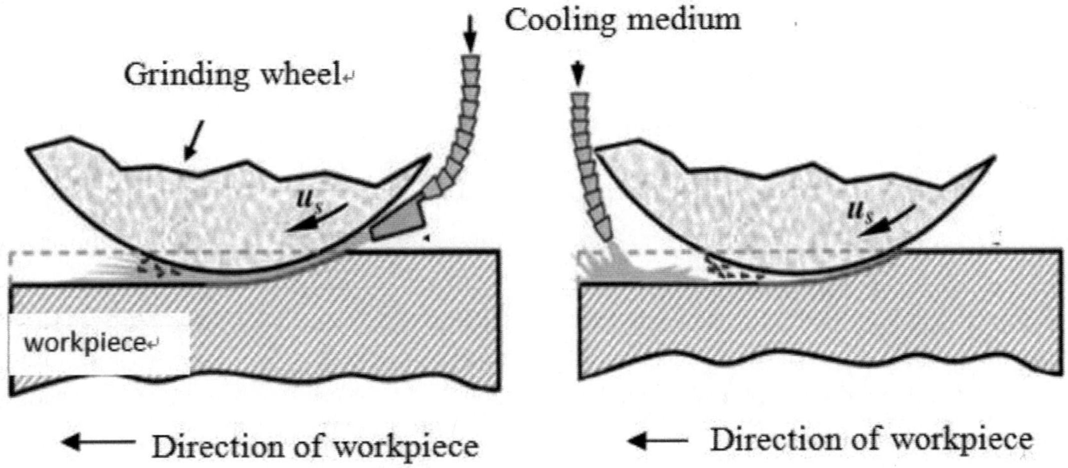

Figure 1. Comprision between jet area of cooling
(a) Ordinary grinding (b) Grind-hardening

The purposes of quenching are to obtain martensite, but also to prevent the decomposition of undercooled austenite.Therefore, greater degree of undercooling is needed to endure that cooling rate greater than the critical cooling rate Vc. The curve C shows that the nose temperature is about 500℃ ~ 600℃. Therefore, there is no need for rapid cooling when the workpiece is cooled below nose temperature of the curve C. In addition, slowly cooling process in the martensitic transformation zone can reduce the structure stress ,contributing to small deformation and fewer cracking [4]. Therefore, hardening medium in grind-hardening should be characterized by rapid cooling at medium temperature region (500-600℃) and slow cooling at low temperature area (Figure 2). In addition, the cooling medium should also meet these requirements that there are no or less harmful oxidation, reduction reactions or other physicochemical reactions between cooling medium and workpiece material, the workpiece surface is not contaminated or less contaminated. It should ensure that it is easy to operate and recycle, non-toxic, low cost. Commonly used mainly . emulsion, dry air and liquid nitrogen are three main kinds of grind-hardening cooling medium.

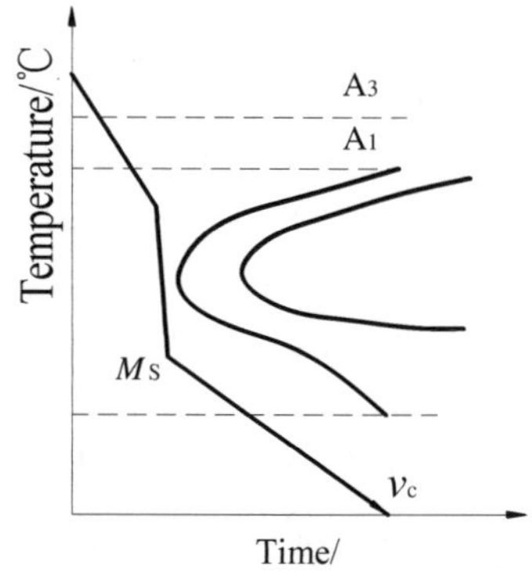

Figure 2. Cooling curve of ideal quenching medium

Emulsion is the mixture of water and emulsified oil, which contribute to lower cooling rate, increasing stability of the steam film and boiling stage happened in the low temperature area. Emulsion is not easy to degenerate and has longer service time and good rust.

Air medium has slower cooling rate and boiling temperature of air is much higher than that of workpiece. Therefore, under normal circumstances, there are no boiling and material changes when cooling, but also, cooling curve shows a steady state. At high temperatures, radiant heat is its main form; convective heat transfer is its predominant form at low temperatures. By using compressed air, cooling rate can be increased several times, better cooling uniformity and hardness uniformity are obtained .

Cooling strength and rate of liquid nitrogen are much greater than that of the emulsion.Cooling process using liquid nitrogen does not result in three stages of thermal shock compare to ordinary medium, and the possibility of workpiece deformation and crack are much lower. In addition, martensitic transformation proceeds more completely when quenching using liquid nitrogen, with minimal retained austenite and martensitic microstructure. Higher hardness,wear resistance and dimensional stability are obtained.

3. Grinding experimental conditions and parameters

3.1. Experimental Materials

AISI1045 with a size of 45×19.5×19.5 mm was selected as the experimental material. The heat treatment method is a anneal above 600 ℃ in vacuum. The initial surface residual stress is -10.00±8.00 MPa (longitudinal direction) and -10.39±8.00MPa (transverse direction), and the hardness is HV0. 5 172.9 ± 11.1.

3.2. Grinding process conditions

Grind-hardening cooling experiments were performed on the Minini Junior 90 CF M286 CNC grinding machine and the workpieces were ground using BWA60MVA1 grinding wheel. The grinding wheel speed is vs=23m/s, the workpiece feed rate is vw=400mm/min and the cutting depth is ap=20μm. Cooling experiments were conducted by using three different cooling method ,including dry air with an ambient temperature of 20℃,liquid nitrogen(Q=2.18L/min) And an emulsion (Noritake SA-02 (1:60), Q=18.8 L/min).

IOP Conf. Series: Materials Science and Engineering **307** (2018) 012080 doi:10.1088/1757-899X/307/1/012080

3.3. Texture and performance testing

The hardness of the grinding surface was measured using a microhardness tester Shimadzu Seisakusho NT-M001. The ground surface topography was observed with a scanning electron microscope (SEM).Gradient residual stress were obtained by using the Movipol-3 electrolytic polishing machine and the X-ray stress analyzer MSF-3M. Transmission electron microscopy (TEM) was used to observe the microstructure of the ground surface.

4. Test results and analysis

4.1. Influence of cooling medium on the surface morphology

Figure 3 shows different surface morphologies hardened with different cooling media. As shown in figure 3a, the workpiece surface cooled by emulsion is a continuous slot with the surface roughness of Ra=0.25μm. The surface was bright and not oxidized. Figure 3b shows grinding surface topography using dry air as cooling medium, workpiece surface appears a darker oxidation color with a surface roughness of Ra=0.98μm. It is obvious that the split groove caused by the lack of lubrication, and spherical particles which is the form of melting grindings in grinding surface , further reducing the grinding surface quality [5]. Grinding surfaces using liquid nitrogen as cooling conditions are showed in Figure 3c, the specimen surface is bright without any oxidation or burns and the surface roughness of Ra=0.27μm, indicating that oxidation is inhibited and liquid nitrogen played an important role to protect the surface.

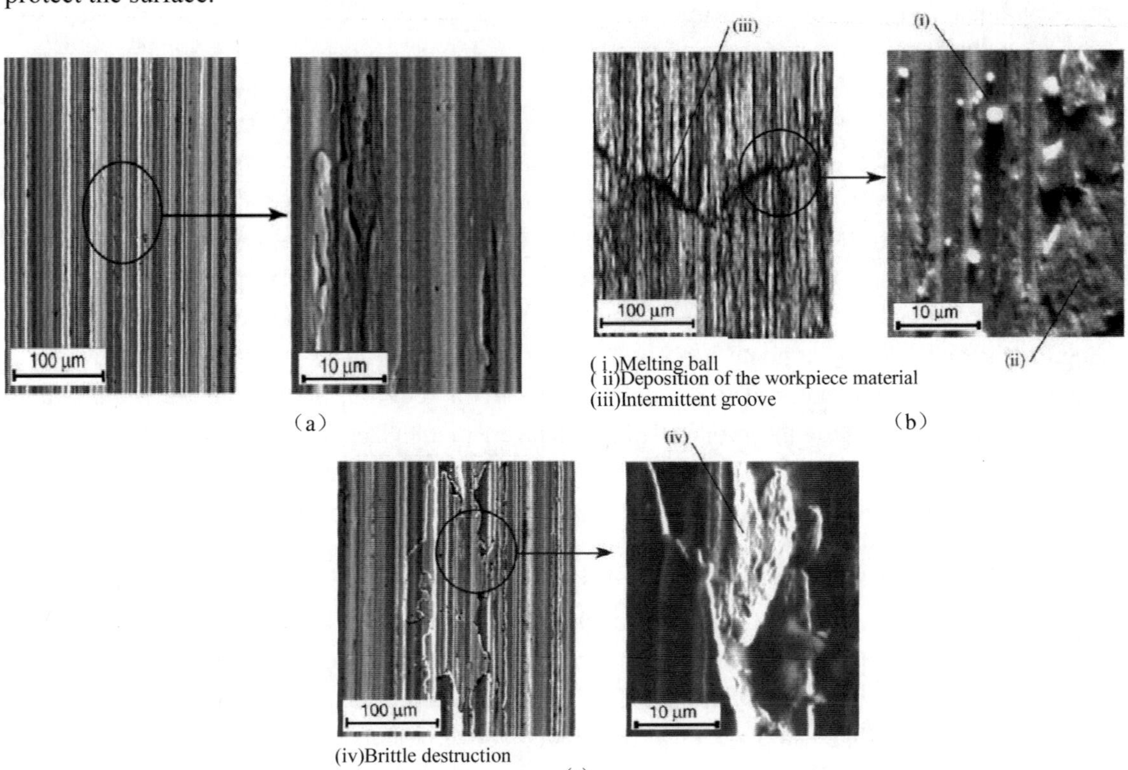

Figure 3. Workpiece surface topograph in different cooling conditions
(a)Emulsion (b)Dry air (c)Liquid nitrogen

IOP Conf. Series: Materials Science and Engineering **307** (2018) 012080 doi:10.1088/1757-899X/307/1/012080

4.2. Influence of cooling medium on the microstructure

Optical micrographs of subsurface regions of the specimen under different cooling conditions were studied. The hardened layer did not occur under emulsion cooling conditions and the material microstruucture is regard as pearlite.

Grind-hardening layers were present in the test pieces under dry air or liquid nitrogen cooling conditions. In the dry air cooling condition, grind-hardened layer is irregular martensite with the depth of 350μm. Compared with the lath martensite produced by ordinary heat treatment, the structure of this martensite layer is finer [6]. This is due to the fact that grinding-hardening is the combined effect of grinding heat and grinding forces during grinding process. As the grinding heat transfer in the sample, the temperature of contact area between grinding wheel and workpiece surface reaches above Ac3 which contribute to austenization,martensite generated during cooling process ,in the meanwhile,workpiece surface deformed due to grinding force.The dislocation density also increases while the grain is shredded, which leads to grain refinement [7]. At the bottom of the hardened layer, there are many ferrite grains with carbides precipitated at the boundaries.

With liquid nitrogen cooling condition, grinding hardened layer will be thinner with depth of only 85μm. The reasion is that the reduction of the surface heating temperature and the increase of the temperature gradient sharply reduce the area at a temperature above AC3, which eventually leads to a corresponding decrease in the depth of the hardened layer. The martensitic structure at the bottom of the hardened layer is smoother and it is difficult to distinguish it from hardened layer under dry air cooling condition using optical microscope.

4.3. Influence of cooling medium on the microhardness

The microhardness of the hardened layer was not changed (HV(500g)=200) under the cooling condition of the emulsion. Under dry air conditions, the microhardness of the grinding hardened layer increased to HV(500g)=750, which is 3.75 times higher than that before processing and 1.15 times of ordinary martensite(HV(500g)=650). The hardness of the grinding hardened layer shows a gradient change. The hardness at the bottom of the hardened layer decreases to HV(500g)=400 due to its changes of microstructure to a ferrite-carbide arrangement. Under the condition of liquid nitrogen cooling, the hardness of the grinding hardened layer is the highest, up to HV(500g)=1100, which is 1.46 times of the microhardness of the sample in dry air. In addition, the microhardness decreases slightly with the thickness of the ground hardened layer, martensite occurs at the bottom of the hardened layer with a hardness as high as HV(500g)=700. Liquid nitrogen cooling has a greater gradient than the air-cooled temperature field, more complete martensitic transformation and very little retained austenite result in higher hardness.

4.4. Influence of cooling medium on the residual stress

The distribution of residual stress on the subsurface of the grinding workpiece is studied. Residual stresses σxx and σyy are respectively measured along the grinding direction and the vertical direction. Under the condition of emulsion cooling, the workpiece is residual compressive stress in both directions within a depth of 100 μm, with a size of 200-400 MPa. In dry air, in the level of the workpiece hardened layer depth (400μm), residual tensile stress occurs in the direction of XX and YY, with the magnitude of 100 ~ 200MPa. The residual compressive stress generated by the liquid nitrogen cooling grinding of the workpiece is 200 MPa, distributed at a depth of 85 μm, which is similar to the emulsion cooling. Grinding hardened layer residual stress is the result of the combined effect of mechanical stress, thermal stress and phase transformation stress [8]. When using the emulsion as cooling medium , the thermal stress decreases, and the surface residual stress value develops toward the direction of compression, which is the main reason that the surface has a large residual compressive stress.

5. Conclusion

(1) Using dry air as the cooling medium, the workpiece surface roughness is much larger and oxidation occurs, and the use of emulsion and liquid nitrogen as a cooling medium, the workpiece surface roughness is smaller and no oxidation occur.

(2) In the condition of emulsion cooling, the workpiece surface structure is pearlite and no hardened layer. in dry air or liquid nitrogen cooling conditions ,the surface structure of the workpiece is martensite and hardened layer occurs. Compared to liquid nitrogen cooling condition,the depth of the hardened layer on the workpiece surface under the condition of dry air cooling is deeper.

(3) The hardness of the surface layer after the workpiece was cooled by emulsion is not changed. The hardness of the workpiece hardened layer under liquid nitrogen cooling was harder than that of the ground hardened layer under the condition of dry air cooling, and the hardness of the former did not change obviously with the depth of the hardened layer , The latter hardness changes obviously with the depth of hardened.

(4) In the emulsion and liquid nitrogen cooling conditions, the residual stress in the hardened layer of the workpiece are residual compressive stress,but tensile stress in dry air conditions,which will lead to the workpiece fracture, fatigue failure , corrosion and wear.

Acknowledgment

This research was supported by the Natural Science Foundation of P. R. China (No.51075192), National key projects of P. R. China（2013ZX04009031）and A Project Funded by the Suzhou City key laboratory of elevator safety technology.

References

[1] J D Liu,G C Wang 2003 Modern Manufacturing Engineering 11pp81-83
[2] J X Ren,D A Hua 2011 Grinding Principle(Beijing:Publishing House Of Electronics Industry
[3] J D Liu,G C Wang 2004 Tooling Engineering 37pp11-14
[4] G C Wang 2015 Grind-Hardening Technology(Beijing:National Defence Industry Press)
[5] L Q Wu,Y Qin,D Qian 2013 Ordnance Material Science and Engineering 36pp138-140
[6] J D Liu,G C Wang 2004 Tool Engineering 38pp67-69
[7] K M Liu,Z Ma 2012 Hot Working Technology 41pp215-217
[8] J D Liu,D P Hou 2008 Heat Treatment of Metals 33pp127-130

IOP Conf. Series: Materials Science and Engineering **307** (2018) 012081 doi:10.1088/1757-899X/307/1/012081

Influence of the sintering temperature on morphology and particle size of silver synthesized by spark discharge

A A Lizunova, A A Efimov, P V Arsenov, V V Ivanov

Research Centre for Functional Materials, Moscow Institute of Physics and Technology, Dolgoprudny 141700, Russia
E-mail: efimov.aa@mipt.ru

Abstract. The morphology and particle size of silver were investigated depending on the sintering temperature in the range from 25 to 1000 °C. Silver nanoparticles synthesized by the multi-spark discharge generator and then sintered in a tube furnace. It is found that the particle size decreases irregularly with increasing sintering temperature, according to TEM and aerosol measurements. Individual spherical nanoparticles are formed at the sintering temperature more than 200 °C. The size of the particles is almost constant of about 45 nm in the temperature range of sintering of 200-900 °C. However, the particle size is significantly reduced to 7 nm at a sintering temperature of about 1000 °C. Thus, it was found that the variation in sintering temperature makes it possible to significantly change the morphology and size of synthesized nanoparticles, which is relevant for a number of applications in optics and medicine.

1. Introduction

It is known that nanoparticles have unique properties that differ from the properties of bulk materials [1]. In this connection, nanoparticles are widely studied and used in various fields of science and technology. For example, silver nanoparticles have unique optical properties due to surface plasmon resonance [2]. This resonance can be tuned to the necessary wavelength by changing the particle size and shape [3]. Such a reorganization is necessary for several interesting applications in Raman spectroscopy [4], optical data storage [5], biosensors [6], plasmon solar cells [7], etc. In this connection, there are requirements for the production of nanoparticles with the desired size and shape. Various methods are used to prepare nanoparticles for optical studies, for example, ion implantation [8], magnetron sputtering [9], wet chemistry methods [10], etc. These methods usually offer very poor and often quite limited control of the size of the spherical nanoparticles. The aerosol synthesis method, such as the synthesis of nanoparticles by a spark discharge, allows a pretty flexible control over the size of the produced particles. However, the shape of the produced particles has a fractal-like structure and a small primary particle size of less than 10 nm [11,12]. In this connection, there is a need to study ways of changing the shape of the particles to spherical and coarsening of particles. Previously, it is known that the shape and size can be changed with a tube furnace [13], but most of the research is not related to spark discharge. In this work we study the change in the size and shape of particles from a spark generator using a tube furnace.

2. Experimental

Figure 1 shows a schematic diagram of the experimental setup. Silver nanoparticles were generated using the multi-spark discharger generator (m-SDG) as a result of electrical erosion of silver electrodes in argon. The mode of the m-SDG was as follows: the energy stored in the capacitor, the repetition rate of discharges and the gas flow rate were 3.9 J, 4 Hz and 1.0 l/min, respectively. After the m-SDG, the aerosol passed through a custom-build tube furnace, which was used for sintering of the nanoparticles. The inner diameter and the length of the tube furnace were 30 and 500 mm, respectively. The particle size distributions, morphology, crystal structure of nanoparticles were investigated using transmission electron microscope (TEM) JEOL JEM2100. The obtained aerosol nanoparticles were deposited on a carbon film TEM grid using an electrostatic sampling device NAS 3089, TSI Inc. Real-time measurements of size and concentration of the nanoparticles in the aerosol flow were determined by an aerosol spectrometer SMPS 3936, TSI Inc.

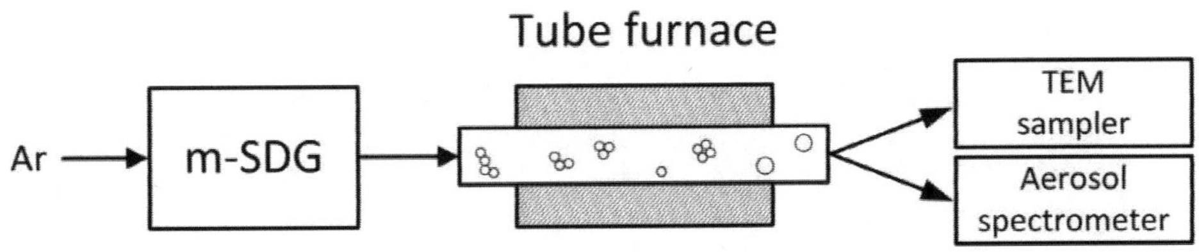

Figure 1. Experimental setup to study the effect of sintering on the amorphology and size of aerosol nanoparticles using a tube furnace.

3. Results and discussions

The received dependences of the effective sizes of aerosol particles on the temperature of the furnace are shown in figure 2. The size of silver particles decreases from 150 to 17 nm with increasing temperature from 25 to 1000 °C, according to the data of the aerosol spectrometer. At the same time, it is observed, that in the temperature range from 250 to 900 °C, the size of nanoparticles is almost constant at about 70 nm and in the temperature range of 900-1000 °C was a sharp decline to 17 nm. The TEM data confirm the character of the temperature dependence of the effective particle size in the range 250-1000 °C. However, at temperatures below 250 °C, there was found a significant discrepancy between the results of the TEM and the aerosol spectrometer, which is caused by the existence in the aerosol of a significant fraction of aggregates of nanoparticles and an estimation of their sizes under the assumption of sphericity in the TEM analysis.

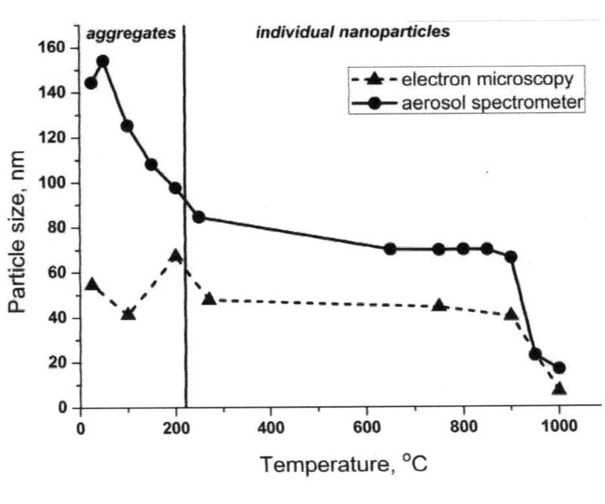

Figure 2. Dependence of the particle size on the sintering temperature, according to data from a transmission electron microscope and an aerosol spectrometer.

Typical TEM images of silver nanoparticles obtained at different temperatures of a tube furnace are shown in Figure 3. At temperatures below 200 °C, there was a prevalence of indestructible aggregates of fractal-like structure nanoparticles with a crystallite size of 10-45 nm, more than 200 °C - individual nanoparticles with a shape close to spherical. Thus, the analysis of TEM images showed that individual particles are absent in aerosols synthesized at temperatures of 25 and 100 °C, at 200 °C the fraction of aggregates is 35%, at 270 °C less than 10%. The average diameter of particles received at temperatures of 270-900 °C is about 45 nm, at 1000 °C - 7 nm. Based on high-resolution electron microscopy image and electron diffraction from the selected area diffraction pattern, the resulting particles are crystalline and match to the cubic phase of silver. Besides the primary large spherical particles and aggregates, a small fraction of silver clusters with dimensions of 2-4 nm is observed at all temperatures, which are mainly localized around the non-spherical particles and in the space between the particles in the aggregates (figure 3b) in the samples synthesized at 200 and 270 °C.

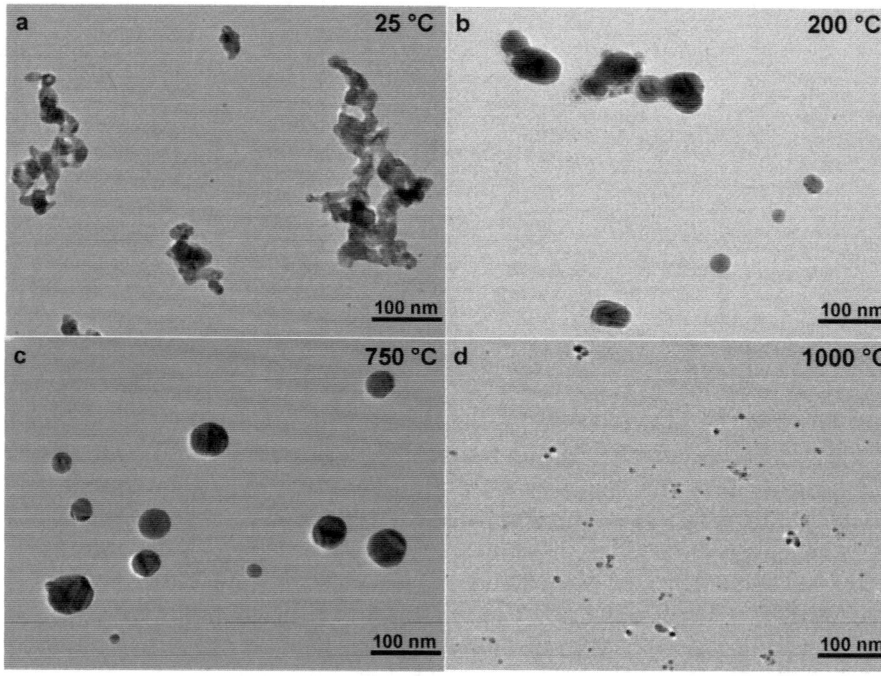

Figure 3 (a, b, c, d). The morphology of silver nanoparticles obtained by a spark discharge at the following sintering temperatures: (a) 25 °C, (b) 200 °C, (c) 750 °C, (d) 1000 °C.

4. Conclusion

The use of a tube furnace to heat the streams of aerosol particles synthesized in a spark generator makes it possible to vary the average dimensions of individual spherical crystalline silver particles from 7 to 45 nm with a temperature change from 1000 to 250 °C. In addition, a change in the sintering temperature in the range of 25-200 °C also makes it possible to produce an aerosol with different morphological composition and a fraction of aggregates of a fractal-like structure with crystallite sizes greater than 15 nm. Thus, it has been found that controlling the sintering temperature makes it possible to change the structure and size of the silver particles in real time and, therefore, to rapidly change their properties. This result is relevant for various applications in the field of optics and medicine.

Acknowledgments

This work was supported by the Russian Science Foundation (project # 15-19-00190).

References

[1] Stark W J, Stoessel P R, Wohlleben W and Hafner A 2015 Industrial applications of nanoparticles *Chem. Soc. Rev.* **44** 5793–805
[2] Iskandar F 2009 Nanoparticle processing for optical applications – A review *Adv. Powder Technol.* **20** 283–9

[3] Kelly K L, Coronado E, Zhao L L and Schatz G C 2003 The Optical Properties of Metal Nanoparticles: The Influence of Size, Shape, and Dielectric Environment *J. Phys. Chem. B* **107** 668–77
[4] Li Y-S, Cheng J and Chung K-T 2008 Surface-enhanced Raman spectroscopy using silver nanoparticles on a precoated microscope slide *Spectrochim. Acta. A. Mol. Biomol. Spectrosc.* **69** 524–7
[5] Hu D, Lin J, Jin S, Hu Y, Wang W, Wang R and Yang B 2016 Synthesis, structure and optical data storage properties of silver nanoparticles modified with azobenzene thiols *Mater. Chem. Phys.* **170** 108–12
[6] El-Dessouky R, Georges M and Azzazy H M E 2012 Silver Nanostructures: Properties, Synthesis, and Biosensor Applications *Functional Nanoparticles for Bioanalysis, Nanomedicine, and Bioelectronic Devices Volume 1* ACS Symposium Series vol 1112(American Chemical Society)pp 359–404
[7] Catchpole K R and Polman A 2008 Plasmonic solar cells *Opt. Express* **16** 21793–800
[8] Vytykacova S, Svecova B, Nekvindova P, Spirkova J, Mackova A, Miksova R and Böttger R 2016 The formation of silver metal nanoparticles by ion implantation in silicate glasses *Nucl. Instrum. Methods Phys. Res. Sect. B Beam Interact. Mater. At.* **371** 245–50
[9] Maréchal N, Quesnel E and Pauleau Y 1994 Silver thin films deposited by magnetron sputtering *Thin Solid Films* **241** 34–8
[10] Iravani S, Korbekandi H, Mirmohammadi S V and Zolfaghari B 2014 Synthesis of silver nanoparticles: chemical, physical and biological methods *Res. Pharm. Sci.* **9** 385–406
[11] Charvet A, Bau S, Coy N E P, Bémer D and Thomas D 2014 Characterizing the effective density and primary particle diameter of airborne nanoparticles produced by spark discharge using mobility and mass measurements (tandem DMA/APM) *J. Nanoparticle Res.* **16** 2418
[12] Efimov A, Lizunova A, Sukharev V and Ivanov V 2016 Synthesis and Characterization of TiO2, Cu2O and Al2O3 Aerosol Nanoparticles Produced by the Multi-Spark Discharge Generator *Korean J. Mater. Res.* **26** 123–9
[13] Harra J, Mäkitalo J, Siikanen R, Virkki M, Genty G, Kobayashi T, Kauranen M and Mäkelä J M 2012 Size-controlled aerosol synthesis of silver nanoparticles for plasmonic materials *J. Nanoparticle Res.* **14**

IOP Conf. Series: Materials Science and Engineering **307** (2018) 012082 doi:10.1088/1757-899X/307/1/012082

Dry aerosol jet printing of conductive silver lines on a heated silicon substrate

A A Efimov[1], P V Arsenov[1], N V Protas[1], K N Minkov[2], M N Urazov[1], V V Ivanov[1]

[1]Department of Physical and Quantum Electronics, Moscow Institute of Physics and Technology, Dolgoprudny 141700, Russia
[2]Moscow Institute of Electronics and Mathematics, National Research University of Higher School of Economics, Moscow, Russia

Abstract. A new method for dry aerosol jet printing conductive lines on a heated substrate is presented. The method is based on the use of a spark discharge generator as a source of dry nanoparticles and a heating plate for their sintering. This method allows creating conductive silver lines on a heated silicon substrate up to 300 °C without an additional sintering step. It was found that for effective sintering lines of silver nanoparticles the temperature of the heated substrate should be about more than 200-250 °C. Average thickness of the sintered silver lines was equal to ~20 μm. Printed lines showed electrical resistivity equal to 35 μΩ·cm, which is 23 times greater than the resistivity of bulk silver.

1. Introduction

In the last decade, aerosol jet printing has been actively investigated to create cheap electronic components and devices. This technology allows creating of printed antennas [1], thin-film transistors [2,3], light-emitting diodes [4,5], solar cells [6], and others. In conventional aerosol jet printing technology, functional ink is used as a source of aerosol particles. The use of ink causes several problems related to their preparation, storage, removal of solvents and etc. In this connection, alternative sources of particles for aerosol jet printing are being investigated. For example, the use of dry silver aerosol particles obtained in a spark discharge for printing conductive lines has recently been reported [7]. As a rule, such deposited lines require sintering to reduce contact resistances between the nanoparticles. The sintering process takes additional time and requires moving the sample from the print zone to the sintering zone, which is not convenient in the production process of a large number of printed samples, for example, the conductive buses of solar cells. Thereby, in this paper, it is proposed to combine the printing process and thermal sintering in order to reduce the time for the formation of functional lines. The proposed approach is realized using dry aerosol printing with silver nanoparticle beam on a heated silicon substrate and described in detail in the paper.

2. Experimental

The scheme of the experiment on the formation of conductive silver lines using dry aerosol printing on a heated silicon substrate is shown in figure 1. Silver nanoparticles in size range from 5 to 150 nm were produced using the multi-spark discharge generator [8] as a result of electrical erosion of silver electrodes. The generator parameters were as follows: energy, repetition rate and flow rate were 3 J, 4 Hz and 2 l/min, respectively. Further, the nanoparticles stream was focused into a narrow beam by means of a coaxial nozzle with an aerosol flow rate and a sheath flow rate of 30 and 90 ml/min, respectively. Nanoparticles beam, after focusing, was deposited on a heated silicon substrate stated on a hot-plate. The temperature of the hot plate was regulated in the range from 25 to 300 °C. Lines of silver nanoparticles were printed at a distance from the nozzle to the substrate, the speed of the substrate moving and the number of printing layers equal to: 0.5 mm, 7 mm/min and 5 layers, respectively.

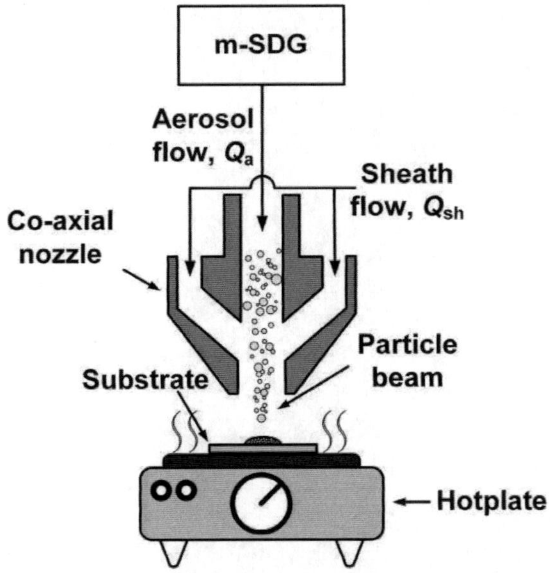

Figure 1. The scheme of the experiment on the formation of conductive silver lines using dry aerosol printing on a heated silicon substrate.

The width and length of the formed lines were measured using an optical and scanning electron microscope. The line profile was measured using an optical profilometer. Electrical resistivity of the printed and sintered lines was measured by four-probe method.

3. Results and discussions

figure 2 shows the dependence of the resistivity of the silver line on the temperature of the substrate at a dry aerosol printing process. Figure 2 shows that within the limits of measurement error and the reproducibility of the printing process the resistivity of the samples does not change significantly and is about 10^8 μΩ·cm for the substrate temperature range from 25 to 100 °C. This result indicates that temperatures of up to 100 °C are not yet sufficient for efficient sintering of nanoparticles. A significant reduction in the resistivity begins to observe at a substrate temperature above 150 °C. For example, at a substrate temperature of 150, 200 and 250 °C, the resistivity of the samples was $3{\cdot}10^5$, $2{\cdot}10^3$, and 35 μΩ·cm, respectively, see figure 2.

Consequently, from the experimental data in figure 2, it can be concluded that the temperature of the start of sintering of silver nanoparticles, received in a spark discharge, exceeds 100-150 °C. This result agrees with the data of other researchers on the sintering of nanoparticles with printing process. It was also found that at a temperature above 250 °C the resistivity of the printed samples did not change significantly. Consequently, the use of high substrate temperatures above 250 °C is not advisable. The inset of figure 2 shows that the cross-section profile of the line printed on heated substrate at 300 °C has a bell-shaped form with average thickness and width equal to ~20 and ~200 μm, respectively.

Figure 3 shows SEM images of the silver lines formed at different temperatures of the silicon substrate at 25, 150 and 250 °C, respectively. Figure 3a shows that the sample formed at a substrate temperature of 25 °C consists of very small particles with a large number of voids between them. These voids limit the number of percolating paths for electric current, therefore, the specific electric resistance of this sample is very high. When the temperature is raised to 150 and 250 °C, as shown in Figures 3b and c, respectively, the particle sizes have grown and the voids between them are minimized.

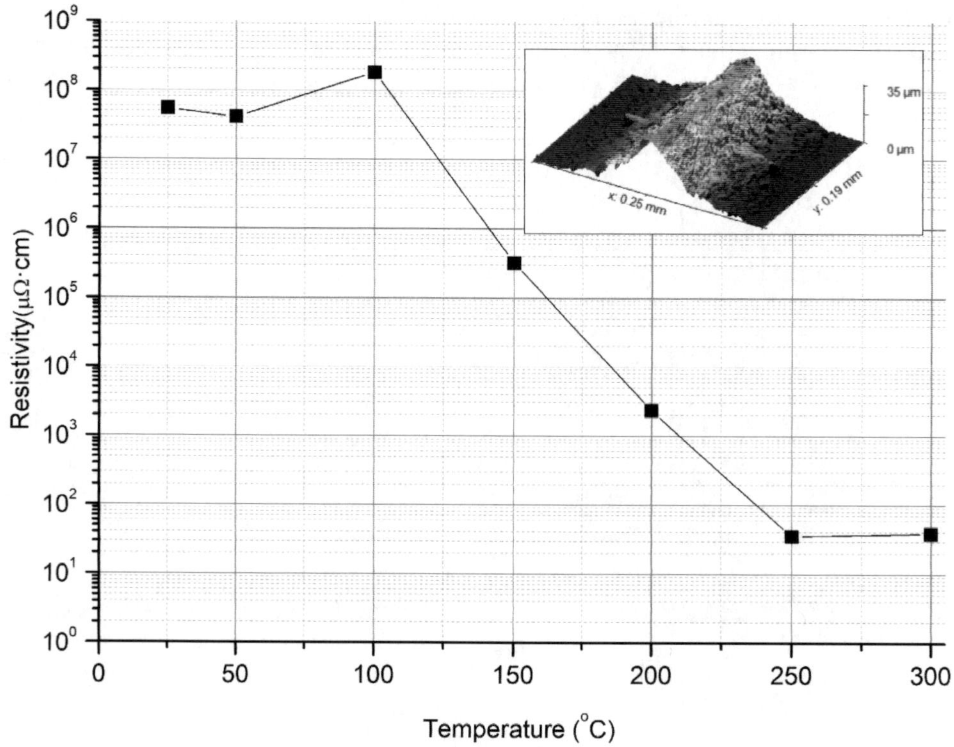

Figure 2. The resistivity of silver lines depending on the silicon substrate temperature at a dry aerosol jet printing. The inset shows a 3D-profile of silver line printed on heated substrate at 300 °C.

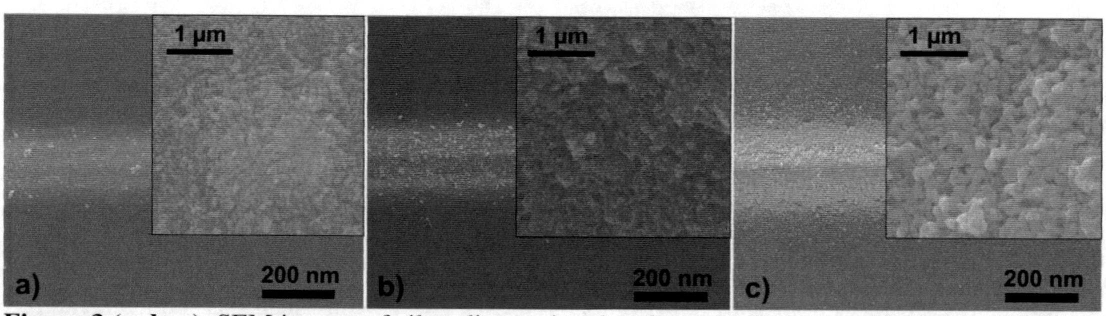

Figure 3 (a, b, c). SEM images of silver lines printed on heated substrate at 25 °C **(a)**, 150 °C **(b)**, and 250 °C **(c)**. The insets show more detail.

Besides, when depositing nanoparticles on a heated substrate, the possibility of their fixing on the surface increases, which means that the resulting particles line is formed from a large number of particles when sintering which forms a more voluminous structure having a low resistance. Based on the results of the experiments, it was also established that the resistivity of the best samples obtained at a substrate temperature of 250 °C is approximately 23 times worse than bulk silver. As we know, from the other works [9] this result is acceptable for some practical applications of e.g. for the production of microheaters, interconnectors, microstrip antennas, gas flow sensors etc. Thus, the considered approach of aerosol printing of conductive lines on heated substrates is quite promising.

4. Conclusion

A new method of dry aerosol printing of conductive silver lines on heated silicon substrates are investigated. This method allows forming conductive lines in one-step without a separate post-sintering operation. It was found that for effective sintering lines of silver nanoparticles the temperature of the heated substrate should be about more than 200-250 °C. Printed silver lines had an average thickness and resistivity equal to ~20 μm and 35 μΩ·cm, respectively. The same microstructures can be used in printed electronics devices.

Acknowledgments

This work was supported by the grant of the President of Russian Federation for young scientists MK-2302.2017.8 and carried out using the equipment of the Center for the collective use of high-precision measuring technologies in the field of photonics (ckp.vniiofi.ru).

References

[1] He Y, Oakley C, Chahal P, Albrecht J and Papapolymerou J 2017 Aerosol Jet printed 24 GHz end-fire quasi-Yagi-Uda antenna on a 3-D printed cavity substrate *2017 International Workshop on Antenna Technology: Small Antennas, Innovative Structures, and Applications (iWAT)* 2017 International Workshop on Antenna Technology: Small Antennas, Innovative Structures, and Applications (iWAT) pp 179–82

[2] Cho J H, Lee J, Xia Y, Kim B, He Y, Renn M J, Lodge T P and Daniel Frisbie C 2008 Printable ion-gel gate dielectrics for low-voltage polymer thin-film transistors on plastic *Nat. Mater.* **7** 900–6

[3] Vaillancourt J et al. 2008 All ink-jet-printed carbon nanotube thin-film transistor on a polyimide substrate with an ultrahigh operating frequency of over 5 GHz *Appl. Phys. Lett.* **93** 243301

[4] Zhou L, Zhuang J Y, Song M S, Su W M and Cui Z 2014 Enhanced performance for organic light-emitting diodes by embedding an aerosol jet printed conductive grid *J. Phys. Appl. Phys.* **47** 115504

[5] Rudorfer A, Tscherner M, Palfinger C, Reil F, Hartmann P, Seferis I E, Zych E and Wenzl F P 2016 A study on Aerosol jet printing technology in LED module manufacturing Fifteenth International Conference on Solid State Lighting and LED-based Illumination Systems vol 9954 (International Society for Optics and Photonics) p 99540E

[6] Yang C, Zhou E, Miyanishi S, Hashimoto K and Tajima K 2011 Preparation of Active Layers in Polymer Solar Cells by Aerosol Jet Printing *Acs Appl. Mater. Interfaces* **3** 4053–8

[7] Efimov A, Potapov G, Nisan A, Urazov M and Ivanov V 2017 Application of the Spark Discharge Generator for Solvent-free Aerosol Jet Printing *Orient. J. Chem.* **33** 1047–50

[8] Efimov A, Lizunova A, Sukharev V and Ivanov V 2016 Synthesis and Characterization of TiO2, Cu2O and Al2O3 Aerosol Nanoparticles Produced by the Multi-Spark Discharge Generator *Korean J. Mater. Res.* **26** 123–9

[9] Perelaer J, Hendriks C E, de Laat A W M and Schubert U S 2009 One-step inkjet printing of conductive silver tracks on polymer substrates *Nanotechnology* **20** 165303

IOP Conf. Series: Materials Science and Engineering **307** (2018) 012083 doi:10.1088/1757-899X/307/1/012083

Impact of the air gap at the casting/water-cooled-mold interface on casting microstructure

Y D Zeng ,X F Li, Y Liu

College of Materials Science and Engineering, Inner Mongolia University of Technology, Huhehaote,010051, China
E-mail: zyidan@sina.com
Abstract. In this paper, we report the results of experiments on gravity castings based on the experimental model of the controllable air gap. On basis of the temperature profiles at different locations of casting and mold were first measured, Beck's non-linear estimation procedure was applied to obtain the interfacial heat-transfer coefficient (IHTC) at the casting/mold interface. Using this as the boundary condition, the microstructures of the casting were simulated under different air-gap widths and formation time, which were compared to the microstructures of the measured casting. Furthermore, the impacts of the width of the air gap and the formation time of the air gap on the microstructure were analyzed. The results indicate that the air gap at the interface reduces the IHTC at the casting/mold interface, alters the solidification process of the casting, and has a great influence on the casting microstructure. Further, when the width of the air gap increases, the zone of the columnar grain of the casting decreases, the zone of the equiaxial grain is enlarged, and the columnar grains and equiaxial grains become coarser. When the formation time of the air gap increases, the zone of columnar grains of the casting is enlarged, the zone of the equiaxial grain is reduced, and the columnar grains and the equiaxial grains become finer.

1. Introduction

It is widely considered that a mold cooled by the circulating water (water-cooled mold) can rapidly exchange heat with a casting during the solidification process of the casting, which has a strong cooling effect on the casting. Thus, the microstructure of the obtained casting is fine, the segregation is slight, and the overall obtained mechanical performance is high. However, during the solidification process, an air gap will form at the casting/mold interface, which strongly hinders the heat exchange between the casting and the mold, thus greatly lowering the cooling effect of the mold and further affecting the solidification process and casting microstructure.

Casting microstructure has been a hot topic in related research, and as computer technology and mathematics have developed, the coupled cellular automata and finite-element method has become one of the major approaches in the investigation of microstructures. In recent years, a significant amount of related research has been reported[1–3].

Currently, for simulation of a microstructure, more studies focus on nucleation[4], dendritic growth [5], grain morphology[6], grain dimension[7,8] and the evolutionary process of microstructure morphology[2,9~10]. The influence of the formation of the air gap at the interface on the microstructure has been ignored. The impact of the air gap at the interface on the microstructure simulation of a sand mold casting may be neglected, but for metal casting and water-cooled casting the thermal resistance at the casting/mold interface caused by the interfacial air gap is far greater than the thermal resistance of the mold itself. Therefore, if the thermal resistance at the interface is neglected, a large deviation can occur in microstructure simulation results.

IOP Conf. Series: Materials Science and Engineering **307** (2018) 012083 doi:10.1088/1757-899X/307/1/012083

Therefore, in this paper we report the results of casting experiments based on the experimental model of the controllable air gap, of adopting the non-linear estimation method to obtain the interfacial heat-transfer coefficient (IHTC), and of determining the impact of the air gap on the IHTC. Based on such work, we performed simulations of a solidification microstructure and experimental studies that will help improve the accuracy of microstructure simulation.

2. Experiment

2.1 Experimental solution

To study the impact of the air gap on the microstructure of a casting, the width of the air gap and the formation time of the air gap were chosen as the control factors in the experiments. The widths of the air gap were set as 0, 0.5, 1, and 2mm and the formation times of the air gap were set as 90 and 140s. Based on measured temperature, the non-linear estimation method was applied to obtain the IHTC. The IHTC was further used as the boundary condition for performing simulations of the solidification microstructure.

2.2 Experimental model

As shown in Figure 1, the experimental model is a combined mold, in which one side is a movable water-cooled mold. The remaining portion is thermal insulation material. When molten metal is poured into the mold, the water-cooled mold moves towards the casting to closely contact the casting, such that no air gap is formed. When the solidification thickness reaches a preset value, the water-cooled mold is moved in the opposite direction, such that an air gap is manually formed at the interface. As a consequence, this model can accurately control the size and formation time of the air gap. A pressure sensor is installed on the push rod of the water-cooled mold, which can be relied on to accurately acquire the contact status between the water-cooled mold and the casting.

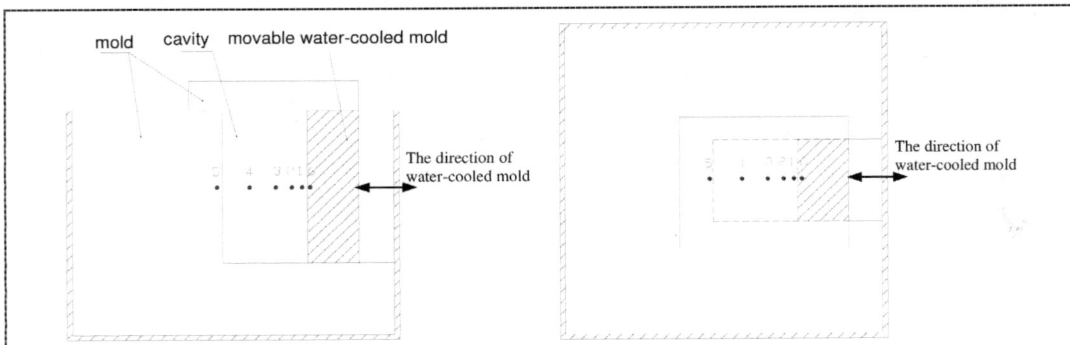

(a) (b)

Figure 1. Experimental apparatus: (a) Front sectional view. (b) Horizontal sectional view.

2.3 Temperature measurement

The distribution of temperature measurement points using thermocouples is shown in Figure 1, where points 1–4 represent points that are 5, 10, 30, and 50 mm, respectively, from the casting/water-cooled mold interface. Points 5 and point 6 are located in the mold and in the water-cooled mold, respectively, for measuring the temperatures of the walls of the mold and the water-cooled mold, respectively. K-type thermocouples with a wire diameter of 0.1mm are used for temperature measurements. In addition, the temperature at each measurement point is measured using an exposed thermocouple. The DAQ-USB-2401 module is utilized to collect data, the sampling frequency is 10Hz, and the temperature variation at each temperature measurement point is recorded during the experimental procedure.

2.4 Preparation of specimens

IOP Conf. Series: Materials Science and Engineering **307** (2018) 012083 doi:10.1088/1757-899X/307/1/012083

The casting material is the industrial pure aluminum, the pouring temperature is controlled within 780–790°C, and the top gating system is applied. To observe the impact of the air gap on the microstructure of the casting, specimens are obtained at locations of the casting as illustrated in Figure 2 for preparation of the metallographic specimens.

3. Experimental results and analysis

3.1 Curves of measured temperature

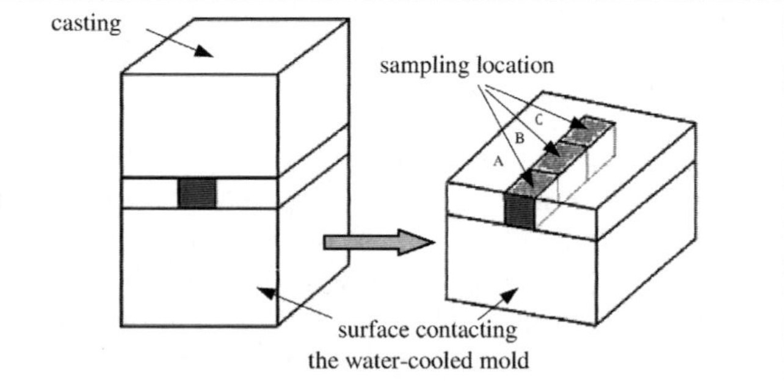

Figure 2. Sampling locations of metallographic specimens

Figure 3 shows the temperature variation curves corresponding to the measurement points 1–6 under conditions of different air-gap formation times and different air gap widths. It can be seen that, without the air gap [Figure 3(a)], the curves of the temperature measurement points 1–4 inside the casting do not exhibit the phenomenon that the temperature decreases first and then increases, and the temperature reducing rate at each point is high. When the air-gap formation time is 90 s and the air gap width is 1 or 2 mm [see Figures 3(b) and 3(c)], the curves of the temperature measurement points 1–4 inside the casting first decrease rapidly and then increase, and the temperature reducing rate at each point is significantly lower than that without an air gap. Further, the wider the air-gap width, the higher the temperature recovery at the casting surface. This indicates that the heat exchange at the casting/water-cooled-mold interface occurs very quickly before the air gap is formed, and that the temperature at the surface of the casting decreases rapidly. After the air gap is formed, the air gap hinders the heat exchange between the casting/water-cooled mold, such that the heat exchange rate decreases, which results in the temperature recovery at the surface of the casting; that is, when the width of the air gap increases, the thermal resistance at the casting/mold interface increases. When the air-gap width is 2 mm, if the air-gap formation time increases from 90 to 140s, the temperature variance at each measurement point can be seen as shown in Figures 3(c) and 3(d). The temperature curves all show that the temperature first decreases, later increases, and then decreases again. As the air-gap formation time becomes longer, the amplitude of the temperature recovery is reduced. This is because the elongated air-gap formation time often indicates an elongated contact time between the casting and the mold, and the heat exchange between the casting and the mold decreases as the time increases, such that the impact of the air-gap formation on the cooling rate of the casting is reduced. Thus, the temperature recovery amplitude of the casting is relatively small.

3.2 Solution of the IHTC

The non-linear estimation method proposed by Beck[11] is used to calculate the IHTC, and the result is shown in Figure 4. As the solidification time increases, the IHTCs under different conditions all display a decreasing trend. When there is no air gap, the IHTC is as shown in Figure 4(a). When the air-gap width is 1 mm, the IHTC decreases relatively slowly [as shown in Figure 4(b)], and when the air-gap width is 2 mm, the IHTC decreases rapidly [as shown in Figure 4(c)]. By comparing the three IHTC variation curves [Figures 4(a)–4(c)], it can be seen that when the air-gap width increases, the IHTC decreases, and the greater the air-gap width, the faster the IHTC decreases. Moreover, as the air-gap formation time increases, the IHTC is reduced [see Figures 4(c) and 4(d)].

3.3 Microstructure simulation and experimental results

The CAFÉ module of PROCAST software is applied to simulate the casting microstructure, which is compared to that obtained by experimental casting.

3.3.1. Microstructure

The formation times of the air gaps are 90 and 140 s and the air-gap widths are 0, 1, and 2 mm. The

IOP Conf. Series: Materials Science and Engineering **307** (2018) 012083 doi:10.1088/1757-899X/307/1/012083

simulated microstructure of the cross-section of the casting is displayed in Figure 5, and, from the surface that contacts the water-cooled mold, four regions are found sequentially, which are the surface

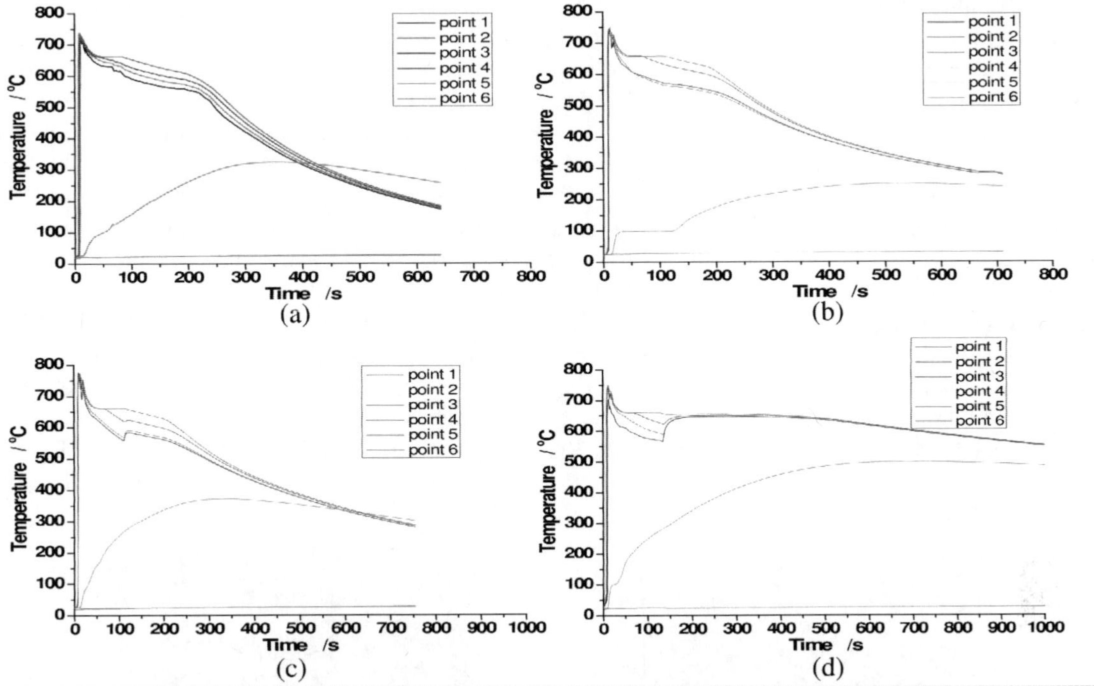

Figure 3. Measured temperature curves under different air gap widths and formation time

(a) no air gap (b) air-gap formation time 90 s and air-gap width 1 mm (c) air-gap formation time 90 s and air-gap width 2 mm (d) air-gap formation time 140 s and air-gap width 2mm

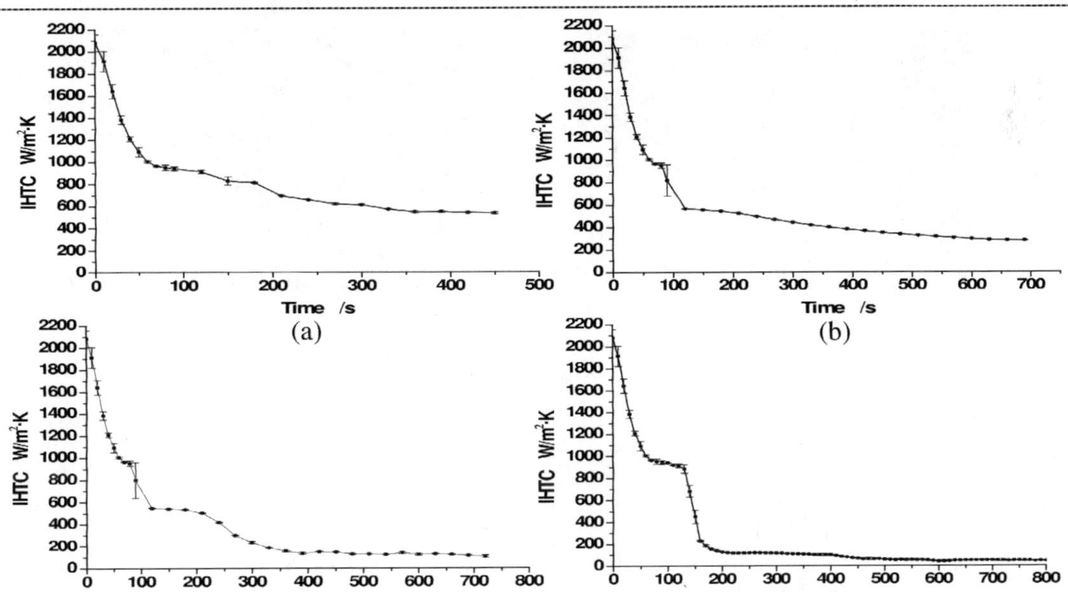

Figure 4. Calculation of IHTCs under different air-gap widths and formation time

(a) no air gap (b) air-gap formation time 90 s and air-gap width 1 mm (c) air-gap 90 s and air-gap width 2 mm (d) air-gap formation time 140 s and air-gap width 2mm fine equiaxial grain zone, fine columnar grain zone, coarse columnar grain zone, and central coarse equiaxial grain zone.

Under the condition of the same air-gap formation time, as the air-gap width increases, the area of the columnar grain zone decreases and the columnar grains become coarser, the central equiaxial grain zone expands, and the grain size of the equiaxial grains increases significantly, as indicated in Figures 5(a)–(c). Under the condition of the same air-gap width, when the air-gap formation time increases, the area of the columnar grain increases and the columnar grains become finer, the equiaxial grain zone is reduced, and the grain size of the equiaxial grains decreases, as shown in Figures 5(c) and 5(d). The cooling effect of the water-cooled mold on the casting is far greater than the other five walls of the casting, and the heat dissipation of the casting mainly relies on the heat exchange between the casting and the water-cooled mold. Accordingly, the direction of heat exchange opposite the growth direction of the columnar grains is the main heat-flow direction.

Before the air gap is formed, the molten metal forms fine columnar grains under the strong cooling effect of the wall, and after the air gap is formed, as the air-gap width increases, the IHTC at the casting/mold interface decreases rapidly. Thus, the cooling effect of the mold on the casting decreases significantly, which weakens the heat exchange in the direction opposite the growth direction of the columnar grains. Consequently, the possibility that the grains grow in the form of equiaxial grains increases, which results in the contraction of the columnar grain zone and the expansion of the equiaxial grain zone.

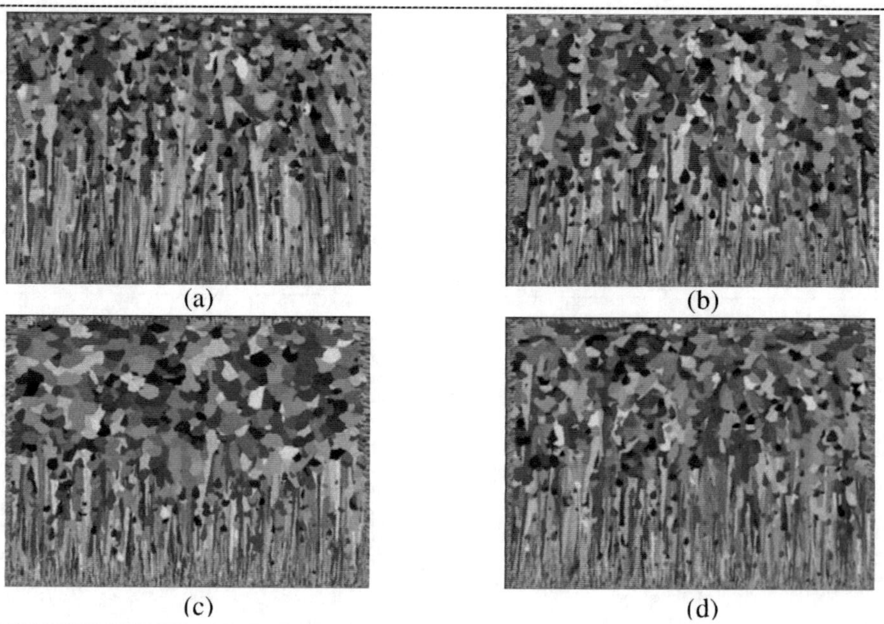

Figure 5. Simulated microstructures of casting cross-section

(a) no air gap (b) air-gap formation time 90 s and air-gap width 1 mm (c) air-gap formation time 90 s and air-gap width 2 mm (d) air-gap formation time 140 s and air-gap width 2mm

As the air-gap formation time increases, the contact time between the casting and the water-cooled mold increases. Specifically, the time that the casting/mold interface remains at a high IHTC is increased, and the heat exchange in a direction opposite the growth direction of the columnar grains weakens relatively slowly. Thus, the growth rate of the columnar grains is relatively fast, which results in an increase of the columnar grain zone and a reduction in the equiaxial grain zone.

3.3.2. Cell spacing and grain size

Given conditions where the air-gap formation time is 90 s and the air-gap widths are 0, 1, and 2 mm, the cell spacing is simulated and calculated results are obtained. When there is no air gap, the cell spacing of fine columnar grains and coarse columnar grains are 32 and 112 um., respectively. After the air gap is formed, the cell spacing of coarse columnar grains increases significantly, the average value of which increases from 137 to 281 um, and the average grain size of the equiaxial grains increases from 261 to 499 um.. The air-gap formation causes the cooling rate of the casting to decrease significantly, and the cell spacing increases rapidly and correspondingly.

When the air-gap width is 2 mm, and the formation time of the air gap is 90 and 140s, the average distance between the coarse columnar grains is 281 and 123 um., respectively, and the average grain size of the equiaxial grains is 499 and 335 um, respectively. The longer the air-gap formation time, the longer the time the casting/mold interface remains at a high IHTC, such that the cooling rate of the casting is fast, which causes the columnar grains to be relatively fine.

In Figure 6, the air-gap width is 1mm and air-gap formation time is 90 s, the microstructures of the casting at the locations marked A, B, and C in Figure 2 are obtained, which indicates that the farther the microstructure is from the water-cooling interface, the coarser it is. Such a conclusion is consistent with the microstructure simulation. Under other conditions of the air-gap width, the microstructure of the casting shows similar behavior. The average cell spacing at locations A and B of the specimen are approximately 40 and 150um, respectively, and the average grain size at location C is 400um.

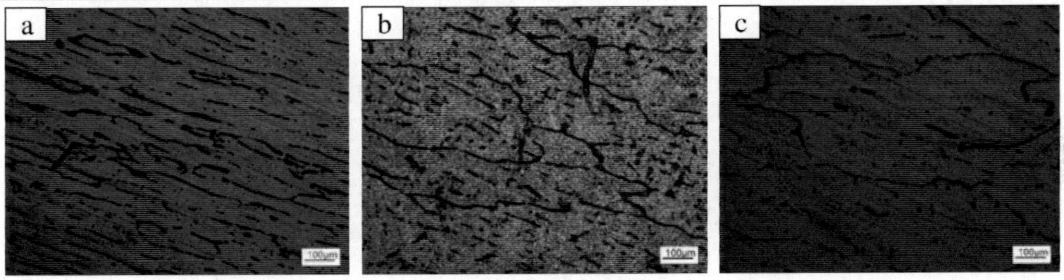

Figure 6. Measured microstructures of casting cross-section
(a) at locations A (b) at locations B (c) at locations C

4. Conclusions

The formation of the air gap at the casting/mold interface decreases the IHTC at the interface, resulting in a decreased cooling rate of the casting, which has a great impact on the casting microstructure. When the air-gap width increases, the IHTC decreases, and the change in the microstructure is that the area of the columnar grain zone decreases, the area of the equiaxial grain zone increases, and the columnar grains and equiaxial grains become coarser. When the air-gap formation time increases, the time that the casting/mold interface remains at the high IHTC becomes longer, the cooling rate of the casting becomes faster, the area of the columnar grain zone increases, the area of the equiaxial grain zone decreases, and the columnar and equiaxial grains become relatively finer.

References

[1] Zaeem M A , Yin H B and Felicelli S D 2013 *Applied Mathematical Modelling* **37** 3495-503
[2] Vandyoussefi M and Greer A L 2002 *Acta Materialia* **50** 1693-705
[3] Gandin C A. and Rappaz M 1997 *Acta mater* **45** 2187-95.
[4] Seo S M, Kim I S , Jo C Y and, Ogi K 2007 *Materials Science and Engineering A* **449–451** 713-6
[5] Eshraghi M, Felicelli S D and Jelinek B 2012 *Journal of Crystal Growth* **354** 129-34
[6] Yin H and Felicelli S D 2010 *Acta Materialia* **58** 1455-65
[7] Omar L B , Uriel M H , José R and Christophe P 2017 *Materials & Design* **113** 369-76
[8] Reyes L A, Páramo P and Zamarripa A S 2015 *Materials & Design* **83** 301-7
[9] Han R H , Dong W C and, Lu S P 2014 *Computational Materials Science* **95** 351-61
[10] Zhang J W, Ou F L, Seufze Wand Taminge K . 2016 *Additive Manufacturing* **11** 32-9
[11] Beck J V and Blackwell B F 1985 *Inverse heat conduction Ill-posed problems*, Wiley, New York.